T0206249

ARTIFICIAL LIFE VII

ARTIFICIAL LIFE VII

Proceedings of the Seventh International Conference on Artificial Life

edited by Mark A. Bedau, John S. McCaskill, Norman H. Packard, and Steen Rasmussen

A Bradford Book

The MIT Press
Cambridge, Massachusetts
London, England

© 2000 Massachusetts Institute of Technology

All rights reserved. No part of this book may be reproduced in any form by any electronic or mechanical means (including photocopying, recording, or information storage and retrieval) without permission in writing from the publisher.

Library of Congress Cataloging-in-Publication Data

International Conference on Artificial Life (7th : 2000 : Portland, Or.)
    Artificial Life VII : proceedings of the seventh International Conference on Artificial Life / edited by Mark Bedau . . . [et al.].
        p.   cm.—(Complex adaptive systems)
    "A Bradford book."
    Includes bibliographical references and index.
    ISBN 978-0-262-52290-8 (pbk.: alk. paper)
    1. Biological systems—Computer simulation—Congresses. 2. Biological systems—Simulation methods—Congresses. I. Title: Artificial life seven. II. Title: Artificial life 7. III. Bedau, Mark. IV. Title. V. Series.
QH324.2.I556   2000
570'.1'13—dc21   00-033942
                                                            CIP

The MIT Press is pleased to keep this title available in print by manufacturing single copies, on demand, via digital printing technology.

Contents

III Evolutionary and Adaptive Dynamics

IV Robots and Autonomous Agents

V Communication, Cooperation, and Collective Behavior

VI Methodological and Technological Applications

Preface

Artificial life is an interdisciplinary research enterprise investigating the fundamental properties of living systems by means of the simulation and synthesis of life-like processes in artificial media. Although isolated specialists in different disciplines had worked in this area for many decades, perhaps the first attempt to collect this disparate work and encourage its cross-pollination was the conference on "Evolution, Games, and Learning: Models for Adaptation in Machines and Nature" organized by Doyne Farmer, Alan Lapedes, Norman Packard, and Burton Wendroff in 1985 at Los Alamos. This conference was followed by another two years later, this one organized by Christopher Langton and baptized "Artificial Life." Since then, the name has stuck and the biennial conferences have continued, moving through Santa Fe, Boston, Nara, and Los Angeles, to Portland. The growth of this collective research enterprise over the past decade and a half, spawning a number of professional journals and specialized conference series in Europe and Asia, is nothing short of astounding.

The artificial life community is still strikingly interdisciplinary. This conference, the seventh in the series, includes authors in biology, physics, chemistry, computer science, mathematics, psychology, economics, robotics, information science, physiology, and philosophy. The interdisciplinary nature of artificial life creates special challenges. It is hard to keep abreast of relevant new work when it uses different specialized vocabularies and methodologies and is published in disparate venues, and it is hard to establish and follow high standards of scientific rigor that different disciplines with unique histories and intellectual conventions will each find acceptable. The coincidence of this year's conference with the birth of a new millennium provides a natural opportunity to address these challenges. Artificial life will remain a vital research activity only if we periodically look backward and reassess our work, so we can continually augment a foundation of solid achievements. We must also periodically look forward and identify the most important open questions so that we can promote fruitful research activities and evaluate their progress over time. Looking backward and forward in this way enables us to renew and redefine our interdisciplinary center of gravity and to reshape the direction of future research. Hence, the theme of this year's conference: *Looking backward, looking forward*.

Over a hundred papers were submitted to the conference, and about half of these will be presented as talks. This volume contains all of the papers to be presented as talks, as well as some of the papers to be presented as posters. All of the papers were reviewed by at least three people on the program committee for quality of science, quality of presentation, and relevance to the conference. The papers fall into seven broad topic areas: (1) the origin of life, self-organization, and self-replication, including astrobiology, artificial chemistry, molecular self-assembly, and molecular information processing; (2) development and differentiation, including multicellular development, gene-regulation networks, and morphogenesis; (3) evolutionary and adaptive dynamics, including modes of selection (natural, neutral, kin, etc.), evolvability, and cultural evolution; (4) robots and autonomous agents, including bio-inspired robots, autonomous and adaptive agents, and evolutionary robotics; (5) communication, cooperation, and collective behavior, including the evolution of social, linguistic, economic, and technical systems; (6) methodological and technological applications, ranging from commerce and industry to medicine; (7) and the broader context, including discussion of the historical origins of artificial life, philosophical analysis of artificial life's distinctive methodologies, and connections between artificial life and artistic creativity.

Creating this conference has crucially depended on many co-organizers. I am especially indebted to my co-editors and Program Committee Co-Chairs: John McCaskill, Norman Packard, and Steen Rasmussen. Their excellent scientific judgment coupled with their hard work and generous spirit made the process of shaping the scientific character of this conference especially rewarding and inspiring. The blizzard of activity that has culminated with this volume has depended on the good will, boundless energy, and technical wizardry of one person more than any other: our Technical Secretary, Titus Brown. Kathleen Stackhouse and Eilis Boudreau admirably oversaw the hundreds of details involved in the conference's local arrangements. Eilis Boudreau and Carlo Maley, Workshops and Tutorials Co-Chairs, shouldered the responsibility for overseeing more than a dozen special-interest gatherings at the conference and producing the *Artificial Life VII Workshops Proceedings*. These workshops and tutorials add immeasurably to the value of the conference experience. Tracy Teal, conference Webmaster, cheerfully and capably created our conference web pages and updated and revised them, sometimes on a daily basis. Ken Willett, conference Treasurer, helped the conference get off on a sound financial footing by preparing and revising our budget. And Peary Brug has helped raise additional support for the conference.

Many people at Reed College made invaluable contributions to the conference. Dean of Faculty, Peter Steinberger, provided crucial early support. Aurelia Carbone created the beautiful conference poster. Mike Raven helped with technical trouble-shooting. And the staffs in the Business office, the Computing and Information Services, the News and Publications office, the Conference and Events Planning office, and Food Services created a wonderfully constructive and cooperative community for organizing the conference.

A very warm thanks is due to all our sponsors, especially Reed College and Intel Corporation, whose early financial support made this conference possible at all. My warm thanks also go out to Bob Prior at MIT Press for his continual support of the artificial life conferences. I am grateful to all my colleagues who agreed to serve on the conference's international Scientific Advisory Board: David Ackley, Chris Adami, Rik Belew, Hughes Bersini, Maggie Boden, Sung-Bae Cho, Dario Floreano, Stephanie Forrest, Inman Harvey, Paulien Hogeweg, Phil Husbands, Tashaki Ikegami, Kunihiko Kaneko, Jozef Kelemen, Christopher Langton, Ju-Jang Lee, Maja Mataric, Jean-Arcady , Melanie Mitchell, Domenico Parisi, Jordon Pollack, Tom Ray, Mitchel Resnick, Masanori Sugisaka, Luc Steels, Charles Taylor, Jon Umerez, Stewart Wilson, Yong Guang Zhang. And a very special thanks goes out to all my colleagues who carefully reviewed the papers submitted to the conference: David Ackley, Chris Adami, Wolfgang Banzhaf, Hugues Bersini, Eric Bonabeau, Sung-Bae Cho, John Collier, Michael Conrad, Michael Dyer, Emmeche, Dario Floreano, Robert French, Inman Harvey, Paulien Hogeweg, Phil Husbands, Takashi Ikegami, Norman Johnson, Kunihiko Kaneko, Brian Keeley, Marc Lange, Kristian Lindgren, Carlo Maley, Paul Marrow, Barry McMullin, Filippo Menczer, J.J. Merelo, Jean-Arcady Meyer, Alvaro Moreno, Chrystopher Nehaniv, Stefano Nolfi, Charles Ofria, Domenico Parisi, Tom Ray, Craig Reynolds, Moshe Sipper, Eugene Spafford, Russell Standish, Luc Steels, Chuck Taylor, Tim Taylor, Guy Theraulaz, Adrian Thompson, Mark Tilden, Jon Umerez, Barbara Webb, Michael Wheeler, Claus Wilke, and Andy Wuensche. The intellectual integrity of a field is protected by the diligence and good judgment exercised by those who participate in the peer review process; we all owe them our appreciation and thanks.

Finally, on a personal note, I want to express my warmest thanks to Kate O'Brien, whose sound instincts, balanced judgment, and positive outlook make her such an extraordinarily valued companion in this and all other aspects of life.

Mark A. Bedau
Conference Chair, Artificial Life VII

Portland, April 2000

I Origin of Life, Self-Organization, and Self-Replication

A Self-Replicating Universal Turing Machine: From von Neumann's Dream to New Embryonic Circuits

Héctor Fabio Restrepo, Daniel Mange, and **Moshe Sipper**

Logic Systems Laboratory, Swiss Federal Institute of Technology
CH – 1015 Lausanne, Switzerland
E-mail: {name.surname}@epfl.ch

Abstract

Borrowing inspiration from von Neumann's dream and the idea of a true cellular automaton, we describe a multicellular universal Turing machine implementation with self-replication and self-repair capabilities. This implementation was made possible thanks to a new "multicellular" automaton developed as part of the Embryonics (embryonic electronics) project. This new automaton, in which every artificial cell contains a complete copy of the genome, is endowed with self-replication and self-repair capabilities. With these properties and by using a modified version of the W-machine, it was possible to realize the mapping of the universal Turing machine onto our multicellular array.

Keywords: self-replication, self-repair, universal Turing machine, cellular automata, Embryonics.

Introduction

The Embryonics (embryonic electronics) project is inspired by the basic processes of molecular biology and by the embryonic development of living beings. By adopting three fundamental features of biology — multicellular organization, cellular division, and cellular differentiation — and by transposing them onto the two-dimensional world of integrated circuits in silicon, we show that properties of the living world, such as self-replication and self-repair, can also be attained in artificial objects (integrated circuits).

Our goal in this paper is to present self-replicating machines exhibiting universal computation, i.e., universal Turing machines. We demonstrate that the dream of von Neumann, the self-replication of such a machine, can be realized in actual hardware thanks to the Embryonics architecture.

In the next section we present a brief reminder of specialized and universal Turing machines. We then survey classical self-replicating automata and loops. The following section introduces the Embryonics architecture based on a multicellular array of cells and describes the implementation of a self-replicating specialized Turing machine. We next present the architecture of an ideal and of an actual universal Turing machine able to self-replicate. A discussion of our results follows in the final section.

Turing machines

In the 1930's, before the advent of digital computers, several mathematicians began to think about what it means to be able to compute a function. The theory of Turing machines was the response to this question. It is important to mention that Alonzo Church and Alan Turing independently arrived at equivalent conclusions; their common definition was: *A function is computable if it can be computed by a Turing machine.*

Turing machines were conceived by Alan Turing in his historic paper, *"On Computable Numbers, with an Application to the Entscheidungsproblem"* [22], which was his response to the Entscheidungsproblem, posed by the German mathematician David Hilbert. Hilbert asked if there existed, in principle, any definite method which could be applied to determine the truth of any mathematical question.

Turing machines are one of the earliest and most intuitive ways to render precise the notion of effective computability. This is now the foundation of the modern theory of computation and computability.

Specialized Turing machines

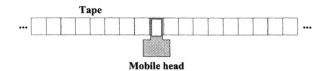

Figure 1: *A specialized Turing machine.*

A specialized Turing machine (Figure 1) is a finite-state machine (*the program*) controlling a mobile head, which operates on a tape (*the data*). The tape, composed of a sequence of squares, contains a string of symbols. The head is situated, at each moment, on some square of the tape and has to carry out three operations to complete a step of the computation. These operations are:

1. reading the square of the tape being scanned;
2. writing on the scanned square;
3. moving the head to an adjacent square.

A Turing machine can be described by three functions f_1, f_2, f_3:

$$Q^+ = f_1(Q, S) \qquad (1)$$

$$S^+ = f_2(Q, S) \qquad (2)$$

$$D^+ = f_3(Q, S) \qquad (3)$$

where Q and S are, respectively, the current internal state and the current input symbol, and where Q^+, S^+, and D^+ are, respectively, the next internal state, the next input symbol, and the direction of the head's next move [13].

The universal Turing machine

Turing had the further idea of the universal Turing machine (UTM), capable of simulating the operation of any specialized Turing machine, and gave an exact description of such a UTM in his paper [22].

A universal Turing machine, U, is a Turing machine with the property of being able to read the description (on its tape) of any other Turing machine, T, and to carry out correctly (one step at a time) what T would have done. The necessary components of the machine U are a finite-state machine (the program of U) controlling a mobile head, which operates on a tape; the data on the tape describe completely the machine T to be simulated (the data of T and the program of T, i.e., the three functions Q^+, S^+, and D^+ describing T).

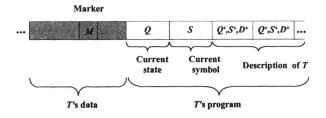

Figure 2: *Universal Turing machine's tape, describing specialized machine T.*

Figure 2 shows the organization of U's tape. To the left is a semi-infinite region, which contains the data of T's tape. Somewhere in this region is a marker M indicating where T's head is currently located. The next region contains the current internal state Q and the current input symbol S of T. The third region is used to record the description of T, i.e., the three functions Q^+, S^+, and D^+ for each combination of Q and S.

Self-replication: A brief survey
Self-replicating automata

The early history of the theory of self-replicating machines is basically the history of von Neumann's thinking on the matter [5, 18, 23]. Von Neumann's cellular automaton [23], as well as all the machines described in this paper, is based on the following general hypotheses:

- the automaton deals exclusively with the flow of information; the physical material (usually a silicon substrate) and the energy (power supply) are given a priori;
- the physical space is two-dimensional and as large as desired;
- the physical space is *homogeneous*, that is comprised of identical *molecules*, all of which have the same internal architecture and the same connections with their neighbors; only the *state* of a molecule (the combination of the values in its memories) and its position can distinguish it from its neighbors;
- replication is considered as a special case of growth: this process involves the creation of an identical organism by duplicating the genetic material of a mother entity onto a daughter one, thereby creating an exact clone.

To avoid conflicts with biological definitions, we do not use the term "cell" to indicate the parts of a cellular automaton, opting rather for the term "molecule". (In biological terms, a "cell" can be defined as the smallest part of a living being which carries the complete blueprint of the being, that is the being's *genome*.)

The molecule of von Neumann's automaton is a finite-state machine with 29 states. The future state of a molecule depends on the present state of the molecule itself and of its four cardinal neighbors (north, east, south, west). The exhaustive definition of the future state, the *transition table*, thus contains $29^5 = 20,511,149$ lines.

In his historic work [23], von Neumann showed that a possible *configuration* (a set of molecules in a given state) of his automaton can implement a universal constructor (Uconst) endowed with the following three properties:

1. universal construction (Figure 3);
2. self-replication of the universal constructor (Figure 4);
3. self-replication of a universal computer (Ucomp), i.e., a universal Turing machine (Figure 5).

According to the biological definition of a cell, it can be stated that von Neumann's automaton is a unicellular organism: its genome is composed of the description of the universal constructor and computer D(Uconst + Ucomp) written in the memory M (Figure 5); as each molecule of this description needs five molecules of the genome [23], it can be estimated that the genome is composed of approximately five times the number of molecules of the

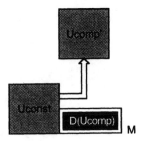

Figure 3: *Universal construction of von Neumann's automaton: a possible configuration can implement a universal constructor Uconst. Then, given the description D(Ucomp) of any one machine Ucomp, including a universal Turing machine, the universal constructor can build a specimen of this machine (Ucomp') in the molecular space.*

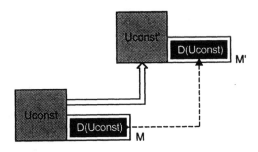

Figure 4: *Self-replication of the universal constructor: given the description D(Uconst) of the constructor itself, it is then possible to build a copy of the constructor in the molecular space: the constructor interprets first the description D(Uconst) to build a copy Uconst' whose memory M' is empty (translation process), and then copies the description D(Uconst) from the original memory M to the new memory M' (transcription process).*

universal constructor and computer.

In summary:

- the dimensions of von Neumann's automaton are substantial (on the order of 200,000 molecules) [6]; it has thus never been physically implemented and has been simulated only partially [4,16,17];
- the automaton implements the self-replication of a universal computer (a universal Turing machine).

Though von Neumann and his successors Burks [3,23], Thatcher [3], Lee [8], Codd [4], Banks [2], and Nourai and Kashef [14] demonstrated the theoretical possibility of realizing self-replicating automata with universal calculation, a practical implementation requires a markedly different approach. It was finally Langton, in 1984, who initiated a second stage in self-replication research.

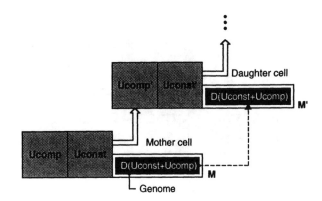

Figure 5: *Self-replication of a universal computer: by attaching to the constructor a universal computer Ucomp (a universal Turing machine), and by placing the description D(Uconst + Ucomp) in the original memory M, the universal constructor produces a copy of itself (Uconst') and a copy of the universal computer (Ucomp') through the mechanism described in Figure 4 (interpretation and then duplication of the description D).*

Self-replicating loops

In order to construct a self-replicating automaton simpler than von Neumann's, Langton [7] adopted more liberal criteria. He dropped the condition that the self-replicating unit must be capable of universal construction and computation. Langton's mechanism is based on an extremely simple configuration in Codd's automaton [4] called the periodic emitter, itself derived from the periodic pulser organ in von Neumann's automaton [23]. The molecule of Langton's automaton is a finite state machine with only 8 states. The future state, as with von Neumann's automaton, depends on the present state of the molecule itself and its four cardinal neighbors. The exhaustive definition of the future state, the transition table, contains only 219 lines, a very small subset of the theoretically possible $8^5 = 262,144$ lines (thanks to the use of default rules and symmetry assumptions).

Langton proposed a configuration in the form of a loop (Figure 6), with a constructing arm (pointing to the north in the left loop and to the east in the right loop) and a replication program, or genome, which turns counterclockwise. After 151 clock periods, the left loop (the mother loop) produces a daughter loop, thus obtaining the self-replication of Langton's loop.

Referring again to biological definitions, we observe that Langton's self-replicating loop is a unicellular organism; its genome, defined in Figure 6, comprises 28 molecules and is a subset of the complete loop which includes 94 molecules.

In summary:

- the size of Langton's loop is perfectly reasonable, since it requires 94 molecules, thus allowing complete simu-

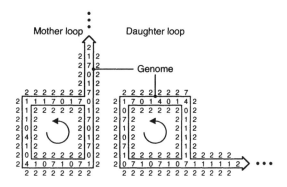

Figure 6: *In Langton's self-replicating loop, the genome, which turns counterclockwise, is characterized by the sequence, read clockwise: 170 170 170 170 170 170 140 140 1111. The signals "1" are ignored, the signals "70" cause the extension of the constructing arm by one molecule, while the signals "40", repeated twice, cause the arm to turn 90° counterclockwise. After 151 clock periods, the left loop (the mother loop) produces a daughter loop, thus obtaining the self-replication of Langton's loop. The genome is both interpreted (construction of a copy at the end of the constructing arm: translation process) and copied (duplication at the junction of arm and loop: transcription process).*

lation;

- there is no universal construction or calculation: the loop does nothing but replicate itself; comparing Figure 4 and Figure 6 reveals that Langton's self-replicating loop represents a special case of von Neumann's self-replication; the loop is a non-universal constructor, capable of building, on the basis of its genome, a single type of machine: itself.

Self-replicating loops with computing capabilities

The loops of the previous section exhibit only rudimentary computing and constructing capabilities, their sole functionality being that of self-replication. Lately, new attempts have been made to redesign Langton's loop in order to embed calculation capabilities. Tempesti's loop [19] is a self-replicating automaton which preserves some of the more interesting features of Langton's loop (in particular, it preserves the structure based on a square loop to dynamically store information, and the concept of a constructing arm); nevertheless, Tempesti introduced important modifications to Langton's design. Tempesti's loop attaches an executable program that is duplicated and executed in each of the copies (Figure 7), a process demonstrated for a simple program that writes out (after the loop's replication) LSL, acronym of the Logic Systems Laboratory [21].

Self-replicating loops with universal computing capabilities

Perrier et al.'s self-replicating loop [15] exibits universal computational capabilities (Figure 8). The system consists of three parts, loop, program, and data, all of which are replicated, followed by the program's execution on the given data. In the figure, P and D denote states belonging to the set of program states and to the set of data states, respectively.

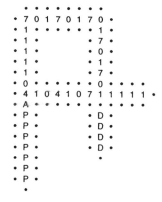

Figure 8: *Perrier et al.'s self-replicating loop. P denotes a state belonging to the set of program states. D denotes a state belonging to the set of data states. A is a state which indicates the position in the program.*

The universal computational model chosen for this work was the W-machine, introduced by Hao Wang [24] and named for him by Lee [9], who explored its relation with finite automata. A W-machine is like a Turing machine with two symbols $S = 0$ and $S = 1$, save that its operation at each time step is guided not by the three functions f_1, f_2, f_3 of a state table but by an instruction from the following list [1]:

PRINT 0, PRINT 1, MOVE DOWN, MOVE UP, IF 1 THEN (n) ELSE (*next*), STOP

The complete program for a Turing machine is a finite ordered list of instructions (a program) equivalent to the state table. After execution of an instruction of the first four types, control is automatically transferred to the next instruction. The conditional jump transfers control to the n-th instruction if the square under scan is a 1 symbol, otherwise it transfers control to the next instruction.

Adding functionality to Langton's loop is, in fact, not possible without major alterations. Perrier et al. developed a relatively complex automaton, in which a two-tape Turing machine was appended to Langton's loop. This automaton exploits Langton's loop as a sort of carrier: the first function of Perrier's loop is to allow Langton's loop to build a copy of itself. The main function

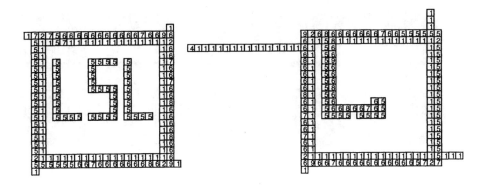

Figure 7: *Tempesti's self-replicating loop has an attached executable program that is duplicated and executed in each copy.*

of the offspring is to determine the location of the copy of the Turing machine. Once the new loop is ready, a messenger runs back to the parent loop and starts to duplicate the Turing machine, a process completely disjoint from the operation of the loop. When the copy is finished, the same messenger activates the Turing machine in the parent loop (the machine had to be inert during the replication process in order to obtain a perfect copy). The process is then repeated in each offspring until the space is filled [15].

The automaton thus becomes a self-replicating universal Turing machine, a powerful construct which is unfortunately handicapped by its complexity: in order to implement a Turing machine, the automaton requires a very considerable number of additional states (63), as well as a large number of additional transition rules. This complexity, while still relatively minor compared to von Neumann's universal constructor, is nevertheless too high to be considered for a hardware application. So once again, adapting Langton's loop to fit our requirements proved too complex to be efficient [21].

Self-replication of specialized Turing machines on a multicellular array: The Embryonics approach

Arbib [1] was the first to suggest a true "cellular" automaton, in which every cell contains a complete copy of the genome, and a hierarchical organization, where each cell is itself decomposed into smaller and regular parts, the "molecules"; unlike all previous realizations, this new architecture is a true "multicellular" artificial organism.

This key idea was the basis of the Embryonics (embryonic electronics) project, under development by Mange and his colleagues since 1993, whose ultimate objective is the construction of large-scale integrated circuits, exhibiting properties such as growth, self-repair (healing), and self-replication, found up until now only in living beings [11, 12, 20].

Embryonics features

Essentially, Embryonics is a modified automata-based approach in which three biologically inspired principles are employed: multicellular organization, cellular differentiation, and cellular division. According to the *multicellular organization* feature, the artificial organism is divided into a finite number of cells (Figure 9), where each cell realizes a unique function, described by a subprogram called the *gene* of the cell.

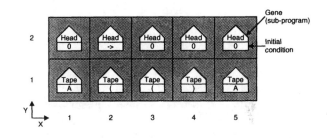

Figure 9: *Multicellular organization of a specialized Turing machine, a parenthesis checker.*

We will confine ourselves to a simple example of a two-dimensional artificial organism (Figure 9): a specialized Turing machine, a parenthesis checker [13], implemented with ten cells and featuring two distinct genes, the *tape gene* and the *head gene*. Each cell is associated with some *initial condition*. In our example the head cells are distinguished by the initial values "0" and "→", the tape cells by "A", "(", and ")" values.

Let us call *genome* the set of all the genes of an artificial organism, where each gene is a sub-program characterized by a set of instructions, by an initial condition, and by a position (its coordinates X, Y). Figure 9 then shows the genome of our Turing machine, with the corresponding horizontal (X) and vertical (Y) coordinates. Let then each cell contain the entire genome (Figure 10): depending on its position in the array, i.e., its place in the organism, each cell can interpret the genome and

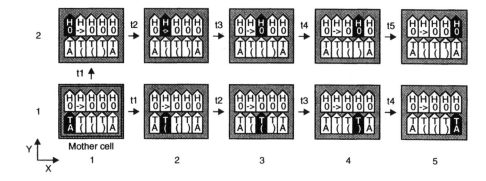

Figure 10: *Cellular differentiation and cellular division of a specialized Turing machine, the parenthesis checker; $t_1...t_5$: five successive divisions.*

extract and execute the gene (with its initial condition) which configures it. According to the *cellular differentiation* feature, it can interpret any gene of the genome (including the initial condition), given the proper coordinates.

At startup, the mother cell or *zygote* (Figure 10), arbitrarily defined as having the coordinates $X, Y = 1, 1$, holds the one and only copy of the genome. At time t_1, according to the *cellular division* feature, the genome of the mother cell is copied into the two neighboring (daughter) cells to the north and to the east. The process then continues until the two-dimensional space is completely programmed. In our example, the furthest cell is programmed at time t_5.

In all living beings, the string of characters which makes up the DNA is executed sequentially by a chemical processor, the *ribosome*. Drawing inspiration from this biological mechanism, we will use a microprogram to compute first the coordinates of the artificial organism, then the initial conditions of each cell, the tape gene and the head gene, and finally the complete genome. The calculation of this microprogram is detailed in [10, 12]. Its software implementation requires basically two kinds of instructions: a *test instruction* (**if** VAR **else** $LABEL$), and an *assignment instruction* (**do** $X = DATA$).

Each artificial cell (called MICTREE for "microinstruction tree") is implemented as an element of a new kind of coarse-grained programmable logic network, which is realized on a special field-programmable gate array (FPGA) circuit. This artificial cell consists basically of a *binary decision machine*, executing the above-mentioned instructions, a random access memory, storing the microprogram of the genome, and several programmable connections linking the cell to its four immediate neighbors (to the north, east, south, and west) [11, 12].

Self-repair and self-replication

In order to demonstrate *self-repair*, we added two spare cells in each row, to the right of the original Turing machine, all identified by the same horizontal coordinate ($X = 6$ in Figure 11). The spare cells may be used not only for self-repair, but also for the example of a Turing machine necessitating growth of the tape of arbitrary, but finite, length.

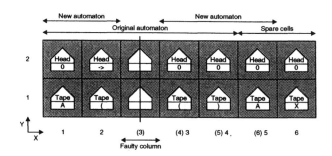

Figure 11: *Self-repair of a 10-cell parenthesis checker in a 14-cell array.*

The existence of a fault is detected by a $KILL$ signal which is calculated in each artificial cell by a built-in self-test realized at the FPGA level. The state $KILL = 1$ identifies the faulty cell and the entire column to which the faulty cell belongs is considered faulty, and is deactivated (column $X = 3$ in Figure 11). All the functions of the artificial cells to the right of the column $X = 2$ are shifted by one column to the right. Obviously, this process requires as many spare columns to the right of the array as there are faulty columns to repair (there are two spare columns in the example of Figure 11). It also implies that the artificial cell has the capability of bypassing the faulty column and shifting to the right all or part of the original cellular array. During such a process, the actual values calculated by the cells are destroyed and the whole calculation should be restarted.

The *self-replication* of an artificial organism rests on two hypotheses:

- there exist a sufficient number of spare cells (unused cells at the upper side of the array, at least ten for our

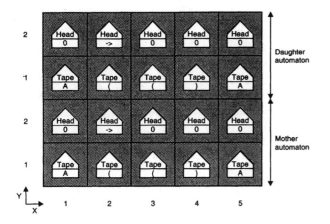

Figure 12: *Self-replication of a 10-cell parenthesis checker in a 20-cell array.*

example);

- the calculation of the coordinates produces a cycle at the cellular level ($Y = 1 \rightarrow 2 \rightarrow 1$ in Figure 12).

As the same pattern of coordinates produces the same pattern of genes (with the initial conditions), self-replication can be easily accomplished if the microprogram of the genome, associated with the homogeneous network of cells, produces several occurrences of the basic pattern of coordinates ($Y = 1 \rightarrow 2$ in Figure 9). In our example, repetition of the vertical coordinate pattern, i.e., the production of the pattern $Y = 1 \rightarrow 2 \rightarrow 1 \rightarrow 2$ (Figure 12), produces one copy, the *daughter automaton*, of the original *mother automaton*. Given a sufficiently large space, the self-replication process can be repeated for any number of specimens in the Y axis (remember that the X axis is reserved for self-repair and/or for a possible growth of the Turing machine).

With a sufficient number of cells, it is obviously possible to combine self-repair (or growth) toward the X direction and self-replication toward the Y direction.

Self-replication of a universal Turing machine on a multicellular array

The preceding section presented a self-replicating two-dimensional artificial organism implementing a specialized Turing machine, the parenthesis checker, which was made of ten MICTREE artificial cells. By using the same type of cells we now show how it is possible to design and build a universal Turing machine (UTM) with self-replication capabilities.

Multicellular architecture of a universal Turing machine

Conventional universal Turing machines [13] consist of a finite but arbitrarily long tape, and a single read/write mobile head controlled by a finite-state machine, which

is itself described on the tape (Figure 2). In order to implement a universal Turing machine in an array of MICTREE artificial cells, we made three fundamental architectural choices (Figure 13):

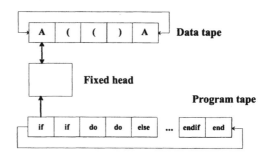

Figure 13: *Universal Turing machine architecture for the parenthesis checker example.*

1. The read/write head is fixed; the tapes are therefore mobile.
2. The data of the given application (the specialized Turing machine to be simulated) are placed in a mobile tape, the *data tape*; this tape can shift right, shift left, or not at all.
3. The finite-state machine for the given application is translated into a very simple program written in a language called PICOPASCAL[1]; each instruction of this program takes place in a square of a second mobile tape, the *program tape*; this tape just needs to shift left.

The fixed head, which is in fact an interpreter of the PICOPASCAL language, has to continuously execute cycles consisting of four operations:

1. reading and decoding an instruction on the program tape;
2. reading a symbol on the data tape;
3. interpreting the current instruction, and writing a new symbol on the current square of the data tape;
4. shifting the data tape (left, right, or none) and the program tape (left).

An application: A binary counter

In order to test our UTM implementation, we used, as a simple but non-trivial example, a binary counter [13], a machine which writes out the binary numbers 1, 10, 11, 100, etc. The counter's state table (Figure 14) has two internal states ($Q \in \{0 \rightarrow, 1 \leftarrow\}$) and two input states ($S \in \{0, 1\}$), S being the value of the current square read

---

[1]PICOPASCAL, itself derived from NANOPASCAL [12], is a minimal subset of PASCAL; the transformation of a state table into such a program is directly inspired by the W-machine with the major advantage of avoiding the jumps necessitated by the **IF 1 THEN** (*n*) **ELSE** (*next*) instructions.

on the data tape. Depending on the present internal state Q and the present input state S, the specialized Turing machine will:

1. write a new binary value $S^+(0,1)$ on the current square of the data tape;
2. move its tape to the right ($Q^+ = 0 \rightarrow$) or to the left ($Q^+ = 1 \leftarrow$), which is equivalent to moving the data tape to the left or to the right, respectively;
3. go to the next state $Q^+(0 \rightarrow, 1 \leftarrow)$.

$Q+,S+$	$S=0$	$S=1$
$0\rightarrow$	$0\rightarrow,0$	$1\leftarrow,1$
$1\leftarrow$	$0\rightarrow,1$	$1\leftarrow,0$
Q		

Figure 14: *State table of the binary counter.*

The PICOPASCAL program equivalent to the state table (Figure 14) is given in Figure 15.

ADDR	DATA	PROGRAM	
00	5	if (Q)	
01	5	if (S)	
02	A	do 0	(S)
03	9	do 1<-	(Q)
04	4	else	
05	B	do 1	(S)
06	8	do 0->	(Q)
07	6	endif	
08	4	else	
09	5	if (S)	
0A	B	do 1	(S)
0B	9	do 1<-	(Q)
0C	4	else	
0D	A	do 0	(S)
0E	8	do 0->	(Q)
0F	6	endif	
10	6	endif	
11	2	end	

Figure 15: *PICOPASCAL program equivalent to the state table of Figure 14.*

An ideal architecture for the universal Turing machine

A universal Turing machine architecture is ideal in the sense that it is able to deal with applications of any complexity, characterized by:

1. a finite, but arbitrarily long data tape;
2. a read/write head able to interpret a PICOPASCAL program of any complexity;
3. a finite, but arbitrarily long program tape.

It must be pointed out that, for any application, the program tape and the read/write head (the PICOPASCAL interpreter) are always characterized by finite dimensions; only the data tape can be as long as desired, as is the case for the binary counter.

An ideal architecture, embedding the current example, but compatible with any other application, is as follows (Figure 16):

1. The data tape, able to shift right, left, or hold, is folded on itself; the initial state is defined by $QL1 : 0, QC, QR0 : 1 = 00100$, where QL are the squares to the left of the central square QC, and QR are squares to the right of QC; the data tape is able to grow to the left of $QC(QL2, QL3, ...)$ and to the right of $QC(QR2, QR3, ...)$.
2. The fixed read/write head, which is not detailed here, is basically composed of a state register Q,S (storing the current values of internal and input states Q,S, respectively, with an initial state $Q, S = 01$) and a stack $ST1 : 3$ characterized by a 1-out-of-3 code (one-hot encoding). At the start of the execution of the PICOPASCAL program (Figure 15, i.e., in address $ADDR = 00$), the stack is in an initial state $ST1 : 3 = 100$; roughly speaking, each **if** instruction will involve a PUSH operation, each **endif** a POP operation, and each **else** a LOAD operation. When $ST1 = 1$, the **do** instructions are executable. The main characteristic of the stack is its scalability: for any program exhibiting n nested **if** instructions, the stack is organized as a $n+1$ square shift register. Both the $ST1 : 3$ stack and the Q, S register are able to grow to accommodate more complex applications.
3. The program tape is folded on itself; it is able to grow to accommodate more complex applications.

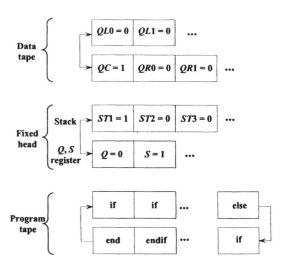

Figure 16: *UTM's ideal architecture.*

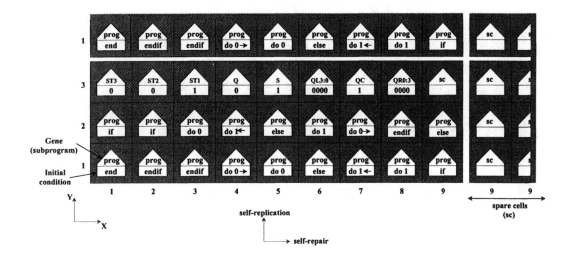

Figure 17: *UTM's actual implementation for the binary counter example on a multicellular array of 27 MICTREE cells.*

An actual implementation of the UTM for the binary counter example

In order to implement the binary counter application with a limited number of MICTREE artificial cells, we have somewhat relaxed the characteristics of the ideal architecture described earlier. Our final architecture is made up of three rows ($Y = 1, 2, 3$) and nine columns ($X = 1...9$) organized as follows (Figure 17):

- The 18 instructions of the PICOPASCAL program (Figure 15) take place in the program tape, using the two lower rows ($Y = 1, 2$) of the array.
- The read/write head is composed of a $ST1 : 3$ stack and the Q, S register ($X = 1...5, Y = 3$), while the data tape is implemented thanks to three cells ($X = 6...8, Y = 3$) displaying 9 bits $QL3 : 0, QC, QR0 : 3$.

In order to demonstrate self-repair, we added spare cells in each row, at the right-hand side of the UTM, all identified by the same horizontal coordinate ($X = 9$ in Figure 17). As previously mentioned, more cells may be used not only for self-repair, but also for a UTM necessitating a growth of the tape of arbitrary, but finite, length.

Self-replication rests on two hypotheses:

- there exist a sufficient number of spare cells (unused cells at the upper side of the array, at least $3 \times 9 = 27$ for our example);
- the calculation of the coordinates produces a cycle at the cellular level (in our example: $Y = 1 \rightarrow 2 \rightarrow 3 \rightarrow 1$).

Given a sufficiently large space, the self-replication process can be repeated for any number of specimens in the Y axis. With a sufficient number of cells, it is obviously possible to combine self-repair (or growth) towards the X direction and self-replication towards the Y direction.

Discussion

In this paper we presented a new, true "multicellular" automaton, in which every cell contains a complete copy of the genome; we have shown that such a multicellular automaton is able to self-replicate and to self-repair.

We then showed that it is possible to embed a universal Turing machine in such a multicellular array, thus obtaining a self-replicating and self-repairing universal Turing machine.

The mapping of the universal Turing machine onto our multicellular array was made possible thanks to the introduction of a modified version of the W-machine, i.e., an interpreter of the PICOPASCAL language. We showed that an ideal architecture was able to deal with applications of any complexity, i.e., with a semi-infinite data tape. We also presented an actual implementation in which we relaxed somewhat the characteristics of the ideal architecture in order to use a limited number of MICTREE artificial cells. We slightly simplified our implementation by presenting the example of the binary counter in which the data are binary (in general we might have discrete values) and where the direction of the head's moves coincides with the internal state (in general functions Q^+ and D^+ are independent).

The property of universal construction raises issues of a different nature, since it requires (according to von

Neumann) that a MICTREE cell be able to implement organisms of any dimension. This challenge can be met by decomposing a cell into molecules and tailoring the structure of cells to the requirements of a given application [11].

Acknowledgments

This work was supported in part by grant 21-54113.98 from the Swiss National Science Foundation, by the Consorzio Ferrara Richerche, Università di Ferrara, Ferrara, Italy, and by the Leenaards Foundation, Lausanne, Switzerland. We thank Barry McMullin and the other two anonymous reviewers for their helpful remarks.

References

[1] M. A. Arbib. *Theories of Abstract Automata.* Prentice-Hall, Englewood Cliffs, N.J., 1969.

[2] E. R. Banks. Universality in Cellular Automata. In *IEEE 11th Annual Symposium on Switching and Automata Theory*, pages 194–215, Santa Monica, California, October 1970.

[3] A. Burks, editor. *Essays on Cellular Automata.* University of Illinois Press, Urbana, Illinois, 1970.

[4] E. F. Codd. *Cellular Automata.* Academic Press, New York, 1968.

[5] G. T. Herman. On Universal Computer-Constructors. *Information Processing Letters*, 2(3):61–64, August 1973.

[6] J. G. Kemeny. Man Viewed as a Machine. *Scientific American*, 192:58–68, April 1955.

[7] C. G. Langton. Self-reproduction in cellular automata. *Physica D*, 10:135–144, 1984.

[8] C. Lee. Synthesis of a Cellular Computer. In J. J. Tou, editor, *Applied Automata Theory*, pages 217–234. Academic Press, London, 1968.

[9] C. Y. Lee. Automata and Finite Automata. *Bell System Tech. Journal*, XXXIX:1267–95, 1960.

[10] D. Mange, D. Madon, A. Stauffer, and G. Tempesti. Von Neumann revisited: A Turing machine with self-repair and self-reproduction properties. *Robotics and Autonomous Systems*, 22(1):35–58, 1997.

[11] D. Mange, M. Sipper, A. Stauffer, and G. Tempesti. Towards robust integrated circuits: The embryonics approach. *Proceedings of the IEEE*, April 2000. to appear.

[12] D. Mange and M. Tomassini, editors. *Bio-Inspired Computing Machines: Towards Novel Computational Architectures.* Presses Polytechniques et Universitaires Romandes, Lausanne, Switzerland, 1998.

[13] M. L. Minsky. *Computation: Finite and Infinite Machines.* Prentice-Hall, Englewood Cliffs, New Jersey, 1967.

[14] F. Nourai and R. S. Kashef. A universal four-state cellular computer. *IEEE Transactions on Computers*, c-24(8):766–776, August 1975.

[15] J.-Y. Perrier, M. Sipper, and J. Zahnd. Toward a Viable, Self-Reproducing Universal Computer. *Physica D*, 97:335–352, 1996.

[16] U. Pesavento. An implementation of von Neumann's self-reproducing machine. *Artificial Life*, 2(4):337–354, 1995.

[17] J. Signorini. Complex Computing with Cellular Automata. In P. Manneville, N. Boccara, G. Y. Vichniac, and R. Bidaux, editors, *Cellular Automata and Modeling of Complex Physical Systems*, volume 46 of *Springer Proceedings in Physics*, pages 57–72. Springer-Verlag, Heidelberg, 1990.

[18] M. Sipper. Fifty Years of Research on Self-Replication: An Overview. In M. Sipper, G. Tempesti, D. Mange, and E. Sanchez, editors, *Artificial Life*, volume 4, pages 237–257, Cambridge, Massachusetts, 1998. The MIT Press.

[19] G. Tempesti. A New Self-Reproducing Cellular Automaton Capable of Construction and Computation. In F. Morán, A. Moreno, J. J. Merelo, and P. Chacón, editors, *ECAL'95: Third European Conference on Artificial Life*, volume 929 of *Lecture Notes in Computer Science*, pages 555–563, Heidelberg, 1995. Springer-Verlag.

[20] G. Tempesti, D. Mange, and A. Stauffer. Self-Replicating and Self-Repairing Multicellular Automata. *Artificial Life*, 4(3):259–282, 1998.

[21] Gianluca Tempesti. *A Self-Repairing Multiplexer-Based FPGA Inspired by Biological Processes.* PhD thesis, Swiss Federal Institute of Technology, Lausanne, Switzerland, 1998.

[22] A. M. Turing. On Computable Numbers, with an Application to the Entscheidungsproblem. *Proceedings of the London Math. Soc.*, 42:230–265, 1936.

[23] J. von Neumann. *Theory of Self-Reproducing Automata.* University of Illinois Press, Urbana, Illinois, 1966. Edited and completed by A. W. Burks.

[24] H. Wang. A Variant to Turing's Theory of Computing Machines. *Journal of the ACM*, IV:63–92, 1957.

Creating a Physically-based, Virtual-Metabolism with Solid Cellular Automata

Alan Dorin

School of Computer Science & Software Engineering
Monash University, Clayton, Australia 3168
aland@cs.monash.edu.au

Abstract

A physically-based system of interacting polyhedral objects is used to model physical and chemical processes characteristic of living organisms. These processes include auto-catalysis, cross-catalysis and the self-assembly and spontaneous organization of complex, dynamic structures constituting *virtual organisms*. The polyhedra in the simulation are surfaced with bonding sites in states akin to those of cellular automata. These bonding sites interact with sites on neighbouring polyhedra to apply forces of attraction and repulsion between bodies and to trigger transitions in their states. Locally controlled assembly of this kind acts without the guidance of an external agent or central control. Such mechanisms, perhaps a defining property of biological construction, are seldom employed to create complex artificial structures. This paper therefore presents a novel model for the construction of complex virtual structures using multiple reactive, virtual elements, acting independently under virtual physical and chemical laws.

Introduction

At the microscopic level, the components of an organism are continually and co-operatively acting to maintain the organism's physical identity. Organisms are self-assembling, parallel machines whose many and varied components maintain a stable organization under perturbation. This is achieved through the local physical and chemical interactions of the individual components. This paper presents a virtual system which operates on the same principles.

Models of some typical biological structures have been created using a physically-based system of interacting elements called *Self-Organizing (Solid) Cellular Automata* (SOCA) [Dorin 98, 99]. Models of chains and polymers, liposome-like clusters and membranes and a model of cluster reproduction by locally-induced fracture were demonstrated in these earlier publications. Using this simulation framework, new models of catalysis and auto/cross-catalysis are presented here. These models are then combined to create a *virtual organism*, which self-assembles from a set of simulated cross-catalytic reactions. This virtual organism has a simple virtual metabolism and reproduces by fracture in a manner reminiscent of the reproduction of single cells.

Various related fields of study, as well as some proposed areas of application for SOCA are briefly listed below. A summary of the SOCA system follows. The next section serves as an introduction to chemical reactions and catalysis. A description of SOCA models of irreversible chemical reactions and simulations of catalysis, auto-catalysis and cross-catalysis follow. Finally a virtual organism which self-assembles via the interactions of SOCA elements engaged in virtual cross-catalysis is presented. Proposals for future work and conclusions are then given. Below is a brief summary of material relevant to the present study.

Physical Simulation - Researchers in computer graphics and robotics model solid/fluid interactions for the purpose of visualizing simulated physical processes (e.g. [Sims 94, McKenna & Zeltzer 90]). Similarly, these techniques are applied in the solid cellular automata simulation framework.

Cellular Automata (CA) - CA's have been widely studied as examples of complex dynamical systems [Gardner 71, Wolfam 84], under the banner of artificial life [Langton 86], and as examples of components in a self-reproducing machine [Burks 70, Gardner 70]. The global behaviour, said to be *emergent* [Cariani 91] from local interactions, shares features with that of the processes which maintain an organism [Dorin 96]. This was considered vital in the implementation of a model of life, hence the inclusion of its properties in the SOCA model.

Philosophy of Biology - The work of [Maturana & Varela 80], [Kauffman 93] and [Prigogine 85] to describe the organization of living things may be summarized thus: A living thing is the matter contained within a space defined by a set of chemical processes which produce the components of which they themselves are constructed. An organism is a network of self-sustaining, self-bounding, auto/cross-catalytic chemical processes. These are taken here to be essential traits of any living thing, and hence necessary features in a model organism with any claim to *virtual* life. The model organism presented below was built with these considerations in mind.

Self-Organization/Assembly – [Penrose 59], [Ingber 98], [Fleischer 95] have presented self-assembling wooden machines, physical models of the mechanical properties of cells and software models of the development of multi-celled organisms respectively. These studies are all similar in spirit to that presented here.

Other authors have explored similar systems including a mechanical system of blocks and magnets [Hosokawa et al 95] and a model of the self-assembly of the T4 bacteriophage [Goel & Thompson 98]. [Steels 95] investigates the development of language treating it as a self-organizing system. [Banzhaf 94] uses binary strings to explore auto-catalysis and metabolic formation. [Saitou & Jakiela 95a,95b] have examined the ordering of sub-assembly processes within larger scale construction tasks.

Reactive, Distributed Artificial Intelligence - Attempts have been made to utilize the emergent properties of interacting reactive agents [Drogoul & Dubreuil 92, Ferber & Jacopin 91]. The agents in these works aim to satisfy their own goals. The result of their local interactions is a global stable state in which all agents are satisfied and a solution to the problem at hand is found.

Molecular Dynamics and Supra-molecular Inorganic Chemistry - Papers in supra-molecular chemistry grapple with self-assembly [Lawrence et al 95]. The interactions of groups of molecules may be visualized according to the shapes they form and the bonding sites these present to their surroundings. Molecules have characteristically arranged bonding sites which link to other molecules to form large *supra-molecules*. It is helpful to visualize molecules as polyhedra whose vertices are bonding sites [Muller et al 95], the resulting supra-molecules are visualized as organized collections of polyhedra[1]. SOCA are an ideal choice for models of these phenomena.

The SOCA System

The elements in the SOCA system are simulated, rigid, convex polyhedra suspended in a fluid. The densities and geometric properties of the elements may be specified. Collisions are detected and analytically-derived impulses prevent bodies from interpenetrating in a manner similar to that of [Baraff 89].

The fluid model incorporates viscosity acting on the bodies as fluid drag. For the purposes of this paper the fluid is stationary at large scales. Small scale Brownian motion helps 'jiggle' the elements in the fluid into stable states. In keeping with the fluid models found in [Dorin 94, Wejchert & Haumann 91], effects of solids on the fluid medium are ignored, only effects of the fluid on solids within it are considered.

In addition to the above properties, the faces of each element have a state visualized as their colour. This is analogous to the state of a CA cell. Lookup tables stored with each element dictate the behaviour of a face depending on its current state and the state of the faces on *other* elements within its vicinity.

Faces may apply forces to the bodies on which they lie in response to the presence or absence of faces on other elements in particular states. The forces act at the center of the face in question and therefore provide linear and

---

[1]The models described in the supra-molecular chemistry literature visualize bonding sites at the vertices of the polyhedra. This scheme was not adopted for the experiments described here although modification of the SOCA software to emulate the chemist's approach is trivial.

angular acceleration to the element on which they lie. The scale of the force generated is determined using a lookup table particular to the element on which the face resides. This value may be scaled according to the surface area of the face and the range over which the neighbouring face is detected.

There exists a special *inert* state similar to the quiescent or background state of CA's. An inert face does not interact with any faces around it.

Here is a sample force table from [Dorin 98] intended to model the interactions of ferrous bar-magnets.

	inert	blue	green
inert	0.0	0.0	0.0
blue	0.0	-1.0	+1.0
green	0.0	1.0	-1.0

Tab1 Sample force table

A positive force indicates a direction *towards* the neighbouring face (attraction), a negative force a direction *away* from the neighbouring face (repulsion). Interactions between inert faces and any other face always result in a force of 0. (Therefore inert faces are usually omitted from force tables altogether.)

When a blue face on a solid element (read down the left column of the table) encounters a green face on a neighbouring element (read across the top row of the table), a force acting at the centre of the blue face is generated (read from the cell where the relevant row and column intersect) in a direction towards the centre of the green face [Fig1] (and vice versa since elements in this example have identical symmetrical transition tables). The strength of the force is attenuated across space according to an inverse-square law.

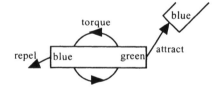

Fig1 Torque on simulated bar magnet

Besides generating a force, a face may undergo a change of state. This is triggered, like the application of forces, by the presence or absence of faces on other elements in particular states and at particular distances. It is not possible for a face on one polyhedron to *control* the behaviour of a face on another, it may only *trigger* a response dictated by the *transition* table of the neighbouring body.

A sample transition table [Tab2] from [Dorin 98] appears below. It was used to construct organized collections of elements which reproduced by fracture. (I.e. when they reach a threshold size they split to form two collections, each of which then continues to grow until reaching the threshold and once again splitting. Details are provided in the earlier paper.)

	green	blue	red
green	8,red,0	-	1,red,0
blue	-	-	-
red	-	-	10,green,0

Tab2 Transition table for fracturing cluster formation
(tuple: threshold, transition, priority)

The table states that if a green face (read down left column) encounters green faces on other elements (read across top row) the interaction specified in the table (in the cell where the relevant row and column intersect) will become active.

This green/green interaction states that if the current green face simultaneously encounters eight green neighbours, it will change state to red. The third value in the cell is a priority for the interaction used to select from a set of interactions whose requirements are met simultaneously. This feature is not used in the experiments here and is omitted from the tables below.

The table above also specifies interactions for changing a green face in the presence of one red neighbour to a red face, and a red face in the presence of ten red faces to a green face.

A force table and transition table is specified for each element in the SOCA system. To date, all experiments have been performed using elements with identical force and transition tables. This simplifies the specification of the transitions required to obtain a desired outcome.

For the following experiments, the face of an element may be in one of the states *yellow, magenta, cyan, blue, green, red* or *grey* (after the colours in which they are drawn) or in the *inert* state. The strength of each applied force is here independent of the surface area of the face on which it acts. The density of all elements is identical unless otherwise stated.

The geometry of the elements used in the following examples was arrived at after some experimentation but is not mandatory. The system is sensitive to changes in dimensions of its elements. Largely this is because short or small elements are less stable than long elements under forces tending to align them in specific directions. This is due to the reduced lever arm available on a short element for the application of torque. Inter-element collisions in these simulations are elastic (the coefficient of restitution, epsilon, equals one).

Modelling Chemical Systems

A simple model reaction may be simulated between 100 randomly placed cubic SOCA elements. Each of these may be given five grey sides and one coloured side which, to begin with, is either red or blue (50% each of red/blue). These elements shall be referred to as blue-faced and red-faced respectively, similarly for other elements with single coloured faces.

The reaction red + blue => cyan + magenta may be simulated using the transition table [Tab3] and an empty force table. The transition table contains entries which convert a blue-faced element to a cyan one in the presence of red, and a red-faced element to a magenta one in the presence of blue.

	grey	red	blue	cyan	magenta
grey	-	-	-	-	-
red	-	-	1,mag.	-	-
blue	-	1,cyan	-	-	-
cyan	-	-	-	-	-
magenta	-	-	-	-	-

Tab3 Transition table: red + blue => cyan + magenta

A graph showing the concentrations of red and cyan-faced elements for this irreversible process is given [Fig2]. Concentrations of red and cyan elements are drawn. Magenta and blue concentrations are symmetrical to these. In this simulation, elements are initiated with random velocities. Their reaction only occurs when the blocks are within each other's predetermined *neighbourhood region*. This depends on chance movements since no inter-element forces act in this run.

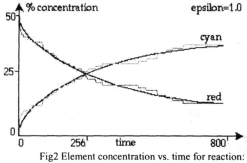

Fig2 Element concentration vs. time for reaction:
red + blue => magenta + cyan

This graph is the expected outcome for a simple irreversible chemical reaction in which the reactants are being gradually consumed [Dorin 99]. The most notable feature is the intersection of the curves after 256 time steps. This point of equal concentration of cyan and red will be compared with the results in the following section.

Catalysis

A catalyst is a substance which binds substrate molecules (reactants) to its surface. The substrate molecules are held proximate to one another and in the correct orientation for them to interact to form the reaction's products [Beck et al. 91, p178]. Once the bound molecules have reacted, the catalyst is freed in an unaltered state allowing it to bind more substrate.

A virtual catalyst to drive the red+blue=>magenta+cyan reaction must bind to red and blue faces, bringing them together so that they can react. This may be achieved using a cubic element with two adjacent (not opposite) binding sites, green and yellow, which attract red and blue faces respectively, but repel cyan and magenta faces so as to free itself once the reaction has occurred. Here are the force and transition tables for the system [Tab4].

	grey	green	blue	red	cyan	mag.	yellow
grey	0	0	0	0	0	0	0
green	0	0	0	+500	0	-500	0
blue	0	0	0	0	0	0	+500
red	0	+500	0	0	0	0	0
cyan	0	0	0	0	0	0	-500
mag.	0	-500	0	0	0	0	0
yellow	0	0	+500	0	-500	0	0

Tab4a Force table for catalyzed reaction
red + blue =>cyan + magenta

	grey	green	blue	red	cyan	mag.	yellow
grey	-	-	-	-	-	-	-
green	-	-	-	-	-	-	-
blue	-	-	-	1,cyan	-	-	-
red	-	-	1,mag.	-	-	-	-
cyan	-	-	-	-	-	-	-
mag.	-	-	-	-	-	-	-
yellow	-	-	-	-	-	-	-

Tab4b Transition table for catalyzed reaction
red + blue => cyan + magenta

The figure [Fig3] shows the randomly placed elements of one virtual catalyst (green/yellow) and reactants (red and blue-faced elements). Frame two shows the red face having been pulled towards the green face of the now re-oriented catalyst. Likewise the blue face is approaching the catalyst's yellow face. Frame three depicts the newly created cyan and magenta products, brought about by the combination of the red and blue model substrate bound to the catalyst. In this frame, the yellow face of the catalyst has rotated out of view due to the repulsion between it and the cyan face. As can be seen also, the green face has rotated away from the magenta face due to the repulsive force between them.

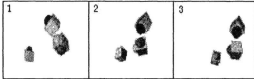

Fig3 Images from a catalyzed irreversible reaction

A run of the reaction commencing with 50 red-faced, 50 blue-faced and 10 yellow/green catalyst elements produced concentration plot [Fig4]. The virtual catalyst clearly speeds up the production of magenta and cyan from red and blue. The point at which equal concentrations of all elements was reached with the catalyst present appears at 150 simulation time units. The un-catalyzed reaction required 256 time units [Fig2] before reaching this point.

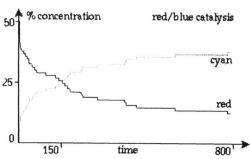

Fig4 Element concentration vs. time for catalyzed reaction red + blue => magenta + cyan

Auto-catalysis

A process of *auto-catalysis* may be illustrated as shown [Fig5] after [Prigogine & Stengers 85, p134]. This represents the production of (more) B from A, in the presence of B. The process is an example of a *positive feedback loop*, in conventional notation: A+B=>B+B.

Fig5 Auto-catalytic loop

Auto-catalysis may be modelled using solid cellular automata elements which commence with two green end caps and four grey sides each. Here are the force and transition tables for the model involving the conversion of such elements into polyhedra with blue end caps replacing the original green ones [Tab5].

	grey	green	blue
grey	-1	-1	-1
green	-1	0	+100
blue	-1	+100	-100

Tab5a Force table for auto-catalysis

	grey	green	blue
grey	-	-	-
green	-	-	1, blue
blue	-	-	-

Tab5b Transition table for auto-catalysis

These tables state that a green face in the presence of a blue face will undergo a transition to create another blue face. The blue catalyst has an affinity for green faces and vice versa. Green faces are not attracted to, nor repelled from each other. But blue catalyst *is* repelled from others of its kind.

The conversion of green faces to blue at first proceeds slowly, but as more catalyst is produced and dispersed by repulsion through the collection of raw green material, the reaction rate increases until the SOCA space contains nothing but dispersed blue-faced elements [Fig5]. (Some of the blue elements have been dispersed beyond the camera's view in frame four).

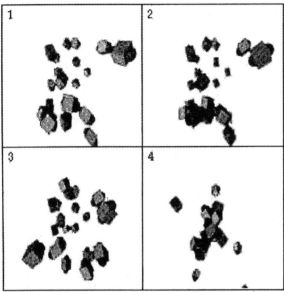

Fig5 Images from an auto-catalytic reaction

Cross-catalysis

The process of cross-catalysis is modelled in this section. Also, what is sought in this section is a virtual chemical process which is not only self-sustaining, but forms and maintains a dynamic, recognizable virtual structure. As indicated in the introduction, this would constitute an example of virtual life.

The SOCA elements of this experiment are cubes with faces in any of the states: inert, grey, green, yellow, blue, red, magenta, cyan or black. The simulation commences with elements randomly oriented and positioned.

Cross-catalysis involves two reactions which occur simultaneously, each producing as a product a catalyst for the other. This may be written in terms of SOCA face-states:

red   + blue => inert + yellow   (catalyzed by green)
magenta + cyan => inert + green   (catalyzed by yellow)

Hence a red face is converted to an inert face in the presence of a blue face. The blue face is simultaneously converted to a yellow face. Similarly, the magenta face is converted to inert and the cyan face which causes this is converted to green. A green face catalyzes the red/blue interaction, and a yellow face catalyzes the magenta/cyan interaction. The catalysis operates as described earlier.

A system commencing with a 25% mixture of elements,

each with one face coloured red, blue, cyan or magenta, will approach a concentration of 50% inert elements, and 25% each of yellow and green-faced elements. The further each reaction goes, the more catalyst it produces for the other reaction in the pair, which in turn produces more catalyst for the first, until the supply of reactants is exhausted. Hence this simulation combines auto-catalysis and catalysis of an irreversible reaction. Each reaction produces catalyst for its own acceleration indirectly by accelerating the other process in the pair.

Cross-catalysis may be established using the SOCA transition and force tables [Tab6]. Unlike the catalyst used with the irreversible reaction, (which had yellow and green faces to bring blue and red faces into proximity), the catalyst in the current reaction has a single face which attracts both reactants. I.e. the green-faced catalyst attracts *both* blue and red, and the yellow-faced catalyst attracts *both* cyan and magenta. This is not a necessary feature of the catalyst, as has been shown, but it does reduce the complexity of the force and transition tables.

	grey	green	blue	red	cyan	mag.	yellow
grey	0	0	0	0	0	0	0
green	0	0	+500	+500	0	0	0
blue	0	+500	0	0	0	0	0
red	0	+500	0	0	0	0	0
cyan	0	0	0	0	0	0	+500
mag.	0	0	0	0	0	0	+500
yellow	0	0	0	0	+500	+500	0

Tab6a Force table for cross-catalysis

	grey	green	blue	red	cyan	mag.	yellow
grey	-	-	-	-	-	-	-
green	-	-	-	-	-	-	-
blue	-	-	-	1,yell.	-	-	-
red	-	-	1,inert	-	-	-	-
cyan	-	-	-	-	-	1,green	-
mag.	-	-	-	-	1,inert	-	-
yellow	-	-	-	-	-	-	-

Tab6b Transition table for cross-catalysis

The virtual cross-catalytic reactions proceed more and more rapidly once the first random collision between reactants creates a catalyst. An amount of catalyst can even be included at the start of the simulation to kick-start the reactions. In either case, the conversion of reactant into product occurs in a manner like that described for auto-catalysis. The process is not quite as rapid as in the earlier case, due to the need for separate elements to be pulled towards the catalyst.

It was also found that the inert by-product of the reactions initially remains near the catalyst where it interferes with the catalysis by colliding with approaching raw material. This may be responsible for slowing down the conversion of raw material to some extent, but the degree to which this is true has not been tested. If interference from inert bodies is seen as a problem, they can be removed from the system by any of the means discussed in [Dorin 99].

This simple cross-catalysis offers little for discussion beyond what has already been presented on catalysis and auto-catalysis. However at a simple level, the cross-catalytic reactions operate in the manner of a virtual organism. They require raw material to be fed into the system (red, blue, cyan, magenta), which is acted upon to produce (inert) waste material and material for the maintenance of the same set of processes, the catalyst (yellow and green). What is lacking is the assembly and maintenance of a recognizable topology for the virtual organism. This is now addressed.

Topology and Cross-catalysis

A suitable topology for the solid cellular automata is a cluster created in [Dorin 98]. This is a robustly assembled and easily maintained formation in which a collection of elements have one of their faces pulled towards the group's center by mutual attraction. The opposite faces are forced by mutual repulsion to fan out radially, away from the group's center. Using the transition table [Tab2], such clusters may be made to fracture like dividing cells. This property will be useful for creating and maintaining a dynamic topology in a virtual organism.

The assembly of a cluster from cross-catalysis by-products like those of the previous section is achieved by introducing coloured faces opposing the blue and cyan faces on the elements which have them. Each blue or cyan-faced element is given a grey face on the side opposite the pre-existing coloured face. The force table given shortly [Tab7] is set up for mutual attraction between grey faces. This attraction does not accelerate the red/blue interaction because the red elements do not have grey back-faces. Similarly, the magenta/cyan interaction is not accelerated by the grey/grey attraction.

The grey/grey attraction does however pull blue and cyan-faced elements together. They usually meet with their grey faces pointing towards one another. When each of these elements meets the other reactant in its pair, as specified by the cross-catalytic reactions, its blue or cyan face is converted to yellow or green accordingly. The grey back-faces are unmodified by this interaction so what remains is a pair of catalysts (one green and one yellow), bound at their back by mutual attraction between grey faces.

If individual blue or cyan-faced elements meet others which have already been converted to catalysts, their grey faces are still attracted to the catalyst's grey back-face. The elements sit beside one another until the blue or cyan face interacts with a red or magenta face to become a catalyst itself. Once again, two catalysts (one yellow, one green), are left back to back in a cluster.

Sometimes several green, yellow, blue or cyan elements form a cluster. Nevertheless, the same process unfolds during which these are converted into a cluster of yellow and green catalysts. The more green catalysts there are in a cluster, the more this cluster will accelerate the local production of yellow catalyst, and vice versa. Hence clusters form with approximately equal numbers of green and yellow-faced catalyst elements.

This then is the formation of a recognizable topology, but it is not yet a dynamic topology. Organisms continually consume material from the environment, convert it to the material they need to build themselves and discard any waste material. The virtual structure as described simply 'grows'. It consumes red, blue, cyan and magenta from the environment, and produces inert waste. But the virtual body of this organism continually accrues yellow and green catalyst. Once material has formed a part of the virtual body it is not removed or exchanged for fresh material.

The final addition to the model is the specification of an additional transition. When a grey face has three grey neighbours, it is converted to an inert face. This breaks the bond of a catalyst to its cluster and it is pulled from the group by any neighbouring reactant, or is left to drift freely about the space. Alternatively, by introducing weak repulsive forces between yellow and green faces (these must not be strong enough to break the bond formed by their opposing grey faces), the yellow or green catalyst with an inert back-face will be forced out of the cluster to be replaced by a fresh yellow or green element with a grey back-face, as it is produced.

Although this scheme gives the desired dynamic topology, it also has an unwanted side-effect. If three blue or cyan elements come into proximity due to attraction between their grey back-faces, these grey faces may be converted into inert faces and the blue or cyan elements can never form a part of a catalyst cluster. Thus additional waste is produced along with the catalyst which participates in the virtual organism.

To counter this, and to answer a question frequently posed by those who believe that reproduction is a necessary feature of life, an additional transition has been added to the model. The transition and force tables for the model incorporating all of the above interactions as well as the new transition are given [Tab7].

The extra transition requires that when a grey element has four grey neighbours, instead of being transformed to an inert face, it is converted to a black face. This black face is repelled from all grey faces, but attracted to other black faces. In addition, the black face will undergo a transition back to grey if it comes within range of five black neighbours.

	grey	green	blue	red	cyan	mag.	yellow	black
grey	+500	0	0	0	0	0	0	-100
green	0	-100	+500	+500	0	0	-100	-100
blue	0	+500	0	0	0	0	0	0
red	0	+500	0	0	0	0	0	0
cyan	0	0	0	0	0	0	+500	0
mag.	0	0	0	0	0	0	+500	0
yellow	0	-100	0	0	+500	+500	-100	-100
black	-100	-100	0	0	0	0	-100	+500

Tab7a Force table for cross-catalysis/dynamic topology

	grey	green	blue	red	cyan	mag.	yellow	black
grey	4,black	-	-	-	-	-	-	-
green	-	-	-	-	-	-	-	-
blue	-	-	-	1,yell.	-	-	-	-
red	-	-	1,inert	-	-	-	-	-
cyan	-	-	-	-	-	1,green	-	-
mag.	-	-	-	-	1,inert	-	-	-
yellow	-	-	-	-	-	-	-	-
black	-	-	-	-	-	-	-	5,grey

Tab7b Force table for cross-catalysis/dynamic topology

When four grey (or five black) faces meet up, an element is expelled from the group to start its own cluster. The asymmetry between the numbers of neighbours required to convert between grey and black (i.e. four grey and five black faces, instead of four of each) prevents repeated oscillations between the black and grey states by a group of elements, each with the requisite number of neighbours.

A rendering of the present experiment is reproduced [Fig6]. Two collections of catalyst are numbered. These clusters of grey and black-backed yellow and green elements are easily spotted amongst the other inert, blue, red, cyan and magenta material.

Fig6 Dynamic, fracturing clusters, formed by cross-catalysis

Clusters one and two arose from the same roots. Initially a single large cluster grew as all elements with grey faces were attracted towards a central location. This group exceeded its quota of grey elements and fractured. The figure clearly shows cluster one undergoing this process again as the number of elements in its body exceeds the quota permitted by the transition table.

Using the transition and force tables as they stand, there is no waste catalyst, it may all participate in the construction of virtual life. The catalyst is regularly expelled from a given structure and replaced by new material. Hence the clusters' topologies are dynamic.

Material is consumed in the production of the yellow and green building blocks. These blocks operate for awhile within a structure before removal and replacement. Waste produced by the bodies is left to float about the space. The structures *metabolize* in a simple (virtual) sense.

Additionally, although it is *not* required for a model of life, the structures are able to reproduce. They do this by fracture during which their virtual metabolism is not interrupted, as is true of real biological reproduction and metabolism. This example of virtual life, a representation of certain aspects of real life, is complete.

Conclusion and Future Work

A physical model of interacting molecules has been presented. The model contains virtual polyhedra with faces in states akin to those of cellular automata. These states may be used to govern the interactions between elements and so simulate various chemical reactions.

Presented in this paper were simulations of irreversible reactions, catalyzed irreversible reactions as well as auto and cross-catalysis. These simulations were amalgamated to produce a model of a simple virtual organism capable of self-assembly and maintenance. The model was also capable of reproduction by fracture whilst maintaining its virtual metabolism.

The construction of more complex static structures using SOCA has been described in [Dorin 98]. Dynamic structures which behave in ways similar to *gliders* and *spinners* from Conway's *Game of Life* have also been produced [Dorin 99]. In the future it is hoped to combine all of these models to construct virtual organisms with more complex physical forms capable of simple locomotion in a manner similar to that of a glider.

Acknowledgments

The author would like to thank Damian Conway for his supervision during the completion of the PhD from which this work arose. Also due thanks are those who provided practical and philosophical suggestions and additional information during the candidature including: David Baraff, Kurt Fleischer, Kevin Korb, Humberto Maturana, Jon McCormack, Craig Reynolds, Karl Sims.

"The chief difficulty Alice found at first was in managing her flamingo: she succeeded in getting its body tucked away, comfortably enough, under her arm, with its legs hanging down, but generally, just as she had got its neck nicely straightened out, and was going to give the hedgehog a blow with its head, it *would* twist itself round and look up in her face, with such a puzzled expression that she could not help bursting out laughing", Carroll, L. *Alice's Adventures In Wonderland* parallel Alan's Adventures with SOCA.

References

Banzhaf W., 1994. Self-Organization in a System of Binary Strings, In Proceedings *Artificial Life 4*, Brooks, Maes (eds), MIT Press, 109-118

Baraff D., 1989. Analytical Methods for Dynamic

Simulation of Non-Penetrating Rigid Bodies. *Computer Graphics*, Vol23, No3, ACM Press, July, 223-232

Beck W.S., Liem K.F., Simpson G.G., 1991. Life, An Introduction To Biology. 3rd edn, Harper Collins

Burks A.W., 1970. Essay 1-Von Neumann's Self-Reproducing Automata. In *Essays On Cellular Automata*, Burks (ed), Univ. Illinios Press, 3-64

Cariani P., 1991. Emergence and Artificial Life. In Proceedings *Artificial Life II, SFI Studies in the Sciences of Complexity*, Langton et al (eds), Vol10, 775-798

Dorin A., Martin J., 1994. A Model of Protozoan Movement for Artificial Life. In Proceedings *Computer Graphics International 1994: Insight Through Computer Graphics*, Gigante, Kunii (eds), World Scientific, 28-38

Dorin, A., 1996. Computer Based Life:Possibilities and Impossibilities. In Proceedings *ISIS:Information, Statistics and Induction in Science*, Dowe, Korb, Oliver (eds) World Scientific Press, 237-246

Dorin, A., 1998. Physically-based, Self-organizing Cellular Automata. In *Multi-Agent Systems- Theories, Languages and Applications*, Zhang, Lukose (eds), Lecture Notes in Artificial Intelligence 1544, Springer-Verlag, 74-87

Dorin, A., 1999. The Philosophy and Implementation of Physically-based Visual Models of Artificial Life. Ph.D. diss., School. of Comp. Sci. and Soft. Eng., Monash Univ., Australia

Drogoul A., Dubreuil C., 1992. Eco-Problem-Solving Model: Results of the N-Puzzle. In Proceedings *3rd European Workshop on Modelling Autonomous Agents in a Multi-Agent World Decentralized AI3*, Demazeau, Werner (eds), North-Holland, 283-296

Ferber J., Jacopin E., 1991. The Framework of ECO-Problem Solving, In Proceedings *2nd European Workshop on Modelling Autonomous Agents in a Multi-Agent World, Decentralized AI2*, Demazeau, Muller (eds), North-Holland, 181-194

Fleischer K.W., 1995. A Multiple-Mechanism Developmental Model for Defining Self-Organizing Geometric Structures, PhD diss., Caltech

Gardner M., 1970. Mathematical Games-On Cellular Automata, Self-reproduction, the Garden of Eden and the Game 'Life'. *Sci. American*, Vol223 No4, 112-117

Gardner M., 1971. Mathematical Games-The Fantastic Combinations of John Conway's New Solitaire Game 'Life'. *Sci. American*, Vol224, No2, 120-123

Goel N.S., Thompson R.L., 1998. Movable Finite Automata (MFA):A New Tool for Computer Modeling of Living Systems,. In *Artificial Life, SFI Studies in the Sciences of Complexity*, Langton (ed), Addison-Wesley, 317-339

Hosokawa K., Shimoyama I., Hirofumi M., 1995. Dynamics of Self-Assembling Systems:Analogy with Chemical Kinetics, *Artificial Life*, Langton (ed), Vol1, No4, MIT Press, 413-427

Ingber D.E, 1998. The Architecture of Life. *Sci. American*, January, 30-39

Kauffman S.A., 1993. The Origins of Order- Self-Organization and Selection in Evolution. Oxford Univ. Press, 289

Langton C.G., 1986. Studying Artificial Life with Cellular Automata, *Physica 22D*, North-Holland, 120-149

Lawrence D.S., Jiang T., Levett M., 1995. Self-Assembling Supramolecular Complexes, *Chemical Review*, Vol95, American Chemical Society, 2229-2260

Maturana H., Varela F., 1980. Autopoiesis, The Organization of the Living. In *Autopoiesis and Cognition, The Realization of the Living*, Reidel, 73-140

McKenna M., Zeltzer D., 1990. Dynamic Simulation of Autonomous Legged Locomotion. *Computer Graphics*, Vol24, No4, ACM Press, August, 29-38

Muller A., Reuter H., Dillinger S., 1995. Supramolecular Inorganic Chemistry: Small Guests in Small and Large Hosts. *Chem. Int. Engl.*, Angew (ed), Vol34, VCH Verlag, 2323-2361

Penrose L.S., 1959. Self-Reproducing Machines. *Sci. American*, Vol200, No6, June, 105-114

Prigogine I., Stengers I., 1985. Order out of Chaos. Harper Collins, Chpts 5-6

Saitou K., Jakiela M.J., 1995a. Automated Optimal Design of Mechanical Conformational Switches, *Artificial Life*, Langton (ed), Vol2, No2, MIT Press, 129-156

Saitou K., Jakiela M.J., 1995b. Subassembly Generation via Mechanical Conformational Switches. *Artificial Life*, Langton (ed), Vol2, No4, MIT Press, 377-416

Sims K., 1994. Evolving Virtual Creatures. In Proceedings of *SIGGRAPH 94*, ACM Press, 15-34

Steels L., 1995. A Self-Organizing Spatial Vocabulary, *Artificial Life*, Langton (ed), Vol2, No3, MIT Press, 319-332

Wejchert J., Haumann D., 1991. Animation Aerodynamics, *Computer Graphics*, Vol25, No4, ACM Press, July, 19-22

Wolfram S., 1984. Universality and Complexity in Cellular Automata, *Physica 10D*, North-Holland, 1-35

Self-Replicating Worms That Increase Structural Complexity through Gene Transmission

Hiroki Sayama

New England Complex Systems Institute
24 Mt. Auburn St., Cambridge, MA 02138
sayama@necsi.org

Abstract

A new self-replicating cellular automata (CA) model is proposed as a latest effort toward the realization of an artificial evolutionary system on CA where structural complexity of self-replicators can increase in some cases. I utilize the idea of 'shape encoding' proposed by Morita and Imai (Morita & Imai 1996b) and make the state-transition rules of the model allow organisms to transmit genetic information to others when colliding against each other. Simulations with random initial configuration demonstrate that it is possible that the average length of organisms and the average frequency of brancing per organism both increase, with decreasing self-replication fidelity, and saturate at some constant level. The saturation is caused in part by the fixation of place and shape of organisms onto particular sites. This implies the necessity of introducing some fluidity of site arrangements into the model for further development of evolutionary models using CA-like artificial media.

Introduction

Research on self-replicating patterns on CA was founded by von Neumann (von Neumann 1966) and now is viewed as one of the origins of artificial life research (Marchal 1998). A number of attempts to embody artificial organisms on CA have been conducted so far in this area. They may be categorized into four groups as follows[1]:

(1) Implementation of universal constructors based on von Neumann's self-reproducing automaton, studied in '50s–'70s (von Neumann 1966; Codd 1968; Vitányi 1973; Pesavento 1995)

(2) Search for a minimal system capable of non-trivial self-replication, studied in '80s–'90s (Langton 1984; Byl 1989; Reggia *et al.* 1993; Sipper 1994; Morita & Imai 1996b)

---

[1]See also (Sayama 1998a, Chap.3) and (Sipper 1998).

(3) Addition of other computational capabilities to self-replicators, studied in '90s–present (Tempesti 1995; Perrier, Sipper, & Zahnd 1996; Chou & Reggia 1998)

(4) Realization of emergence and evolution of self-replicators, studied in '90s–present (Lohn & Reggia 1995; 1997; Chou & Reggia 1997; Sayama 1998a; 1998b; 2000)

(1), (2), and (3) are efforts to implement regulated behavior (e.g. construction, self-replication, computation) manually designed according to the designer's idea, while (4) strives to obtain unexpected behavior (e.g. emergence of self-replicators or evolution) that may arise from robust or random state-transition rules.

In terms of category (4), the evolutionary processes so far attained using CA were just a change of the size of self-replicating loops, i.e. either increase (Chou & Reggia 1997) or decrease (Sayama 1998a; 2000) in size of the loops. Whether the complexity-increasing evolution of artificial organisms (the evolution of their ability to do more complicated things) is attainable using CA is an open question originally posed by von Neumann at the beginning of this area (von Neumann 1966; Marchal 1998), which still has been unsolved. One of the reasons for this is that the idea of Langton's self-replicating loop (Langton 1984) that has formed the basis for many succeeding studies requires a simple, square (or rectangular) shape of organisms to enable their replication. To remove this restriction, it is necessary to employ a model much more flexible in terms of the shape of self-replicating organisms.

The work introduced in this article extends this effort to realize a new CA model where complexity of virtual organisms can increase along time. I mainly focus on the possibility of increase in structural complexity of artificial organisms, based on the assumption that any other aspects of complexity such as the function of artificial organisms should stem from their structure. To construct a new CA model, I employ the shape-encoding mechanism proposed by Morita and Imai (Morita & Imai 1996b) that makes a variety of patterns capable of self-replication. I then make the state-transition rules of the model allow organisms to transmit genetic information to others when colliding against each other, which may give rise to their variation.

In the following sections I introduce the design of the new model and demonstrate through simulations that the average length and the average frequency of brancing per organism can both increase in this model. Such processes take place with the decrease of self-replication fidelity due to overcrowding, and always saturate at some constant level. It is suggested that such saturation is caused in part by the fixation of place and shape of patterns onto particular sites, which is an intrinsic limitation of CA that prevents us from creating open-ended evolution there.

Model

The shape-encoding mechanism

The main property to be added to the CA-based self-replication model is the variety in shape of self-replicators. For this purpose, I utilize the shape-encoding mechanism proposed by Morita and Imai (Morita & Imai 1996b), which is a mechanism to let an organism dynamically generate genetic codes from its own phenotypical pattern by self-inspection (Laing 1977). An example of their self-replicating automata is shown in Fig. 1. This worm performs a unique form of self-replication, in which its shape is continuously being encoded into genotype at its tail, and these encoded genes are conveyed to the head and decoded there for construction of its offspring. This is different from other prevalent systems in which genetic information is described in a static form and information flows only from genotype to phenotype.

With the shape-encoding mechanism, a great variety of patterns can self-replicate, which is quite useful for the goals of this study. Since Morita and Imai did not consider collisions among organisms, however, it is necessary to incorporate the state-transition rules for such situations.

Space

The proposed model uses a two-dimensional discrete space, where a virtual organism is represented as a contiguous region composed of mutually connected 'structure cells[2].' A structure cell is a square that has one

[2]Note that the word 'cell' is used here for a particular state as a part of a virtual organism represented on the CA space, while 'site' is used for a substrate unit that composes the space itself.

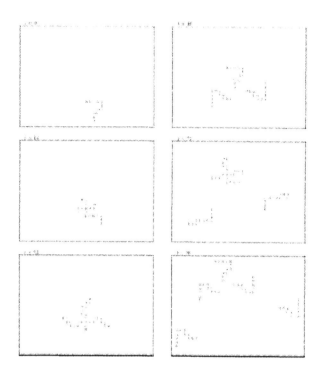

Figure 1: A self-replicating worm with the shape-encoding mechanism by Morita and Imai (from (Morita & Imai 1996b) by courtesy of the original authors).

input port and three output interfaces on its edge, and also three internal 1-bit registers inside itself, as shown in Fig. 2. The internal registers hold information about the existence of Central, Left, and Right genes that sequentially describe how the shape of that organism is formed. They are continuously being updated to have new values coming from the input port, while the old values they previously had are sent out through the output interfaces. The values of internal registers are conveyed successfully if and only if the cell's input port is correctly rooted to one of the output interfaces of another adjacent structure cell. In addition, each entire cell takes either active or passive mode. A passive cell simply conveys genetic information, while an active cell plays more important roles in growth and dissolution of the structure patterns.

The above-mentioned cell property is implemented using sixty-five-state CA with von Neumann neighborhood (five-site neighborhood), where the state each site will take after one update is determined locally according to the states it and its four adjacent sites (upper, lower, right, and left neighbors) have at present. The design of states used here is shown in Fig. 3. The states

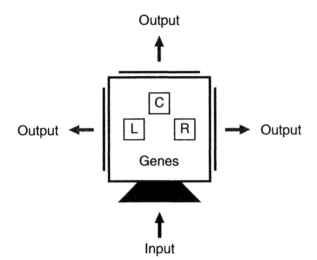

Figure 2: Schematic illustration of a structure cell in the new model. It has one input port, three output interfaces, and three internal 1-bit registers that hold information about the existence of Central, Left, and Right genes. The entire cell takes either active or passive mode.

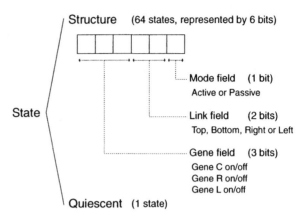

Figure 3: Design of states in the new model.

consist of one quiescent state and sixty-four structure states. The latter are composed of three parts: Mode field (1 bit), Link field (2 bits) and Gene field (3 bits). This implementation is based on the idea of 'multi-data-field CA' (Chou & Reggia 1997) (also known as 'partitioned CA' (Morita & Imai 1996a)) in which the bits that constitute one state are divided into some fields and treated separately. The Mode field stores the cell's current mode, the Link field does its direction, and the Gene field does the values of its internal registers, respectively.

State-transition rules

The state-transition rules used in this model are defined similarly to those of the original model by Morita and Imai, except for modifications made so as to keep the organisms working even in a situation of collision. These rules can be described in words as follows[3]:

- **For quiescent state:**

  - If stimulated by one of the adjacent active structure cells, it will turn into a blank structure cell rooted to the stimulating cell.

  - Otherwise it will remain quiescent.

---

[3]Contact me (sayama@necsi.org) for the complete rule set.

- **For structure states:**

  - **Passive:**
    * If rooted to another structure cell, it will copy that cell's internal register values into its own, then become active if it received at least one gene and if there is no cell rooted to itself, i.e. it is the head of a pattern.
    * If not rooted to any structure cell, i.e. it is the tail of a pattern, it will encode which output interface is linked to other cells into its internal registers and become active.
    * In either case of the above two, if stimulated by one or more of the adjacent active structure cells, it will superimpose these cells' internal register values onto its own, using a logical 'OR' operation.

  - **Active:**
    * If rooted to another structure cell and there is no cell rooted to itself, i.e. it is the head of a pattern, it will become passive and copy the root cell's internal register values into its own. Then, if stimulated by one or more of the adjacent active structure cells, it will superimpose these cells' internal register values onto its own, using a logical 'OR' operation.
    * Otherwise it will become quiescent.

In the above description, 'rooted' means that the cell's input port is correctly linked to one of the output interfaces of another structure cell, and 'stimulated' means that there is an adjacent active structure cell that attempts to make a new structure cell onto that site according to direction by genes.

Note that there is no special rule provided here for cutting off the construction arm of self-replicating

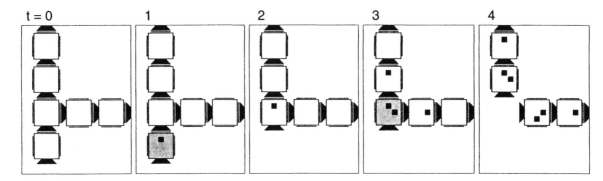

Figure 4: Shape encoding process at the tail of a pattern. The lowermost cell that has no root becomes active (indicated by gray in figures) and encodes which output interface is linked to other cells into its internal registers ($t = 1$). This active cell disappears while the encoded gene is conveyed upperward at the next time ($t = 2$). Such a process takes place repeatedly ($t = 3, 4$).

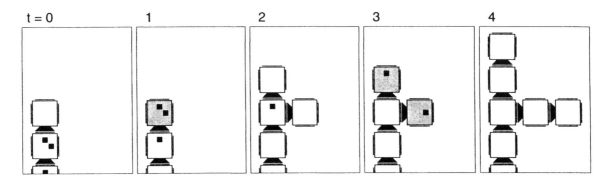

Figure 5: Shape decoding process at the head of a pattern. The uppermost cell to which no cell is rooted becomes active at the arrival of genes ($t = 1$). This active cell stimulates its adjacent quiescent sites to let them turn into blank structure cells according to direction by the genes ($t = 2$). Such a process takes place repeatedly ($t = 3, 4$).

loops that were often added to state-transition rules in earlier models. Since the primary motivation of this study is to extend the flexibility in shape of self-replicators, eliminating the constraint of loop structures is essential.

Behavior

Microlevel behavior

The shape-encoding/decoding behavior under the rules defined above are shown in Fig. 4 and 5. Shape encoding or decoding takes place in a cell which momentarily becomes active and determines the appearance of newly growing structure cells or its own disappearance.

The most interesting innovation introduced by this model, compared to earlier models, is that collisions of organisms lead to gene transmission beyond their boundaries, instead of irregular behavior or structural

dissolution. This process is depicted in Fig. 6. Such interaction of phenotype, which directly affects the genotype, is not found for sophisticated life forms such as eukaryotic organisms including human beings. Bacteria Concerning the beginning of life, however, it may have been an important source of variation in driving evolution of primitive life forms born with so small complexity that there was no distinct line between genotype and phenotype.

Self-replication

A variety of patterns can replicate themselves in the proposed model due to the shape-encoding mechanism. One of the simplest self-replicating organisms in this model is shown in Fig. 7. It occupies only two sites in the space. It is remarkable that even such small organisms self-replicate through the interaction of genotype

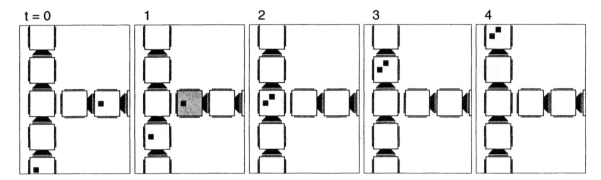

Figure 6: Gene transmission process occuring when a pattern collide into another. The central cell becomes active at the arrival of a gene ($t = 1$). However, the place onto which it wants to create a new structure cell is occupied by another existing pattern. Then the gene coming from the right pattern is transmitted beyond the boundary of two patterns and superimposed onto the left pattern's gene information using a logical 'OR' operation ($t = 2$).

and phenotype, involving transcription and translation of genes. Thus this process is non-trivial according to Langton's classification (Langton 1984). Other examples of small self-replicators are shown in Fig. 8.

Macrolevel behavior

This model displays interesting evolutionary behavior at the macro level. Simulations were conducted with initial configuration in which blank structure cells were randomly distributed at some specified density. All the results shown in this section are obtained using a square space of 200×200 sites with cutoff boundary conditions applied to its edges.

The typical outline of macrolevel behavior in this model is the following: (1) At first there is a short transient period when most of 'junk' patterns contained in the initial configuration are screened out of the space. (2) If some self-replicating organisms survive the initial transient period without being extinct, they begin to generate their respective colonies. (3) When the growing colonies crash against each other (or against sterile patterns remaining), in some cases one of the competing clusters absorbs the other; in other cases the crash happens to generate a new kind of cluster and it overcomes its 'parent' clusters. (4) Eventually, the space falls into one of the following types of final states:

Type I Static or periodic state with no self-replicator.

Type II Dynamic state in which the space is filled like a mosaic with clusters each of which is made of self-replicators of the same kind.

Type III Dynamic state in which the space is filled with a dense cloud made of a jumble of complicated self-replicators with low self-replication fidelity.

Type IV Static (or almost static) state in which the space is filled with infinitely growing static webs that originated in looped patterns included in the initial configuration.

Examples of these four final states are shown in Fig. 9. Which type the system finally falls into is dependent on the number and the kind of self-replicating patterns that survive the initial transient period. In general, larger space and higher initial density of structure cells make the initial population more diverse, which leads to the greater probability of the appearance of type III or IV final state. How the final states depend on initial density of structure cells is roughly shown in Fig. 10.

The most interesting behavior is the process of evolution toward the type III final state, which is principal behavior of this model for a regime with the initial density between 0.15 and 0.3. Fig. 11 shows a typical example. In this case, at first two colonies of simple self-replicators are formed after initial screening period ($t = 0–108$). The collisions among them and other sterile patterns give rise to appearance of some other self-replicators ($t = 330$). Once a dense cloud of more complicated self-replicators is formed ($t = 500$), it gradually proliferates ($t = 760–1000$) and finally fills up all the space ($t = 1500–2000$). This process looks like an evolutionary process in real biology performed by variation and natural selection. However, we should note that, since the organisms in this model change their genotype or phenotype through direct interaction very frequently, it is no longer possible to trace their lineages and consider their reproduction and selection dynamics to be the same as that in real biology.

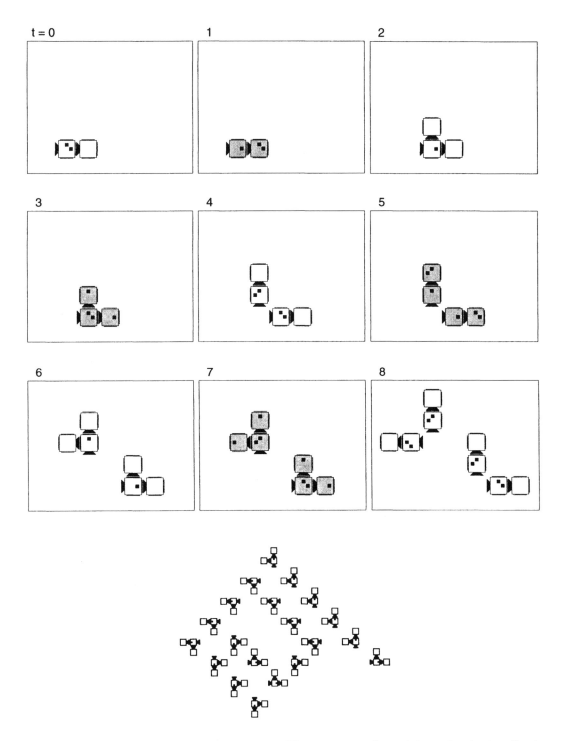

Figure 7: One of the simplest self-replicating organisms. This worm travels straight and emits its offspring to the left side repeatedly. The lower figure is a growing colony of them.

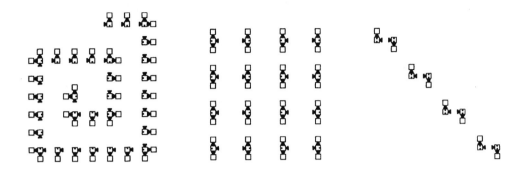

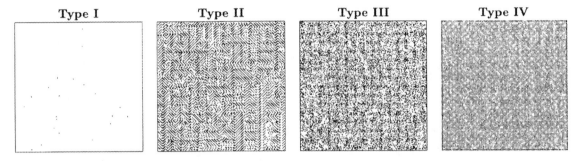

Figure 8: Examples of other kinds of simple self-replicators.

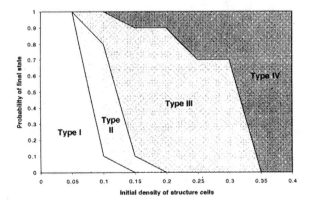

| Type I | Type II | Type III | Type IV |

Figure 9: Four types of the final states of evolution. For simplicity, only the direction of structure cells is plotted. In the initial configurations are randomly directed structure cells set onto 10% of all the sites for type I, II, III, and 30% for type IV. Type I: At $t = 200$ with random number seed 1234. Type II: At $t = 1500$ with random number seed 4321. Type III: At $t = 2000$ with random number seed 12345. Type IV: At $t = 1000$ with random number seed 1234. (With these seed numbers you can simulate them again using a Java applet intoduced at the end of this article.)

Increase of structural complexity

In the evolution toward the type III final state, a kind of increase of structural complexity is observed. Temporal development of structural complexity of the organisms in Fig. 11 is characterized in Fig. 12 using the average of length of organisms, the average frequency of branching per organism, and the average number of genes per organism. These graphs show that all the measured quantities are increasing after the appearance of the dense cloud of complicated worms around $t = 500$. The emergence and fixation of more complicated worms implies that this model has a nature to favor those with such complicated shape to some extent.

However, such increase of complexity always saturates at some constant level, i.e. the evolution in this model is definitely restricted. Fig. 13 shows an example of such a saturated situation taken from the final state in Fig. 11. It is observed in this figure that each

Figure 10: Dependence of final states upon initial density of structure cells. Ten simulation results are shown for each initial density. The space is of 200×200 sites.

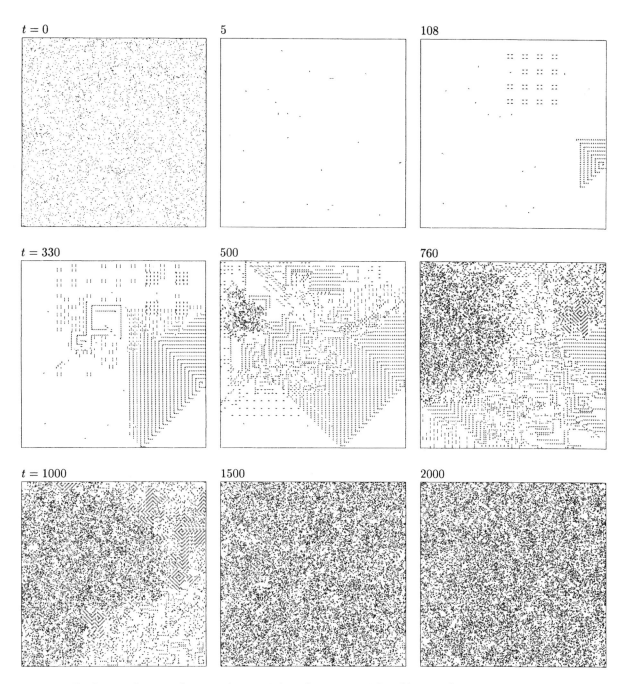

Figure 11: Evolution of worms from random initial configuration with 10% sites of structure cells. For simplicity, only the direction of structure cells is plotted. The random number seed used is 12345.

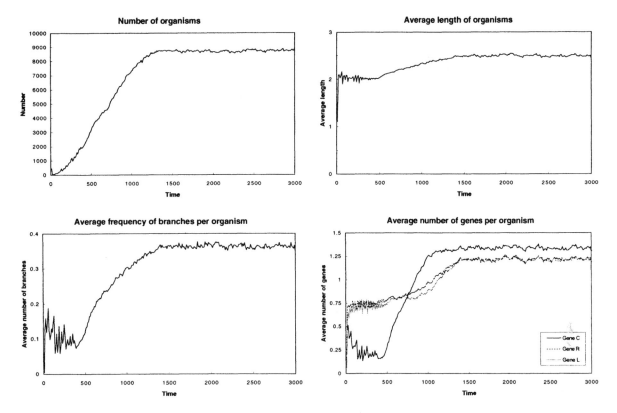

Figure 12: Evolution of structural complexity characterized by the average length of organisms (upper right), the average frequency of branching per organism (lower left), and the average number of genes per organism (lower right) in the case shown in Fig. 11.

organism has many gene information accumulated but most of them are lost without being translated to phenotype. The self-replication fidelity is thus diminishing there due to overcrowding, where the shape of offspring is determined more significantly by the environmental constraints, i.e. the availability of room around the organism for its growth.

One of the possible reasons for this saturation is the shortage of local room necessary for full translation from genotype to phenotype due to the fixation of place and shape of organisms onto particular sites. We find in Fig. 13 that there are still some areas of unused sites left near the crowded organisms. They cannot make use of such empty areas in their vicinity because their location and shape are strictly fixed to particular sites, which is one of the intrinsic features of CA. In contrast, biochemical polymers in real cells such as DNA can move and change shape adaptively reacting upon external forces, which enables many ribosomes simultaneously translate genetic information into proteins in a very compact local area. To real-

ize open-ended evolution on CA-like artificial media, it would be necessary to develop and use a newer model of space that has some fluidity in terms of place and shape of virtual organisms.

Conclusion

In this article, I introduced a new self-replication model with shape-encoding mechanism on a sixty-five-state CA space with von Neumann neighborhood, where virtual organisms transmit genetic information over their boundaries through collisions of phenotypes. An interesting result was obtained from simulations that there were some cases where the structural complexity of organisms characterized by the average length of organisms and the average frequency of branching per organism increased as the population evolved. However, an unlimited increase of structural complexity could not occur in this model. This limitation should be caused in part by an intrinsic problem of CA that the location and the shape of virtual organisms are strictly stuck to particular sites so that there is no local room

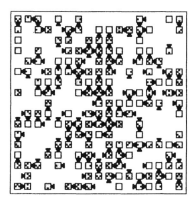

Figure 13: Enlargement of a central part of the saturated population in Fig. 11 at $t = 2000$.

to allow further complicated structures to evolve. It would be necessary to introduce some fluidity of site arrangements into the model for further development of artificial evolutionary models using CA-like space.

For the readers who may want to see the dynamic behavior of the worms introduced, a Java applet for simulating them was developed and is available at http://necsi.org/postdocs/sayama/worms/.

Acknowledgments

I would like to thank Kenichi Morita and Katsunobu Imai for permission to use their figure in this article. Thanks are also due to Yaneer Bar-Yam, Barry McMullin, and two anonymous reviewers for valuable comments. This research was supported in part by a grant from the Murata Overseas Scholarship Foundation.

References

Byl, J. 1989. Self-reproduction in small cellular automata. *Physica D* 34:295–299.

Chou, H. H., and Reggia, J. A. 1997. Emergence of self-replicating structures in a cellular automata space. *Physica D* 110:252–276.

Chou, H. H., and Reggia, J. A. 1998. Problem solving during artificial selection of self-replicating loops. *Physica D* 115:293–312.

Codd, E. F. 1968. *Cellular Automata*. ACM Monograph Series. New York: Academic Press.

Laing, R. 1977. Automaton models of reproduction by self-inspection. *Journal of Theoretical Biology* 66:437–456.

Langton, C. G. 1984. Self-reproduction in cellular automata. *Physica D* 10:135–144.

Lohn, J. D., and Reggia, J. A. 1995. Discovery of self-replicating structures using a genetic algorithm. In *Proceedings of the 1995 IEEE International Conference on Evolutionary Computation (ICEC'95)*, 678–683. Piscataway, New Jersey: IEEE.

Lohn, J. D., and Reggia, J. A. 1997. Automatic discovery of self-replicating structures in cellular automata. *IEEE Transactions on Evolutionary Computation* 1(3):165–178.

Marchal, P. 1998. John von Neumann: The founding father of artificial life. *Artificial Life* 4(3):229–235.

Morita, K., and Imai, K. 1996a. Self-reproduction in a reversible cellular space. *Theoretical Computer Science* 168:337–366.

Morita, K., and Imai, K. 1996b. A simple self-reproducing cellular automaton with shape-encoding mechanism. In Langton, C. G., and Shimohara, K., eds., *Artificial Life V: Proceedings of the Fifth International Workshop on the Synthesis and Simulation of Living Systems*, 489–496. Nara, Japan: MIT Press.

Perrier, J. Y.; Sipper, M.; and Zahnd, J. 1996. Toward a viable, self-reproducing universal computer. *Physica D* 97:335–352.

Pesavento, U. 1995. An implementation of von Neumann's self-reproducing machine. *Artificial Life* 2(4):337–354.

Reggia, J. A.; Armentrout, S. L.; Chou, H. H.; and Peng, Y. 1993. Simple systems that exhibit self-directed replication. *Science* 259:1282–1287.

Sayama, H. 1998a. *Constructing evolutionary systems on a simple deterministic cellular automata space*. Ph.D. dissertation, Department of Information Science, Graduate School of Science, University of Tokyo.

Sayama, H. 1998b. Introduction of structural dissolution into Langton's self-reproducing loop. In Adami, C.; Belew, R. K.; Kitano, H.; and Taylor, C. E., eds., *Artificial Life VI: Proceedings of the Sixth International Conference on Artificial Life*, 114–122. Los Angeles, California: MIT Press.

Sayama, H. 2000. A new structurally dissolvable self-reproducing loop evolving in a simple cellular automata space. *Artificial Life* 5(4):343–365.

Sipper, M. 1994. Non-uniform cellular automata: Evolution in rule space and formation of complex structures. In Brooks, R., and Maes, P., eds., *Artificial Life IV: Proceedings of the Fourth International Workshop on the Synthesis and Simulation of Living Systems*, 394–399. Cambridge, Massachusetts: MIT Press.

Sipper, M. 1998. Fifty years of research on self-replication: An overview. *Artificial Life* 4(3):237–257.

Tempesti, G. 1995. A new self-reproducing cellular automaton capable of construction and computation. In Morán, F.; Moreno, A.; Merelo, J. J.; and Chacón, P., eds., *Advances in Artificial Life: Proceedings of the Third European Conference on Artificial Life (ECAL'95)*, volume 929, Lecture Notes in Artificial Intelligence of *Lecture Notes in Computer Science*, 555–563. Granada, Spain: Springer-Verlag.

Vitányi, P. M. B. 1973. Sexually reproducing cellular automata. *Mathematical Biosciences* 18:23–54.

von Neumann, J. 1966. *Theory of Self-Reproducing Automata*. Urbana, Illinois: University of Illinois Press. edited and completed by A. W. Burks.

Self-Organisation in Micro-Configurable Hardware

Uwe Tangen
GMD – German National Research Center for Information Technology
Biomolecular Information Processing
Schloß Birlinghoven
53754 St. Augustin, Germany

Abstract

This paper reports on a model situated between the biological and the artificial scenario using a second generation, massively parallel reconfigurable computer – POLYP – based on the Xilinx micro-reconfigurable field programmable gate arrays (FPGAs, 6200 series) with additional distributed SRAM memory circuits under local control and broad-band dynamically reroutable optical interconnect technology.

It is shown that self-organisation of clocked Boolean networks indeed is possible where logical description and execution evolve concomittently. Self-replicators emerge, evolution over long timescales show the rise and fall of certain species and an inherent mutation control is established.

Introduction

The evolution of programs (Rasmussen *et al.* 1990; Ray 1991; Adami & Brown 1994) can be perceived as an intermediate step towards self-organising systems. There, non evolvable register machines determine the artificial physics evolving programs are subjected to. With the invention of the micro-configurable field programmable gate arrays (Xilinx 1997) the old question of bootstrapping, see fig. 1, can now also be tackled in artificial silicon environments. It becomes possible to evolve not only the description or the genotype but additionally the functionality or phenotype. Further types of artificial evolution systems have been investigated using cellular automata (Banzhaf 1993; Garis 1996; Perrier, Sipper, & Zahnd 1996; Chou & Reggia 1997; 1998), evolving Turing-machines (McCaskill 1988; Thuerk 1993) or λ-calculus based programs (Fontana 1991). These models are still abstract models even if configured into hardware. None of these models directly evolve configuration strings for hardware and thus a cycle of description and function in real hardware is not established.

Though still artificial, the physical distinction between functionality and description promises a deeper insight into emergence of self-organising systems with increasingly complex inner-organisational structures. As a necessary step towards increasing complexity it has to be shown that viable self-replicators do exist in the hardware used. The next question which has to be addressed is the evolvability of these self-replicators. Both questions are tackled in this contribution at least to a preliminary extent.

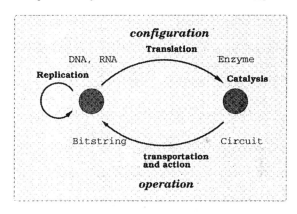

Figure 1: The bootstrapping problem
Nature has solved the problem of how to store and replicate information. DNA and/or RNA is replicated and translated via dozens of enzymes and factors which are themselves products of these translation processes. This is a bootstrapping problem – high fidelity replication couldn't take place before enzymes were invented and specific enzymes couldn't be evolved before replication had become commonplace.

The typically drastic effects of a single bit-flip in the configuration memory or program (Holland 1986; Rasmussen *et al.* 1990) seems to decrease the probability of emerging self-replicators. In nature, though the currently known biologically active substances react in a much less brittle way when their genomes

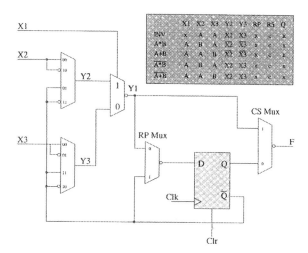

Figure 2: Inside the micro-configurable computer
Eight custom designed boards are assembled via broadband optical interconnects into a two-dimensional dataflow topology. Each of these 18 layered double-sided surface mounted boards host four Xilinx XC4000 series chips for interfacing and control and eight dynamically microconfigurable Xilinx XC6264 chips (termed as agents). Additionally, 58 fast 4 MBit SRAM devices are soldered on each board, allowing for huge populations of bit-strings flowing through the evolving system.

Figure 3: An elementary function cell (Xilinx 1997)
Including additional local routing elements, all possible configurations of a function cell can be realized by specifying 25 bits. Restricting the functionality to certain types further reduces this number. Currently, in contrast to Thompson (1996), the flip-flops are always activated – high frequency oscillators should not have a chance to disturb evolution or destroy the chips. Only clocked Boolean logic is investigated with this setup. Each agent has 16384 function cells to be configured.

are mutated (Martinez *et al.* 1996) – there always exist very crucial spatial configurations (c. f. active sites of enzymes) which must be maintained to retain the functionality of this entity (Spada, Honegger, & Plückthun 1998). When networks of many interacting enzymes and genes are considered, these mutations may also destroy in an on-off manner the functionality of the whole system (Simon 1962; Kauffman 1986). This would seem to prevent the creation of self-replicating construction machines as investigated by von Neumann (1966). So far, current models of self-organisation, e.g. (McCaskill 1988; Ikegami & Kaneko 1990; Fontana 1991; Thuerk 1993; Tangen 1994; Bedau, Snyder, & Packard 1998), represent intermediate steps towards open-ended evolution of self-replicating construction systems. The work reported here is no exception of this observation.

POLYP

Already a number of remarkably powerful computers have been built using reconfigurable silicon chips – two of the world's largest in our department, NGEN (McCaskill *et al.* 1994; 1997) built in 1993 and POLYP (Tangen & McCaskill 1998). In the dynamically reconfigurable machine POLYP, each of the boards, see

fig. 2, connected into a two dimensional topology with periodic boundary conditions, has eight dynamically reconfigurable chips (called agents). Two conventional FPGAs (termed as distributors) serve to control the online-evolution. In this work, a small microcontroller – adapted from (Chapman 1994) and configured into each distributor – initiates a new configuration at each of the four attached agents, sets control registers and reads out collected bit-strings which are then interpreted as new configuration strings.

The model

Bit-strings – equivalent to DNA or RNA, see fig. 1 – flow systolically (McCaskill *et al.* 1997) through the machine. Evolution is realized in "reactors" – defined areas in the micro-configurable chips subjected to evolution, see fig. 5. Four such reactors are configured into each agent. Input of new bit sequences is on the left and output on the right side. In the default configuration, sequences are shifted bit-serially from left to right with a clock frequency of 16 MHz. Two filters on the right side of each reactor serve to observe 16 single outputs for the occurrence of valid configuration

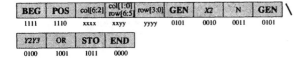

Bin.	Mem.	Comment
0000	END	End marker and nil sequence
0001	NXL	Go left after next writing
0010	NXR	Go right after next writing
0011	RDP	Reverse direction pointer immediately
0101	GEN	General configuration function
1011	STO	Stop codon
1110	POS	Position of configuration pointer
1111	BEG	Begin of sequence

Table 1: Four bit codons
Currently, out of 16 possible codons eight are used. Sequences containing undefined codons are not recognised as syntactically correct configuration strings. Some codons have additional parameter sections attached.

Figure 4: A valid configuration bit-string
Every box is called a codon with a length of four bits. Commands are written in capitals and parameters printed with a smaller font. The sequence is read from left to right and starts with the BEG-codon, bits shown underneath. With the POS-codon and its parameter an absolute position of the function cell to be configured is defined. Illegal values of the position pointer are masked out. Arbitrary many GEN-codons can be concatenated to completely define the 25 configuration bits of a function cell. In the example shown, the input **X2** will be configured with a north direction – input is taken from south. The **Y2Y3** multiplexers are set to realize an OR-function. The STO-codon then realizes the configuration and moves the configuration pointer one step ahead – in this case to the northern function cell.

strings (also called individuals). From these eight filters per chip one output lane is collected and locally stored. 64 pick-up points per agent are realized. The microcontroller in each distributor collects these configuration strings and reconfigures locally the sequence generating agents.

Bit-strings are called self-replicating if they are able to reach these pick-up points and reconfigure the agents configuration so that replica of these strings are also able to reach pick-up points.

In principle, the mapping (or basic code) between valid configuration strings and the hardware bound configuration bits is arbitrary. Different mappings tested (data not shown) revealed that the specific impact of point mutations, deletions or insertions should be as small as possible. A four bit code with parameter sections appears to be suitable. In this implementation, eight different four-bit codons are used – all others result in syntax errors. Some codons have parameter sections attached, tab. 1.

A valid configuration bit-string, e. g. see fig. 4, starts with a BEG (begin codon) and stops with an END codon. Four codons, NXL (next left), NXR (next right) and RDP (reverse direction) in conjunction with STO (stop codon) determine the next cell to be configured. After each STO, the configuration of a cell is done and the next neighbouring cell becomes the new configuration site. Configuration commences northbound at a usually random position in the evolvable area of the agents. A POS (position codon) with its three parameter codons determines the absolute loca-

tion of the next cell to be configured. The codon GEN, with it's parameters, tab. 2, defines which bits in the configuration memory of the agents are set or reset. Several GEN codons can be concatenated and thus each of the 25 bits of each function cell can be configured, fig. 3. To ensure activation of the flip-flop in the function cells the **CS**-bit is always forced to zero by the distributor.

Experiments soon showed that constants or oscillators easily dominate the evolving system and thus disrupt the flow of bit sequences completely. Several means have to be invented to reduce the probability of being trapped in such trivial fixed-points. One is to subdivide the reactors into small boxes. Configuration bit strings (also named as individuals) usually are restricted to one box. Which box to be taken is a random decision. Only the POS-codon in a configuration bit string is able to absolutely address a certain function cell to be configured. Actually, in the experiments reported, only the first column of boxes per reactor is used for evolution. This means that evolution has not to obey the need for cooperation between boxes. Boxes are not chained in the default regime. Bit-strings usually pass only one box before leaving a reactor.

Experimental setup

Experiments are conducted according to the following procedure. Initially, all reactors are reset to the default configuration. This default configuration guarantees that bits are able to flow from the left input boundary to the right outputs. Each agent is accompanied by a small population in the UNIX-host of 128 configuration bit-strings and is reconfigured 16 times by the distributor. The UNIX-host workstation is then interrupted and the board is read out and analysed. Thereafter, one configuration bit-string is written into an on-board SRAM and the board is launched again. This procedure is done asynchronously, interrupt controlled, with

Bin.	Memn.	Comment
0000	VAL	Set value of register
0001	X1	Input definition of X1 variable
0010	X2	Input definition of X2 variable
0011	X3	Input definition of X3 variable
0100	Y2Y3	Definition of Y2 and Y3 muxes
0101	MCSRP	Definition of M, CS and RP muxes
0110	S	Southern output multiplexer
0111	W	Western output multiplexer
1000	N	Northern output multiplexer
1001	E	Eastern output multiplexer
1010	CX1	Change X1-input
1011	CX2	Change X2-Input
1100	CX3	Change X3-Input
1101	CY2Y3	Change Y2Y3-Muxes
1110	CNEWS	Change NEWS-output directions
1111	CX123	Change X1,X2,X3-input directions

Table 2: Parameters of the command GEN
The parameter field of GEN is two codons wide. The first four bits determine which section in the configuration cell should be changed and the second four bits specify the according change. For example, see fig. 3, the code 0010 determines the three bits necessary to determine unambiguously the input **X2**. The **CS** multiplexer bit is always forced to low, see also fig. 4.

all eight boards.

To avoid early fixed-points in the evolutionary process, random $2*2$ boxes in the reactors are reset to the default configuration at a slow rate $d \sim 3\%$.

The time measure is defined as the number of reconfigurations – in this case the number of board invocations, which amounts to 128 agent reconfigurations per time step. Every 100 board invocations, all hitherto gathered configuration bit-strings are analysed by the controlling C-program in the host.

The experiments reported lasted about one hour CPU- or two hours real-time each with 10^6 board invocations – which amounts to $1.3 * 10^8$ chip reconfigurations. With a continuous 16MHz clock about $6 * 10^{10}$ rising clock edges were employed per experiment.

Mutation is realized at two levels. Intrinsic mutation is realized by the evolving logic. As an extrinsic mutation rate, valid configuration bit-strings at the host level are mutated bit-wise. Additionally, at a slow rate, new genes are inserted or deleted. Mutation at the host level always produces valid configuration strings and is in fact a mutation in the phenotype space. Because of the online-evolution procedure, valid configuration bit-strings are able to reconfigure the reactors 15 times before a host mutated configuration bit-string enters the configuration apparatus situated in the distributors. These internal configuration bit-strings are subjected to evolutionary intrinsic changes due to the

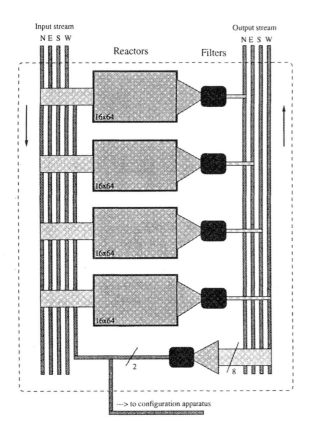

Figure 5: Local reactors competing for reconfiguration resources
Four reactors of size $16 * 64 = 1024$ function cells are shown which are situated in each agent. With a total of 256 reactors in the system $262,144$ function cells are subjected to evolution. Actually, every reactor is subdivided into small squares of $2*2$ called boxes. Without the aid of POS-codons configuration bit strings are restricted to these small areas. Configuration pointers leaving a box are mapped to the opposite boundary. Which box is to be configured is typically chosen at random – if no POS-codon forces the configuration of a certain function-cell of a reactor.

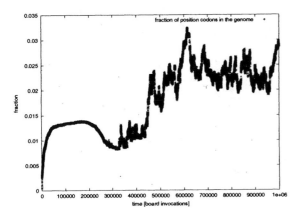

Figure 6: Evolution of reconfiguration control
The concentration of POS-codons of an experiment with $p = 0.016$ (see text) is shown. Initially, no sequences contain this positioning command. Sequences with a POS-codon are able to control the location where reconfiguration is going to start at the chip. The first species, dominating until about $t = 20000$, then becomes instable and is replaced or outnumbered by a competition or association of at least two species at $t = 30000$.

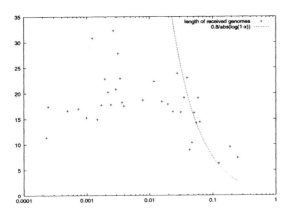

Figure 7: Error threshold for self-organisation
The average length of valid reconfiguration strings, y-axis, is shown. The configuration bit-string mutation rate p, x-axis is varied from 10^{-4} to 0.5. A decay rate of about 3% per board invocation is used. Every dot averages the length of received valid configuration bit-strings of 10^6 board invocations or 128 million chip reconfigurations. 35 experiments are plotted. A $\nu = 0.8/|ln(1-x)|$ curve is plotted to show what the behaviour in a quasi-species (Eigen 1971) would be. A $ln\sigma = 0.8$ seems to be a good choice in the high error regime – which is in the same order as the values found in (Tangen 1994).

logic configured into the reactors.

The complexity measure $c = (n_o/n_t) * n_d$ used here, only serves as an indication of the inner properties of the system. All 64 different populations are merged and sorted lexically by their genome. It is counted how frequent sequences are, n_t, and how many agents, n_o, host this specific sequence. The quotient of n_t/n_o says something about the proliferation capability of this configuration sequence. Multiplied with the number of different sequences in the whole system, n_d, two conflicting quantities are multiplied. Thus the value is low for bit noise in the system and for one dominating species as well.

Results

Introducing correct bit-strings into the evolving regions of some agents ignites self-organisation and evolution. These hand-coded initial bit-strings are designed as self-replicators which are bit-serially fed into a few lanes of two of the 64 agents. Almost always these ignition sequences are replaced in the course of evolution.

In order to allow for self-organisation to work a sufficient large number of syntactically correct bit-strings has to be maintained. Bit-strings which passed the agents' filters are reintroduced several times at the reactor inputs. Additionally, it turned out that a minimum decay rate of $d = 1$ is necessary to allow for self-organisation. With a decay rate of zero, the system is almost immediately trapped in fixed-points. Constants and oscillators disrupt the data-flow.

Typically, the location where configuration bit-strings change the configuration of the chips is chosen at random – in between the limits given by the reactors. An exception from this rule is when evolution itself uses the POS-codon to specifically determine a single function-cell to start configuration. Because the initial strings do not contain any POS-codons the concentration of POS-codons in the course of time gives insight into the evolutionary processes. In fig. 6, it is shown that individuals are indeed able to use this method of specifying the starting location of the reconfiguration procedure. Mutations alone would not be able to maintain such a large number of POS-codons in the system. Until time-step 20000, a single species or system of species seems to dominate the system. The decline between $t = 20000$ and $t = 30000$ is a frequent observation with these evolving systems. At about $t = 30000$ other species are able to dominate space. Due to the stochastic behaviour of concentration at least two species are competing for reconfiguration resources. This might also be a predator/prey cycle. An interesting question is what the emerging fitness

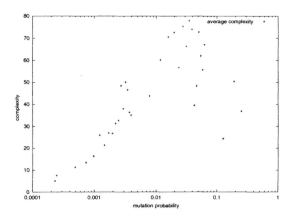

Figure 8: A simple complexity measure
The average complexity measure $c = (n_o/n_t) * n_d$, y-axis, is shown. With n_o the number of chips where the sequence occurred, n_t the total number received and n_d the number of different sequences in the system. The mutation rate p, x-axis is varied from 10^{-4} to 0.5. A decay rate of about 3% per board invocation is used. Every dot accumulates 10^6 board invocations or 128 million chip reconfigurations. The results of 35 experiments are plotted.

landscapes in the system look like and whether this hardware evolution system behaves as a quasi-species (Eigen 1971). In fig. 7, the average length of picked-up sequences is plotted against the mutation rate. Thirty five experiments with 10^6 board invocations were conducted. The average length of received sequences in each experiment with a varying mutation probability p is taken over the whole time span. At least in the high error regime, the system shows quasi-species behaviour. For comparison, a $\nu = 0.8/|ln(1-x)|$ curve is plotted to show what the behaviour in a quasi-species (Eigen 1971) would be. A maximum value of the simple complexity measure can also be observed, fig. 8, but at a different mutation rate p.

Discussion

The emergence of bit-strings specifying the absolute position of cells to be configured is the strongest hint on a working evolutionary system. Additionally, a simple measure of complexity could give increasing values with time. The length of generated sequences can increase in the course of evolution, too. The succession of different species dominating the system shows the operation of evolutionary processes. The underlying fitness landscapes present a special challenge and seem to have high peaks sparsely located in sequence and phenotypic space. At least, connected neutral networks

are not abundant in the system. A special property of this approach is that there are no external fitness landscapes involved. No one is telling a configuration bit-string whether it has good properties or not. This could be done of course, but the aim of the current work is to find out which are the viable configuration bit-strings and how do they interact. The only selection criterion is to survive in the system and this means that configuration bit-strings have to reach these pickup points in the agents without being destroyed or changed by the logic they have to pass. This not only allows us to study the emergence of certain chemistries but also to look for higher information processing capabilities, e. g. immune responses, facilitation of species specific sequence replication.

Usually only a small part of a configuration bit-string is at any time instance located in one reactor. For example in fig. 4, the simple valid configuration bit-string which only configures one single function cell, is $13 * 4 = 52$ bits long. A reactor in its default configuration would therefore be able to store $16/52 = 30.8\%$ of this string. It is not very probable that much more bits are stored along a data-path of a valid configuration bit-string – even in evolved reactors. Calculating this for typical sequences of length 40 codons only 10% of a configuration string reside at any time in the reactor. This strong discrepancy between the length of a sequence and the resulting power of configuration changes is a consequence of really specifying each detail of the functionality. Using predefined logic blocks would be helpful but would rise the question of the very logic modules to be used.

For self-organisation to emerge, it was crucial to amplify artificially valid configuration bit-strings. In biology this does not necessarily mean sequence amplification because RNA- or DNA-strands might be processed several times before degradation.

Though the pressure to evolve self-replicators is decreased, sequence-replication is done that easy in silicon (a sequence is flowing along two inputs of buffers, it is not that easy in biology (Mullis *et al.* 1986; Sievers & von Kiedrowski 1994) but conceivable) that with an artificial amplification not too high self-replicators should emerge.

Further experiments pointed to a possible difference between mutation in the configuration space (phenotype) and in the sequence space of the configuration bit-strings (genotype). Intrinsic variation has been observed more in the sense of many repeats, which seem to be the main source for increasing sequence lengths.

The preliminary conclusion is that artificial evolving systems in a binary world with a closed bootstrapping loop are viable and may help to understand natural biological processes. This Boolean ansatz has the advantage that no complex sequential register machines are necessary as is the case with Tierra (Ray 1991) or Avida (Adami & Brown 1994). Boolean properties can be found in nature everywhere and enzyme networks naturally will have properties of Boolean networks. Additionally, gene-regulation has a lot to do with Boolean decisions. No matter how complex the emerging configuration bit-strings of synchronous digital silicon hardware evolution are – they are constituted by simple clocked Boolean functions.

Acknowledgements

The help and support of J. S. McCaskill and the German Ministry of Science (BMBF) is greatly acknowledged.

References

Adami, C., and Brown, C. T. 1994. Evolutionary learning in the 2d artificial life system "avida". In Brooks, R., and Maes, P., eds., *Artificial Life IV*. Cambridge, MA: MIT Press. 377–381.

Banzhaf, W. 1993. Self-replicating sequences of binary-numbers - foundations general. *Biol. Cyber.* 69:269–274.

Bedau, M. A.; Snyder, E.; and Packard, N. H. 1998. A classification of long-term evolutionary dynamics. In Adami, C.; Belew, R. K.; Kitano, H.; and E., T. C., eds., *Artificial Life VI*. Cambridge, Massachusetts: MIT Press. 228–237.

Chapman, K. 1994. Dynamic microcontroller in an xc4000 fpga. *Xilinx Application Note*.

Chou, H.-H., and Reggia, J. A. 1997. Emergence of self-replicating structures in a cellular automata space. *Physica D* 110:252–276.

Chou, H.-H., and Reggia, J. A. 1998. Problem solving during artificial selection of self-replicating loops. *Physica D* 115:293–312.

Eigen, M. 1971. Selforganization of matter and the evolution of biological macromolecules. *Naturwissenschaften* 58:465–523.

Fontana, W. 1991. Algorithmic chemistry: A model for functional self-organization. In Langton, C. G., ed., *Artificial Life II*. Reading, Massachusetts: Addison-Wesley. 159–202.

Garis, H. d. 1996. Cam-brain: The evolutionary engineering of a billion neuron artificial brain by 2001 which grows/evolves at electronic speeds inside a cellular automata machine (cam). *Lect. Not. Comp. Sci.* 1062:76–98.

Holland, J. H. 1986. Escaping brittleness: the possibilities of general-purpose learning algorithms applied to parallel rule-based systems. In Michalski, R. S.; Carbonell, J. G.; and Mitchell, T. M., eds., *Machine Learning II*. Los Altos, CA: Morgan Kaufman. 593–623.

Ikegami, T., and Kaneko, K. 1990. Computer symbiosis - emergence of symbiotic behavior through evolution -. *Physica D* 42:235–243.

Kauffman, S. A. 1986. Autocatalytic sets of proteins. *J. Theor. Biol.* 119:1–24.

Martinez, M. A.; Pezo, V.; Marliere, P.; and Wain-Hobson, S. 1996. Exploring the functional robustness of an enzyme by in vitro evolution. *EMBO Journal* 15:1203–1210.

McCaskill, J. S.; Chorongiewski, H.; Mekelburg, K.; Tangen, U.; and Gemm, U. 1994. Ngen - configurable computer hardware to simulate long-time self-organization of biopolymers (abstract). *Physical Chemistry* 98:1114–1114.

McCaskill, J. S.; Maeke, T.; Gemm, U.; Schulte, L.; and Tangen, U. 1997. Ngen a massively parallel reconfigurable computer for biological simulation: towards a self-organizing computer. *Lec. Note Comp. Sci* 1259:260–276.

McCaskill, J. S. 1988. *Polymer Chemistry on Tape: A Computational Model for Emergent Genetics*. Max-Planck–Society. Report.

Mullis, K.; Faloona, F.; Scharf, S.; Saiki, R.; Horn, G.; and Erlich, H. 1986. Cold spring harb. symp. *Quant. Biol.* 51:263.

Perrier, J.-Y.; Sipper, M.; and Zahnd, J. 1996. Toward a viable, self-reproducing universial computer. *Physica D* 97:335–352.

Rasmussen, S.; Knudsen, C.; Feldberg, R.; and Hinsholm, M. 1990. The coreworld: Emergence and evolution of cooperative structures in a computational chemistry. *Physica D* 42:111–134.

Ray, T. S. 1991. An approach to the synthesis of life. In Langton, C. G.; Taylor, C.; Farmer, J. D.; and Rasmussen, S., eds., *Artificial Life II*. New York: Addison-Wesley. 371–408.

Sievers, D., and von Kiedrowski, G. 1994. Self replication of complementary nucleotide-based oligomers. *Nature* 369:221–224.

Simon, H. A. 1962. The architecture of complexity. *Proc. Amer. Phil. Soc.* 106:467–482.

Spada, S.; Honegger, A.; and Plückthun, A. 1998. Reproducing the natural evolution of protein structural features with the selectively infective phage (sip) technology. the kink in the first strand of antibody kappa domains. *J. Mol. Biol.* 283:395–407.

Tangen, U., and McCaskill, J. S. 1998. Hardware evolution with a massively parallel dynamically reconfigurable computer: Polyp. In Sipper, M.; Mange, D.; and Perez-Uribe, A., eds., *ICES '98 Evolvable Systems: From Biology to Hardware*, volume 1478. Heidelberg: Springer. 364–371.

Tangen, U. 1994. *The Extension of the Quasi-Species to Functional Evolution*. Jena: PhD Thesis.

Thompson, A. 1996. An evolved circuit, intrinsic in silicon, entwined with physics. *Lect. Not. Comp. Sci.* 1259:390–405.

Thuerk, M. 1993. Ein modell zur selbstorganisation von automatenalgorithmen zum studium molekularer evolution. *Uni. Jena.* Diss.

von Neumann, J. 1966. *Theory of Self-Reproducing Automata*. Urbana: Burks, A. W. University of Illinois Press.

Xilinx. 1997. Xc6200–field programmable gate arrays data sheet. *Xilinx* 1:1–73.

Reaction Mechanisms in the OO Chemistry

Hugues Bersini
IRIDIA – Université Libre de Bruxelles
CP 194/6
50, av. Franklin Roosevelt
1050 Bruxelles – Belgium
bersini@ulb.ac.be

Abstract

The work presented in this paper continues the work presented in the previous Alife (Bersini, 1999) conference by improving the connections between the computerized object-oriented chemistry and the real natural chemistry. This improvement is the result of a clearer definition of the type of computational structure which codes for a molecule, of the different reaction mechanisms between two molecules to obtain one or several new ones, and of the way the dynamics (i.e. the molecular concentration change in time) and the metadynamics (i.e. the appearance of new molecules) simultaneously make the whole chemical system to evolve in time. What makes a molecule computationally unique, in terms of a strictly ordered tree, will be deeply described. Simulation results are presented for a simple chemical reactor composed of four atoms with different valence, and allowing molecules to deterministically or randomly interact according to two mechanisms: the chemical single-link-crossover and the open-bond reaction. How the chemical kinetics is called into play will be shown for very elementary reactions.

Introduction

An "existential problem" inherent in a lot of works being developed under the Alife banner is their exact positioning between the natural sciences from which they originate: biology or chemistry, and some very abstract scheme, appealing at a pure computational level, but too much abstract for the results obtained to be of any feedback on the inspiring natural sciences. This position is so unstable that the temptation is great to escape this neutral situation by joining one of the two extremes, either to become a pure natural scientist manipulating test-tubes, or just dealing with software in a close formal or engineering perspective. As a consequence of this instability, this work attempts to slightly fill in the gap between the computer simulations of the chemical components and the way they interact on one side and their natural counterpart on the other side.

Physics naturally strides towards the "dematerialization" of objects by restricting their study to the laws of behavior and the laws of interaction. The "physical nature" of an objet reduces to a set of variables, the measure of which is experimentally possible, and that can easily be manipulated in a computer. Chemistry on the other hand fully integrates material object in its study, and the challenge is great for whatever computational approach, proposed in Alife, to see how the borders of materiality can be crossed. Several times in this paper, this difficulty will recur. While computational biology is similarly confronted with the limits of materiality, it might be judicious to attack this limitation one level below by allowing the capture of a reality with is known to be easier to model. That is while chemistry appears to be a key target for attempts of computational replica, and entails artificial chemistry to strategically precede artificial biology.

No doubt that we are today very far to be able to replicate in software and in all details a real molecule and the simplest interaction between two of them. It turns out to be pretty laborious to integrate in whatever computer simulation capital aspects as the 3-D shape, the energetic and electronic properties, both of a single molecule and of the way they interact. What is possible however is to capture in software, and in a preliminary attempt, some basic scholar chemistry like the molecular composition and combinatorial structure, basic reaction mechanisms as chemical crossover or bonds opening/closing and simple dynamical aspect as first or second order reactions.

A lot of chemical aspects is still however left aside (which reaction takes place, which link is involved, what is the value of the reaction rate,...) but in such a way that it should be easy for a more informed chemist to parameterize the simulation in order to account for the influence of these aspects.

The choice of object-oriented (OO) programming naturally follows from this attempt to reduce the gap between the computer simulations and its natural counterpart. It is intrinsic to OO programming that by thinking about the problem in real-world terms, you naturally discover the computational objects and the way they interact. In contrast, procedural programming forces the conversion of the real-world problem into the computer basic language. Like in our precedent paper, the next section will recall the UML class diagram, recapitulating the seven main classes of our chemical software and the way they relate to each other. UML diagrams (Eriksson and Penker, 1998) allow understanding the central structure of the program, and the interaction of its modules, without resorting to the code. As a matter of fact, the section will plead for an increased use, in our community, of tools of analysis like UML. UML eases the development of the model but, first of all, helps the communication of this model with no need to get into a lot of implementation details.

In this work a molecule is neither a lambda-calculus expression (Fontana, 1992), any kind of operator (Dittrich, 1999; Dittrich and Banzhaf, 1998) nor a string (Farmer, Kauffman, and Packard, 1986; Holland, 1995), but is taken to be a computational tree, whose vertical and horizontal organization is strictly defined, due to the symmetry in the way atoms are connected. These organizational rules prevent two identical molecules, obtained as respective outcome of different reaction histories, to co-exist at any moment in the system. This is a replica, for our molecular computational tree, of the notions of "sameness" and "normal form" stressed by Walter Fontana is his lambda calculus framework (Fontana, 1992; Fontana and Buss, 1996). Those rules will be given and illustrated by numerous examples of molecules in section three. Also the section will precise the kind of chemical isomers that can be differentiated when represented as computational tree, and the isomers that this representation can't allow to differentiate.

In chemistry tutorials, it is frequent to find different names for different types of chemical reactions like: "decomposition" (when one molecule breaks down into two or more other simpler ones) or "combination" (when two molecules combine to form a new one) or "single and double replacement" (involving exchange of partners), etc. Based on the precise computational definition of a molecular identity given in the previous section, the section three will propose a new nomenclature for several reaction mechanisms, each of them being precisely described: the single-link-crossover, the multiple-links-crossover, the open-bond and closed-bond reactions, the changing-variance and the re-organizational reactions. Section four will show the results of some simulations obtained, departing from four atoms of valence 1,2 and 4, and which compose four elementary molecules of two atoms. These four diatomic molecules will initiate a chain of reactions where two molecules picked randomly will be able to interact according to only two of the reactions: the single-link-crossover and the open-bond reaction. The last section will show what happens to the simulation when we allow, besides the metadynamics (leading to an artificial chemistry said "strongly constructive" (Fontana, 1992; Dittrich, 1999)), the molecule to change their concentration in time. A simple first-order reaction will be simulated. Various algorithmic alternatives for the whole simulation loop will be discussed. Artefactual effects resulting from different algorithmic choices will be illustrated and discussed. Whereas in a lot of Alife simulations, the dynamics and the metadynamics are kept separated, a clear interest of this work is their simultaneous consideration.

The OO Chemistry Class Diagram

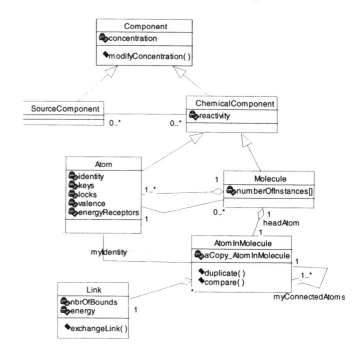

Figure 1: The UML class diagram of the OO Chemistry

It is not OO programming language (java, c++, smalltalk,) whose use is advocated here (although java contributes a lot to the spreading of Alife applets through the Internet). Everything programmed in an OO language can be alternatively done in a procedural way. It is rather the exploitation for communication ends of what OO languages, by increasing their use, have made more and more necessary, useful and prevalent in the computer world. That is, the development of high-level graphical notations to describe what the program is doing and how the reality is actually simulated without the need to resort to the code. The Unified Modeling Language (UML) is a major opportunity for our community to improve the deployment, the better diffusion and better understanding of the Alife models, and especially when addressed to researchers not feeling comfortable reading programs.

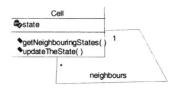

Figure 2: The Cellular Automata UML class diagram

Take the class diagram whose main function is to identify the classes and how they do interact, and see in fig.2 the simplest example of what a cellular automata basically reduces to. Each cell interacts with n neighbors by just getting their state and, on the basis of this value, updating their own state.

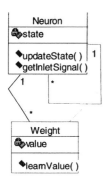

Figure 3: The UML neural net class diagram

A neural net (fig 3) has an additional class, the "Weight" that can learn its value. One neuron is associated with n weights and n neurons, since updating the state of a neuron needs the value of the connected weights and neurons. Additionally, each weight links together two neurons. Class diagrams are eventually more useful in simulations involving a large number of classes. A typical example is ecosystem simulations that generally contain different sorts of predators, preys, and resources (with neurons in these creatures and sometimes genes…). On the other hand, sequence diagrams can be useful when it is the sequence of events that counts to identify the process. For instance, a critical debate, very vivid in our community, addresses the choice of synchronous or asynchronous updating in CA kind of models. A sequence diagram easily helps to pinpoint the difference between these two updating mechanisms.

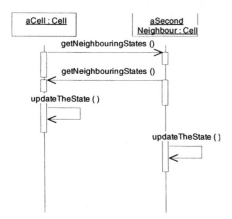

Figure 4: A UML sequence diagram helping to differentiate synchronous from asynchronous updating

Fig.4 represents the sequence diagram in the synchronous case for two cells of the cellular automata. For the asynchronous case, it is enough to shift the second "getNeighbourState()" message below the first "updateState()".

In the same line of thought, the use of UML is this paper aim at easily describing to computer scientists and chemists the components of the chemical simulation (fig.1) and how they do interact. A lot of similarities are found with the UML class diagram presented in an earlier paper (Bersini, 1999). The **Component** super class includes as main attributes the *concentration*. This concentration changes in time by natural decrease or increase and as a result of the reactions and their specific rate, in a way that will be discussed in section five. The **Chemical Component** is a subclass of Component and the super-class of two further sub-classes: Atoms and Molecules. The **Atom** is the first sub-class of the Chemical Component and describes the basic objects of the whole system. The fundamental attribute is the *valence*, which indicates in which proportion this atom will connect with another one to form a molecule. For instance an atom with valence 4 (for instance with identity "1" for reasons to appear later) will connect with four atoms of valence 1 (for instance with identity "4") to form the molecule: *1(4 4 4 4)*. Connections between atoms are perfectly symmetric.

The second major attribute is the *identity*, which, in our simulation, relates to the value of the variance. Atom with a high variance will be given a small identity value. This identity simply needs to be an ordered index ("1", "2",…) for the organization rules shaping the molecular tree be possible. The way this identity is defined depends on what we take to be unique to any atom. Actually, in chemistry this identity is given by the atomic mass i.e. the number of protons and neutrons.

Molecule is the second sub-class of Chemical Component. The following section will show how the molecules are structured in a unique way. As shown in the class diagram, molecules are compounds of atoms. An attribute called *numberOfInstances* is a vector of integers whose elements are the number of times one specific atom appears in the molecule (i.e. four "4" and one "1" in the molecule *1(4 4 4 4)*). Molecules are trees that are computationaly structured with pointers of class **AtomInMolecule**. Each molecule possesses one and only one AtomInMolecule pointer called the *headAtom* and which can be seen as its "front door" (it would be the "1" in the molecule *1(4 4 4 4)*). As soon as an atom enters into a molecule, it is transformed into an AtomInMolecule object. AtomInMolecule relates to atom since the identity of such an object is the same as its associated atom. AtomInMolecule are responsible for coding the tree structure (Aho and Ullman, 1995) of the molecule since they possess pointer attributes (called *myConnectedAtoms*) pointing to a vector of AtomInMolecule objects.

An addition with respect to the previous work is the class **Link**. An object "Link" connects two AtomInMolecule. It has a given *energy* so that the weakest link is the first to break, and a *number of bonds*. For instance two atoms of valence 4 will connect (to form a diatomic molecule *4(4)*) with a link containing 4 bonds, and one atom of valence 4 will connect with four atoms of valence 1 (*1(4 4 4 4)*), each link containing one bond. Link objects intervene in the unfolding and the coding of the reaction mechanisms. For instance, one major method associated with the class Link is *"exchangeLink"* involved in crossover molecular reactions.

The Molecular Unique Computational Structure

Whatever chemical notation you adopt, for instance the "line-bond" or Kékulé structure (for instance the butane molecule in fig.5), one way of reproducing the connectivity pattern of a molecule is by means of a computational tree.

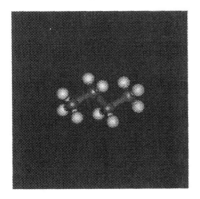

Figure 5: The "butane" molecule

For facility and space, the following linear notation will be adopted to describe a molecular computational tree. One example will be enough to understand it. Take the following molecule:

$$1 (1 (4 4 4) 2 (1 (3 3 3)) 2 (2 (3)) 2 (4))$$

"1" is an atom with valence 4, "2" is an atom with valence 2, and "3" and "4" are atoms with valence 1. The graphic tree version is given in fig.6. In our linear notation, the "butane" molecule of fig.5 would become:

$$C(C (C (H H H) H H) C (H H H) H H)$$

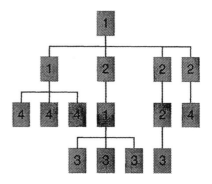

Figure 6: The computational tree structure for a molecule

if (the identity of n < the identity of m) { the smaller is n }
else
if (the identity of n = the identity of m)
{ if (n has no connected atom and m has no connected atom)
{the smaller is n}
else
if (the number of connected atoms of n > the number of connected atoms of m) {the smaller is n}
else
if (the number of connected atoms of n = the number of connected atoms of m)
{for all j connected atoms of n and m
{if (the identity of the jth connected atom of n < the identity of the jth connected of m) {the smaller is n , break-the-loop}
else
{ for all connected atoms of n and m
{ redo recursively the same testing procedure}}}}

Table 1: Which is the smaller between the atomInMolecule n and m.

In a first approximation, the connectivity shows symmetry both vertically and horizontally. The following rules need to be respected in order to shape the molecular tree in a unique way:

Vertically: The highest node, i.e. the front door of the molecule (the initial "1" in our example of fig.6) must be the smallest of all the AtomInMolecule objects composing the tree.

Horizontally: Below any node (i.e. any AtomInMolecule) the sub-nodes are arranged from left to right in an increasing order, the smallest to the left, the greatest to the right.

Clearly these two rules depend on the definition of "smaller" between two AtomInMolecule nodes. It is defined in a way described in table 1 for two AtomInMolecule "n" and "m". You can see that the example given in fig.6 indeed comply with these rules: the highest "1" is connected first to a "1" then to a "2" whereas the other "1", although first connected to a "1", are afterwards connected to atoms with higher identity. The horizontal rule is equally verified for all connected atoms.

Organizing the tree is such a way allows differentiating two structural isomers, i.e. molecules that contain the same number of the same atoms but in a different arrangement (like the well-known chemical examples of the butane and the methylpropane molecules, both C_4H_{10} but with distinct connectivity patterns). However such a re-organisation can miss the differences still remaining between molecules showing a same connectivity pattern but with different spatial organisations i.e. the geometrical isomers. These different spatial organisations can induce different optical properties, which is of no relevance for the remaining of the paper, but can also play a major role in the type of reactions these molecules are subject of. This consists of a first major obstacle coming from the "materiality" of the chemical object (briefly commented in the introduction) and that deserves a future deep investigation.

The Different Reaction Mechanisms

Several reaction mechanisms called "decomposition", "combination", "replacement" are repeatedly described in the chemical literature. Based on our syntactical definition of what is the molecular identity, a new nomenclature will be used here for the possible chemical reactions, with a simple illustration for each reaction mechanism. Every time a new molecule is created as a result of the combination of two molecular reactants, this new molecule needs to be reshaped according to the organization rules previously defined (transformed into its normal form (if borrowing Fontana's words (1996)).

- *Single-link crossover*: the weakest links of each molecule are exchanged (remember that "1" has valence 4, "2" has valence 2 and "3" and "4" have valence 1, and suppose the link "2-3" is weaker than the link "2-4").

$$1(1) + [4] \ 2 \ (3 \ 4) \ \rightarrow \ 1(3 \ 3 \ 3 \ 3) + 1(2(4) \ 2(4) \ 2(4) \ 2(4))$$

The bold values between brackets are *stoichiometric coefficients* needed in order to balance the chemical equations

- *Multiple-link crossover*: several weak links are homogeneously exchanged between the two molecules.

$$1(4 \ 4 \ 4 \ 4) + [2] \ 2(2) \ \rightarrow \ 1(2 \ 2) + [2] \ 2(4 \ 4)$$

The linear notation cannot account for the number of bonds characterizing the links. Here the 2-2 and the 1-2 links are double bonds.

- *Open-bond reactions*: in the first molecule, a link containing i bonds opens itself to make j links of i/j bonds (here the first link "1-2" of the first molecule opens itself (i.e. frees one bond) and the first link of the second molecule breaks)

$$1(2 \ 2) + 2(3 \ 3) \ \rightarrow \ 1(2(3) \ 2(3) \ 2)$$

- *Close-bond reactions*: it is the reciprocal of the previous reaction - in the first molecule, the two first links of one bond merge to form one link of two bonds:

$$1(2(3) \ 2(3) \ 2) \ \rightarrow \ 1(2 \ 2) + 2(3 \ 3)$$

- *Changing-variance reactions*: in the first molecule, the first atom of variance i increase or decrease its variance (here decrease)

$$1(4 \ 4 \ 4 \ 4) \ \rightarrow \ 1(4 \ 4) + 4(4)$$

- *Re-organizational reactions*: it is just a re-organization of the links in the molecule:

$$1(1(2 \ 3) \ 3 \ 3 \ 3) \ \rightarrow \ 1(3 \ 3 \ 3 \ 3) + 1(2)$$

This re-organisation is often accompanied by a change in the variance of one of the atom (here the "1").

Simulating the Single-Link-Crossover and the Open-Bond Reactions

We now show the results obtained by running the chemical simulators with the four atoms "1" (valence 4), "2" (valence 2), "3" (valence 1), "4" (valence 1) and the four basic diatomic molecules: *1(1), 2(2), 3(3), 4(4).*

The simulator runs in the following way:

1. Take randomly two molecules
2. Make them interacting according to either the single-link-crossover or the open-bond reaction (random choice). In both cases, the link involved in each molecule (either

exchange or open) is the weakest link (each link receives a random energetic value).

3. Generate the new molecules in their normal form only if they do not already exist in the system.

A list of some firstly obtained molecules is given below:

1 (3 3 3 3) , 1 (2 (2 (3)) 3 3 3) , 1 (1 (3 3 3) 3 3 3) , 1 (4 4 4 4) , 1 (3 4 4 4) , 2 (2 (4) 3) , 1 (1 (4 4 4) 3 3 4) , 1 (1 (3 3 3) 1 (3 3 3) 1 (3 3 3) 1 (3 3 3)) , 1(2 (1 (3 3 3)) 3 3 3) , 1 (2 (4) 3 3 4) , 1 (2 (1 (4 4 4)) 4 4 4) , 1 (3 3 4 4) , 2 (2 (4) 4) , 1 (1 (3 3 3) 1 (4 4 4) 1 (4 4 4) 1 (4 4 4)) , 1 (1 (3 3 3) 1 (3 3 3) 1 (3 3 3) 2 (2 (1 (3 3 3)))) , 1 (2 (2 (2 (1 (3 3 3)))) 3 3 3) , 1 (1 (4 4 4) 1 (4 4 4) 1 (4 4 4) 2 (1 (3 3 3))) , 1 (1 (2 (4) 2 (4) 2 (4)) 2 (3) 2 (3) 2 (3)) , 1 (1 (1 (4 4 4) 2 (1 (3 3 3)) 2 (3)) 1 (1 (4 4 4) 2 (1 (3 3 3)) 2 (3)) 1 (1 (4 4 4) 2 (1 (3 3 3)) 2 (3)) 1 (1 (4 4 4) 2 (1 (3 3 3)) 2 (3))).

It is worth verifying the correct vertical and horizontal organization of all the molecular trees to see the rules presented in the previous section in action.

After several seconds of simulation you easily get molecules long of 47 atoms like the following one:

1 (1 (2 (1 (1 (1 (2 (1 (1 (1 (1 (1 (1 (1 (2 (1 (1 (1 (2 (1 (1 (1 (2 4) 3 3) 2)) 2) 2) 2)) 2) 2) 4 4) 2) 2) 2)) 2) 2) 2)) 2) 1 (2 4) 3 3)

or 93 atoms like:

1 (1 (1 (4 4 4) 1 (4 4 4) 2 (1 (1 (4 4 4) 1 (4 4 4) 1 (4 4 4)))) 1 (1 (4 4 4) 1 (4 4 4) 2 (1 (1 (4 4 4) 1 (4 4 4) 1 (4 4 4)))) 1 (1 (4 4 4) 1 (4 4 4) 2 (1 (1 (4 4 4) 1 (4 4 4) 1 (4 4 4)))) 1 (1 (4 4 4) 1 (4 4 4) 2 (1 (1 (4 4 4) 1 (4 4 4) 1 (4 4 4)))))

You could substitute "1" with "C", "2" with "O" and "3" or "4" with "H" in order to obtain realistic organic molecules.

The Dynamics comes into play

Every chemical component, and thus molecule, is assigned an important attribute: its concentration (between brackets in the following) which can change in time. In the next simulations to be shown, like for classical chemical kinetics, the molecular concentration will change as the sole effect of the reactions. For instance in a first-order reaction whenever an interaction takes place between two molecules A and B to give two new products E and F:

$$A + B \rightarrow E + F$$

$$[E] = [F] \leftarrow k*[A]*[B]$$

if E and F don't exist already. k is called the reaction rate.

$$[E] \leftarrow [E] + k*[A]*[B]$$

if E already exists (the same for "F")

And in all cases:

$$[A] \leftarrow [A] - k*[A]*[B]$$
$$[B] \leftarrow [B] - k*[A]*[B]$$

Take the elementary case of a chemical reaction involving only two one-valence atoms "3" and "4" composing two diatomic molecules 3(3) and 4(4), and interacting by single-link-crossover:

$$3(3) + 4(4) \rightarrow [2]3(4) \qquad \text{(the [2] is the stoichiometric coefficient)}$$

Below (figure 7), you can see two plots showing the concentration evolution in time of the three molecules 3(3), 4(4) and 3(4): The first plot in the case of a reversible reaction, the second one in the case of an irreversible reaction.

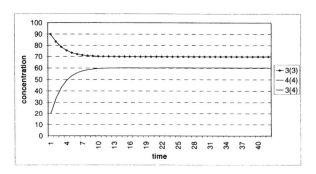

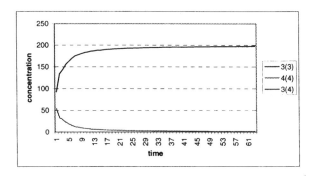

Figure 7: Time evolution of the concentration of the three interacting molecules: up – in the reversible case, down – in the irreversible case.

These two plots experimentally obtained could be analytically obtained by resolving in this case the simple differential equations. While the analytical road is possible for very simple reaction mechanisms, it is far to be the case for much more complicated reaction schemes involving a lot of intermediaries. For these reactions, rapidly leading to a formidable analytical complexity, one is forced to resort to computer simulations. The complete simulation of our chemical systems can take three forms. The first is *random interaction* shown in table 2.

```
Ad infinitum do {
- time = time + 1
- Select randomly one molecule A
- Select randomly one molecule B
- Make the reaction (A,B) according to a specific reaction
scheme
 - If the products of the reaction already exists, increase
their concentration, if not add them in the system with
their specific concentration.
- Decrease the concentration of A and B
}
```

Table 2: The random interaction algorithm

In the case of several interacting molecules, figure 8 shows the evolution in time of 5 molecules among others.

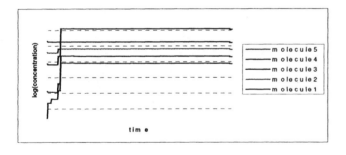

Figure 8: Time evolution of the concentration of the interacting molecules for the random interaction algorithm.

The flat line presence is due to the long period where nothing changes in the concentration of a specific molecule due to the birth of new molecules. The 5 molecules selecting in the plot are indeed the ones that re-enter in a reaction after a long period following their birth. Now the random interactions, although commonly found in a lot of artificial chemistry schemes (Fontana, 1992; Dittrich, 1999) does not make a lot of sense if the molecules are not single chemical objects but are concentration of chemical objects. In this case, the reaction rate "k", to some extent, already account for the randomness of the molecular collisions. It becomes

more natural to resort to a deterministic type of simulation such as indicated in table 3.

```
Ad infinitum do {
- time = time + 1
- For all molecules i of the system
 For all molecules j (going from 1 to i) of the system
 { - Make the reaction (i,j) according to a specific
reaction mechanism
  - If the products of the reaction already exist, increase
their concentration, if not add them in the system with their
specific concentration.
  - Decrease the concentration of i and j }}
```

Table 3: The deterministic interaction algorithm

The presence of the second loop going from "1 to i" in the loop maintains everything deterministic and just accounts for the symmetry of the reactions ("A reacts with B" is equivalent to "B reacts with A"). Although the best algorithmic version, because naturally leading to the analytical solution, the problem with such an algorithm is that everything rapidly exponentially explodes. After 4 or 5 time increments, the simulation is captured in a nearly infinite loop (at time t=0: 10 reactions, at t=1: 55 reactions, at t=2: 1540 reactions etc.).

A final algorithmic compromise is still deterministic but takes the time increment inside the double loop such that a reaction occurs at every time step, with nevertheless the sequence of molecular interaction kept fixed. The time evolution of some molecules is shown in figure 9.

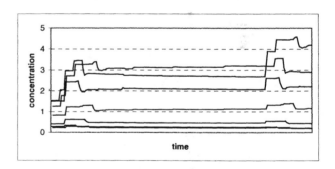

Figure 9: Time evolution of the concentration of seven molecules for the deterministic interaction algorithm

Interestingly enough, this version of the algorithm allows the detection of chaining and self-maintaining reaction (which is related with the Fontana's level 1 of self-maintaining molecular system (1992)). As a matter of fact, the concentration of certain molecules seems to oscillate in time. This is only possible if some reaction intermediaries are first consumed then regenerated as products of subsequent reactions. Fully self-maintaining molecular loops in which all molecules are the products of some interaction involving other molecules can be distinguished from weaker self-maintaining reactions in which only some of the molecules are obtained as a consequence of reactions.

However the observed cycles are nothing but artifactual effect of the version of the algorithm. They would naturally disappear to give fixed point in the more exact version of the algorithm (table 3). Indeed, the figure 7, in the reversible case, the smallest version of a molecular self-maintaining system, shows clearly the appearance of fixed point. This computer artifact reminds a similar effect and related discussion concerning "the asynchronous versus synchronous updating" found in the cellular automata literature (Bersini and Detours, 1994).

In order to clarify further this point, let's suppose two trivial examples of chemical reaction. In the first example, the reactions are the following (let's suppose two simple first order reactions with reaction rates k_1 and k_2 and three initial conditions $[A] = [A]_0$ and $[B] = [C] = 0$)

$$A \rightarrow B \text{ and } B \rightarrow C$$

The plot in fig.10 shows how the molecule concentrations evolve in time in our simulation. There is no visible difference with the analytical solution given by:

$$[A] = [A]_0 \exp(-k_1 t)$$
$$[B] = k_1 [A]_0 (\exp(-k_1 t) - \exp(-k_2 t))/(k_2 - k_1)$$

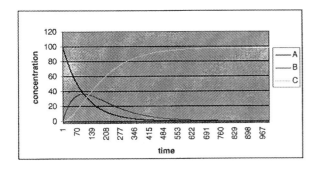

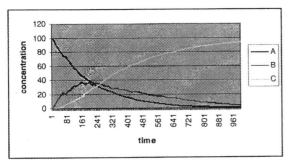

Figure 10: the concentration evolution in time of the molecules A,B,C - up): for the deterministic algorithm and down): for the random one

You can also see the plot obtained by a random choice between the two reactions and observe some noisy fluctuations that nevertheless still keep the global trends. Therefore for this simple reaction and, as a rule, as far as the

deterministic algorithmic is computationally tractable, the analytical solution is obtained for sure. Randomness introduces noise in the plotting, and the amount of degradation is dependent on the complexity of the reaction scheme.

Let's suppose now a second more complex example consisting in a loop of 5 reactions (the last reaction closing the chain):

$$A \rightarrow B, B \rightarrow C, C \rightarrow D, D \rightarrow E, E \rightarrow A.$$

The "exact" plotting of the deterministic algorithm is given in fig.11. Adopting the random interaction algorithm described in table 2 gives the noisy plot shown in the same figure. Here the randomness induces a much severe degradation of the plotting.

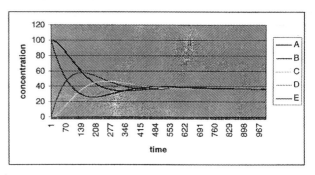

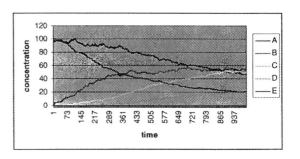

Figure 11: the concentration evolution in time of the molecules A,B,C,D,E - up): for the deterministic algorithm and down): for the random one

The intermediary "compromising" algorithm that could indeed appear as a form of time dilatation gives the most interesting plotting. The looping structure of the reaction induces the same type of artificial oscillation discussed previously. As a matter of fact this oscillatory pattern is the direct result of the loop and could indeed be exploited to detect this loop. In figure 12 you can see the evolution in time of the concentration of the molecule A for the three versions of the algorithm: determinist, "compromising" and the random. The oscillatory solution is the highest one since the concentration of A decreases slower than for the fully deterministic algorithm. During the first time steps of the reaction, the two alternative simulations gives result very different form the analytical result.

Which version of the algorithm to adopt becomes a difficult question depending on the objective pursued by the simulations. A computer experiment aiming at the highest fidelity to the chemical reality should adopt the basic deterministic version. In such a case, it is capital to limit the number of possible reactions by a deeper analysis of which reaction turns out to be possible.

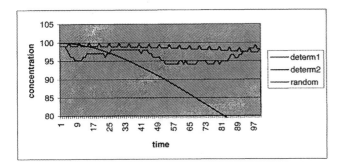

Figure 12: the concentration of the molecule A evolving in time for the three versions of the algorithm.

So while chaining reactions and self-maintaining molecular systems are certainly common in the whole reactor, their precise detection could be harder among the multitude of fixed points that would allow the simulation of the whole system, provided much computing power is available. Easily detecting these self-maintaining structures could rather privilege the use of the second compromising version of the algorithm.

Finally the random version common in some artificial chemistry simulation already known in Alife aims at compensating for the absence in these works of the concentration variable. Introducing this variable make useless the random version all the more as it can considerably degrade the concentration evolution in time.

Conclusions

The kind of computational chemistry being investigated in this work lies at an abstraction level that might allow the simultaneous fulfillment of two of the main objectives of Alife. A first one is to offer chemists a computational platform, which following an adequate parameterization could help them to model and understand the evolution of a particular chemical system. Indeed a large number of key aspects remain to be explored and tuned in the simulation among which:

- the value of the reaction rates
- whenever two molecules collide, which reaction mechanism takes place and which links in each molecule are involved in this reaction

- the kind of relationship that exists between the reaction rates and the composition of the molecules involved in the reaction.
- the dependence of the reactions on the energy sources

Getting all this missing knowledge, it might be possible to simulate in a distant future the historical Miller's experiment aiming at replicating the origin of life. This could shed some new light on the length of the organic molecules obtained by Miller, of surprising complexity but still so far from the complexity of a DNA.

On the other hand, Alife simulations can stand on their own and lead to the discovery of generic laws characterizing the behavior of complex systems. When concerned with chemistry, the best representative of this research road is Walter Fontana (1992, 1996) who is trying to develop a pure mathematical abstraction allowing to better formalize the appearance of interesting dynamic and self-maintaining structures. His assimilation of molecules with operators of the lambda calculus, and reactions with molecules operating one on another seems to be unrelated with this work.

However, it is possible that the algorithmic choices made in our simulation (molecules = computational trees and reactions = symmetric combinatorial exchanges or combinatorial re-organization of the trees), closer to real chemistry, could still appear as a possible instance of Fontana's mathematical developments. If this is the case, we could have for the same price a chemical computational platform benefiting from his mathematical solid progress while able to present a friendly interface to chemical practitioners.

References

1. Aho V. A., Ullman, J.D. (1995) : Foundations of Computer Science – Computer Science Press - W.H. Freeman and Company – New York.
2. Bersini, H. (1999): Design Patterns for an Object-Oriented Computational Chemistry – In proceedings of the 5th European Conference on Artificial Life (ECAL'99) – Eds : Floreano, Nicoud, Mondada – Springer – Verlag - pp. 389-398
3. Bersini, H. and V. Detours: Asynchrony induces stability in cellular automata based models – In proceedings of Artificial Life V – Eds : Brooks and Maes – MIT Press – pp. 382 – 387.
4. Detours, V., Bersini, H., Stewart, J. and Varela, F. (1994): Development of an Idiotypic Network in Shape Space – Journal of Theor. Biol. (170), 401-404 (1994)
5. Dittrich, P. and Banzhaf, W. (1998): Self-evolution in a constructive binary string system. Artificial Life, 4(2):203-220.
6. Dittrich, P. (1999): Artificial Chemistries – Tutorial held at ECAL'99 – European Conference on Artificial Life 13-17 September, 1999 – Lausanne, CH.

7. Farmer, J.D., Kauffman, S.A. and Packard, N.H. (1986) Autocatalytic reaction of polymers. Physica D, 22:50-67.

8. Fontana, W. (1992): Algorithmic Chemistry. In Artificial Life II: A Proceedings Volume in the SFI Studies in the Sciences of Complexity (C.G. Langton, J.D. Farmer, S. Rasmussen, C. Taylor, eds.), vol. 10. Addison-Wesley, Reading, Mass (1992)

9. Fontana, W. and L.W. Buss (1996). The barrier of objects: From dynamical systems to bounded organization in Casti, J. and Karlqvist, A., editors, Boundaries and Barriers, pages 56-116. Addison-Wesley.

10. Holland, J.H. (1995): Hidden Order – How adaptation builds complexity – Helix Books – Addison Wesley Publishing Company (1995)

11. Kauffman, S. (1993): The Origins of Order: Self-Organization and Selection in Evolution – Oxford University Press

12. Eriksson, H-E, Penker, M. (1998): UML Toolkit – John Wiley and Sons

A Less Abstract Artificial Chemistry

Pietro Speroni di Fenizio

COGS, Sussex University, Brighton, BN1 9QG, UK
Now at: ICD, Joseph-von-Fraunhofer-Str. 20, 44227 Dortmund, Germany
e-mail: pietro.s@altavista.net

Abstract

We start with AlChemy system [Fontana, 1992], and with four further modification we move into an Artificial Chemistry system where for each element is present an atomic structure, elements follow the conservation laws on the number of atoms, and the operation permits to two elements to generate more than one element. The resulting system still generates Organisations (self-sustaining, closed sets), but a wider variety of them are easily reachable without imposing any filter rules on the results of the operation [Fontana, and Buss, 1993]. The resulting Organisations are nearly equally divided into two classes (A and B), one (A) which contains Organisations that metabolise external elements to keep themselves 'alive', while the other (B) tends to expand until no place is left in reactor, without any need on external elements.

Introduction

Artificial Chemistries study the interaction of many elements, in a system where their interaction can generate new elements of the same kind. In this respect they are similar to an abstraction of chemistry where chemical elements interact generating other chemical elements.

Artificial Chemistries have been used to show how a self-organising set (organisation) naturally arises in a system of interacting elements. [Fontana, 1992]. Other researchers have developed his work further [Dittrich, and Banzhaf, 1998], [Suzuki, and Tanaka, 1998].

With Organisations, we intend [Fontana, 1992] a set of elements such that each element can be generated by the conjunct action of the others, and such that given any two elements in the system their interaction will always generate only elements of the set.

In many systems studied either the number of possible elements were finite (as in Suzuki), or the number of elements in the system is fixed (as in Fontana). Systems where the number of possible elements is not fixed and the number of elements inside the system is not fixed are rare. We present a system of this kind.

In AlChemy, and in general when the number of elements in the system is fixed, a standard operation is adopted. At each time step three elements are randomly chosen (A, B and X). A function is applied that associates to the first two elements (A and B) another one (C_{AB}), if this new element is acceptable the X element is eliminated and the C_{AB} element is instead inserted in the system.

A different operation is here proposed. Two elements are randomly chosen (A and B) and a multi set (that is a set where the same element can appear more than once) of N elements is generated. That is

$$A+B \rightarrow C_1+C_2+\ldots+C_{n_{AB}} \qquad (n_{AB}>=1)$$

The system using this new operation can model interactions more similar to the chemistry ones. Like reactions of the form $A+B \rightarrow C+D$.

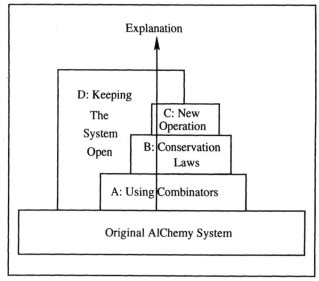

Figure 1. This figure shows the modifications to the alchemy system presented. Each one builds upon the previous. We could have stopped after each modification, but the most interesting results seem to be present only when all the modifications are working together.

Modifications Applied

To build a system where the number of elements is not fixed, the system doesn't explode, and the number of possible elements is not limited either, we need to change the original Fontana AlChemy system in some ways.

a. Combinators are used.
b. Conservation laws are applied.
c. The operation is changed.

d. The system is kept as an open system.

As a first modification the whole system is rewritten using combinators as base elements instead than lambda-terms. We then apply the conservation laws on the number of simple combinators (1-term combinator) used. This is made possible by the atomic structure of combinators. Then the operation is changed, moving to a more chemical-like operation. From now on the number of elements will not be fixed , yet the whole system will be bounded by the conservation laws. Yet the system in those conditions will tend to stop very soon, and to prevent it from stopping we add the last modification. The system is inserted in a small flux of elements that sometimes will produce new elements and sometimes will randomly eliminate one of the existing ones.

A. The Algebra of Combinators

In AlChemy each element is a lambda-term that can be seen equally well as an operator or as data [Fontana, 1992][Hindley and Seldin, 1986]. At each time step two elements are randomly chosen and the first one is applied to the second to generate the new one. In our system each element is a combinator rather than a lambda-term.

A combinator is a string whose elements are either elements from a finite alphabet or combinators themselves [Hindley and Seldin, 1986]. Combinators contained in other combinators are represented with a parenthesis before ('(') and after (')'). The alphabet we normally use contains 6 basic element: I, K, S, B, C, W. Any sequence of those letter is a valid combinator, such as: KI, KIBBBW, BSWWB, but also, K(BBSKI)S is a valid combinator with BBSKI as a sub-combinator. Or even, BBB(SK(KSSI))I is a more complicated combinator with two levels of sub-combinators inside.

As with lambda-terms, each combinator can be seen both as an operator and as data. As an operator it will transform a string (combinator) into another string (combinator).

When a combinator is applied to another the second is added to the first as a sub-combinator, so for example if we apply SSK to KI the result would be SSK(KI). In any combinator the first element is never enclosed by parenthesis, if they are present they can be omitted, so the combinator (SI)KKS is the same as the combinator SIKKS.

Every letter of the alphabet, as an operator, act upon the remaining elements by rearranging them, occasionally making copies of some elements or destroying others. So for example the combinator S, applied to 3 elements x1, x2, x3 (S $x_1x_2x_3$) acts to generate the combinator x1x3(x2x3). This is represented as:

$$S\ x_1x_2x_3 \rightarrow x_1x_3(x_2x_3)$$

The → operation is called 'combinatorial reduction' and each 1-term combinator (element of the alphabet) rearranges the data in different ways (see table 1 for

details). Each element just rearranges the first elements leaving the rest unaltered.

A.	Effect	A.	Effect
I	I $x_1s_0 \rightarrow x_1s_0$	B	B $x_1x_2x_3s_0 \rightarrow x_1(x_2x_3)s_0$
K	K$x_1x_2\ s_0 \rightarrow x_1\ s_0$	C	C $x_1x_2x_3s_0 \rightarrow x_1x_3x_2\ s_0$
S	S $x_1x_2x_3s_0 \rightarrow x_1x_3(x_2x_3)s_0$	W	W $x_1x_2\ s_0 \rightarrow x_1x_2x_2\ s_0$

Table 1. Basic Atom Types. Note: A. stands for Atom, s_0 is just the remaining sub-string and can be equal to $\varnothing$.

Not all combinators can be reduced, a combinator whose first element is not followed by enough terms, and where in each set of parenthesis the first term is always followed by too few other sub terms, cannot be reduced. For example: KK(SKI) can be reduced to K and K(SKKI) can be reduced to K(KI(KI)) which can be further reduced to K(I)=KI. While K(SKI) cannot be further reduced. A combinator which can't be further reduced is said to be in its normal form. Thus when we speak of a combinator we are really speaking of a whole equivalence class of strings, and two combinators will be equivalent if can be reduced to the same combinator. This definition is well posed since [Hindley and Seldin, 1986] if a combinator possesses a normal form this is unique (or in other words by normalising one part of a combinator before another we will never reach two different normal forms). This does not mean that every combinator does have a normal form. There are in fact infinite combinator that either enter into cycles or continue to get longer and longer as they 'reduce'. For example WWW reduces to itself.

The normal form of a combinator is equivalent to the normal form of a lambda-term in Fontana's program AlChemy. In our system our elements will be combinators expressed in their normal form.

B. Adding the Conservation Laws

Having used combinators instead of lambda terms we can now see each element as being constructed from smaller atomic parts. We can thus impose the conservation laws upon the total number of atoms (that in our case will be the 1-term combinator). To permit this a list of all free atoms that are available at each moment is kept. This list will in general be called 'the Pool'. At each time step, as two elements interact, and as the resulting element is reduced and normalised, every atom that has to be added is taken from the Pool, and every atom that is used up is dropped back into the pool. At the end we will consider as acceptable only the operations that don't require the presence of more atoms than the ones present in the Pool.

An example of the use of the Pool could be:

Suppose the Pool contains 20 B, 20 C, 20 I, 20 K, 20 S, 20 W. The two elements SKI and KW are randomly chosen.

The first is applied to the second and the combinator SKI(KW) is generated. 1 S, 2 K, 1 W, and 1 I are taken from the Pool (20 B, 20 C, 19 I, 18 K, 19 S, 19 W). Then the normalisation process begins:

SKI(KW)→K(KW)(IKW); 1 S is added to the Pool and 1 K and 1 W are subtracted from it (20 B, 20 C, 19 I, 17 K, 20 S, 18 W).

K(KW)(IKW)→KW; 2 Ks, 1 I, and 1 W are added to the Pool (20 B, 20 C, 20 I, 19 K, 20 S, 19 W).

KW is a combinator in its normal form. KW is then added to the list of elements and another element is randomly chosen, and deconstructed to its basic atoms that are then added to the Pool. At each time step the number of elements in the Pool is checked and if in any moment there are not enough free atoms the whole reaction is considered elastic (in this environment) and the reaction never to have happened.

C. A Different Operation

Our next modification on the system will be to change the operation that permits the generation of new elements after having chosen the two interacting ones. We want the operation to be such that given two elements we generate a whole family of N elements, with N dependent only on the two interacting elements. We want an operation such that given a pair of elements the generated family is uniquely defined. This will be possible if we just redefine the action of one of the combinators. The key of the new operation will be the combinator K and its action upon other combinators. From table 1 we can see that K keeps the first combinator following it (x_1) and destroys the second one (x_2). We can, instead, consider K as merely detaching the second combinator x_2, leaving it as an independent unit to float away. To make things more similar to the actual chemistry we will eliminate, at each time step, the two elements that interacted, keeping only their product. We are thus using up the elements in the operation, as molecules are effectively used up in real chemical interaction.

The original operation could be written as:

A+B+X→A+B+A(B),

while the new one would be written as:

A+B→ C_1+C_2+...+$C_{n_{AB}}$ (n_{AB}>=1)

Of course every C_i will have to be normalised and the whole operation will happen only if all the element reach a normal form, without exhausting all the free atoms in the Pool.

D. Keeping the System Open

To prevent the system from stopping and to keep a flow of energy through the system we added, as a last rule, that new elements (using the free atoms from the pool) would be inserted, when the number of random elements dropped

under a Critical Minimum Value (CMV). For each empty position under the CMV there was a fixed probability (p1) that at each time step the position would be filled by a new random element. Also each element would have a fixed probability (p2) to be destroyed. This didn't modify the behaviour of organisations (where many copies of each element is in general present), but would prevent the system from being stopped by having too many huge unusable elements.

Results

We made 180 runs with parameters: starting number of elements=300, CMV 300, p1=0.1, p2=0.0001, number of atoms=2000 for each type, max length accepted 100 atoms, max depth accepted 20 levels of parenthesis.

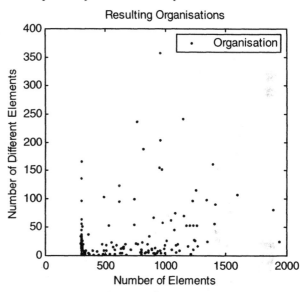

Figure 2. Resulting Organisations. Each resulting Organisation is drawn as a point. Organisations either have around 300 Elements (Class A, Organisations that balance the input of new elements with the destruction of old ones) or are scattered around (Class B, Organisations that expand until all the atoms are used)

Running the system we immediately noticed some results.
1.   Even though the number of element is not fixed from the beginning, after a transient phase it tends to stabilise itself around an average value with little or no oscillation around that value.
2.   Some rare times the value does indeed generate bigger fluctuation around the chosen value. Those experiments are always very rare.
3.   Each time the experiment is run, with a different random seed, we reach a different organisation. We can thus notice that the space of all the possible organisations

is very rich. This makes it possible (as a possible future work) to search for an organisation with certain characteristics.

4. Some of the Organisations hold a finite number of elements, some an infinite number (even if, of course, at any one moment only a finite number can be present in the experiment). Interestingly sometimes an infinite Organisation is generated that holds as sub-Organisations both a finite and an infinite one. The two Organisations are kept in balance by the conservation laws. In fact the Organisations tend to be based on different atoms, and would not compete one with the other, as they do in the original AlChemy program.

5. Two different kinds of organisations tend to be generated

a. Organisations that get smaller until new random elements are inserted. When this happens they 'use' those new elements to expand and the cycle begins again.

b. Organisations that gets bigger and bigger until they end the available atoms. Those organisations, under our starting conditions, tend to have between 350 and 1000 elements.

We present an example of organisation for each type.

Organisation of Type A

Our first example is an organisation generated by a single element.

C(C(K(CKK))(WC))(C(K(CKK))(WC))=α.

This element applied to any other generates:

α * a$\rightarrow$$\alpha$(a) $\rightarrow$$\alpha$, K, a.

α applied to 'a' releases a copy of K and a copy of 'a'. K element applied to any element 'a' generates K(a).:

K * a$\rightarrow$K(a)

thus K(α), K(K(α)), …, $K^n(\alpha)$. are generated, as well as KK, K(KK), …,$K^n(KK)$.

This organisation contains an infinite number of elements. Since $K^n(b)$ applied to any element 'a' generates:

$K^n(b)$ * a $\rightarrow$ $K^n(b)(a)$ $\rightarrow$ $K^{n-1}(b)$, a   (for n>1).

K(b) * a $\rightarrow$ K(b)(a) $\rightarrow$ b, a.

thus no new elements are generated and O≡{α, K, KK, $K^n(\alpha)$, $K^n(KK)$} is an organisation.

O tend to keep its number of elements fixed. Only α can increase, through a reaction, the number of elements present in the reactor. Yet every new element, K, generates a complementary reaction (K *a) that decreases the number of elements present. O tends to remain of a fixed size until some elements are randomly destroyed. When the number of elements decreases too much new elements are inserted and the organisation metabolises them increasing its total number of elements.

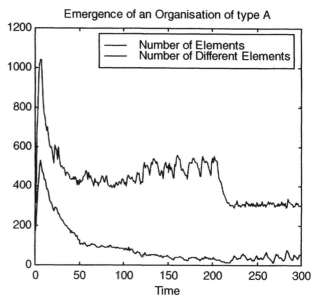

Figure 3. Organisation of type A. After a transition phase of about 210 generations the organisation emerges. The number of elements is always between 296 and 330. The number of different elements present has wider oscillation, since the organisation uses external elements and has itself an infinite number of elements.

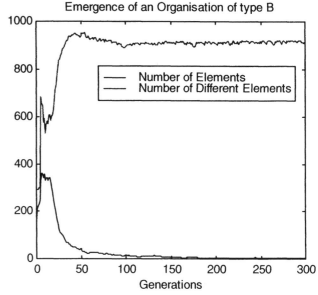

Figure 4. Organisation of Type B. A much shorter transition phase is necessary. The end Organisation has about 900 elements of between 4 and 6 different types.

Organisation of Type B

The second organisation we present is generated by 4 different elements:

C(K(WC))(K(WC))=α;

S(SS(S(W(S(SWB)))))(B(K(S(SS(S(W(S(SWB)))))))))=β;

S(K(SSK))(K(K(SSK)))=γ;
B(WW)(W(B(WK)W))=δ

The relative reactions are:

α * a= α, a.

γ * a= γ, a, a.

δ * a= aaa(aaa)(aaa), Wa, Wa, Wa.

δ * α = 9 α, 3 W(α)

δ * γ = 17 γ, 3 W(γ)

β * a= Out of memory.

The dynamic of the organisation is driven by the γ element that pushes the whole system into having more and more elements. The δ element tend to have no effect since it requires an enormous effort in terms of free atoms. While the α has no effect whatsoever.

Organisations as Abstract Metabolisms

In this section we will show how those organisations differ from the organisations previously built by being a metabolism with respect to the insertion of new elements.

A general definition of metabolism, fit for our work can be found in [Bagley and Farmer, 1992, p. 127]:

"A metabolism takes in material from its environment and reassembles it in order to propagate the form of the host organism.... Another function of a metabolism is to extract free energy from the environment and make it available for the functions of the organism...".

This definition is totally abstract and can also be applied to non physical systems. There are two different functions that are accomplished from a metabolism: an energetic function and a physical one.

An Organisation can act as a metabolism in two different ways. Directly if are the interactions between its elements and the environment that produce the transformation in the external elements. Indirectly if it generates the conditions under which the external elements cannot survive, without directly interacting with the external elements.

Organisations of the first type (type A) tend to contract until new elements are inserted. As new elements are inserted they interact with those new elements using them to produce more elements of the former organisation.

For example an element of an organisation could be:

a such that a * b = a, b.

Those elements regenerate themselves all the time.

If they are invaded by an element S, chances are that S is applied to 'a' for three consecutive times. When this happen Sa, Saa, and Saaa are subsequently produced. Saaa is then reduced to aa(aa)→ ... → a, a, a, a. So three copies interact with S but four are generated.

Organisations of the second type tend to have a rather different reaction to external elements. They don't need external elements to make copies of themselves. Yet, since they tend to expand until they use up all the atoms of a particular kind, an organisation is then 'interested' in destroying an external atom to free its elements. This cannot be done directly, yet it is possible to prove that the organisation effectively induces an environment where those external elements are easily destroyed. This happens as the system reaches a state where all the free atoms of many different kinds are used up and many reactions are inhibited.

We believe that this work provide a good base to continue the exploration of artificial chemistries and artificial metabolism. We see as a next important step the study of system where sub-elements can both be released and be destroyed. We also see as an important future step the study of a system were such a metabolism is inserted in a metric space to provide the possibility for boundaries to evolve and as a last aim for evolution to start.

Acknowledgements

I am deeply grateful to Inman Harvey for his support in writing this paper. I wish to thank Mark Bedau, Chris Adami and Peter Dittrich for the many helpful comments that inspired this work. A last thanks goes to the Sussex University for its important support.

References

Bagley J. R. and Farmer J. D. 1992. Spontaneous Emergence of a Metabolism. In Proceedings of The Second International Conference on Artificial Life, 94-140. Edited by Langton C. G., Taylor C., Farmer J. D., and Rasmussen S., Redwood City, Addison-Wesley.

Dittrich, P., and Banzhaf W. 1998. Self-Evolution in a constructive Binary String System. Artificial Life 4(2):203-220.

Fontana W. 1992. Algorithmic Chemistry. In Proceedings of The Second International Conference on Artificial Life, 159-209. Edited by Langton C. G., Taylor C., Farmer J. D., and Rasmussen S., Redwood City, Addison-Wesley.

Fontana W., and Buss L. W. 1993. "The arrival of the Fittest": toward a Theory of Biological Organization. Bulletin of Mathematical Biology.

Hindley J. R., and Seldin J. P., 1986. Introduction to Combinators and λ-Calculus. Cambridge University Press.

Kauffman, S. 1993. The Origins of Order: Self-Organisation and Selection in Evolution. Oxford University Press.

Suzuki Y., and Tanaka H. 1998. Order Parameter for Symbolic Chemical System. In Proceedings of The Sixth International Conference on Artificial Life, 130-139. Edited by Adami C., Belew R., Kitano H., and Taylor C., MIT Press.

Chemical evolution among artificial proto-cells

Yasuhiro Suzuki and Hiroshi Tanaka

Bio-Informatics,
Medical Research Institute,
Tokyo Medical and Dental University
Yushima 1-5-45, Bunkyo, Tokyo 113 JAPAN

Abstract

We develop an Artificial Cell System (ACS) based on an abstract chemical system. ACS consists of a multiset of symbols, a set of rewriting rules (*reaction rules*) and membranes. Throughout simulations, we find that chemical evolution like behavior emerges and cells evolve to a structure consisting of several cell-like membranes. We investigate the correlation among the type of reaction rules (rewriting rules), characteristic of a membrane and the evolution of cells, and then find the characteristics of a membrane effects on its evolution and obtains a parameter to describe the correlation. Furthermore, we introduce a genetic method to the system, and we attempt to apply it to Genetic Programming.

Introduction

A membrane is an important structure for living systems. It distinguishes "self" from its environment and hierarchical structures inside the system (like cells, organs and so on) are composed by membranes. Membranes change their structure dynamically and constitute a system. We are interested in their dynamical structure in terms of computation.

A membrane is composed of "chemical compounds (denoted by symbols)" which are generated through chemical reactions in the cell. In each cell there is some chemical compounds and these chemical compounds interact with each other according to the rewriting rule (*reaction rules*).

Based on the principles outlined above, we develop an "*Artificial Cell System*" (*ACS*). It consists of a multiset of symbols, a set of rewriting rules (reaction rules) and membranes. *ACS* consists of an abstract chemical system, "*Abstract Rewriting System on MultiSets (ARMS)*" that is a multiset transform system,(Suzuki 1996). It consists of a multiset of symbols and a set of rewriting rules. Although not many alife researches have tackled this topic previously (McMullin and Varela 1997), (Mizaro, Moreno, et. al. 1999), the focus of these researches is on the formation of a membrane. The aim of this study is to investigate the role of membrane in terms computation, thus we do not treat its formation.

Since *ARMS* is not a strings rewriting system but a (multi)-*set* rewriting system, it can deal with many degrees of freedom such as a *chemical solution* and the *concentration* of *chemical compounds*. We confirmed that it could simulate phenomena that we find in real bio-chemical reactions such as the "*Belouzov-Zhabotinsky reaction (BZ-reaction) (Suzuki 1998)*" the *BZ-reaction* is a spontaneous chemical oscillation, and is considered as the basic mechanism of bio-chemical systems.

ARMS

We will introduce the multiset rewriting system, "*Abstract Rewriting system on MultiSets*" in this section. Intuitively, *ARMS* is like a chemical solution in which *molecules* floating on it can interact with each other according to reaction rules. Technically, a chemical solution is a finite multi-set of elements denoted by $A^k = \{a, b, \ldots, \}$; these elements correspond to *molecules*. Reaction rules that act on the molecules are specified in *ARMS* by rewriting rules. As to the intuitive meaning of *ARMS*, we refer to the study of chemical abstract machines (Berry 1992). In fact, this system can be thought of as an underlying "*algorithmic chemistry* (Fontana 1994)."

Let A be an *alphabet* (a finite set of abstract symbols). The set of all strings over A is denoted by A^*; the empty string is denoted by λ. (Thus, A^* is the free monoid generated by A under the operation of concatenation, with identity λ.) The length of a string $w \in A^*$ is denoted by $|w|$.

A *rewriting rule* over A is a pair of strings (u, v), $u, v \in A^*$. We write such a rule in the form $u \rightarrow v$. Note that u and v can also be empty. A *rewriting system* is a pair (A, R), where A is an alphabet and R is a finite set of rewriting rules over A.

With respect to a rewriting system $\gamma = (A, R)$ we define over A^* a relation $\Longrightarrow$ as follows: $x \Longrightarrow y$ iff $x = x_1 u x_2$ and $y = x_1 v x_2$, for some $x_1, x_2 \in A^*$ and $u \to v \in R$. The reflexive and transitive closure of this relation is denoted by $\Longrightarrow^*$. A string $x \in A^*$ for which there is no string $y \in A^*$ such that $x \Longrightarrow y$ is said to be an *dead* one (in other words, from a dead string no string can be derived by means of the rewriting rules).

From now on, we work with an alphabet A whose elements are called *objects*; the alphabet itself is called a *set of objects*.

A *multiset* over a set of objects A is a mapping $M : A \longrightarrow \mathbf{N}$, where $\mathbf{N}$ is the set of natural numbers, 0, 1, 2,.... The number $M(a)$, for $a \in A$, is the *multiplicity* of object a in the multiset M. Note that we do not accept here an infinite multiplicity. The set $\{a \in A \mid M(a) > 0\}$ is denoted by $supp(M)$ and is called the *support* of M. The number $\sum_{a \in A} M(a)$ is denoted by $weight(M)$ and is called the *weight* of M.

We denote by $A^{\#}$ the set of all multisets over A, including the empty multiset, $\emptyset$, defined by $\emptyset(a) = 0$ for all $a \in A$.

A multiset $M : A \longrightarrow \mathbf{N}$, for $A = \{a_1, \ldots, a_n\}$, can be naturally represented by the string $a_1^{M(a_1)} a_2^{M(a_2)} \ldots a_n^{M(a_n)}$ and by any other permutation of this string. Conversely, with any string w over A we can associate a multiset: denote by $|w|_{a_i}$ the number of occurrences of object a_i in w, $1 \le i \le n$; then, the multiset associated with w, denoted by M_w, is defined by $M_w(a_i) = |w|_{a_i}, 1 \le i \le n$.

The union of two multisets $M_1, M_2 : A \longrightarrow \mathbf{N}$ is the multiset $(M_1 \cup M_2) : A \longrightarrow \mathbf{N}$ defined by $(M_1 \cup M_2)(a) = M_1(a) + M_2(a)$, for all $a \in A$. If $M_1(a) \le M_2(a)$ for all $a \in A$, then we say that multiset M_1 is included in multiset M_2 and we write $M_1 \subseteq M_2$. In such a case, we define the multiset difference $M_1 - M_2$ by $(M_2 - M_1)(a) = M_2(a) - M_1(a)$, for all $a \in A$. (Note that when M_1 is not included in M_2, the difference is not defined).

A rewriting rule such as

$$a \to a \ldots b,$$

is called a *heating rule* and denoted as $r_{\Delta > 0}$; it is intended to contribute to the stirring solution. It breaks up a complex *molecule* into smaller ones: *ions*. On the other hand, a rule such as

$$a \ldots c \to b,$$

is called a *cooling rule* and denoted as $r_{\Delta < 0}$; it rebuilds *molecules* from smaller ones. In this paper, reversible reactions, i.e., $S \rightleftharpoons T$, are not considered. We shall not formally introduce the refinement of *ions* and *molecules* though we use refinement informally to help intuition (on both types of rules we refer to (Berry 1992)).

A *multiset rewriting rule* (we also use to say, *evolution rule*) over a set A of objects is a pair (M_1, M_2), of elements in $A^{\#}$ (which can be represented as a rewriting rule $w_1 \to w_2$, for two strings $w_1, w_2 \in A^*$ such that $M_{w_1} = M_1$ and $M_{w_2} = M_2$). We use to represent such a rule in the form $M_1 \to M_2$.

An *abstract rewriting system on multisets* (in short, an *ARMS*) is a pair

$$\Gamma = (A, R)$$

where:

(1) A is a set of objects;

(2) R is a finite set of multiset evolution rules over A;

With respect to an *ARMS* Γ, we can define over $A^{\#}$ a relation: $(\Longrightarrow)$: for $M, M' \in A^{\#}$ we write $M \Longrightarrow M'$ iff

$$M' = (M - (M_1 \cup \ldots \cup M_k)) \cup (M_1' \cup \ldots \cup M_k',)$$

for some $M_i \to M_i' \in R, 1 \le i \le k, k \ge 1$, and there is no rule $M_s \to M_s' \in R$ such that $M_s \subseteq (M - (M_1 \cup \ldots \cup M_k))$; at most one of the multisets $M_i, 1 \le i \le k$, may be empty.

With respect to an *ARMS* $\Gamma = (A, R)$ we can define various types of multisets:

- A multiset $M \in A^{\#}$ is *dead* if there is no $M' \in A^{\#}$ such that $M \Longrightarrow M'$ (this is equivalent to the fact that there is no rule $M_1 \to M_2 \in R$ such that $M_1 \subseteq M$).

- A multiset $M \in A^{\#}$ is *initial* if there is no $M' \in A^{\#}$ such that $M' \Longrightarrow M$.

How *ARMS* works

Example In this example, an *ARMS* is defined as follows;

$$
\begin{aligned}
\Gamma &= (A, R), \\
A &= \{a, b, c, d, e, f\}, \\
R &= \{a, a, a \to c : r_1, b \to d : r_2, c \to e : r_3, \\
&\quad d \to f, f : r_4, a \to a, b, b, a : r_5, f \to h : r_6, \}.
\end{aligned}
$$

The set of the rewriting rules, R is $\{r_1, r_2, r_3, r_4, r_5, r_6\}$. We assume the maximal multiset size is 4 and the initial state is given by $\{a, a, b, a\}$. In *ARMS*, rewriting rules are applied in parallel. When

$$\{a, b, a, a\} \quad \subseteq a, \ a, \ a, \ a, \ b$$
$$\downarrow \qquad \text{(the left hand side of } r_1, r_2, \ r_5)$$
$$\{c, d\} \quad \subseteq c, \ d \text{ (the left hand side of } r_3, r_4)$$
$$\downarrow$$
$$\{e, f, f\} \quad \subseteq f \text{ (the left hand side of } r_6)$$
$$\downarrow$$
$$\{e, h, h\} \quad \text{There are no rule to apply, it reaches}$$
$$\text{the } death \text{ state}$$

Figure 1: Example of rewriting steps of $ARMS$

there are more than two applicable-rules, then one rule is selected randomly. Figure 1 illustrates an example of rewriting steps of the calculation from the initial state.

At the first step, the left hand side of rule of r_1, r_2 and r_5 are included in the initial state. In the next step, r_3 and r_4 are applied in parallel and $\{c, d\}$ is rewritten into $\{e, f, f\}$. In step 3, by using r_6, $\{e, f, f\}$ is transformed into $\{e, h, h\}$. There are no rules that can transform the multiset any further so, the multiset is in a *dead* state.

Artificial Cell System

In this section, we introduce the basic structural ingredients of ARMS, membrane structures and how ACS works.

The membrane structure (MS)

To describe the membrane and its structure in $ARMS$, we first define the language MS over the alphabet $\{[,]\}$ whose strings are recurrently defined as follows:

(1) $[,] \in MS$

(2) if $\mu_1, ..., \mu_n \in MS$, n $\geq$ 1, then $[\mu_1, ..., \mu_n] \in MS$;

(3) there is nothing else in MS.

The most outer membrane M_0 corresponds to a container such as a test tube or reactor and it never dissolves.

Consider now the following relation on MS: for $x, y \in MS$ we write $x \sim y$ if and only if we can write the two strings in the form $x = [_1...[_2...]_2[_3...]_3...]_1, y = [_1...[_3...]_3[_2...]_2...]$, i.e., if and only if two pairs of parentheses which are not contained in one other can be interchanged, together with their contents. We also denote by $\sim$ the reflexive and transitive closures of the relation $\sim$. This is clearly an equivalence relation. We denote by $\overline{MS}$ the set of equivalence classes of MS with respect to this relation. The elements of $\overline{MS}$ are called *membrane structures*.

It is easy to see that the parentheses [,] appearing in a membrane structure are matching correctly in the

usual sense. Conversely, any string of correctly matching pairs of parentheses [,], with a matching pair at the ends, corresponds to a membrane structure.

Each matching pair of parentheses [,] appearing in a membrane structure is called a *membrane*. The number of membranes in a membrane structure μ is called the *degree* of μ and is denoted by deg(μ). The external membrane of a membrane structure μ is called the *vessel;* membrane of μ. When a membrane which appears in $\mu \in \overline{MS}$ has the form [ ] and no other membranes appear inside the two parentheses then it is called an *elementary* membrane.

ACS and ACSE

We will define two types of ACS;

(1) ACS and

(2) ACS with an Elementary membrane (ACSE).

ACSE is different only in the way of dissolving and dividing from ACS.

Descriptions of ACS

A transition ACS is a construct

$$\Gamma = (A, \mu, M_1, ..., M_n, R, MC, \delta, \sigma),$$

where:

(1) A is a set of objects;

(2) μ is a membrane structure (it can be changed throughout a computation);

(3) $M_1, ..., M_n$, are multisets associated with the regions 1,2, ... n of μ;

(4) R is a finite set of multiset evolution rules over A.

(5) MC is a set of membrane compounds;

(6) δ is the threshold value of dissolving a membrane;

(7) σ is the threshold value of dividing a membrane;

μ is a membrane structure of degree n, n $\geq$ 1, with the membranes labeled in a one-to-one manner, for instance, with the numbers from 1 to n. In this way, also the regions of μ are identified by the numbers from 1 to n.

Rewriting rules are applied in following manner:

(1) The same rules are applied to every membrane. There are no rules specific to a membrane.

(2) All the rules are applied in parallel. In every step, all the rules are applied to all objects in every membrane that can be applied. If there are more than two rules that can apply to an object then one rule is selected randomly.

(3) If a membrane dissolves, then all the objects in its region are left free in the region immediately above it.

(4) All objects and membranes not specified in a rule and which do not evolve are passed unchanged to the next step.

Rewriting rule R is a finite set of multiset rewriting rules over A. Both the left and the right side of a rule are obtained by sampling with replacement of symbols. A set of reaction rules is constructed as the overall permutation of both sides of the rules.

Input and Output Chemical compounds are supported from outside of the system to M_0 and some compounds are exhausted from M_0. All chemical compounds are transformed among cells, a randomly selected chemical compound is transformed into the membrane just above or below it. Although a membrane does not allow specificity of transport across the membrane, a cell can control its chemical environment by chemical reaction.

Dissolving and dividing a membrane of ACS A membrane is composed of a "membrane compound" which is in fact a symbol. To maintain a membrane, it needs to have a certain minimal volume. A membrane disappears if the volume of membrane compounds decreases below the needed volume to maintain the membrane. Dissolving the membrane is defined as follows:

$$[_h a, ...[_i b, ...]_i]_h \rightarrow [_h a, b, ...]_h,$$

where the ellipsis $\{...\}$ illustrates chemical compounds inside the membrane. Dissolving takes place when

$$\frac{|w_i|_{MC}}{|M_i|} < \delta$$

where δ is a threshold value for dissolving the membrane. All chemical compounds in its region are then set free and they are merged into the region immediately above it.

On the other hand, when the volume of membrane compounds increases to a certain extent, then a membrane is divided. Dividing a membrane is realized by dividing it in multisets random sizes. The frequency at which a membrane is divided is decided in proportion to its size. As the size of a multiset becomes larger, the cell is divided more frequently. Technically, this is defined as follows;

$$[_h a, b, ...]_h \rightarrow [_h a, ...[_i b, ...]_i]_h$$

Dividing takes place when

$$\frac{|w_h|_{MC}}{|M_h|} > \sigma$$

where σ is a threshold for dividing the membrane. All chemical compounds in its region are then set free and they are separated randomly by new membranes.

Description of ACSE

ACSE is different only in the way of dividing and dissolving cells from ACS. Dissolving the membrane is defined as follows:

$$[_h a, b, ...]_h \rightarrow [_0 a, b, ...,]_0$$

Dissolving takes place when

$$\frac{|w_h|_{MC}}{|M_h|} < \delta$$

where δ is a threshold value for dissolving the membrane. All chemical compounds in its region are then set free and they are merged into the region of M_0.

Dividing is defined as follows;

$$[_h a, b, ...]_h \rightarrow [_h a, ...]_h [_i b, ...]_i.$$

Dividing takes place when

$$\frac{|w_h|_{MC}}{|M_h|} > \sigma,$$

where σ is a threshold for dividing the membrane. All chemical compounds in its region are then set free and they are separated randomly in the old and new membranes. Hence, in ACSE, a structured cell such as $[a, b[c, [d, e]]]$ does not appear.

A Cell like Chemoton Because the components of a membrane diminish with the lapse of a certain time, a cell has to generate the components to maintain the membrane through chemical reactions in the cell in ACS and ACSE. Hence, all survived cells in ACS and ACSE become cells like a *chemoton*(Gánti 1975). We confirmed this through simulations.

Evolution of Cells When a cell grows and the cell exceeds the threshold value for dividing, it divides into parts of random sizes. This can be seen as a kind of *mutation*. If a divided cell does not have any membrane compounds, it must disappear soon.

Furthermore, to maintain the membrane through chemical reactions inside the cell can be seen as *natural selection*. If cells can not maintain the membrane compounds, it must disappear soon.

Thus, both dividing membranes and dissolving membranes produce evolutionary dynamics. These correspondences are summarized into;

Natural Selection	Dissolving a membrane,
Mutation	Dividing a cell into parts of random size.

Behavior of *ACSE* and *ACS*

We will show some experimental results of ACSE and ACS in this section.

ACSE The evolution of elementary cells can be regarded as an approximate model of the chemical evolution in the origin of life.

The following *ACSE* was simulated;

$$\Gamma = (A = \{a,b,c\}, \mu = \{ [,]_0, ...[,]_{100}\} M_0 = \{[a^{10}, b^{10}, c^{10}]^{100}\}, R, MC = \{b\}, \delta = 0.4, \sigma = 0.2),$$

where:

(1) R, the length of the left- or right-hand-side of a rule is between one and three. Both sides of the rules are obtained by sampling with replacement of the three symbols a, b and c;

(2) Membrane structures are assumed to be ($\mu = \{[_1]_1...[_{100}]_{100}\}$).

Through the simulation we discovered that the strength of a membrane affects the behavior of cells. The strength of a membrane is defined as the frequency of decreasing membrane compounds.

When a membrane is strong When a membrane is strong, the most stable cell consists of only one membrane, cells of this type become " mother" cells and they produce "daughter" cells.

In order to display a state of a cell we transform the state of a cell to a number by using the transformation function; $f(M(a), M(b), M(c)) = 10^2 \times M(a) + 10^1 \times M(b) + 10^0 \times M(c)$. For example, the state $\{a, a, b, c, c\}$ is transformed into $10^2 \times 2 + 10^1 \times 1 + 10^0 \times 2 = 212$.

The figure 2 illustrates the evolution of cells when a membrane is strong. The cells that are close to the horizontal axis are mother cells. Some daughter cells depart from the group and evolve different types of cells, even though almost all cells are in the group. In this case, dissolving a membrane compound takes place per 100 steps.

The figure 3 is focused to the mother cells. At first there are about ten groups, and some of them become extinct: after 200 steps there remain about four groups.

When a membrane is weak Figure 4 illustrates the case when a membrane is weak, a membrane dissolves every 3 steps. In this case, the system can not form a group of mother cells such as in the previous case. Since the group of cells drifts to more stable cells, the cells grow larger. Even if a large cell divides into parts of random sizes, the probability of including

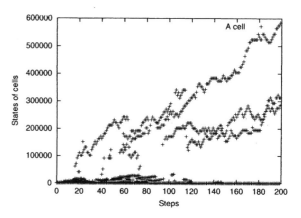

Figure 2: When a membrane is strong. The lines illustrate the regions where cells exist and points correspond to the state of cells.

enough membrane compounds to maintain its membrane are larger than a small cell.

We believe that this behaviors of evolution is similar to the evolution of viruses (Tanaka 1999). The settings of this simulation are so rough, however, that the possibility remains open that chemical evolution in origin of life is similar to virus evolution. This will be addressed in future research.

The correlation between the behavior of an ACS and the characteristics of rewriting rules

Description of a simulation Next, the following *ACS* was simulated;

$$\Gamma = (A = \{a,b,c\}, \mu = \{ / 0\}, M_0 = \{a^{10}, b^{10}, c^{10}\}, R, MC = \{b\}, \delta = 0.4, \sigma = 0.2),$$

where:

(1) R, the length of the left- or right-hand-side of a rule is between one and three. Both sides of the rules are obtained by sampling with replacement of the three symbols a, b and c;

(2) Membrane structures are not assumed.

λ_e **parameter** In order to investigate the correlation between the rewriting rule and the behavior of the model, we will introduce the λ_e parameter (Suzuki 1998). This parameter is introduced as an order parameter of *ARMS*.

Let us define the λ_e parameter as follows:

$$\lambda_e = \frac{\Sigma r_{\Delta S > 0}}{1 + (\Sigma r_{\Delta S < 0} - 1)} \tag{1}$$

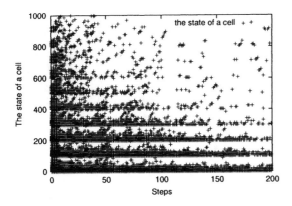

Figure 3: Evolution of mother cells. The points correspond to cells.

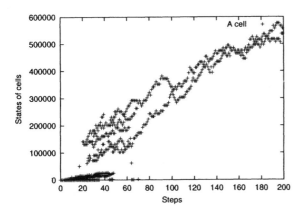

Figure 4: When a membrane is weak

where $\Sigma r_{\Delta S > 0}$ corresponds to the number of *heating rules*, and $\Sigma r_{\Delta S < 0}$ to the number of *cooling rules*. This parameter is well-defined when the number of rules is greater than 1.

When the *ARMS* only uses rules of the type $r_{\Delta S < 0}$, λ_e is equal to 0.0. On the contrary, if the *ARMS* uses rules of the type $r_{\Delta S > 0}$ and $r_{\Delta S < 0}$ with the same frequency, λ_e is equal to 1.0. Finally, when the *ARMS* only uses rules of the type $r_{\Delta S > 0}$, λ_e is greater than 1.0.

λ_e indicates the degree of reproduction in a cell. When λ_e close to 0.0, the degree is quite low and as λ_e is getting larger than 1.0, the degree becomes higher.

λ_e parameter and system's behavior The behavior of *ACS* is classified into four classes by this param-eter as follows;

- **Type I**: A cell does not evolve and disappears;

- **Type II**: The period of dividing membranes and dissolving membranes appears cyclically.

- **Type III**: A cell evolves to a complex, hierarchically structured cell;

- **Type IV**: All chemical compounds inside a cell increase rapidly. However, cells hardly divide;

where each λ_e value is not important very much. Although they change in the different environments, these four classes are unchanged.

Type I (λ_e close to 0.0)   When λ_e is close to 0.0 the *cooling rule* is mainly used, and cells will hardly grow up. Thus membranes will hardly be divided (figure 5).

Type II (λ_e in between 0.5 and $\sim$ 1.0)   When λ_e is in between $0.5 \sim 1.0$, membranes are more likely to be divided than when λ_e is close to 0.0. But, since *cooling rules* are likely to be used, the membrane compounds do not increase very much. Thus, when a cell evolves to a certain size the membrane compounds of each cell decrease and they are dissolved (figure 6). Thus there emerges a *cell cycle* like behavior.

step	state
0.	$[a^{10}, b^{10}, c^{10}]$
1.	$[a^9, b^5, c^{10}]$
2.	$[a^{10}, b^2, c^7]$
3.	$[a^{11}, b^3, c^7]$
4.	$[a^6, b^4, c^5]$
	....
10.	$[a^1, b^4, c^2]$
	....
16.	$[a^1, b^4]$

Figure 5: An example of state transition of *ACS*: Type I (λ_e close to 0.0)

Type III (λ_e in between 1.05 $\sim$ 2.33)   When λ_e exceeds 1.0 and the *heating rule* is likely to be used, a cell grows up easier and the frequency of cell division becomes high. Thus a cell evolves into a large complex cell (figure 7).

Type IV (λ_e more than 2.5)   When the λ_e parameter becomes much larger than 1.0, membranes will hardly be divided, because the number of compounds of all kinds in the cell increase and it is difficult to specifically increase the number of only membrane compounds. Consequently, a cell will hardly divide (figure 7).

step	state
0.	$[a^{10}, b^{10}, c^{10}]$
1.	$[a^6, b^9, c^9]$
...	
91.	$[a^4, b^4, c^1]$
92.	$[[b^2][b^3, c^1]]$
93.	$[[b^2][b^1, c^1]]$
94.	$[a^6, b^1, c^1]$
...	
114.	$[a^2, b^5]$
115.	$[[a^1, b^2][a^1, b^3]]$
116.	$[[a^1, b^2][b^2][a^1, b^1]]$
...	
140.	$[[[a^1, b^1][b^1, c^1]][[b^2][a^1, b^1]][b^1, c^1]]$
141.	$[[a^3, c^3][a^3, c^2][a^1, c^2]]$
142.	$[a^6, c^5]$

Figure 6: An example of state transition of ACS: Type II (λ_e in between 0.5 and ~ 1.0)

step	state
0.	$[a^{10}, b^{10}, c^{10}]$
1.	$[a^{10}, b^{10}, c^9]$
...	
41.	$[a^2, b^7, c^5]$
42.	$[[b^2][b^6, c^1]]$
43.	$[[b^4][b^3, c^1]]$
44.	$[[b^4][b^2][b^2, c^1]]$
45.	$[[b^4, c^2][b^2, c^2][b^1, c^3]]$
46.	$[[[b^3, c^1][b^1, c^1]][b^2, c^2][b^1, c^3]]$
47.	$[[[[b^2][a^1, b^1]][b^1, c^1]][a^1, b^3, c^2]]$ $[a^1, b^2, c^3]]$
140.	$[[[a^1, b^1][b^1, c^1]][[b^2][a^1, b^1]][b^1, c^1]]$
214.	$[[[[a^{16}, b^3, c^2][a^5, b^5, c^1][a^3, b^2]]$ $[a^4, b^3]][[[b^2][b^2][a^1, b^1]][a^3, b^1]]$ $[a^3, b^1][b^4][a^3, b^1]]]$

Figure 7: An example of state transition of ACS: Type III (λ_e in between 1.05 $\sim$ 2.33)

We believe that the division in four classes describes the basic behavior of the system. However, when δ or σ are changed the λ_e value that corresponds to each type is also changed. Thus a deeper investigation is needed with respect to the correlation between δ, σ and λ_e more precisely.

Genetic ACS (GACS)

Since a rewriting rule promotes a reaction, it can be regarded as an enzyme. Here we extend ACS with evolutionary mechanism We called the system Genetic ACS (GACS).

Descriptions of $GACS$

A transition in $GACS$ is a construct

step	state
0.	$[a^{10}, b^{10}, c^{10}]$
1.	$[a^{12}, b^5, c^9]$
2.	$[a^{14}, b^4, c^{10}]$
3.	$[a^{13}, b^6, c^{11}]$
4.	$[a^{14}, b^8, c^{12}]$
5.	$[a^9, b^9, c^{10}]$
6.	$[a^7, b^{10}, c^1 0]$
...	
87.	$[a^{17}, b^{12}, c^{21}]$
...	
147.	$[a^{14}, b^{20}, c^{29}]$
...	
242.	$[a^{17}, b^{50}, c^{44}]$
...	
300.	$[a^3, b^{56}, c^{58}]$

Figure 8: An example of state transition of ACS: Type IV (λ_e more than 2.5)

$$\Gamma = (A, \mu, M_1, ..., M_n, R, \delta, \sigma),$$

where:

(1) A is a set of objects;

(2) μ is a membrane structure (it can be changed throughout a computation);

(3) $M_1, ..., M_n$, are multisets associated with the regions 1,2, ... n of μ;

(4) R is a finite set of multiset evolution rules over A.

(5) δ is the threshold of dissolving membrane;

(6) σ is the threshold of dividing membrane.

The way of applying rewriting rules, the way of dissolving and dividing and input and output are the same as ACS and ACSE.

An enzyme We denote a set of reaction rules as follows;

	a	b	c
a	x_{aa}	x_{ab}	x_{ac}
b	x_{ba}	x_{bb}	x_{bc}
c	x_{ca}	x_{cb}	x_{cc},

where x_{ij} means the number of compounds i which are transformed from j. For example, 2_{ab} means a rewriting rule, $b \rightarrow a, a$. We call the table a transformation table.

Transmission of an enzyme When a membrane is divided, the enzyme which is inside the membrane is copied and passed down to a new divided cell. At that time, a point mutation occurs only in the copied enzyme and it is passed down to the new cell. The enzyme remain in the old membrane as well as the new one. Point mutations occur every time a membrane divides. When a membrane is dissolved the enzyme which is inside the membrane loses its activity. A point mutation is a rewriting of the number of x_{ij}. Thus, it changes the number of transforming compounds to i. We assumed $x_{ij} \in \{0, 1, 2\}$. In ACS and ACSE the system have only one rewriting rule, however, in GACS, each membrane has a set of rules.

An experimental result of a GACS

We will show experimental results of GACS. At first, the following *GACS* was simulated;
$$\Gamma = (A = \{a, b, c\}, \mu = \{/\ 0\}M_0 = \{a^{10}, b^{10}, c^{10}\}, R, MC = \{c\}, \delta = 0.4, \sigma = 0.2),$$

where the transformation table (R) is set as follows in the initial states;

	a	b	c
a	0_{aa}	0_{ab}	1_{ac}
b	1_{ba}	0_{bb}	0_{bc}
c	0_{ca}	1_{cb}	0_{cc}.

The productivity of membrane compounds p is defined as the ratio of the total number of non-membrane compounds to be produced to the total number of membrane compounds to be produced;

$$p = \frac{\Sigma_{j=a}^{j=b} x_{cj}}{\Sigma_{j=a}^{i=b} \Sigma_{j=a}^{j=c} x_{i,j}},$$

When $p = 0$ the enzyme does not produce any membrane compounds, when $p = 1$, it produces the same number of membrane compounds to the non-membrane compounds, and when $p > 1$, it produces more membrane compounds than non-membrane compounds. Figure 9 illustrates the time series of productivity, where the vertical axis illustrates the productivity, the horizontal axis illustrates the steps and each dot is an enzyme. It shows that at first almost enzymes evolve to $p > 1$. However, after 100 steps, the productivity of the enzymes decrease.

After 100 steps, both the number of cells and the sizes of cells increase exponentially. Furthermore, the structure of cells becomes complicated. Figure 10 illustrates the correlation between the number of cells, the size of cells and the number of steps, where each dot corresponds to a cell.

Figure 11 illustrates the internal nodes of the whole system. If we regard M_0 as the root and other cells

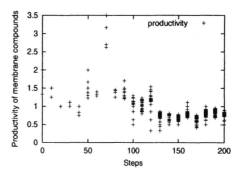

Figure 9: Productivity of enzymes.

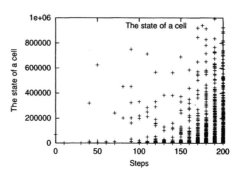

Figure 10: State transition of the system

as internal nodes and leaves, we can regard the whole system as a tree. In order to indicate the complexity of the tree, we use the number of internal nodes in the tree. Figure 11 illustrates that the number of internal nodes increases exponentially, after 150 steps.

It is interesting that when a cell grows into a hierarchical cell, the enzyme evolves to a low productivity one.

The reason of this behavior can be considered as follows; the enzyme whose productivity is high always suffers from mutations, because it promotes membrane division thus it generates more mutations than low productivity ones. If every cell is an elementary cell, an enzyme have to keep producing membrane compounds at a high rate. However, when the cell forms structure, it is not necessary one with high productivity, because, in a structured cell, if an inside cell dissolves, the cell that includes the dissolved cell obtains its membrane compounds.

Therefore, during evolution of a cell into a structured cell, the cell needs a high productivity enzyme. However, once it evolves a structured cell, high pro-

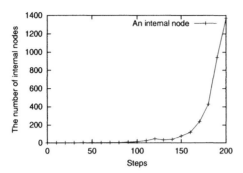

Figure 11: Internal nodes of the cell

ductivity enzymes are weeded out. This is the role of membranes in terms of computation.

Genetic Programming by using GACS

We attempted to generate a program by using a GACS. In the GACS, a program corresponds to an enzyme, thus we breed an enzyme which can solve a particular problem. We apply it to a simple problem *doubling*; calculate the double value of the number of a and b then show the result as the number of c ($c = 2(a + b)$).

Description of GACS A *GACS* is defined as follows;

A transition *GACS* is a construct $\Gamma = (A = \{a, b, c\}, \mu = \{[_1]_1 \cdots [_{100}]_{100}\}, M_0 = \{[a^2, b^2, c^0]^{100}\}, R)$. In the initial state, all transformation tables R are

	a	b	c
a	0_{aa}	0_{ab}	1_{ac}
b	1_{ba}	0_{bb}	0_{bc}
c	0_{ca}	1_{cb}	0_{cc},

and one hundred of elementary cells are assumed inside M_0. No compounds are transformed among cells and no input and output are assumed. Although we performed simulation by using different type of cells in the initial state, the experimental results are same, thus we will address only when each cell is $[a^2, b^2, c^0]$ in the initial state.

Dissolving and dividing a membrane The way of dissolving and dividing are the same as in ACSE. After n rewriting steps, if the number of c is smaller than 7, the membrane is dissolved and the enzyme inside it loses its activity, for example,

$$[_h a^3, b^5, c^8]_h [_i a^4, b^5, c^1]_i \to [_h a^3, b^5, c^8]_h.$$

In the above example, the number of c inside the membrane i is smaller than 7, so the membrane is dissolved.

After n rewriting steps, if the number of c is larger than 9, the membrane is divided and a point mutation takes place in its enzyme. Furthermore, a new enzyme is passed down to a new cell. When a cell divided, the inside multiset of the divided cell and its parent cell are set to $\{a^2, b^2, c^0\}$ again, so they try to solve the problem again. The results in:

$$[_h a^3, b^5, c^{10}]_h \to [_h a^2, b^2, c^0]_h [_i a^2, b^2, c^0]_i$$

In the example, because the number of c inside the membrane h is larger than 9, the membrane h is divided and a new membrane i emerges. Then compounds which are inside both membranes i and h are set to $\{a^2, b^2, c^0\}$. In this GACS, cells continue to solve the problem.

The fitness of the GACS The fitness of an enzyme is defined as the number of steps to reach the solution. By using this fitness, good enzymes are there that can solve the problem within a smaller number of steps than the others are selected.

Experimental result At first, all enzymes are set to,

$$
\begin{aligned}
a &\to b, \\
b &\to c, \\
c &\to a.
\end{aligned}
$$

After 5000 rewriting steps, the enzymes that reach the solution within 8 steps are selected;

$$
\begin{array}{llll}
a &\to a, c, & a &\to a, b, c, c, \\
b &\to a, b, c, c, & b &\to a, b, c, c, \\
c &\to c, c; & c &\to c, c; \\
& & a &\to a, c, c, \\
& & b &\to b, \\
& & c &\to c, c.
\end{array}
$$

For each rule, one hundred of elementary cells that are $[a^2, b^2, c^0]$ are set again and calculations performed again. Next, the enzymes that can solve the problem within 5 steps are selected. Then, there remains only one enzyme;

$$
\begin{aligned}
a &\to a, c, c, \\
b &\to c, c, \\
c &\to c, c.
\end{aligned}
$$

it is a solution of this simulation. In fact a solution of this problem is

$$
\begin{aligned}
a &\to c, c, \\
b &\to c, c, \\
c &\to c;
\end{aligned}
$$

Thus the survived enzyme evolved to a similar enzyme in the solution.

By using this method, we have attempted to treat the system as an artificial living system of computation. In the future we plan to create a GACS as an artificial living system of computation that can solve more complicated problems. In such a system, in order to obtain the result, we observe their output (behaviors) and change the condition, do not stop their computations. In other words, we steer them in the right direction by using the selection and mutation, and lead them to our settled goal. Although GACS may not fit to make optimizer, we believe this method can apply to design artificial living things such as robots.

Discussion

P systems

The above ACSes correspond to a class of P systems, a parallel molecular computing model proposed by G. Păun(Păun 1998) that is based on the processing of multisets of objects in cell-like membrane structures. A P system is a multiset transformation system. However, it is different from (Berry 1992), since it includes a "*membrane*" in its computing mechanism. In P systems *cells* are structured like living cells. The system itself is composed of several cells that are delimited from the neighboring cells.

The computing power of P systems is equal to that of a Turing machine, as proved in (Păun 1998), and an algorithm to compute the SAT problem in linear time (Păun 1999) was proposed. Various P systems have also been proposed and their mathematical properties have been investigated. On the computing power of ACSes that are treated in this paper is open problems.

Conclusion

It is obvious that living systems never do calculation only by using *pen* and *paper*. Thus, if we could abstract *computation* in terms of living systems then we might obtain a new computational world.

We have not found the new world yet, however, we have already obtained some *hints* throughout the observations on living systems, such as "parallelism" and "membranes." In a living system, every cell and every organ is acting in parallel and they are consisted of membranes.

To implement our system on a parallel computer and investigating its mathematical properties are our future work. We expect that these studies lead us to a new computing paradigm which goes beyond the Turing Machine based computing paradigm.

Acknowledgment

We thank for fruitful discussions with Dr. Steen Rasmussen, Prof. Dr. W. Banzaf gave us important comments and encouraged us, long discussions with P. Dittrich were useful. F. Pepper and SOMA research group gave us important suggestions. And the authors would like to express many thanks to Dr. Gheorghe Păun for his useful comments, discussions and mathematical refinements. This research is supported by Grants-in Aid for Scientific Research No.11837005 from the Ministry of Education, Science and Culture in Japan.

References

Fontana, W. and L.W. Buss, The arrival of the fittest: Toward a theory of biological organization. *Bulletin of Mathematical Biology* 56: 1–64. 1994.

Berry, G. and G. Boudol, The chemical abstract machine. *Theoretical Computer Science* 96: 217–248. 1992.

Gánti, T., Organization of chemical reactions into dividing and metabolizing units: the chemotons. *Biosystems* 7: 189–195. 1975.

McMullin, B. and F. Varela, Rediscovering Computational Autopoiesis, *ECAL'97*. 1997.

Mirazo, K., A. Moreno, F. Moran, et. al., Designing a Simulation Model of a Self-Maintaining Cellular System *ECAL'97*. 1997.

Nicolis, G. and I. Prigogine. *Exploring Complexity, An Introduction*. San Francisco: Freeman and Company. 1989.

Păun, G., Computing with Membranes, *Turku Center for Computer Science TUCS Technical Report No. 208* (submitted, also on http://www.tucs.fi). 1998.

Păun, G., P Systems with Active Membranes: Attacking NP Complete Problems, Center for Discrete Mathematics and Theoretical Computer Science CDMTCS-102 (also on http://www.cs.auckland.ac.nz/CDMTCS). 1999.

Suzuki, Y. and H. Tanaka. Order parameter for a Symbolic Chemical System, *Artificial Life VI*:130-139, MIT press. 1998.

Suzuki, Y. S., Tsumoto and H. Tanaka. Analysis of Cycles in Symbolic Chemical System based on Abstract Rewriting System on Multisets, *Artificial Life V*: 522-528. MIT press. 1996.

Tanaka, H., F. Ren, S. Ogishima, Evolutionary Analysis of Virus Based on Inhomogeneous Markov Model, ISMB'99, p 148, 1999.

A search for multiple autocatalytic sets in artificial chemistries based on boolean networks

Harald Hüning

Electrical Engineering, Imperial College, London SW7 2BT, UK

Abstract

Populations of strings which interact in ways defined by an artificial chemistry can self-organise spontaneously into an autocatalytic set. This paper considers populations of binary strings with fixed length and a reaction scheme that uses strings as both data (or tape) and machine. Here the machine is a boolean network where some parameters are determined by a string from the population. The input to the machine is given by a second string drawn from the population. In the artificial chemistry based on boolean networks, simulations have revealed a high sensitivity on a probabilistic rate that filters out trivial patterns. By variation of the rate parameter, multiple stable sets have been found. Short string lengths are used here, in order not to rely only on simulations, but also to keep the reaction graph small enough to be able to search for possible autocatalytic sets. A search method has been developed that finds all closed subgraphs of the reaction network, which indicate to a high degree what autocatalytic sets are possible. While simulations most often give only one as a result, the search saves many simulation runs, because it is independent of the initial populations. The resulting number and size of autocatalytic sets gives information about any freedom of the system to adapt, e.g. when coupling such a system to an environment that can impose constraints. So this description of the behaviour of artificial chemistries appears useful for further artificial studies of molecular evolution and the origin of life.

Introduction

Artificial chemistries make abstractions from the natural chemistry in the representation of molecules, populations, and the implementation of reactions. Although artificial, they have already provided simulations of the cooperation of molecules in metabolic or catalytic networks (Farmer, Kauffman & Packard, 1986), and they may also serve to discover laws of evolution (Bagley, Farmer & Fontana, 1992).

The study of artificial chemistries has the advantage that in computer simulations all details can be made observable, and more than one trajectory of the dynamics can be investigated. The general questions are

first, when do self-replicating units appear, starting only from the existence of simpler elements, and given some abstract chemical rules? Secondly, how can replicating units evolve, and do any effects broadly share laws with evolution and the origin of life in the chemistry of nature?

Farmer, Kauffman & Packard (1986) have first simulated interacting strings with concatenation and cleavage reactions. They have found an *autocatalytic set*, i.e. a set of strings which are all produced from reactions within the set. Another way of defining the chemistry is based on the notion of using strings as data and machines, where the machines are automata that transform strings. Ikegami & Hashimoto (1995) have used different types of strings for data and machine, while Dittrich and Banzhaf (1998) have used just one type of string for both data and machine. Replication arises from closed subgraphs of the random catalytic chemistry (Stadler, Fontana & Miller, 1993), where data and machine correspond to two molecules catalysing a reaction.

Regarding the evolution of autocatalytic sets, Bagley, Farmer & Fontana (1992) have studied evolutionary modifications of reaction graphs, and have found that autocatalytic sets can change with varying degrees. In the data-machine type of simulation, Dittrich and Banzhaf (1998) have observed an exchange of string segments in one example. This is reminiscent of a mutation operator, but it acts on individual strings rather than on replicating units like autocatalytic sets.

All these studies suggest that autocatalytic sets typically involve only a small part of the reaction network, and in the large networks defined by artificial chemistries no other method than simulations of the dynamics (or fixed-point solutions) seems to give information about these sets.

Motivation

The current work addresses the question of how many different autocatalytic sets are possible in a given artificial chemistry. Simulations give only one set as a stable result, and sometimes more sets can be recognised by their competition. When there would be an abundance of possible sets in a system, then one could regard the particular strings in a set as somewhat arbitrary, or having degrees of freedom. This is important for further studies of interactions with an environment, e.g. imposing constraints on the survival of strings or on selected reactions, or injecting strings into the population considered. If there is a degree of freedom in the strings that can exist in a dynamically stable population, no matter in which autocatalytic set, then such interactions with the environment can lead to decisions that affect the convergence of the population. In contrast, if there would be always only one autocatalytic set for one artificial chemistry, then there is not much freedom for decisions. However, as Bagley, Farmer & Fontana (1992) have studied, the coming into existence of a single molecule can change an autocatalytic set dramatically.

Most examples of autocatalytic sets in the literature are lacking a characterisation of the number and size of the sets. The sets reported for example by Farmer, Kauffman & Packard (1986) and Bagley, Farmer & Fontana (1992) are very large, and it is very hard to gain further insight into the dynamics. So it seems desirable to study smaller systems, where for example the difference between the connected graphs of reactions and the typically smaller autocatalytic sets can be investigated.

Here smaller string lengths are chosen to reduce the size of the reaction graph. The system uses strings as both data and machine (Dittrich & Banzhaf, 1998), because this type of chemistry admits in principle the construction of arbitrary new strings. Dittrich and Banzhaf (1998) mention a phase where the the system explores the space of strings. However, the automaton (like a computer processor) that they have used, does not seem appropriate for use with much shorter strings, since bit-words are translated into machine instructions.

A different scheme for transforming bit-strings is given by boolean networks (Kauffman, 1969), sometimes also called weightless neural networks (Aleksander & Morton, 1990). The artificial chemistry is here defined by means of boolean networks for the following reasons. Boolean networks allow relatively general functions depending on the number of bits that each node reacts to. They implement discrete functions like Dittrich and Banzhaf's (1998) automaton, so it is expected that the system has autocatalytic sets with similar characteristics. Most importantly, small string lengths can be chosen in a systematic way, since

the strings need not be translated into a series of instructions.

Thus the particular questions addressed in this paper are:

1. What is the number and size of the autocatalytic sets?

2. How do autocatalytic sets relate to properties (subgraphs) of the reaction network?

3. Do the boolean networks have similar characteristics as Dittrich and Banzhaf's (1998) automaton, even with small string lengths?

The behaviour of artificial chemistries has been mainly investigated by simulations and not analytically, due to the large number of different strings involved. For small reaction networks the dynamics can often be analysed deterministically (Eigen & Schuster, 1978). For very large reaction networks stochastic simulations are used, unless a restricted reaction network can be extracted (Bagley, Farmer & Fontana, 1992). The disadvantage of simulations is their dependence on the initial populations, so usually many simulation runs have to be carried out.

This paper aims at characterising the behaviour of artificial chemistries independently of the initial populations. By searching for potentially stable sets in the reaction graph, behaviours may be found that are missed by simulation studies alone. The search can be made quite exhaustively in the simple artificial chemistry with short binary strings, and the applicability to longer strings is discussed below.

Boolean networks have much been studied with the output feeding back directly on the input. The global behaviour of boolean networks has been illustrated as basins of attraction fields (Wuensche, 1997). These are diagrams with a node for every distinct state, and a line for every possible state transition. For any initial state, one can follow the states (trajectory) in the diagram, until one encounters a state again. So the attractors are either a state with a transition to itself, or a state cycle. Kauffman (1969) has studied the cycle lengths of these attractors.

The search in a reaction graph of an artificial chemistry is comparable to, but more complex than searching for state cycles, since boolean networks on their own have deterministic behaviour, but the simulated reactions are stochastic and involve the collisions of two strings.

Dittrich and Banzhaf (1998) have used the following reactor algorithm that carries out the stochastic collisions:

1. Randomly select two strings from a population

2. If the reaction $s_1 + s_2 \Rightarrow s_3$ exists according to a filter, replace a randomly chosen string from the population with s_3

Their automaton uses one string as input or data string, and the other one as encoded machine instructions. The instructions are carried out to process the input string sequentially. For each bit in the input string one bit is written into the output string, so all strings have the same length. A filter condition is applied to avoid self-replication of strings, and the all-0 string is disallowed as well. The results are characterised by an exploration phase with high diversity of the strings in the population, until typically a small number of strings suddenly rises in their population number and takes over the population. For example, an autocatalytic set of eight strings has been reported, where all reactions of any two of the eight strings produce another string of the set. Similar autocatalytic sets are expected using the boolean networks, where shorter string lengths allow searching in the reaction graphs.

The next section introduces the artificial chemistry with boolean networks. In simulations the filter condition is varied. Then follows a graph-based global search for the possible stable sets, which is more efficient than the simulations.

Artificial chemistry based on boolean networks

The simulation system consists of a population of binary strings and a machine that transforms strings, implementing a reaction scheme (Fig. 1). A boolean network is used as the machine, or in other words to implement a reaction scheme for the population of binary strings.

Fig. 1 illustrates the reactor algorithm mentioned in the introduction. At every time step two strings are chosen randomly (with replacing) from the population, and used by the machine, in this case a boolean network (Fig. 2). One of the strings acts as input to the machine, the other is used to specify parameters of the machine itself, very much like setting the weights of a standard neural network. So the machine can be different in every time step. Processing of an input string gives an output string which is put back into the population if it passes a filter. If the string is put back, it replaces another randomly chosen string. This is the only operation that changes the population in the current simulations.

The filter in Fig. 1 can prohibit certain strings from being put back. Here, the filter is chosen as putting back the strings with a probability that depends on some property of the string, e.g. the number of '1's in

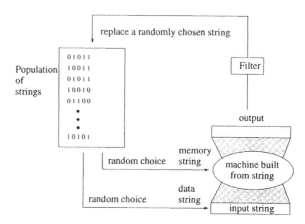

Figure 1: Reactor algorithm with a population and machine. The 'memory string' sets parameters of the machine, which is a boolean network.

it. These probabilities are given directly in a look-up table. Such a filter has similar effects on the dynamics as kinetic rate constants in chemical models.

Details of the machine are shown in Fig. 2. There are as many Random Access Memories (RAMs) as there are bits in the strings. Each RAM has the same number n of random, but fixed, connections to the input string. The bits conveyed by the connections form a bit-word that addresses a memory location in the RAM. The content of each RAM comes from the memory string as in Fig. 2. In the assignment from the memory string to the 2^n memories in each RAM, every bit is used several times, again in a random but fixed way.

The connections and the string-to-memory assignment are the fixed parameters of the boolean network, while its function changes with the memory string at every time step. The outputs of all RAMs are combined to the output string, and if it is passed through the filter, it is put back into the population.

Simulation

At the beginning of a simulation, the population is initialised with random strings. Then the reactor algorithm is run. Some examples of reactions from the first simulations are shown in Table 1.

The last three reactions in Table 1 exemplify how the all-0 and all-1 strings are produced from other strings. The all-1 string eventually takes over the whole population, so it 'wins' in this first simulation (see Fig. 3).

When the whole population has converged to the same string, no more changes can happen. The all-0 and all-1 strings are trivial, because they are repro-

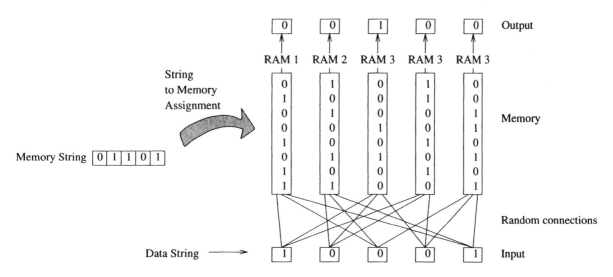

Figure 2: Boolean network that implements the machine with parameters taken from the memory string. For each output bit there is a random access memory (RAM), in this example 5 with $n = 3$ connections each. Each RAM has 2^n memory locations that are filled with bits from the memory string. Fixed parameters are the random but fixed connections, and the sting-to-memory assignment, where each bit is used several times.

```
01101 + 10110  =>   00010
10110 + 10011  =>   10111
10111 + 10101  =>   10001

10101 + 10110  =>   11111
00000 + 11111  =>   00000
00000 + 00011  =>   00000
```

Table 1: Reactions from a first simulation. The first string is the memory string, the second is the input. The boolean network has five RAMs with 3 connections each.

duced independently of the connections and memory assignment of the boolean network. How much has the system to be changed in order to stabilise at any other of the $2^n - 2$ (in this case 30) strings? Other authors have disallowed any self-replication, and the creation of the all-0 string altogether. Here a more moderate filter condition has been found sufficient.

A first global measurement

The following method shows that the trivial patterns have a much higher chance to be produced from random initial conditions. Globally counting the results of all possible reactions reveals, that for the chosen fixed paramters (connections, memory assignment) there is a much higher number of reactions that produce the trivial strings than others. In particular, all pairwise

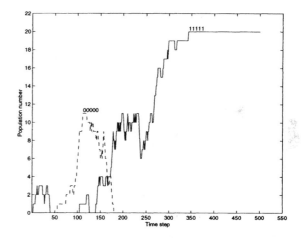

Figure 3: Population of the strings 00000 and **11111** in the first simulation. RAMs and connections as in Table 1. The population size is 20.

combinations of all possible strings are presented to the boolean network, and the occurrences of each of the possible output patterns are counted. Then these are averaged for all strings with the same number of '1's, see Fig. 4.

The histogram shows the average times a string with a given number of '1's is produced over the whole reaction scheme. The higher number of producing all-0

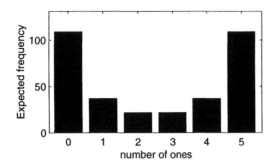

Figure 4: Histogram of the number of '1's, averaged over the $\binom{5}{k}$ strings with the same number of '1's.

and all-1 strings has been found typical for boolean networks with different random connections and string-to-memory assignments. Assuming an even distribution in the initial string population, the histogram shows the bias towards producing the trivial strings at the beginning of a simulation.

The filter rate

In contrast to filtering out trivial strings completely, here a probability like a rate or fitness is applied to whether putting back a string into the population or not. For each string that is output from the machine, the number of ones is counted, and the filter probability is looked up in Table 2.

Number of ones	0	1	2	3	4	5
Rate (probability)	x	1	1	1	1	x

Table 2: Modified filter condition (probability of replacing). For simplicity, only the filter rate for all-0 and all-1 patterns x is chosen to be variable.

The filter rate x has been varied in simulations, and surprisingly it is not necessary to fully neutralise the bias in Fig. 4. Merely by changing x from 1 to 0.95 in the system from Fig. 3 makes the population converge to a different string (11011). For another example of a boolean network (with other connections and string-to-memory assignment), very many changes of the attractor with the variation of the rate x have been found, see Table 3.

The many different results depending on the rate may be interpreted on the one hand as a lack of robustness of the results, but on the other hand there appear to be several stable results, such that small changes in a parameter can switch between them. There appears to be a potential to add further constraints to the system

rate x	stable	steps
0	01110	48700
0.05	10001	1800
0.1	10001	10200
0.15	10001	600
0.2	10001	600
0.25	01110	32200
0.3	10001	2500
0.35	10001	1200
0.4	10001	8700
0.45	10001	11000
0.5	- 0 -	17600
0.55	10001	5300
0.6	01110	1000
0.65	01110	700
0.7	01110	2100
0.75	- 1 -	1300
0.8	- 0 -	600
0.85	- 1 -	1100
0.9	01110	500
0.95	- 1 -	300
1	- 1 -	400

Table 3: Results of simulations where the filter rate x is varied. Trivial strings are denoted by '- 1 -' or '- 0 -' respectively. The boolean network has five RAMs as before, but now 2 connections each. The population size is 20.

that decide about the attractor. Before the range of possible attractors is investigated, another simulation with more than one resulting string is presented.

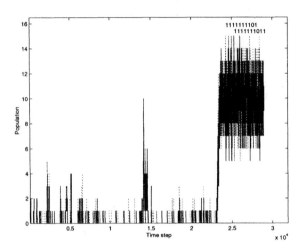

Figure 5: An autocatalytic set. The boolean network has 10 RAMs with 2 connections each. The population size is 20.

An autocatalytic set

The simulation in Fig. 5 shows a set of strings that co-exist in a stationary way. The population of the two strings keeps fluctuating, but both stay at a high population number, suggesting that they stabilise each other. Fig. 6 shows the reaction pattern underlying this simulation in the stationary state. Clearly none of the strings can reproduce alone. Deterministic equations for this reaction scheme are

$$\dot{x_1} = x_1 x_2 + x_2 x_2 - x_1 f$$
$$\dot{x_2} = x_1 x_1 + x_1 x_2 - x_2 f \qquad (1)$$

with

$$f = 2 x_1 x_2 + x_1^2 + x_2^2$$

Linearisation around the fixed point $x_1 = x_2 = 0.5$ gives the eigenvalues $\lambda_1 = -2$ and $\lambda_2 = -1$, so the system is locally stable. For more details see Hüning (2000) and Eigen & Schuster (1978).

A search method

The previous section has shown that simulation results depend on the filter rate x, and that autocatalytic sets can be found in artificial chemistries based on boolean networks. This section aims at finding the total number of stable sets.

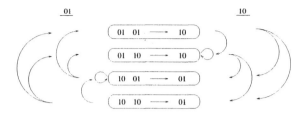

Figure 6: Reactions involved in the autocatalytic set. Nodes are reactions. Arrows show that the two reaction partners can always be produced within the set.

Like attractor cycles can be found by a global search for boolean networks, here cycles are searched for in the reaction graph. However, there are always two strings meeting in a reaction, so the graph is more complex. The reactions requiring the collision of two strings give every single string many possible successors in the graph. So the cycles in the graph can be heavily interconnected, and the dynamics is not confined to a cycle like for recurrent boolean networks.

In a simulation, pairs of strings are chosen randomly, so the reactions have a stochastic order. Over time, the population of strings can be imagined to home in onto a stable autocatalytic set, i.e. a small number of strings closely connected by their reactions.

There are properties of the reaction graph that are indicating the existence of autocatalytic sets:

1. all members of a set are produced from reactions within the set

2. all reactions between the members produce members of the set

The second property may not hold strictly, when dynamic competition reduces the size of the set (see parasites in the discussion). Due to above properties of the reaction graph, possible autocatalytic sets can be searched for, avoiding simulations and their dependence on initial conditions. The search only gives possible sets, because their size can still be reduced by dynamical properties, e.g. competition between subsets. Simulations or analysis of the dynamics can be used after the search to investigate the stability of the sets.

The search for stable sets involves keeping a list of strings as set members, and evaluating all possible reactions between them. All reaction results are added to the list, then the reactions are evaluated again, until no more strings are added. In many cases it may be sufficient to start the list with two strings, but in general more combinations may have to be tried exhaustively. A case where the search has to start from

Set	Type
0	A
0 … 31	
0, 31	B
14	A
14, 17	B
14, 31	B
17	A
17, 31	B
31	A

Table 4: Results of the search for possible sets. The members of a set are shown as integers, 14 corresponds to 01110, and 17 corresponds to 10001. The second set from the top denotes the whole reaction scheme, while the reaction schemes A and B are given in Table 5. The boolean network is identical to the simulation in Table 3.

Type	Reactions	Stability
A	31 + 31 -> 31	stable
B	17 + 17 -> 17	unstable
	17 + 14 -> 17	
	14 + 17 -> 14	
	14 + 14 -> 14	
C	17 + 17 -> 1	stable
	17 + 1 -> 17	
	1 + 17 -> 1	
	1 + 1 -> 17	

Table 5: Types of reaction patterns and their stability.

more than two strings is for example a singly connected catalytic chain of self-replicating strings.

Results of the search

With this search method, the sets in Table 4 have been found for the same fixed parameters of the boolean networks as for the simulation in Table 3. Many sets share the same types of reaction graph, so only the type of graph is given in the Table 4, and Table 5 gives the reaction patterns. The search shows that the strings 01110 and 10001 can be expected from simulations, and indeed only these strings, apart from the trivial strings (cf. Table 3).

A similar fixed-point analysis than for equation (1), but for reaction pattern B of Table 5 proves that the pattern of type B is unstable (in contrast to the pattern in Fig. 6). Only one of the strings takes over the whole population.

Set	Type
0	A
0 … 31	
0, 31	B
1, 17	C
31	A

Table 6: Results of the search for possible sets using an example network with 5 RAMs and 3 connections each.

With different random choices of the connections and string-to-memory assigments of the boolean network, different stable sets are found. There are possible sets with more than two members, where only a subset may be dynamically stable, and more symmetric and unsymmetric reaction patterns for two-member sets.

For example, the sets in Table 6 have been found for one random choice of 3 connections per RAM. Here appears a stable reaction pattern C (see Table 5) of a two-member set, like in Fig. 6 and equation (1).

The search shows that typically there are a number of stable sets, such that small variations in a simulation (like the filter rate) can change the result (which of the sets the simulation converges to). This can be interpreted on the one hand as there being room free to apply constraints to the system. However, the strings that can remain in a stationary population are not arbitrary. The search shows that the total number of different stable sets is rather small. Although one string changes the boolean network (machine) at every step, the whole system has a rather small number of stable sets, corresponding to only a few possible machines made from the strings in a set.

Discussion

The results of the search may not totally coincide with all sets that can possibly be found by simulation, because:

1. For large sets the dynamics may further reduce the set (e.g. by competition within the closed subgraph)

2. An autocatalytic set may be stable, although it has parasites (see below). Parasitic branches are not detected by the current search.

A parasite is a species that is supported (catalysed) by the set, but does not give back any support (does not have a catalytic link back into the set). The current search method cannot detect parasitic branches, because it adds all reaction results to the set. Parasites cannot simply be detected as dead-end paths, because here a product exists for every two partners. Filter rates greater than zero do not reduce the graph of reactions. Assuming an autocatalytic set with a parasite, the search algorithm would find a larger set, and it may fail to satisfy the conditions for the set to be closed after all.

In general, even sets with parasitic branches may be dynamically stable, depending on the coupling strengths (Eigen & Schuster, 1978). In this case, detecting parasites would give better results of a search. For parasites that are not only dead-end subgraphs it is still not solved, however, how a graph-based search could be extended to discriminate host and parasite, and possibly to evaluate which of the two the competition is likely to favour. The detection of parasitic branches may become more relevant when some strings are fully filtered out.

What can be expected for longer string lengths, where the exhaustive search even for small starting sets cannot be carried out, because the number of reactions grows with the square of the number of strings? One may still be able to use a sampling version of the search. For longer strings or variable lengths, one may sample the possible strings and follow the reaction graph for a certain depth.

Wuensche (1997) has implemented a similar sampling method for the attractor basins of random boolean networks. This method plots a histogram of the cycle lengths, which may already be significant, although only a small proportion of the state transitions has been searched. Similarly, here a histogram of stable sets could be made. The sampling version of the search may still have the advantage over simulations that it is independent of the initial conditions.

Conclusion

The artificial chemistries with boolean networks allow the study of self-replicating sets. Since no qualitative differences between Dittrich and Banzhaf's (1998) automaton reaction and the boolean networks have been found, the use of boolean networks has allowed to use shorter strings. Simulations have shown a high sensitivity to the filter rate, and that different stable sets can result as a consequence of such little changes. The search finds the different stable sets more efficiently, at least for short strings, because it is independent of the initial population and the filter rate.

In addition, the search method may be extendable to a reverse characterisation of initial conditions when the stable set is given. Initial conditions may be characterised by containing sufficient support for a set, or the decision boundary of two competing sets may be expressed by the sizes of certain subpopulations.

The current results suggest the following answers to the questions raised in the introduction:

1. (Number and size of autocatalytic sets?) The search gives the number of potential sets and their reaction subgraphs, which can be classified into types. All potential sets including the all-string set require testing for their dynamical stability. Finding the number of autocatalytic sets was motivated by the possibility that artificial chemistries could be considered as adaptive systems. However, a small number of autocatalytic sets indicates too little freedom for adaptation.

2. (Autocatalytic sets related to subgraphs?) The closed subgraphs are a good indicator, but not necessary for the existence of autocatalytic sets (see above discussion of parasites). However, according to the two criteria in the section "A search method", the number of sets that are found is only limited by the number of combinations in the starting sets. It appears that only very peculiar types of autocatalytic reaction patterns would require starting sets much greater than three strings.

3. (Boolean nets compared to Dittrich and Banzhaf's automaton?) As in Dittrich and Banzhaf's (1998) simulation, there is first an exploration phase before any autocatalytic set is found. The sets that are found can have more than one member, so the the boolean network does not give rise to different characteristics. For some boolean networks very many potential sets are found, which seems to be an interesting property regarding the short string length of five bits. However, the very many types of reaction patterns resulting from some searches still require more investigation. It may be possible that trends are found that hold for longer string lengths as well.

Since the behaviour has been found rather fixed, it seems not to be universal enough for attempting adaptation to external constraints. Applying some external fitness can only select from the rather small number of possible stable sets.

Rather than attempting universal adaptation, this type of system may serve for studying evolution. Boolean network may be selected that exhibit bigger sets where the dynamics is not stationary (e.g. like the waves observed by Dittrich, Ziegler & Banzhaf (1998)), or other influences like mutations or interactions with an environment may be added. The description of the system resulting from the search can facilitate a quick comparison of different systems (as characterised by RAM connections, memory assignment *and* initial populations according to the possible sets found).

A subject of future research may be to inject new species into the population, to consider several interacting spatial compartments, or to study if predictions can be made about consequences of any changes of the boolean network paramters (RAM connection, memory assignment). In the case of injecting species into the population, the current results can help to characterise the *reachability* of stationary states from given initial conditions.

Acknowledgements

Many thanks to W. Banzhaf, P. Dittrich, and J. Ziegler for helpful discussions.

This work was supported by the European Commission, in the programme Training and Mobility of Researchers.

References

Aleksander, I. & Morton, H. *An Introduction to Neural Computing,* Chapman & Hall 1990.

Bagley, R. J., Farmer, J. D. & Fontana, W. (1992). Evolution of a Metabolism. In C. G. Langton, C. Taylor, J. D. Farmer & S. Rasmussen (Eds.) *Artificial Life II.* Santa Fe Institute Proceedings Vol. X, (pp. 141-158). Redwood City, CA: Addison-Wesley.

Dittrich, P. & Banzhaf, W. (1998). Self-Evolution in a Constructive Binary String System. *Artificial Life* 4(2), 203-220.

Dittrich, P., Ziegler, J. & Banzhaf, W. (1998). Mesoscopic Analysis of Self-Evolution in an Artificial Chemistry. Proc. of *Artificial Life VI,* Los Angeles, June 26-29, 1998; C. Adami, R. Belew, H. Kitano, and C. Taylor (Eds.), MIT Press, pages 95-103.

Eigen, M. & Schuster, P. (1978). The Hypercycle. A Principle of Natural Self-Organization. Part B: The Abstract Hypercycle. *Die Naturwissenschaften* **65**, 7-41.

Farmer, J. D., Kauffman, S. A. & Packard, N. H. (1986). Autocatalytic replication of polymers. *Physica D* **22**, 50-67.

Hüning, H. (2000). Convergence Analysis of a Segmentation Algorithm for the Evolutionary Training of Neural Networks. To appear at: *The First IEEE Symposium on Combinations of Evolutionary Computation and Neural Networks.* May 11-12, 2000, San Antonio, TX, USA.

Ikegami, T. & Hashimoto, T. (1995). Coevolution of Machines and Tapes. In F. Moran et al. (Eds.) *Advances in Artificial Life.* LNAI 929, pp. 234-254.

Kauffman, S. A. (1969). Metabolic stability and epigenesis in randomly constructed genetic nets. *J. Theoret. Biol.* **22**, 437-467.

Stadler, P. F., Fontana W. & Miller, J. H. (1993). Random catalytic reaction networks. *Physica D* **63**, 378-392.

Wuensche, A. (1997). Attractor Basins of Discrete Networks. PhD Thesis, The University of Sussex, Brighton, UK.

Searching for Rhythms in Asynchronous Random Boolean Networks

Ezequiel A. Di Paolo

GMD—German National Research Center for Information Technology
Schloss Birlinghoven, Sankt Augustin, D-53754, Germany
Ezequiel.Di-Paolo@gmd.de

Abstract

Many interesting properties of Boolean networks, cellular automata, and other models of complex systems rely heavily on the use of synchronous updating of the individual elements. This has motivated some researchers to claim that, if the natural systems being modelled lack any clear evidence of synchronously driven elements, then asynchronous rules should be used by default. Given that standard asynchronous updating precludes the possibility of strictly cyclic attractors, does this mean that asynchronous Boolean networks, cellular automata, etc., are inherently bad choices at the time of modelling rhythmic phenomena? In this paper we focus on this subsidiary issue for the case of Asynchronous Random Boolean Networks (ARBNs). We find that it is rather simple to define measures of *pseudo-periodicity* by using correlations between states and sufficiently relaxed statistical constraints. These measures can be used to guide an evolutionary search process to find appropriate examples. Success in this search for a number of cases, and subsequent statistical studies lead to the conclusion that ARBNs can indeed be used as models of coordinated rhythmic phenomena, which may be stronger precisely because of their built-in asynchrony. The methodology is flexible, and allows for more demanding statistical conditions for defining pseudo-periodicity, and constraining the evolutionary search.

Introduction

Cellular automata, coupled-map lattices, Boolean networks, and a number of variants of these classes have been the centre of copious research effort. The aims of this research have been mainly focused on two distinct directions: a piece of work will, in general, study these entities in themselves, as classes presenting interesting properties from a formal viewpoint, and also possibly from a practical one, or it will *use* some version of these entities as a modelling tool applied to a scientific end. One may be interested in the properties of cellular automata *per se* as a universal class, or one may use cellular automata to model some biological phenomenon such as morphogenesis. We hurry to stress that there is much overlap and cross-fertilization between these two directions, but that the distinction holds nonetheless.

Significant contributions from the first route to the second sometimes occur when the limitations of a class of formal entities are exposed. Modellers can assimilate this knowledge, and choose their modelling tools and techniques accordingly. In recent years, much evidence has been gathered suggesting that many of the initially interesting features of the above mentioned classes for modelling complex systems have depended crucially on the use of a synchronous rule for updating the atomic elements. In contrast, the implementation of asynchronous updating rules has tended to produce trivial, rather than complex, behaviour.

In this paper we will be concerned with a subsidiary aspect of one such demonstration of the effects of asynchrony, (Harvey & Bossomaier, 1997). However, the scope of our worries is more general and the results presented here could be applied more widely. As will be described below, Harvey and Bossomaier (1997) show that asynchronous updating introduces radical changes in the long term behaviour of random Boolean networks. In the asynchronous case, the typical number of different attractors per network is smaller than in the synchronous case, and these seem to be mainly of the fixed point type, suggesting much larger basins of attraction. As a further difference, the expected number of point attractors will not depend on the size of the networks. These findings cast doubts on some interpretation made when using random Boolean networks to model genetic regulatory networks, (e.g., Kauffman, 1969, 1993). In passing, the authors draw the correct conclusion that it is not possible for such networks to exhibit non-stationary cyclic long term behaviour due to the randomness of the updating scheme.

A large number of biological systems produce rhythms which arise from the complex interaction of many elements, and are not due to the existence of an external clock that orchestrates the behaviour of these basic constituents, (a variety of examples can be found in Winfree, 1980). For instance, patterns of global rhythmic activity have been observed in ant nests, (Franks, Bryant, Griffiths, & Hemerik, 1990; Cole, 1991b), while the behaviour of individual ants in isolation is not rhythmic in itself (Cole, 1991a). This phenomenon has been successfully modelled using continuous maps that interact asynchronously, (Sole, Miramontes, & Goodwin, 1993).

If, as has been rightly argued, asynchronous updating should be the modeller's default choice, should we conclude that the impossibility of exhibiting cyclic attractors means that asynchronous Boolean networks are inappropriate for modelling rhythmic phenomena such as the above? This question has not been addressed explicitly by Harvey and Bossomaier (1997), or by authors drawing similar conclusions for cellular automata, (e.g., Ingerson & Buvel, 1984; Bersini & Detours, 1994). It seems that the limitations of asynchronously driven systems regarding cyclic behaviour should prompt the modeller to discard them at an early stage as good tools for studying rhythm in biological, and other complex systems.

It will be shown that this would be a hasty conclusion, and that the long term behaviour of some asynchronous random Boolean networks can be characterized by marked rhythms. In order to do this, we will provide a simple way of defining and measuring pseudo-periodic behaviour, and use this measure to guide an evolutionary algorithm in the search for cases exhibiting this pseudo-periodicity.

Asynchrony as the default modelling choice

The work by Nowak and May (1992) on spatial patterns in a population of players of the Prisoner's Dilemma is by now almost a classic in the growing literature on the role of artefacts in simulation models. The complex spatial patterns obtained in their model, which suggest interesting implications regarding the polymorphic conviviality of cooperators and defectors, depend critically on the synchronous updating scheme they use. When asynchrony is introduced no spatial pattern appears, and the much gloomier picture of global defection as the stable strategy results, (Huberman & Glance, 1993)[1]. But this is not the only example from which a similar lesson can be learned.

Abramson and Zanette (1998) show that asynchronous updating in globally coupled logistic maps suppresses much of the complexity of the synchronous case. The state of each site results from a sum of the application of a logistic map to the previous state and a global coupling term reflecting the mean activities of all the other sites. In the case of synchronism, as the strength of the coupling increases so does the complexity of the global behaviour from partially ordered to turbulent regimes. With asynchronous updating, complexity *decreases* as the coupling between maps is made stronger. Abramson and Zanette (1998) argue that this is a significant difference, and concur with Rolf, Bohr, and Jensen (1998) in that, unless one can advance sufficient reasons to the contrary, asynchronous updating is "more physical".

Analogous results have been reported for the case of cellular automata, (e.g., Ingerson & Buvel, 1984; Bersini & Detours, 1994), and Boolean networks (Harvey & Bossomaier, 1997). The latter authors suggest that the assumption of synchrony is perhaps made "most dangerous" when it is associated with the idealisation that individual elements can be safely modelled as having discrete states, (the "Boolean idealisation" as they call it). This is interesting, although the above example of coupled maps (in which elements are "non-Boolean") suggest that the dangers may in fact be more widespread.

The methodological lesson we can derive from these and other cases is that, in the absence of knowledge about specific time delays, orchestration by an external clock should not be the default choice when modelling complex systems of many interacting elements. This is especially relevant to studies addressing phenomena related to local or global synchronization, or entrainment in such multicomponent systems.

Attractors in asynchronous random Boolean networks

A Boolean network is an array of nodes, each of which can have any one of two states (0 or 1). Each node is connected to other nodes in the network. By computing a Boolean function of their states, a new state for the node is determined. Random Boolean networks form a class of networks in which the links between nodes and the Boolean function are specified at random. They are divided into subclasses depending of the total number of nodes (N), and the number of links that influence each node (K), which is assumed here to be the same for all nodes[2].

Boolean networks have been used as models of different biological phenomena including morphogenesis,

---

[1] In (May, Bonhoeffer, & Nowak, 1995) the original choice of synchronous updating is defended by saying that it may be appropriate for some biological situations. This is, no doubt, true, although they fall short of justifying that such is indeed the case for the situation they are modelling.

---

[2] More general classes are obtained when K indicates the average number of connections to each node.

and immune response. The most well known use of random Boolean Networks has been as models of genetic regulation (Kauffman, 1969, 1993). They have also served to model idealised developmental processes, (Dellaert & Beer, 1994). In the majority of cases, synchronous methods have been used to update the network.

Harvey and Bossomaier (1997) have studied asynchronous random Boolean networks (ARBNs) by exploring the nature of their attractors using numerical experiments, and by presenting some general arguments about what can be expected from ARBNs as a class. In contrast with their synchronous cousins, ARBNs have a significant trend to evolve towards fixed point attractors suggesting that these attractors have much larger basins of attraction in the asynchronous case, a finding in accordance with previous observations by Ingerson and Buvel (1984), and Bersini and Detours (1994) on cellular automata (which in particular cases may be thought of as a special sub-class of Boolean networks). The average number of point attractors in an ARBN tends to be small when compared with the synchronous case, and does not depend on the size of the network. These observations would invalidate, if asynchrony were to be used, Kauffman's (1969, 1993) conclusions about the significance the supposedly intrinsic order of genetic regulatory networks. Kauffman has argued that different cell types in multicellular organisms correspond to different attractors of the genetic regulatory network, and that the number of cell types is roughly related to the size of the genome in the same way as the number of different attractors in a Boolean network is related to its size N for low K (roughly, $\sqrt{N}$ for $K = 2$), and, consequently, possibly for the same reasons. A similar comparison is made between the length of cell division cycles and the typical length of attractors. But these analogies rely critically on the applicability of synchronous updating to the real case which remains to be justified.

In short, Harvey and Bossomaier (1997) contribute to the methodological lesson drawn in the previous section. Asynchronous updating should be the default choice when Boolean networks are used as modelling tools, unless there is sufficient justification for considering synchrony as characteristic of the phenomenon being modelled.

Not all the attractors found in ARBNs are of the fixed point type. Those that are not have been termed "loose attractors", which can be broadly defined as the sub-set of states of the network with more than one element such that, if a given state belongs to this sub-set, then the state that follows after asynchronous updating will also belong to the sub-set. Cyclic attractors, like those found using synchronous updating, cannot be found in ARBNs. The proof is simple for standard random updating (random selection with replacement of the node to be updated, i.e., each node is updated at random with uniform probability independently of previous updates). If we suppose that the network has a cyclic attractor which is not a fixed point, then there must be two consecutive states in this attractor differing in at least the state of one single node. Since a time step is *defined* as N random node updates, there is a non-zero probability that the node that should have changed its state remains without being updated. Therefore, the two consecutive states will not differ in the state of this node as required. Notice that the proof does not work for other forms of asynchronous updating which guarantee that all nodes will be updated after N single node updates[3].

Autocorrelation and pseudo-periodicity

Using standard random updating (as will be used in this paper), is it possible for these 'loose' attractors to show marked rhythms? This question must be answered on two fronts simultaneously. On the one hand, one should specify what is meant by 'rhythm' in this case, and, on the other, one should try to find a way of using this criterion to search for cases that qualify.

Strictly speaking, the observation that ARBNs cannot produce cyclic attractors is true but only of relative significance for the researcher interested in rhythmic behaviour, for instance, in biology. This is mainly because the definition of periodicity for deterministic systems does not conform well with the relaxation of the assumption of an external driving clock. Effectively, in order to say that cyclic attractors cannot be found in ARBNs, one must take back the discarded external clock, this time not as a driving, but as a measuring device. This is achieved by using a system-independent time scale for defining when a new state of the whole network has occurred. We must therefore adopt a view of rhythmic behaviour which focuses more on the operational relationships between the states of the system — for instance, by noticing regularities in the ordering and/or statistical properties of patterns — and less on the externally measured individual duration of the states.

---

[3]As a simple counterexample, suppose that all nodes in the network but one fixate on a given state, and will remain unchanged independently of how the update is performed, and suppose that the remaining node is connected to a number of the other nodes, and to itself with a rule that specifies that, whatever the value of the other nodes, its own value must change. Since the updating scheme guarantees that the node will be updated, it will flip its value every time step.

Self	Self - 1	Self + 1	New Value
0	0	0	0
0	0	1	0
0	1	0	1
0	1	1	0
1	0	0	1
1	0	1	0
1	1	0	1
1	1	1	1

Table 1: Update rules for hand-built example of pseudo-periodic ARBN.

The real measure of rhythmic behaviour in this case will be a measure of how patterns occurring at different instants in the history of a system relate to one another. For the case of ARBNs in particular it is possible to devise a variety of simple measures based on correlations between states occurring at different points during the evolution.

In this paper, perhaps the simplest of these possible measures will be used because it will provide us with the case most similar to deterministic periodic behaviour. Other measures are imaginable, and the methodology used to search for cases that rank high under these measures is, in principle, equally applicable. The chosen measure indicates the degree to which a given state in an ARBN of N nodes *approximately* recurs after *approximately* $P \times N$ single node updates. Networks ranking high on the scale defined by this measure will be called *pseudo-periodic*[4]. The meanings of 'approximately' must be made clear in both cases. We first define an order index j which is incremented by one unit after N random updates to single nodes, but we will not equate periodicity with strict recurrence of states using this index. Instead, the correlation between two states of the network will be used to that end. The state at time j is denoted by $S(j)$, a vector whose components $s_i(j)$ correspond to the state of each node i in the network. The correlation between the states at two different times j and j' is:

$$C(j, j') = \frac{1}{N} \sum_{i=1}^{N} s_i^*(j) s_i^*(j'),$$

where $s_i^*(j)$ is the linear scaling of $s_i(j)$ into $[-1, 1]$. Highly correlated states will be taken to mean also highly similar states from the point of view of the sys-

[4]This term should not be confused with "*quasi-periodicity*" as used to refer to toroidal attractors with an irrational ratio of frequencies in continuous deterministic systems.

Figure 1: Asynchronous evolution of hand-built example. Initial condition: all nodes but one set to 0. $N = 32$, $K = 3$, time increases from left to right; 1000 updates. Black corresponds to 1 and white to 0.

tem's operation or its global significance. This is an assumption that need not be true in general, as discussed in the last section. A more global measure of the behaviour of the network is given by the correlation function between a state and its k'th successor, averaging over M successive states:

$$AC(k) = \frac{1}{M} \sum_{j=1}^{M} C(j, j + k),$$

with $k = 0, 1, 2, \ldots$. For sufficiently large values of M this function will give an idea of how well correlated, on average, is any given state with a state occurring k time steps afterwards. In this case, the function will be called simply *autocorrelation*. Notice that a given network may possess different autocorrelation functions depending on how many attractors it has and how much they differ in their statistical properties. A *sufficient* condition for ensuring non-stationary pseudo-periodic behaviour with pseudo-period P will be to ask that at least one of the autocorrelation functions has distinct peak values for k close to P which means high similarity between $S(j)$ and $S(j + P)$.

Let's consider the following hand-designed example for $N > 3$ and $K = 3$. Each node is connected to the nodes which immediately precede and follow it as indexed (with wrapping-up at the end), and to itself. The Boolean functions are the same for all nodes[5] and are shown in table 1. What these rules say is that if a node is in state 0, then it must remain 0 unless there is a 1 on the preceding node and a 0 in the following node. If a node is in state 1, then it must remain 1 unless there is a 0 in the preceding node and a 1 in the following. With synchronous updating, these rules and connectivity produce travelling waves with period $P = N$. For the case of asynchronous updating we find that the evolution of the network can also be characterized by waves with a pseudo-period $P \cong N$.

Figure 1 shows an example of the asynchronous evolution of this network, and figure 2 shows the corresponding autocorrelation function averaged over 10000

[5]This example is a particular case of a homogeneous Boolean network. In the general case, as in the rest of the paper, Boolean functions differ from node to node.

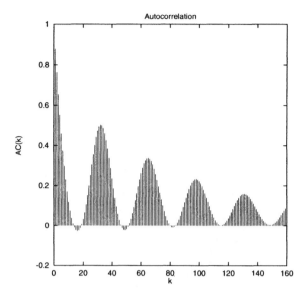

Figure 2: Autocorrelation corresponding to hand-built example as averaged over 10000 consecutive states and over 10 runs. $N = 32$, $K = 3$; peaks in $k = 0, 32, 65, 99, 132$.

consecutive states and over 10 runs. Rhythmic behaviour is apparent both by direct observation of the evolution, and from the autocorrelation function. The evolution of the network can be interpreted as a soliton wave with varying width but having a distinctly low variability in speed. The autocorrelation function shows clear peaks near multiples of a pseudo-period of $P \cong N$. It is important to notice that each successive peak is a bit lower than the previous one, showing an effect of 'memory decay'. This is mainly due to the fact that a highly correlated state will recur after *about* P time steps, and therefore actual recurrence becomes more uncertain the further upstream one moves.

Evolving rhythms

Is the above example an unique case? This question can be answered by devising a method for searching for more examples. The idea is to use a target autocorrelation function and a search method that provides us with networks which will approximate this target. We have used a simple genetic algorithm as a search tool. ARBNs are described using a binary genotype which encodes their connectivity and the rules governing how each node is updated. Any network with parameters N and K can be encoded in a genotype of length G_L:

$$G_L = N(K \log_2 N' + 2^K),$$

(a)

(b)

Figure 3: Asynchronous evolution of evolved ARBN with $N = 16$, $K = 2$ and a target period $P = 16$, (see also figure 4). 1000 consecutive states starting from a random initial condition are shown in (a). These can be compared with another 1000 consecutive states occurring later in the same run, (b). The interval between the two figures is of 8000 steps.

where N' is the first power of 2 greater than or equal to N. The factor in parentheses corresponds to the number of binary loci necessary to encode K connections plus one Boolean function of K arguments. Other encodings are possible.

Individual networks are run for between 500 and 1000 time steps (each time step being equivalent to N node updates), and for a number of trials (usually between 4 and 10) starting from different random initial conditions. After each trial, the autocorrelation $AC(k)$ is calculated for $k = 0, 1, ..., 2N - 1$ by averaging for all the states in the run (except the last $2N$).

The fitness of a network is calculated for each trial as $1 - D$, where D is the normalized distance between the network's autocorrelation, and the target autocorrelation. Fitness scores are averaged over the trials, and the value of one standard deviation is deducted to benefit low variability between the trials. Point mutation, uniform crossover, and a rank-based selection scheme are used. The rate of mutation per loci is chosen in accordance to the genotype length so as to have a probability of no mutation in a given genotype of about 80%. The size of the population is of 90 networks.

The choice of an adequate target autocorrelation function is crucial for the success of the search. This target need not correspond to any realisable network. Instead, its definition has been guided by considerations of evolvability. Our choice in all the cases presented here has been to define a target autocorrelation using steps between the values of 0 and 1. States will

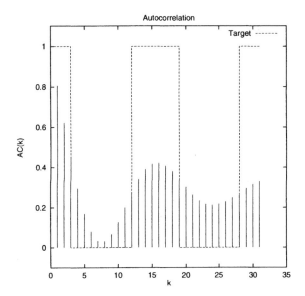

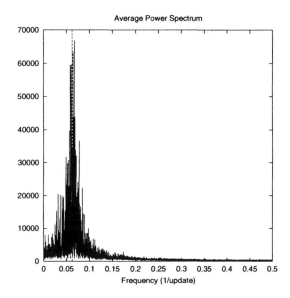

Figure 4: Autocorrelation for an evolved network with $N = 16$, $K = 2$, and $P = 16$, as averaged over 10000 consecutive states. The dashed line shows the target autocorrelation used in the fitness function.

Figure 5: Power spectrum for evolved network corresponding to figures 3 and 4, as averaged over variations in the state of individual nodes. The frequency corresponding to the target period is shown with a vertical dashed line.

be highly correlated around the chosen pseudo-period P, so that a value of 1 is assigned to values of k in $[nP - e, nP + e]$, with $n = 0, 1, 2, ...$ For any other value of k the autocorrelation is 0. The width of the square peak $2e$ is carefully chosen so as to strike a balance between the number of values of $AC(k)$ equal to 0 and those equal to 1. This balance is important in order not to bias the search process, and not because we can expect typical pseudo-periodic ARBNs to necessarily exhibit such balance in their autocorrelation functions.

So far, ARBNs have been successfully evolved for $N = 16, 32, 64$ and $K = 2, 3, 4$ using target periods of $P = N/2, N, 2N$. A few trials with shorter target periods and larger N have been attempted only with minor success. The number of generations has oscillated between 1000 and 5000, often obtaining good results after about 500 generations[6].

Figures 3, 4, 5, and 6 correspond to an evolved network with $N = 16$, $K = 2$, and $P = 16$. Figure 3 shows the first and the last 1000 steps corresponding

to a 10000-step run. Although some nodes are frozen most of the time, the remaining ones form distinct patterns which appear with a marked rhythm. The form of the pseudo-periodic attractor is stable as can be appreciated by comparing both figures. This result is not directly implied in the constraints used for performing the evolutionary search, which condition only the shape of the autocorrelation.

The autocorrelation function is shown in figure 4, together with the target autocorrelation, the range of which goes from $k = 0$ to $k = 31$ (dashed line). This function has been calculated by averaging over 10000 steps corresponding to 10 different runs starting from different random initial conditions. It shows a clear, though wide, peak around $k = 16$.

Further evidence of the rhythmic behaviour of this network can be obtained by calculating the power spectrum (using Fast Fourier Transform) for each node in the network. This is shown in figure 5 where the N spectra have been averaged to give an idea of the behaviour of the whole. There is a marked peak corresponding to a frequency near $1/P = 0.0625$.

Figure 6 presents a histogram showing the percentage of recurrence of individual states corresponding to the same network, again over a 10000-step run. States

---

[6]The search method implicitly selects networks with rapid transients. Harvey and Bossomaier (1997) have observed that standard asynchronous updating produces the shortest transients, so we do not consider this to be a major problem. However, simple modifications to the search algorithm could avoid this if necessary.

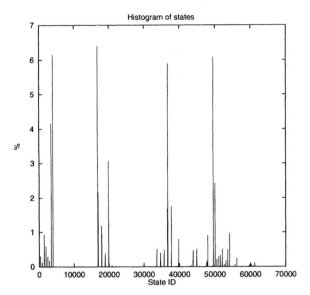

Figure 6: Histogram of states corresponding to the run shown in figure 3. States are labeled by using the integer number that they encode as a binary string. Only about 12 states recur with a frequency greater than 1%.

are identified using the integer number that they encode as binary strings. The number of states that recur in the attractor with a frequency greater than 1% is about 12.

The rhythmic behaviour of an evolved ARBN with $N = 32$, $K = 2$, and $P = 32$ is shown in figure 7, and the corresponding autocorrelation in figure 8 where a more defined peak around P can be observed.

Finally, one interesting observation we have gathered from all the cases tested so far is that all the evolved pseudo-periodic ARBNs behave as strictly periodic if they are updated synchronously with periods corresponding to the evolved pseudo-periods. We do not know how generalised is this observation.

Conclusion

It has been shown that, although they cannot reproduce strictly cyclic behaviour, ARBNs need not be discarded *a priori* as possible models of rhythmic phenomena since they may be able to capture many features of interest of such phenomena. Few people would hesitate in calling a natural system 'rhythmic', or even 'periodic', if its behaviour exhibited a power spectrum such as the one shown in figure 5.

Precisely because they do not incorporate synchrony

Figure 7: Asynchronous evolution of evolved ARBN with $N = 32$, $K = 2$, and a target period $P = 32$, for 1000 time steps.

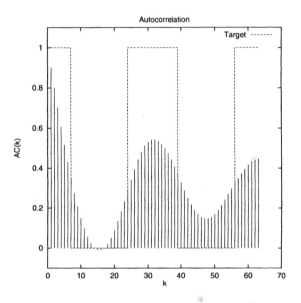

Figure 8: Autocorrelation for ARBN corresponding to figure 7, averaged over 10000 consecutive states. The dashed line shows the target autocorrelation.

by default, ARBNs, and similar asynchronous systems, may constitute quite strong models if they are able show spontaneous rhythmic behaviour as compared with models using built-in synchrony from the start.

It is possible to provide simple ways for recognizing pseudo-periodicity, and use them to guide an evolutionary search process to find examples. Probably a variety of search processes could have worked equally well or better for this task, but success using a genetic algorithm is suggestive of ways natural selection could have acted on natural systems with analogous properties, if rhythmic or coordinated behaviour happened to be of some functional value.

Before drawing conclusions about the relatively minor success in evolving ARBNs with other pseudo-periods or values of N, it may be necessary to understand better the evolutionary task itself, and check whether the choices for encoding and mutation rates could not be improved. As it stands, the evolutionary landscape exhibits a high degree of neutrality (there are many ways of producing functionally equivalent ARBNs), and choice of search parameters would perhaps need some re-consideration in view of this fact.

As briefly mentioned, our definition of pseudo-periodicity relies on similarity between states using a correlation measure. Such similarity need not correspond to *significant* similarity in the context of a natural system. It is easy to think of many (not necessarily pathological) cases in which alterations to the strict order of single events may produce radically different results. In those cases, our definition does not work, and it is an open question whether others would. Two speculative solutions could perhaps be of some use in such cases. One is the utilization of *weighted correlation* as a measure of similarity. If it is functionally important that certain states recur pseudo-periodically more reliably than others, then they could be assigned a higher weight in the calculation of correlations. The other possible solution is the use of *single element autocorrelation* on which stricter statistical demands could be made on particular nodes (like less variability in pseudo-period) if it were necessary. It is not clear yet whether these solutions would work in general.

Finally, we have also mentioned that the impossibility of reproducing strictly cyclic behaviour depends on the use standard random updating. Other schemes which guarantee that all nodes be updated in a same time step admit some cases of strict cycles. This is not important in itself, but should warn us that these latter forms of updating are perhaps as artificial as synchronous updating, and should not constitute a default choice for the modeller either. Observations like this one open the question of which is the most adequate way of defining the passage of time in a way that accords naturally with the type of systems involved. In this work, a time index has been defined in a intuitive manner in accordance to the size of the system so as to simulate statistically parallel (uniform time delays) and independent updating for all nodes. We believe that such a choice is appropriate since we did not rely heavily upon it to define pseudo-periodicity (variability in the pseudo-period is allowed to be large). It remains an open issue whether other, 'more physical' time indexes, appropriate for the particular systems in question, could be defined.

References

Abramson, G., & Zanette, D. H. (1998). Globally coupled maps with asynchronous upating. *Phys. Rev. E*, **58**, 4454–4460.

Bersini, H., & Detours, V. (1994). Asynchrony induces stability in cellular automata based models. In Brooks, R., & Maes, P. (Eds.), *Artificial Life IV, Proceedings of the Fourth International Conference on Artificial Life*. Cambridge, MA: MIT Press.

Cole, B. J. (1991a). Is animal behaviour chaotic? Evidence from the activity of ants. *Proc. R. Soc. Lond. B*, **244**, 253–259.

Cole, B. J. (1991b). Short-term activity cycles in ants: generation of periodicity by worker interaction. *Am. Natur.*, **137**, 244–259.

Dellaert, F., & Beer, R. (1994). Toward an evolvable model of development for autonomous synthesis. In Brooks, R., & Maes, P. (Eds.), *Artificial Life IV, Proceedings of the Fourth International Conference on Artificial Life*. Cambridge, MA: MIT Press.

Franks, N. R., Bryant, S., Griffiths, R., & Hemerik, L. (1990). Synchronization of the behavior within nests of the ant *Leptothoras acervorum* I. *Bull. Math. Biol.*, **52**, 597–612.

Harvey, I., & Bossomaier, T. (1997). Time out of joint: Attractors in asynchronous random Boolean networks. In Husbands, P., & Harvey, I. (Eds.), *Proceedings of the Fourth European Conference on Artificial Life*, pp. 67–75. Cambridge, MA: MIT Press.

Huberman, B. A., & Glance, N. S. (1993). Evolutionary games and computer simulations. *Proc. Natl. Acad. Sci. USA*, **90**, 7715–7718.

Ingerson, T. E., & Buvel, R. L. (1984). Structure in asynchronous cellular automata. *Physica D*, **10**, 59–68.

Kauffman, S. (1969). Metabolic stability and epigenesis in randomly constructed genetic nets. *J. theor. Biol.*, **22**, 437–467.

Kauffman, S. A. (1993). *The Origins of Order*. Oxford: Oxford University Press.

May, R. M., Bonhoeffer, S., & Nowak, M. A. (1995). Spatial games and the evolution of cooperation. In Moran, F., Moreno, A., Merelo, J. J., & Chacon, P. (Eds.), *Proceedings of the Third European Conference on Artificial Life*, pp. 749–759 Granada, Spain. Berlin: Springer.

Nowak, M. A., & May, R. M. (1992). Evolutionary games and spatial chaos. *Nature*, **359**, 836–829.

Rolf, J., Bohr, T., & Jensen (1998). Directed percolation universality in asynchronous evolution of spatio-temporal intermittency. *Phys. Rev. E*, **57**, R2503–R2506.

Sole, R. V., Miramontes, O., & Goodwin, B. C. (1993). Oscillantions and chaos in ant societies. *J. theor. Biol.*, **161(3)**, 343–357.

Winfree, A. T. (1980). *The geometry of biological time*. New York: Springer Verlag.

Levels of Compartmentalization in Alife

S. Kazadi,* D. Lee, R. Modi, J. Sy , W. Lue

Jisan Research Institute

Abstract

This paper addresses the use of particular encoding schemes in evolutionary systems. We define three paradigms of DNA encodings: *non-compartmentalized*, *partially compartmentalized*, and *fully compartmentalized*. We demonstrate that there is a significant and increasing advantage to the use of *partially* and *fully compartmentalized* models as the complexity of a structure increases. Implications for the design of evolutionary systems including biological systems are discussed.

keywords: compartmentalization, transposons, artificial evolution, natural evolution

1 Introduction

Transposons (MacPhee, 1991; Finnegan, 1994) are fascinating pieces of DNA which have the surprising ability to move around, create duplicate copies, and excise themselves over successive generations and during duplication. They are units of DNA which encode complete functions not dependent on other pieces of DNA for their expression. Transposons are nearly ubiquitous in Nature and are thought to make up over 90% of the DNA found in human beings. Their general capability to provide functions, generate new functions of a cell, create genetic defects, and correct them is a rich set of effects that has yet to be generated by the artificial life community.

Their power comes from the fact that DNA that can encode whole functions can be reconnected to produce novel systems of fundamental importance with great efficiency. They are a subject of intense study in the genetic algorithm and evolutionary computation communities and should become important in the artificial life community beyond the scope of the building block hypothesis and related issues.

How transposons arose is also intellectually interesting. Perhaps the simplest question centers around the nature of the origin of these structures. Transposons certainly

*To whom correspondences should be addressed. email: sanza@caltech.edu

appear in natural systems and their analogs appear in some evolutionary systems. However, how much of the structure of transposons depends on the nature of the replicating system, and how much depends on the requirement to build efficient evolutionary systems? If transposons are simply a result of the way in which life historically arose on earth, then they have no natural place in artificial systems. However, if they are a result of an evolutionary requirement, then it would seem to be advantageous to understand that requirement and to build them into our evolutionary models.

Many evolutionary models exist in the recent literature, each one with its own general structure. Of these, genetic algorithms are the most common and they have been used to do a variety of things including designing hardware and optimizing functions. Genetic algorithms are powerful due to their ability to share information that is beneficial. This is the basis of the *building block hypothesis* and forms the basis for a great deal of genetic algorithm literature (Forrest and Mitchell, 1993). One important problem that has been recently addressed is the identification of *linkages* between parts of the genome and identifications of recodings of the search space in such a way that these linkages are minimized (Kazadi 1997; Kazadi 1998; Munetomo and Goldberg, 1999).

A particularly exciting application of genetic algorithms has recently been undertaken by Adrian Thompson (Thompson 1999a(b,c), 1998a(b)) in studying hardware evolution of field programmable gate arrays (FPGA), and is becoming more popular as a practical means of generating useful hardware designs. In these studies, a genetic algorithm is used to generate circuits in the FPGA. However, this work does not seem to deal with the use of advanced genetic operators nor take advantage (or even seem to be aware) of techniques of compartmentalization.

Tierra (Ray 1992; Ray 1994; Adami 1995a) and Avida (Adami 1994; Adami 1995b; Ofria and Adami 1999) are examples of artificial ecosystems in which no explicit fitness function has been designed. The evolving agents are self-replicating strings of DNA modelled to function in a virtual environment using an analog of the 80x86 processor. The main difference between these two systems is that Tierra is built on a single completely connected soup, while Avida is designed on a two-dimensional spatially separated toroidal lattice. Each agent may replicate itself to a neighboring cell during its processor time, may compete with agents currently occupying the cell for the space, and has a nonzero probability of a number of addition and alteration mutations at every replication. While these systems exhibit fascinating dynamics, they evolve at a rate which has a power law distribution of epochs in time, a somewhat slower evolution rate than one might wish to be restricted to.

Genetic programming (Koza 1992; Koza 1994) contains no explicit limitation in the length of the genome. Functional trees are built up which serve to solve a practical design problem. The ability of the trees to be used in a modular way allows subfunctions to be built up and to be incrementally added to a functional tree system. This is an example of a *partially compartmentalized* model, which has advantages we clarify in this study.

The rest of the paper is organized as follows. Section 2 discusses compartmentalization and the expected time of construction of differing paradigms. Section 3 presents the theoretical distribution of epochs in time, providing motivation for application of this theory to existing models of evolutionary systems. Finally, Section 4 offers some concluding remarks.

2 A Mathematical Model of Compartmentalization

In this section, we introduce the concept of DNA compartmentalization and discuss a mathematical justification for its requirement in evolutionary systems. We focus on evolutionary systems which are built up incrementally, weighing both the addition of new elements and the modification of connections between elements. We assume that DNA is a string made up of *building blocks*, which are units of DNA that may take on one of several possible settings. On top of these building blocks are connections which must be properly configured between independent blocks or groups of building blocks. We assume that the length of the DNA is not constant and that new building blocks may be added to the structure with some probability.

2.1 The nature of compartmentalization

Definition 1: A **compartment** is a subset of a DNA string consisting of elements which code for the structures which carry out a specific function, group of functions, or no function at all. Such a grouping must exclude elements which in whole or in part define other functions. An **elemental compartment** is a compartment which is not capable of being broken up into smaller compartment, with retention of all functions initially encoded.

Note that an elemental compartment can have multiple functions.

Definition 2: A **non-compartmentalized encoding** is an encoding in which there exists no natural structure within the DNA which codes a given device. Any specific element of DNA is equally probable of being a part of a given substructure in the evolving system.

This includes, among others, infinite length genetic algorithms. Unless explicitly introduced in the system, there are no self-contained functional units within such implementations.

Definition 3: A **partially compartmentalized encoding** is an encoding in which there exist natural substructures which may be completely duplicated without alteration of function.

Genetic programming algorithms are examples of partially compartmentalized encodings.

Definition 4: A **completely compartmentalized encoding** is an encoding in which connectivity and functionality may at once be specified by the DNA or be a natural consequence of some "natural" laws governing the behavior of the structures built by the encoding.

Each change in the design of a system is of one of two fundamental types. One changes the fitness of the design, while the other does not. This is an important distinction, and we formally define these.

Definition 5: We define an **evolutionary step** to be to be a change in a design that also changes its fitness. We define a **null step** to be a change in a design that produces no change in fitness.

The aftermath of an evolutionary step is an *epoch*, in which the new, more functional structure takes over a population of less functional structures. By successfully competing against the others, the fitter organism can create multiple copies of the new structure. New designs are then based on this model and will then have a lower improvement time. A null step, on the other hand, produces no such epoch and so all alterations must be *built on top of* the change in question thereby further increasing the improvement time.

2.2 Rates of evolution

Let us assume that we are working with linear DNA made up of building blocks of some kind and a method of encoding the connection between the building blocks. These building blocks may generally encode a physical entity or a computational entity.

Let us assume that we have N possible assignments for each building block. First, note that the probability of adding a specific building block is typically of $O\left(\frac{1}{N}\right)$. This, of course, assumes that the addition of each possible building block is equally likely. The probability of connecting any two building blocks correctly is $O\left(\frac{1}{m(m-1)}\right)$ where m represents the number of elements currently in the genome. So, the probability of correctly adding and connecting a building block to a device with $m - 1$ elements is

$$p_a = O\left(\frac{1}{Nm(m-1)}\right) . \tag{1}$$

Thus, the probability of a particular design containing M elements is

$$p_d = O\left(\frac{1}{N}\prod_{i=2}^{M}\left(\frac{1}{Ni(i-1)}\right)\right) = O\left(\frac{1}{N^M M!(M-1)!}\right) \tag{2}$$

yielding an expected time to completion of the algorithm of

$$\tau \propto O\left(N^M M!(M-1)!\right). \tag{3}$$

If one does not require a particular order to the nodes being designed, then the probability of success is conditioned by the number of possible identical solutions. In this case

$$p_d \propto O\left(\frac{M!}{N^M M!(M-1)!}\right) \tag{4}$$

giving the expected time to completion

$$\tau_n \propto O\left(N^M (M-1)!\right). \tag{5}$$

Now let us work out the analogous evolution time for the compartmentalized approach. We assume that we have k compartments, each of length L of the same building blocks. In general, we assume that each compartment, as shown above, has a probability of being formed as given in (2.4). The probability of connecting each of these compartments correctly is typically of order

$$p_{pc} \propto O\left(\frac{1}{k(k-1)L^{2k}}\right) \tag{6}$$

making the total probability

$$p_p \propto O\left(\frac{1}{L^{2k}k(k-1)N^L(L-1)!}\right) \tag{7}$$

The expected time of construction in the partially compartmentalized model is

$$\tau_p \propto O\left(k^2 N^L (L-1)! L^{2k}(k-1)\right). \tag{8}$$

Removing the time due to the connection of compartments yields the expected time to build in the compartmental model. Thus,

$$\tau_c \propto O\left(kN^L(L-1)!\right). \tag{9}$$

This means that the ratio of the times is

$$\frac{\tau_p}{\tau_n} = \frac{N^L(L-1)!L^{2k}k^2(k-1)}{N^M(M-1)!}. \tag{10}$$

If we then let $M = kL$, we have

$$\frac{\tau_p}{\tau_n} = \frac{N^L(L-1)!L^{2k}k^2(k-1)!}{N^{kL}(kL-1)!}. \tag{11}$$

The corresponding ratio for the compartmental model is

$$\frac{\tau_c}{\tau_n} = \frac{kN^L(L-1)!}{N^{kL}(kL-1)!}. \tag{12}$$

Example 1: Suppose that we have $k = 2$ and $L = 4$. Then equation (2.11) is

$$\frac{\tau_p}{\tau_n} = \frac{2N^4 3! 4^4}{N^8 7!} = \frac{6N^4 256}{1260N^8} = \frac{1536}{420N^4} \approx 3.66N^{-4}. \tag{13}$$

If we let $N = 10$, a reasonable choice for many evolutionary algorithms, the ratio is

$$\frac{\tau_p}{\tau_n} \simeq 3.66 \times 10^{-4}. \tag{14}$$

Clearly, this is a significant advantage with a very small level of compartmentalization. The corresponding calculation for the compartmentalization model is

$$\frac{\tau_c}{\tau_n} = \frac{N^4 3!}{N^8 7!} = \frac{1}{24N^4} \approx 4.2 \times 10^{-6} \tag{15}$$

an even more significant advantage.

2.3 Types of compartmentalization

The use of compartmentalization in generating more efficient evolution algorithms is of paramount interest in this work, and we now turn to implications of this formalism.

The ability of a device to be built up incrementally from discrete modules allows one to evolve the modules individually. This greatly reduces the expected time of completion from factorial in the number of building blocks to factorial in the number of building blocks per node multiplied by a linear term in the number of compartments. This also increases the time of completion by a factor that is factorial in the number of building compartments, but quadratic in the number of building blocks per compartment. Thus, as long as the number of building blocks is significantly larger than the number of compartments, the development time will improve greatly. We also note that this represents a *worst case scenario*; the situation is significantly easier if several of the building blocks are similar and may be duplicated.

These mathematical definitions lead us back to our previous definitions. We can see that there are three fundamental types of DNA encodings, coupled strongly to the physical system in which they live. Non-compartmentalized encoding is characterized by equation (2.5). These types of systems make use of absolutely no structural information and have no well-defined functional groupings. Often times the data concerning functional groupings is scattered and must be dealt with across the genome. Infinite dimensional or incremental genetic algorithms, in which the number of elements in a vector is unbounded, are such algorithms. Algorithms which are not include genetic programming, and the Avida and Tierra systems by Adami et. al. and T. S. Ray, respectively.

The second class of structures which are partially compartmentalized include those previously mentioned: Tierra, Avida, and genetic programming. These models have implicit structure, and have elements evolving which are partially compartmental. In Tierra and Avida these might be evolving subroutines whose combinations produce the functional creature. In genetic programming these are the subtrees, which may be moved about intact and combined with other trees. These algorithms quickly create good behaviors, but have a hard time creating a large number of functional subgroups. Rather, these paradigms should generate progressively larger subgroups as the genome increases in length in order to keep the connection term of equation (2.8) small. As the length of the compartments rises, the probability of finding a useful compartment decreases factorially leading to a convergence in the compartmentalization and size of compartment. This leaves the algorithm to develop compartments of high quality and may lead to stagnation of the paradigm. These types of algorithms, while providing a great advantage over the completely non-compartmentalized model, can converge due to the second term of equation (2.8). If the number of compartments increases, so too does the difficulty in generating the design, owing only to the correct interaction between compartments.

For some algorithms, though, it might be possible to build which remove this term. These algorithms would have the interactions built as part of their structure, so that a particular compartment in the design would only be able to interact with a handful of other compartments in the design. This would allow the number of interactions to be severely curtailed, yielding a nearly linear computation time in the number of compartments. This is what biological life is capable of accomplishing through the use of proteins and embedding of this level in the physics of the universe. Such fully compartmentalized models represent a great advantage to any type of evolution.

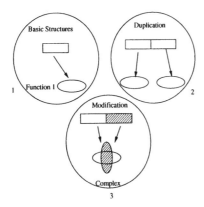

Figure 1: The fully compartmentalized model.

As one can see, Figure 1 illustrates

this model. Compartmentalization in the evolutionary process organizes the DNA string based on functionality. By doing this, the number of interactions between compartments and the time for completion are reduced to a minimum while the DNA string's full capabilities for adaptability and survival are exploited. The product of the new DNA element is automatically incorporated properly with the previous one, without the need for evolution of this compatibility. The way in which this is done is model specific and embedded in the system.

Thus, we provide this table of paradigms and computation times.

Paradigm	Computation Time
Non-compartmentalized	$\tau_n \propto O\left(N^{kL}\,(kL-1)!\right)$
Partially Compartmentalized	$\tau_p \propto O\left(k^2 N^L\,(L-1)!L^{2k}\,(k-1)\right)$
Fully Compartmentalized	$\tau_c \propto O\left(kN^L\,(L-1)!\right)$

Table 2.1: Table of computational design expressions for different paradigms

3 Compartmentalization in the Generation of Epochs

An evolutionary system represents a balance between two competing requirements, the ability to shield existing systems from mutation and the need to adapt to the environment. Evolutionary systems have a tendency to leave existing designs unchanged, as the set of useful designs for a particular task are typically sparse in the space of all possible designs. Recklessly changing a design would be unlikely to produce a feasible design, resulting in the extinction of the species in question. The competing tendency, of modifying existing designs, also exists in order in to produce more capable offspring. This tendency to change things requires a method of changing DNA without significantly negatively altering the functionality of it. Thus, the problem for an evolutionary design system is to preserve the function of evolving agents while still improving.

If we again assume that DNA is essentially a string of building blocks, we may estimate the probability that an improvement will occur. Let us also assume that we have some method for copying segments of DNA into new additional segments of DNA. This copying mechanism must then depend on the choice of the beginning and ending segments of DNA. We begin by investigating the non-compartmental model.

First, we assume that we have a length-independent mutation rate $0 < r < 1$ which gives the likelihood that at each iteration a given building block will mutate. Then, if we have a sequence of L building block elements and N ways to mutate each one, the probability of choosing and correctly mutating S of them, given that each mutation is independent, is

$$p_{n_M} \propto O\left(\left(\frac{r}{N}\right)^S (1-r)^{L-S}\right).$$

Now, the probability of correctly copying any given DNA sequence

$$p_{n_c} \propto O\left(\frac{1}{M(M-1)}\right) \tag{16}$$

for each pair of cutting/copying points. So, the total probability of correctly copying the chosen DNA set and correctly mutating it (creating a new functional piece of DNA) is

$$p_{n_{cr}} \propto O\left(\frac{1}{M(M-1)} \sum_{i=0}^{M-S-1} (M-S-i)\left(\frac{r}{N}\right)^S (1-r)^i\right) \tag{17}$$

$$= \frac{1}{M(M-1)S^2}\left(\frac{r}{N}\right)^S \frac{(1-r)\left[(1-r)^{M-S}-1\right]+r(M-S)}{r^2} \tag{18}$$

Once the correct changes have been made, it is necessary to incorporate these changes into the genome. This involves both correctly removing and adding the linkages between the existing and new structures respectively. In non-compartmentalized models, these are done independently so that in general, the probabilities are of order

$$p_{n_{inc}} \propto O\left(\left(\frac{1}{M\langle L\rangle}\right)^S\right) \tag{19}$$

where $\langle L\rangle$ is the expected length of the advantageous mutation. The product of these probabilities gives us the total probability of an epoch-generating event. Thus, the total probability is of order

$$p_n = O\left(\frac{r^{S-2}(1-r)\left[(1-r)^{M-S}-1\right]+r^{-1}(M-S)}{M(M-1)S(S-1)(MN\langle L\rangle)^S}\right) \tag{20}$$

yielding a time of order

$$\tau_n \propto O\left(\frac{M(M-1)S(S-1)(MN\langle L\rangle)^S}{r^{S-2}(1-r)\left[(1-r)^{M-S}-1\right]+r^{-1}(M-S)}\right) \tag{21}$$

If $r \ll 1$ then

$$\tau_n \propto O\left(\frac{rM(M-1)S(S-1)(MN\langle L\rangle)^S}{r^{S-1}(M-S)r+(M-S)}\right) \tag{22}$$

In compartmentalized models, creating copies of DNA pieces can be done simply by choosing the compartment of interest and directly copying it. This means that the probability of choosing the correct compartment is

$$p_{c_c} = O\left(\frac{1}{k}\right) \tag{23}$$

where k is again the number of compartments. The probability of modifying the copy correctly is given as

$$p_{c_m} \propto O\left(\frac{r^S (1-r)^{L-S}}{N^S}\right) \tag{24}$$

giving the expected time to correct modification as

$$\tau_{c_m} \propto O\left(\frac{N^S k}{r^S (1-r)^{L-S}}\right) \tag{25}$$

which becomes, if $r \ll 1$,

$$\tau_{c_m} \propto O\left(\frac{N^S k}{r^S (1-(L-S)r)}\right) \tag{26}$$

A partially compartmentalized model requires that the compartment of interest be correctly connected, and the previous connection be correctly disconnected. Thus, the probability of disconnecting and reconnecting the appropriate compartments is given by

$$p_{p_c} \propto O\left(\left(\frac{1}{kL}\right)^S\right) \tag{27}$$

making the time

$$\tau_c \propto O\left((kL)^S\right). \tag{28}$$

Thus, the total time is

$$\tau_p \propto O\left(\frac{N^S k (kL)^S}{r^S (1-r)^{L-S}}\right) \tag{29}$$

Finally, a completely compartmentalized model has the connection taken care of by the underlying structure. Thus, the expression (3.15) is altered, becoming

$$\tau_c \propto O\left(\frac{N^S k}{r^S (1-r)^{L-S}}\right) \tag{30}$$

If we again take $kL = M$, we may form another comparative table of the three evolutionary models.

Paradigm	Improvement Time
Non-Compartmental	$\tau \propto O\left(\frac{r(kL)((kL)-1)S(S-1)((kL)N\langle L\rangle)^S}{r^{S-1}((kL)-S)r+((kL)-S)}\right)$
Partially Compartmental	$\tau_{cm} \propto O\left(\frac{N^S k^{S+1} L^S}{r^S (1-r)^{L-S}}\right)$
Fully Compartmental	$\tau_{cm} \propto O\left(\frac{N^S k}{r^S (1-r)^{L-S}}\right)$

Table 3.1: A summarizing table of the different expected computation times for improvements in the different models.

Typical performances of the three models are as given in Figure 2.

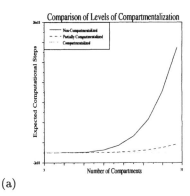

(a)

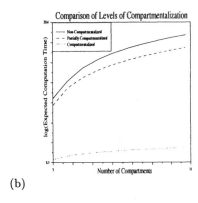

(b)

Figure 2: This gives the typical expected performance of the three compartmental models in linear (a) and log (b) format. The advantages of the partially and fully compartmentalized models are clearly visible.

Table 3.1 and Figure 2, illustrate the advantage of a compartmental model in generating *epochs*, or sweeping displacements of the current dominant species in an artificial evolutionary system.

Evolutionary time is measured in epochs, rather than years, as these are the only important time measures. Because of the existence of epochs, the time required for a design or adaptation is *additive* rather than multiplicative. Any evolutionary system which is capable of generating evolutionary change which results in an epoch will come to be the dominant system in use. Clearly compartmentalization exhibits a significant and increasing advantage in the generation of any given improvement in any particular trait. Thus, it would seem that compartmentalization is a significantly advantageous design paradigm for evolutionary systems, both natural and artificial.

4 Implications of the Theory

In this Section, we present some of the implications and paradigm modifications that seem to result from this theoretical exposition.

4.1 The Rise of Transposons and Viruses

Each of these computation times in Tables 2.1 and 3.1 indicates a dependence on the number of different tries or trial steps one might expect to take in an evolutionary model. Clearly, if one had a large number of elements in a population of possible designs, this task would occur more rapidly, in terms of numbers of iterations. However, the mere duplication of a piece of unconnected DNA does not allow one to make multiple alteration trials, as it has no significant improvement over any other improvement. This means that no successful copy will trigger an epoch, during which elements in the population are replaced by a new dominant species. In other words, each of these events can be viewed as an isolated set of events set into motion by a single mutation. How then might one be able to try many different designs without independently discovering the correct copy mutation each time?

The answer is surprisingly simple. Rather than making a single copy of any given copied DNA, one might create multiple copies of the DNA segment in question (Robertson and Martos, 1997). This allows each mutation step in the paradigm to affect multiple copies of the DNA segment in question. The probability of getting the correct mutation increases linearly with the number of copies. Thus, the ability to create a new element is significantly improved by using multiple copies of compartments.

In the linear model, since the probability of disconnecting and reconnecting the correct elements is a $4L$th order function of the number of elements in the genome, progressively longer genomes lead to an extremely fast increase in the length of time. In this case, creating copies leads to a significantly longer computation time.

The case is similar in the case of partially compartmentalized DNA models. The effect comes from the increase in connection time, which increases as L^{2k} where, as before L is the compartment size, and k is the number of compartments.

Finally, a fully compartmentalized DNA model has its assembly taken care of by the physical system underlying the scheme. No increase in assembly time is necessary, in general, making the ability of the paradigm to parallelize unbounded. Thus, it is in the best interests of the system to build large redundant DNA strains, as the probability of evolving more quickly increases with larger and more redundant DNA.

As an aside, we consider for a moment, biological DNA. In biological DNA, the machinery that leads to making copies is not provided by any source other than the DNA making the copies itself. This means that the DNA must provide for the making of copies, and its own activation, which typically means movement of the DNA segment in question into an active region.

Since it is advantageous for an evolutionary system which makes use of transposons to have them, it is reasonable to expect that the proliferation of transposons and retrotransposons have more to do with the nature of evolution than with the details of biological life. These are nothing more than the compartments of fully compartmentalized DNA. Once discovered, it is expected that these would become a fully integrated part of the genome.

This leads one to speculate about the connection between transposable elements and viruses. It is certainly a possibility that the apparent similarity between transposable elements and viruses is not accidental. It may be that transposons arose as a result of the need for their inclusion in the evolutionary process, and that these almost independent transposable elements became (and perhaps are still becoming) independent of the cell in the form of viruses and retroviruses. If this is the case, then it would imply that, while transposons and transposable elements are results of the requirement of efficient evolutionary systems, viruses are a byproduct of biological life, essentially a global or population-based transposable unit.

This also raises the alarming possibility that transposons may continually give rise to beneficial viruses. If it is indeed possible for transposons to transform themselves into viruses by combining with appropriate parts of existing cell DNA, it may be possible for these viruses to leave the host and infect multiple other hosts. The most dangerous result of this might be the creation of a virus which carries antibiotic resistance in bacteria. In close proximity to other bacteria, this would provide a very effective method of passing helpful resistance on to other organisms not created in reproduction. In fact, it is currently known that such infections do exist. It is not presently known whether or not a transposon can give rise to a virus in a bacteria.

4.2   Lessons for Alife

Alife, of course, is the study of lifelike systems. What makes this discussion important for alife is both its motivation and its implications. This study was originally motivated out of a desire for an understanding about the existence of transposons and retrotransposons (and related structures such as viruses and retroviruses). The existence of transposons can be either an evolutionary necessity, or a fluke provided by carbon-based biology. This paper has implied that the existence of transposons is indeed an evolutionary necessity. It is not known whether this extends to viruses, as it is not clear that the use of viruses is much different evolutionarily from an epoch.

In order to more fully understand these systems, a significant amount of simulation work needs to be done. We have seen that at small levels of design complexity, the models are nearly interchangeable, while at large levels of complexity, there is a significant difference between the performance of different evolutionary models. We have also indicated several

reasons that one might expect an evolutionary system which is properly equipped to create compartmentalization, if the capabilities exist. However, in order to fully investigate this, new simulations need to be constructed which have the capability of creating both types of DNA structures.

Previously a number of very specific, and somewhat contrived, simulations have been built, in which agents were allowed to grow and interact in simulated environments. However, no simulation has yet been built which has allowed for the creation of complex, compartmental DNA. Yet in order to understand how evolutionary systems may be used efficiently to complete a variety of useful tasks, one must incorporate evolution's most powerful tools. This would seem to require abandonment or modifications of particular algorithms which are currently in use. Alternatively, one might construct fitness functions which reward small changes in the a design, creating multiple epoch events, and essentially making the three paradigms rest on nearly equal footing. However, this is difficult to do in practice, and unlikely to be a viable possibility without regular human intervention.

Genetic algorithms are the most often exploited algorithm for undertaking genetic studies, and studies of population dynamics, although the encoding is often a perturbation of the standard GA paradigm[1]. Most GAs, however, do not employ connections. This is because GAs are primarily studied as optimization routines. Thus, the probability of a good solution is $p = N^{-m}$ rather than that given in equation (2.4). However, this requires that the details of the evolution be built into the system, rather than discovered by a search through the system. As such, a system like this would seem to be unsuitable for alife studies, as they do not allow the study of such novel structures.

These may be modified quite easily to incorporate compartmentalization. This requires recoding the DNA. The first requirement is that there must be no limit on the length of the DNA in question (Harvey, 1997). This allows functionality to be added when necessary. The second is that there must be a **copy** operator added to the basic three operators. The third is that the functionality must be derivable from the DNA string, allowing the string to designate active and inactive parts and section activation and deactivation markers. Finally the fourth requirement is that crossover be performed paying strict attention to the way in which it is carried out; disruption of functional groups must be avoided.

Genetic programming is already partially compartmental, but suffers strongly from interaction connection time as the genome increases in size. It is often unclear how to move an intact compartment, or how to know whether or not a compartment is intact in the first place. Adding a number of *capping nodes* which indicate

1. A connection family, and

2. The end of a structure

---

[1]We view a genetic algorithm as any evolutionary algorithm with a fixed string length and a finite nonzero population size. This is the generalization of the binary string genetic algorithm.

would allow the fully compartmental model to develop. This would assume that each connection might only be compatible with other connections of the same family. Thus, if one subtree attempted to connect to another subtree containing only connections of different family connectors, it would not be possible. This would *by design* reduce the second term of the numerator in equation (2.8). One would also be required to implement a *physics* in which all possible legal combinations of connections are attempted for each genome before evaluating it. However, allowing the units to have any unlimited connection labels allows an adaptive algorithm to be implemented in which the connection labels would be renamed if structures that were sufficiently different used the same labels.

What makes these modifications useful is only that they empower the algorithm to produce compartmental structures. The tendency of these algorithms to use these structures depends on the details of how mutations are done, and the dynamics of the systems. However, if the ability exists, it will be in the best interests of the algorithm to eventually make use of it. Whether or not this is useful in real time depends only on the proportionality constants developed earlier.

5 Conclusion

In this paper, we've motivated the existence of compartments in the DNA of evolutionary systems by examining the probability that a given structure may be built, given that it is composed of a specific number of building blocks of a finite number of types and equal probabilities. Three general types of evolutionary paradigms have been formulated based on their use of compartmentalization. These are *non-compartmentalized*, *partially compartmentalized*, and *fully compartmentalized* models. The main difference in the models derives from the creation of linkages between building blocks which model the system. The gains in using full compartmentalization in the design of complex structures are large when compared to the use of standard non-compartmentalized DNA.

Though no known natural or artificial system completely satisfies the assumptions of this model in that building blocks are normally not independent in their impact on the design of a piece of hardware or a biological system, the model is general enough that its main implications may be viewed as representative of other systems. Thus, we believe that these results are generally applicable to current paradigms, and may be used as a motivation for the alterations of these paradigms.

More importantly, we have understood that the removal of the difficulty in producing correct linkages is of fundamental importance when considering building complex structures. The single most difficult part of designing an improvement to an existing structure resides in this term. Models that partially or completely remove this term may be expected to perform many orders of magnitude better than those that do not. The design of new paradigms that incorporate this

"physical" term into the design of the system rather than the structure may be expected to be useful tools in the design of new and interesting structures.

6 Acknowledgements

This research was funded by and conducted at the Jisan Research Institute, a pioneering institution dedicated to training our nation's young people in the art of science. Research Mentors from many computational fields create research groups made up of bright young scientists and conduct research yielding a variety of exciting new developments in those fields of science.

References

[1] Adami C (1994). *On Modeling Life*. **Artificial Life IV, proceedings of the Fourth International Workshop on the Synthesis and Simulation of Living Systems**, R.A. Brooks and P. Maes eds. (MIT Press, Cambridge, MA), p. 269; *Artificial Life* 1. 429-438.

[2] Adami C. (1995a). *Learning and Complexity in Genetic Auto-Adaptive Systems*. **Physica D**, 80: 154-170.

[3] Adami C, Brown CT, Haggerty M. (1995b). *Abundance Distributions in Artificial Life and Stochastic Models: "Age and Area" revisited*. **Proc. of ECAL 95**.

[4] Burns N. et al. (1992) *Symmetry, flexibility, and permeability in the structure of yeast retrotransposon virus-like particles*. **The EMBO Journal**. 11(3) 1155-1164.

[5] Dittrich P., Banzhaf W.(1998) *Self-Evolution in a Constructive Binary String System*. **Artificial Life** 4(2): 203-220.

[6] Forrest S. and Mitchell M. (1993) *Towards a stronger building-blocks hypothesis: effects of relative building-block fitness on GA performance*. **Proceedings of a Workshp on Found. of Genetic Algorithms**. Los Altos, CA: Morgan Kaufmann Publishers.

[7] Harvey I. (1997) *Open the Box*. **Workshop on Evolutionary Computation with Variable Size Representation. Seventh Int. Conf. on Genetic Algorithms**. San Mateo, CA: Morgan Kaufmann Publ..

[8] Kazadi S. (1997) *Conjugate Schema in Genetic Search*. **Proc. of Seventh Int. Conf. on Genetic Algorithms**. San Mateo, CA: Morgan Kaufmann Publ., 10-17.

[9] Kazadi S. (1998) *Conjugate Schema and the Basis Representation of Crossover and Mutation*. **Evolutionary Computation**. 6 (2): 129-160.

[10] Keymeulen D., Iwata M., Kuniyoshi Y., Higuchi T.(1999)."Online Evolution for Self-Adapting Robotic Navigation System Using Evolvable Hardware," *Artificial Life*, Massachusetts: MIT Press.

[11] Koza J. (1992) **Genetic Programming**. Cambridge, MA: MIT Press.

[12] Koza J. (1994) **Genetic Programming II**. Cambridge, MA: MIT Press.

[13] MacPhee D. (1991) *The significance of deletions in spontaneous and induced mutations associated with movements of transposable DNA elements: possible implications for evolution and cancer*. **Mutation Research**, 250: 35-47.

[14] Munetomo M. and Goldberg D. (1999) *Linkage identification by non-monotonicity detection for overlapping functions*. **Evolutionary Computation**. 7 (4), 377-398.

[15] Ofria C and Adami C. (1999). *Evolution of Genetic Organization in Digital Organisms*. **Proc. of DIMACS. Workshop on "Evolution as Computation"**, L. Landweber and E. Winfree eds., Princeton, Springer-Verlag.

[16] Ray, T. (1992) *An Approach to the Synthesis of Life*. **Artificial Life II**. Langton C., Taylor C., Farmer J., and Rasmussen, S., eds. Redwood City, CA: Addison-Wesley Publishing Co., 371-408.

[17] Ray, T. (1994) *Evolution and Complexity*. **Complexity: Metaphors, Models, and Reality**. Cowan G., Pines D., Meltzer D. (Eds.). Santa Fe: Addison-Wesley.

[18] Robertson H. and Martos R. (1997) *Molecular evolution of the second ancient human* mariner *transposon, Hsmar2, illustrates patterns of neutral evolution in the human genome lineage*. **Gene** 205: 219-228.

[19] Stork D, Jackson B, and Walker S, (1991) *'Non-Optimality' via Pre-adaptation in Simple Neural Systems*, **Artificial Life II**. Langton C., Taylor C., Farmer J., Rasmussen S. (Eds.) Santa Fe: Addison-Wesley, 409-430.

[20] Thompson A., Layzell P., and Zebulum, R. (1999a) *Explorations in design space: Unconventional electronics design through artificial evolution*. **IEEE Transactions on Evolutionary Computation**. 3(3) 167-196.

[21] Thompson A. and Layzell, P. (1999b) *Analysis of Unconventional Evolved Electronics*. **Comm. of the ACM**. 42 (4): 71-79.

[22] Thomspon A. (1999c) *Evolutionary design for novel technologies*. **IEEE Colloquium on Evolutionary Hardware Systems**. p.4.

[23] Thompson A. (1998a) *On the automatic design of robust electronics through artificial evolution*. **Proc. 2nd Int. Conf. on Evolvable Sys: From biology to hardware (ICES98)**. M. Sipper, D. Mange, & A. Peres-Uribe (Eds.), Springer Verlag, pp. 13-2.

[24] Thompson A. (1998b) *Exploring beyond the scope of human design: automatic generation of FPGA configurations through artificial evolution*. **Proc. 2nd Int. Conf. on Evolvable Systems**. M. Sipper, D. Mange, &A. Peres-Uribe (Eds.), Springer Verlag, 13-24.sss

Exploring Gaia Theory: Artificial Life on a Planetary Scale

Keith Downing
The Norwegian University of Science and Technology
Trondheim, Norway
keithd@idi.ntnu.no

Abstract

Gaia theory, the view that the biota can both affect their environment and do so in a manner that benefits life in general, is an extremely controversial interpretation of the complex relationships between the biota and biosphere. Since individual Gaian phenomena can span spatial scales from cellular to planetary, they evade thorough analysis and empirical validation. Consequently, a good deal of Gaian thinking revolves around an abstract computer model, Daisyworld [24]. However, this model fails to properly account for natural selection's role in Gaian emergence. Although we propose an alternate scheme that offers some improvement - one based on evolutionary computation and individual-based simulation - the field remains wide open for investigations from the alife perspective. This paper reviews both models along with a few natural Gaian phenomena before generalizing a set of common primitive features and emergent properties from the real and artificial examples. These shared characteristics will hopefully provide a backbone for a much-desired "Gaia-logic" and assist other alife researchers in the search for additional Gaian models.

Introduction

In the 1960's, James Lovelock, a NASA atmospheric chemist, analyzed infrared spectrometer readings of the Martian atmosphere to assess the probability of life on Mars. He found an atmosphere very near chemical equilibrium, a telltale sign of a dead planet [14, 15]. Since Earth's atmosphere is very far from equilibrium, Lovelock argued that the biota are the key to maintaining this dissipative, low-entropy state. At the time, the idea that living organisms could affect large-scale environmental change was highly controversial, but considerable evidence of life's influence, particularly on geochemistry, has persuaded many earth scientists to accept the biota as a driving force of planetary change.

However, Lovelock and Lynn Margulis [16] felt that the biota had an even stronger role: they could not only affect the planet, but could do so in a manner

that was beneficial for life. In short, the biota indirectly regulate planetary conditions within a window of survival that is largely defined by their own physiologies. Many Gaians expand this definition of the basic Gaia theory to include all situations in which the biota modify the environment to the benefit of life itself.

This radical viewpoint has evoked cries of "teleology" and "pop ecology" from a host of renowned scientists, but putting aside ancient prejudices and stale dogma, one sees a variety of interesting examples where the biota appear to play a major role in making the planet more liveable. These include the regulation of local climate by algae [4, 17], the control of global temperatures by photosynthetic organisms [26], the maintenance of relatively constant marine salinity [8] and nitrogen-phosphorus (N:P) ratios [12, 18, 23] by aquatic biota, and the emergence of efficient recycling loops among diverse microbial species [23, 25].

As evidence of these phenomena accumulate, many natural scientists have accepted Gaia's essence: life begets life. However, many Neo-Darwinians remain skeptical, since the evolutionary origins of ecosystem-level homeostatic loops with biotic components are difficult to envision, and are in fact counter to the competitive, survival-of-the-fittest views of natural selection. Gaian interactions involve the coordination of many biological, chemical and physical activities, as illustrated by the networks of diverse microbes involved in the global chemical cycles. Reconciling the emergence of coordinated, multi-species, distributed controllers with Neo-Darwinian evolution is no simple task, even when coordinated strategies are clearly the best for all organisms.

The problem is greed: populations evolve in directions that garner the highest fitness for their individual members, and as countless game-theoretic analyses and alife simulations indicate, when greedy individual behaviors yield higher payoffs than cooperation, global coordination is unstable or never even emerges.

Consequently, one of the key challenges to Gaia scientists is to show the compatibility of Gaia theory and individual-level natural selection. Artificial life research can greatly assist this effort through simulations of environmental and evolutionary dynamics that produce emergent Gaian phenomena. A handfull of these systems already exist, but none has convincingly shown the inevitability of Gaian behavior on planets that support life. In short, nobody has exposed an underlying "Gaia-logic" that could apply to Earth and other actual planets or possible worlds.

This paper briefly describes two contemporary Gaian models: a) Daisyworld, the central Gaian metaphor since 1983 [24], and b) Guild, our own model that extends some of the key Daisyworld concepts to more closely address the issue of natural selection [6]. We then attempt a basic classification of the key components of Gaian systems, both natural and artificial, in hopes of a) forming the foundation for a "Gaia-logic", and b) providing a starting point for other alife researchers to delve into this fascinating area.

Proposed Gaian Phenomena

Gaia scientists (a.k.a. Gaians) search for interesting examples of the bi-directional interactions between organisms and their environments. A few of the more popular instances indicate the nature and scope of postulated Gaian behavior.

Algae are a major focus of Gaians, since they appear to be key components in homeostatic loops that control temperature over both small and large spatiotemporal scales. A small-scale effect involves algae and dimethylsulfide (DMS). Algae release DMSP, a DMS precursor, into the atmosphere. Through various chemical reactions, DMSP leads to DMS and on to sulphate aerosols, which serve as condensation nucleii for cloud formation, and clouds reduce temperature via shading [4]. In nutrient-rich waters, temperature normally has a positive influence on algae growth. So a decrease in algae density reduces DMSP emissions, causing fewer clouds to form, leading to increased local irradition and temperature, thus promoting algae growth. Algal density and local temperature are therefore mutually regulated by their Gaian bond.

A more recent twist on the algae-DMS theory exploits the positive relationship between cloud cover and wind to argue that when algal density becomes too great, the increased DMS and cloud cover leads to higher winds, which then blow the algae to areas of less nutrient competition [7]. The presence of airborne algae supports this theory, but the quantitative significance of this phenomena is highly speculative.

On a larger spatiotemporal scale, algae photosynthesis (like that of land plants) takes up CO_2 to build carbon-based biomass. Although most of this carbon is quickly returned to the oceans and atmosphere via the respiration processes of algae or higher organisms, a small percentage is not immediately recycled and sinks to the ocean depths, where it is eventually respired by microorganisms but where mixing rates are extremely low and currents move very slowly. Hence, the released CO_2 may not return to surface waters for several centuries, often via equatorial outgassing. Thus, the net effect of the marine food chain is atmospheric CO_2 reduction over large timescales, with algae as the biota-biosphere interface [21, 26]. Reductions in atmospheric CO_2 will then lead to temperature decreases via a weakened greenhouse effect, and the lower temperature will inhibit plankton growth, thus reducing CO_2 consumption. Once again, the plankton appear to be pivotal links in a homeostatic network.

A particularly intriguing Gaian possibility is the invariance of the nitrogen:phosphorous (N:P) ratio in marine environments at a value near 6.7, the "Redfield ratio" (1958). With few exceptions, Redfield ratios are found both in water concentrations and in the biomasses of algae and zooplankton in all the world's oceans. The big question is whether the biota have adapted their internal concentrations to the biosphere, or whether they have caused global changes in N:P ratios to suit their needs for the magical 6.7, which may represent some physiological optimum. As [23] points out, if the biota are passively adapting to the oceanic concentrations, then why don't other chemical elements have matching ratios in marine organisms and waters? He argues that negative feedbacks between populations of nitrogen-fixing and denitrifying bacteria could lead to emergent control of the Redfield ratio. Furthermore, [12] uses difference-equations for these 2 populations within coupled atmospheric-oceanic compartmental models to show that these regulatory loops actually do produce water-column N:P ratios near 6.7 - thus indicating that, indeed, local biotic mechanisms can lead to emergent chemical regulation.

In these and other proposed Gaian phenomena, many of the individual links (e.g., DMS to cloud cover) in the feedback loops have been confirmed to some qualitative degree, and in several cases, the relationship in a link becomes the focus of hundreds of life-science research papers (e.g. the algae-CO_2-climate connection). But putting all the links together into a loop and then proving the quantitative significance of the myriad interactions across various spatiotemporal scales is a gargantuan task. This complex intertwining of many "biogeochemical" [21]) factors can quickly overwhelm the naive computer modeller.

Models of Gaia

Fortunately, one need not master all the biogeochemical vagaries to do Gaia research. To wit, [24] used a brilliantly simple computer model, Daisyworld, to thwart many of the major attacks upon Gaia theory. The fact that Daisyworld remains the centerpiece of the Gaian movement after 17 years illustrates both a) the difficulty of incorporating any real natural data into a convincing Gaian model, and b) the need of Gaian researchers for a set of basic principles (i.e. a Gaia-logic) to explain a wide variety of Gaian phenomena. Daisyworld has filled that need for a long time, but its shortcomings are becoming more evident as further natural data accumulates and the required scope of Gaia-logic expands.

Daisyworld

Watson and Lovelock's classic Daisyworld model [15, 24] has served as Gaia theory's prototype for over a decade. Hence, understanding Daisyworld is a key prerequisite to understanding Gaia theory. Since most proposed Gaian phenomena are very complex, involving geophysical factors and population dynamics spread over an entire planet, operational computer models of Gaian activity are scarce. Daisyworld is the notable exception, since it illustrates the emergence of distributed planetary temperature regulation via an extremely simple differential-equation model of two competing species. Unfortunately, Daisyworld has done little to reinforce the tenuous connection between Gaia and evolution.

In the classic Daisyworld, two species of daisy, black and white, are grown on a simulated planet. Both species have the same preferred temperature, 22.5°C, at which their growth rates are maximal. The black variety have low albedo (i.e. reflectivity) and therefore create higher local temperatures than the ambient, whereas the white daisies and their high albedo create local temperatures below the ambient. The growth rates of the daisies are directly influenced by their local temperatures, which are a function of both the local albedo and the global temperature, which in turn depends upon solar intensity, the cumulative albedo of all daisies, and the albedo of the uncovered (by daisies) ground. This simple combination of two daisy types with the same optimal temperature leads to self-organized regulation of global temperature in the face of a slow but steady increase in solar input, similar to that from our own sun over the past 4.5 billion years.

Briefly, at low temperatures, black daisies proliferate due to their ability to raise local temperatures closer to 22.5°. A globe nearly covered in low-albedo black daisies absorbs a high percentage of the incoming solar energy, thereby increasing global temperature to 22.5° at a much faster rate than could the increasing solar input alone. Global temperature is then maintained near 22.5° even as the solar input soars, due to the effect of the white daisies, whose high albedo enables them to create local temperatures near 22.5° and outcompete the black daisies, whose local temperatures are too high. Eventually, however, the solar input is too great and even the white daisies succumb to heat death, at which point global temperatures rise unabated. However, across a wide range of solar inputs, one of the two daisy populations keeps the planetary temperature nearly constant, thus providing key evidence that simple local interactions among the biota can have global regulatory consequences. The invisible hands of teleology are unnecessary, and Gaia theory thereby takes a huge step away from the mysticism that was unduly attributed to it by a host of renowned Neo-Darwinian biologists.

However, Gaia theory is still very vulnerable on the evolutionary flanks, and Daisyworld provides only minor fortification. From the evolutionary standpoint, the main critique of Daisyworld is the lack of significant genetic diversity for color (i.e. albedo) and preferred temperature. [20] shows that if black and white daisies have preferred temperatures of 27.5°C and 17.5°C, respectively, then the regulatory range (of solar inputs) is reduced. He further claims that if preferred temperature were under genetic control, then cold-loving species would dominate under reduced solar input but quickly give way to heat-loving species as solar forcing increased. In short, the species whose combination of preferred temperature and albedo best matched the current solar trend would dominate. This "greedy" acceptance of the ambient conditions destroys the global regulatory behavior. The modified Daisyworld simulations of [19] verify Saunders' claim.

The Guild System

Our Guild system [6] was designed to show the basic compatibility of Gaia theory and natural selection by simulating the evolutionary emergence of a diverse interacting set of species whose combined activity could control some aspect of the physical environment. We take a few steps beyond Daisyworld by a) including a wide range of diverse genotypes and b) incorporating a new metric for Gaian activity: nutrient recycling.

Biogeochemical Motivations Another intriguing example of life's ability to create favorable conditions for more life involves the creation of efficient recycling pathways for poorly-supplied nutrients. As detailed in [23], the external supplies of critical elements such as carbon, nitrogen and phosphorus to terrestrial and aquatic ecosystems are far below the amounts actually required by the biota. The deficit is filled by recycling processes wherein C, N and P atoms are shuttled among different compounds that are ingested and expelled by various organisms.

For example, carbon is taken up by photosynthesizing plants as CO_2 and used to build organic carbon compounds such as carbohydrates, which are then transferred to herbivores or detritus-consuming microorganisms, only to be returned to the atmosphere as carbon dioxide by respiration in plants, animals, bacteria and fungi. A small percentage of the carbon sinks out of aquatic and terrestrial ecosystems as organic detritus and the calcium-carbonate shells of buried microorganisms, returning millions (to hundreds of millions) of years later via geophysical processes such as volcanism, deep-sea thermal ventilation, and rock weathering [21]. In the case of weathering, the biota have been shown to significantly accelerate this key step of the carbon, nitrogen and phosphorus cycles [22], so many links of the circuit feel the biotic presence.

The net result of these recycling loops is that the biota annually consume 200 times more carbon, 500-1300 times more nitrogen, and 200 times more phosphorus than is supplied by external fluxes [21, 23]. These numbers represent the "cycling ratios" for the three elements (computed as the intra-biota transfer rate divided by the external flux). Without this amplification, Earth's biota would be restricted to a fraction of their current total biomass; and without the biota, there would be no amplification.

In short, the coordination of biochemical processes across a diverse range of organisms enables life to thrive to a degree that dwarfs that of an uncoordinated, low-recycling environment. Furthermore, the abundance of critical nutrients adds stability to the environment, enabling the biota to endure periods of fluctuating external inputs. Once again, life begets life via its effects upon the environment.

If we classify organisms by the chemicals that they consume and produce (i.e., by their metabolisms), then each group constitutes a "biochemical guild" [23]. The formation of recycling loops is therefore dependent upon the emergence of the proper complement of biochemical guilds such that the waste products of one guild become the resources of another.

Simulated Emergence of Recycling and Control
Our Guild system combines abstract models of chemistry, biological growth and natural selection to simulate the emergence of both a) nutrient recycling networks, and b) the regulation of global chemical ratios. We borrow one key mechanism from Daisyworld: the ability of organisms to create local buffers against the global environment, where the combined buffering effects of many organisms can then exert an influence on the global situation. However, we avoid much of Daisyworld's hard-wiring by providing a large genotype space, defined by a genetic-algorithm chromosome.

The simulations are seeded with a single species, so all additional genotypes must arise by mutation and crossover. Furthermore, the regulatory task is one involving the coordinated effort of a wide range of temporally co-existent genotypes; a single dominant species cannot do the job alone. Hence, it is the biotic community as a whole that regulates global conditions, and these heterogeneous communities emerge from a homogeneous seed population that is subjected to nothing more than competition for resources, reproduction (by splitting) of successful resource gatherers, and genetic operators. This emergence of coordinated group regulation via standard individual selection in a large genotype space significantly fortifies Lovelock and Watson's rebuff of the Neo-Darwinians. In addition, our model illustrates the emergence of recycling networks. So both perspectives on Gaian activity: recycling and regulation, are commensurate with Neo-Darwinism.

The Guild system employs a standard genetic algorithm (GA) along with a simple model of chemical interactions. The environment consists of n nutrients/chemicals, $N_1..N_n$, with input and output fluxes I_k and O_k, respectively, and internal stores E_k for k = 1..n. An organism's genome determines both the chemicals that it feeds on and those that it produces during metabolism; an organism cannot consume and produce the same chemical. Organisms reproduce by splitting; the genetic operators are mutation, during splitting, and crossover, via gene swaps between organisms. The growth, reproductive and genetic dynamics are intended to mirror those of bacteria, which are the basis of Earth's primary biochemical guilds.

In addition, the organisms are assumed to be most active (i.e. have the highest feeding rates) when the relative fractions of the environmental chemicals, E_k, in the organism's immediate vicinity are near a particular user-defined optimal ratio. By producing and consuming chemicals, the organisms can create local ratios that differ from the global values, thus providing a semi-protective buffer against their surroundings. For example, organisms that consume N_1 and produce N_2 will have, respectively, lower and higher local amounts than the global values. Conceptually, the preferred ratio is analogous to an ambient factor such as pH, whose value is dependent upon many different chemical concentrations. So individual growth is governed by both

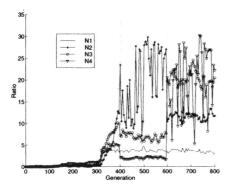

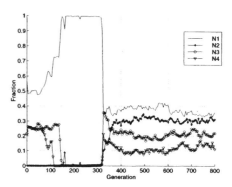

Figure 1: The evolution of cycling ratios (i.e., total inter-organism transfer / external flux) for 4 nutrients in a run of the Guild system. The gradual increase indicates the emergence of recycling in a community of biochemical guilds. Dashed vertical lines denote time points of extreme change in external input fluxes.

Figure 2: The evolution of the global fractions of 4 nutrients in a run of the Guild system, where all guilds have optimal growth when the proportions are (.4, .3, .2, .1) for nutrients $N_1..N_4$, respectively, as denoted by the dashed horizontal lines. Note the relative stability of optimal growth fractions despite major fluctuations in the external input fluxes in generations 400 and 600.

the availability of food resources and the degree of satisfaction with the resource ratios within one's buffer.

As shown in [6], a variety of Guild runs exhibit the emergence of both a) nutrient recycling (i.e. high cycling ratios), and b) control of the global nutrient ratios at levels near the optima. Control is particularly evident when the environmental input fluxes are drastically perturbed, and yet the biota maintain nearly optimal nutrient ratios.

Figures 1 and 2 shows the results of a typical Guild run in which the initial population of size 100 consists of a single phenotype that produces N_1 and consumes N_2. The environment is initially devoid of nutrients, with input fluxes of (20, 20, 20, 20) units/timestep for $N_1 ..N_4$, respectively, and output fluxes of 1% of the standing amounts, E_i, i = 1..4. At generation 400, the input fluxes change to (5, 10, 25, 50), and then to (50 25 10 5) at generation 600. The biota have optimal growth with ambient nutrient ratios (i.e. normalized E_i values) of (.4, .3, .2, .1).

Figure 1 illustrates the gradual rise in cycling ratios as phenotypic diversity rises and the recycling loops form, while Figure 2 shows the approach of nutrient ratios to their biota-preferred values (dotted horizontal lines) and their persistence in the face of the two large disturbances at generations 400 and 600.

As a brief causal explanation, competition drives the initially homogeneous biota toward greater trophic diversity (i.e., diversity of consumed nutrients), and since each organism must produce at least one non-consumed chemical as waste, a diversity of outputs also emerges. This increasing biotic heterogeneity re-

sults in the fortuitous formation of recycling networks. When all of the pieces (i.e., guilds) of these networks fall into place, previously under-consumed (and thus accumulating) nutrients are taken into the food chain, fueling a population explosion and an increase in cycling ratios. The elevated nutrient transfer within the recycling network then facilitates further population growth within each guild. The high transfer fluxes between these large interconnected guilds dwarf the environmental input and output fluxes, thus reducing the biota's sensitivity to external perturbations.

Competition within and between this diverse collection of well-populated guilds results in a frequency-dependent selection that enables the guilds to effectively control global chemical ratios via their cumulative production and consumption. For example, consider a guild G that consumes nutrient N_k. The local environment of G will have a lower proportion of N_k than the global environment, so changes in the total biomass of G will decrease E_k and hence the global proportion of N_k. Now if the global ratio is near optimal, then any major increase in G's biomass will push the global ratio away from optimal and give a selective advantage to the members of other guilds, namely, those that produce N_k. This will then push E_k back up toward the optimal proportion. In short, any deviations from optimal (i.e. error) of the global ratios will create an environment that favors guilds that (fortuitously) decrease the error. Importantly, the selective advantage of these guilds stems not from their global influence, but from their ability to create pleasant local environments for their own growth. Thus, the interplay between the guilds, orchestrated by standard natural selection, achieves and maintains a stable optimal nutrient ratio.

In summary, natural selection operating on a collection of diverse competing biochemical guilds leads to the emergence and stability of both a) a self-sustaining nutrient recycling network and b) a distributed controller. For an in-depth description of the Guild model, the parameter settings (such as mutation, crossover, metabolism and feeding rates) for particular test cases, and full details of simulation results, see [6].

Clearly, models of this simplicity cannot fully explain complex biogeochemical phenomena, but they can often illustrate the sufficiency of particular mechanisms for deriving similar patterns. This work shows that simple local interactions, under the scrutiny of natural selection, can lead to interesting cooperative arrangements. Since these particular cooperative results, efficient cycling networks and distributed global chemical regulation, are both viewed as fundamental examples of Gaia in action, our simulations lend support to the basic compatability of Gaia and evolution. The Guild system's use of a legitimately large genotype space and true evolutionary simulations makes a somewhat stronger argument for this compatability than Daisyworld's small set of hard-wired genotypes.

Unfortunately, the Guild model, like the original Daisyworld, side-steps the evolving-preferences issue [20], since all guilds are assumed to have the same constant preferred chemical ratios. When included in the Guild genomes, these preferences inhibit regulation, as individuals simply evolve preferences to the current conditions. These "regulatory parasites" are clearly a problem.

An improved chemistry model in the Guild system could help tackle the evolving-preferences dilemma. Given a set of chemical and energetic primitives, certain preferences of ambient chemicals and physical factors would arise to match the restrictions imposed by the set of metabolic possibilities. Hence, the constraints of chemical principles upon ambient preferences (and their ease or difficulty of change) would shed some light on the true range of freedom that organisms actually have in "breaking from the regulatory ranks". This agrees with Williams [25], who contends that Gaia and the global chemical cycles can be best understood from the cellular level of enzymes, their production and regulation.

Characteristics of Gaian Processes

Williams' focus on the cellular level motivates his contention that Gaia is a basic property of living systems.

Thus, in great similarity to alife researchers' mutual quest for a "bio-logic" [10], Williams and other Gaians have begun looking for a "Gaia-logic". The two pursuits are intimately related, since both involve the emergence of distributed phenomena from local interactions, with adaptivity occuring on both lifetime and evolutionary time-scales. As the Guild system illustrates, the basic alife techniques can aid this search for Gaia-logic.

Our own vague conception of "Gaia-logic" spans many levels, from the metabolic activities of single cells to population and ecosystem dynamics, with natural selection playing a key role. Our search for a concrete theory begins with an analysis of several Gaian systems, both natural and artificial.

Although the sample space of theorized natural Gaian phenomena is rather small, and the space of Gaian computer models is even smaller, there are a few key features of the natural and artificial systems that appear to be vital components of Gaia. Alife models of Gaia will need to embody many of these characteristics, either as explicit primitive constraints or as emergent properties.

Primitive Factors

The first basic mechanism is trivial: organisms must be affected by their physical environments. This is often modelled as relationships between ambient factors and growth rates. An apparent implicit assumption in Gaia theory is that these ambients should consist of something other than available food resources, such as temperature, pH, salinity, etc. The daisies in Daisyworld have optimal growth at 22.5° C, while the Guild organisms grow best at particular ratios of ambient chemicals (which serve as crude abstractions of pH). The forms of the functional relationships between ambients and growth can have significant effects upon the emergent ecosystem dynamics.

A second basic principle is frequency-dependent population growth. The standard situation involves negative frequency-dependent selection in a typical Malthusian manner: higher population density entails more competition for resources and lower overall fitness. However, since Gaia embodies the notion of organisms making the planet more liveable for each other, some positive relationships between population size and average fitness will often occur, but these are often more emergent phenomena than the negative connections, which are typically implicit in resource limited situations.

The third property is the inverse of the first: organisms must have some causal means of altering their physical environment. Normally, this influence occurs

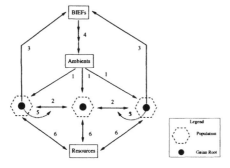

Figure 3: Summary of the basic causal mechanisms involved in many proposed Gaian phenomena: 1) ambient effects upon growth rate, 2) frequency-dependent population growth, 3) direct biotic effects upon an environmental factor (BIEF), 4) BIEF effects (possibly indirect) upon ambient factors, 5) local selective advantage of the Gaian root, and 6) biotic production and consumption of resources.

only at the aggregate level. For example, one daisy in Daisyworld cannot exhibit a significant influence on the planetary temperature, but a large population can. Similarly, in Guild, one organism cannot change environmental chemical ratios by more than a miniscule amount, but the combination of all guilds constitutes a powerful force. In nature, a single plankton cannot produce enough DMSP to form a cloud, but many million plankton can.

Let us define the "Gaian root" as the substance or physical condition produced by an organism that has some causal effect upon the environment. In the three examples above, the Gaian roots are local albedo, local metabolic inputs and outputs, and DMSP, respectively.

The fourth characteristic is a set of causal links relating the biologically-influenced environmental factors that are directly influenced by the biota (call them BIEFs) to those ambients that govern biological growth rates. In some cases, the causal chain may be short or trivial (i.e. the BIEFs are the ambients), while others involve a complex environmental model. Daisyworld has a relatively short pathway from local albedo to aggregate global albedo to global temperature and back to local temperature (a function of the global temperature and local albedo), which then controls growth. In the plankton-DMS example, the chain involves DMSP, DMS, cloud formation, shading effects upon temperature and wind, etc. Finally, in Guild, the chain loops from local chemical fluxes to global chemical ratios to effective local ratios (a function of the global ratio and the local fluxes) to growth.

The fifth factor is local selective advantage. Since the global environmental effects of individual organisms are microscopic, if standard individual-based natural selection is to guide the evolution of Gaian phenomena, then the bases for these small environmental effects (i.e. the Gaian roots) must also produce some immediate local selective advantage. Otherwise, organisms that did not expend the energy to contribute to environmental change, i.e. "cheaters", would have a selective advantage over the contributors.

For example, [3] estimates that the metabolic cost for plankton to produce DMSP (a dimethyl-sulphide precursor) exceeds the climatic benefit (accrued by the collective of millions of plankton) by 9 orders of magnitude. However, [13] find that internal DMSP concentrations are critical to preventing osmotic water loss in phytoplankton. So clearly, DMSP provides an immediate local selective advantage, while the wider influence is merely a pleasantly coincidental side-effect that connects phytoplankton to climate in a manner that gives little or no benefit to the individual organisms but can play a key role in the long-term stability of populations, communities and ecosystems.

In the DMSP example, the local mechanism relating DMSP to dehyradation prevention is quite different from the factors that connect DMS to cloud formation and temperature change. The DMSP plays two different roles. However, in other situations the Gaian root will have a similar local and global role. The local buffering effects in both Daisyworld and Guild are typical examples. In the former, a daisy's individual albedo allows it to experience a different temperature than the ambient, which is largely determined by the collective albedo of all daisies. Still, the Gaian root, albedo, functions similarly at the local and global levels. In Guild, an organism's consumption and production have a strong influence upon the effective local chemical ratio, which often differs from the global ratio, but the mechanism is the same at both levels: production and consumption (either individual or net) govern chemical ratios. By allowing individual organisms to essentially bathe in a local environment that differs from the global, these buffering mechanisms allow the Gaian root to incur a local selective advantage while contributing a small piece to a more global influence. Without the local selective advantage, the global effect accrued by a small population of colonizing individuals would be so insignificant as to make genes for the Gaian root selectively neutral or deleterious.

Finally, Gaian phenomena involve the consumption and production of resources by the biota. Later Daisyworld extensions include resources, and they are also a key ingredient in Guild. Similarly, a reasonable simulation of the plankton-DMS connection will include algal nutrient supplies and DMSP release. This not only supports the emergence of one Gaian phenomena, nutrient recycling, but generally renders biological simulations more realistic.

Figure 3 summarizes these six primitive characteristics, many of which appear in natural and artificial systems that exhibit Gaian behavior. Although some of these primitives could also be emergent in particular simulations or real-world phenomena, they are generally taken as givens in Gaian discussions. A Gaia-logic will presumably build on factors such as these and proported relationships between them such that emergent Gaian dynamics can be predicted from instantiations of these primitives in different biogeochemical contexts.

Emergent Properties

The hallmark emergent property of Gaian systems is biotic-abiotic feedback, with the emphasis being on negative feedback (i.e. homeostatic) circuits. Essentially, the biota, BIEFs and ambients are connected in a homeostatic loop such that perturbations to population density or the environment are regulated back toward their normal values. Thus, the biota participate in the regulation of factors such as climate via population density, and perhaps diversity. Furthermore, if the biota have optimal growth under particular ambient conditions, the indirect biotic influences often combine with natural selection to push the ambients toward those prefered values.

In many cases, the causal link from the biota to the ambients is indirect. For example, under a relatively cold climate regime, a homeostatic loop involving plants and global temperature might work as follows: an increase in temperature (the ambient) would lead to increased photosynthesis and plant growth, which would draw down more CO_2 from the atmosphere. This would reduce the absorption of the earth's infrared heat by the atmosphere (the greenhouse effect) and thereby reduce temperature. Unfortunately, in a warm climate regime, the loop could become positive, leading to runaway warming: increasing the ambient temperature of a plant species that is already living near its optimal growth temperature would lead to reduced plant growth, reduced CO_2 draw-down (and increased CO_2 output to the atmosphere from organic decay), and hence, increased greenhouse warming.

The proported dominance of similar positive feedbacks at the end of glacial periods (along with the mere fact that the biota "permit" the large temperature swings of the glacial cycles) leads many to question the importance of homeostasis in Gaia theory. Furthermore, the central Gaian notion of the biota making the planet more favorable for life does not entail purely homeostatic connections to abiotic factors, since the explosive growth enabled by positive feedback is also important for the rapid colonization of environments following migrations, catastrophic environmental changes, etc. Hence, biotic-abiotic feedback loops in general, whether positive or negative, appear to be key emergent Gaian patterns.

In Daisyworld, the initial explosive growth of black daisies results from a simple positive feedback: during the cold period of low solar irradiance, black daisies reduce planetary albedo, which drives temperatures up toward the optimal, which increases the growth rate for daisies (black and white). Then, when the global temperature is near the optimum, the competition between the black and white daisies manifests homeostatic control over a wide range of solar fluxes. When the sun becomes too hot, positive feedback takes over again: the death of white daisies causes planetary albedo to drop, making the effects of solar luminance on temperature even greater, killing more daisies, etc. Incidentally, Lovelock [15] views many such drastic changes in environmental factors as signs of a once-present Gaian homeostasis that eventually succumbed to an uncontrollable ambient.

In Guild, any increase in a particular strategy can alter the global chemical ratio so as to favor a different strategy. Thus, the populations of different strategists/guilds are homeostatically controlled via their influence upon the environment. The biota becomes a robust distributed controller of itself and the environment via the forces of natural selection upon diverse biochemical guilds.

A second emergent property is Lenton's [11] frequency-dependent selective feedback. He differentiates between the effects of population density upon growth versus selective advantage. The former is the second primitive factor above, while the latter is more emergent and specifically characteristic of Gaia: if a population of organisms can incur environmental change, then these alterations may raise or lower their selective advantage over other populations.

For example, in Daisyworld, the proliferation of dark daisies warms the planet, which increases the selective advantage of the light over the dark varieties. In Guild, if one metabolic strategy dominates the population, it will often decrease its own selective advantage by altering global chemical concentrations so as to favor other strategies, as detailed earlier.

In general, a negative relationship between population density and selective advantage is important in Gaian systems such as Guild, where the emergence of diversity is a key prerequisite to Gaian control. In

Daisyworld, the relationship is present but not as vital, since any single type of daisy can control global temperature over a given interval of solar irradiance. Regardless, the concept of frequency-dependent selective feedback probably constitutes a key aspect of Gaia-logic.

A third emergent property, recycling, is gaining increasing popularity as a fingerprint of Gaian dynamics [23, 25]. As discussed earlier, the ability of diverse species (often microorganisms) to coordinate their metabolic inputs and outputs such that few resources go to waste can provide both a) a resource base that supports several orders of magnitude more biomass than do the external nutrient fluxes alone, and b) considerable insurance against wild fluctuations in those exogenous flows.

These conditions support Gaian control, since a) the populations become large enough that their aggregate environmental effects can be felt, and b) as the competition for food abates, natural selection can begin to operate on characteristics other than those directly related to resource gathering, such as the side-effects of the Gaian root (e.g. dehydration prevention, local albedo or chemical-ratio alteration, etc.).

Discussion

Although the inclusion of several of the above 6 primitive features would appear to "hard-wire" feedback into the system, they function more as necessary than sufficient groups of mechanisms. Feedback formation depends not only on relationships between entities such as population densities, chemical concentrations, and ambient physical conditions, but the quantitative amounts of those entities as well. A model with a proper set of necessary mechanisms can be seeded with just the right amounts of each entity such that significant feedbacks "emerge" at time 0, but more interesting simulations begin with rather uninteresting combinations of entities that eventually self-organize and evolve into active causal loops.

In general, Gaia research has focused on uncovering these necessary mechanisms and showing that they fit into nice little causal circuits. The key issue now becomes one of emergence: given these primitive relationships, do the forces of nature actually drive systems toward biotic and environmental conditions that manifest powerful feedback states. Daisyworld and Guild indicate that under various conditions, they can, but additional field work and computer models are needed.

To date, few researchers have investigated Gaian issues. The EUZONE model [5] of the evolution of aquatic ecosystems is motivated by Gaian thinking and achieves the emergence of one species, vertically-migrating photosynthesizers, which creates a niche for another species, aerobic bottom-feeders, via its effects upon the chemical environment. However, environmental regulation does not arise in EUZONE.

The Guild system parallels research into the emergence of autocatalytic-sets [9], metabolic-systems [1] and hypercycles [2], except that we focus on (a) interaction pathways involving both organisms and chemicals, and (b) the self-organized regulation of the environment by the evolving biota. In general, a whole host of alife systems involve populations of genotypes that encode for feeding, mating and other strategies, but questions regarding recycling throughput and whether or how the phenotypes regulate the surroundings are normally not addressed.

While Williams [25] claims that the basis of Gaia-logic lies in the cells themselves, we feel that Gaia theory is not so easily reducible and must encompass many levels, from enzymes and cells to populations and ecosystems; and with Neo-Darwinian natural selection guiding the way, due to (not in spite of) the natural tendency of organisms to selfishly maximize their own fitness. Our alife perspective to the problem leads to the obvious question: could Gaia be a basic emergent phenomena in a wide variety of ecosystems? Since a good many of the biological influences on the environment are chemical, the alife spirit motivates a follow-up query: is Gaian emergence possible across a wide variety of biochemistries, from carbon-based to silicon-based (in the sense of silicon being the key structural molecule in organic chains) to chemistries based on other periodic tables, to those involving strings, s-expressions, etc.?

Answers to these questions call for an approach similar to [9], wherein many sets of randomly generated primitive interactions are simulated, and the emergent patterns categorized. To wit, we are working on a Guild extension in which random chemistries are generated; they differ with respect to sanctioned reactions and their energy yields. Biota have genetically-determined metabolisms with energy-production capabilities determined by the chemistry. Different chemistries will presumably lead to different fitnesses for particular metabolisms, and hence to the evolution of different chemical-transfer networks at the ecosystem level. The frequency of emergent Gaian phenomena such as recycling and control across these possible biochemical worlds will hopefully shed some light on whether Gaia constitutes "order for free" [9] derived from a Gaia-logic whose generality encompasses more than "life as we know it" [10].

While many biologists deride Gaia theory, many others view it as a potentially revolutionary breakthrough on par with Darwinism itself. This promise, along with the fact that a good deal of Gaian thinking is based on the rather trivial, yet enlightening, Daisyworld model, should be read as an open invitation to all natural-science modellers to join in the pursuit of Gaia-logic.

References

[1] R. Bagley and J. Farmer. Spontaneous emergence of a metabolism. In C. Langton, C. Taylor, J. Farmer, and S. Rasmussen, editors, *Artificial Life II*, volume 10, pages 93–140. Addison-Wesley, 1992.

[2] M. Boerlijst and P. Hogeweg. Self-structure and selection: Spiral waves as a substrate for prebiotic evolution. In C. Langton, C. Taylor, J. Farmer, and S. Rasmussen, editors, *Artificial Life II*, volume 10, pages 255–276. Addison-Wesley, 1992.

[3] K. Caldeira. Evolutionary pressures on planktonic production of atmospheric sulphur. *Nature*, 337:732–734, 1989.

[4] R. Charlson, J. Lovelock, M. Andreae, and S. Warren. Ocean plankton, atmospheric sulfur, cloud albedo and climate. *Nature*, 326:655–661, 1987.

[5] K. Downing. Euzone: Simulating the emergence of aquatic ecosystems. *Artificial Life*, 3(4):307–333, 1998.

[6] K. Downing and P. Zvirinsky. The simulated evolution of biochemical guilds: Reconciling gaia theory and natural selection. *Artificial Life*, 5(4), 2000.

[7] W. Hamilton and T. Lenton. Spora and gaia: How microbes fly with their clouds. *Ethology, Ecology Evolution*, 10:1–16, 1998.

[8] G. Hinkle. Marine salinity: Gaian phenomenon? In P. Bunyan, editor, *Gaia in Action*, pages 75–88. Floris Books, Edinburgh, Scotland, 1996.

[9] S. Kauffman. *The Origins of Order*. Oxford University Press, New York, 1993.

[10] C. Langton. Artificial life. In C. Langton, editor, *Artificial Life: Proceedings of an Interdisciplinary Workshop on the Synthesis and Simulation of Living Systems*, pages 1–49. Addison-Wesley, 1989.

[11] T. Lenton. Gaia and natural selection. *Nature*, 394:439–447, July 1998.

[12] T. Lenton. Redfield revisited: Regulation of nitrate, phosphate and oxygen in the ocean. *Global Biogeochemical Cycles*, 1999.

[13] P. Liss, A. Hatton, G. Malin, P. Nightingale, and S. Turner. Marine sulphur emissions. *Philosphical Transactions of the Royal Society of London*, 352:159–169, 1997.

[14] J. Lovelock. *Gaia: A New Look at Life on Earth*. Oxford University Press, 1979.

[15] J. Lovelock. *The Ages of Gaia: A Biography of Our Living Earth*. Oxford University Press, Oxford, England, 1995.

[16] J. Lovelock and L. Margulis. Atmospheric homeostasis by and for the biosphere: The gaia hypothesis. *Tellus*, 26:2–10, 1974.

[17] R. Monastersky. The plankton-climate connection. In C. Barlow, editor, *From Gaia to Selfish Genes: Selected Writings in the Life Sciences*, pages 25–29. The MIT Press, Cambridge, Massachusetts, 1991.

[18] A. Redfield. The biological control of chemical factors in the environment. *American Scientist*, 46:205–221, 1958.

[19] D. Robertson and J. Robinson. Darwinian daisyworld. *Journal of Theoretical Biology*, 195(1):129–134, 1998.

[20] P. Saunders. Evolution without natural selection: Further implications of the daisyworld parable. *Journal of Theoretical Biology*, 166:365–373, 1994.

[21] W. Schlesinger. *Biogeochemistry: An Analysis of Global Change*. Academic Press, Boston, 1997.

[22] D. Schwartzman and T. Volk. Biotic enhancement of weathering and surface temperatures on earth since the origin of life. *Palaeogeography, Palaeoclimatology, Palaeoecology*, 90:357–371, 1991.

[23] T. Volk. *Gaia's Body: Toward a Physiology of Earth*. Copernicus, New York, 1998.

[24] A. Watson and J. Lovelock. biological homeostasis of the global environment: The parable of daisyworld. *Tellus*, 35B:284–289, 1983.

[25] G. Williams. *The Molecular Biology of Gaia*. Columbia University Press, New York, 1996.

[26] P. Williamson and J. Gribbin. How plankton change the climate. *New Scientist*, pages 48–52, March 1991.

II Development and Differentiation

Complex organization in multicellularity as a necessity in evolution

Chikara Furusawa and Kunihiko Kaneko

Department of Pure and Applied Sciences
University of Tokyo, Komaba, Meguro-ku, Tokyo 153, JAPAN

Abstract

By introducing a dynamical system model of a multi-cellular system, it is shown that an organism with a variety of differentiated cell types and a complex pattern emerges through cell-cell interactions even without postulating any elaborate control mechanism. Such organism is found to maintain a larger growth speed as an ensemble by achieving a cooperative use of resources, than simple homogeneous cells which behave 'selfishly'. This suggests that the emergence of multicellular organisms with complex organization is a necessity in evolution. According to our theoretical model, there appear multipotent stem cells initially, which exhibit stochastic differentiation to other cell types. With the development and differentiation, both the chemical diversity and complexity of intra-cellular dynamics are decreased, as a general consequence of our system. Robustness of the developmental process is also confirmed.

Introduction

Multicellular organisms consist of various differentiated cell types. Although a life cycle of a multicellular organism always starts from a single cell or a few stem cells without cellular diversity, through the course of development, cells differentiate into several types, to form a complex organism. Recent advances in molecular biology have clarified a molecular basis of the mechanisms in the developmental process (1). However, it is still not clear why multicellular organisms should have such complexity, nor why such inhomogeneities in cell types and patterns commonly evolve. These are not trivial problems (2), since a simple cell system with identical cells could replicate fast, and a simple intra-cellular biochemical network would be sufficient (or more fit) to produce identical cells rapidly and faithfully. In this study, we give a novel standpoint for the problem of why multicellular organisms in general have diverse cell types with complex patterns and dynamics.

Some theoretical studies, based on appropriate dynamical systems of biochemical networks, have revealed the possibility for the emergence of inhomogeneous patterns with diverse cell types, starting from homogeneous cells. A pioneering study for the cellular inhomogeneity is due to Turing, where instability of a homogeneous state arising from reaction and diffusion leads to pattern formation (3). The authors and Yomo have extended this dynamical system approach to construct a logic of cell differentiation, in which a robust developmental process with various cell types emerges, based on the interplay between intra-cellular dynamics and cell-cell interactions (4; 5; 6). Differentiation into several cell types has also been studied by using random genetic networks (7). However, it is still unknown how such biochemical or genetic networks that allows for diversity are evolved.

It should be noted here that any theory accounting for the cellular diversity requires a specific choice of the genetic and biochemical reaction networks and parameters. Then, to explain the ubiquitous nature of cellular diversity in multicellular organisms, we must understand how such reaction dynamics can emerge only through the struggle for life, without postulating a finely tuned mechanism.

Fossil records suggest that multicellular organisms appear several times independently, such as in fungi, plants, animals and so on (2). Nevertheless, existing multicellular organisms have common characteristics, such as the diversity of cell types. In addition, although an organism consists of various cell types, its reproduction process always starts from a single cell or a few homogeneous cells, such as a fertilized egg. Although the biochemical dynamics are complicated and there are large molecular fluctuations, the developmental process is, in general, rather robust, even against macroscopic perturbations, such as removal of some cells. Although the initial cell types have ability for differentiation to other types, the loss of multipotency of cells is also a general feature through the developmental process. The existence of these universal characteristics in existing multicellular organisms suggests that there underlies some universal logic for surviving organisms, which is not explained just by accidental accumulations of lucky mutations.

The purpose of this study is to understand what type of inter- and intra- cellular dynamics in multicellular organism evolves through competition, as a universal structure for multicellularity. To achieve this purpose, we need to clarify the relationship between the growth of multicellular organisms and some characteristics of intra-cellular dynamics, by constructing a model for a cellular system with internal reaction process and interaction. In this modeling, we cannot rely on detailed knowledge of specific cellular processes in existing organisms, since the question we address is rather general, and should be answered independently of such specific examples. Hence, we adopt dynamical systems models that display only essential features of the developmental process. As a minimal model of cellular dynamics, we include only intra-cellular biochemical reactions with enzymes, simple cell-cell interactions through diffusion, and cell division as a result of biochemical reactions within each cell.

We have performed numerical experiments of a class of models with this general and minimal content, using thousands of different biochemical reaction networks generated randomly. As a result of this study, it is shown that complex multicellular organisms with a variety of differentiated cell types and a complex pattern can emerge even within the minimal model, with randomly chosen reaction networks. Furthermore, we find a strong evidence that the evolution to such multicellular organisms is inevitable, since they achieve a cooperative use of resources and have a larger growth speed as an ensemble than simple systems with homogeneous cells. Stochastic differentiation of a multipotent stem cell, decrease of complexity in intra-cellular dynamics along the development, and stability of developmental process against macroscopic perturbations are also confirmed as a general characteristics of our model system, that are also observed universally in existing multicellular organisms.

model

The basic strategy of the modeling follows the previous works (6). Our model consists of the following three parts:

- Internal biochemical reaction dynamics in each cell
- Cell-cell interaction through media
- Cell division and cell death

In Fig.1, we show the schematic representation of this model. Cells are assumed to be completely surrounded by a one-dimensional medium. The state of cell i is assumed to be characterized by the cell volume and $c_i^{(m)}(t)$, the concentrations of k chemicals($m = 1, \cdots, k$). The concentrations of chemicals change as a result of internal biochemical reaction dynamics within each cell and cell-cell interactions communicated through the surrounding medium. The corresponding chemical concentration in the medium is denoted by $C^{(m)}(x,t)$, where x denotes the position along the one-dimensional axis. We use a one-dimensional model only for its tractability; conclusions we draw in this case are consistent with the result of preliminary simulations of a two-dimensional model.

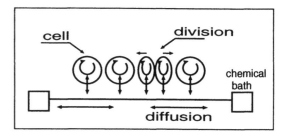

Figure 1: Schematic representation of our model.

Internal reaction dynamics: For the internal chemical reaction dynamics, we choose a catalytic network among the k chemicals. Each reaction from the chemical i to j is assumed to be catalyzed by some chemical ℓ, determined by a matrix (i, j, ℓ). To represent this reaction-matrix, we adopt the notation $Con(i, j, \ell)$ which takes unity when the reaction from the chemical i to j is catalyzed by ℓ, and takes 0 otherwise. We assume that the rate of increase of $c_i^{(m)}(t)$ (and decrease of $c_i^{(\ell)}(t)$) through this reaction is given by $c_i^{(\ell)}(t)(c_i^{(j)}(t))^\alpha$, where α is the degree of catalyzation($\alpha = 2$ in most simulations here). Each chemical has several paths to other chemicals, and thus a complex reaction network is formed. The network represents intra-cellular metabolic process in a broad sense, that can also include reactions associated with genetic expressions, signaling pathways, and so on.

Of course, the biochemical mechanisms of cells are very much complicated. We do not take into account of the details here, since our purpose is to show a general relationship between the growth of a cell society and cellular dynamics. What we need here is essentially the biochemical reaction dynamics that is complex enough to exhibit nonlinear oscillatory dynamics, including chaos. It should be noted that in the real biological systems, the intra-cellular reaction dynamics are complex enough to exhibit such oscillatory behavior, that are observed in some chemical substrates as Ca, cyclic AMP, and so on (8; 9; 10).

Besides the change in chemical concentrations, we take into account the change in the volume of a cell. The volume is now treated as a dynamical variable, which increases as a result of transportation of chemicals into the cell from the environment. As a first approximation, it is reasonable to assume that the cell volume is proportional to the sum of chemicals in the cell. We note that the concentrations of chemicals are diluted as a result of increase of the volume of the cell. With the above assumption, this dilution effect is identical to impose the restriction $\sum_\ell c_i^{(\ell)} = 1$, that is, the normalization of chemical concentrations at each step of the calculation, while the volume change is calculated from the transport, as will be given later.

Cell-cell interaction Cells interact with each other through the transport of chemicals out of and into the surrounding medium. Here we consider only indirect cell–cell interactions via diffusive chemical substances, as a minimal form of interaction. We assume that the rates of chemicals transported into a cell are proportional to differences of chemical concentrations between the inside and the outside of the cell.

The diffusion of a chemical species through cell membrane should depend on the properties of this species. In this model, we consider the simple case in which there are two types of chemicals, one that can penetrate the membrane and one that cannot. We use the notation σ_m, which takes the value 1 if the chemical $c_i^{(m)}$ is penetrable, and 0 otherwise. Also for simplicity, we assume that all the chemicals capable of penetrating the membrane have the same diffusion coefficient.

With this type of interaction, corresponding chemicals in the medium are consumed. To maintain the growth of the organism, the system is considered to be immersed in a bath of chemicals through which some (nutritive) chemicals are supplied to the cells. This situation is realized by imposing boundary conditions at each end in which the nutritive chemical's concentrations are kept constant.

To sum up all these process, the dynamics of chemical concentrations in each cell is represented as follows:

$$
\begin{aligned}
dc_i^{(\ell)}(t)/dt = & \sum_{m,j} Con(m,\ell,j)\, e_1\, c_i^{(m)}(t)\, (c_i^{(j)}(t))^\alpha \\
& - \sum_{m',j'} Con(\ell,m',j')\, e_1\, c_i^{(\ell)}(t)\, (c_i^{(j')}(t))^\alpha \\
& + \sigma_\ell D(C^{(\ell)}(p_i^x,t) - c_i^{(\ell)}(t)) \\
& - c_i^{(\ell)}(t) \sum_{m=1}^{k} \sigma_m D(C^{(m)}(p_i^x,t) - c_i^{(m)}(t))
\end{aligned}
$$

where the terms with $\sum Con(\cdots)$ represent paths coming into and out of ℓ, respectively. The third term describes the transport of chemicals out of and into the surrounding medium, where D denotes the diffusion constant of the membrane and p_i^x denotes the location of the i-th cell. The last term gives the constraint of $\sum_\ell x_i^{(\ell)}(t) = 1$ due to the growth of the volume.

The diffusion of penetrable chemicals in the medium is governed by a partial differential equation for the concentration of chemical $C^{(\ell)}(x,t)$. For each chemical $C^{(\ell)}$, at a particular location:

$$
\begin{aligned}
\partial C^{(\ell)}(x,t)/\partial t = & -\tilde{D}\nabla^2 C^{(\ell)}(x,t) \\
& + \sum_i \delta(x - p_i^x,)\sigma_\ell D(c_i^{(\ell)} - C^{(\ell)}) .
\end{aligned}
$$

We assume the following boundary condition:

$$
C(0,t) = C(x_{max},t) = \begin{cases} const. & \text{for nutrients} \\ 0 & \text{otherwise} \end{cases}
$$

where $\tilde{D}$ is the diffusion constant of the environment, x_{max} denotes the extent of the lattice, and $\delta(x)$ is Dirac's delta function.

Cell division and cell death Each cell takes penetrable chemicals from the medium as the nutrient, while the reaction in the cell transforms them to unpenetrable chemicals which construct the body of the cell such as membrane and DNA. As a result of these reactions, the cell volume changes. In this paper, we assume that a cell divides into two when the cell volume becomes double the original.

The chemical composition of two divided cells are almost identical with their mother's, with slight differences between them due to random fluctuations. In other words, each cell has $(1 + \epsilon)c^{(l)}$ and $(1 - \epsilon)c^{(l)}$, respectively, with a small "noise" ϵ given by a random number with a small amplitude, say from $[-10^{-6}, 10^{-6}]$. Although the existence of this imbalance is necessary to the cellular diversity that is essential to our scenario for multicellularity and discussed later, the mechanism or the degree of imbalance is not important. The important feature of our model is that microscopic differences between the cells can be amplified through the instability of the internal dynamics.

After cell division, two daughter cells appear around their mother cell's position, and the positions of all cells are adjusted so that the distances between adjacent cells are constant. As a result, the total length of the chain of cells increases. As the initial state, a single cell, whose chemical concentrations are determined randomly, is placed in the medium. According to the process described above, cells divide to form a chain.

In some cases, a cell starts to release penetrable chemicals to the medium rather than taking them. As a result, the volume of the cell becomes smaller. We assume that a cell dies when the cell volume is less than a given threshold. After the cell death, the positions of all cells are adjusted to keep the distance between adjacent cells to be constant.

Result

Of course, the behavior of the model depends on each intra-cellular reaction network. To examine how the nature of cell growth and dynamics are correlated, here we have carried out simulations of the model by considering 1000 different reaction networks, generated randomly. Throughout the paper, the number of chemical species k is 20, and each chemical has 6 reaction paths to other chemicals, chosen randomly. Each chemical reaction path is catalyzed by some other (or the same) chemical, again chosen randomly. Among the 20 chemicals, there are 5 chemicals capable of penetrating cell membranes, while 3 chemicals are supplied as the nutrition. The parameter are fixed at $e_1 = 1.0, D = 0.15, \tilde{D} = 1.0$, and $x_{max} = 500$.

Classification of growth behavior: exponential growth and linear growth

In Fig.2 some examples of the growth curve of cell number are plotted for different reaction networks. It is demonstrated that the growth can be classified into two classes:

(I) 'fast' growth in which the increase of cell number grows exponentially in time t, and (II) 'slow' growth in which the cell number grows linearly in time. These two classes, indeed, are also distinguished by the nature of the corresponding cellular dynamics.

In case (II), the chemical compositions and dynamics of all cells are identical. These dynamics fall into either a fixed state (fixed point) or a regular oscillation (a limit-cycle attractor). In this situation, only a few cells around the edges of the chain can divide. Since cells are not differentiated, chemicals required for cell growth are identical for all cells. Thus, once the cells at the edges consume the required resources, the remaining cells can no longer grow. This is the reason for the linear growth.

Case (II) can be further classified into two cases according to the diversity in chemicals present (which may also be regarded as the diversity in expressed genes). In one case, (IIa), only a few dominant chemicals exist, and all other chemicals vanish in each cell. Here, only a small number of reaction paths are used. In the other case, (IIb), a variety of chemicals coexists, and a large number of the reaction paths are used. These reactions are balanced and lead to fixed-point dynamics (or, rarely, periodic oscillations).

In case (I) with faster growth, it is found that the chemical reaction dynamics of the cells are more complex than in case (II). In this case, cells in the organism take various different states. The microscopic differences between the chemicals in two daughter cells are amplified through the internal biochemical dynamics and the cell-cell interaction. This leads to chaotic chemical dynamics, which make the cells to take various different phases of intra-cellular dynamics.

For most such cases, the cells differentiate into various cell types that are defined by distinct chemical dynamics and compositions. In Fig.3, an example of developmental process with various cell types and larger growth speed is shown. In this example, a single cell, introduced as initial condition, shows the internal dynamics plotted in Fig.3(b). The cell with this internal dynamics is represented as a "white" cell in the figure. At the first stage of development, the "white" cells reproduce the same type cells and a cluster of the "white" cells is formed. With further increase of the cell number, some of the "white" cells start to exhibit different types of chemical dynamics

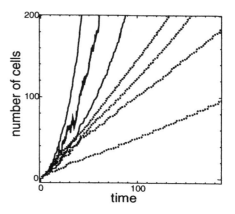

Figure 2: Growth curves of cell numbers. Temporal evolutions of cell number are plotted. Each growth curve was obtained by using a different chemical reaction network chosen randomly. The solid curves correspond to exponential growth, where cells are differentiated, while the dotted curves correspond to linear growth without cell differentiation.

(Fig.3(c),(d)), due to some influence of the cell-cell interaction on the intra-cellular dynamics. These different dynamics are clearly distinguishable as "digitally" separated states. Hence, this phenomenon is regarded as cell differentiation. With this differentiation, the capability for cell division is also differentiated as the development progresses. The total increase in cell number is due to the division of certain type(s) of cells, and most other cell types stop dividing or shrink their volume.

It should be noted that, this differentiation process is caused neither by the tiny differences introduced at the cell division, nor by any external mechanisms. This phenomenon is caused by instability in the dynamical system consisting of all the cells and the medium. When the instability of the total cell system exceeds some threshold due to the increase of the cell number, the differentiations start. Then, the emergence of another cell type stabilizes the dynamics of each cell again. Indeed, this differentiation is a general feature of a system of interacting units that each possesses some non-linear internal dynamics (11; 12) as has been clarified by isologous diversification theory (4; 5; 6).

In case (I), the division of cells is not restricted to the edge of the chain. With differentiation, cells begin to play different roles and come to require different chemical as nutrition. For this reason, chemical flow into the inside of the chain takes place, and internal cells are supplied with the nutritive chemicals they require. For this reason, even the internal cells are able to grow. This flow is sustained by the diffusion process between cells possessing different chemical compositions and exhibiting different phases of chemical oscillations. In Fig.4, typical examples of the flow of chemicals along the chain

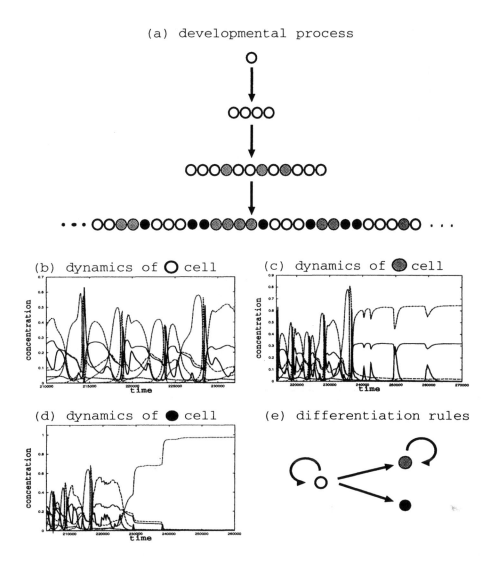

Figure 3: An example of the developmental process with larger growth speed. (a) The developmental process of the spatial pattern with differentiated cells, starting from a single cell. Up to several divisions, a single type of cell reproduces, maintaining its characteristic type of dynamics. This type is represented as "white" cells. The time series of the chemical concentrations of this type of cell are plotted in (b), where the time series of 8 chemical concentrations among the 20 are overlaid. With further increase of cell number, some of the "white" cells start to exhibit different types. The transitions from the "white" cells to two distinct cell types are represented by the plots of "gray" and "black" cell types in (c) and (d) in the time series of concentration.

In figure (e), automaton-like representation of the rules of differentiation is shown. The path back to the own cell type represents the reproduction of the same type, while the paths to other types represent the potential of differentiation to the corresponding cell type. Here, the "white" cells are regarded as the stem cells that have the potential both to reproduce themselves and to differentiate into other cell types, while the differentiated ("gray" and "black") cells have lost such potential. The "black" cells lose the potential both to reproduce and to differentiate, and their cell volume gradually becomes smaller toward to the cell death.

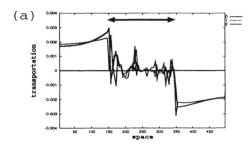

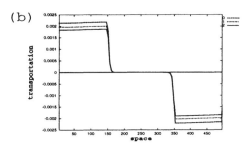

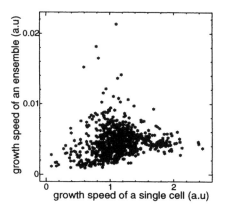

Figure 5: Relationship between the growth speed of a single cell and that of an ensemble. The ordinate corresponds to the growth speed of an ensemble, measured as the inverse of the time for the cell number to reach 200, starting from a single cell, while the abscissa represents the inverse of the time for an isolated cell to divide. Each point is obtained by using a different chemical reaction network. The points around the peak of the growth as an ensemble correspond to case (I) with cellular heterogeneity.

Figure 4: Examples of the flow of chemicals along the medium. The horizontal axis indicates the space along the one-dimensional medium, while the vertical axis shows the transportation of chemicals in the medium, measured as a amount of chemicals that pass the corresponding site during a certain period. Figure (a) and (b) show typical examples in case (I) and case (II) with 200 cells, respectively. The arrow in the figure indicates the region that cells exist in the medium.

of cells are plotted for both cases (I) and (II). Fig.4(b) corresponds to case (II), in which no flow of chemicals exists at the inside of the organism. In this case, nutritive chemicals from the outside are consumed out by the cells around the edges of the chain. On the other hand, Fig.4(a) shows the transport of chemicals in case (I) sustained by the complex cellular dynamics, that supply the nutritive chemicals with the cells at the inside.

Growth speed of a single cell and that of an ensemble

It should be noted that, the faster growth in case (I) is based on the interplay between complex cellular dynamics and cell-cell interaction. Therefore, this faster growth cannot be elucidated by the dynamics of a single cell, but is a property as a whole system. To clarify the relationship between the dynamics of a single cell and that of an ensemble, we compare the growth of an isolated cell with that of an ensemble of cells.

In Fig.5, the growth speeds of a single cell and an ensemble of cells are plotted. Here each point corresponds to a different reaction network, generated randomly. The growth speed of a single cell is regarded as the time required for an isolated single cell to divide, and that of an ensemble is computed from the time for a single cell

to reach a given number of cells (here 200). As shown in the figure, the growth speed of an ensemble is not monotonically related to that of a single cell.

The points around the peak of the growth for an ensemble correspond to case (I) of exponential growth. Here, the growth speed of a single cell is not large. In each cell, a variety of chemicals coexists, supporting complex reaction dynamics and cell differentiation.

There are much simpler intra-cellular reaction processes, with only a few auto-catalytic reactions used dominantly (i.e. case (IIa)), that can produce the rapid replication of a single cell. In this case, the growth speed of a single cell is often large, while the growth speed of an ensemble always remains small. Although such simple cells with low diversity of chemical species can exhibit large growth speeds as single cells, they cannot grow cooperatively, and their growth speeds as ensembles are suppressed. In some sense, simple cells with rapid growth are 'selfish': Although they may grow faster as single cells, the growth speed is suppressed as the number of cells increases, since they experience strong competition for resources (see Table 1).

Complex cells belonging to case (I), which have slower growth speeds as single cells, come to surpass the simpler cells in growth rate as they form ensembles, by differentiating into distinct types. Note that no elaborate mechanism is required to form such differentiated ensembles. Some fraction of the randomly chosen biochemical networks we considered exhibit dynamics sufficiently complex to allow for spontaneous cell differentiation. Hence, it is natural to conclude that complexity of multicellular

Class	Growth of an ensemble	Growth of a single cell	Intra-cellular chemical dynamics	Diversity of chemicals	Cell differentiation
I	Rapid (exponential)	Intermediate	Chaotic (for stem cells)	High	from stem cells
IIa	Slow (linear)	Rapid/Intermediate/Slow	Fixed Point	Low	None
IIb	Slow (linear)	Intermediate	Fixed Point or Periodic	High	None

Table 1: Classification of growth behavior

organisms with differentiated cell types is a necessary course in evolution, once a multicellular unit emerges from cell aggregates. In fact, by carrying out the evolution experiment numerically, with mutation to reaction networks and selection of the cell ensembles with higher growth speed, we have confirmed that cells of the case (I) emerge and survive through evolution.

In the following part of this paper, we discuss the characteristics of such complex organisms in detail. Note that these characteristics are observed commonly to case (I).

Differentiation from stem cell

In the developmental processes in case (I), the initial cell type exhibit transitions to other types. For most such cases, some of cells of the initial cell type remain to be of the same type during the developmental process. That is, cells of the initial cell type have the potentiality either to proliferate and to switch to other types. Thus, they can be regarded as a stem cell. On the other hand, the type of differentiated cells maintain their type after division, while rarely they act as stem-type cells over limited types of cells at lower hierarchy. Here, a specific rule of differentiation emerges form the dynamics, even though it is not explicitly implemented (see Fig.3 as a example).

The differentiation of the stem cell of our model is "stochastic", in the sence that the choice for a stem cell whether to replicate or to differentiate cannot be determined completely by the condition of the medium, but is determined in a stochastic manner arising from chaotic intra-cellular chemical dynamics. The probability of the differentiation can be regulated depending on the states of surrounding cells and the medium. This regulation of differentiation can sustain the robustness of organisms, as will be discussed. In a real biological system, the stochastic differentiation of stem cells and the robustness through regulation of that differentiation are widely observed (e.g. hematopoietic system (13)), and are expected to be a general feature of the spontaneous differentiation.

Decrease of complexity in cellular dynamics with developmental process

The dynamics of a stem-type cell of our model show irregular oscillation with orbital instability like chaos, and keep a variety of chemicals. The cells with this complex dynamics have a potential to differentiate into several cell types with simpler cellular dynamics, for example, with a fixed-point and regular oscillation (see Fig.3 as an example). The differentiated types of cells with simpler dynamics lose the potential of differentiation, and produce cells of the same type or stop dividing. Here, we can see a correspondence between the loss of multipotency and the decrease in some complexity of intra-cellular dynamics. So far, it is still unclear which is the most relevant quantity to characterize this direction of development, that plays the similar role as the entropy for thermodynamic irreversibility. In our model simulation, with the direction of development (loss of multipotency), following tendencies can be confirmed quantitatively:

(i) **Diversity of chemicals decreases**

(ii) **Intra-cellular dynamics become less chaotic.**

First, we focus on the diversity of chemicals in cellular dynamics. The diversity of chemicals of the i-th cell can be measured as $S_i = \sum_{j=1}^{k} p_i(j) \log p_i(j)$, with

$p_i(j) = < \frac{c_i^{(j)}}{\sum_{m=1}^{k} c_i^{(m)}} >$, with $< .. >$ as temporal average.

In Fig.6, the diversity of chemicals in the initial cell type and the growth speed of an ensemble of cells are plotted. As in Fig.5, results from 1000 randomly generated reaction networks are shown in the figure. The chemical diversity in the initial cell type S_{ini} is measured as the average over cells when the number of cells is small (here 10), and the growth speed of an ensemble is computed with the same way as in Fig.5. In Fig.6, the points with a high growth speed of an ensemble around $S_{ini} > 2$ correspond to case (I) with cellular heterogeneity. This figure shows that, the faster growth in case (I) requires higher diversity of chemicals in the cellular dynamics of the initial cell type.

In the course of development, the diversity of chemicals in the cells changes. In cases (IIa) and (IIb), cells with the initial cell type reproduce themselves without differentiation, so that the diversity of chemicals in all cells remain almost same. On the other hand, In case (I) with differentiations, the diversity of chemicals of each cell takes several distinct values corresponding to differentiated cell types. An important point here is that the differentiated cell types always have lower diversity of

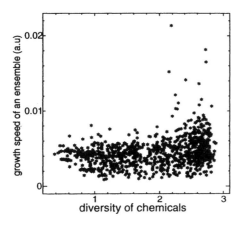

Figure 6: Relationship between the growth speed of a single cell and the chemical diversity of the initial cell type. The ordinate corresponds to the growth speed of an ensemble with the same way as in Fig.5. The horizontal axis represents chemical diversity in the intra-cellular dynamics of the initial cell types. Each point is obtained by using a different chemical reaction network, as in Fig.5. Note that the maximum value of chemical diversity in the intra-cellular dynamics S_i is $\log 20 \simeq 3.0$ with 20 chemicals in a cell.

chemicals than that of the cell type before the differentiation. In Fig.7, we show some examples of the development of the chemical diversity in case (I) organisms. Each graph of Fig.7 shows a histogram of the number of cells as for the chemical diversity in cellular dynamics. The upper figures give the distribution when organisms consist of a small number of cells. In this stage, cells have not been differentiated yet, and the dynamics of cells belong to the same type. As mentioned above, this initial cell type has high diversity of chemicals. The lower figures show the chemical diversity of organisms after development, that contain several cell types. Each peak in the figures corresponds to distinct cell types differentiated from the initial cell type. These examples clearly show the tendency of decrease in the diversity of chemicals along the direction of development.

Next, to confirm the decrease in complexity of dynamics with the loss of multipotency, we have measured the Kolmogorov-Sinai(KS) entropy, that is known to be a good measure of the variety of orbits and the degree of chaos. Although the KS entropy is defined using all degrees of freedom in the system including all cells and the medium, here we measure only the complexity of the intra-cellular dynamics by neglecting the instability arising from cell-cell interactions. Hence we compute the "KS entropy" of the intra-cellular dynamics by a sum of positive Lyapunov exponents in the restricted phase space for intra-cellular dynamics of each cell.

We have found that the KS entropy of differentiated cells are always smaller than that of stem-type cells before differentiation. For example, the KS entropy of the

stem-type "white" cells in Fig.3 is 6×10^{-3}, while it is 5×10^{-4} for "gray", and is less than 1×10^{-6} for "black" cells. The results strongly suggest that the multipotency of stem-type cell is sustained by complexity of intra-cellular dynamics of the cell.

Stability of cell society

In case (I), an organism with various cell types has the stability against macroscopic perturbations, such as the removal of some cells, as its intrinsic property.

As mentioned above, the differentiations of stem cells obey a specific rule with some probability for each path of differentiation. In this differentiation process, the probability of differentiation is not fixed, but is regulated depending on the interactions among the surrounding cells. For example, in Fig.3, the differentiation rate of "white" cells to other cell types is controlled by the cell-type distribution of neighboring cells.

It should be noted that the information on the distribution of cell types in the cell society is embedded in each internal dynamics. Each attracting state of internal dynamics, corresponding to a distinct cell type, is gradually modified in the phase space with the change of the distribution of other cells. This modification of internal dynamics is much smaller than the differences between different cell types. Thus, there are two types of information in internal dynamics; the analogue one which embeds the global distribution of cell types and the digital information to give each cell type. This analogue information controls the rate of the differentiation, because the probability of the differentiation from the stem cell to other cell types depends on their modification. On the other hand, change of the distribution of distinct cell types, introduced by the differentiation, again brings about the change in the analogue information.

As a result of this interplay between two types of information, the higher level dynamics emerges, which controls the rate of the division and differentiation depending on the number of each cell type. This dynamics of differentiation allows for the stability at the level of ensemble of cells. As an example, let us consider an organism with three cell types ("white","black","gray" in Fig.3). When the "black" cells are removed to decrease their population, the frequency of differentiations from "white" to "black" is enhanced, and the original distribution is roughly recovered, following the modification of dynamics in "white" cells.

It should be stressed that dynamical differentiation process in case (I) is always accompanied by this kind of regulation process, without any sophisticated programs implemented in advance. In this process, the differentiations occur when the instability of the system exceeds some threshold through the increase of the cell number, and the emergence of differentiated cells stabilize it. Therefore, a large perturbation such as removal of cells

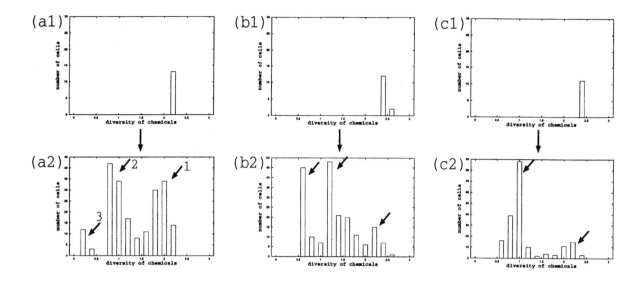

Figure 7: Three examples of changes in distribution of intra-cellular chemical diversity along the developmental process. Each graph shows a histogram of the number of cells as for the chemical diversity in cellular dynamics. The upper figures (a1), (b1), and (c1) give the distribution when the organisms consist of a small number of cells. The lower figures (a2), (b2), and (c2) show the chemical diversity in cells after various cell types have emerged through the development. Each peak, that is marked by an arrow, corresponds to a distinct cell type. The developmental process (a1)→(a2) shows the result corresponding to Fig.3, in which the peaks marked "1", "2", and "3" correspond to "white", "gray", and "black" cells, respectively, while (b) and (c) are obtained from different reaction networks.

makes the system unstable again, and then the differentiations toward the stable state occur. In other words, only the cell types, that have this regulation mechanism to stabilize the coexistence with other cell types, can appear in the developmental process in case (I).

Summary and discussion

In the present paper, we have studied a dynamical system model of developmental process, to make clear the relationship between the growth of multicellular organisms and some characteristics of intra-cellular chemical dynamics.

Our results are robust against the change in model parameters, as long as the internal reaction and intra-cellular diffusion terms have a comparable order of magnitude. The results are also independent of the details of the model, in particular, of the choice of catalyzation degree as we have confirmed for $\alpha = 1$ or 3. The same two classes and the same relation between the growth and cellular diversity are obtained, as long as the number of chemicals is sufficient (say for $k > 10$) and number of the reaction paths is in the medium range (e.g., 3∼9 for each chemical at $k = 20$; otherwise, the intra-cellular dynamics fall onto a fixed point for most cases without differentiation).

To sum up, our study has provided evidence that an ensemble of cells with a variety of dynamics and sta-

ble states (cell types) has a larger growth speed than an ensemble of simple cells with a homogeneous pattern, because of its greater capability to transport and share nutritive chemicals. Since no elaborate mechanism is required for the appearance of this complex cell system, our results suggest that complexity of multicellular organisms is a necessity course in evolution, once a multicellular unit emerges from cell aggregates.

We have also discussed some characteristics observed commonly to such complex cell societies with larger growth speed (i.e. case (I)).

In developmental process in case (I), the multipotent stem cells appear at the initial stage, which provide cells with several distinct cell types through the differentiations. This appearance of both multipotent stem cells and differentiated cells can give a primitive separation between germ cells and somatic cells, that is observed generally in existing multicellular organisms. In fact, by using a model on a 2-dimensional medium and appropriate rules of cell-cell adhesion, we have observed the emergence of a life-cycle of a multicellular replicating unit. The next generation is formed by releasing stem-type cells to the medium (14).

In case (I), along the direction of developmental process, from the multipotent stem cells to differentiated cells, some complexity in intra-cellular dynamics decreases. Here, the reaction dynamics of a multipotent stem cell show complex chaotic oscillation with a variety of chemicals, while a cell differentiated from the stem cell has always simpler dynamics, such as a fixed-point or a limit cycle. We have quantitatively confirmed this tendency along the developmental process, as the decrease of chemical diversity and KS-entropy of intra-cellular dynamics.

The differentiation of a stem cell is determined stochastically, that is sustained by the chaotic intra-cellular reaction dynamics. The probability of differentiation is regulated depending on the cell-type distribution of surrounding cells, through modification of cellular dynamics. This regulation of differentiation allows for the stability at the level of organism, without any sophisticated mechanisms implemented in advance.

It should be stressed that, these characteristics in the complex cell society are a natural consequence for a multicellular organism to acquire a larger growth speed. Therefore, we predict that all existing multicellular organisms have these characteristics commonly, even if their multicellularity might have evolved independently. Our goal in the present approach is to elucidate a universal structure in a multicellular system, and to provide a novel viewpoint to understand the developmental process, such as its stability and irreversibility, independent of detailed information of an organism.

Our result concerning the relationship among the heterogeneity in cell types, complexity in dynamics, and the growth speed of a single cell and ensemble provides experimentally testable predictions. Since even primitive organisms such as *Anabena* (15; 16) and *Volvox* (17) exhibit differentiation in cell types and some spatial pattern, the relationship can be verified. In fact, for a mutant of *Volvox* that possesses only homogeneous cells, the growth becomes slower in comparison to the wild-type (18). The negative correlation between the growth speed and the cell diversity is also suggested in the culture of a hematopoietic system (19).

Cancer organisms are also a candidate to verify the relationship. They can be regarded as a novel multicellular organism that tries to evolve independently during the lifetime of their host. As is well known, there are two types of cancer clusters, those that are subject to a growth limit and those that are not. Although this difference is often attributed to external factors, such as the existence of blood vessels, we expect from the present study that the heterogeneity in cellular states is also relevant.

The relationship between the loss of multipotency and the decrease of complexity in reaction dynamics are also experimentally verifiable. In real organisms, multipotency is generally lost through the development. In terms of molecular biology, the decrease of chemical diversity can be interpreted as the decrease of the number of activated genes. We predict that a stem cell with multipotency, such as an ES cell and a hematopoietic stem cell, includes a variety of weakly activated genes, while a terminal differentiated cell has a smaller number of strongly activated genes. In addition, we predict that the temporal variation in intra-cellular dynamics is larger for the stem cell. Dependence of Ca oscillation on the cell type, for example, has already been measured (20). It is relevant to study how the complexity in dynamics changes with the loss of multipotency beginning from the ES cell.

We would like to thank Tetsuya Yomo for stimulating discussions. This research was supported by Grants-in-Aid for Scientific Research from the Ministry of Education, Science, and Culture of Japan (Komaba Complex Systems Life Science Project).

References

1. B. Alberts, D. Bray, J. Lewis, M. Raff, K. Roberts, & J. D. Watson, *The Molecular Biology of the Cell*(Garland, New York, ed. 3, 1994)
2. J. Maynard-Smith and E. Szathmary, *The Major Transitions in Evolution* (W.H.Freeman, 1995)
3. A. M. Turing, *Phil. Trans. Roy. Soc. B*, **237**, 5 (1952)
4. K. Kaneko and T. Yomo, *Bull.Math.Biol.*, **59**, 139 (1997)
5. K. Kaneko and T. Yomo, *J. Theor. Biol.*, **199** 243 (1999)
6. C. Furusawa and K. Kaneko, *Bull. Math. Biol.*, **60**, 659 (1998)
7. S. A. Kauffman, *J. Theo. Biol.*, **22**, 437 (1969)
8. A. Goldbeter, *Biochemical oscillations and cellular rhythms* (Cambridge University Press, New York, 1996)
9. B. Goodwin, *Temporal Organization in Cells* (Academic Press, London, 1963)
10. B. Hess and A. Boiteux, *Ann. Rev. Biochem*, **40**, 237 (1971)
11. K. Kaneko, *Physica* **41D**, 137 (1990)
12. K. Kaneko, *Physica* **75D**, 55 (1994)
13. M. Ogawa, *Blood*, **81**, 2353 (1993)
14. C. Furusawa and K. Kaneko, *Artificial Life*, **4**, 79 (1998)
15. J. W. Golden, S. J. Robinson, and R. Haselkorn, *Nature*, **314**, 419 (1985)
16. H-S Yoon and J. W. Golden, *Science*, **282**, 935 (1998)
17. D. L. Kirk and J. F. Harper, *Int. Rev. Cytol.*, **99**, 217 (1986)
18. L. Tam and D. L. Kirk, *Development*, **112**, 571 (1991)
19. J. Suda, T. Suda and M. Ogawa, *Blood*, **64(2)**, 393 (1984)
20. M. Dolmetsch, A. Xu, and K. Lewis *Nature*, **392**, 933 (1998).

Sympatric Speciation from Interaction-induced Phenotype Differentiation

Kunihiko Kaneko [1] and Tetsuya Yomo[2]

[1] Department of Pure and Applied Sciences,University of Tokyo, Komaba, Meguro, Tokyo 153, JAPAN
[2] Department of Biotechnology, Faculty of Engineering, Osaka University 2-1 Suita, Osaka 565, JAPAN

Abstract

A novel viewpoint for evolution is presented, by taking seriously into account the relationship between genotype and phenotype. First, as a consequence of dynamical systems theory, phenotypes of organisms can be differentiated into distinct types through the interaction, even though they have identical genotypes. Then, with the mutation in genotype, it is shown that the genotype also differentiates into discrete types, while maintaining the 'symbiotic' relationship between the types. This process is robust against sexual recombination, because offspring with intermediate genotypes are less fit than their parents. Accordingly, a plausible scenario for sympatric speciation is presented. Relevance of our scenario to the historical evolution as well as to artificial evolution is discussed.

Introduction

The question why organisms are separated into distinct groups, rather than exhibiting a continuous range of characteristics(1), originally raised by Darwin(3), has not yet been fully answered, in spite of several attempts to explain sympatric speciation (see also Maynard-Smith and Szathmary(23; 2)). Here, we provide an answer to this question, by presenting a novel and plausible mechanism for the sympatric speciation, based on dynamical systems theory.

Difficulty in stable sympatric speciation, i.e., process to form distinct groups with reproductive isolation, lies in the lack of a known clear mechanism how two groups, which have just started to be separated, coexist in the presence of mutual interaction and mixing of genes by mating. So far people try to propose some mechanism so that the two groups do not mix and survive independently, as is seen in sexual isolation by mating preference (e.g., (24; 22; 27; 11; 21; 4)). However, this type of theory cannot answer how such mating preference that is 'convenient' for sympatric speciation, is selected. In addition to this drawback, there lies another serious problem. In the conventional theory of sympatric speciation, if one group may disappear by fluctuations due to finite-size population, the other group does not necessarily reappear. Coexistence of one group is not necessary for the survival of the other. Hence the speciation process is rather weak against possible fluctuations that should exist in a population of finite size.

Of course, if the two groups were in a symbiotic state, the coexistence would be necessary for the survival of each. However, the two groups have little difference in genotype in the beginning of speciation process, and it might be hard to imagine such a 'symbiotic' mechanism. Accordingly, it is generally believed that sympatric speciation, robust against fluctuations, is rather difficult. As long as we assume that the phenotype is a single-valued function of genotype for a given environment, this conclusion will be plausible and general.

Let us recall the standard standpoint for the evolution in the present biology(8; 1). (i) First, each organism has genotype and phenotype. (ii) Then, the fitness for survival is given for a phenotype, and Darwinian selection process acts for the survival of organisms, to have a higher fitness (iii) Only the genotype is transferred to the next generation (Weissman's doctrine) (iv) Finally, there is a direct flow only from a genotype to phenotype, i.e., a phenotype is determined through developmental process, given a genotype and environment (the central dogma of molecular biology). Although there may be some doubt in (iii) (and (iv)) for some cases, we follow this standard viewpoint here.

Note, however, that (iv) does not necessarily mean that the phenotype is 'uniquely determined'. In the standard population genetics, this uniqueness is assumed, but it is not necessarily postulated within the above standard framework. Indeed, there are three reasons to make us doubt this assumption of the uniqueness, one theoretical and two experimental.

First, we have previously proposed isologous diversification theory, where two groups with distinct phenotypes appear even from the same genotype(6; 7; 15; 16; 17). In this theory, due to the orbital instability in developmental process, any small difference (or fluctuation) is amplified to a macroscopic level, so that the dynamical state of two organisms (cells) can be different, even if they have a same set of genes. The organisms are differentiated into discrete types through the interaction, where existence of each type is necessary to eliminate the dynamic instability in developmental process, which underlies when the ensemble of one of the types is isolated. Hence, existence of each type is required for the survival of each other, even though every individual has identical, or slightly different genotypes.

Second, it is well known experimentally that in some mutants, various phenotypes arise from a single genotype, with some probability(10). This phenomenon is known as low or incomplete penetrance(25). Although the existence of an organism with low penetrance is a 'headache' in genetics, it is observed even in *C elegans*. Although it might sound strange, it is an established fact that the uniqueness of phenotypes is not always true.

Last, the interaction-induced phenotypic diversification is clearly demonstrated in an experiment reported by one of the authors and his colleagues, for specific mutants of *E. coli*. In fact, they show (at least) two distinct types of enzyme activity, although they have identical genes. These different types coexist in a well stirred environment of a chemostat (20; 19), and this coexistence is not due to spatial localization. Here, coexistence of each type is supported by each other. In fact, when one type of *E. coli* is removed externally, the remained type starts to differentiate again to recover the coexistence of the original two types. It is now demonstrated that distinct phenotypes (as for enzyme activity) appear, according to the interaction among the organisms, even though they have identical genes. Indeed, the mechanism for this differentiation is understood theoretically by the above 'isologous diversification theory'.

Hence, we take this interaction-induced phenotypic differentiation from a single genotype seriously into account and discuss its relevance to evolution. We will show that this phenotypic differentiation is later fixed to genotypes through mutation to genotypes, in spite of the fact that we have assumed only the flow from genotype to phenotype, and will give a general mechanism for the sympatric speciation. Last, we will discuss relevance of our theory to biological evolution, as well as to artificial life studies.

Model

To study phenotypic and genotypic relationship, we have to consider a developmental process that maps a genotype to a phenotype. As an illustration, we consider an abstract model consisting of several biochemical processes. Each organism possesses such internal dynamic processes which transfer external resources into some products depending on the internal dynamics. Through this process, organisms mature and eventually become ready for reproduction.

Here, the phenotype is represented by a set of variables, corresponding to biochemical processes. Each individual i has several cyclic processes $j = 1, 2, \cdots, k$, whose state at time n is denoted by $X_n^j(i)$. With k such processes, the state of an individual is given by the set $(X_n^1(i), X_n^2(i), \cdots, X_n^k(i))$, which defines the phenotype. This set of variables can be regarded as concentrations of chemicals, rates of metabolic processes, or some quantity corresponding to a higher function. The state is not fixed in time, but changes temporally according to a set of deterministic equations with some parameters.

Genes, since they are nothing but information expressed on DNA, could in principle be included in the set of variables. However, according to the central dogma of molecular biology (requisite (iv) in *Introduction*), the gene has a special role among such variables. Genes can affect phenotypes, the set of variables, but the phenotypes cannot change the code of genes. During the life cycle, changes in genes are negligible compared with those of the phenotypic variables they control. In terms of dynamical systems, the genes can be represented by control parameters that govern the dynamics of phenotypes, since the parameters in an equation are not changed through the developmental process, while the parameters control the dynamics of phenotypic variables. Accordingly, we represent the genotype by a set of parameters. Only when an individual organism is reproduced, this set of parameters changes slightly by mutation.

To be specific we consider the following model consisting of the processes (i)-(iii) given below.

(i) Dynamics of the phenotypic state:

The dynamics of the variables $X_n^j(i)$ consist of a mutual influence of cyclic processes { $X_n^\ell(i)$ } and interaction with other organisms ($X_n^\ell(i')$). First, as a simple model we split the state variable $X_n^\ell(i)$ into its integer part $R_n^\ell(i)$ and the fractional part $x_n^\ell(i) \equiv mod[X_n^\ell(i)]$. The integer part $R_n^\ell(i)$ is assumed to give the number of times that the cyclic process has passed since the individual's birth, while the fractional part $x_n^\ell(i)$ gives the phase of oscillation in the process.

As a simple model, we assign a phase of oscillation to each cyclic process and assume that there are mutual influences depending on the phase state of processes. The ℓ-th process has a flow from other processes, while there is a flow from the process to the other processes. As a simple example, the internal dynamics of the cyclic process is assumed to be represented by $\sum_m \frac{a^{\ell,m}}{2} sin(2\pi x_n^m(i))$. Hence the internal dynamics are given by

$$X_{n+1}^\ell(i) = X_n^\ell(i) +$$
$$\sum_m \frac{a^{\ell m}(i)}{2} sin(2\pi x_n^m(i)) - \sum_m \frac{a^{m\ell}(i)}{2} sin(2\pi x_n^\ell(i)).$$

Next, the interaction between individuals is introduced through competition for resources, with which each cyclic process progresses. The ability to obtain resources generally depends on the internal state of the unit $x_n^j(i)$. Again, we choose our model so that only the phase is relevant to the interaction and take $psin2\pi(x_n^\ell(j))$ as the ability to obtain resources. Assuming that all elements (whose number is N_n) compete for resources s^j for each step, we take the following interaction term:

$$Interaction^\ell(i) = psin(2\pi x_n^\ell(i)) + \frac{s^\ell - \sum_j psin2\pi(x_n^\ell(j))}{N_n}.$$

Here, the second term comes from the constraint that $\sum_i Interaction^\ell(i) = s^\ell$, due to the condition that units compete for a given resource s^ℓ at each time step.

Now, by summing up the two processes, the developmental dynamics of phenotypes in our model is given by

$$X_{n+1}^\ell(i) = X_n^\ell(i) +$$
$$\sum_m \frac{a^{\ell m}(i)}{2} sin(2\pi x_n^m(i)) - \sum_m \frac{a^{m\ell}(i)}{2} sin(2\pi x_n^\ell(i))$$

$$+ psin(2\pi x_n^\ell(i)) + \frac{s^\ell - \sum_j psin2\pi(x_n^\ell(j))}{N_n}. \quad (1)$$

(ii) Growth and death: Each individual splits into two when a given condition for the growth is satisfied. Taking into account that the cyclic process is required for reproduction, we assume that a unit replicates when the accumulated number of cyclic processes goes beyond some threshold. As a specific example, the condition for the reproduction is given by $\sum_\ell R_n^\ell(i) \geq Thr$. The rotation number $R_n^\ell(i)$ is reset to zero when the corresponding individual splits.

To introduce the competition for survival, death of an individual has to be included. Here each individual is eliminated both by random removal of organisms at some rate as well as by a given death condition based on their state. The latter condition is given by the elimination of such individual that satisfies $R_n^\ell(i) < -10$. In other words, if the cyclic process of an individual progresses reversely too much, it dies.

(iii) Genetic parameter and mutation:

Next, genotypes are given by a set of parameters $a^{m\ell}(i)$, representing the relationship between the two cyclic processes ℓ and m ($1 \leq \ell \neq m \leq k$). Following the above argument, genes, represented by parameters in the model slowly mutate by reproduction. With each division, the parameters $a^{m\ell}$ are changed to $a^{m\ell} + \delta$, with δ as a random number over $[-\epsilon, \epsilon]$. Here the small parameter ϵ corresponds to the mutation rate.

Scenario for Symbiotic Sympatric Speciation

We have carried out several simulations of the model with $k = 3, 4$, and 5, and some other variants (18). Also, we have carried out the simulations of a model consisting of a metabolic process of autocatalytic networks (26). Since a common speciation is obtained for all the models,

we basically describe numerical results of the model in the last section, to demonstrate our general scenario for the sympatric speciation. An example of the speciation process in the model is shown in Fig.1 and Fig.2. The evolution of the genotype-phenotype relationship, plotted by (a^{23}, R^2) at every reproduction event, is given in the sequence of Fig.1, while the values of $a^{23}(i)$ are plotted at every reproduction in Fig.2. Note the change in the scale for the genotype axis by figures in Fig.1, which illsutrates the progress of separation in genotype parameters. As shown in Fig.1a, the phenotype differentiates initially into two groups and this phenotypic change is subsequently fixed to the genotypes. The scenario for the speciation here is described as follows (18). (see Fig.3 for schematic representation).

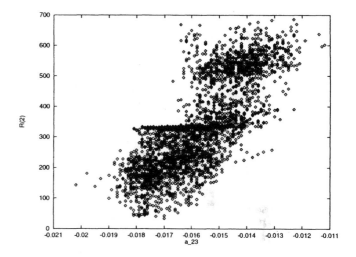

Fig. 1a)

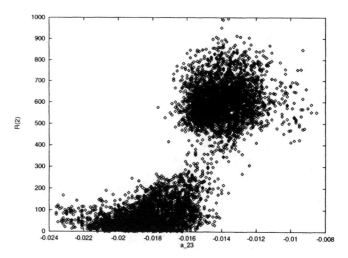

Fig.1 b)

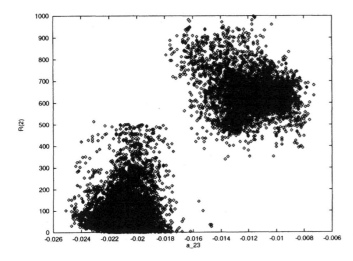

Fig.1 c)

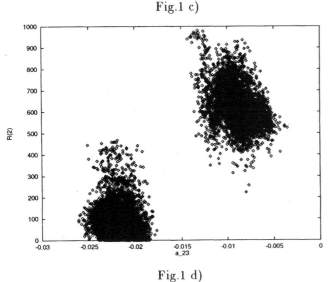

Fig.1 d)

Figure 1: Evolution of genotype-phenotype relationship. As a phenotype, the integer part $R^2(j)$, the number of cyclic process 2 used for reproduction, is adopted, while a^{23} is used as a genetic parameter. (a^{23}, R^2) is plotted for every division of individuals. The 501-3500th divisions are plotted in (a), 6000-10000 divisions in (b), 10000-20000 divisions in (c), and 20000-30000 divisions in (d). Initially, phenotypes are separated, even though the genotypes differ only slightly, as shown in (a). Later, the genotypes are also separated, according to the difference in phenotypes, as shown in (b) and (c). In the simulation shown in Fig.1. and Fig.2, the parameters are set at $p_k = 1.5/(2\pi)$, $s^1 = 10, s^2 = 8, s^3 = 5$, while threshold number Thr for the reproduction is set at 1000, and the mutation rate of the parameters is 0.001. The population size fluctuates around 500, after an initial transient. (Hence the generation number is given roughly by dividing this division number by 500.) Initially, the genotype parameters are set as $a^{ij} = \frac{-0.1}{2\pi}$.

Stage-1: Interaction-induced phenotypic differentiation

When many individuals interact competing for finite resources, the phenotypic dynamics start to be differentiated even though the genotypes are identical or differ only slightly. This differentiation generally appears if nonlinearity is involved in the internal dynamics of some phenotypic variables. Slight differences in variables between individuals are amplified by the internal dynamics (e.g., metabolic reaction dynamics). Through interaction between organisms, the difference in phenotypic dynamics are amplified and the phenotype states tend to be grouped into two (or more) types. The dynamical systems mechanism for such differentiation was first discussed as clustering (12), and then extended, to study the cell differentiation (6; 7; 13; 16; 17). In fact, the orbits of $(x_t^1(i), x_t^2(i), \cdots, x_t^k(i))$ lie in a distinct region in the phase space, depending on each of the two groups that the individual i belongs to. Note that the difference at this stage is not fixed in either the genotype or the phenotype. The progeny of a reproducing individual may belong to a distinct type from the parent. If a group of one type is removed, then some individuals of the other type change their type to compensate for the missing type. To discuss the present mechanism in biological terms, consider a given group of organisms faced with a new environment and not yet specialized for the processing of certain specific resources. Each organism has metabolic (or other) processes with a biochemical network. As the number of organisms increases, they compete for resources. As this competition becomes stronger, the phenotypes become diversified to allow for different uses in metabolic cycles, and they split into two (or several) groups. Each group is specialized in processing of some resources. Here, the two groups realize a differentiation of roles and form a symbiotic relationship. Each group is regarded as specialized in a different niche, which is provided by another group.

Stage-2: Co-evolution of the two groups to amplify the difference of genotypes

At the second stage of our speciation, difference in both genotypes and phenotypes is amplified. This is realized by a kind of positive feedback process between the changes in geno- and phenotypes.

This process consists of two parts. The first part, essential to the genetic fixation, is genetic separation due to the phenotypic change. This occurs if the parameter dependence of the growth rate is different between the two phenotypes. In other words, there are (one or) several parameters such that the growth rate increases with them for the upper group and decreases for the lower group (or the other way around).

Indeed, such parameter dependence is not exceptional.

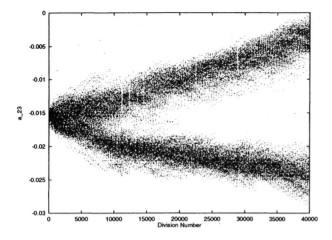

Figure 2: The evolution of the genotypic parameter, corresponding to Fig. 1. The parameter $a^{23}(i)$ is plotted as a dot at every division (reproduction) event, with the abscissa as the division number.

As a simple illustration, assume that the use of metabolic processes is different between the two groups. If the upper group uses one metabolic cycle more, then the mutational change of the parameter $a^{\ell m}$ to enhance the use of the cycle is in favor for the upper group, while the change to reduce it may be in favor for the lower group. Indeed, several numerical results (given e.g., in Fig. 1 and Fig. 2) support that there always exist such parameters. This dependence of growth on genotypes leads to genetic separation of the two groups.

Although the above process is most essential to speciation, the genetic separation is often accompanied by the second process, the amplification of phenotypic difference due to the genotypic difference. In the situation of Fig. 3, as the parameter $a^{\ell m}$ is increased, the phenotype variable R^j tends to increase and vice versa. This is possible if $\partial R^j / \partial a^{\ell m}$ is larger for the upper group. In a typical and clear example, as in Fig. 1 and Fig. 3c, $\partial R^j / \partial a^{\ell m}$ is positive for the upper group and negative for the lower group. With this process, the separation of the two groups is amplified both in genotypes and phenotypes. We again emphasize that the existence of such parameter(s) that satisfy the two conditions is not unusual.

With this separation of two groups, each phenotype (and genotype) tends to be preserved by the offspring, in contrast with the first stage. Now, distinct groups with recursive reproduction have been formed. However, up to this stage, the two groups with different phenotypes cannot exist by themselves in isolation. When isolated, offspring with the phenotype of the other group start to appear. The developmental dynamics in each group, when isolated, are unstable and some individuals start to be differentiated to recover the other group. The dynamics, accordingly each phenotype, is stabilized by

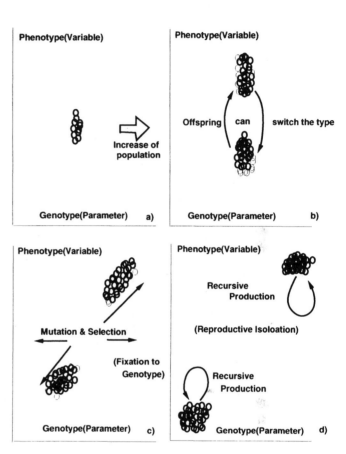

Figure 3: Schematic representation of speciation process, plotted as phenotype-genotype relationship. (a) Initially, there is a group of organisms with distribution centered around a given phenotype and genotype. (b) Then, with the increase of population, phenotype is differentiated into discrete types. (c) Then according to the difference of phenotype, genotype is also differentiated. (d) Finally, the two groups differentiate both in genotypes and phenotypes, and form distinct species. Indeed, these two groups are separated also by sexual recombination, since the hybrid offspring cannot produce its progeny.

each other through interaction. Hence, two groups are in a symbiotic state, and the evolution of one group is related with that of the other. To have such stabilization, the population of each group has to be balanced through the interaction. Even under random fluctuation by finite-size populations and mutation, the population balance of each group is not destroyed. Indeed, the growth speed of each group remains of the same order at this stage. With this co-evolutionary process, the phenotypic differentiation is fixed to the genotype.

Accordingly, our mechanism of genetic diversification is stable. This is why our mechanism works as a stable sympatric speciation, as will be shown in the next section.

Stage-3 Genetic Fixation and Isolation of Differentiated Groups

Complete fixation of the diversification to genes occurs at this stage. Here, even if one group of units is isolated, the offspring of the phenotype of the other group are no longer produced. Offspring of each group keep their phenotype (and genotype) on their own. This is confirmed by numerically eliminating one group of units.

Now, each group has one phenotype corresponding to each genotype, even without interaction with the other group. Hence, each group is a distinct independent reproductive unit at this stage. This stabilization of a phenotypic state is possible since the developmental flexibility at the first stage is lost, due to the shift of genotype parameters. The initial phenotypic change introduced by the interaction is now fixed to genes. Genetically distinct groups with independent reproduction are formed with this genetic fixation.

To check the third stage of our scenario, it is straightforward to study the further evolutionary process from only one isolated group. In order to do this, we pick out some population of units only of one type, after the genetic fixation is completed and both the geno- and phenotypes are separated into two groups, and start the simulation again. When the groups are picked at this third stage, the offspring keep the same phenotype and genotype. Now, only one of the two groups exists. Here, the other group is no longer necessary to maintain stability.

Speciation as Reproductive Isolation in the Presence of Sexual Recombination

The speciation process is defined by both genetic differentiation and by reproductive isolation (Dobzhansky 1937). Although the evolution through the stages I-III lead to genetically isolated reproductive units, one might say that the term 'speciation' should not be used unless the process shows isolated reproductive groups under the sexual recombination. In fact, one may wonder if the present differentiation scenario works under sexual recombination, since the two genotypes from parents are mixed in the offspring by recombination.

On the other hand, since the present scenario is robust against perturbations, such as the removal of one phenotype group, it may be expected to be stable against sexual recombination, which mixes the two genotypes and may bring about a hybrid between the two genotypes. To examine this stability, we have extended the previous model to include this mixing of genotypes by sexual recombination. To be specific, when the reproduction occurs for two individuals i_1 and i_2 that satisfy the threshold condition $(\sum_\ell R_n^\ell(i_k) > Thr)$, the two genotypes are mixed, by producing two offspring $j = j_1$ and j_2, as

$$a^{\ell m}(j) = a^{\ell m}(i_1)r_j + a^{\ell m}(i_2)(1 - r_j) + \delta \qquad (2)$$

with a random number $0 < r_j < 1$ to mix the parents' genotypes, while the term by δ represents the random mutation. (Asexual reproduction is not included in the simulation). We have made several simulations of this type by choosing the same setting as the previous model.

In Fig.4, we have plotted the evolution of the parameter $a^{12}(i)$ by each reproduction event. As shown in Fig.4, the two distinct groups are again formed in spite of the above mixing of genotypes by sexual recombination.

Even if two separated groups may start to be formed according to our scenario, the above recombination can form 'hybrid' offspring with intermediate parameter values $a^{\ell m}$ between the two group. We have again plotted the evolution of the parameter $a^{12}(i)$ by each reproduction event. As shown in Fig.4, the two distinct groups are again formed in spite of the above mixing of genotypes by sexual recombination.

Of course, the mating between the two groups can produce an individual with the parameters in the middle of the two groups, according to eq.(2). However, an individual with intermediate parameters between the two groups starts to have a lower reproduction rate. Such individual requires much longer time to reach the threshold condition for reproduction whatever phenotype it takes. Before the reproduction condition is satisfied, the individual dies with a higher probability. Hence, an individual from the parents of two distinct groups becomes harder and harder to produce their offspring, with time.

To demonstrate this post-mating isolation, we have also measured the average offspring number of individuals over given parameter (genotype) ranges and over some time span, in Fig.5. As the two groups are formed with the split of the parameter values, the average offspring number of an individual having the control parameter between those of the two groups starts to decrease. Soon the number goes to zero, implying that the hybrid between the two groups is sterile. In this sense, sterility (or low reproduction) of the hybrid appears as a result. Hence it is proper to call the process I-III as speciation, since it satisfies genetic differentiation and reproductive isolation (under the sexual recombination).

Note that we have not assumed any preference in mating choice. Rather, it is natural, according to the present scenario, that mating preference in favor of similar phenotypes evolves, since it is disadvantageous for individuals to produce a sterile hybrid. In other words, the present mechanism also provides a basis for the evolution of sexual isolation through mating preference(24; 22; 27; 11; 21; 4). For example, assume that each individual has a tendency to prefer a mating partner with a closer phenotype, or to avoid an individual with too much different phenotype. With this mating preference, the rate to form a sterile hybrid is reduced. Hence, the pre-mating isolation will evolve as a consequence of post-mating isolation. At any rate, the mating preference can strengthen the speciation process of our mechanism, but never hinders the mechanism from working.

It should be noted that our mechanism for the speciation can work in asexual and sexual reproduction in the same way. The phenotype (R^j) separates into two groups first in the present case with sexual recombination as in the previous asexual case. Later the change is mapped onto the parameters $a^{\ell m}$. The speciation process progresses following the three stages given in §3. Indeed the stability of the speciation against sexual recombination is naturally expected, since the coexistence of two distinct phenotype groups is supported by the isologous diversification, i.e., differentiation to distinct phenotypes under the same genotypes. Even though the genes are mixed, the phenotypes are tended to be separated into distinct groups. Hence the separation into distinct groups is not blurred by the recombination.

Relevance to Biology

According to our scenario, the speciation is a result of interaction-induced phenotypic differentiation. To check the condition for speciation, we have performed numerical experiments of our model, by choosing parameters so that differentiation into two distinct phenotype groups does not occur initially. In this case, separation into two (or more) groups with distinct pheno/geno-types is never observed, even if the initial variance of genotypes is large, or even if a large mutation rate is adopted.

Next, the genetic differentiation always occurs when the phenotype, (represented by the rate of each cyclic process R^k), differentiates into two (or more) distinct groups. After the initial separation into two groups, the fixation into parameters *always* follows, as long as mutation exists. Hence, phenotypic differentiation is a necessary and sufficient condition for the speciation process, in a standard biological situation, i.e., a process with reproduction, mutation, and a proper genotype-phenotype relationship.

Note that the interaction-induced phenotypic differ-

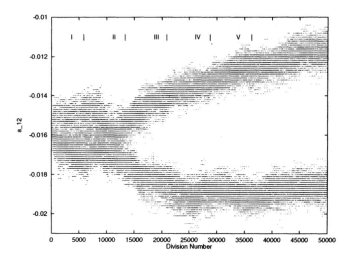

Figure 4: An examples of the speciation process with sexual recombination. The parameter a^{13} of divided units is plotted with the division event. The parameters are $p_k = 1.6/(2\pi)$, $s^1 = s^2 = s^3 = 2$, with initial parameters $a^{ij} = (-.1)/(2\pi)$. The total population fluctuates around 350 in this example.

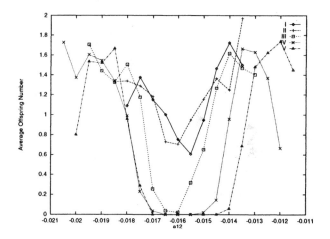

Figure 5: The average offspring number before death is plotted as a function of the parameter (genotype), for simulations of a model with sexual recombination. We have measured the number of offspring for each individual during its lifespan, for the data of Fig.4. By taking a bin width 0.005 for the genotype parameter a^{12}, the average offspring number over a given time span is measured to give a histogram. The histogram over the first 7500 divisions (about 20 generations) is plotted by the solid line (I), and the histogram for later divisions is overlaid with a different line, as given by II (over 7500-15000 divisions), III (1.5-2.25 $\times 10^4$), IV(2.25-3 $\times 10^4$), and V(3.75-4.5 $\times 10^4$). The stages I-V are also represented in Fig.4. As shown in the increase of the dip in the middles, a hybrid offspring will be sterile after some generations.

entiation is deterministic in nature. Once the initial parameters of the model are chosen, it is already determined whether such differentiation will occur or not. Hence, the speciation process is also deterministic in nature, in spite of stochastic mutation process therein. In fact, the speciation process (e.g., the time required for it or the population ratio between the two groups) changes only little between runs that adopt a different random number.

This speciation process is rather fast, once the condition for phenotypic differentiation is satisfied. In the simulations shown in the figures 1 and 2, the speciation is completed around the first 50 generations. Hence, our scenario may give a new and plausible explanation on the variation of time scales on which evolution proceeds, e.g., punctuated equilibrium(9).

In our speciation process, the potentiality for a single genotype to produce several phenotypes decreases. After the phenotypic diversification of a single genotype, each genotype again appears through mutation and assumes one of the diversified phenotypes in the population. Thus the one-to-many correspondence between the original genotype and phenotypes eventually ceases to exist. As a result, one may expect that a phenotype is uniquely determined for a single genotype in wild types, since most organisms at the present time have gone through several speciation processes. One can also expect that a mutant tends to have a higher potentiality to produce various phenotypes from a single genotype. Hence our theory explains why low or incomplete penetrance(10; 25) is more frequently observed in mutants than in a wild type.

Our theory is expected to shed a new light on a possible relationship between developmental process and speciation. For example, consider the questions why the insect has often a larger number of species, or why there are some species that stop evolution, such as *living fossils*. These questions are thought to be tightly related with the developmental flexibility. Our theory predicts that the evolution in phenotype and genotype is accelerated when the developmental process to map between the two types is flexible.

Finally, it should again be stressed that *neither any Lamarckian mechanism nor epigenetic inheritance is assumed* in our theory, in spite of the genetic fixation of the phenotypic differentiation. Only the standard process allowing from genotype to phenotype is included in our theory. Note also that genetic 'takeover' of phenotype change was also proposed by Waddington as genetic assimilation, in possible relationship with Baldwin's effect (28). Using the idea of epigenetic landscape, he showed that genetic fixation of the displacement of phenotypic character is fixed to genes. In our case the phenotype differentiation is not given by 'epigenetic landscape', but rather, the developmental process forms different char-

acters through the interaction. Distinct characters are stabilized through the interaction. With this interaction dependence, the two groups are necessary with each other, and robust speciation process is possible.

Relevance to Artificial Life

Discussion of the mechanism involved in evolution often remains vague, since no one knows for sure what has occurred in history, within limited fossil data. Most important in our scenario, on the other hand, lies in experimental verifiability. As mentioned, isologous diversification has already been observed in the differentiation of enzyme activity of *E. coli* with identical genes(20; 19). We have already started an experiment of the evolution of *E. coli* in the laboratory(29), controlling the strength of the interaction through the population density. With this experiment we can check if the evolution on the genetic level is accelerated through interaction-induced phenotypic diversification, and can answer if our scenario really occurs in nature. In this sense, our evolution theory is testable in laboratory, in contrast with many other speculations. In the same sense, our study is relevant to the field of artificial life (AL), since AL attempts to understand some biological process such as evolution, by constructing an artificial system in laboratory or in a computer from our side.

A problem in most of the present AL studies lies in that it is too much symbol-based. They generally assume some rule for a biological system, represented as manipulation over symbols. Such process will eventually be written by a universal Turing machine. Hence it generally faces with the problem that the emergence may not be possible in principle in such system, since the emergence originally means a generation of a novel, higher level that is not originally written in a rule. The same drawback lies in the symbol-based study of evolution (i.e., a study starting from the evolution of symbols corresponding to genes), and indeed, the AL study on the evolution is often nothing but a kind of complicated optimization problem.

According to our theory, first the phenotype is differentiated, given by continuous (analogue) dynamical system, which is later fixed to genes that serve as a rule for dynamical systems. Now, rules written by symbols (genetic codes) are not necessarily the principal cause of the evolution(14).

In the present paper, we have mainly studied the speciation process from one species to two. However, our theory is straightforwardly extended to study further speciation processes. By including a larger number of processes, one can study successive speciation, as is relevant to the adaptive radiation. There, it is important to study the evolution process in an "open" phase space, i.e., with an increasing number of variables and parameters (phenotypes and genotypes). With this extension, the origin and evolution of diversity will be understood, that is one of the focus issues in the artificial life studies.

We thank M. Shimada and J. Yoshimura for illuminating comments. This work is supported by Grant-in-Aids for Scientific Research from the Ministry of Education, Science and Culture of Japan (11CE2006;Komaba Complex Systems Life Project; and 11837004).

References

1. Alberts B. et al., *The Molecular Biology of the Cell*(Garland, New York ed.3. 1994),

2. Coyne J.A. and H.A.Orr, "The evolutionary genetics of speciation", Phil. Trans. R. Soc. London **B 353** 287-305 (1998).

3. Darwin C., *On the Origin of Species by means of natural selection or the preservation of favored races in the struggle for life* (Murray,London,1859).

4. Dieckmann U. and M.Doebeli, "On the origin of species by sympatric speciation", *Nature* **400**,354(1999).

5. Dobzhansky T.,*Genetics and the Origin of Species* (Columbia Univ. Press.,New York,ed.2,1951).

6. Furusawa C. and K.Kaneko, "Emergence of Rules in Cell Society: Differentiation, Hierarchy, and Stability" *Bull. Math. Biol.***60**,659(1998).

7. Furusawa C. and K. Kaneko, "Emergence of Multicellular Organism: Dynamic differentiation and Spatial Pattern", Artificial Life 4, (1998), 79-93

8. Futsuyma D.J., *Evolutionary Biology* (Sinauer Associates Inc., Sunderland, Mass,ed.2,1986).

9. Gould, S. J. and N. Eldredge. 1977. "Punctuated equilibria: The tempo and mode of evolution reconsidered", *Paleobiology* 3: 115.

10. Holmes L.B., "Penetrance and expressivity of limb malformations", *Birth Defects. Orig. Artic.Ser.* **15**,321(1979).

11. Howard D.J. and S.H.Berlocher,Ed., *Endless Form: Species and Speciation*(Oxford Univ. Press., 1998)

12. Kaneko K., 'Clustering, Coding, Switching, Hierarchical Ordering, and Control in Network of Chaotic Elements" *Physica***41D**,137(1990).

13. Kaneko K., "Relevance of Clustering to Biological Networks", *Physica***75D**,55(1994).

14. Kaneko K., "Life as Complex Systems: Viewpoint from Intra-Inter Dynamics", Complexity, 3 (1998) 53-60

15. Kaneko K. and C.Furusawa, "Robust and Irreversible Development in Cell Society as a General Consequence of Intra-Inter Dynamics", Physica A, (2000) in press

16. Kaneko K. and T.Yomo, "Isologous Diversification: A Theory of Cell Differentiation ", *Bull. Math. Biol.***59**,139(1997).

17. Kaneko K. and T.Yomo, 'Isologous Diversification for Robust Development of Cell Society ", *J. Theor. Biol.***199**,243(1999).

18. Kaneko K. and T.Yomo, "Symbiotic Speciation from a Single Genotype", submitted to Proc. Nat. Acad. Sci. USA

19. Kashiwagi A.,T.Kanaya,T.Yomo,I. Urabe, "How small can the difference among competitors be for coexistence to occur", *Researches on Population Ecology***40**, (1999), in press.

20. Ko E.,T.Yomo, and I.Urabe, "Dynamic Clustering of bacterial population", *Physica***75D**,81(1994)

21. Kondrashov A.S. and A.F.Kondrashov, "Interactions among quantitative traits in the course of sympatric speciation", *Nature***400**,351(1999).

22. Lande R., "Models of speciation by sexual selection on phylogenic traits", *Proc. Natl. Acad. Sci. USA***78**,3721(1981).

23. Maynard-Smith J. and E.Szathmary, *The Major Transitions in Evolution* (W.H.Freeman,1995).

24. Maynard-Smith J., " Sympatric Speciation", *The American Naturalist***100**,637(1966).

25. Opitz J.M., "Some comments on penetrance and related subjects", *Am-J-Med-Genet.***8**,265(1981).

26. Takagi H. and K. Kaneko, "Evolution of genetic code through isologous diversification of cellular states", in these proceedings.

27. Turner G.F. and M.T.Burrows, " A model for sympatric speciation by sexual selection", *Proc. R. Soc. London***B 260**,287(1995).

28. Waddington C.H., *The Strategy of the Genes* (George Allen & Unwin LTD., Bristol, 1957)

29. Xu W. A.Kashiwagi, T.Yomo, I.Urabe, "Fate of a mutant emerging at the initial stage of evolution", *Researches on Population Ecology***38**,231(1996).

Evolution of genetic code through isologous diversification of cellular states

Hiroaki Takagi and Kunihiko Kaneko
Department of Pure and Applied Sciences
University of Tokyo, Komaba, Meguro-ku, Tokyo 153, JAPAN

Abstract

Evolution of genetic code is studied as the change in the choice of enzymes that are used to synthesize amino acids from the genetic information of nucleic acids. We propose the following scenario: the differentiation of physiological states of a cell allows for the different choice of enzymes, and this choice is later fixed genetically through evolution. To demonstrate this scenario, a dynamical systems model consisting of the concentrations of metabolites, enzymes, amino acyl tRNA synthetase, and tRNA-amino acid complex in a cell is introduced and numerically studied. It is shown that the biochemical states of cells are differentiated by cell-cell interaction, and each differentiated type takes to use different synthetase. Through the mutation of genes, this difference in the genetic code is amplified and stabilized. Relevance of this scenario to the evolution of non-universal genetic code in mitochondria is suggested.

Introduction

The protein synthetic system adopted in today's living organisms has a very large and complex network. It consists of over 120 kinds of molecules, such as tRNA, ARS(aminoacyl tRNA synthetase), mRNA, 20 kinds of amino acids, lybosome, ATP, etc. In this system, genetic code plays an important role to link genetic information in DNA to phenotypic functions, where genetic code was considered to be stable. From such considerations and experimental results, the genetic code was once considered to be universal, and "frozen accident theory" was proposed by F.Crick(1), in which the genetic code is assumed to be fixed by frozen accident in the early history of life. From recent studies, however, several non-universal genetic codes were found, for example, in mitochondrial DNA. Now it is recognized that genetic code is not universal and can change in a long term. Considering these stability and flexibility of the genetic code, it is important to study the evolution of genetic codes with these two aspects, which might look like contradicting superficially.

To discuss the evolution of genetic codes, it is necessary to point out two basic features of genetic codes.

The first point concerns about the relationship between genetic codes and the molecular structure. Although tight chemical coupling between codon and amino acid was initially pursued such as a key-keyhole relationship, it is now believed that there is no specific interaction between codon and amino acid(2). Ueda et al. have recently discovered "polysemous codon" in certain Candida species, where two distinct amino acids are assigned by a single codon(3). Now looseness in genetic code is seriously studied.

Second, the evolutionary change of genetic codes has also been studied after the discovery of non-universal genetic code. Among these studies, "codon capture theory", proposed by Osawa & Jukes is most popular(4). The essence of the theory is as follows: If some change to genetic code occurred without any intermediate stage, a sense codon would be changed to a nonsense one, which would cause vital damage to the survival. Therefore, it is necessary to pass through some intermediate stage in evolution, during which the change of genetic code is not fatal. If genetic code is degenerate and some specific triplet is hardly used, tRNA and ARS that correspond to the specific triplet can change their coding without fatal damage.

With these two points in mind, we consider the problem of evolution of genetic codes. First, we expect that genetic code must have passed through the stage with some ambiguity or looseness in the course of the evolution, since otherwise it is hard to imagine that the genetic code has evolved without having a fatal damage to an organism. Then, how is such looseness supported? How is a different coding for the translation supported biochemically? If the difference in genetic codes were solely determined by a genetic system all through the evolutionary process, it would be difficult to consider how the change from one code to another could occur smoothly, without a fatal damage to a cell. Instead, we propose here that the difference in the translation is not solely determined by the nucleus, but also influenced by the physiological state of a cell, at least at some stage of evolution. Indeed, as will be shown, it is rather plausible in an intracellular biochemical process that a cell with identical genes can have different physiological states. Such differentiation, indeed, is expected to occur according to the "isologous diversification theory", proposed for cell differentiation(7; 8; 9; 10; 11; 12; 13). Since the translation system from nucleic acid to amino acid is influenced by several enzymes within a cell, the difference in the physiological state can introduce some change in the translation also. By constructing a model including several biochemicals, we will give an example with non-unique correspondences from nucleic acids to amino acids.

Choosing such biochemical dynamics allowing for differentiation in physiological states and ARS, we then take into account the mutation into a genetic system, and study how different genetic codes are established through the evolution. Through the extensive simulation, we propose the following scenario for the evolution of genetic code: first, phenotypic differentiation occurs for metabolic dynamics through cell-cell interaction. Then each differentiated group of cells starts to use different ARS, and adopt a different way in translating nucleic acid to protein (enzyme). Then, through evolutionary process with competition for reproduction and mutation to genes, this difference in physiological state results in a difference in genes, and one-to-one correspondence is established between differentiated phenotype and mutated genes, so that each group can clearly be separated both in phenotype and genotype. After this evolutionary process, the difference in the translation is fixed. Each group finally achieves a different genetic code, that is now fixed in time, and the initial ambiguity or looseness in coding is reduced.

Background of our modeling

Here, we construct an abstract model to demonstrate the scenario for the evolution of genetic code. Of course, it is almost impossible to describe all factors of complex cellular process. Furthermore, even if we succeeded in it, we could not understand how the model works, since the model is too much complicated. Rather, we extract only some basic features of a problem in concern, and construct a model to understand a general aspect of the evolution of the genetic code. Taking this standpoint, Kaneko, Yomo, and Furusawa proposed "isologous diversification" for the cell differentiation and development, based on the study of coupled dynamical systems. Due to the dynamic instability and cell-cell interaction, it is shown that the biochemical states of cells differentiate to take distinct types(7~13). It is also demonstrated experimentally by Yomo that a type of E.coli with identical genes can take distinct groups with different enzymic activities(6). Evolution starting from such interaction-induced phenotypic differentiation is also investigated, by including mutation and competition for the survival. There, it is shown that the genotypes are also differentiated and fixed according to differentiated phenotypes, although only the flow from genotype to phenotype is

assumed as the standard evolution theory requires.

Here we follow this line of studies, and make a minimal model to understand the evolution of the genetic code. In particular, we show how differentiated "phenotypes" are organized, that adopt a different coding in the translation from nucleic acids to amino acids, following the isologous diversification theory. Then, with the evolution with mutation of genes, the different translations will be shown to be established.

Our model

A cell has a set of variety of biochemicals. Considering the metabolic and genetic process, at least four kinds of basic compounds are necessary, namely, metabolic chemicals, enzymes for metabolic reaction, chemicals for genetic information, and enzymes to make translation of genetic information to protein. In the present paper, these four kinds of chemicals are chosen as the metabolites (metabolic chemicals), enzymes for metabolites, tRNA-amino acid complexes, and ARS, respectively for this set of chemicals. Now, as a state variable characterizing the cell, we introduce

$c_i^{(j)}(t)$: concentration of jth metabolic chemical in ith cell
$a_i^{(j)}(t)$: concentration of jth enzyme for metabolites in ith cell
$e_i^{(j)}(t)$: concentration of jth ARS in i's cell
$x_i^{(j)}(t)$: concentration of jth tRNA-amino acid complexes in ith cell

As for the dynamics of these chemicals, we consider the following processes.

- intra-cellular chemical reaction network

- inter-cellular interaction

- cell division and mutation

Now, we describe each process. See Fig.1 and Fig.2 for schematic representation of our model.

intra-cellular chemical reaction network

In general, each biochemical reaction in cells is catalysed by some enzymes. Here, each metabolic reaction is assumed to be catalyzed by each specific enzyme, and a simple form of reaction rate is adopted given by just the product of the concentrations of the substrate and enzyme in concern. (This specific form is not essential, and the same qualitative results are obtained by using some other form, such as Michaelis-Menten's one.) Here we choose a network consisting of reactions from some metabolite j to other metabolite k catalyzed by the enzyme k. The network is chosen randomly, and represented by a reaction matrix $W(j,k)$, which takes 1 if there is a reaction path, and 0 otherwise. The network is

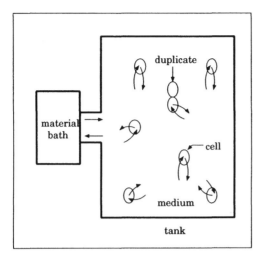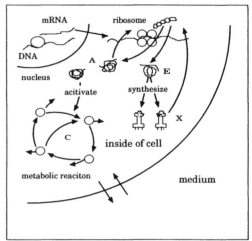

Figure 1: Schematic representation of our model

fixed throughout the simulation. Of course, the dynamics can depend on the choice of the reaction network. Here we choose such network that allows for some oscillatory dynamics. The oscillation is rather commonly observed, as long as there is a sufficient number of autocatalytic paths.

Next, each enzyme, including ARS, that for the synthesis for tRNA, is synthesized from amino acids. This synthesis is again catalyzed by some enzyme. This synthesis is given by a resource table. For the enzyme $a^{(j)}$ and the ARS $e^{(j)}$, the tables are given by $V(j,k)$ and $U(j,k)$ respectively. We also set all entries of $V(j,k)$, $U(j,k)$ at random, by keeping a normalization $\sum_k V(j,k) = \sum_k U(j,k) = 1.0$.

Third, we assume that ARS produces tRNA-amino acid complexes in proportion to its amount. The correspondence between the two is given by a reaction matrix $T(j,k)$, which is 1 if ARS $e^{(j)}$ produces tRNA-amino acid $x^{(k)}$, and 0 otherwise. To include ambiguity, we allow one to many correspondence between e to x. The matrix $T(j,k)$ is again chosen randomly.

Accordingly the concentration change of chemicals by intracellular process is given by

$$dc_i^{(j)}(t)/dt = D_1 \sum_{k=0}^{P-1} W_i^{(k,j)} a_i^{(k)}(t) c_i^{(k)}(t)$$
$$- D_1 \sum_{k=0}^{P-1} W_i^{(j,k)} a_i^{(j)}(t) c_i^{(j)}(t)$$

$$da_i^{(j)}(t)/dt = D_3 \left(\sum_{k=0}^{Q-1} V_i^{(j,k)} x_i^{(k)}(t) \right) a_i^{(j)}(t) c_i^{(l(j))}(t)$$

(where $l(j):j \rightarrow l$ gives a one-to-one mapping.)

$$de_i^{(j)}(t)/dt = D_4 \left(\sum_{k=0}^{Q-1} U_i^{(j,k)} x_i^{(k)}(t) \right) a_i^{(m(j))}(t) c_i^{(n(j))}(t)$$

(where $m(j),n(j):j \rightarrow l$ give one-to-one mappings.)

Finally, tRNA-amino acid complexes, which are the materials of all enzymes, are assumed to change with a faster time scale than the above three types of chemicals. Hence, we adiabatically eliminate its concentration to give the equation for it. By setting

$$dx_i^{(j)}(t)/dt = D_5 \sum_{k=0}^{R-1} T_i^{(k,j)} e_i^{(k)}(t)$$
$$- D_3 x_i^{(j)}(t) \sum_{k=0}^{P-1} V_i^{(k,j)} a_i^{(k)}(t) c_i^{(l(k))}(t)$$
$$- D_4 x_i^{(j)}(t) \sum_{k=0}^{R-1} U_i^{(k,j)} a_i^{(m(k))}(t) c_i^{(n(k))}(t) = 0$$

and we obtain

$$x_i^{(j)}(t) = D_5 \sum_{k=0}^{R-1} T_i^{(k,j)} e_i^{(k)}(t)$$
$$/ \{ D_3 \sum_{k=0}^{P-1} V_i^{(k,j)} a_i^{(k)}(t) c_i^{(l(k))}(t)$$
$$+ D_4 \sum_{k=0}^{R-1} U_i^{(k,j)} a_i^{(m(k))}(t) c_i^{(n(k))}(t) \}$$

Note that the translation process of genetic information to proteins is given by the process between X

(tRNA-amino acid complexes) and E (ARS). One can discuss the difference in coding by examining which species of E has nonzero concentration, and acts in the translation process. We first see how the difference in physiological states given by C affects in the choice of E, for our purpose of the problem.

Throughout the paper, we set the number of metabolites and enzymes (C and A) as $P = 16$, that of ARS (E) as $R = 12$, and that of tRNA-amino acid (X) as $Q = 6$ (see table 1).

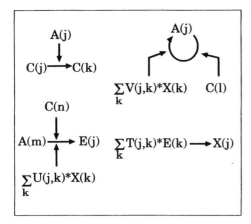

Figure 2: Schematic representation of intra-cellular reaction network

cell-cell interaction

According to the isologous diversification theory, cell-cell interaction is essential to establish distinct cell states. Here we consider the interaction as diffusion of some chemicals through the medium. In this model, we assume that only metabolic chemicals (c) are transported through the membrane, which is rather plausible biologically (see table 1). Assuming that cells are in a completely stirred medium, we neglect spatial variation of chemical concentrations in the medium. Hence we need only another set of concentration variables

$C^{(j)}(t)$: concentration of jth metabolic chemical in the medium

Therefore, all the cells interact with each other through the same environment.

As a transport process we choose the simplest diffusion process, i.e., a flow proportional to the concentration difference between the inside and outside of a cell.

Of course, the diffusion coefficient depends on the metabolic chemical. Here, for simplicity we assume that all the chemicals c are classified into either penetrable or impenetrable ones. The former has the same diffusion coefficient D_2, while for the latter the coefficient is set

to 0. Here we define the index σ_m, which takes 1 if a chemical $c^{(m)}$ can penetrate the membrane, and otherwise 0. Each cell grows by taking in penetrable chemicals from the medium and transforms them to other impenetrable chemicals.

factor	cell-cell interaction	number
metabolic chemicals (C)	exists	P=16
enzyme (A)	not exists	P=16
ARS (E)	not exists	R=12
tRNA-amino acid (X)	not exists	Q=6

Table 1: compositions of the reaction network

Accordingly, the term for the diffusion

$$\sigma_j D_2(C^{(j)}(t) - c_i^{(j)}(t)) \qquad (1)$$

is added to the equation for $dc_i^{(j)}(t)/dt$, while the concentration change in the medium is given by

$$dC^{(j)}(t)/dt =$$
$$D_6(\overline{C^{(j)}} - C^{(j)}(t)) \quad - \quad \sigma_j D_2 \frac{(\sum_{k=0}^{N-1} C^{(j)}(t) - c_k^{(j)}(t))}{Vol},$$

where the parameter Vol is the volume ratio of a medium to a cell, and N is the number of cells. Since these chemicals in the medium are consumed by a cell, we impose a flow of penetrable chemicals from the outside of the medium, that is proportional to the concentration differences. This term is given by the term $D_6(\overline{C^{(j)}} - C^{(j)}(t))$, where the external concentration of chemicals $C^{(j)}$ is denoted by $\overline{C^{(j)}}$.

The variables $c_i^{(j)}$, $a_i^{(j)}$, $e_i^{(j)}$, and $x_i^{(j)}$ stand for the concentrations. Since the volume of a cell can change with a flow of metabolites, its change should be taken into account. Here, we compute the increase of the volume from the flow of chemicals by the sum of the term in eq.(1). The concentration is diluted in accordance with this increase of the volume. With this process the sum

$$\sum_{j=0}^{P-1} c_i^{(j)}(t) + \sum_{j=0}^{P-1} a_i^{(j)}(t) + \sum_{j=0}^{R-1} e_i^{(j)}(t)$$

is preserved through the development of a cell, and the sum is set at 1 here.

cell division

Each cell gets resources from the medium and grows by changing them to other chemicals. With the flow into a cell, the chemicals are accumulated in each cell. As mentioned, this leads to the increase of the volume of a cell. We assume that the cell is assumed to divide, when the volume is twice the original. After the division, the volume of each cell is set to be half. In the division process, a cell is divided into two almost equally, with some fluctuations. Hence, the concentrations of chemicals $b_i^{(j)}$ (where b represents either c, a, or e) are divided into $(1 + \eta)b_i^{(j)}$ and $(1 - \eta)b_i^{(j)}$, with η as a random number over $[-10^{-2}, 10^{-2}]$. As will be shown later, this fluctuation can be amplified to lead to the cell differentiation. The amplitude is not essential, but the existence of fluctuation itself is relevant to have the differentiation.

mutation

To discuss the evolution of genetic codes in a long run, we need to include mutation to genes. In our model, the genetic information is translated from DNA into amino acid. Here both U and V are changed by the mutation to the table of enzyme. At each division, each element of the matrix U or V is mutated by a random number κ with the range of $[-\varepsilon, \varepsilon]$, where ε corresponds to the amplitude of the mutation rate, which we later set at 10^{-3} for most simulations.

Note that this matrix corresponds to genotype, while other chemical concentrations give biochemical states of the cell. Since in our model, there is no direct process to change the matrix from the concentrations, the "central dogma" of the molecular biology is satisfied, i.e., genotype changes phenotype, but not otherwise. We also assume that mutation to genes affects only to the catalytic abilities of enzymes a, e, and not to the specificity of catalytic reactions.

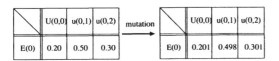

Figure 3: example of mutation process

Recall that the difference in the genetic code is represented by which kinds of ARS are used in a cell, depending on the physiological state of the cell. Here, we are interested in how this difference is fixed genetically through the evolution. With the change of the matrix element of U corresponding to the ARS, the use of specific kind of ARS may start to be fixed, with the increase of some matrix element (to approach unity). If this is the case, specific mappings between ARS and tRNA-amino acid complex are selected, to establish a different coding system. We will pursue the possibility of this scenario in the following sections, based on the simulation results of our model.

Isologous diversification

First, we discuss the behavior of the present cell system, without mutation. We assume that intracellular

chemical dynamics for a single cell system, show oscillation. Since there are many oscillatory reactions in real cells such as Ca^{2+}, cAMP, NADH, and the oscillation is easily brought about by autocatalytic reactions (as also given by the hypercycle (5)), the existence of oscillation is a natural assumption(14; 15).

We have carried out several simulations by taking a variety of reaction networks that produce oscillatory dynamics. In many of such examples, we have found the differentiation process to be discussed, and here we focus on such case, mostly using one example, by fixing a given network. The oscillation of chemical concentrations in this adopted example is shown at the first stage (type 0) in Fig.4. This oscillation of chemical concentrations is observed for most initial conditions, although for rare initial condition there is also a fixed point solution whose basin volume is very small. Note that the oscillation of chemicals, and accordingly the expressions of genes, show on/off type switching, as is true in realistic cell systems.

Now we discuss the behavior of cells with the increase of cells. As cells reproduce, the chemical states start to be differentiated, in consistency with the "isologous diversification", originally discussed for the development of a multicellular organism(7 ~13). First, the phase coherence of oscillations is lost in the intra-cellular dynamics with the increase of the number of cells. Then, the chemical states of cells differentiate into 2 groups. Each group has a different composition in metabolites and also in other enzymes. In the example shown in Fig.4, the type 1 cell is differentiated from the type 0 cell. Here the type 0 cell has a higher activity with a larger metabolite concentrations, than the type 1. In order for a cell to grow, metabolites, enzyme, and ribonucleic acids are necessary. The growth speed of a cell depends on the balance among the concentrations of chemicals $c_i^{(j)}$, $a_i^{(j)}$, $e_i^{(j)}$, and $x_i^{(j)}$ in our model. Hence the growth speed of a cell also differentiates, depending on the concentration of metabolites. Since the dynamic states of chemicals are stabilized by the cell-cell interaction, these states, as well as the number ratio between the two types of cells, are stable against fluctuations.

The differentiation itself has already been studied in the earlier models (7~13). With the introduction of the transcription from tRNA, we can discuss the difference in the use of genetic codes here. Depending on the different metabolic states, use of ARS is also differentiated. In the present example, the type 0 cell uses $e^{(1)}$, $e^{(5)}$, $e^{(7)}$, and $e^{(9)}$, and the concentrations of other ARS are zero. On the other hand, the type 1 cell uses $e^{(5)}$ and $e^{(7)}$ (see Fig.5). Therefore, each cell type has a different phenotype-genotype mapping. Accordingly, we have found that different coding for the translation is adopted depending on the physiological state of the cell.

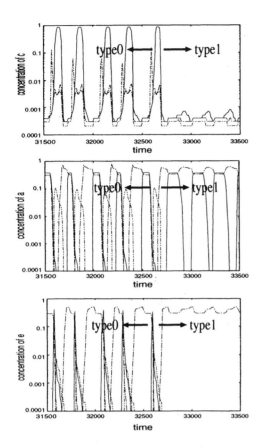

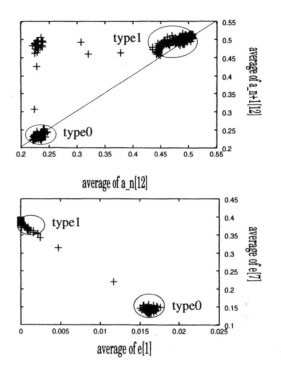

Figure 5: The return map of the average concentration of $a^{(12)}$ (the upper panel), and the average concentration of $e^{(1)}$ and $e^{(7)}$ (the lower one) of each cell, plotted at every division event. In the return map, the chemical average of a mother cell as abscissa and that of its daughter cell as ordinate are plotted. As shown, the type 1 cell keeps its type after division, while the type 0 cell either proliferates or differentiates to type 1.

Figure 4: Oscillations of chemicals. The time series of some $c_i^{(j)}$, $a_i^{(j)}$, $e_i^{(j)}$ are plotted by semi-log scale. The parameters are set at $D_1 = 3.0$, $D_2 = 0.050$, $D_3 = 100.0$, $D_4 = 100.0$, $D_5 = 1.0$, $D_6 = 0.050$, $Vol = 100.0$, $\overline{C^{(j)}} = 0.010$ (for all j) in all the simulations shown in the present paper.

When the type 1 cells are isolated, (i.e., by removing the type 0 cells), their state switches to another type with distinct chemical composition. This type is called type 2. The type 2 cell uses $e^{(7)}$ only in all ARS. In other words, the use of $e^{(5)}$ that is common to type 0 and type 1 cells is abandoned, when the type 0 cells do not coexist. This suggests that the adopted coding system may change depending on cell-cell interaction(see Fig.6 for schematic representation).

Evolutionary process introducing mutation to genes

Now we consider the evolutionary process of the genetic code, by introducing the change of the matrices U and V, giving the translation from nucleic acids to amino acids. At every division a small noise is introduced to U, V as mutation to genes. This noise corresponds to the

fluctuation to the mapping between genotype and phenotype, and our purpose is to see how the evolution of coding progresses in the presence of isologous diversification. To include the selection process, cells are removed randomly, so that the total number of cells is kept within a certain limit. Since cells continue to divide competing for resources, the selection process works as to the division speed of a cell. Here the limit is set to be 150 in this simulation.

First, we study the case when the phenotypes are not differentiated in our model. We choose a reaction network W so that the chemical dynamics fall onto a fixed point. Then, no differentiation in cell types is observed. In this case, even if the mutation to U and V is added, no important change is observed. The values of matrix elements and chemical concentrations are distributed with the variance given from the mutation rate, but no differentiation to different groups of chemical states and matrix elements (genotype parameters) is observed. All the cells keep adopting the same translation code from nu-

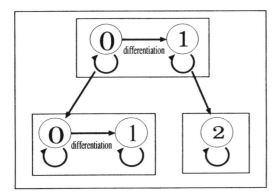

Figure 6: Schematic representation of the differentiation to the types observed in our model.

cleic acid to amino acid and this coding does not change in time.

Since we are interested in the evolution of the code, we do not adopt such network without differentiation, but adopt the case in which the chemical states of a cell are differentiated to allow for different uses of ARS to synthesize amino acids. Here, we choose the matrix W adopted in the previous section, which brings about the differentiation. As will be shown, existence of discrete cell types by isologous diversification is a necessary condition to have a distinct group in the genotype (see also (10; 11)).

Of course, the evolutionary process depends on the mutation rate, which is given by the amplitude of noise added to the matrix elements ε. First, if ε is larger than 10^{-2}, differentiation produced initially is destroyed, and the types are not preserved by cell divisions. With such high mutation rate, the distribution of matrix elements by cells is broader, and both the genotypes and phenotypes are distributed without forming any distinct types. Then, the initial loose coupling between genotype and phenotype remains. No trend in the evolution of codes is observed.

When the mutation rate is lower, the genotypes, i.e., the matrix elements also start to differentiate. Each group with different compositions of metabolites starts to take different matrix element values. An example of the time course of some matrix elements is shown in the upper panel of Fig.8. Two separated groups are formed according to the differentiated chemical states of metabolites given in Fig.5. With the mutation and selection process, the genotype is also differentiated following the phenotypic differentiation. This differentiation, originally brought about by the interaction among cells, is embedded gradually in genotypic functions.

Not all the elements of U,V parameters, but only some of them show this splitting process. In fact, metabolites or enzymes having higher concentrations are often re-

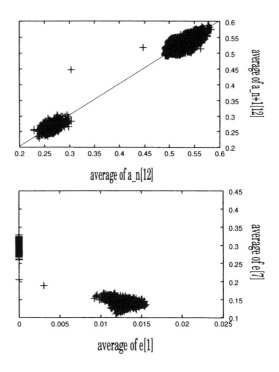

Figure 7: The upper figure is return map of the average concentration of $a^{(12)}$, plotted in the same way as Fig.5, while the lower figure shows the temporal change of $e^{(1)}$ and $e^{(9)}$, plotted at every division event. Each cell keeps recursive production.

sponsible for the differentiation. To estimate the splitting speed in genotype space, we have plotted the distance between the two types in the elements of the matrices U and V. We have measured the following distance between the averaged values of a given matrix element of each type, i.e.,

$$d^{(j,k)} = |\ 1/N_0 \sum_{i \in type0}^{N_0} S_i^{(j,k)} - 1/N_1 \sum_{i \in type1}^{N_1} S_i^{(j,k)}\ |$$

where S represents either V or U, and N_0 and N_1 are the number of type 0 and 1 cells respectively. As shown in the lower panel of Fig.8, the separation progresses linearly in time, although the mutational process is random. In this sense, this separation process is rather fast and deterministic in nature, once the phenotype is differentiated. Furthermore, the slope in the figure is different by chemicals, although the same mutation rate is adopted for all elements. For some of other matrix elements, no separation occurs, of course.

With this mutation process, the difference in chemical states is also amplified as shown in Fig.8. With this evolutionary process, the differentiation starts to be more rigid. In Fig.7, we have plotted the return map of the

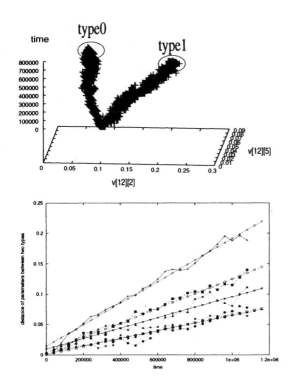

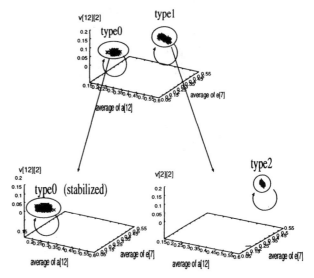

Figure 9: Genotype-phenotype relation after the transplant experiment. The set of values $(a^{(12)}, e^{(7)}, V^{(12,2)})$ is plotted at every division event.

Figure 8: The upper figure gives the temporal change of $V^{(12,2)}$, and $V^{(12,5)}$, namely, the activity of $a^{(12)}$ for the composition $x^{(2)}$, and $x^{(5)}$. The parameter values are plotted at each division event. The lower figure shows the temporal change of the distance between the averages for the matrix elements of each type.

chemical states. Now, the frequency of the differentiation event from type 0 to type 1 is decreased in time. Each type keeps recursive production.

Next, we survey this separation process by "transplant" experiment, to see if each group of cells exists on its own. At the initial stage of evolution, when type 0 cells are extracted, some of them spontaneously differentiate to type 1 cells. Type 0 cells cannot exist by themselves. With the evolution to change the genes, the rate of differentiation to type 1 cells from isolated type 0 cells is reduced. Later at the evolution ($\sim$ 600 generation), transplanted type 0 cells no more differentiate to the type 1, and the type 0 cell exists on its own with stability.

On the other hand, as already mentioned, the type 1 cells, when transplanted, are transformed to the type 2 cells, where different ARS are used in the translation process (see Fig.9). This characteristic feature does not change through the evolution.

With this type of fixation process, the difference in the correspondence between nucleic acid and protein (en-

zyme) is fixed. For matrix U, one of the elements $U^{(i,j)}$ for given j is larger through the evolution. As shown in Fig.10, $U^{(7,0)}$ increases for the type 1 cell, with the decrease of $U^{(7,5)}$, implying that the correspondence between $x^{(0)}$ and the ARS $e^{(5)}$ is stronger. In other words, the loose correspondence between the nucleic acid and amino acid is reduced in time, to establish a rigid relationship between them.

Due to the evolution of matrix elements, the difference in the correspondence between the type 0 and 1 cells gets amplified. Hence, the difference in the correspondence initially brought about as a different metabolic state is now fixed into genes, and each type of cell, even after isolation, keeps a different use of ARS for the translation of the genetic information.

Summary and discussion

In the present paper, we have studied how different correspondences between nucleic acid and enzymes are formed and maintained through the evolution. To discuss this problem we have adopted a model of a cell consisting of
(a) intracellular metabolic network
(b) ambiguity in translation system.
(c) cell-cell interaction through the medium
(d) cell division
(e) mutation to the correspondence between nucleic acid and enzyme

Our scenario for the evolution of genetic code is summarized as follows.

(1) First, due to the intracellular biochemical dynam-

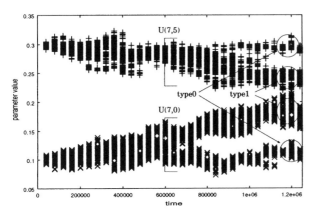

Figure 10: The change of the matrix U through the evolution. The parameter values $U^{(7,0)}$ and $U^{(7,5)}$ are plotted at each division event. First, each type starts to take different $U^{(7,0)}$ values and later $U^{(7,5)}$ values. For example, the type 1 cell starts to use more $x^{(0)}$ accordingly for $e^{(7)}$.

ics with metabolites, enzymes, tRNA, and ARS, distinct types of cells with distinct physiological states are formed (for example, denoted by type 0 and type 1 cells, respectively). Each cell type has different chemical composition and also uses different species of ARS for the protein synthesis. Hence, different types of cells utilize different correspondences between nucleic acid and enzymes. The differentiation at this stage is due to cell-cell interaction. For example, the type 1 cell is differentiated from type 0 cell, and can maintain itself only in the presence of type 0 cells. The difference in the correspondence, however, is not fixed as yet, and by each cell division, each cell can take a different metabolic state, and the correspondence is changeable.

(2) Next, by adding mutation as a small noise to catalytic ability of enzyme by each division of a cell, each distinct cell type starts to be stable, and keeps its type after the division. The difference in chemical states is now fixed to parameters that characterize the catalytic ability. Accordingly, each cell type with distinct metabolic states is fixed also to the catalytic ability of enzymes represented by genes.

(3) After the fixation of distinct types is completed both in phenotype and genotype, these types are maintained even if each type of cells is isolated. Each type uses different ARS for the translation between nucleic and amino acids, and this difference in the usage is amplified through the evolution. At this stage, one can say that different coding, originally introduced as distinct physiological states of cells through cell-cell interaction, is established genetically.

The presented result here is rather general, as long as

the differentiation of cellular states occurs in the general setup of the model with the processes (a)-(e). Although the network we have adopted is randomly chosen, it is expected that the same scenario for the evolution of genetic code is obtained as long as this general setup with (a)-(e) is satisfied.

We note again that the evolutionary process here is based on the standard Darwinian process without any Lamarckian mechanism, although the genetic fixation occurs later from the phenotypic differentiation. It should also be mentioned that our process differs from Baldwin's effect (16) in the point that the interaction-induced phenotypic differentiation from a same genotype is essential to from distinct genotypes later.

The origin of mitochondrial non-universal genetic code

The results in this paper can give new insights to the non-universal genetic code of mitochondria. From recent studies in the molecular biology, it is suggested that mitochondria had used almost the same code as universal one before "symbiosis in cell"(17), and its genetic code was deviated after symbiosis(4).

According to our scenario for the evolution of genetic code, the coding system can depend on cell-cell interaction. The isolated type 1 cell, i.e., the type 2 cell, has a different use of ARS than the cell in coexistence with the type 1 cell. With the interaction, the cells take a different coding system. Furthermore, this difference in the coding is established through the evolution. In this sense, it is a natural course of evolution that mitochondria, which starts to live within a cell and has strong interaction with the host cell, will establish a different coding system through the evolution.

Although the evolution to switch to a different coding might look fatal to an organism, a cell can survive via the loose coupling between genotype and phenotype. The loose coupling produced by the cell-cell interaction is essential to the evolution to non-universal genetic codes.

It should also be stressed that the coding is not necessarily solely determined by the genetic system. In a biochemically plausible model, we have demonstrated that the change in the physiological state of a cell can lead to difference in coding. Based on our theory we believe that this dependence on the physiological state is essential to the study of non-universal coding in mitochondria and others. Furthermore, such possibility of the difference in coding may not be limited to the phenomena at the early stage of evolution. It may be possible to pursue such possibility experimentally, by changing the nature of interaction among cells or intracellular organs keeping genetic information.

We would like to thank T.Yomo, C.Furusawa for stimulating discussions. This work is supported by Grant-in-Aids for Scientific Research from the Ministry of Education, Science and Culture of Japan (11CE2006;Komaba Complex Systems Life Project; and 11837004).

References

1. Crick,F.H.C. (1968) The Origin of the Genetic Code, *J.Mol.Biol.*, 38, 367-379.

2. Smith,J.M., Szathmary,E. (1995) The Major Transitions in Evolution, W.H.Freeman Spektrum.

3. Suzuki,T., Ueda,T., and Watanabe,K. (1997) The 'polysemous' codon -a codon with multiple amino acid assignment caused by dual specificity of tRNA identity, *TheEMBOJournal*,16,1122-1134.

4. Osawa,S. (1995) Evolution of the Genetic Code, Oxford Univ.Press.

5. Eigen,M., Shuster,P. (1979) The Hypercycle: A Principle of Natural Self-Organization,Berlin:Springer-Verlag.

6. Ko,E., Yomo,T., Urabe,I. (1994) Dynamic clustering of bacterial population, *PhysicaD*,75,81-84.

7. Kaneko,K., Yomo,T. (1994) Cell Division, Differentiation, and Dynamic Clustering,*PhysicaD*,75,89-102.

8. Kaneko,K., Yomo,T. (1997) Isologous Diversification: A theory of cell differentiation, *Bull.Math.Bio.*, 59, 139-196.

9. Kaneko,K., Yomo,T. (1999) Isologous Diversification for Robust Development of Cell Society, *J.Theor.Biol.*, 199, 243-256.

10. Kaneko,K., Yomo,T. (1999) Symbiotic Speciation from a Single Genotype, submitted to *Proc.Nat.Acid,Sci.*

11. Kaneko,K., Yomo,T. (2000) Sympatric Speciation from Interaction-induced Phenotype Differentiation, in these proceedings.

12. Furusawa,C., Kaneko,K. (1998) Emergence of Rules in Cell Society: Differentiation, Hierarchy, and Stability, *Bull.Math.Bio.*, 60, 659-687.

13. Furusawa,C., Kaneko,K. (1998) Emergence of Multicellular Organisms with Dynamic Differentiation and Spatial Pattern,*ArtificialLife*4,79-93.

14. Hess,B., Boiteux,A. (1971) Oscillatory Phenomena in Biochemistry, *Ann.Rev.Biochem.*,40,237-258.

15. Tyson,J.J., Novak,B., Ordell,G.M., Chen,K., and Thron,C.D. (1996) Chemical Kinetic Theory: Understanding Cell-cycle Regulation, *Ternd.Bioch.Sci.*, 21, 89-96.

16. Baldwin, J.M. (1902) Development and evolution, New York: Macmillan.

17. Margulis,L. (1981) Symbiosis in Cell Evolution, W.H.Freeman and Company.

Evolution of Differentiation in Multithreaded Digital Organisms

Thomas S. Ray and **Joseph F. Hart**

ATR Human Information Processing Research Laboratories
2-2 Hikaridai, Seika-cho, Soraku-gun, Kyoto 619-02, Japan
ray@hip.atr.co.jp jhart@hip.atr.co.jp
http://www.hip.atr.co.jp/~ray

Abstract

Examination of code execution patterns in different threads of multithreaded digital organisms reveals the clustering of threads into tissues which differ in patterns of code execution. Evolution in a network environment has led to the differentiation of new tissues, which through a division of labor, work together to accomplish the sensory function of a single tissue type of the ancestral organism.

Introduction

(Ray and Hart, 1998) stated:

"The central objective of this project is to study the conditions under which evolution by natural selection leads to an increase in complexity of the replicators. For the purpose of this study, the primary quantitative measure of complexity is the level of differentiation of the multicellular organism. The study begins with the most primitive level of differentiation: two cell types. There are two milestones in the study:

1. The differentiated state persists through prolonged periods of evolution.

2. The number of cell types increases through evolution."

"In the work reported here, only the first of these two milestones has been achieved. There has been no sign of an increase in the number of cell types."

The current study reports the first evidence of an increase in the number of cell types. This evolutionary increase in the level of differentiation of the multicellular digital organisms was detected through the development of new tissue analysis tools.

The digital organisms of this study are self replicating multithreaded machine code organisms living on a network of computers. This system is generally known as Tierra (Ray, T. S. 1991.; Ray, T. S. 1994a.; Ray, T. S. 1994b.; Ray, T. S. 1995.). This study does not address the transition from single threaded to multithreaded organisms as addressed by previous work (Ofria, Adami, Collier and Hughes, 1999.; Ofria, Adami, 1999.), but rather evolution from an ancestral multithreaded state. The intention here is to observe thread differentiation as a function of time rather than environmental complexity, which is addressed by (Ofria, Adami, Collier and Hughes, 1999.).

The digital organisms are able to gather information about other machines on the network, process that information, and use it in making decisions about movements between machines. The digital organisms move between machines in order to improve their access to resources, primarily CPU time, which is their energy resource.

Analogies

Here we are making analogies between some features of digital organisms and organic organisms. The objective of making these analogies is not to create a digital model of organic life, but rather to use organic life as a source of ideas on how to create a richer evolutionary process in the digital medium.

In organic organisms, the "genome" is the complete DNA sequence, of which a copy is found in each "cell". Each cell is a membrane bound compartment, and requires its own copy of the DNA, as the genetic information is not shared across the cell membranes. The entire genome includes many "genes", which are segments of DNA that code for specific functions such as individual proteins.

While each cell contains a complete copy of the genome, each individual cell expresses only a small subset of the genes in the entire genome. The specific subset of genes that are expressed in a cell determine the "cell type". Groups of cells of the same type form a "tissue". Different tissues are composed of cells that have "differentiated" in the sense that they express different subsets of the genes in the genome.

In the digital organisms of this study, the genome consists of the complete sequence of executable machine code of the self replicating computer program. Each thread of a multithreaded process is associated with its own virtual CPU. These threads (CPUs) are considered analogous to the cells. However, the threads of a process all share a single copy of the genome, because they operate in a shared memory environment where the genetic information can easily be shared between CPUs. Duplication of the genome for each thread would be redundant, wasteful and unnecessary. In this detail, our digital system differs quite significantly from the organic system.

The genome of the digital organism includes several segments of machine code with identifiable functions, which are coherent algorithms or subroutines of the overall program represented by the entire genome. These individual algorithms can be considered analogous to the genes. Each thread (CPU) has access to the entire genome, yet each thread will execute only a subset of the complete set of genes in the genome. The specific subset of genes executed by a single thread determine its cell type. Groups of threads of the same cell type form a tissue. Different tissues are composed of threads that are differentiated in the sense that they execute different subsets of the algorithms (genes) in the genome.

Another difference between the organic and digital systems is that in the digital system there is no spatial or geometric relationship between cells. However, there exist logical relationships between threads. For example, the odd threads of a tissue may behave differently from even threads.

Developmental Behavior of the Network Ancestor

The work reported here is focused on the evolution of the differentiated multicellular condition. The multithreaded digital organisms live in a networked environment where spatial and temporal heterogeneity of computational resources (most importantly CPU time) provides selective pressure to maintain a sensory system that can obtain data on conditions on various machines on the network, process the data, and make decisions about where to move within the network.

The experiment begins with a multithreaded ancestral seed program (0960aad) that is already differentiated into three cell types: a sensory tissue, a reproductive tissue, and a copy tissue. The entire seed program includes about 320 bytes of executable machine code. However, no single thread executes all of this code, just as no cell in the human body expresses all of the genes in the human genome.

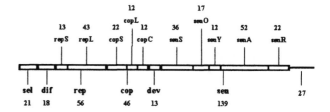

Figure 1: Lower labels indicate the six major genes and their sizes in bytes. Upper labels indicate subdivisions of the major genes, and their sizes.

The network ancestor genome has been somewhat arbitrarily labeled as composed of six genes, some of which have been further subdivided (Figure 1). Two of the genes are executed only during the development from the unicellular to the mature ten celled form (**sel**, **dif**). One gene is executed only by the reproductive tissue (**rep**), and one gene is executed only by the sensory tissue (**sen**). Two genes are executed by the sensory, reproductive, and copy tissues (**cop**, **dev**).

Figure 2 illustrates early development in the ancestor. The ancestor is born as a single undifferentiated thread which begins by executing the **sel** gene to perform the self examination. Execution then flows into the **dif** gene which causes the embryonic cell to split into two cells, which then differentiate into the sensory and reproductive tissues.

The mechanism of differentiation is a conditional jump. When a thread splits into two threads, the value of the dx register is altered such that the two daughter threads have different values (The value in the dx register is copied from the mother thread to the daughter thread, then both values are shifted left. The value in the daughter thread is then incremented by one). By making conditional jumps based on tests of the values of the dx registers, one thread begins executing the **sen** gene, while the other thread begins executing the **rep** gene.

There exists a mechanism of gene promotion, in which genes promote the expression of other genes, through function calls. In a function call, execution jumps from the location of a call, to the code of the called function (another gene), and when the function is complete, execution returns to the instruction following the location of the call. The **rep** and **sen** genes promote the expression of the **cop** gene, which in turn prompts the expression of the **dev** gene. The **dev** gene is also directly promoted by the **sen** gene (Figure 3).

The function of the **dev** gene is to split a single thread into a tissue of 2^n threads, without differentia-

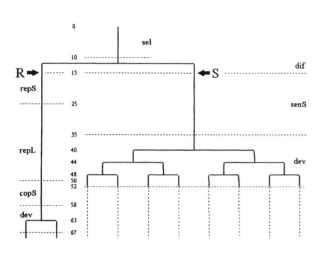

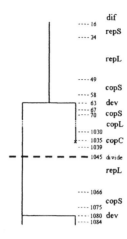

Figure 2: Early development and gene expression in the ancestor, 0960aad. The column of numbers indicates time, in parallel instructions executed by the organism. Horizontal dashed lines show the points of transition between expression of different genes in the threads crossed by the dashed lines. The labels (i.e., **sel**, **dif**, **senS**, etc.) indicate the gene expressed by a thread in a specific time interval. R arrow and S arrow show the point of differentiation into the sensory and reproductive threads.

Figure 4: The single reproductive thread calls the **cop** gene to effect the replication of the genome with two threads, after which the daughter is born and sent to another machine on the network. The other machine is chosen based on data processed by the sensory system. The reproductive process repeats until the death of the organism.

tion. The value of n is determined by data in a CPU register.

The **cop** gene performs a string copy function, copying a block of memory from one location to another. This is the means by which the genetic information is replicated from mother to daughter. It is also the means by which sensory data is copied during sensory processing. The **cop** gene parallelizes the string copy function, by calling the **dev** gene and then dividing the data among the 2^n threads. After the data has been copied, all but the original thread halt.

The **cop** gene has three subgenes. The **copS** subgene (copy tissue setup), sets up register values and calls the **dev** gene to parallelize the function. The **copL** subgene (copy loop) copies the data. The **copC** subgene (copy tissue cleanup) halts all but the original copy thread and restores register values before returning from the copy function.

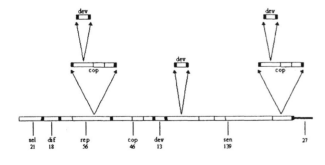

Figure 3: Genes can promote the expression of other genes by making function calls.

The reproductive thread expresses the reproductive gene, **rep**, which contains two subgenes (Figure 4). The **repS** subgene (reproduction setup) prepares some register values. The **repL** subgene (reproduction loop), is an infinite loop that calls the **cop** gene, spawns a daughter, and repeats the reproduction process. The **cop** gene parallelizes the reproductive tissue into two threads, each of which copies half of the genome.

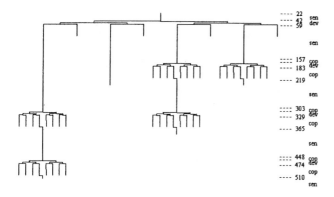

Figure 5: Sensory processing in the ancestor (0960aad).

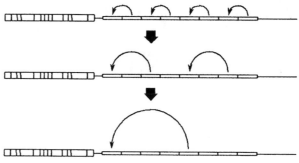

Figure 6: Sensory processing in the ancestor, 0960aad.

In each reproductive cycle after the genome has been copied, the **cop** gene returns control to the **rep** gene, and the reproductive thread executes the **divide** instruction, which spawns the daughter as an independent process. Before giving birth to the daughter, the reproductive thread reads data that has been processed by the sensory system. By this time, the sensory system has examined data from fifteen machines on the network, and placed the IP address of the "best" looking machine at a certain location. At birth the reproductive system sends the daughter to the machine whose IP address is stored at that location. Meanwhile the sensory thread expresses the **sen** gene, beginning with **senS**, which calls the **dev** gene causing the sensory thread to divide into a tissue of eight sensory threads. The **dev** gene then returns control to **sen** and the eight sensory threads begin the complex process of gathering and processing sensory data (Figure 5).

Adjacent to the 320 byte genome is a 512 byte data area used by the sensory system. Each of the eight sensory threads reads a 64 byte TPing data structure into a block of the data area. This initial gathering of sensory information is followed by a series of three pairwise comparisons, which result in the "best" looking data being placed in the leftmost block. It is from this location that the reproductive tissue will find the IP address of the machine to which the daughter will be sent (Figure 6).

After the eight parallel sensory threads have each gathered their TPing data, half of the threads (the odd threads) halt, leaving the four remaining threads to perform the first round of four pairwise comparisons. If the data on the right is "better" than the data on the left, the sensory thread calls the **cop** gene to copy the sensory data from the right block to the left block. The **cop** gene calls the **dev** gene to parallelize the copy

process into eight threads. After the sensory data has been copied, seven of the eight copy thread halt before returning control to the **sen** gene.

Two of the four sensory threads then halt, and the remaining two threads perform the second round of two pairwise comparisons. One of the two remaining sensory threads then halts, and there is a third and final comparison leaving the best data in the leftmost position. Note that the **sen** gene only calls the **cop** gene (and thus **dev** gene) if the comparison indicates that the data needs to be copied from right to left. For this reason the pattern of gene expression in the sensory tissue is dependent on the data gathered from the environment.

The **sen** gene is divided into five subgenes. The **senS** subgene (sensory tissue) setup, sets up some register values, calls the **dev** gene to split the sensory thread into a tissue of eight threads, and collects the TPing data. The **senO** subgene (sensory processing coordination), helps the different sensory threads to coordinate the division of the data among themselves, and recognizes the completion of the sensory cycle. The **senY** subgene (sensory system synchronization) causes the odd threads to halt after gathering the TPing data, and forces synchronization of the threads (threads may lose their synchronization because some threads must copy TPing data while others may not). The **senA** subgene (sensory data analysis) compares two sets of TPing data, and decides if it is necessary to copy the data from left to right. The **senR** subgene (sensory data report) calls the **cop** gene if necessary to copy the TPing data from left to right.

Tissue Analysis Methods

The rationale for the current experiment was originally presented by Ray (Ray, T. S. 1995.). Technical details of the implementation have been reported in Charrel (Charrel, Agnes, 1995.) and Ray (Ray, T. S.

1997.; Ray, T. S. 1998.). Further details are available on the web at:

http://www.hip.atr.co.jp/~ray/tierra/
netreport/netreport.html.

Thus only those experimental methods new to the current report will be presented here. These new methods are those involved in the analysis of tissue types.

If new tissues do in fact evolve in Tierra, they may differentiate gradually, starting with only slight differences which then gradually increase through evolution. To help us get a clear and realistic picture of this process, we must classify threads into tissues through a careful multistep step process. This process will be described through the example of the analysis of the ancestral organism (0960aad).

In Tierra, a tissue can be strictly and objectively defined as the set of threads which execute exactly the same code. Any two threads which execute a different set of Tierran machine code instructions will be classified into different tissues. We will refer to tissues defined in this manner as "strict tissues".

However, this strict definition is only a starting point for the grouping of threads into tissues. There may exist slight difference in the code executed by two threads that will be of little significance. For example, when a single thread divides into a multi-thread tissue, the last threads to be formed generally execute slightly fewer instructions than the threads which formed earlier. We can recognize groups of trivially different tissues by doing a cluster analysis on the code expression patterns of the strict tissues. Strict tissues that are only trivially different should be grouped together as a single tissue type.

Even two threads which execute exactly the same algorithm may execute different code due to conditional branches or halts, based on variable data gathered from the environment. Due to such conditional behavior, the same thread could execute different code at different times, depending on the data that it gathers from the environment. We will refer to differences in code execution due to conditional behavior as "behavioral differences". Strict tissues which are behaviorally different can be recognized only through a careful examination of their algorithms. Strict tissues that are behaviorally different should be grouped together as a single tissue type.

The classification of threads into strict tissues is an automated process, based on data gathered during a finite period of observation. Termination of the period of observation generally causes some threads to terminate before fully completing their algorithm. These incomplete threads are spuriously classified as distinct tissues. Tissues whose distinction is due to such "observational differences" should be grouped with tissues which have completed the same algorithm.

When two or more strict tissues are grouped together as a single tissue because they are trivially, behaviorally, or observationally different, we call the grouping a "narrow tissue". The common theme in grouping strict tissues into narrow tissues is to recognize the unity of threads which execute the same algorithm.

The definition of narrow tissues on the basis of algorithms can also lead to the need to split a single strict tissue into two narrow tissues. This situation arises uniquely in the case of undifferentiated embryonic threads which later differentiate. These threads first execute the algorithm for embryonic development, and after differentiation execute the algorithm of a mature tissue, such as the reproductive tissue.

Narrow tissues can be further grouped into "broad tissues", based on their functional actions. The broad tissues recognized in this study are embryonic, copy, reproductive, and sensory. This study will focus on the evolution of new narrow tissue types in the sensory broad tissue.

The first step in the tissue analysis process is to group threads into strict tissues on the basis of their patterns of code execution, a process that can be completely automated by recording the code executed by each thread of sample organisms living in a network environment. This leads to the groupings shown in Table 1.

The execution patterns of the ten strict tissues are then compared in a similarity matrix (Table 2) so that we can perform a cluster analysis.

Because the amount of code executed by two tissues may be different, the similarity may not be symmetric. For example, tissues 0 and 6 show a nearly symmetric relationship, with a 5% and 6% similarity. Tissues 7 and 9 on the other hand are highly asymmetric, with 100% and 12% similarities (meaning that tissue 7 executes all the code of tissue 9, while tissue 9 executes only 12% of the code of tissue 7).

The similarity matrix is then subjected to a cluster analysis using the PC ORD 4 software from MjM Software (http://www.ptinet.net/~mjm/pcordwin.htm). For the ancestral organism, 0960aad, it produced the cluster diagram shown in Figure 7.

A glance at the cluster diagram suggests some natural groupings, however, these must be considered in the light of a careful analysis of the functions of the

strict	# threads	sel	dif	repS	repL	copS	copL	copC	dev	senS	senO	senY	senA	senR
0	1	sel	dif	repS	repL	copS	copL	copC	dev					
1	1		dif			copS	copL	copC	dev	senS	senO	senY	senA	senR
2	17					copS	copL	copC	dev	senS	senO	senY	senA	senR
7	3					copS	copL		dev	senS	senO	senY	senA	senR
3	28								dev	senS	senO	senY		
6	1								dev	senS	senO	senY	senA	
4	155					copS	copL	copC	dev					
5	114					copS	copL	copC	dev					
8	9					copS	copL		dev					
9	12					copS	copL		dev					

Table 1: A listing of the ten strict tissues identified in the ancestor, 0960aad, the number of threads found in each tissue, and the genes expressed in each tissue.

	T0	T1	T2	T3	T4	T5	T6	T7	T8	T9
T0	1.00	0.43	0.39	0.04	0.21	0.22	0.05	0.29	0.17	0.16
T1	0.24	1.00	0.81	0.20	0.12	0.12	0.52	0.74	0.10	0.09
T2	0.27	0.99	1.00	0.26	0.14	0.15	0.64	0.91	0.12	0.11
T3	0.11	0.94	1.00	1.00	0.11	0.11	1.00	0.97	0.11	0.11
T4	0.95	0.95	0.95	0.19	1.00	1.00	0.19	0.71	0.71	0.71
T5	0.95	0.95	0.95	0.18	0.95	1.00	0.23	0.73	0.73	0.68
T6	0.06	0.98	0.99	0.40	0.04	0.06	1.00	0.98	0.06	0.04
T7	0.21	0.99	1.00	0.28	0.12	0.13	0.70	1.00	0.13	0.12
T8	1.00	1.00	1.00	0.25	0.94	1.00	0.31	1.00	1.00	0.94
T9	1.00	1.00	1.00	0.27	1.00	1.00	0.27	1.00	1.00	1.00

Table 2: Strict tissue similarity matrix for the ancestor 0960aad. Table entries are computed by dividing the number of instructions executed in common by the two tissues by the total number of instructions executed by the tissue of the row.

Broad	Narrow	Strict	Function
Embr/Repr	A	0	embry/reprod
Copy	B	4, 5, 8, 9	copy
Sensory	C	1, 2, 3, 6, 7	full sensory

Table 3: Grouping of strict tissues into narrow tissues.

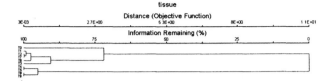

Figure 7: cluster diagram - 0960aad

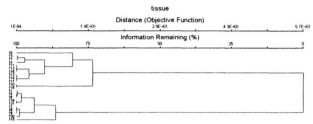

Figure 8: cluster diagram - 3414aaj

different strict tissues, and the role of trivial, behavioral, and observational differences. These considerations lead to the groupings given in Table 3.

Narrow tissue A consists of the original embryonic thread, which eventually differentiates into a reproductive thread. This one thread strict tissue is unique by virtue of being the only thread to execute the embryonic code (**sel**, **dif**). However, it also expresses the **rep** gene and the **cop** gene.

Narrow tissue B expresses the copy gene (**cop**), which is promoted by the reproductive code to copy the genome, and by the sensory code to copy TPing sensory data. The threads of tissue B are created by the **dev** gene for the parallelization of the copy function, and halt when the data has been copied, thus they express only the **dev** and **cop** genes.

Narrow tissue C expresses the sensory gene, **sen**. The algorithm of the sensory gene includes a data reduction loop in which the odd threads halt, the even threads compare the data on the left and right, and depending on the results of the comparison, may or may not promote the copy gene to copy the TPing data.

Within this sensory gene we find two kinds of behavioral difference: tissue 6 does not copy the data while tissues 1, 2, and 7 do copy the data; tissue 3 differs from tissues 1, 2, 6, 7 in that it halts early in the first sensory cycle. We also find observational differences: tissue 7 does not finish the copy loop during the observation period while tissues 1 and 2 do finish. We also find trivial differences: tissue 1 is the original sensory thread, originating in the **diff** gene, and thus differs slightly from the rest of the sensory tissues in that it executes a small part of the **diff** gene before entering the sensory code.

Developmental Behavior of an Evolved Organism

The tissue analysis procedure was applied to an organism captured after five days of evolution (3414aaj). An examination of homology between the genome of **3414aaj** and the genome of the ancestor revealed the existence of all thirteen ancestral genes in the evolved genome. These thirteen genes accounted for the entire evolved genome; there were no new genes added or lost.

Classification of the threads of 3414aaj into strict tissues produced the pattern shown in Tables 5. The cluster analysis (Figure 8) together with consideration of trivial, behavioral, and observational differences leads to the classification into narrow and broad tissues shown in Table 4.

The grouping of strict tissues into narrow tissues in 3414aaj follows the same procedures detailed for the ancestor in the methods section above. However, in the end we are left with two narrow sensory tissues which differ both in their code execution pattern and their function.

Narrow tissue D loads the TPing data into the sensory data buffers using the **senO** gene. Narrow tissue E does not execute **senO**, but does perform the comparison of the left and right data using the **senA** gene, and uses the **senR** gene to call the **cop** gene if it is necessary to copy the data.

After loading the TPing data buffers, the threads of narrow tissue D go on to execute the **senA**, **senR**, and possibly the **cop** gene. The comparisons and copies made by tissue D have no effect, however, because the same data is later operated on by the threads of tissue E.

It should be noted that tissue E is a sub-set of tissue D, in the sense that tissue D executes all the code of tissue E, while tissue E executes only a portion of the code of tissue D. Where the execution of code overlaps between the two, in the genes **senA**, **senR**, **copS**, and **copL**, the code has no effect in tissue D. The functional division of labor between the two tissues is thuse absolute, even where there is overlap of code execution.

Broad	Narrow	Strict	Function
Embr/Repr	A	0	embry/reprod
Reproductive	B	3,4,7,12,14,17	reprod
Copy	C	10,11,13,16,18	copy
Sensory	D	1,2,5,6,8,9	sensory load
Sensory	E	19,20,21	compare, cpy
Vestigial	F	15	vestigial

Table 4: Grouping of strict tissues into narrow and broad tissues in a five day evolved organism, 3414aaj.

strict	#T	sel	dif	repS	repL	copS	copL	copC	dev	senS	senO	senY	senA	senR
0	1	sel	dif	repS	repL	copS	copL		dev					
3	1				repL	copS	copL	copC	dev					
4	1				repL	copS	copL		dev					
7	1				repL	copS	copL	copC	dev					
12	2				repL	copS	copL	copC	dev					
14	5				repL	copS	copL	copC	dev					
17	8				repL	copS	copL	copC	dev					
11	2					copS	copL	copC	dev					
13	5					copS	copL	copC	dev					
16	8					copS	copL	copC	dev					
10	38					copS	copL		dev					
18	48					copS	copL		dev					
1	1		dif			*copS*	*copL*		dev	senS	senO	*senY*	senA	*senR*
2	1					copS	copL		dev	senS	senO	*senY*	senA	senR
5	2					copS	copL		dev	senS	senO	*senY*	senA	senR
6	1					*copS*	*CopL*		dev	senS	senO	*senY*	senA	*senR*
8	2								dev	senS	senO	*senY*	senA	
9	1								*dev*	*senS*	*senO*	*senY*	*senA*	
19	3					copS	copL		dev				senA	senR
20	4					copS	copL		dev				senA	senR
21	1					*copS*	*copL*		*dev*				*senA*	*senR*
15	8										*senO*	*senY*		

Table 5: A listing of the twenty-two strict tissues identified in the evolved organism, 3414aaj, the number of threads found in each tissue, and the genes expressed in each tissue. Genes named in italics are expressed in the tissue, but have no effect (generally because another thread overwrites the same data area).

Discussion

The ancestor (0960aad) has a single sensory tissue type which performs all of the sensory functions. In the organism resulting from five days of evolution, 3414aaj, these sensory functions have been divided into two separate tissue types, with a functional division of labor. The threads of tissue D load the TPing data into buffers, and the threads of tissue E compare the data and copy from left to right if necessary.

While the functional division of labor is complete and clear cut, the difference in code execution patterns is not as clean. Tissue D also executes the code executed by tissue E, but in tissue D that common code has no effect.

The data presented here reflect only our first observations of the evolution of higher levels of differentiation in multithreaded digital organisms. We expect that further observations are likely to reveal additional examples, and possibly examples of yet greater and more clear cut differentiation.

References

Charrel, Agnes. 1995. Tierra network version. ATR Technical Report TR-H-145.
http://www.hip.atr.co.jp/~ray/pubs/charrel/charrel.pdf

C. Ofria, C. Adami, T.C. Collier, and G.K. Hsu. 1999 Evolution of Differentiated Expression Patterns in Digital Organisms, Lect. Notes Artif. Intell. 1674, 129-138 Proceedings ECAL'99
http://www.krl.caltech.edu/avida

C. Ofria, C. Adami. 1999 Evolution of Genetic Organization in Digital Organisms Proc. of DIMACS workshop on Evolution as Computation, Jan 11-12, Princeton University, L. Landweber and E. Winfree, eds. Springer Verlag, p. 167
http://www.krl.caltech.edu/avida

Ray, T. S. 1991. An approach to the synthesis of life. In: Langton, C., C. Taylor, J. D. Farmer, and S. Rasmussen eds., *Artificial Life II, Santa Fe Institute Studies in the Sciences of Complexity*, vol. XI, 371-408. Redwood City, CA: Addison-Wesley.
http://www.hip.atr.co.jp/~ray/pubs/tierra/tierrahtml.html

Ray, T. S. 1994a. Evolution, complexity, entropy, and artificial reality. *Physica D* 75: 239-263.
http://www.hip.atr.co.jp/~ray/pubs/oji/ojihtml.html

Ray, T. S. 1994b. An evolutionary approach to synthetic biology: Zen and the art of creating life. *Artificial Life* 1(1/2): 195-226.
http://www.hip.atr.co.jp/~ray/pubs/zen/zenhtml.html

Ray, T. S. 1995. A proposal to create a network-wide biodiversity reserve for digital organisms. ATR Technical Report TR-H-133.
http://www.hip.atr.co.jp/~ray/pubs/reserves/reserves.html

Ray, T. S. 1997. Selecting Naturally for Differentiation. In: Koza, John R., Kalyanmoy Deb, Marco Dorigo, David B. Fogel, Max Garzon, Hitoshi Iba, and Rick L. Riolo eds. *Genetic Programming 1997*: Proceedings of the Second Annual Conference, July 13–16, 1997, Stanford University, 414-419. San Francisco, CA: Morgan Kaufmann.
http://www.hip.atr.co.jp/~ray/pubs/gp97/gp97.html

Ray, T. S. 1998. Selecting Naturally for Differentiation: preliminary evolutionary results. *Complexity*, 3(5): 25-33. John Wiley and Sons, Inc.

Ray, T. S. and Joseph Hart. 1998. Evolution of Differentiated Multithreaded Digital Organisms. In: *Artificial Life VI* proceedings, C. Adami, R. K. Belew, H. Kitano, and C. E. Taylor eds., 295-304. The MIT Press, Cambridge.

Design by Morphogenesis

Cefn Hoile[1,2] and Richard Tateson[2]

[1]COGS, University of Sussex, Brighton BN1 9QH, UK.   cefn@newscientist.net

[2]BT Labs, Martlesham Heath, Ipswich IP5 3RE, UK.   richard.tateson@bt.com

Abstract

We present a simulation of morphogenesis and cell interaction which allows us to address embryogenesis as an engineering problem. A space of cell control functions is defined using a cybernetics approach by identifying the controlling variables and the degrees of freedom of each individual cell. These functions receive inputs which are dependent on a cell's context, and determine its responses to that context. Multiple cells are coupled through a simulation of a 3d reaction-diffusion fluid matrix in order to generate interactive behaviour. A fitness function is designed which characterises the desired behaviour of the cellular collective. We tune the parameterised control function for the individual cells using a genetic algorithm to maximise the fitness of their collective behaviour.

Introduction

Over the last decade of artificial life research there have been several impressive approaches to simulating morphogenesis (see below). Morphogenesis, literally "creation of shape", is the process by which a fertilised egg changes into the form of the adult organism. It involves cell division and movement. Of course, to achieve the correct pattern of cell movements the cells must change their behaviour. For example genes will be turned on, new proteins will be expressed, perhaps resulting in two neighbouring cells adhering. The understanding of these underlying molecular processes is the realm of developmental biology.

Some of the previous simulations of morphogenesis include simulation of the molecular processes. In the work reported here we aim to show morphogenesis at the cellular level and are not concerned with a biologically plausible set of developmental mechanisms within the simulated cells.

Looking forward to artificial life in the coming decade, we see the simulation of morphogenesis as an approach to system design. As artificial life techniques are increasingly applied to real industrial problems, the issue of how to "engineer" emergent behaviour will become ever more pressing.

In the next section we review earlier work relevant to simulation of morphogenesis. We then explain our motivation for work in this area and give details of our simulation and results to date. Finally we look ahead to future work.

Related Research

Several simulation approaches have been employed to study different classes of cellular behaviour.

Furusawa and Kaneko simulate cell-differentiation using catalytic networks of different chemical species to describe the internal dynamics of cells in a shared fluid (Furusawa and Kaneko 1997). They identify cell-type attractors, and characterise their *potency*. The combinations of cells which can coexist in stability within the same matrix was also found to have attractors. Their catalysis matrix is the basis of the reaction matrix in Design by Morphogenesis (DBM), although the binary values have been replaced by continuous values.

Peter Eggenberger (Eggenberger 1997) uses a model of gene expression to generate symmetrical 3D morphologies from a single seed cell. The interaction of cell products, regulatory molecules and genetic material is prescribed by an affinity function. A genetic algorithm is used to tune the affinities of the genetic material. Triangulation of position is made possible by three sources of diffusible morphogens which are placed in the environment. Forms with bilateral symmetry emerged from the epigenetic interactions of the cells and these diffusible morphogens.

Nicholas Savill and Paulien Hogeweg's (Savill and Hogeweg 1997) models are based on earlier work (Glazier and Graner 1993). Their techniques focus on the energetics of membrane surfaces, and may be extended to explore the generation of tissue types with different mechanical properties, physical behaviour, and complex membrane topologies, such as those found in neural tissues. Their classic work is the simulation of slug formation, cell sorting and slug motility in the slime mould *Dictyostelium*.

Geometrical approaches to the simulation of cell membranes in 2D have been explored (Weliky and Oster 1990). However, these techniques would be challenging to scale to 3D.

Detailed biological models are reported in a manifesto for the computer simulation of cell biology (Kitano et al 1997). The focus of their Virtual Cell Laboratories project is to replicate the known dynamics of individual cells at the biochemical level, as a potential replacement for cell culture. Kitano does not envisage the simulation of multi-cellular tissues in this simulator.

Motivation for Design by Morphogenesis

Multi-cellular organisms have a number of features which are desirable in artificial systems. They can adapt to changing circumstances, through learning or evolution. "Faults" arising from internal errors or injury can often be rectified. Irredeemable faults are seldom the cause of catastrophic failure because the redundancy of the system allows most or all functions to continue.

This contrasts markedly with most current artificial systems. Here the human designer takes on the burden of solving all these problems. The designer must gather a large amount of information about the operating conditions and specification of the system to be designed. Having described the problem, the designer must provide an explicit, practical and comprehensible solution.

This design process usually results in a system which only operates under previously anticipated conditions, and which employs a centralised, hierarchical, non-redundant control system. Such control systems are applauded as transparent and efficient. However, many solutions are unavailable to human designers not because they are poor solutions in terms of performance in the system of interest but merely because they are not comprehensible by humans.

As systems become more complex, by virtue of increasing size, interconnections and dynamism, the shortcomings of rational human design will be exposed. This is already happening in the computation and communication fields. For example there have been failures of large software control systems for air (Lyu 1996) and space travel (Ward and Seligsohn 1990), and of highly complex terrestrial data networks (Andrews 1991).

These facts have led to a great increase of interest over recent years in the potential for non-rational design through evolutionary algorithms (Thompson and Layzell 2000). The DBM work was motivated by two complementary desires. Firstly to simulate the process of morphogenesis. Secondly to exploit morphogenesis as part of a design process for artificial systems. The importance of the genotype to phenotype mapping is well recognised by researchers who use evolutionary algorithms (Shipman et al 2000). We aim to mimic nature's "mapping" mechanism, and employ its principles in human design problems.

Multi-cellular Dynamics and 'DBM' Dynamics

The dynamics of an isolated cell, or an isolated point in a reaction-diffusion soup can be treated as a closed phase space, in which the rate of change of the system variables is determined entirely by their current values. However, in a multi-cellular organism, these are not independent systems. Each infinitesimal point in a fluid is affected by its neighboring points. Each cell embedded in the fluid is affected by the chemistry taking place around it.

Separating the multi-cellular organism into subsystems with variables and parameters is somewhat artificial given these interdependencies, but the definition of fundamental components is demanded by the formalism of computer simulation. In a discrete time simulation, we can then prescribe how the phase spaces of subsystems intersect, setting the rate of change of variables at intersections to the sum of the contributing vectors of the intersecting spaces.

DBM navigates the multi-cellular repertoire by exploring the range of possible vector fields for each of the coupled sub-units in the simulation. Each sub-system's trajectory through its phase space can be contributed to by other sub-systems, according to their coupling. In each case a parameterised function is employed which can describe a wide range of possible vector fields. The parameters of these functions represent the genome of the individual.

A discrete-time simulation of these coupled units offers a good approximation of the global dynamic qualities of a continuous system. The repeated assessment of interdependent variables over small time segments closely parallels the analytic treatment over infinitely small intervals of a deterministic, fixed-dimensionality, dynamical equation by calculus.

Simulation

The spherical cells are simulated as volumes of fluid with a semi-permeable boundary, embedded at a point in a fluid space.

Genomic response is captured by a feed-forward sigmoidal neural network - theoretically capable of representing any input to output mapping. This has the advantage of representing continuous response functions which cannot be represented by discrete gene-switching models such as boolean networks.

To simulate the chemical medium we employ a 3D cellular automata-style reaction-diffusion model with a fixed set of chemical species.

The parameters specified by the candidate genome are:
- Weights of neural network employed in control of individual cells.
- Reactivity and diffusivity of each chemical species in the system.

Each of these values has an upper and lower bound. The remaining parameters are hard coded to suitable values.

The implementation is broken up into subsystems:

- **Cell Control Mechanism**
  Each cell takes the concentrations of chemicals in its cytoplasm, its membrane, the membranes of cells which it contacts, and its local fluid environment. These values constitute the *input vector* for the neural network whose weights are specified by the genome.
  The resulting *output vector* dictates the cell's behaviour which includes:
  - rate of emission of chemicals into its cytoplasm and membrane
  - rate of active transport of chemicals from its cytoplasm into the local fluid environment
  - change of membrane permeability for each chemical

- chemotaxis (movement in the direction of maximum concentration gradient for each chemical)
- cell division
- cell death

- **Cell Update Mechanism**
  The output states from the Cell Control Mechanism, the positions of each cell, and the concentrations and gradients of each chemical in the local fluid environment form the *input vector* for this subsystem.
  The *output vector* specifies the actual exchanges between the cell and the reservoirs with which it interacts, and the actual movement of the cell within the simulation space.

- **Grid Cube Update Mechanism**
  (Elements of reaction-diffusion CA)
  The *input vector* is the concentration of each chemical within the subcube and within its facing neighbours.
  The *output vector* specifies the chemical exchange with each of its facing subcubes and the rate of transformation from each chemical to another within this subcube.

Simulation Runs

Each simulation run begins with a single cell placed at the centre of a nil concentration 3D space, analogous to the initial state of the zygote. The cell and its daughters then alter and respond to their chemical environment for a fixed period of simulation time. If the number of cells reduces to zero or increases above a prescribed threshold, the simulation is called to an early halt, and the candidate receives a penalty fitness. Otherwise, the candidate is given a fitness which captures its degree of competence at producing a desired 3d form.

Candidates and Evaluation

In our approach, a genome of floating point numbers represents the candidate for assessment. These numbers are used as parameters for the control functions of the coupled subsystems. The dynamics of these coupled systems result, after a number of timesteps, in a multi-cellular form.

The points occupied by cells are read into a matrix. Principal components analysis is undertaken to establish the axes of maximal variance. The original matrix is then re-oriented onto the axes of the principal vectors.

This re-orientation is necessary because cell divisions take place according to a random vector. It is not predetermined in which direction the original spherical symmetry and its descendants will break. Using the arbitrary x, y and z axes of the simulation space would penalise most distributions even if they had the perfect shape, since they would have the wrong orientation. Transforming the axes according to the distribution itself eliminates this problem.

The initial population comprises thirty individuals. In each generation, thirty new individuals are generated by crossover, employing a probabilistically biased rank to select the parents. All sixty individuals are then trialed, and rank selection used to select the thirty individuals which are carried forward to the next generation.

In each simulation run, we employ a fitness function which selects those candidates whose re-oriented distributions best approximate to a target form. Candidates are free to use whatever "affordances" are offered by the simulation dynamics to maximise their fitness.

The strategies they adopt to achieve target structures could offer insights into multi-cellular dynamics in a real fluid environments. They could also indicate design principles for robust self-organising artificial systems.

Results

A two segment target structure was used to test the repertoire of the DBM system. Segmentation is a form achieved very early in the morphogenesis of natural organisms. It is revealed by the organisation of the early embryo into alternating stripes (with the stripes distinguished either by morphology or by some less visible differentiator such as gene expression state). Hence these trials were intended to explore early symmetry breaking and structural formation of an artificial "embryo".

To mimic this process of natural segmentation it was decided to generate segments with differential densities of cells. The ideal candidate would have a high density of cells in specified sections and low densities of cells in between. In a real organism this would correspond to sections of cells of a certain differentiated type, separated by cells of another differentiated type.

Since the transformed co-ordinates always have their greatest variance in the z-axis, this was used to represent a longitudinal axis, along which the segments should form. Individual cells were given credit in proportion to the value of a negative cosine function centred on the mean containing the same number of peaks as the desired number of segments. The wavelength of the function for two segments was, therefore, twice the standard deviation in the z axis. For any size or orientation of distribution, this function will contain the appropriate number of peaks in the longitudinal axis. If the z co-ordinate of an individual cell fell somewhere on the peak of this function, the candidate was credited. If it fell in a trough, it was penalised.

During this run, a penalty score of -20 was assigned to candidates producing more than 40 cells, or whose cells all committed cell suicide. This corresponds to a distribution with the average permissible number, (twenty cells), whose cells are all maximally penalised for incorrect positioning.

The function was used to encourage the development of two segment distributions. Since the distance between any two cells is exactly twice the standard deviation from the mean, both cells are located at the peak of the negative cosine wave. This provides a guaranteed score of 2.0. However, producing such a distribution is not a trivial matter, and it is a simple demonstration that DBM cells can regulate their own activity. Cells during this run were able to divide to produce up to 64 cells within the number of timesteps allotted. At some point during the simulation, the input to the cell control system must have generated an output to trigger division in order to generate two cells. However, further division must have been inhibited by some change which the cells initiated to themselves or their environment, leading to a change in their input vector.

Later solutions in this run were similar, although they contained a larger number of cells in each cluster. The initial division is followed by a migration by the cells away from each other. The daughter cells of these two initial cells, however, did not migrate, maintaining their proximity to the peaks of the negative cosine wave.

The initial migration increases the variance in the z axis of the transformed coordinates. Hence a larger number of cells can share the peaks of the distribution. Below, this strategy is shown for full-term cell distributions of 2, 4, 6, 10, 11, 25 and 37 cells from generation 450.

2 and 4 cell distributions Fitness 2.0 and 3.917

6 and 10 Cell Distributions Fitness 5.502 and 6.382

11 and 25 cell Distributions Fitness 6.211 and 11.233

37 cell Distribution Fitness 12.482

Unfortunately, as the number of cells increases, it seems to become harder to maintain the separation of the two clusters and their symmetry. Although originally a candidate may score very highly, the next evaluation introduces an element of noise for which the cell collective is not able to compensate.

The 19 cell distribution below scored only 1.320, through the apparent merging of the two cell clusters.

19 and 25 cell distributions Fitness 1.320 and –4.883

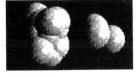

The two clusters of the 25 cell distribution above are asymmetrical. This causes the standard deviation to be significantly reduced, as the mean tends towards the largest cluster. As a result of the repositioned mean the fitness function penalises many of the cells.

Although the genetic algorithm is able to drive the distributions towards large numbers of cells in each segment, the instability that this introduces causes more candidates to achieve negative scores. Where the maximum fitness is driven up, the mean is often driven down. This has the overall result of causing selection to favour cells with stable configurations such as the two-cell system.

Conclusion and Future Work

In this paper we have presented an argument for the use of abstract biological models to explore multi-cellular dynamics. We have detailed a specific approach to the simulation of complex multi-cellular behaviour, and presented an early optimisation run demonstrating that this approach may be used to replicate phenomena in early embryological development.

Future work is to include exploring somatic optimisation by multi-cellular systems, such as neural organisation. The neurons of two cloned rats with identical DNA produce different neural structures during embryological development (Edelman 1992). Nevertheless, both structures "design themselves" in order to process information and participate in the response mechanisms of the rat. In the absence of a single global controller, this self-organisation must take place according to local interactions.

The exploration of more challenging optimisation based on natural systems will be carried out alongside use of the simulation in abstract design problems. For example, an individual's morphology could be evaluated as a solution to a telecommunications network topology problem.

Future implementations may include the use of a spatially distributed genetic algorithm in which only neighbouring individuals in a population space may mate. Innovations may then emerge in particular neighbourhoods, without being diluted or disrupted by genetic material from other solutions. Combinations of benefits can still occur at the borders between competitive neighborhoods.

It is also possible to employ an "evolutionary strategy" algorithm (Bäck et al 1997), in which real-valued genetic information is not mutated randomly within a specific range, but according to variances prescribed at other locations in the genome. This could encourage the gradual variation of network weights and chemical coefficients towards an optimal solution, avoiding large changes in dynamics which can accompany large changes in parameters in complex coupled systems.

Future developments are also planned to improve the performance of the simulator. Parallel processing will permit larger populations, more generations and a finer grained discrete-time simulation, hopefully allowing more complex behaviours to emerge.

Acknowledgments

Many thanks to Brian Goodwin, Inman Harvey, Alan Steventon, Mark Shackleton, Erwin Bonsma, Alex Penn and the reviewers for discussions and critical reading.

A pre-release version of Eos, an evolutionary and ecosystem platform developed at BT Labs, was used to marry the cellular simulation with a genetic algorithm. For Eos information see http://www.labs.bt.com/projects/ftg/

References

Alt, A., Deutsch, A. and Dunn, G. (eds.) 1997. Dynamics of Cell and Tissue Motion. Birkhauser Verlag, Basel.

Andrews, E. L. 1991.Computer Maker Says Tiny Software Flaw Caused Phone Disruptions. *N.Y. Times*. 10th July.

Bäck, T., Hammel, U., and Schwefel, H-P. 1997 Evolutionary Computation: Comments on the History and Current State, IEEE Transactions on Evolutionary Computation, Vol. 1, No. 1

Edelman, G. M. 1992. Bright air, brilliant fire: on the matter of the mind. BasicBooks, New York.

Eggenberger, P. 1997. Evolving Morphologies of Simulated 3D Organisms Based on Differential Gene Expression. In Proc. 4th European Conf. Artificial Life. pp 205-213. Husbands, P. and Harvey, I. (eds), MIT Press.

Furusawa, C. and Kaneko, K. 1997. Emergence of Differentiation Rules leading to Hierarchy and Diversity. In Proceedings of the Fourth European Conference on Artificial Life. pp 172-181. Husbands, P. and Harvey, I. (editors), MIT Press.

Furusawa, C. and Kaneko, K. 1998. Emergence of Multicellular Organisms with Dynamic Differentiation and Spatial Pattern. Artificial Life IV pp79-93 MIT Press.

Glazier, J. A. and Graner, F. 1993. Simulation of the differential adhesion driven rearrangements of biological cells. *Physical Review* E (47) Number 3 (March)

Gleick, J. 1988. Chaos – Making a New Science. Heinemann.

Goodwin, B 1994. How The Leopard Changed Its Spots Weidenfeld and Nicolson

Harvey, I. and Bossomaier, T. Time out of Joint: Attractors in Asynchronous Random Boolean Networks. In Proceedings of the Fourth European Conference on Artificial Life. pp 67-75. Phil Husbands and Inman Harvey (editors), MIT Press.

Kauffman, S. 1993. The Origins of Order. Oxford University Press, New York.

Kitano, H., Hamahashi, S.,Kitazawa, J., Takao, K. and Imai, S. 1997. Virtual Biology Laboratories. In Proceedings of Fourth European Conference on Artificial Life. pp 274-283 Phil Husbands and Inman Harvey (editors), MIT Press.

Lyu, M. R. (ed), 1996. Handbook of Software Reliability Engineering. Institute of Electrical & Electronic Engineers, McGraw-Hill Book Company, New York.

Painter, K. J. 1997 Chemotaxis as a Mechanism for Morphogenesis. DPhil Thesis, University of Utah.

Pittenger M.F., Mackay A.M., Beck S.C., Jaiswal R.K., Douglas R., Mosca J.D., Moorman M.A., Simonetti D.W., Craig S. and Marshak D.R. 1999 Multilineage potential of adult human mesenchymal stem cells. *Science* Vol. 284 pp 143-147

Savill, N. J. and Hogeweg, P. 1997. Modelling morphogenesis: from single cells to crawling slugs. *J. Theor. Biol.* Vol. 184 pp 229-235

Shipman, R., Shackleton, M., Ebner, M. and Watson, R. 2000. Neutral search spaces for artificial evolution: a lesson from life. In Artificial Life VII: Proceedings of the Seventh International Conference. Bedau, M., McCaskill, J., Packard, N. and Rasmussen, S. (editors)

Slack, J. M. W. 1991. From Egg to Embryo. Cambridge University Press.

Slotine, J-J. E. 1994 Stability in adaptation and learning. Animals to Animats 3. pp 30-34 MIT Press.

Thompson, A. and Layzell, P. 2000 Evolution of Robustness in an Electronics Design. In Proc. 3rd Int. Conf. on Evolvable Systems, Springer Verlag. (in press)

Turing, A. The Chemical Basis of Morphogenesis. 1952. *Phil. Trans. R. Soc.* Vol B237 pp 37-72.

Ward, M. and Seligsohn, D. 1990. Missing tests sent Ariane on path to doom. *New Scientist*. p 10, 27th July.

Weliky, M. and G. Oster. 1990. The mechanical basis of cell rearrangement 1. Epithelial Morphogenesis during Fundulus Epiboly. *Development* Vol. 109 pp 373-386.

Evolutionary Neural Topiary[†]:
Growing and Sculpting Artificial Neurons to Order

Alistair G Rust[1,2], Rod Adams[1] and Hamid Bolouri[2,3]

[1]Department of Computer Science, University of Hertfordshire, UK
[2]Science & Technology Research Centre, University of Hertfordshire, UK
[3]Division of Biology, California Institute of Technology, USA
http://strc.herts.ac.uk/NSGweb/   {*a.g.rust, r.g.adams, h.bolouri*}*@herts.ac.uk*

Abstract

Designing artificial systems with ever more biologically-plausible 'brains' continues apace and permits investigations into the computational capabilities of engineered systems. Creating artificial neurons with biologically-realistic morphologies is however a non-trivial problem. This paper addresses *growing neurons to order*, neurons with morphologies exhibiting strong biological traits. A biologically-inspired simulator of neural development is coupled with a genetic algorithm to evolve 3-dimensional neuron morphologies. The morphology of a biological neuron provides the exemplar target against which the developmental evolution process is gauged.

Realising the Potential of *Brain Building*

During the infancy of Artificial Life, the goal of building artificial organisms and systems with artificial brains – *Brain Building* (deGaris, 1990) – was often a mooted topic. As the domain of Artificial Life has matured, so has *Brain Building*, such that a number of leading-edge, research corporations currently invest substantial resources into developing artificial brains (RIKEN, 2000; NASA, 2000; ATR-HIP, 2000). Much of the maturation has come through the constant increase in computational resources which provide the means with which to simulate artificial neural systems to ever higher levels of biological-plausibility (Hines and Carnevale, 1997).

With biological-plausibility comes the opportunity to fully exploit the complexities of neural computation. Single neurons, let alone networks and systems, possess adaptive, dynamic computational capabilities (Koch and Segev, 1998). One of the key determinants of these capabilities is the relationship between a neuron's morphology and its function (Mainen and Sejnowski, 1996). However, only a handful of computer simulations have explored the relationship between function and form in biologically-plausible terms, e.g. (Mel, 1994).

Our research explores the feasibility of evolving developmental programmes that create biologically-plausible structured neural systems (Rust, 1998). Previously we have used simulated neural development to grow artificial neurons that were functionally evaluated against biological neurons (Rust and Adams, 1999). Experiments showed a strong relationship between function and form in artificial neurons where for example, implausible morphologies possessed inappropriate functionality. Even the final neurons, although functionally similar, had morphologies which still differed from their biological targets. So to mimic the computational capabilities of real neurons and to speed up the search for artificial equivalents, we argue that a method of creating biologically-cogent artificial neurons is required. Namely we are interested in *growing neurons to order*. In this paper the evolution of neuron morphology is explored.

Modelling Neural Development

Modelling of neural morphology using the re-writing rules of L-systems (Lindemayer, 1968) has been explored by a number of groups (Burton et al., 1999; Ascoli, 1999). (Ascoli, 1999) in particular has been using L-system rules whose grammar is derived from observations of biological neural morphology (e.g. typical branching angles and rate of dendritic diameter reduction at branch points). L-system neuron models however, do not typically allow interactions between a growing neuron and its environment. The development of a neuron and hence its visual form are dependent on the implementation of the hand-crafted re-writing rules.

Fleischer and Barr developed an extensive simulator which incorporated many self-organising bio-physical phenomena expressed as differential equations (Fleischer, 1995). Attempts were made at evolving neural morphology but these were computationally intensive due to the large parameter search space involved.

We have implemented a 3D model of neuro-biological development, in which neuron-to-neuron connectivity is created through interactive self-organisation (Rust et al., 1998; Rust, 1998). Development occurs as a number of overlapping stages, which govern how neurons extend axons and dendrites, collectively termed neurites. Neurons grow within an artificial embryonic environment, into which neurons and their neurites emit local chemical gradients. The growth of neurites is influenced by

---

[†]*topiary*: a branch of gardening, the clipping of trees into imitative and fantastic shapes.

the local gradients and the sets of interacting, developmental rules. The *interactive rules* enable neurites to navigate and branch in response to local developmental conditions, and to prune unwanted connections.

The developmental rules are controlled by parameters, much in the same way as gene expression levels can be thought of as parameters for biological development. By varying these parameters, a variety of neuron and network morphologies can be achieved. Examples of individual neurons are illustrated in Figure1.

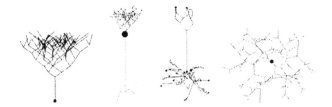

Figure 1: Examples of 3 dimensional neurons grown using the developmental simulator.

Evolving the developmental model for a specific network then becomes equivalent to the search for optimal sets of developmental parameters using, for example a genetic algorithm (GA). Previously we have used a GA to evolve developmental parameters which lead to the creation of an edge-detecting retina (Rust et al., 1998). In this paper the simulator is used to grow single neurons within developmental environments rich with gradients of chemical attractants. The single neurons are then evaluated in terms of their morphological characteristics against the dendritic structure of a biological neuron.

Analysing Neural Morphology
Numerical Evaluation

In order to verify the match between an artificial neuron and a desired biological counterpart, means of characterising biological neurons are required. However, no one set of benchmark measures (quantitative, topological and/or qualitative) exists within neuroscience literature, with which separate classes of neurons can be reliably characterised or compared. Hence in the experiments reported in this paper, as with complementary work (Ascoli, 1999), a minimal set of characteristics is specifically selected. The chosen set aims to reliably represent the geometrical and spatial characteristics of the target biological neuron in the least terms.

Some approaches seek to find generic solutions for classes of neurons by averaging measures from neuron databases. In this paper we aim to gain an understanding of how the morphological characteristics of individual neurons arise before attempting more generic, classbased approaches. Consequently we seek a methodology to clone artificial neurons from a given target neuron. The chosen exemplar neuron for this paper, a layer 5 *pyramidal* neuron, is shown in Figure 2.

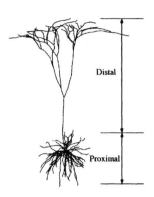

Figure 2: The dendritic morphology of the target, layer 5 *pyramidal* neuron, where the cell soma is located at the centre of the proximal dendritic tree. Morphology obtained from (Mainen and Sejnowski, 1996).

Visual Evaluation

One proviso in using only representative numerical characteristics is, that although particular artificial solutions may satisfy the selected criteria, they may not necessarily be solutions which are visually satisfying. For example, although a neuron may be evolved to have the correct number of tips (terminal segments), it does not mean that the overall geometry of a solution is necessarily consistent with its biological counterpart. In this paper we therefore examine evolving artificial neurons to criteria based upon numerical characteristics alone as well as criteria based on visual appearance alone.

Simply using a GA to traverse the parameter space looking for visually desirable morphologies is not however an effective method. In early generations due to the large diversity in initial populations, visual discrimination will be slow since most morphologies will be discarded until relevant, morphological search sub-spaces are found. A more directed search, where the paths between generations of morphologies could be more directly traversed, would be desirable in this instance.

One such directed approach was used by Dawkins to evolve his artificial *biomorphs* (Dawkins, 1991). An analogous approach towards the evolution of neuron morphology based on visual discrimination is used here. Developmental parameters are selected to form a genome. Each generation of neuron is created by selecting each gene in turn and creating 2 mutated copies of the gene by simply incrementing and decrementing its current value by a pre-determined delta. The values of the remaining genes in the genome remained fixed. Genes are given minimum and maximum values which they can not exceed and a start value is randomly selected from within this range. For each generation, a neuron for each genome is grown and displayed on the screen. Based on a visual comparison between the evolved neuron morphologies and the target biological neuron, a morphology is selected for the next generation.

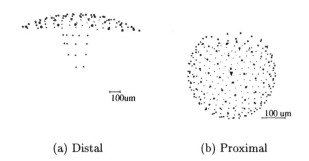

(a) Distal   (b) Proximal

Figure 3: Arrangement of chemical attractants for the 2 developmental environments. The soma of the pyramidal neuron is the lowest object in (a) and the cone at the centre of the sphere of attractants in (b).

Developmental Strategies

Self-Organising Interactions

A feature of the developmental simulator is the ability to influence or sculpt the growth of neurons by exploiting the potential of the interactive, self-organising mechanisms. This was achieved by using the placement and temporal expressions of chemical attractants to guide dendritic growth. For example, the placement of attractants can allow asymmetrical growth, orientated along one particular axis. Being able to specify characteristics of the attractants, provides a more intuitive approach to growing neurons to order. This can avoid relying upon adapting developmental rules and parameters, which can often feel like tinkering with a *black box*.

Developmental environments of attractants were therefore constructed to interact and sculpt the neurons as they grow. These are illustrated in Figure 3. In the case of the proximal dendritic tree, its morphology is approximately radial and symmetrical, hence an environment of attractants placed on the surface of a sphere is used (Figure 3(b)). The dendritic tree is not however perfectly symmetrical as the upper dendrites have shorter lengths compared to their neighbours. To sculpt the artificial neurons to this subtle degree of asymmetry, a small proportion of the attractants are positioned marginally closer to the centre of the sphere.

The morphology of the distal dendrites is directly reflected in the arrangement of attractants (see Figure 3(a)). Intermediate attractants are placed to guide dendrites vertically before the arc of attractants at the top of the environment are encountered. In both sets of developmental environments the attractants were placed to reflect the spatial dimensions (μm) occupied by the biological neuron.

Phased Development

Neurons are known to respond to different environmental cues in accordance with their spatial and temporal locations (Hall, 1992). Trying to reproduce the shape of a complex dendritic tree is therefore unlikely to succeed using an algorithmic model with global settings.

One potential solution to this problem is to split the development of the morphology into different phases and optimise these phases separately. Criteria for subdividing the morphology can then be based on such properties as branching frequency and the orientation of growth. For each phase of development growth control parameters can be given different ranges of potential values at different times. Branching for example, may then be controlled by inhibitory values when it is not required and excitatory values when it is required.

The development of the artificial pyramidal neuron is divided into 3 such phases. The first phase consists of the initial vertical development of the dendritic tree away from the soma, where preliminary branches are established. The effects of attractants are time dependent during this phase to induce interactive branching at the desired times. The end points of phase 1 are then used as the starting points for the second phase of branching, directed towards the arc of uppermost attractants. The final phase produces the development of the proximal dendrites into the environment of attractants arranged on the surface of a sphere (see Figure 3(b)).

A complete pyramidal neuron is achieved by combining all 3 phases of the developmental process.

Results

Since the proximal dendrites are basically 'spherical', visual selection is neither useful nor necessary. Instead, we used automated evolution based on numerical measures of morphology. Visual selection was used to sculpt the asymmetrical distal dendritic tree.

Visual Selection: Distal Dendrites

For both phases 1 and 2, 4 developmental parameters were evolved. Any number of parameters could have been chosen but 4 parameters which have key effects on branching interactions were selected. Specifically the parameters adapted were: (i) the probability of branching as a function of local attractant gradients, (ii) inhibition of branching following a previous branch, (iii) branch inhibition due to saturation caused by nearby attractants and (iv) local repulsion between growing dendrites.

Figure 4 shows the various stages of the visual selection process. Since the genome of the neuron contained 4 genes, 8 morphologies are grown for each generation (each gene is mutated both positively and negatively whilst the other genes remain fixed). A typical screenshot of an intermediate generation of phase 1 is shown in Figure 4(a). The finally selected morphology is illustrated in Figure 4(b). The morphology for phase 2 was evolved in the same manner.

Numerical Selection: Proximal Dendrites

Six morphological properties of the proximal dendrites were chosen to be the criteria upon which evolved neu-

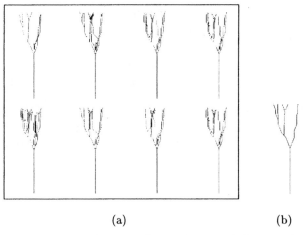

(a)   (b)

Figure 4: Typical images from phase 1 of the visual selection process. (a) A screen-shot of one generation of 8 neurons during an intermediate stage (25 generations). (b) The final, visually selected neuron (52 generations).

rons were evaluated. The criteria were: (i) total length of the dendritic tree, (ii) number of dendritic segments (i.e. the number of dendritic lengths between branch points (iii) number of tips, (iv) average length of tips, (v) length of the longest tip and (vi) the centre of gravity. (vi) was aimed at providing a spatial perspective.

These properties were extracted from data from the neuron and were chosen to effectively describe the morphology of the neuron in the least possible terms. The error value of a neuron was calculated using:

$$e = \sum_{i=1}^{6} \frac{|a_i - t_i|}{t_i} \qquad (1)$$

where a_i is the value of a property of the evolved neuron and t_i is the target value for the proximal dendrites.

The GA used in these experiments was GENESIS (Grefenstette, 1990). 14 developmental parameters were encoded in the GA using 44 bits. The encoded parameters controlled the times at which dendrites could branch and how the growing tips would interact with the attractants in its environment. The population size was 50 and each population was randomly initialised. The crossover rate was 0.6 using dual point crossover. The mutation rate was set such that at each generation approximately 15% of the population would undergo a bit mutation. The role of the GA was to minimise the error from the evaluation function (1).

Ten simulations were performed on a 400MHz Pentium PC running Linux. Each population of neurons was evolved for 150 generations where each population, on average, took 50 minutes to grow and to be evaluated. The best individual neuron evolved had a fitness value of 0.195, which is equivalent to an average error of 3.25% per evaluated property. The average error for the 10 experiments was 0.296 (4.93% per property).

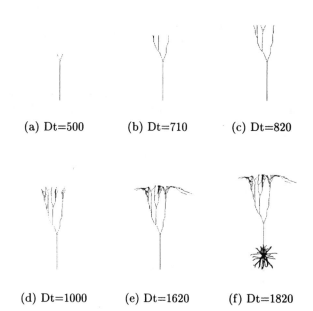

(a) Dt=500   (b) Dt=710   (c) Dt=820

(d) Dt=1000   (e) Dt=1620   (f) Dt=1820

Figure 5: Snapshots of the growth of the pyramidal neuron in all 3 developmental phases, at various developmental times (Dt). (c) is the end of phase 1, (e) is the end of phase 2 and (f) shows the final phase.

Combining the Phases

Figure 5 shows the complete developmental process combining all 3 phases of the developmental evolution process. The morphology of the evolved artificial neuron is compared against its biological target in Figure 6.

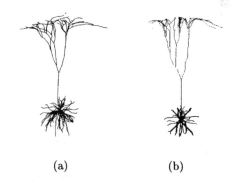

(a)   (b)

Figure 6: A comparison of morphologies between (a) the target biological neuron and (b) the best evolved neuron.

Discussion

The evolution of the proximal dendrites demonstrated that if visual discrimination of a neuron is not critical and its dendritic tree can be assumed to be approximately symmetrical, then numerical measures alone can be sufficient. Due to the radial, symmetrical morphology of the proximal dendrites a regular spacing of attractants was adequate. Growing neurons in this way can then be a blind, automatic process.

Where the morphology of the dendritic tree is more complex, then a blind process may fail. To evolve the morphology of the distal dendrites using numerical selection alone, would require a larger set of numerical characteristics to be extracted and used as guides for development. Choosing representative measures is a non-trivial task and there is no guarantee that these measures can be effectively interpreted by the evolutionary process to produce desired morphologies. (Evolutionary computation (EC) literature contains many examples of algorithms adapting to unwanted behaviours and functions.) Evolving using visual examples of morphologies enables a more directed approach, somewhat more intuitive than a *black box* method. Greater control of the developmental process was also afforded by splitting the morphology of the pyramidal neuron into separate, tractable sub-problems.

The modelled mechanisms of interaction encapsulate the self-organising principles inherent in bio-physical processes. These interactions are shared between a growing neuron entity and its developmental environment. Due to these interactions, fewer parameters which control the growing behaviour of the neuron need to be encoded and evolved in the genome. This thereby leads to a reduction in the size of the genome under evolution, which has 2 potential benefits. Firstly, it reduces the reliance on the chosen EC tool alone to identify optimal solutions by modifying the developmental control parameters. The *evolutionary workload* can be more evenly divided between the bio-physical interactions of the developmental model and the EC method. Secondly, reduced genome sizes should lead to shorter evolutionary searches.

Future Work

As well as validating the model further on other exemplar neurons, the coupling of evaluation of evolved morphologies with dynamical membrane models needs to be explored. Previous work (Rust and Adams, 1999) explored the relationship between function and morphology but without directing the growth process to produce more visually biological-like morphologies. The question to be addressed is, having selected a morphology based on its similarity to a biological neuron, does functionality (e.g. the pattern of spike trains) come for 'free'?

During previous experiments evolving artificial neurons with functional characteristics akin to biological examples (Rust and Adams, 1999), the major bottleneck was the computational cost of evaluating evolved neurons with the chosen compartmental membrane model. On-going work aims to identify a computationally less-intensive membrane model to permit faster evaluations.

Conclusions

A biologically-inspired developmental simulator subject to evolutionary adaptations, was presented in this paper which was capable of generating biologically-realistic neuron morphologies. Growing dendritic trees were sculpted through self-organising interaction with chemical attractants positioned within the developmental environment. In-conjunction with this, dividing the search for the morphology of a complex neuron into separate and solvable phases proved to be a beneficial strategy. Such a combination of techniques and strategies provides a jumping off point to explore even more dynamic artificial neurons and networks.

References

Ascoli, G. A. (1999). Progress and perspectives in computational anatomy. *Anatomical Record*, 257(6):195–207.

ATR-HIP (2000). ATR, Human Information Processing Laboratories. www.hip.atr.co.jp/.

Burton, B. P., Chow, T. S., Duchowski, A. T., Koh, W., and McCormick, B. H. (1999). Exploring the brain forest. *Neurocomputing*, 26-27:971–980.

Dawkins, R. (1991). *The Blind Watchmaker*. Penguin, London, 1st edition.

deGaris, H. (1990). Building brains with GenNets. *Proc. of the Int. Neural Network Conf.*, pages 1036–1039, Dordrecht, Netherlands. Kluwer Academic Publishers.

Fleischer, K. (1995). *A Multiple-Mechanism Developmental Model for Defining Self-Organizing Geometric Structures*. PhD dissertation, California Institute of Technology.

Grefenstette, J. J. (1990). GENESIS Version 5. ftp.aic.nrl.navy.mil:/pub/galist/src.

Hall, Z. W. (1992). *An Introduction to Molecular Neurobiology*. Sinauer Associates, Sunderland, MA, 1st edition.

Hines, M. L. and Carnevale, N. T. (1997). The NEURON simulation environment. *Neural Computation*, 9(6):1179–1209.

Koch, C. and Segev, I. (1998). *Methods in Neuronal Modeling: From Ions to Networks*. MIT Press, 2nd edition.

Lindemayer, A. (1968). Mathematical models for cellular interactions in development. *Journal of Theoretical Biology*, 18:280–299.

Mainen, Z. F. and Sejnowski, T. J. (1996). Influence of dendritic structure on firing pattern in model neocortical neurons. *Science*, 382:363–366. www.cnl.salk.edu/~zach/patdemo.html.

Mel, B. W. (1994). Information-processing in dendritic trees. *Neural Computation*, 6(6):1031–1085.

NASA (2000). NASA, Jet Propulsion Laboratory, Revolutionary Computing Technologies: Biological Computing. cism.jpl.nasa.gov/program/RCT/BioCompUD.html.

RIKEN (2000). Riken Brain Science Institute. www.riken.go.jp/.

Rust, A. G. (1998). *Developmental Self-Organisation in Artificial Neural Networks*. PhD thesis, Department of Computer Science, University of Hertfordshire.

Rust, A. G. and Adams, R. (1999). Developmental evolution of dendritic morphology in a multi-compartmental neuron model. In *Proc. of the 9th Int. Conf. on Artificial Neural Networks (ICANN'99)*, vol. 1, pages 383–388, IEE.

Rust, A. G., Adams, R., George, S., and Bolouri, H. (1998). Developmental evolution of an edge detecting retina. In *Proc. of the 8th Int. Conf. on Artificial Neural Networks (ICANN'98)*, pages 561–566, London. Springer-Verlag.

III Evolutionary and Adaptive Dynamics

Connectivity and Catastrophe - Towards a General Theory of Evolution

David G. Green, David Newth and **Michael G. Kirley**

School of Environmental and Information Science
Charles Sturt University, Albury NSW 2640 AUSTRALIA

Abstract

Here we show that connectivity and catastrophe play a key role in driving species evolution within a landscape. They also form a special case of a more general process, which occurs widely in natural and artificial systems. In this process, catastrophes cause a temporary phase change in the connectivity of a system. Different mechanisms (selection and variation) predominate in each phase. The system passes through the critical point without being poised there. The "chaotic edge" associated with the phase change may be an important source of variety in biological and other systems. In species evolution (and landscape ecology) the process is mediated by cataclysmic events, which fragment widely distributed species and trigger population explosions of new species. The proposed mechanism explains a number of biological and physical phenomena, as well as certain algorithms used for optimisation and evolutionary programming.

Introduction

One of the great challenges facing science is to understand the processes by which complex systems adapt and change (Bossomaier and Green 2000). An important lesson of complexity theory is that superficially different systems show deep similarities in their structure and behaviour (Green 1992). The existence of these universal properties raises the possibility that deep similarities exist also in the mechanisms underlying self-organization and other processes of system change. In other words, can we identify universal processes that operate in a wide range of different systems and circumstances?

The concept of evolution has been a central idea in science for over 150 years. From protein structure and function to taxonomic hierarchies, the theory underlies most of modern biology. More recently the idea has crept into other fields, especially computing. For example, genetic algorithms and other evolutionary methods now play an increasingly important role in problem solving. Given the importance of evolution, it is an obvious question to ask how widespread are evolutionary processes?

The aim of this discussion is to explore the potential for generalizing the concept of species evolution to encompass a much wider range of phenomena. In particular we will show here that certain aspects of species evolution form a special case of a general process of self-organization, which acts in many different kinds of systems.

Models of Evolution and Self-Organization

In the sense that they involve systems changing through time, many processes are akin to species evolution.

An important theme in research into complexity and artificial life has been to understand how systems evolve, adapt and change (Langton 1989; Kauffman 1992; Depew and Weber 1997). For example, Prigogine (1980) introduced the idea of dissipative systems, which are thermodynamic systems that produce energy internally and exchange it with an external environment. Dissipative systems include living organisms as well as many physical systems, such as certain chemical reactions and interstellar gas clouds. In such systems, which are far from equilibrium, global patterns can emerge out of minor, local events.

Holland (1995), who introduced the genetic algorithm, has stressed the role of adaptation in self-organization. He suggested that seven basic elements are involved in the emergence of order in complex adaptive systems. These include four properties - aggregation, nonlinearity, flows, and diversity - and three mechanisms - tagging, internal models, and building blocks.

The role of criticality in system change is well known (eg. Bak and Chen 1991). Several authors, including Kauffman (1992) and Langton (1990, 1992), have suggested that the critical region between ordered and chaotic behaviour plays a central role in evolution. Systems that are in a state lying close to this "edge of chaos" have the richest, most complex behaviour. Natural systems with these properties would adapt better and therefore have a selective advantage over other systems. Thus these authors suggest that many systems tend to evolve so that they lie close to the edge of chaos. These ideas are consistent with conclusions of others (eg. Freeman 1975), based on experimental observations, that chaos appears to be an important source of novelty and variety in living systems.

Another important model is that of the "fitness landscape" (Kauffman 1992; Depew and Weber 1997). In this model the entire range of potential genotypes that a population may possess is mapped onto an imaginary landscape in which the axes denote particular factors or properties of the organisms (eg. size, drought tolerance). The entire population can then be viewed as a scatter of points in the landscape, with each point being an individual in the population. Processes such as mutation and selection can be seen as forces pushing the cloud in particular directions. *It is important not to confuse these imaginary fitness landscapes with the real landscapes that we discuss below.*

Evolution in a Landscape

The neo-Darwinian theory of evolution is based around three main elements:

- Populations with genetic structure;

- Competition and selection of the fittest individuals;

- Mutation and variation.

The assumption is that populations change by slowly accumulating genetic changes, with natural selection weeding out unfit individuals. "Fitness" refers to how well an individual's adaptation to environmental conditions helps it to survive and reproduce. Environmental factors that affect survival impose selective pressures on a population.

The slow accumulation of new characters mentioned above is often termed a gradualistic theory. In the Nineteenth Century it replaced earlier theories of catastrophic evolution. However catastrophism underwent a revival in the late Twentieth Century as the mounting fossil evidence revealed a picture of intermittent mass extinctions (Raup 1986; Raup and Jablonski 1986; Kauffman and Walliser 1990), often followed by a burst of speciation. These observations are inconsistent with a purely gradualistic view, which would predict a more or less constant turnover of species.

Punctuated Equilibria

Seeking to explain the evident stop-start nature of the fossil record Eldredge and Gould (1972) proposed the theory of punctuated equilibria. According to this theory, species tend to remain stable for long periods of time. The equilibrium is punctuated by abrupt changes in which existing species are suddenly replaced. Many biologists have queried the theory, arguing that speciation could not occur so rapidly. They also point to the incompleteness of the fossil record to explain missing transitional forms.

Perhaps the best known explanation for sharp boundaries in the fossil record is the idea that comets and other cataclysmic events in the past caused sudden mass extinctions. This idea had been around for a long time but was not taken seriously until Alvarez et al. (1980) found evidence that a massive cometary impact had coincided with the extinction of the dinosaurs. Their evidence was a thin layer rich in the metal iridium. This layer coincided precisely with the Cretaceous-Tertiary (K-T) boundary. Subsequent research has identified that this iridium layer is worldwide. It seems to have resulted from an impact on the Yucatan Peninsula in central America.

Landscape Connectivity

It has always been acknowledged that the spatial distribution of plants and animals plays a role in evolution. In particular the popular expression "speciation by isolation" sums up the key role that landscapes play in the origins of new species. However, the exact mechanism by which isolation occurs is a much neglected question.

In previous studies (eg. Green 1989, 1994a, 1994b) we have shown that spatial distributions play an important role in the dynamics of natural communities. In particular, landscape connectivity plays a crucial role. "Connectivity" here means processes that affect genetic "communication" within a population. Examples include animal migration, seed dispersal, disturbances, and the distribution of potential habitats. Sites in a landscape are "connected" if there are patterns or processes that link them in some way. These links arise either from static patterns (eg. landforms, soil distributions, and contiguous forest cover) or from dynamic processes (eg. dispersal, fire). Note that a particular landscape may have radically different degrees of connectivity with respect to different processes. For instance following a major fire that clears (say) 90% of forest within a region, the forest remnants are likely to be genetically isolated from one another. However to a population invading the same landscape, the sites that are available for colonization are highly connected.

The key result to emerge from studies of connectivity (Green 1994a) is that landscapes can exist in two different phases: connected and disconnected (Fig. 1). Sites in a landscape are "connected" if the local populations interbreed with each other (i.e. share genetic information). Dispersal between sites is essential to maintain genetic homogeneity within populations. Should this dispersal connectivity fall below a critical level, then a regional population effectively breaks up into isolated subpopulations (Fig. 2).

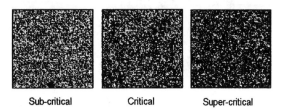

Figure 1: Phase change in the connectivity of a cellular automaton grid as the proportion of "active" cells increases (after Green 1994a). Grey denotes active cells; white denotes inactive cells. The proportion of active cells increases from left to right. The black areas indicate connected patches of active cells. Notice that a small change in the number of active cells produces a phase change in the system: from many small patches, isolated from one another, to essentially complete connectivity of the entire system.

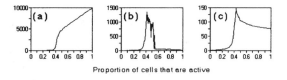

Figure 2: Critical changes in connectivity of a CA grid (cf Fig. 1) as the proportion of "active" cells increases (x-axes). (a) Average size of the largest connected subregion (LCS). (b) Standard deviation in the size of the LCS. (c) Traversal time for the LCS. Each point is the result of 100 iterations of a simulation in which a given proportion of cells (in a square grid of 10,000 cells) are marked as active. Note that the location of the phase change (here ≈ 0.4) varies according to the way we define connectivity within the model grid.

Landscape Phases and Evolutionary Processes

We can briefly summarize the process of evolution as follows. For most of the time the system - that is populations of plants and animals in their environment - sits in an undisturbed state. In this state, individual species exist in either of two states: connected and fragmented.

For species that consist of a single, connected population, genetic information is constantly being circulated throughout the entire population. The effect of this constant genetic mixing is to inhibit variation. Wild mutations are culled out and the entire population is kept within narrow bounds (Fig. 3). Under these conditions only natural selection can produce change. Poorly adapted members of the population are culled out and well-adapted individuals produce most of the offspring. The nett result is that the average adaptation of the population gradually increases through time.

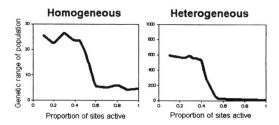

Figure 3: Simulated genetic drift in a landscape. The figure shows the range of gene values (G), after 10,000 "generations" of a population that initially is: (left) homogeneous (G = 0 everywhere); and (right) heterogeneous ($-100 < G < 100$), in response to the proportion P of active sites. In each case, the range $0.4 < P < 0.6$ forms a critical region.

By way of contrast, neutral variations are bound to increase in species that consist of fragmented populations (Fig. 3). Isolated subpopulations may be small enough that random mutations can become fixed. Also there is nothing to stop different subpopulations from randomly drifting apart in their genetic makeup. This is true even in the absence of selective pressure. When selective pressure does operate, its strength and direction usually varies from place to place. So under these conditions selection acts to accelerate and magnify the differences between subpopulations.

The Role of Cataclysms

The above account summarizes what happens in the environment most of the time. However the slow, steady accumulation of changes is often interrupted by disturbances. The history of life is peppered with cataclysms, both great and small. Most discussion of cataclysms has focused on the impact of comets and other events with the potential to disrupt the entire planet. However the biosphere is continually subjected to cataclysms of all sizes. Great events, such as the impact of a comet, are as rare as their effects are vast. Smaller events are more common. So common are small disturbances that every year the Earth's surface is marked by thousands of fires, storms, volcanic eruptions, and innumerable other events. On a smaller scale these small, but common cataclysms, add up to have a similar effect to that of really large events.

The most important effect of a cataclysm is to clear large tracts of land at a single stroke. In doing so the cataclysm plunges an ecosystem into a different phase. Suddenly the normal rules, described above, no longer apply. The absence of restrictions has two main effects. First it carves up widespread species into isolated subpopulations. Sometimes those subpopulations will again spread and recombine. However the opposite happens too. That is, cataclysms often act to break subpopulations for long periods of time. When this happens speciation begins.

The other effect of cataclysms is to free up large tracts of land that were formerly occupied. This sets all the species back to square one. Species that were formerly dominant have to compete against newcomers. In normal times, dominant species exclude competitors for the simple reason that they already occupy territory. After a cataclysm they are set back to the same state as their competitors. If climatic conditions have changed, or if some new, superior competitor appears, then they may lose their dominant position.

The final point to note about these cataclysms is that after they clear territory, the surviving populations expand to fill the void. In doing so they compete and rearrange both their territories and their relationships to each other. When this resorting dies down the system re-enters the "normal" phase that we described first.

Phase Changes and Punctuated Equilibria

The above effects help to explain the phenomenon of punctuated equilibrium. The normal conditions effectively prevent speciation in common, widespread species. On the other hand they promote speciation in fragmented populations. Importantly they also prevent new species from spreading. So new species, which form in small, isolated populations, remain as small isolated populations.

Cataclysms act to release the above restraints. A consequence is that any new species that have formed are no longer suppressed. If conditions favour them, they spread rapidly, perhaps even replacing the species that spawned them. So the proliferation of new species after a cataclysm is not necessarily a speeding up of evolution, but rather an unleashing of new species that had already formed.

Several lines of evidence exist to support the above theory. One is the phenomenon of hybridization (Barber 1970; Levin 1970), which occurs when genetically distinct, but related populations come into contact. Field studies (Briggs 1986) reveal that the extent of hybridization is much greater in disturbed environments, which allow much greater movement of plants. Another line of supporting evidence arises from the dynamics of long-term vegetation change, which we address in the next section. It provides not only an analogy, but also a possible fine scale mechanism.

Phase Changes in Vegetation History - An Analogue for Evolution

The above model for species evolution emphasizes the role of landscapes and disturbances. Similar mechanisms within landscapes also exist on finer scales. An excellent example is provided by vegetation history. Studies, using preserved pollen, of postglacial vegetation history in Europe and northeastern North America (Davis 1976) reveal that the sequences of forest changes during the last 10,000 years were remarkably uniform over vast areas. It was assumed that these "pollen zones" represent periods of more or less constant forest composition. Before the advent of radiocarbon dating, pollen zones were used to establish the relative chronologies between sites. Subsequent research has shown that these zones are associated with post-glacial migrations of tree populations (Davis 1976, Webb 1981). Most significantly, the zone boundaries, which are usually defined by invasions and other sudden changes in plant populations (Fig. 4), often coincide with major fires (Green 1982). Pollen and charcoal records (Green 1987, 1990) show that competition from established species suppresses invaders. By clearing large areas, major fires remove competitors and trigger explosions in the size of invading tree populations.

The parallels between vegetation change and evolution are striking: pollen zones versus geologic eras, sudden changes in community composition versus mass extinctions, and major fires versus cometary impacts. This correspondence is so striking that it implies some fundamental process underlies the similarities (Green 1994b). Simulation studies imply that biotic processes in landscapes are responsible. In the case of forest change seed dispersal acts as a conservative process (Green 1989). Because they possess an overwhelming majority of seed sources, established species are able to out-compete invaders. By clearing large regions, major fires enable invaders to compete with established species on equal terms. Conversely seed dispersal also enables rare species to survive in the face of superior competitors by forming clumped distributions. This process also provides a mechanism that maintains high diversity in tropical rainforests (Green 1989).

A Model for Catastrophe Induced Phase Changes

One of the strongest indicators of the possibility of a universal theory of evolution is the existence of common properties underlying the structure and behaviour of all complex systems (Green, 1992, 1994b, 1994c, 1999). In any complex system, connectivity is best expressed as a directed graph (X, E) ("digraph"). This is a set X of "nodes", of which some or all are joined by a set E of "edges". We represent elements of the system as nodes and interactions by edges. The universal nature of digraphs is assured by the following theorems (Green 1994c).

Theorem 1 *The patterns of dependencies in matrix models, dynamical systems, cellular automata, semigroups and partially ordered sets are all isomorphic to directed graphs.*

Theorem 2 *In any array of automata the state space forms a directed graph. If both the array and the number of states are finite, then so is the resulting set of directed states.*

The first theorem shows that digraphs are inherent in all of the ways we represent complex systems. So, assuming the models are valid, digraphs are present in the structure of virtually all complex systems. The

second theorem shows that we can also regard the behaviour of complex systems as directed graphs.

The most important consequence of the above theorems is that properties of directed graphs explain many phenomena, such as criticality, in systems that had previously been treated as distinct (Green 1992; 1994b). Most prominent of these properties is the "connectivity avalanche". Erdos and Renyi (1960) examined what happens if one takes a set of nodes and adds edges progressively to pairs of nodes chosen at random. At first the set of connected nodes are very small. But at a certain point in the procedure, a "connectivity avalanche" occurs. Adding just a few extra edges suddenly joins virtually all of the nodes into a single "giant component". This amounts to a phase change in the system - from essentially disconnected to fully connected.

The above theorems imply that the connectivity avalanche is responsible for many kinds of phase changes in complex systems (Green 1992; 1994b). For example, if we represent a landscape as a grid of cells (using the formalism of cellular automata), and represent the distribution of (say) a plant species by cells in a particular state, then we find that as the occupied proportion of the landscape increases, a phase change occurs in the size of the largest "patch" (Fig. 2).

We can regard this phase change as an elementary form of chaos (a "chaotic edge"). Because of the sudden change from disconnected to connected, the system is highly sensitive to initial conditions at the phase change. Also, because of the extremely high variance the size and composition of patches in any two systems are likely to be quite different from one another.

The universality of graphs in the structure and behaviour of complex systems suggests that phase changes may play a role in system evolution. In particular it enables us to generalize the model proposed above for species evolution in a landscape to identify a potentially universal mechanism based on phase changes triggered by disturbances. We suggest that the inherent variability of phase changes in connectivity (Fig. 2b) provides a source of novelty in many systems (Green 1994b). Taken in the broadest sense, we can understand variation (c.f. mutation) to mean changes within a system's components or its connectivity. We can interpret selection as constraints that either prevent variation or else push it in a particular direction. Adopting the above idea, we suggest that many systems flip-flop backwards and forwards across a "chaotic edge" associated with a phase change in their structure or behaviour. This phase-shift mechanism (Fig. 5) works as follows:

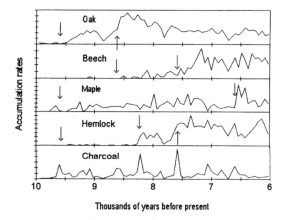

Figure 4: Cataclysmic change in postglacial forests (after Green 1990). Pollen and charcoal records (from Everitt Lake, Nova Scotia) show that competition from established species suppresses invaders. Major fires clear large tracts of land, remove competitors and trigger explosions (arrows) of invading tree populations.

- The system can exist in either of two phases - a connected phase, wherein selection predominates; and a disconnected phase, wherein variation predominates.

- Most of the time the system rests in the connected phase. Selection maintains the system in a more or less steady state.

- External stimuli may disturb the system, forcing its structure to shift across the phase change. Whilst the system is in the disconnected phase, variation has free rein.

- Following a disturbance, connectivity gradually builds up again within the system until it reorganizes itself ("crystallizes") into a new stable structure. Because of the variability associated with the phase change, this new structure is likely to be quite different from the structure prior to the disturbance.

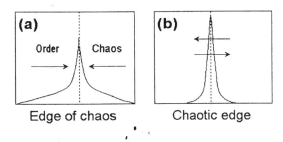

Figure 5: Contrasting the edge-of-chaos and phase-shift models of evolution in complex systems. The x-axis represents a connectivity "order" parameter appropriate to the system concerned. The spike represents the critical point where a phase change occurs. (a) In the edge-of-chaos model complex systems evolve to lie near or at the critical point (the spike) between ordered and chaotic phases. (b) In the phase shift model, which is described here, external stimuli flip the system across the chaotic edge into the phase where variation predominates. The system then gradually returns, crystallizing into a new structure or behaviour as it does so. See the text for further explanation.

Application of the Model to Evolution in Natural and Artificial Systems

In this section we look at a number of cases where the above model appears to apply.

Feedback Networks

Competition, predation, herbivory and other interactions among species are common types of connectivity within any ecosystem. One issue is to assess the effects of suites of interactions that can arise in complex ecosystem (eg. rainforests). Levins (1970; 1977) introduced the idea of "loop analysis": by knowing the sign of particular interactions we can trace whether feedback loops are positive (i.e. destabilizing) or negative (i.e. stabilizing). Theoretical studies of population dynamics indicate that random assemblages usually form non-viable systems (Tregonnig and Roberts 1979). That is, the interactions are highly likely to form positive feedback loops, which lead to the extinction of one or more species. In a system with few interactions, such instabilities are less likely to occur than in a richly connected one (May 1972; 1974). Thus gradually adding new species (eg. by migration) to a viable system increases the overall connectivity until the phase change occurs and positive feedback becomes inevitable. The system would then lose species and collapse back to a new, sub-critical state. Further immigrant species would then set the process off again.

Neural Function

Experimental evidence (eg. Freeman, 1975) implies that some neural systems exploit phase transitions as a source of novelty and flexibility, which allows them to continually adapt to new stimuli. In particular, brain function may involve changes in sifts in connectivity between neurons. Freeman (1992) has shown that, unlike current artificial neural networks, living neural systems (eg. cat's brains) exploit chaos as a source of novelty in creating memory patterns. We suggest that the above novelty arises from the inherent unpredictability of connectivity patterns associated with phase transitions.

Socio-economics and Cultural Change

We expect that our model will be relevant to human social and economic systems. For instance, changes in communications, such as the introduction of the Internet, changes the patterns of connectivity between groups and individuals (Bossomaier and Green 1998). More generally, any new technology disturbs the prevailing socio-economic framework, which eventually settles down into a new pattern (Toffler 1970). The precise application of the theory to social systems awaits the development of models that define appropriate connectivity models for socio-economic systems. However some authors have pointed to empirical evidence suggesting that connectivity models may provide fresh insights. For instance Dunbar (1996) has pointed out that different forms of social interaction (speech versus grooming) led humans to have larger natural group sizes than apes or monkeys.

Geomorphology

The notion of gradualism first arose in the context of geomorphology. In the early 1800s Lyell showed that geological change can be attributed to the action of erosion and other slow continual processes. In truth, however, geological change includes both gradual and catastrophic processes. For instance, volcanism builds mountains and lava plains, which erosion then shapes over time. Likewise erosion slowly shapes the course of streams, but intermittent floods can create new channels overnight.

Optimization

The notion of a fitness landscape, mentioned earlier in connection with species evolution, can be extended to optimization problems. We can imagine all possible solutions to a given problem as being mapped onto a hypothetical landscape (in some cases the existence of orthogonal parameters makes this possible literally), in which the elevation represents the value of the object function that we wish to maximize (or minimize). Optimization algorithms therefore seek to locate the hilltops (or valleys) in this solution landscape. The risk with simple hill-climbing, and other local optimization procedures, is that we may wind up stuck (say) on a minor foothill without finding the tallest mountaintop. To counteract this prolbme, many algorithms make use of the phase-change mechanism we have propsed here. For instance simulated annealing, uses a pseudo-temperature measure, which effectively allows solutions to drop downhill by a certain amount. At the start of the cooling schedule this temperature drop is large enough to allow the algorithm to wander anywhere in the landscape. In other words the solution landscpae is fully connected. However as the cooling procedes the system eventually passes a point where the connectvity breaks down and the algorithm becomes trapped on a particular hill. Likewise, the "great deluge" algorithm follows a random walk around the solution landscape, which it gradually floods until individual hills become isolated from one another.

In our own work on optimization, we have exploited the flip-flop, phase change mechanism more explicitly in the cellular genetic algorithm (see below).

Evolutionary Algorithms

Evolutionary Algorithms (EA's) are biologically inspired models for solving computationally hard problems. Evolutionary Algorithms are abstracts models that typically ignore key features of biological and population dynamics. We argue that mimicking nature more closely leads to more robust algorithms.

The Cellular Genetic Algorithm (CGA) is a parallel genetic algorithm that mimics adaptation in a landscape by mapping the population of solutions onto a pseudo landscape (Kirley et al. 1998). It raises the prospect of exploiting the population's spatial organisation to solve multi-objective and other novel problems. Our experiments show that random "disasters", which clear space, can break down the connectivity, leading to isolated subpopulations whose genetic makeup rapidly drifts apart (Fig. 6). Intermediate rates of disturbance are essential to maintain a high diversity of genetically distinct individuals.

Conclusion

In essence the theory we have developed here proposes that many systems develop and change by a mechanism involving phase changes. Left to itself such a system will remain in a more or less constant state. However external events may disturb the system, flipping it into a different phase in which variation, rather than selection, dominates. The phase transition is an essentially chaotic phenomenon that perturbs the systems in unpredictable ways, and acts as a source of novelty. Following the phase change the system gradually drifts back into its original phase, but can settle into a completely new steady state.

This model differs from other theories that have been proposed. First, it differs from the kinds of critical collapse described by Bak and Chen (1991). In a critical collapse, a system collapses spontaneously when connectivity exceeds the critical threshold. In our model, the system is normally static. It is an external disturbance that triggers the change. The two theories describe different aspects of critical behaviour and are complementary. Most importantly, the system does not keep returning to its original state. Instead it

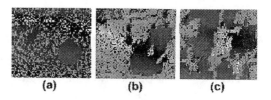

Figure 6: Exploiting connectivity phase shifts in the cellular genetic algorithm. Lighter colours indicate fitter individuals. Black indicates a disaster zone. (a) The landscape view at generation 3, after a disaster has struck. (b) Additional disasters have hit the landscape, patchy subpopulations are beginning to form. Parts of the disaster zone shown in (a) have been reclaimed. (c) The cumulative affect of many disasters across the landscape. (After Kirley et al. 1998).

evolves via a sequence of disturbances and crashes.

The present theory differs too from the "edge of chaos" model. Rather than settling in the critical region, the phenomena we describe here exhibit jumps through the critical region. The systems do not settle and remain in a critical state. More generally, the two models were developed to describe different things. The edge of chaos model arose from considerations of the behaviour of automata, with the relevant critical region lying within the system's state space. In contrast the model developed here derives from considerations of system structure, such as the connectivity within a landscape.

In this account we have described a particular evolutionary mechanism. In itself, this is not a universal theory of evolution. Inevitably, our general theory is an abstraction. And in the abstraction some important details are lost. For instance we say nothing about the nature of the processes of selection and mutation. These details do matter. For example, in considering landscapes, these processes can also be affected by the spatial patterns and processes (Green 1997).

A truly universal theory would need to encompass all of the above issues, as well as many others besides. Nevertheless, we feel that the mechanisms we have identified point to the practicality of setting species evolution in the context of a general theory of evolutionary processes. Certainly recent reviews of evolutionary theory (eg. Depew and Weber, 1997) take on board work on complexity and related phenomena. Developing a general evolutionary theory is surely one of the great scientific challenges for the new Millennium.

References

Alvarez, L.W., Alvarez, W., Asaro, F. and Michel, H.V. 1980. Extraterrestrial Cause for the Cretaceous-Tertiary extinction. *Science* 208:1095-1108.

Bak, P. and Chen, K. 1991. Self-organized criticality. *Scientific American* 265:26-33.

Barber, H.N. 1970. Hybridization and the evolution of plants. *Taxon* 19:154-160.

Bossomaier, T.J. and Green, D.G. 1998. *Patterns in the Sand - Computers, Complexity and Life*. Sydney. Allen and Unwin.

Bossomaier, T.R.J. and Green, D.G. eds. 2000. *Complex Systems*. Cambridge University Press.

Briggs, B.G. 1986. Alpine Ranunculi of the Kosciusko Plateau: habitat change and hybridization. In B. Barlow. ed. Flora and Fauna of Alpine Australasia - Ages and Origins. 401-412. Canberra. CSIRO.

Davis, M.B. 1976. Pleistocene biogeography of temperate deciduous forests. *Geoscience and Man* 13:13-26.

Depew, D.J. and Weber, B.H. 1997. *Darwinism Evolving*. Cambridge Mass. MIT Press.

Dunbar, R. 1996. *Grooming, Gossip and the Evolution of Language*. London. Faber and Faber.

Eldredge, N. and Gould, S.J. 1972. Punctuated equilibria: an alternative to phyletic gradualism. In T.M. Schopf ed. Models in Paleobiology. San Francisco. Freeman, Cooper.

Erdos, P. and Renyi, A. 1960. On the evolution of random graphs. *Mat. Kutato. Int. Kozl.* 5:17-61.

Freeman, W.J., 1975. *Mass action in the nervous system*. New York, Academic Press.

Freeman, W. 1992. Tutorial on neurobiology: from single neurons to brain chaos. *International Journal of Bifurcation and Chaos* 2(3):451-482.

Green, D.G. 1982. Fire and stability in the postglacial forests of southwest Nova Scotia. *J. Biogeog.* 9:29-40.

Green, D.G. 1987. Pollen evidence for the postglacial origins of Nova Scotia's forests. *Can. J. Bot.* 65:1163-1179.

Green, D.G. 1989. Simulated effects of fire, dispersal and spatial pattern on competition within vegetation mosaics. *Vegetatio* 82:139-153.

Green, D.G. 1990. Landscapes, cataclysms and population explosions. *Mathematics and Computer Modelling* 13(6):75-82.

Green, D.G. 1992. Emergent behaviour in biological systems. In D. G. Green, and T. J. Bossomaier eds. *Complex Systems - From Biology to Computation*. 25-36. Amsterdam. IOS Press.

Green, D.G. 1994a. Connectivity and complexity in ecological systems. *Pacific Conservation Biol.* 1(3):194-200.

Green, D.G. 1994b. Connectivity and the evolution of biological systems. *J. Biological Systems* 2(1):91-103.

Green, D.G. 1994c. Evolution in complex systems. In R. Stonier and X. H. Yu eds. *In Complex Systems - mechanism of adaptation*, 25-31. Amsterdam. IOS Press. Reprinted as Green, D.G. 1995. *Evolution in complex systems*. Complexity International Vol. 2.

Green, D.G. 1997. Complexity in ecological systems. In . Klomp and I. Lunt eds. *Frontiers of Ecology - Making the Links*. N. 221-231, London. Elsevier.

Green, D.G. 1999. Self-Organization in complex systems. T.J. Bossomaier and D.G. Green eds. *Complex Systems*. 7-41. Cambridge University Press.

Holland, J.H. 1995. *Hidden Order: How Adaptation Builds Complexity*. Ann Arbor. Univ. Michigan Press.

Kauffman, E.G. and Walliser, O.H. 1990. Extinction Events in Earth History. *Lecture Notes in Earth Sciences*. Berlin. Springer-Verlag.

Kauffman, S.A. 1992. *The Origins of Order: Self-Organization and Selection in Evolution*. Omford. Oxford University Press.

Kirley,M., Li, X. and Green, D.G. 1998. Investigation of a cellular genetic algorithm that mimics evolution in a landscape. In X. Yao et al. eds. *Simulated Evolution and Learning. SEAL 98. Lecture Notes in Artificial Intelligence, 1585*. 93-100. Berlin. Springer.

Langton, C.G. ed. 1989. *Artificial Life*. Reading Massachusetts. Addison-Wesley.

Langton, C.G. 1990. Computation at the edge of chaos: phase transitions and emergent computation, *Physica D* 42(1-3):12-37.

Langton, C.G. 1992. Life on the edge of chaos. *Artificial Life II*, 41-91, New York, Addison-Wesley.

Levin, D.A. 1970. Hybridization and evolution - a discussion. *Taxon* 19:167-171.

Levins, R. 1970. Complex systems. *Towards a Theoretical Biology. 3. Drafts* Waddington, C. H. ed. 73-88, Edinburgh University Press, Edinburgh.

Levins, R. 1977. The search for the macroscopic in ecosystems, In G. S. Innes ed. New Directions in the Analysis of Ecological Systems II. 213-222, La Jolla. Simulation Councils.

May, R.M. 1974. *Stability and Complexity in Model Ecosystems*. Princeton. Princeton University Press.

May, R.M. 1972. Will a large complex system be stable? *Nature* 238:413-414.

Prigogine, I. 1980. *From Being to Becoming*. San Francisco. W. H. Freeman and Co.

Raup D.M. 1986. Biological extinction in Earth history. *Science* 231:1528-1533.

Raup, D.M. and Jablonski, D. 1986. *Patterns and Processes in the History of Life*. Berlin, Springer-Verlag.

Toffler, A. 1970. *Future Shock*. New York. Bantam Books.

Tregonning, K. and Roberts, A. 1979. Complex systems which evolve towards homeostasis, *Nature* 281:563-564.

Webb, T. III 1979. The past 11,000 years of vegetational change in eastern North America. *BioScience* 31:501-506.

Neutral search spaces for artificial evolution: a lesson from life

Rob Shipman[1], Mark Shackleton[1], Marc Ebner[2], Richard Watson[3]

[1] BT Labs at Adastral Park, Admin 2-5, Martlesham Heath, Ipswich, IP5 3RE, UK. rob.shipman@bt.com, mark.shackleton@bt.com

[2] Universität Würzburg, Lehrstuhl für Informatik II, Programmiersprachen und Programmiermethodik, Am Hubland, 97074 Würzburg, Germany. ebner@informatik.uni-wuerzburg.de

[3] Brandeis University, Volen Center for Complex Systems, Mail Stop 18, Waltham MA 02454-9110, USA. richardw@cs.brandeis.edu

Abstract

Natural evolutionary systems exhibit a complex mapping from genotype to phenotype. One property of these mappings is neutrality, where many mutations do not have an appreciable effect on the phenotype. In this case the mapping from genotype to phenotype contains redundancy such that a phenotype is represented by many genotypes. Studies of RNA and protein molecules, the fundamental building blocks of life, reveal that this can result in neutral networks - sets of genotypes connected by single point mutations that map into the same phenotype. This allows genetic changes to be made while maintaining the current phenotype and thus may reduce the chance of becoming trapped in sub-optimal regions of genotype space. In this paper we present several redundant mappings and explore their properties by performing random walks on the neutral networks in their genotype spaces. We investigate whether the properties found in nature's search space can be engineered into our artificial evolutionary systems. A mapping based on a random boolean network was found to give particularly promising results.

1. Introduction

Natural evolution differs in many respects from the evolutionary algorithms typically employed today. One such difference is highlighted by the neutral theory of evolution. According to this theory a considerable fraction of all mutations are neutral and only a minute fraction of the remainder are actually beneficial (Kimura 1994). This results in a redundant genotype-phenotype mapping with typically many genotypes representing any given phenotype. The redundancy manifests itself in a number of different ways and at a number of different levels - from the genetic code, consisting of 64 codons mapping into only 20 amino acids, to the complex interplay of molecules forming an organism. A particularly important and

thorough study of the effects of such redundancy was performed in the context of the folding of RNA, and to a lesser extent, protein molecules (Huynen 1996, Huynen, Stadler, and Fontana 1996). These studies revealed a number of interesting properties in the nature of the secondary structures that the primary structures folded into. There were a number of common secondary structures each represented by a very large set of primary sequences i.e. there was large-scale redundancy in the genotype-phenotype mapping. The density was such that these sets were often connected by single-point mutations forming so-called *neutral networks*. Thus, it was possible to traverse the set of genotypes through the simplest of mutations without changing the represented phenotype.

This brought about the possibility of neutral drift allowing larger areas of genotype space to be explored in search of more adaptive secondary structures. However, during such a process there is no pressure influencing movement to areas of genotype space in which more adaptive phenotypes can be found and there is thus a danger of prolonged periods of random drift. It is important, therefore, that the neutral networks representing each of the phenotypes are intertwined with many access points between them. This encourages beneficial transitions from one network to another and minimizes the amount of time spent drifting randomly. Studies of RNA folding suggested that this was the case. As the neutral networks were traversed a relatively high, and roughly constant, number of new structures were discovered at each step (Huynen 1996).

These properties may have a significant impact on the evolvability of a system. Instead of becoming trapped in sub-optimal regions of genotype space, adaptation is able to continue through genetic changes that do not alter the phenotype but enable movement in genotype space to areas that are closer to genotypes representing potentially more adaptive phenotypes. Following on from previous studies (Ebner 1999, Shipman 1999) this work explores a number of redundant genotype-phenotype mappings with a view to ascertaining whether these fundamental properties of living systems can be encouraged in our artificial systems. For related work that aims to exploit the developmental process in engineered systems see (Hoile and Tateson 2000).

The structure of this paper is as follows - section 2 details the four mappings that were studied, section 3 details the methods that were used to ascertain the properties of these mappings, section 4 presents our findings, section 5 gives a discussion of these results and section 6 concludes.

2. Redundant mappings

A number of different mappings were constructed for this work, this paper reports on four of them. In order to allow the calculation of statistics, the phenotype space was fixed at 8 bits, giving $2^8 = 256$ possible phenotypes. Both the length of the genotype and the number of alleles varied between the mappings.

2.1 Voting mapping

The first mapping, shown in figure 1, is based on a voting approach where each bit of the phenotype is influenced by several bits from the genotype. Each phenotype bit is determined by looking at all the bits of the genotype to which it is linked. A bit of the phenotype is set to one if the majority of connected bits in the genotype "vote" in favor of this. Thus, depending on the values of the other relevant bits, a point mutation may or may not have an effect on the phenotype. It is important to note that the set of genotype bits linked to a particular phenotype bit will typically *overlap* with the sets corresponding to other phenotype bits. It is this aspect that permits multiple phenotype bits to potentially be changed simultaneously by a single point mutation. Together with the redundant "majority voting" aspect of the mapping, this permits the scene to be set for future transitions to another phenotype, without actually changing the current phenotype encoded in the genotype. The links between the genotype bits and the phenotype bits are determined in the following way. For each bit of the phenotype we select a number of bits of the genotype which will vote for that phenotype bit, which is typically a constant odd number. For each of the voting bits, we randomly choose whether a set bit will vote in favor of the corresponding phenotype bit being set, or against it being set. Thus each gives either a positive or a negative vote. For instance, in the results reported later, a genotype of 24 bits was used with sets of 17 genotype bits being chosen for each of the 8-phenotype bits. There is thus significant overlap among the sets of voting bits.

In order to confirm that the above mapping was introducing

"useful" redundancy, we compared it with a trivial voting mapping in which each phenotype bit was linked to 3 genotype bits *without overlap*. This mapping exhibits redundancy, but in other respects behaves exactly like a direct encoding where exactly one genotype represents each phenotype. The redundancy in this case was not useful.

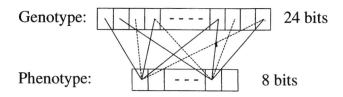

Figure 1. Illustration of the voting map; each of the phenotype bits receives input from an odd number of genotype bits. These bits vote either positively (solid lines) or negatively (dashed lines) for the corresponding genotype bit to be turned on. A positive sum results in a 1 for the genotype bit and a negative sum a 0.

2.2 Cursor based mapping

The second mapping uses a developmental approach where the phenotype is interpreted as a linear sequence of commands similar to the genetic programming paradigm (Koza, 1992). The commands control a write-head, which moves over a write area as shown in figure 2.

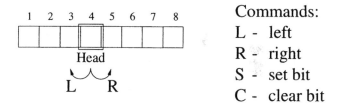

Figure 2. Illustration of the cursor based mapping, the write head can move over the write area and set or clear bits. Four commands are used: L - moves the write head one bit to the left, R – moves the write head one bit to the right, S – sets the bit under the write head, C – clears the bit under the write head. If the write head moves out of the write area it reappears on the other side.

Four different commands are used, two to control the movement of the head and two to set and clear bits on the 8-bit write area. The genotype bits thus consist of 4-alleles to represent each of these commands. All bits on the write area are initially cleared and then all commands represented in the genotype are executed sequentially. The resulting bit string is interpreted as one of the 256 phenotypes. This process is illustrated in figure 3 for a genotype length of 11. This work used a genotype of 32 commands.

Genotype   Phenotype

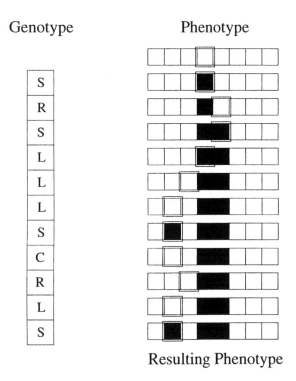

Resulting Phenotype

Figure 3. Development of the phenotype using a cursor based mapping. The genotype is interpreted as a sequence of commands, which either move the write head or set or clear bits on the write area. Initially all bits of the write area are cleared, the binary string following execution of all commands represents one of the 256 phenotypes.

2.3 Cellular automaton mapping

The third mapping also uses a developmental approach in which the genotype specifies a cellular automaton (Wolfram 1984) that is used to form the phenotype. In this work a non-uniform one-dimensional cellular automaton was used (Sipper 1997), which consists of a linear array of cells and a rule table for each of the cells specifying how the state changes over time. The state of the cell and its two immediate neighbors are used to form an index into the rule table for that bit, which specifies whether the next state should be 1 or 0. Thus, 8 entries per phenotype bit are required to fully specify the behavior of the automaton. The initial state is also encoded in the genotype resulting in a total of 72 bits for the 8-bit phenotype used in this work.

To perform the mapping from genotype to phenotype the state of the cellular automaton was initialized with that specified in the genotype. The automaton was then run for a fixed number of time steps, 20 in this work, and the resultant binary string interpreted as one of the 256 phenotypes. This is illustrated in figure 4 for an example cellular automaton and 16 updates.

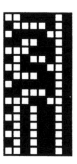

Figure 4. Development of a cellular automaton with the following rule table: 11011100, 00011001, 11100100, 11101010, 01110110, 01001110, 10101001, 10110100. The starting state (encoded in the genotype) is 00111101. The development is shown for 16 time steps after which the cellular automaton has settled into a periodic cycle. A snapshot of the automaton at a fixed time step is interpreted as one of the 256 phenotypes.

2.4 Random boolean network mapping

The fourth mapping is a generalization of the cellular automaton mapping. In this case the wiring of the automaton is also specified in the genotype together with the rule tables and initial state. Thus, the neighborhood of each cell is no longer fixed as the cell itself and its two immediate neighbors but can be any of the phenotype bits as specified by the genotype. This is illustrated in figure 5. More information on random boolean networks can be found in (Kauffman 1993).

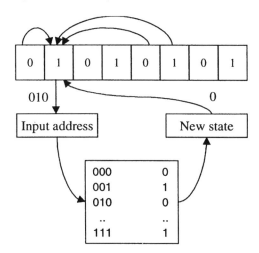

Figure 5. Illustration of the RBN mapping; the genotype specifies the rule table for each cell, the initial state of each cell and the inputs to each cell. For the example bit, the inputs specify that the third rule is used to give its next state.

In this work the number of inputs per cell was fixed at 3 and thus 9 bits were required to specify the inputs for each cell. This resulted in a total of 72 bits to specify the wiring of the 8-bit boolean network, which together with the 72 bits to specify the rule tables and initial state gave a genotype of length 144. To perform the mapping from genotype to phenotype the network was constructed and initialized using the information encoded in the genotype. As for the cellular automaton, it was then run for 20 time steps and the resulting binary string interpreted as one of the 256 phenotypes.

3. Evaluation of the mappings

The main tool used to evaluate the properties of the mappings described above was a random neutral walk (Huynen 1996). This procedure allowed a measure to be made as to the benefit of neutral drift through assessing the number of new phenotypes encountered. The walk began by choosing a genotype mapping into a given phenotype at random. All one-point mutants of this genotype were assessed and the number of new phenotypes encountered was logged. These are termed innovations. In addition a list of neutral neighbors, i.e. one-point mutants mapping into the same phenotype, was formed. One of these neutral neighbors was chosen at random and the procedure was repeated for a given number of steps, 100 steps were used in this work. If no neutral neighbors are found, the walk remains in the same position and no further innovation is possible. This process is illustrated in figure 6. A number of statistics were calculated using this procedure (see Ebner et al. 2000), the following two are reported on in this paper:

3.1 Total number of innovations

For each of the 256 phenotypes, a genotype mapping into that phenotype was chosen at random. A random neutral walk consisting of 100 steps was performed for each of these genotypes and the number of new phenotypes encountered at each step was logged. This procedure was repeated for four independent walks for each of the phenotypes. This resulted in a total of 1024 walks and the averaged cumulative number of innovations was plotted at each step of the walk for each mapping.

3.2 Phenotypic accessibility

For the same set of 1024 walks described in the previous section an accessibility plot was formed that showed which phenotypes were encountered on the neutral walks for all 256 phenotypes. This resulted in a 256 by 256 plot with one axis showing the phenotype that neutral walks were being performed for and the other axis the phenotypes encountered on those walks. This plot gave some impression of the connectedness of phenotype space via the neutral pathways in genotype space.

Genotype Space

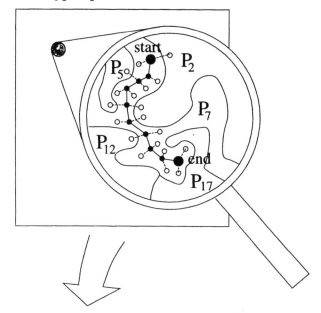

Different phenotypes encountered along random neutral walk:

$$P_2 \qquad P_5 \qquad P_{12}$$

Set of 10 equivalent genotypes found:

Figure 6. A random neutral walk in genotype space. Starting from a randomly chosen genotype, all innovations i.e. previously unseen phenotypes, are logged. A list of neutral neighbors, i.e. neighboring genotypes that map into the same phenotype, is also formed. One of these neutral neighbors is then chosen at random and the procedure repeated for a number of steps.

4. Results

This section shows the statistics that were calculated for each of the redundant mappings. In addition, the statistics for a direct encoding with no redundancy are included in order to allow comparison.

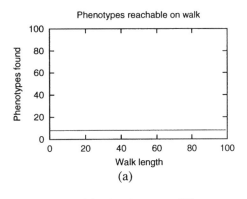

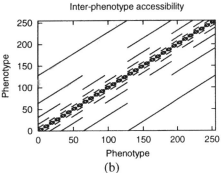

Figure 7. Results for a direct binary encoding. (a) The number of phenotypes found remains constant at 8, reflecting all possible one-point mutations. There are no neutral neighbors and thus there can be no innovation through neutral drift. (b) The accessibility is very sparse reflecting the fact that only 8 new phenotypes are accessible from each phenotype.

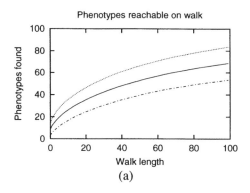

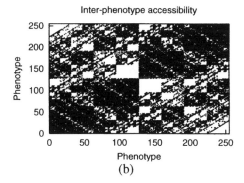

Figure 8. Results for the voting mapping. (a) The number of phenotypes found continues to increase throughout the walk indicating continual innovation. The number eventually reaches a total of 68. The dashed lines indicate +/- one standard deviation. (b) The accessibility plot is denser than for the direct encoding although some transitions were not possible.

4.1 Direct binary encoding

Figure 7 shows results for a direct encoding in which an 8-bit genotype directly specified the phenotype. In effect there was no genotype-phenotype mapping as is commonly the case in artificial evolution.

4.2 Voting mapping

The results in figure 8 show that neutral walks for a voting mapping allow access to a greater number of phenotypes than for a direct encoding. The number of innovations continues to rise after 100 steps indicating an increased probability of finding more adaptive phenotypes through neutral drift. The accessibility plot is also less sparse indicating the discovery of more phenotypes on the neutral walks. Statistics were also gathered for a trivial voting map as discussed in section 2.1. As expected, these results were identical to those for a direct encoding.

4.3 Cursor based mapping

Figure 9 again shows that neutral walks for a cursor based mapping allow the discovery of more new phenotypes than a direct encoding. However, the number of phenotypes encountered at the end of the walk was smaller than that for the voting mapping. The curve is beginning to level off after 100 steps and thus the number of phenotypes found is not likely to increase much further even with longer walk lengths. The accessibility plot is again denser than that for a direct encoding however a number of phenotype transitions were not found on the neutral walks.

4.4 Cellular automata mapping

Figure 10 shows that the number of phenotypes found is greater for the CA than for the previous mappings. The curve is still relatively steep after 100 steps indicating remaining potential for discovering further phenotypes with increased walk length. The accessibility plot is also much denser indicating that many more transitions between phenotypes are possible for the CA mapping than the previous mappings.

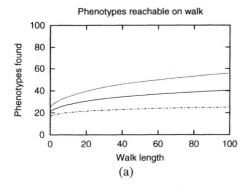

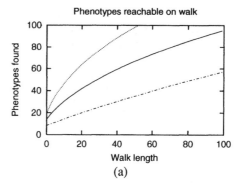

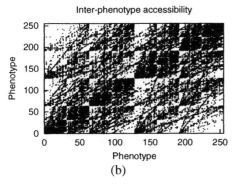

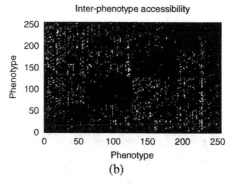

Figure 9. Results for the cursor based mapping. (a) The number of phenotypes found is an improvement over the direct encoding but not as high as for the voting mapping. The final number reached is 40. The dashed lines indicate +/- one standard deviation. (b) The accessibility plot is denser than a direct encoding but some transitions were not found on the neutral walks.

Figure 10. Results for the CA mapping. (a) 95 phenotypes were found after 100 steps on the neutral walk. The curve is still relatively steep indicating continuing innovation. The dashed lines indicate +/- one standard deviation. (b) The accessibility plot is much denser than for the previous mappings indicating greater accessibility between the neutral networks associated with each of the phenotypes.

4.5 Random boolean network mapping

Figure 11 shows that the ability to modify the wiring of an automaton increases the number of innovations on a neutral walk. The number of phenotypes found after 100 steps is approximately 50% greater than for the cellular automaton mapping and the steep curve indicates the potential for further innovations. The density of the accessibility plot is also greater indicating even greater accessibility between the phenotypes.

5. Discussion

It is common practice in artificial evolution to use a direct one-to-one mapping between genotype and phenotype. An example of such a mapping is the direct binary encoding, the results for which were presented in section 4.1. In such a scenario the number of differing phenotypes accessible from any given genotype is restricted to the length of the genotype i.e. all one-point mutants. In many situations it may be common for none of these phenotypes to be better adapted than the current one and thus adaptation will effectively halt at a local optimum.

This is a very different scenario to the seemingly open-ended innovation found in natural systems. The mappings explored in this work show that continuing innovation can be achieved with a suitable mapping between genotype and phenotype. Introducing the same kind of redundancy found in natural evolutionary systems into our artificial systems may increase the efficacy of artificial evolution through reducing the possibility of entrapment at local optima. However, the type of redundancy is crucial. This is highlighted by the voting mapping; the trivial voting mapping can be made to contain very high degrees of redundancy by increasing the number of genotype bits that are used to vote for a given phenotype bit. However, this redundancy will be of no benefit, regardless of how much is introduced. The redundancy is only beneficial if it increases the accessibility between phenotypes, i.e. if it allows more new phenotypes to be discovered than would be the case for a direct encoding. This is not the case for the trivial voting mapping as all mutations effect only one genotype bit and thus neutral mutations cannot enable moves closer to new phenotypes while maintaining the current phenotype. In order to allow for such a scenario a genotype bit must be able to effect more than one phenotype bit, which is the case when the sets of genotype bits voting for each of the phenotype bits are allowed to overlap. The results presented in section 4.2 show that this enables the discovery of many more phenotypes than is the case for the direct encoding.

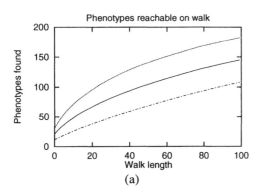

(a)

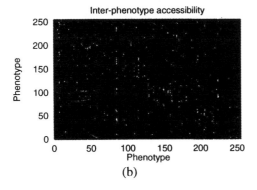

(b)

Figure 11. Results for the RBN mapping. (a) The number of phenotypes found is higher than all the other mappings. A total of 145 different phenotypes were found after 100 steps along the neutral walk. The curve continues to rise indicating continuing innovation. (b) The accessibility plot is the densest of those presented in this paper with most transitions between phenotypes found.

In order to introduce beneficial redundancy, therefore, it is important not only to create a mapping in which there are large connected sets of genotypes, representing each phenotype of interest but also to create a mapping in which the boundaries between the sets are intertwined. In the case of trivial redundancy the sets only come within a point mutation of each other in one position, as is the case for a direct encoding. However, the addition of overlap both increases the number of boundary points and introduces new boundaries with different phenotypes. The cursor-based mapping has a similar effect but on the evidence presented here, is less effective at increasing and expanding the boundary points of the sets. This is highlighted by the smaller number of phenotypes encountered during the neutral walk.

The cellular automaton mapping and, in particular, the random boolean network mapping are especially effective at introducing the right kind of redundancy. The accessibility plots are quite dense indicating that the sets of genotypes representing each of the phenotypes have become very intertwined with a high number of boundary points between them. The number of innovations indicates that boundary points with other sets continue to be found throughout the walk. These properties could considerably aid an artificial evolutionary system.

The success of a neutral mapping is dependent on the balance between structure and randomness. In order to increase the possibility of discovering a given phenotype, it would be desirable for many genotypes mapping into that phenotype to be randomly scattered throughout genotype space. Thus, from any point in that space it is likely that a required genotype will be in relatively close proximity. However this scattering cannot be entirely random, as it is important to maintain a relatively high number of neutral neighbors in order to encourage the formation of connected neutral networks and allow substantial neutral drift. The good results of the RBN mapping reflect a good balance between these two aspects. A large number of different phenotypes were discovered on neutral walks whilst an average of approximately 50% neutral neighbors was maintained.

6. Conclusion

This work has explored the properties of four redundant genotype-phenotype mappings that were constructed in an attempt to mimic the desirable properties found in nature's own redundant search space, evidenced by the work on RNA folding (Huynen, Stadler, and Fontana 1996) for example. In all four cases the redundancy was found to be beneficial, in that movement on the resulting neutral networks allowed for the discovery of a larger number of phenotypes than would be the case for a direct encoding. Thus, the probability of entrapment at local optima when using these mappings would be reduced. One mapping, based on a random boolean network, was found to have particularly good properties and may be of real benefit in an artificial evolutionary system.

The results presented in this work used only a small number of phenotypes and a relatively small sample of the genotype space. With sizeable genotypes exhaustive enumeration of the space is impossible and some form of sampling is required. An intention of future work is to explore other statistics that can help to further reveal the properties of the spaces created by these and other mappings. The performance of the mappings on larger phenotype spaces and in the context of an adaptive fitness walk is also being explored (Shackleton et al. 2000).

Acknowledgments

Part of this work was supported by the Santa Fe Institute, Santa Fe, NM. Thanks go to Susan Ptak from Stanford University, Department of Biological Sciences for helpful discussions and comments. The authors would also like to thank their respective institutions for their continuing support.

References

Ebner, M. 1999. On the search space of genetic programming and its relation to nature's search space. In Proceedings of 1999 Congress on Evolutionary Computation, Volume 2, 1357-1361. IEEE Press.

Ebner, M., Shackleton, M., Shipman, R., and Watson, R. 2000. How redundant mappings influence the ability to adapt to a changing environment. Under submission.

Hoile, C., and Tateson, R. 2000. Design by Morphogenesis. To appear in Proceedings of the Seventh International Conference on Artificial Life. MIT Press.

Huynen, M. A. 1996. Exploring phenotype space through neutral evolution. *Journal of Molecular Evolution,* 43:165-169.

Huynen, M. A., Stadler, P. F., and Fontana, W. 1996. Smoothness within ruggedness: The role of neutrality in adaptation. *Proc. Natl. Acad. Sci. USA,* 93:397-401.

Kauffman, S. A. 1993. *The Origins of Order. Self-Organization and Selection in Evolution.* Oxford University Press, Oxford.

Kimura, M. 1994. Population Genetics, Molecular Evolution, and the Neutral Theory: Selected Papers. The University of Chicago Press, Chicago.

Koza, J. R. 1992. *Genetic Programming, On the Programming of Computers by Means of Natural Selection.* The MIT Press, Cambridge, Massachusetts.

Shackleton, M., Shipman, R., Ebner, M. 2000. An investigation of redundant genotype-phenotype mappings and their role in evolutionary search. To appear in Proceedings of the 2000 Congress on Evolutionary Computation.

Shipman, R. 1999. Genetic Redundancy: Desirable or Problematic for Evolutionary Adaptation? In Proceedings of the 4th International Conference on Artificial Neural Networks and Genetic Algorithms, 337-344, New York. Springer-Verlag.

Sipper, M. 1997. *Evolution of Parallel Cellular Machines: The Cellular Programming Approach.* Springer-Verlag, Berlin.

Wolfram, S. 1984. Cellular Automata as models of complexity. *Nature,* 311(4):419-424.

Mutualism, Parasitism, and Evolutionary Adaptation

Richard A. Watson[1]   **Torsten Reil**[2]   **Jordan B. Pollack**[1]

[1] Dynamical and Evolutionary Machine Organization
Volen Center for Complex Systems – Brandeis University – Waltham, MA – USA
[2] Department of Zoology – University of Oxford – Oxford – UK.
richardw@cs.brandeis.edu

Abstract

Our investigations concern the role of symbiosis as an enabling mechanism in evolutionary adaptation. Previous work has illustrated how the formation of mutualist groups can guide genetic variation so as to enable the evolution of ultimately independent organisms that would otherwise be unobtainable. The new experiments reported here show that this effect applies not just in genetically related organisms but may also occur from symbiosis between distinct species. In addition, a new detail is revealed: when the symbiotic group members are drawn from two separate species only one of these species achieves eventual independence and the other remains parasitic. It is nonetheless the case that this second species, formerly mutualistic, was critical in enabling the independence of the first. We offer a biological example that is suggestive of the effect and discuss the implications for evolving complex organisms, natural and artificial.

1 Introduction

The phrase "survival of the fittest", ubiquitous in our thoughts about evolution, is often taken to mean mutually exclusive competition. Accordingly, mutually beneficial relationships are generally treated as a curio. But biological evidence suggests that mutualism is an important enabling mechanism in evolutionary innovation. In its strongest form, symbiosis can lead to *symbiogenesis*: the genesis of new species via the genetic integration of symbionts [Khakhina 1992, Kozo-Polyansky 1921, Margulis 1992, Merezhkovsky 1909]. For example, eukaryotes, which include all plants and animals, have a symbiogenic origin [Margulis 1992].

Such 'genetic integration' may occur via direct genetic mechanisms such as horizontal gene-transfer, but our earlier work [Watson & Pollack 1999a] provided a simple model of a relatively subtle, indirect mechanism whereby the genetic characteristics of one organism may be acquired by another. Our model parallels Hinton and Nowlan's work, "How Learning Can Guide Evolution" [1987]. Their paper demonstrates the Baldwin effect [Baldwin 1896], a phenomenon whereby learned, or plastic, characteristics can induce equivalent innate, or non-plastic, characteristics. In our adaptation of their model we replaced learning with symbiosis; or more generally, replaced the lifetime plasticity of an organism with lifetime interaction between organisms. This enabled us to show how the characteristics of one organism can be induced in another symbiotic organism. This shaping effect enables the evolution of organisms that would otherwise be unobtainable—or at least, would be very unlikely to occur.

Non-genetic variation guides genetic variation

Our simulation of the effect can be described in two phases. First, symbiotic groups find the solution to a problem (a set of abilities that confers high reproductive fitness) more quickly than the solution can be found by a single organism. This occurs simply because the combinations of abilities formed via lifetime interaction of organisms samples a much larger set of variations than the relatively slow genetic variation from mutation. In the second phase, after an ecosystem of mutually beneficial organisms has become established, the evolution of the individual organisms therein operates in a different environment. Where previously an organism that exhibited some fraction of the necessary abilities, but not all the necessary abilities, would fail, now symbionts may *fill-in* for this organism's inadequacies. Moreover, the greater the fraction of necessary abilities it exhibits the less filling-in is required—i.e. the less it depends on symbionts and the more reliably successful it is. This provides a gradient to guide genetic search toward an organism that can ultimately perform independently. Without the support of symbionts this gradient does not arise and therefore the occurrence of an independent organism exhibiting the solution would require an improbably fortunate random mutation.

Thus, the abilities first discovered by the symbiotic group become encapsulated in the heritable traits of a single individual. We call this effect *symbiotic scaffolding*: the symbionts support each other as partially able organisms, and enable the gradual accumulation of abilities, until ultimately, when their abilities are complete, the scaffolding is not required.

One way to interpret how this effect operates is as a 'smoothing' of the fitness landscape. The dotted curve in Figure 1 shows an arbitrary rugged fitness landscape. Each point on the horizontal axis represents a phenotype (or set of traits), and the dotted curve indicates the fitness of the phenotypes when evaluated independently. Now imagine that when an organism is evaluated, its own characteristics may be supplemented with those of other organisms. This modified set of characteristics will still contain the characteristics of the original organism and will therefore be somewhere in the neighborhood of the original organism's phenotype. If the organism interacts with many other organisms during its lifetime then its fitness will reflect the fitness of a set of points sampled from the neighborhood of its own characteristics. Points on the solid curve in Fig. 1 are an average of the original landscape weighted using a Gaussian. This averaging over the local neighborhood acts so as to 'smooth' the fitness landscape and provide a kind of 'look-ahead' about phenotypes in the nearby vicinity. This modified fitness landscape enables genetic search to escape from local optima in the original landscape and move towards fitness peaks that were formerly unobtainable.[1]

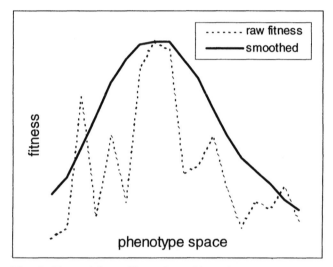

Fig. 1. The guiding effect of symbiotic interaction can be interpreted as a smoothing of the fitness landscape.

Previous experiments with one population

In our previous work we illustrated this effect using an adaptation of Hinton and Nowlan's model of the Baldwin effect. We evaluated individuals in the context of many randomly constructed groups of individuals and, instead of assigning fitness based on their individual ability, we gave them a fitness that reflected the average success of groups they formed. Without the group evaluation individuals could not be evolved to solve the problem. But with the benefit of group evaluation the symbiotic scaffolding effect enabled the evolution of initially mutualist groups that solved the problem together, and ultimately independent organisms that solved the whole problem by themselves. With this effect, we illustrated a

---

[1] This interpretation of scaffolding is quite similar to Hinton and Nowlan's explanation of the Baldwin effect, as the reader may recognize—especially, when we apply the smoothing to Hinton and Nowlan's problem in Figure 2 (which we will introduce shortly).

mechanism whereby the formation of mutualist groups enabled the evolution of organisms that would not otherwise have occurred. In so doing, this effect also shows how the characteristics of an organism can be induced in another symbiotic organism without direct transfer of genetic material. These are important principles for understanding the role of symbiosis in evolutionary adaptation.

Our model does not use the exchange of genes between organisms—the symbionts may be distinct species—and our original experiments did show some evidence for genetic divergence in the population. However, we used only one population of organisms and thus, in general, the symbionts may have been closely genetically related (by inheritance). This weakened the interpretation of the model as symbiosis between different species, and brought into question whether the effect would be seen in an ecosystem of truly separate species.

Mutualism between separate species

In the new experiments we present in this paper we sought to verify that the phenomenon was reproducible when the symbionts were reproductively isolated species. This would show the effect to be relevant to symbiosis between genetically unrelated organisms and therefore more widely applicable in nature.

Thus, in this paper, we form groups by selecting individuals from two separate fixed-size populations each of which reproduces independently. (In all other respects the experimental setup we describe here is the same as that used in our earlier experiments.) As expected, the scaffolding effect is still observed; confirming our hypothesis that mutualist groups consisting of genetically separate species can be instrumental in catalyzing the evolution of complex independent organisms.

But there is an interesting difference from our earlier results. When the symbiotic group members are drawn from two separate species, only one of these species achieves eventual independence. The pressure for the second to become independent falls off and it becomes a parasite—it gains benefit from its perfect partner but provides nothing in return.

The remaining sections of this paper are organized as follows. In the next section we describe our experimental setup, and Section 3 gives results. Section 4 introduces an example from nature that is suggestive of symbiotic scaffolding. Finally, we discuss the implications for artificial life research, and conclude.

2 Experimental set-up

Hinton and Nowlan provide a simple and elegant abstract model of the Baldwin effect which has been replicated and extended many times [Belew 1989, Harvey 1993, Mayley 1996]. Our experiments use an adaptation of their 'extreme and simple scenario'. The model is deliberately abstract so that the combinatorics involved in the effect are clear. We consider a problem that consists of a large number of variables all of which must be correctly specified by an organism in order for that organism to receive any reproductive fitness. In such cases an organism that is partially correct, even one that specifies all but one of the variables correctly, is not rewarded at all. This worst-case scenario is the extreme case of irreducible complexity, in which solutions can only be found by trying possibilities at random.

As an example scenario, we may imagine a metabolic chemical cycle with 20 steps. Each of the 20 steps must be performed by an organism correctly in order to get the chemical cycle going and to thereby confer reproductive fitness.

As stated thus far, this 'needle on a plateau' fitness landscape provides no gradient to lead search towards the solution – an organism with 19 correct steps is not favored over an organism with, say, only 1 correct step. But when we introduce lifetime interaction between organisms this will enable a gradient that leads genetic variation toward the solution. The smoothing afforded by group interaction modifies the fitness of points near the solution so that they are preferred over points farther from it, as depicted in Fig. 2.

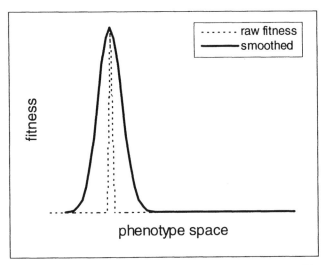

Fig. 2. An illustration of the simple problem space used in the following experiments (and in those of Hinton and Nowlan), depicting the 'smoothed' landscape effected by symbiotic interaction.

Evaluating Groups

To model lifetime interaction we test groups of organisms instead of individual organisms, and we test an organism in many groups during one lifetime. Each organism may prohibit, enable, or have a neutral effect on a step in the chemical process. If an organism is neutral with respect to a step then this step may (or may not) be completed by some other organism (of a different species). That is, an organism can gain the benefit (or penalty) of chemical byproducts created by the processes of other organisms in the ecosystem (for those steps where the organism itself is neutral).

We may crudely represent the relevant traits of an organism in 20 genes where each gene has three alleles: *correct*, *incorrect* or *neutral* corresponding to enabling, prohibitive (preventing completion of the cycle), or neutral interaction with a step in the cycle. This unrealistic simplification enables us to see the mechanisms of interest more clearly but it is not integral to the results that follow.

We use the happenstance co-location of organisms to form groups since it makes minimal assumptions about the nature of symbiont interactions. The use of a more sophisticated model of symbiotic relationship-forming will illustrate the scaffolding effect more strongly—we stress that our model of organism interaction is deliberately trivial so as to prevent details from obscuring the essence of the effect. In our model we may imagine that organisms are randomly distributed in the environment and perpetually mixed. At any one instant there will be some number of other organisms in the immediate vicinity of the organism in question. Thus every organism is tested by combining its abilities with those of several other randomly selected organisms of the other species. Fig. 3 shows how the abilities of organisms are combined. In this figure there are two species, A and B, and we prevent solutions being provided by mutualism within a species by using only organisms from the complementary species to fill-in groups.

```
 first organism, A1: 00-0-11---01-1--0--1
second organism, B1: -1-0-1-001-01-1--0--
 third organism, B2: 1--10-11-0----10---1
fourth organism, B3: -01---0-0----10-0-11
 fifth organism, B4: --111----0101--0-0--
combined abilities: 00100110010111100011
```

Fig. 3. Combining the abilities of organisms. The 20 traits of each organism may take one of three forms: correct, incorrect or neutral shown as 1, 0 and "-", respectively. The neutral abilities of an organism from species A may be filled-in by the abilities of organisms from species B. Notice that the traits of the first organism take priority over all others; for consistency, the traits of the second organism take priority over all but the first, and so on. In some groups, some organisms may be redundant since every trait is specified by at least one of the preceding organisms, as illustrated by the fifth organism shown here. Note that this combination of abilities does not represent the formation of a new organism - it is merely a representation for the result of different species acting in concert.

Since both the selection and ordering of the organisms are random, the details of this filling-in mechanism are largely inconsequential to the results that follow. One important feature, however, is that the fitness of the combined traits is awarded to the first organism only, and that the traits of the first organism are not over-ruled by any other. However, since the first organism will likely fill-in for other organisms in their turn, this asymmetry is reciprocated. Alternate models of interaction and reward distribution may be equally valid; however, the current model is sufficient for our purposes.

A key feature of the mechanism we are modeling is that the search for combinations of abilities via lifetime interaction is much more rapid than that arising from genetic variation alone. Hence we test each organism in 1000 random groups during its lifetime. Each group is formed from a different random selection of organisms drawn from the other species.

Our earlier experiments showed that when appropriate symbionts are reliably available, and incur no additional overheads, then there is no pressure to be independent. But naturally, if there is some cost to relying on symbionts then independent organisms are preferred. In our earlier experiments we added an implicit cost by limiting the availability of symbionts. Implementationally, we limit group sizes probabilistically with the limit randomly selected from an exponential distribution for each group formed. Specifically, the probability of there being exactly **k** members in a group is 2^{-k}, $k \geq 1$. In this way it is most likely that an organism will be evaluated on its own; next most likely it will be evaluated with one other organism, and so on.

Finally, the fitness of an organism is given by **f=1+n,** where **n** is the number of groups (out of the 1000 groups tested) in which the organism in question forms a successful group.

Experiments

The genetic model, the method of interaction, and the evaluation described above are iterated in a genetic algorithm (GA) [Holland 1975]. Hinton and Nowlan chose the population size, number of lifetime trials, and number of variables in the problem carefully so as to make it most unlikely that genetic variation alone would find the solution but very likely that lifetime variation would. We continue to follow the experimental parameters of Hinton and Nowlan where applicable for the same reasons. We use 1000 individuals but here they are divided into two populations of 500 representing two species. Each species, say **A** and **B**, reproduces independently, and as indicated in Figure 3 (above), each member of **A** is evaluated by filling-in its missing abilities with members from **B**, and vice versa. When an individual is being evaluated the group will not contain other members of that individuals' species.

Fitness-proportionate reproduction is applied generationally [Holland 1975] to each population separately. In this way there is no competition between the members of population **A** and the members of population **B**, but the specific configuration of individuals in each population does have an effect on the fitness of individuals in the other population through the group evaluation in which they are involved.

In these experiments we use mutation as our only source of genetic variation. Mutation is applied with a bitwise probability of 0.05 of assigning a new random value. New values are randomly selected to be correct, incorrect or neutral genes with probability 0.25, 0.25, 0.5 respectively. These same proportions are used to construct the initial population so that, on average, an individual will have half its genes neutral and half non-neutral (as illustrated in Fig. 3).

We should emphasize which parts of the results that follow are expected and which parts offer new insight. As stated, the size of the problem (number of steps in the cycle), ratio of alleles in the initial population, number of organisms, and number of groups tested for each organism are deliberately chosen to make the discovery of a successful individual most unlikely and the discovery of a successful group most likely. The interesting part of the effect is what happens *after* successful groups are formed. Specifically, we are looking for how the trends in the make-up of individuals change in the context of symbiotic groups. We will see that before the formation of successful groups, increases in the number of correct alleles are not selected for. However, after groups are formed, there is a trend towards more correct alleles. This trend eventually results in some organisms becoming independent which would not have happened without the presence of the (eventually redundant) mutualists.

3 Results

Fig. 4 shows the number of each allele per organism averaged over all organisms (in both populations) at each generation. We see that the proportion of alleles at the start of the experiment is as per the mutation probabilities, i.e. approximately 0.25,0.25,0.5 for correct, incorrect and neutral respectively. Around the 225th generation a quite dramatic change takes place: the proportion of incorrect alleles falls close to zero whilst the number of correct alleles rises. (The exact generation at which these sharp changes occur varies from run to run due to the stochastic nature of the experiment.) This is the point where symbiotic organisms become established and incorrect alleles are purged from the gene pool. Thereafter we see a clear upward trend in the number of correct alleles in subsequent generations. Unlike our original experiments there is no clear trend towards 20 correct alleles in the following generations. But we will see in a moment that, although the average number of correct alleles over both populations does not continue to rise, some individuals do find all 20 correct alleles.

In Fig. 5 we see that the dramatic changes around 225 generations coincide with the establishment of groups that solve the complete cycle.[2] Then we see, at about 350 generations, the establishment of individuals that are self-

---

[2] Although there are a few instances of successful groups in the first 200 generations they do not take hold in the population.

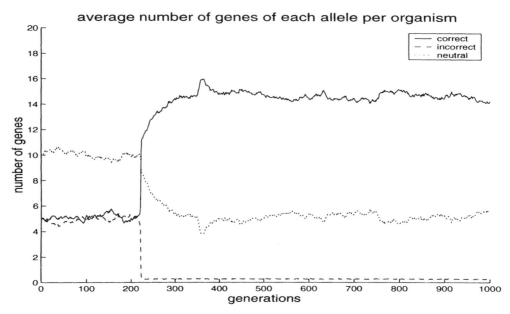

Fig. 4. Number of genes of each allele per organism, averaged over all 1000 organisms at each generation.

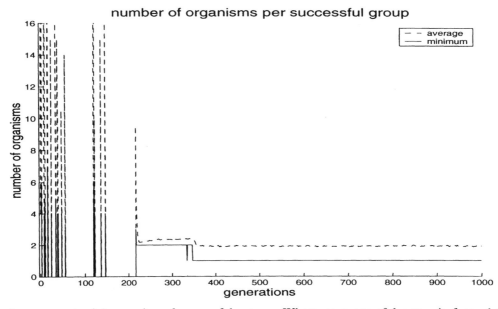

Fig. 5. Average and minimum size of successful groups. Where no successful group is formed (as in some of the first few generations) both the average and minimum group sizes are shown as zero.

sufficient (i.e. we see a minimum group-size of 1).

The effects shown in Figures 4 and 5 have some qualitative similarity with our previous work using a single population. This is as expected. Separating the organisms into two reproductively isolated populations makes no difference to the combinatorics and probabilities involved, and the establishment of complementary mutualists does not depend on genetic relatedness. However, we need to explain why the average number of correct alleles remains lower than in our single population model, and when we examine the make-up of the two populations separately in Fig. 6 (and also Fig. 7) some interesting features are revealed.

From Fig. 6 we can see that in the initial stages the maximum number of correct alleles shows no significant trends but begins to increase when mutualist groups are established. The trends in both populations are identical thus far. But, as one of the species approaches independence the other species starts to decline. There is no asymmetry between the two populations in the set-up

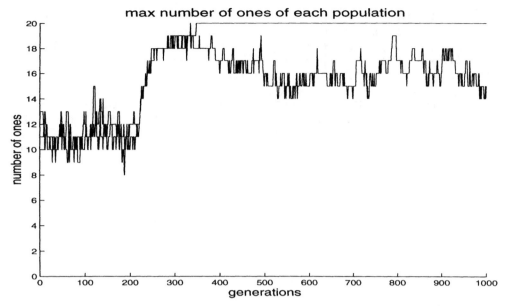

Fig. 6. Maximum number of correct alleles shown separately for each population.

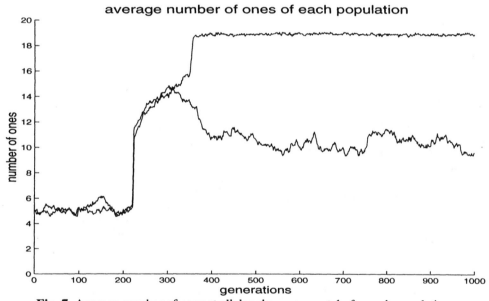

Fig. 7. Average number of correct alleles shown separately for each population.

of the experiment – the asymmetry that develops depends on which of the two populations happens to find independent organisms first. The maximum number of correct alleles in the second population now starts to decline. These trends are seen more clearly in Figure 7 which shows the average number of correct alleles. There is a clear intermediate stage where organisms in both populations are increasing in ability and a clear separation in the latter stage where one species becomes independent. The second species is not contributing anything to the first but does receive benefit from it – it has transformed from a mutualist to a parasite. The rise of

this parasitic species can be understood as follows. Consider two species **A** and **B**. When there are imperfections in members of **A** there is some advantage in increased independence of **B**. But if **A** is perfect then members of **B** with say, one or ten neutral abilities are equally likely to be correctly supported by **A**. Thus, there is no fitness differential for organisms in **B**. The only exception being the preference for an organism that does not need **A** at all. In fact, in a context of perfect hosts, organisms are faced with a 'needle on a plateau' problem space that is the same as the original space without symbionts. So, as one species approaches independence and thereby becomes a nearly perfect host, the pressure for the second species to exhibit correct bits declines.

The exact changes in the second species will be determined by the balance between the remaining selection pressure for correct bits (since the hosts are not quite perfect, on average), and the pressure of mutation which, when individuals have a high number of correct bits, will usually reduce the number of correct bits.

Since the introduction of new alleles via mutation includes both neutral and correct bits[3] in the ratio 2:1 (or 0.5:0.25) respectively, we might expect that the population will likely drift to reflect this ratio, i.e. show an average of approximately 33% correct bits. In practice, it seems that the ratio is reliably higher—not going outside 40-60% in all 17 (of 20) runs of the experiment that exhibited a host/parasite split. The reason for this is not clear to us but may possibly reflect the fact that there is still some slight pressure for correct bits in the parasites since the hosts are not quite perfect (about 19 correct bits on average).

Note that the fitness of the parasites is considerably lower than the fitness of the hosts—recall that most of the time an individual is evaluated on its own given the probabilistic group-size of 2^{-k}, and thus, in most trials, a parasite will get a fitness of zero. In contrast, a perfect host gets maximum fitness in all trials. However, since the hosts and parasites are in separate populations the parasites will not be replaced by hosts (as they were in our original experiments). And, unless some member of the parasite population by chance makes the transition to full independence, the fitness of all parasites is the same: when they are evaluated alone they receive no fitness contribution, when they are evaluated with a perfect host they receive full-fitness. Their average fitness in the context of perfect hosts is only a function of the probability that they are evaluated in a group, and has no dependence on the number of correct traits they exhibit.

In these experiments we have focused on models of two species. Preliminary experiments with three or more species indicate that the likelihood of a species becoming independent decreases with the number of other species that are already independent. The last species finds itself drifting in the same manner as the second species in these experiments.

To summarize, in this overall effect we see an ecosystem of unrelated organisms that become mutualist organisms, scaffolding for each others inabilities. During this mutualist period the organisms increase in their individual ability until one of the species becomes independent. The characteristics formerly exhibited only by mutualist groups have been encapsulated in the characteristics of single organisms. As the first species approaches independence the second species transitions from being a mutualist to being a parasite. Although it now confers no benefit to its independent host, it was

essential in enabling the evolution of the host's independence.

In other runs of the experiment there is considerable variation in the exact generation in which symbiotic organisms become established, and in the generation which exhibits the first self-sufficient individual. Nevertheless, the discovery of an independent species and a residual parasitic species is quite reliable. 17 out of 20 runs of the experiment resulted in an independent species with a parasite. As stated, the average number of correct alleles in the second species (through 1000 generations) remained around 50% in these 17 runs. In the other three runs the second species also found independent organisms. In most runs the intermediate stage, where mutualism is increasing the number of correct alleles, occurs very quickly, but without the mutualism (i.e. without group evaluation) the effect does not occur at all, and an independent organism is never found.

4 An example from nature

The search for examples of symbiont scaffolding in the natural world is complicated by the transient nature of the phenomenon. Evidence supporting our original, one-species model was particularly problematic since the end result expected to find nothing but independent organisms. However, in this two-species model we expect a residual symbiotic partner to be more common and this type of relationship is easier to find.

For the process to be identified as such, a symbiotic relationship has to be recognised in which at least one partner has acquired at least one trait that was previously exclusive to the other partner. There must furthermore be evidence that this particular trait would have been unavailable for the partner in question through conventional evolution. We suggest here that these conditions are met by the bivalve species *Solemya reidi*.

Members of the genus *Solemya* are characterised by a close symbiotic relationship with sulfide oxidising bacteria [Cavanaugh 1994] enabling them to inhabit sulfur rich environments such as deep thermal vents, which would otherwise be inhospitable due to the high toxicity of sulfur [Grieshaber & Völkel 1998]. In addition to providing a detoxification mechanism the bacteria supply the clam with organic compounds from CO_2 fixation.

Intriguingly, one member of the genus, *S. reidi*, exhibits the ability to oxidise sulfide by itself – a trait previously unique to its symbiotic bacteria, which it still harbours it its gills [Powell & Somero 1986]. We suggest that this partnership represents an example of symbiont scaffolding as we have modelled. The scaffolding process may be complete, in which case the bacteria are now parasitic. Alternatively, *S. reidi* may represent a transitory stage of the process, that is, the clam has acquired one characteristic from the scaffolding bacteria but the bacteria still provide other characteristics essential to the clam. Which of these scenarios is the case is not yet clear.

The possibility of a transitory condition is supported by

---

[3] Incorrect bits are still selected against.

the fact that *S. reidi* lives in anthropogenic habitats, namely sewage outfalls and pulp mill effluents, making it likely to be an evolutionary novelty. As such, there may have been insufficient time for symbiont scaffolding to come to an end, i.e. to reach a point where the bacteria are no longer of benefit to the clam and are therefore entirely parasitic.

In summary, we propose that the case of *S.reidi* illustrates symbiotic scaffolding as follows: a) The toxicity of the sulfur-rich environments presents an impossibly hard transition for the adaptation of the clam via conventional evolution; the correct adaptation being analogous to our 'needle on a plateau'. b) The mutualist clam-bacteria symbiosis, as exhibited by other *Solemya* species, was an ancestral stage for *S. reidi* before its own ability to oxidise sulfur. c) The provision of sulfide oxidation by the bacteria smoothed the problem landscape presented to the clam. Given some cost in reliance on symbionts, or some benefit in independent sulfur oxidation, this relationship induced *S. reidi* to gradually evolve the trait itself. In this case, the main adaptive advantage in being less reliant on the bacteria was energy production (sulfide oxidation in *S. reidi* is coupled to ATP synthesis in the mitochondria). The extent of the latter is demonstrated by the absence of both mouth and gut in *S. reidi* [Cavanaugh 1994], indicating total independence from conventional feeding mechanisms. d) Thus, the clam has acquired a characteristic from the bacteria via symbiotic scaffolding. *S. reidi* is now less dependent on the bacteria that catalysed its own abilities, at least in respect of sulfide oxidisation.

While the case of *S. reidi* is suggestive, one caveat must be pointed out. It is possible that the clam's ability to oxidise sulfide has arisen as a consequence of gene transfer from the bacteria to the bivalve's mitochondria. If such inter-species gene transfer has occurred then symbiont scaffolding, as described in the previous sections, would not be required to explain the characteristics of *S. reidi*. In order to clarify the issue, the sequence of the sulfide oxidising enzyme in both symbiotic partners must be determined and tested for homology.

5 Discussion and Future Work

In symbiogenesis, the creation of new species via the genetic integration of pre-adapted symbionts provides a different source of innovation from the Darwinian gradual accumulation of random mutations. And the indirect acquisition of symbiont characteristics that we have modeled in this paper also suggests a different perspective on adaptation. The scaffolding effect provides a mechanism by which non-genetic variation can guide genetic variation. Although in this indirect mechanism it is still the accumulation of random mutations that implements the effect, the mutations now occur under selection which is educated by a form of 'look ahead'. This look ahead is provided by the exploration of phenotypic characteristics afforded by the formation of symbiotic groups that include the organism in question.

Though there are many computational models that explore the evolution of cooperative behavior there are none that use mutualism as an integral part of adaptive innovation. As Artificial Life (ALife) researchers we are well advised to understand the sources of innovation in natural evolution. Changes in the way we look at organism interaction, mutualism, and evolutionary adaptation in general, open new directions for ALife research.

For example, existing ALife models that are based on genetic evolution utilize some fixed representation of genes, of one kind or another, and evolutionary search uses fixed variation operators to explore the space this representation affords. A failing that is therefore common to all such systems is their inevitable *complexity ceiling*. That is, after some initial promise, further innovation is not forthcoming since most variations are detrimental [Kauffman 1993]. In light of this, perhaps the most interesting interpretation of the symbiogenic and symbiont scaffolding mechanisms is that they provide natural evolution with an escape from this problem by creating new 'units of variation'. In the models we have illustrated, the important discovery takes place by shuffling combinations of *organisms*, not by mutation of the genes. Genetic mutation merely follows in its footsteps. At first, sets of abilities are explored by shuffling groups of simple organisms. Then successful groups are encapsulated into composite individuals that exhibit the characteristics formerly exhibited by the group. Now sets of abilities may be explored by shuffling groups of these more complex organisms – variation is now operating on larger units. The potential for the mechanisms to recurse in this way provides the opportunity to scale-up the representation in which search takes place. This, we believe, has potential for over-coming complexity ceilings in ALife models, just as it has been instrumental in enabling major transitions in natural evolution [Maynard-Smith & Szathmary 1995]. Other ongoing research is directed at hierarchical problem solving using these ideas [Watson et al 1998, Watson & Pollack 1999b].

6 Conclusions

This paper has developed previous work investigating the guiding effect of mutualism in evolutionary adaptation. This effect, which we call symbiotic scaffolding, enables the characteristics of mutualist groups to become encapsulated in a single individual and thereby enable the evolution of independent complex organisms that would not otherwise occur. We have shown here that this effect may occur between genetically unrelated species and the new set-up has revealed an interesting new feature. When mutualists of two different species scaffold one-another only one achieves eventual independence and the other remains as a parasite. Finally, we introduced a biological example that is suggestive of this model.

Acknowledgments

Though several audiences suggested re-implementing our previous work with separate species, Peter Todd, of the Max Planck Gesellschaft, suggested directly that when the two populations were separated they may exhibit some interesting asymmetry – which indeed they did. Thanks also to Steven Frank and Vladimir Kvasnicka for sharing comments. Thanks, as always, to the members of DEMO at Brandeis for assisting us in the research process.

References

Baldwin JM, 1896, "A New Factor in Evolution," *American Naturalist*, 30, 441-451.

Belew RK, 1989, "When Both Individuals and Populations Search," in Schaffer JD, ed., *ICGA3*, San Mateo California: Morgan Kaufmann.

Cavanaugh, CM, 1994, Microbial Symbiosis: Patterns of Diversity in the Marine Environment, *Amer. Zool.*, 34: 79-89.

Grieshaber, MK, Völkel, S, 1998, Animal Adaptations for Tolerance and Exploitation of Poisonous Sulfide, *Anuu. Rev. Physiol.*, 60: 33-53.

Hinton GE, & Nowlan SJ, 1987, "How Learning Can Guide Evolution," *Complex Systems*, 1, 495-502.

Harvey I, 1993, "The Puzzle of the Persistent Question Marks: a Case Study of Genetic Drift," in S. Forrest (ed.), *Genetic Algorithms: Proc. of the Fifth Intl. Conference*, pp. 15-22, San Mateo CA, Morgan Kaufmann.

Holland JH, 1975 *Adaptation in Natural and Artificial Systems*, Ann Arbor, MI: The University of Michigan Press.

Kauffman, S, 1993, *The Origins of Order*, Oxford University Press.

Khakhina LN, 1992 *Concepts of Symbiogenesis: Historical and Critical Study of the Research of Russian botanists*, eds. Margulis L, McMenamin M, (translated by Merkel S, Coalston R), Yale University Press.

Kozo-Polyansky BM, 1921, "The Theory of Symbiogenesis and Pangenesis, Provisional Hypothesis" in *Journal of the first All-Russian Congress of Russian Botanists* in Petrograd in 1921. Peterograd (in Russian, see Khakhina 1992).

Mayley G, 1996, "Landscapes, Learning Costs and Genetic Assimilation," in *Evolution, Learning, and Instinct: 100 Years of the Baldwin Effect, special issue of Evolutionary Computation*, Vol 4, No. 3; P.Turney, D. Whitley and R. Anderson (eds).

Maynard-Smith, JM & Szathmary, E, 1995, "*The Major Transitions in Evolution*", WH Freeman and Co.

Margulis L, 1992, *Symbiosis in Cell Evolution: Microbial Communities in the Archean and Ptroterozoic Eons.* 2d ed., W.H. Freeman, New York.

Merezhkovsky KS, 1909 "The Theory of Two Plasms as the Basis of Symbiogenesis, a New Study or the Origins of Organisms," *Proceedings of the Studies of the Imperial Kazan University*, Publishing Office of the Imperial University. (In Russian, see Khakhina 1992).

Powell, MA, and Somero, GN, 1986, Hydrogen-Sulfide Oxidation is Coupled to Oxidative-Phosphorilation in Mitochondria of *Solemya reidi*, *Science*, 233: 563-566.

Watson RA, Hornby GS, & Pollack JB, 1998, "Modeling Building-Block Interdependency," *PPSN V*, Eds. Eiben, Back, Schoenauer, Schweffel: Springer.

Watson, RA & Pollack, JB, 1999a, "How Symbiosis Can Guide Evolution". Procs. of Fifth European Conference on Artificial Life, Floreano, D, Nicoud, JD, Mondada, F, eds., Springer.

Watson, RA, & Pollack, JB, 1999b, "Incremental Commitment in Genetic Algorithms," In Banzhaf, W, Daida, J, Eiben, AE, Garzon, MH, Honavar, V, Jakiela, M, & Smith, RE eds. *GECCO-99*. San Francisco, CA: Morgan Kaufmann.

Coevolving Mutualists Guide Simulated Evolution

Michael L. Best

Media Laboratory, MIT, Cambridge, MA 02139
mikeb@media.mit.edu
www.media.mit.edu/~mikeb

Abstract

We show that the mutual coevolution of cooperating traits amongst interacting populations permit the solution of a matching problem. This solution, within a highly uncorrelated fitness landscape, is difficult in the absence of coevolution or other powerful agencies. We start with the GA environment of Hinton and Nowlan (1987) who originally showed that in the absence of individual learning evolving agents are not able to solve a related problem. While a number of researchers have demonstrated that the coevolution of cooperating and (in particular) competing populations of agents can improve simulated evolution, we argue that coevolved mutualists can help evolution find a solution it otherwise could not solve (namely, selection of some particular single bit string in an uncorrelated landscape). We posit that coevolved mutualists succeed at this problem because they are able to benefit from genetically stored solutions to sub-problems. This result suggests that perhaps natural problems such as wasp/fig tree signaling, or gene-culture coevolution of vocal learning in songbirds or human natural language may be guided by coevolutionary mutualism.

Introduction

Symbiosis is a prevalent phenomena; indeed probably most of the earth's biomass depend on some form of symbiotic relationship (Begon, Harper & Townsend, 1996). Symbiosis exists when two agents from differing species interact in a way that increases the fitness of at least one of those agents. Under *mutualism*, the phenomenon under study here, both agents enjoy a fitness advantage. *Coevolution* occurs when cooperating (or indeed competing) populations interact over trans-generational time.

Related coevolutionary GA work

The Genetic Algorithm (GA) has been used to explore a variety of ecological interactions, in particular forms of competition and cooperation. This research has both aimed at improving the GA's performance on certain optimization and search problems as well as studying directly these particular ecological interactions.

The majority of coevolutionary studies with the GA have concentrated on competitive interactions between populations where an increase in one population's fitness correlates with a decrease in the other. In an influential early experiment, Hillis (1990) coevolved a population of sorting networks against a population of test cases and showed that this improved the quality of the solutions. Similarly, Paredis (1994) has explored coevolutionary interactions between neural networks and their training data. Sims (1994) evolved the morphology and behavior of populations of artificial creatures. Fitness was measured by paring off agents in one-on-one competitions over some common resource. Other researchers have explored, in particular, the predator/prey competitive relationship between a pursuer and evader population (Cliff & Miller 1996). These researchers' most interesting result is a negative one: Coevolution is not a panacea and the emergence of non-trivial adaptively unpredictable behavior is hardly guaranteed.

A number of researchers, describing their work as competitive coevolution, have used the GA to evolve *single* populations of agents who interact under a competitive fitness regime. In these experiments an agent somehow competitively interacts with other agents chosen from within its own population (sometimes across generations) and its fitness is based on this interaction. Reynolds (1994) evolved with competitive fitness a population of agents that act as both pursuers and evaders in a game of tag. Juillé and Pollack (1996) used genetic programming with competitive fitness to evolve solutions to a neural network classification problem.

While the bulk of GA studies have concentrated on competitive interactions some researchers have explored symbiotic relationships. Bull, Fogarty and Pipe (1995), for instance, were influenced by Lynn Margulis' work on cellular complexity to study endosymbiotic relationships. Here, one species is contained symbiotically within a host species. Bull, Fogarty and Pipe employed Kauffman's (1993) NKC coevolutionary model to study when this form of species interaction might emerge. Potter and DeJong (1994) coevolved separate populations that cooperated to solve a function optimization problem. Each population worked on a distinct sub-problem and fitness was proportioned back to these agents based on a credit assignment scheme.

Finally, Smith and Cribbs (1996) explicitly compared cooperating and competing interactions with the GA. Here, competition versus cooperation is with respect to choosing a solution to elements of a control system. It was not clear if these populations interacted ecologically via the fitness function.

Most of these experiments demonstrated that the coevolution of populations, in particular with predator/prey type competitive interactions, increased the quality of and speed to some solution.

The paper

We have now introduced the concept of coevolved mutualists and glossed the related GA work. In the next section we introduce the GA due to Hinton and Nowlan and describe how they showed that individual learning allowed evolution to solve a difficult problem. In the next section we describe a substantial modification to their environment where, instead of individual learning, populations mutually coevolve a solution to the problem of identifying a particular single bit pattern. In our analysis of this simulation we argue that coevoled mutualists solve the problem because they benefit from genetically stored partial solutions. Finally, we end with our conclusions and motivate these findings by considering natural cases of coevolved mutualists: wasp/fig signaling, and gene-culture coevolution of vocal learning in songbirds and human natural language.

Hinton and Nowlan's Simulation

Hinton and Nowlan used a simple GA simulation to demonstrate the effects of individual learning on genetic evolution (Hinton & Nowlan, 1987). In particular their simulation gave support to the Baldwin effect (Baldwin, 1896). This phenomenon, as understood by Hinton and Nowlan, suggests that individual lifetime learning can alter the fitness landscape searched by genetic evolution.

In Hinton and Nowlan's simulation each agent is represented by a string of twenty characters, this is the agent's genotype. Each locus along the string can take on one of three values, namely, "0", "1", or "?". The values, "0" and "1", represent genotypic codings which are described as connecting relay switches for a neural net with 20 potential connections. In the simulation a population of 1000 agents were initially seeded randomly, each locus having a 25% chance of being assigned a "0", 25% chance a "1", and 50% chance a "?".

The "?" value is what allows for lifetime behavioral plasticity amongst the agents. Each agent enjoys during its lifetime a series of learning rounds. During each round the agent runs a simple individual learning algorithm which for each "?" in the genotype assigns either a "0" or "1" as the expressed value. Note that the values assigned to the question mark positions during a learning round are never written back into the genotype nor passed directly on to progeny.

The individual learning algorithm itself is quite simple. For each learning round each agent queries a random number generator once for each question mark. With 50% probability a "0" is assigned to the locus and with 50% probability a "1" is assigned. Thus the agent performs during the learning rounds a random walk of a subspace of the boolean hypercubic landscape (as described by the loci marked by question marks). This search clearly is random with respect to the resultant fitness; it is not biased towards adaptive behaviors but has equal probability to assign a zero or one. For each round of genetic evolution every agent engages in 1000 learning rounds during which each of its "?'s" are subject to individual learning.

The reproductive rounds proceed using, more or less, the traditional genetic algorithm. The fitness of each agent is evaluated according to an external function and agents are selected (with replacement) with probability proportional to that fitness. The selected agents are paired at random and the crossover operator is applied to their genotypes. The mutation operator is not used.

In their simulation the external fitness landscape is unusual, indeed it describes a rather cruel space. The function is low for all dichotomies of the genotype except for the string of all ones for which it is high. Thus it can be described as an impulse function. This fitness landscape is highly uncorrelated and provides a difficult space indeed for genetic evolution to search — a needle-in-a-haystack search. During the selection stage of the GA an agent that is all "1's" will be assigned a high fitness and an agent with *any* "0's" will receive a low value.

But, given the general sketch above, how exactly do they evaluate each agents fitness during the selection phase of the GA? The agent's fitness is a function of its successes during the learning rounds as well as its initial genetic makeup. If during the set of learning rounds the agent happened upon the winning dichotomy, namely the string of all ones, then the agent notices that it has found the co-adaptive set of behaviors and will stop its learning. After all 1000 learning rounds the fitness for an agent is given by, $(1 + 19n / 1000)$, where n is the number of learning rounds that remain after the agent has learned the correct bit string.

The principal measurement of the state of the simulation is the proportion of loci occupied by "0's", "1's", or "?'s". Should genetic evolution make any progress one would expect the percentage of "0's" to drop to zero and the percentage of "1's" to rise. Since the fitness function defined above penalizes agents who have to spend time learning the all-one configuration, one expects there to be selective pressure towards the genetic assimilation of "1's" and the removal of "?'s" from the genotype. We have re-implemented the Hinton and Nowlan algorithm and summarize their main result in Figure 1. Here we see the percentage of "0's" quickly drop to zero and the percentage of "?'s" slowly decline.

The important finding of Hinton and Nowlan's original work is revealed when one compares the results of the simulation in the presence of individual learning against the exact algorithm run without any learning. Without learning there is no significant variation in the frequency of "0's", "1's", and "?'s" from their original random assignment, even after runs of 1,000's of generations. Thus the conclusion is that in the absence of individual learning genetic evolution is lost; it will never find and hold onto the needle-in-the-haystack. But when coupled with the adaptive learning algorithm, genetic evolution is able to find and keep hold of the all-one configuration, bringing along the entire population of agents. The learning mechanism not only increases the *likelihood* of discovering the needle due to searching. It also allows agents which are *close* to the needle to receive some fitness (and therefore reproductive) benefit since they are likely to find the needle during their learning trials. This in effect smooths-out the spiked landscape creating a virtual slope through the fitness landscape; and it is this slope which the genetic algorithm is then able to hillclimb. Thus individual learning "guides" evolution to the adaptive goal.

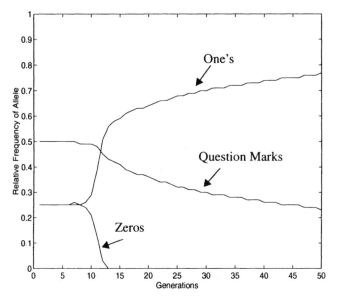

FIGURE 1. Original Hinton and Nowlan result shows that the number of "0's" quickly falls off and the "1's" are genetically assimilated slowly over time. Without individual learning the graph never changes from its initial condition.

A number of researchers have analyzed or enhanced the findings of Hinton and Nowlan. For instance, Belew (1990) studied a culturally-enriched version of the original model. Inman Harvey (1993) studied the persistence of question marks within the evolving population of agents. Through a probabilistic diffusion model he argues that the persistent question marks are due to genetic drift in the system. And I (Best, 2000), have explored social learning algorithms and contrasted them with the original individual learning mechanism.

Coevolution of Mutualists

We have modified Hinton and Nowlan's simulation in order to explore whether the coevolution of mutualists can also assist organic evolution towards an adaptive goal. This adaptive problem is difficult because it is described by a completely uncorrelated landscape. The agents need to arrive at a particular configuration of traits without the aid of helpful partial solutions (in other words, without any hills in the landscape to climb). The essence of the problem is to find some particular single bit-string, given that no fitness hints will be offered as one gets close.

In the original simulation the adaptive goal is described by a string of all "1's". However, any 20-bit string would be equally difficult to arrive at as the all-one string, as long as partial solutions are not rewarded by the fitness function. We consider a type of coevolution where one population is evolving the "answer" while the other population evolves the "solution". In other words, one population evolves an arbitrary 20-bit string and the other population tries to arrive at this particular single 20-bit string in response.

This sort of coevolutionary mutualism is suggestive of many phenomena within the natural world which we discuss in the conclusion.

In our simulation the initial population of 1000 agents is similar to the original population of Hinton and Nowlan. However, the loci are seeded with a "1" 50% of the time and a "0" 50% of the time. There are no plastic "?" loci and no learning rounds. Instead of learning rounds, the population is divided into two sub-populations of 500 each. One population tries via the GA to arrive at a "solution" and the other population tries via the GA to evolve the "answer". The two populations are independent with respect to crossover and selection.

The GA rounds proceed identically to the original simulation except for a modification to the fitness function. Rather then computing fitness from the number of rounds of individual learning it takes to arrive at the all-one dichotomy, an agent's fitness is computed as the number of *matches* between its configuration and the configurations of the agents in the coevolving population (this number is then plugged into the fitness equation above). Two agents match if they have identical bit strings.

The original fitness proportionate selection is used such that agents that match more with agents from their coevolving sub-population are selected with greater frequency. Two-point crossover and no mutation, exactly as is the case with the original simulation, finish off the GA implementation.

Recall that in the original simulation in the absence of learning rounds, the GA is not able to arrive at and maintain the winning 20-bit string. Genetic evolution was not capable of solving this difficult problem without some additional adaptive agency, such as individual learning. Moreover, it is not at all obvious that under the new scheme, with a coevolved 20-bit string as the adaptive goal, the population will be able to converge on and maintain a particular solution. First off, we have the same difficult highly uncorrelated landscape — an impulse function. But now the impulse is moving rather then staying fixed. It is worth emphasizing how this is similar to the original problem. We are attempting to arrive at a particular single 20-bit string in an otherwise completely uncorrelated landscape. In the original experiment, this bit string is fixed. Here, the bit string is allowed to move about.

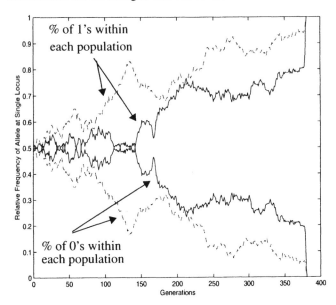

FIGURE 2. Percentage of "0"'s and "1"'s for both populations at a single locus. Both populations steadily fixate on 1's for this position.

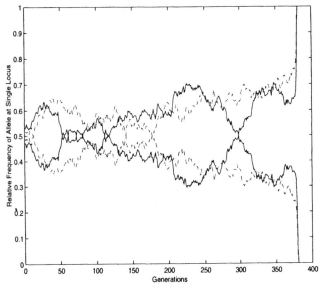

FIGURE 3. Percentage of "0"'s and "1"'s for both populations at a different locus. Here, both populations vacillate between a majority of "0"'s and "1"'s until they finally fixate on a particular setting.

We find that the GA is able to converge to and maintain a solution for the problem when enhanced with mutual coevolution. By "convergence", we mean that all agents of the two populations settle on the same bit string. However, the convergence to a 20-bit solution, given coevolution, is slower (takes more genetic rounds) then it is with individual learning.

In Figure 2 we see a typical example of a single locus for the two sub-populations as they proceed through generations. Each sub-population is represented by a pair of mirror-image lines; they are the percentage of "0"'s and "1"'s at this particular locus across the entire sub-population for each generation. For the locus represented in Figure 2 we see a slow exploration with time of the binary space but with the gradual fixation towards a particular setting. This gradual fixation continues until the knee of the curve, about at generation 350, when the final 20-bit string is converged upon in a handful of final rounds. At this point of final convergence, all agents from the two populations have settled on the same particular bit string.

In Figure 3 we see a slightly different exploration of the binary space for a different locus. Here, rather then a steady gradual fixation towards one particular setting the sub-populations spend the first 150 generations fairly evenly distributed amongst "1"'s and "0"'s at this locus. Indeed, cases when the lines cross each other at the 0.5 mark (where "0"'s and "1"'s are equally distributed across the sub-populations) demonstrate a swap from accumulating "0"'s to "1"'s or vice-versa. For the final 150 generations there is a gradual move toward fixation and, again, the rapid convergence on the final solution during the last rounds.

Note that we performed a large number (over 20) runs of this simulation. Variations from run to run were not significant.

Analysis

Why would the mutual coevolution of a 20-bit string allow the GA to solve this problem while the evolution to a fixed 20-bit string (in the absence of something like learning) is not possible? Figure 4 begins to shed some light on the answer. As the rounds of the GA proceed, the total number of agents that match perfectly across the sub-populations are steadily (if not monotonically) increasing. (I simply sum up all pairwise perfect matches.) Some of this, no doubt, represents progress towards fixation at a particular string. A similar graph of the original Hinton and Nowlan result does not show such steady progress towards the all-one string.

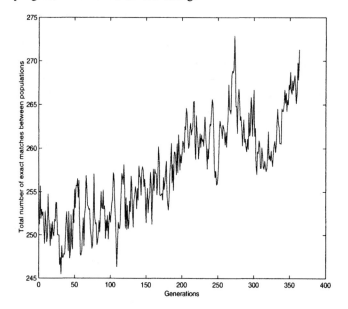

FIGURE 4. Total number of pairwise exact matches between agents of both populations. We see a steady increase in the number of matches with time.

However, this gradual increase in matches is the result of, but not the reason why, coevolution works. A number of researchers have argued for the importance of genetic diversity within evolution (Bedau, 1995), and indeed within coevolution (Cliff & Miller, 1995). Bedau defines two measurements of genetic diversity, between-locus and within-locus. Conceptually, within-locus diversity measures the average distance between each agent and the consensus or wild-type sequence. In contrast, between-locus diversity measures the distance between the consensus sequence and the scalar average of the consensus sequence taken across each locus. Thus (adapted from Bedau, 1995), if we let n be the number of agents in the population, k be the genome length, and w_{ij} be the value of the ith agent at the jth locus, then

$$D_{within} = \frac{1}{nk} \sum_{i=1}^{n} \sum_{j=1}^{k} \left(w_{ij} - \bar{w}_j^i\right)^2,$$

$$D_{between} = \frac{1}{k} \sum_{j=1}^{k} \left(\bar{w}_j^i - \bar{w}^{ij}\right)^2,$$

where

$$\bar{w}_j^i = \frac{1}{n} \sum_{i=1}^{n} w_{ij} \quad \text{and} \quad \bar{w}^{ij} = \frac{1}{nk} \sum_{i=1}^{n} \sum_{j=1}^{k} w_{ij}.$$

Simply put, the within-locus diversity is a measure of how close to convergence the two populations are. As this diversity decreases the populations are matching more and more. In contrast, the between-locus diversity is sensitive to the ratio of "0's" to "1's".

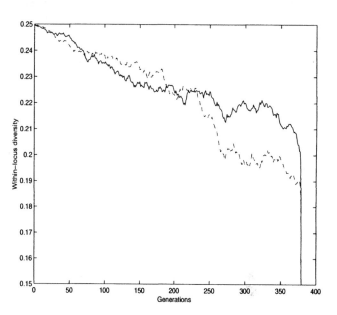

FIGURE 5. The average diversity within a particular loci is decreasing with time for both populations. This is due to the general fixation towards a solution.

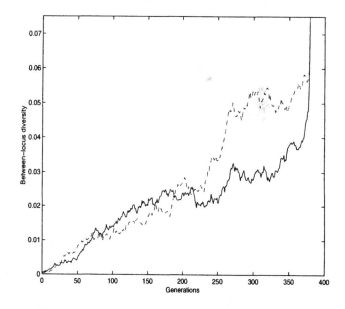

FIGURE 6. The diversity between loci increases with time for both populations. As a particular solution is arrived at it is not evenly distributed between "0's" and "1's".

In Figure 5 we see the within-locus diversity for both populations over time, for a given run, and in Figure 6 we see the between-locus diversity. These graphs show that slowly over generations the average diversity within loci across the populations is slowly diminishing; that is, a solution to the sub-problems (of convergence at any single locus) is slowly accumulating. In contrast, the diversity across all loci across the entire population is increasing; that is, the accumulating weight towards some particular solution does not have an even number of "1's" and "0's".

We believe that these graphs illustrates the reason that mutual coevolution is effective against this problem. By chance, sets of agents across the two sub-populations occasionally match. Those agents enjoy an increase in their fitness and thus have a higher probability of reproducing. But the cross-over operator, with high probability, tears apart these matching agents. Thus, they most likely will not match again in the subsequent round, given that they probably do not recombine again into a match. However, for both of the sub-populations, particular *subsequences* of the loci are retained in the populations as the cross-over operator splits up agents. Under the original Hinton and Nowlan simulation, sub-sequences with the all-one configuration are quickly dispersed into the population of evolving agents and are never (with high probability) reconnected in numbers sufficient to self-sustain. But given the mutual co-evolution of a solution, these subsequence partial solutions accumulate with time. And, unlike the case of a fixed all-one configuration, *any* chance reassembly of partial solutions that occurs to both sub-populations will produce fit agents.

As shown in Figure 5, as the GA rounds continue, the diversity within single positions diminishes. This is due, we believe, to the accumulation of a number of subsequence solutions that are shared across the populations (and the increase in occasional matches confirms this). This pool of available partial solutions grows and grows until finally, in a striking transition point often seen with the GA, a quick chance assembly of a particularly auspicious bit string — one whose set of subsequences are well represented now within the populations — fixates the simulation onto a string that is quickly assembled.

Conclusions

We have shown that coevolved mutualists can match a moving 20-bit string within a highly uncorrelated landscape. Hinton and Nowlan had shown that the unaided GA cannot converge on a fixed 20-bit string with a similar landscape.

An important motivating question is whether this type of phenomenon exists in nature amongst coevolving populations. We believe that a number of cases exist of symbiotic relationships between species in which specific, and indeed arbitrary, signalling is critical to their successful interactions.

One such case is nicely treated by Richard Dawkins (1997). Dawkins describes the relationship between varieties of plants and pollen bearing insects. A tree that pollinates its female flowers with pollen from its own male flowers might, almost as well, not bother to pollinate at all; it would be more cost-effective to simply produce a vegetative clone of itself. As an alternative strategy, plants and insects come into a relationship where the plant gives up something (generally energy-rich nutrients) to coax the insect to act as a DNA vector. If the plant allows just any old insect to act as its DNA transport, for instance any ordinary bee, then that costly pollen is distributed rather broadly over all sorts of plants, only some of which are of the desired species. In order to waste less pollen, the plant can recruit a special species unto itself — if I am an *X*-plant than I want an *X*-bug that transports my pollen like a guided missile only to other *X*-plants. From the wasp's vantage there are similar pressures to getting this right, as this is how they rendezvous with mates and fertilize their own species' eggs.

Dawkins describes the rich set of relations between the variety of fig trees, each of which with its own specialized air-force of pollen transporting wasps. There are 900 species of fig trees and so the species-specific wasp must fly to just the right tree out of these 900 species. The wasp navigates to exactly the right tree thanks to smell signatures which are coevolved to a specific odor amongst the sequence-space. (Of course things are much more complex then we describe here — with cheaters and freeloaders and mimics and this just adds to the evolutionary mix.)

It is this coevolution that may have in its nature the sort of dynamic described by our simulation. The key to the smell signature is that it be quite distinct within the potential space (assuming that the various competing species smell somewhat similarly) and being off by a bit can be as bad as being completely off. If we control for the relative costs of producing one smell versus another, then all signatures are of equal value as long as they have been coevolved between fig and wasp to be identical. Thus the actual outcome is arbitrary within the space of smells but an exact match is critical. Our simulation suggests that this sort of problem may be hard for evolution to solve without the additional affordances of coevolving mutualistic populations.

We are particularly interested in the coevolution of genes with culture (Durham, 1991; Feldman & Laland, 1996). It has been argued that mutualistic coevolution of cultural traits along with genetic ones could account for phenomena within the natural world, for instance, the vocalizations of songbirds (Lachlan & Slater 1998) and human natural language (Deacon, 1997). Human natural language and bird songs, like fig odor signatures, are often arbitrary within the space of vocalizations (Suassurian symbols). In other words, within the space of possible words it is an arbitrary choice we make such that "bat" rhymes with "cat". What is central, is that we all mutually agree. Thus, while humans evolved certain language instincts (Pinker, 1994), language coevolved a symbiotic relationship with humans. In other words, humans evolved to learn and remember a host of words, words evolved (and continue to evolve) to be particularly learnable and memorable.

Our simulation suggests that this sort of coevolution may not be simply a happy enhancement to the mutualistic gene-culture interaction but may be an integral component.

References

Baldwin, J.M. (1896). A new factor in evolution. *American Naturalist,* Vol 30, 441-451.

Bedau, M.A. (1995). Three illustrations of artificial life's working hypothesis. In W. Banzhaf & F.H. Eeckman (Eds.), *Evolution and biocomputation: Computational models of evolution* (pp. 53-68). Berlin: Springer.

Begon, M., Harper, J.L. & Townsend, C.R. (1996). *Ecology: Individuals, populations, and communities (3rd ed.).* Oxford, UK: Blackwell Science.

Belew, R.K.(1990). Evolution, learning, and culture: Computational metaphors for adaptive algorithms. *Complex Systems,* Vol 4, 11-49.

Best, M.L. (2000). How culture can guide evolution: An inquiry into gene/meme enhancement and opposition. To appear, *Journal of Adaptive Behavior.*

Bull. L., Fogarty, T.C. & Pipe, A.G. (1995). Artificial endosymbiosis. In F. Morán, A. Moreno, J.J. Merelo & P. Chacón (Eds.), *Advances in Artificial Life: Third European Conference on Artificial Life* (pp. 273-289). Berlin: Springer.

Cliff, D. & Miller, G. F. (1995). Tracking the red queen: Measurements of adaptive progress in co-evolutionary simulations. In F. Morán, A. Moreno, J.J. Merelo & P. Chacón (Eds.), *Advances in artificial life: Proceedings of the third European conference on artificial life (ECAL95)* (pp. 200-218). Berlin: Springer.

Cliff, D. & Miller, G.F. (1996). Co-evolution of pursuit and evasion II: Simulation methods and results. In P. Maes, M.J. Mataric, J. Meyer, J. Pollack, & S. Wilson (Eds.), *From animals to animats 4: Proceedings of the fourth international conference on simulation of adaptive behavior* (pp. 506-515). Cambridge, MA: MIT Press.

Dawkins, R. (1997). *Climbing mount improbable.* New York: Penguin Books.

Deacon, T.W. (1997). *The symbolic species: The co-evolution of language and the brain.* New York: W.W. Norton & Company.

Durham, W.H. (1991). *Coevolution: Genes, culture, and human diversity.* Stanford, CA: Stanford University Press.

Feldman, M.W. & Laland, K.N. (1996). *TREE,* Vol 11(11), 453-457.

Harvey, I. (1993). The puzzle of the persistent question marks: A case study of genetic drift. In S. Forrest (Ed.), *Proceedings of the fifth international conference on genetic algorithms* (pp. 15-22). San Mateo, CA: Morgan Kaufmann.

Hillis, W. D. (1990). Co-evolving parasites improve simulated evolution as an optimization procedure. *Physica D,* Vol 42, 228-234.

Hinton, G.E. & Nowlan, S.J. (1987). How learning can guide evolution. *Complex Systems,* Vol 1, 495-502.

Juillé, H. and Pollack, J. (1996). Co-evolving intertwined spirals. In L.J. Fogel, P.J. Angeline & T. Bäck (Eds.), *Proceedings of the Fifth Annual Conference on Evolutionary Programming* (pp. 461-468). Cambridge, MA: MIT Press.

Kauffman, S. (1993). *The origins of order: Self-organization and selection in evolution.* Oxford: Oxford University Press.

Lachlan, R.F. & Slater, P.J.B. (1998). The maintenance of vocal learning by gene-culture interaction: The cultural trap hypothesis. Unpublished manuscript.

Paredis, J. (1994). Steps towards co-evolutionary classification neural networks. In R.A. Books & P. Maes (Eds.), *Artificial Life IV* (pp. 102-108). Cambridge, MA: MIT Press.

Potter, M.A. & DeJong, K.A. (1994). A cooperative coevolutionary approach to function optimization. In Y. Davidor, H. Schwefel & R. Männer (Eds.), *Parallel Problem Solving from Nature - PPSN III: International Conference on evolutionary Computation* (pp. 245-259). Berlin: Springer.

Pinker, S. (1994). *The language instinct.* New York: W. Morrow and Company.

Reynolds, C.W. (1994). Competition, coevolution and the game of tag. In R.A. Books & P. Maes (Eds.), *Artificial Life IV* (pp. 59-69). Cambridge, MA: MIT Press.

Sims, K. (1994). Evolving 3D morphology and behavior by competition. In R.A. Books & P. Maes (Eds.), *Artificial Life IV* (pp. 28-39). Cambridge, MA: MIT Press.

Smith, R.E. & Cribbs, H.B. III (1996). Cooperative versus competitive system elements in coevolutionary systems. In P. Maes, M.J. Mataric, J. Meyer, J. Pollack, & S. Wilson (Eds.), *From animals to animats 4: Proceedings of the fourth international conference on simulation of adaptive behavior* (pp. 497-505). Cambridge, MA: MIT Press.

Evolution of Cooperation in Social Dilemma — Dynamical Systems Game Approach

Eizo Akiyama[1] and **Kunihiko Kaneko**[2]

[1] KEIO University, Shin-Kawasaki-Mitsui Building West 3F, 890-12 Kashimada saiwai-ku, Kawasaki-shi 212-0058, Japan
[2] Graduate School of Arts and Sciences, The University of Tokyo, Komaba 3-8-1, Meguro-ku, Tokyo 153-8902, Japan

Abstract

Social dilemma, problems in the formation and maintenance of cooperation among selfish individuals, are of fundamental importance for the biological and social sciences. To consider the spontaneous formation of cooperation in society under social dilemma, dynamical systems game is adopted. To be specific, the 'Lumberjacks' Dilemma (LD) game' is studied, which includes the concrete description of the dynamics of resources depending on players' actions. With the numerical study of the evolution of the strategies in the LD game, it is shown that the cooperation is formed and maintained, through the formation of articulation of strategies based on bifurcation in dynamical systems. Evolution of the cooperative society is found to occur successively with the dynamical change of game structure and of norms adopted in the society. In contrast with the models of the traditional game theory, the cooperation is shown to be sustained even if the number of the members sharing the social dilemma is increased.

Social dilemma

Construction or maintenance of cooperation among a group of people often brings about dilemma, as is seen for example in the problem of garbage disposal, where those who do not care the public good get relatively high utility. This type of dilemma occurs in the following situation: all peoples' collective profits are maximized when they maintain cooperation, but each of them will get a larger personal profit by behaving selfishly. However, if all behave selfishly, this will cause them loss and the society will eventually fall apart. The problem of maintenance of cooperation in a social group is generally called *social dilemma*, The classical story in the social dilemma is 'the tragedy of the commons,' presented by Garret Hardin in 1968 (10). Because of its applicability to a variety of environmental issues, it was frequently adopted in many fields.

Theoretical studies on social dilemma

In socio-biology, emergence and maintenance of cooperation in society is thought to be a result of kin selection (9) or altruistic reciprocity. Study of cooperation based on 'reciprocity' has been started from the experimental and theoretical study of the Iterated Prisoners' Dilemma (IPD) game by Axelrod (3). He showed that the simple strategy TIT-FOR-TAT is evolutionarily stable against other strategies, such as ALL DEFECT, and cooperative society is brought about.

This explanation based on the IPD is applied to a variety of social phenomena. However, many researchers have come to believe that a direct application of the result of the IPD to the problem of cooperation in a group is difficult, because the interaction in a society usually involves more than two individuals. Therefore, the necessity of game models with more than two players is recognized. For example, Axelrod and Dion have pointed out the formulation of the social dilemma as an n-person Prisoners' Dilemma ($n > 2$) (5).

Boyd and Richerson (6), and Joshi (12) have analyzed the n-person Prisoners' Dilemma with evolutionary games, and proved that the condition for the evolutionary stability of (n-person version) TIT FOR TAT is harder to be satisfied with the increase of n. Let us call this *the effect of the number of players*. That is, 'reciprocity' is not sufficient to explain the maintenance of cooperation. Hence, the following question is raised: "If reciprocity is not sufficient to explain the maintenance of cooperation in a social group, how is the cooperation maintained?"

With respect to this question, Boyd and Richerson have considered an additional strategy, 'sanction' against the non-cooperators in the n-person Prisoners' Dilemma. They have also investigated the evolution of sanction (e.g., (8)). Axelrod has also introduced a model that involves a 'metanorm,' which involves incentive to punish not only betrayals but also those who have not punished the betrayals (4). On the other hand, Boyd and Richerson considered a cultural effect to maintain the cooperation (7). All of the above researches deal with the problem of how cooperation in a society can be maintained by the factors other than mere reciprocity between agents. Summing up, there is a recent trend that 'reciprocity' is not sufficient for the maintenance of cooperation in social dilemma, and that sanction or some other strategies are necessary, based on institutions or norms.

Problems in modeling social dilemma

Here we discuss the problem associated with the model study of the social dilemma so far.

From the viewpoint of sociology, it is interesting to study how the cooperation is formed by introducing additional rules, such as the sanction on betrayers. However, with the inclusion of additional strategies, the game itself changes to lose the original dilemma at the level of payoff matrix.[1] In addition, the cooperation in the society may sometimes be achieved by the introduction of external institutions out of interacting agents, such as the government. Although it is also important to study the role of the external institution, it should be noted that the players can form cooperative society even without external institutions.[2] The problem we deal with here is the formation and maintenance of cooperation within interacting players under dilemma.

Dynamical systems game

Here we point out an important characteristic of social dilemma in the real world that has not been treated in traditional game models. Most of the real examples of social dilemma, whether they might result in tragedy or not, have characteristic dynamics, such as the decrease of petroleum resources, fluctuation of livestock resources, that of marine resources, and the change of players' economic condition. Players' behaviors necessarily affect these dynamics. Cooperation among the players cannot be taken into consideration apart from such dynamics. For example, the stability of real cooperation may be deeply related with the stability of the dynamics in the cooperative state. In the present paper, we study social dilemma from a dynamical point view, by adopting a framework, *dynamical systems (DS) game* (1), where the game, conducted repeatedly, changes with time, depending on the players' actions. In the DS game, the game

itself can be affected and changed by players' behaviors or states. In other words, the game itself is described as a 'dynamical system.'

Lumberjacks' Dilemma as a DS Game

As an application of the DS Game framework, we present in this paper what we call the 'Lumberjacks' Dilemma (LD) Game.' Let us consider the following situation: There is a wooded hill where several lumberjacks live. The lumberjacks fell the trees for their living. They can maximize their collective profit if they cooperate in waiting until the trees have fully grown before felling them, and sharing the profits. However, any lumberjack who fells a tree earlier will take the entire profit on that tree. Thus, each lumberjack can maximize his personal profit by cutting trees earlier. If all the lumberjacks do this, however, the hill will go bald and there will eventually be no profit. This situation inevitably brings about a dilemma.

This LD Game can be categorized into the social dilemma. In other words, it can be represented in the form of an n-person version Prisoners' Dilemma if we project it onto static / traditional games. However, there are several important differences: Dynamics of the size of the trees should be expressed explicitly in this LD Game. The yield of a tree, and thus the lumberjacks' profit, differs by the timing when the tree is felled. The profits have a continuous distribution, because the yield of a tree can have a continuous value. A lumberjack's decision today can affect the future game environment through the growth of a tree.

Outline of the LD game model

Let us describe the Lumberjacks' Dilemma Game, which we adopt here.

In the game world, the ecology of the Lumberjacks' Dilemma Game, there are several species of lumberjacks and also several wooded hills. Here we define a set of the lumberjack species as $S = \{1, 2, \cdots, s\}$ and a set of the hills as $H = \{1, 2, \cdots, h\}$.

In each hill, several lumberjacks (players) live and a single tree grows.[3] We define a set of the lumberjacks in a hill as $N = \{1, 2, \cdots, n\}$. n lumberjacks of each hill are randomly selected from S respectively. A lumberjack of species $p(\in S)$ behaves based on the decision-making function f^p. (Several lumberjacks of the same species can live in the same hill.) These n lumberjacks play a game repeatedly until their death (**"Lumberjacks' Dilemma (LD) game"**). Competing with others in the same hill, a lumberjack fells the tree that grows in time. In each repetition (round), he acquires his *score* of the round, which is defined in the following subsection. At the conclusion of the LD game (at the maximum

---

[1] Assume that each player has a strategy for sanction. If the sanction toward players who defy the norm is easy, it is a matter of course that the society gets cooperative. If the cost for punishing betrayers is slight, sanctions to betrayers are advantageous in a long run, since betrayers are eliminated by the sanctions. As a result, the society will be cooperative. In this case, however, it is not appropriate to regard this as "the emergence of cooperative society under the dilemma." Rather one should better say that "cooperation occurs because the dilemma is lost as a result of the change of the rule." In fact, if the rule of game explicitly permits punishment of betrayers, the game has already changed into a different game without social 'dilemma,' at the level of payoff matrix.

[2] Of course, there is no doubt that players cannot avoid tragedy if they cannot communicate with each other, for example, when they live in a vast village. However, a norm may be formed through the players' communications or interactions, even without external institution. For example, negotiations or struggles among nations may sometimes form cooperation, even without external 'meta-power' except for the occasional intervention of the United Nations.

---

[3] Extension to the case with multiple trees is straightforward, as is discussed in (2)

round T), each player's average score over T rounds is measured.

In each of the h hills, an LD game is conducted once, while the processes simultaneously ongoing on all the hills (i.e., the hills as a whole) are called *one generation* of the game. The *fitness of a species* is given by the average of the average scores that all the players of the species have taken in the hills where they have lived respectively. Before the next generation enters the game, the species with the lowest k fitness eliminated from the game world as a selection process. The surviving ($s - k$) species can leave their descendant species to the next generation, and these participants will have the same decision-making function as their respective ancestors. The extinct species are replaced by new k species, mutants of the k species randomly selected from among surviving ($s - k$) species. The same procedure is repeated in the next generation, without the memory of the previous generation. (Throughout all experiments in this paper, the parameters h, s, k and T are commonly set to $h = 60$, $s = 10$, $k = 3$ and $T = 400$, respectively.)

LD game in a hill

We explain in detail the game by the n lumberjacks (players) on each hill, the LD game.

Let us denote the 'state' of the resource of the hill at time t by $x(t)$, the size of the tree. Each lumberjack has a 1-dimension variable that represents his *state* (e.g. the state of satisfaction), which corresponds to the *score* of the round, and has a decision-making function. For example, the state of the player i, who belongs to lumberjack species $S(i)$ in the game world, is denoted by $y^i(t)$, and the decision-making function by f^i, denoted by $y(t) = (y^1(t), y^2(t), \cdots, y^n(t))$, $f = (f^{S(1)}, f^{S(2)}, \cdots, f^{S(n)})$ respectively. Players decide their next actions respectively by referring to the sizes of the tree, $x(t)$, and the states of players, $y(t)$. All of the players' actions are denoted by $a = (a^1(t), a^2(t), \cdots, a^n(t))$. Each player's individual action can be either waiting (doing nothing) or cutting the tree. The set of these two feasible actions is denoted by A.

T-times repetition of the map g make the LD game dynamics $((x(t + 1), y(t + 1)) = g(x(t), y(t)))$. In the simulation of this paper, for the first round of the game in each hill, x is set at 0.10, and $y_j (j \in N)$ is chosen from random numbers from the normal distribution with the mean 0.10 and the variance 0.10. g is composed of approximately three steps — [A] **natural law** [B] **decision making of players** (f) [C] **effects of actions:** [A] The natural law makes the game dynamics but has nothing to do with the decision-making of players. In LD games, the growth of the size of the tree, $x(t + 1) = u_\Xi(x(t))$, and the decrease of the players' states, $y^i(t)' = u_N(y^i(t))$ ($i \in N$). In the LD games in this paper, we set $u_N(y) = 0.8y$. As for u_Ξ, we use two types of maps: [1] $u_\Xi(x) = 0.7x^3 - 2.4x^2 + 2.7x$, and

[2] $u_{\Xi'}(x) = \min(1.5x, 1.0)$. We call the former map, u_Ξ, the *convex map* and the latter, $u_{\Xi'}$, the *(piecewise) linear map* by the shapes of their graphs.[4] [B] A player i's decision-making function, $f^{S(i)}$, **decides his action**, $a^i(t)$, based on the states of his surroundings, which are denoted by $x'(t)$ and $y'(t)$: that is, $a^i(t) = f^{S(i)}(x(t)', y(t)')$. $f^S(i)$, which varies throughout the evolution, is the inner structure of the player i and is invisible by other players. To implement concretely the decision making function, we introduce the *motivation map, mtv_r*[5] for each feasible action r ($r \in A$), which means a player's incentive to take the action r.

[C] Players' actions **affect the state of the resource** in the hill. To be specific, the size of tree i cut by the players is assumed to be reduced according as $x(t + 1) = (1/3)^\nu x(t)'$, where ν is the number of players who cut the tree. The lumber cut from the tree will be divided equally among all the players who cut the tree, which increases the states of the players.

Attractor of game dynamics and the 'AGS diagram'

Let us briefly survey the basic natures of the LD game for the simplified LD game with only a single player. Here we investigate the effect of the change of the decision-making function upon the attractor of game dynamics. We make two simplifications. First, the player never refers to his or her state, y, that is, the player makes his or her decision only by referring to the size of the tree. Second, the player cuts the tree if the size of the tree exceeds a certain value, called the *decision value*, x_d. The decision value uniquely decides the time series of (x, y). The attractor of the time series can be a fixed point, periodic, quasi-periodic, or chaotic motion, which depends on the dynamical law (including the natural law) given

---

[4] What is important here is the fact that we adopted two example natural laws that have the common nature but have different ways of description of the dynamics of x. In either case, the tree grows well during the early rounds and the size of the tree eventually approaches 1.0 (as long as it is not felled). Both the two natural laws make Lumberjacks' Dilemma, in that players can maximize their collective profits by mutual cooperation in waiting for the tree to have fully grown before felling them, while any player who fells a tree earlier can monopolize the entire profit on that tree.

[5] The structure of mtv used in this paper is simply as follows: $mtv_r : (x, y) \mapsto \eta_r x + \sum_{l \in N} \theta_{lr} y^l + \xi_r$ Here, $\{\eta_r\}$ and $\{\theta_{lr}\}$ are real number matrices and $\{\xi_r\}$ is a real number vector. These coefficients, which can change a little by mutation, determine the player's strategy. Each player selects the action whose motivation has the largest value among the set $\{mtv_r\}$. Each element of $\{\eta_r\}$, $\{\theta_{lr}\}$ and $\{\xi_r\}$ of the initial lumberjack species in the game-world are generated by random numbers from the normal distribution with the mean 0 and variance 0.1. Every coefficients of the new species is chosen as random numbers from the normal distribution with the variance 0.1 around the mean value at the parameters of parent species.

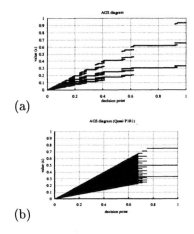

(a)

(b)

Figure 1: AGS diagram of simplified 1-person 1-tree convex LD games — (a) Convex LD game: Change of the attractor with the strategy is plotted. A set of values of x at the attractor (all the values that x takes between the 200th and 400th rounds), is plotted against the change of the decision value, x_d. For example, the two parallel straight segments around $x_d = 0.8$ show that the dynamics of x are attracted to a period-2 cycle taking the values around 0.3 and 0.6, alternately. (b) Linear LD game: Quasi-periodic attractors appear for $x_d \leq 2/3$.

to the system.

As in the 'bifurcation diagram', we have plotted, in Fig. 1-(a), the change of the set of x values at the attractor against the change of x_d. The figure gives a diagram showing how the attractor of game dynamics changes with a parameter in the decision-making function. Let us call such a figure the *AGS diagram* — the transition of the **A**ttractor of the **G**ame dynamics versus the change of the **S**trategy. With the AGS diagram, one can study how the nature of game dynamics shifts among various states (fixed point / periodic / chaotic game-dynamics, and so forth) with the change in decision-making. The following two characteristics in Fig. 1-(a) are noted: [1] For each decision value, its corresponding attractor is always a periodic cycle. [2] There are infinite numbers of 'plateaus,' in which the attractors are completely the same over some range of decision values.[6]

As for the 1-person *linear* LD game, the AGS diagram shows that the dynamics are attracted to a quasiperiodic motion if $x_d \leq 2/3$, otherwise to a periodic motion (Fig. 1-(b)).[7]

---

[6] For examples of such plateaus, see the period-2 and 3 plateaus in Fig. 1-(a).

[7] Detailed study of the evolutionary one-person LD games with these two AGS diagrams is given in (1). In multiple-person LD games, the dynamical structure between strategies gets more influential than the minute structure between a player and the game environment.

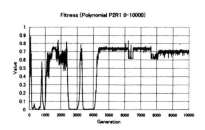

Figure 2: Fitness chart of a 2-person convex LD game.

Evolution in a two-person one-tree convex LD game

Outline of the evolutionary process

In this section, results of the simulation of the evolution of the 2-person 1-tree convex LD game are surveyed. Fig. 2 shows the *fitness value* of each generation, defined as the fitness of the fittest species at each generation, is plotted with the generations as the horizontal axis. We will call this type of figure the *fitness chart*. At the early stage of generation (roughly up to 1000th), lumberjacks try to outfox others by cutting the tree more frequently. As a result, hills become barren and the fitness value gets lower. However, as the generation passes, lumberjacks start to make rules of cooperation. The rules for cooperation adopted in the society change time to time. In our LD game, there are many possible norms on the cooperation agreed among the players. For example, cutting a tree only higher than 0.7 might be called a cooperative action in one society while it might be deemed a selfish action in another. The state of cooperation sometimes collapses completely and the state of non-cooperation resumes. In this manner, the cooperative society is created, changes, and collapses repeatedly over generations, up to approximately the 4,000th generation. After the generation, the cooperative state is established completely, and the non-cooperative society no more appears. Through the cooperative state, the transitions occur among several sets of rules of cooperation that the society mainly adopts.

Formation and transition of rules for cooperation

First we focus on the cooperative societies appeared from the 900th to the 1,800th generation (Fig. 3).

As one can see, from about the 1,000th generation, the fitness value of the generation begins to rise step-by-step. At each epoch with a plateau of stairs, a type of game dynamics specific to each epoch appears. At each step of the stairs, the society changes drastically. Let us take a close look at the epochs A, B and C, where distinct game dynamics dominant at each epoch are called as type A,

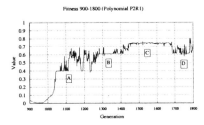

Figure 3: The fitness chart from the 900th generation to the 1,800th generation (an blowup of Fig. 2): A, B, C, and D in this figure represent epoch A (1,100th generation —), epoch B (1,250th generation —), epoch C (1,450th generation —), and epoch D (1,700th generation —), respectively.

B, and C, respectively. The dynamics are identical for all hills in some case, while they are nearly identical in other cases. For instance, the type A game dynamics are observed for more than a half of the 60 hills, in a certain generation of epoch A, while they are used on almost all hills in other generations.

Fig. 4 shows the type A game dynamics (about the actions, the tree size, and the states of the players) which are dominant in **epoch A**. As indicated in Fig. 4-(a) and Fig. 5-(a), the players exhibit the period-5 action cycle of "wait, cut, wait, cut, and wait." In addition, the two players' actions are identical. From the dynamics of tree size (Fig. 4-(b)), one can see that the lumberjacks collect the lumbers while raising appropriately the tree. The actions of type A are considered as a *a norm for cooperation* for lumberjacks living in the game-world. In this period, the game is set up so that the mean height of the tree is approximately 0.12.

Fig. 5-(b) shows the dominant action dynamics in **epoch B**. In this epoch, the mean height of the tree is set at approximately 0.27, while the action in this game is given by the period-3 sequence of "wait, cut, and wait", as shown in Fig. 5-(b). The frequency of cutting is lower than in the case of type A, and thus a more productive environment is provided. The actions of two players are also identical in the type B dynamics. The dynamics similar to those in epoch B are seen in epoch D.

The game in **epoch C** is set up so that the average tree height is approximately 0.45. As seen in Fig. 6 and Fig. 5-(c), each player exhibits the period-4 action, with a sequence of "wait, wait, wait, and cut." The most salient feature in this type of dynamics is that the two players are not synchronized in the action. The action sequence is performed with out of phase, and the two players alternately raise the tree and gather lumbers. As a result, a more productive game environment than epoch A or B is organized.

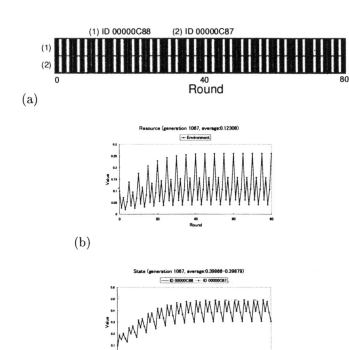

(a)

(b)

(c)

Figure 4: epoch A: (a) Action chart. (b) Resource chart. (c) State chart. In each figure, the horizontal axis shows the round. The black tile represents the action 'wait,' while the white tile represents 'cut.'

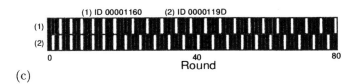

(c)

Figure 6: Action chart of epoch C

Stability of game dynamics

We have observed that lumberjacks form rules to manage the dynamics of the resources cooperatively in the 2-person 1-tree convex LD game. Those dynamic rules of cooperation change with generations. How are the formation and change of such rules possible? How can those forms of cooperation remain stable in spite of the dilemma underlying in our model? To answer these questions, we will analyze the relationship between the evolution of strategies and the game dynamics by the *AGS diagram*.

Articulated structure in the AGS diagram Here, we discuss the stability of game dynamics in a cooperative society. First, let us look at epochs C and D (Fig. 3). The dominant action dynamics of the former are shown in Fig. 5-(c) (1,501st generation), while the

Figure 5: Action chart: (a) epoch A. (b) epoch B. (c) epoch C. In (a)-(c) the action chart at the attractor after the transients is plotted. The black and white circles indicate the time series of actions by the two lumberjacks.

dynamics for epoch D is identical to the pattern of epoch B, Fig. 5-(b). Thus the characteristic game dynamics are completely different between epochs C and D. Nevertheless, the decision-making functions in those ages are quite similar. The difference lies in the values of θ_{11} and θ_{20}. For example, θ_{11} is approximately -0.65 for species C and -0.15 for species D.

Next, let us look at AGS diagrams. The LD game between a lumberjack of the fittest species (lumberjack C) and one of the 2nd-fittest species of the 1,501st generation, results in the period-4 dynamics seen in Fig. 5-(c). Here, AGS diagrams for the game between two lumberjacks are drawn in Fig. 7 with θ_{11} as the parameter of lumberjack C.

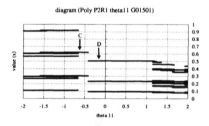

diagram (Poly P2R1 theta11 G01501)

Figure 7: AGS Diagrams regarding θ_{11} (by two lumberjacks in the 1,501st generation): Two segments with the arrow C are a set of attractors of the type C dynamics (the tree size). The three segments with the arrow D are a set of attractors of the type D dynamics.

As seen in the figure, around $\theta_{11} \approx -0.5$, there is a period-2 attractor as shown in two parallel segments. These segments correspond to the dynamics of type D.[8] On the other hand, the three parallel segments ranging from about -0.4 to $+1.0$ correspond to the period-3 action dynamics of type C. Let us call this parallel segment area the *plateau*. The θ_{11} value of lumberjack C (about -0.65) lies within the period-2 plateau. With the change of θ_{11} to the period-3 plateau (about -0.15), the player becomes a type D. Change between epochs C and D is produced by this difference of the parameter for the strategy. Here, dynamical structure of this 2-person convex LD game allows for such change.

When a stable cooperative society in the 2-person convex LD game is achieved as seen in Fig. 7, almost all of the AGS diagrams of the lumberjacks are composed of plateaus of the periodic attractors. In the plateau, the game dynamics do not change regardless of the change of the strategy. Take for example of Fig. 7, where θ_{11} of lumberjack C is approximately -0.65, within the plateau of type C. In this figure, even if the θ_{11} value of this player has a slight deviation due to mutation, the game dynamics to be observed will remain the same. When θ_{11} exceeds -0.4, the game dynamics will jump into type D immediately. Afterwards, even if θ_{11} is further increased, the game dynamics will remain type D.

From the viewpoint of lumberjacks' strategies, the structure of game dynamics is clearly articulated with many plateaus in the AGSs of the lumberjacks when the mutual cooperation is formed among them.

Plateaus in the AGS diagram and their role on the evolution of cooperation In this LD game simulation, several cooperative societies, which have their characteristic game dynamics respectively, last stably over generations. Those game dynamics are also found among the plateaus in the players' AGS diagrams. We call those dynamics *Strategic Metastable game Dynamics (SMD)*. In the SMD, lumberjacks live in the cooperative society that can have stable game dynamics **against the change of some players' decision-making**. In other words, plateaus in the AGS are the candidates for SMD to make cooperative societies.[9]

Using the articulated structure, mutual agreements are made among players as to the formation of cooperation in each SMD. In the cooperative phase in this social dilemma, the game dynamics do not gradually change as the strategy changes. Rather, they suddenly jump to the next game dynamics at a certain phase. In the cases of epochs A, B, C, and D, certain kinds of game dynamics are dominant in the game-world. They are all plateaus in AGS diagrams of lumberjacks living in those ages, and these different plateaus give different SMD. In other words, those plateaus give standards of the society or the rules that most players follow. Their very existence is a *necessary condition* for establishing and stabilizing a cooperative society, while for other conditions see the

---

[8] Although players' actions are attracted to the period-4 dynamics, the period of the dynamics of tree size is 2. This is because the tree dynamics depend solely on whether it is 'cut' or 'not cut,' independent of the player who cuts the tree.

[9] In fact, treasonable game dynamics in competition societies that last long in this LD game should be called SMD, but in this paper, we regard only the dynamics to form the plateaus of the stable cooperation as SMD.

Table 1: Score table of lumberjack A and B; C and D: The values in the table are approximate average score of lumberjacks in the LD game of 400 rounds.

A	0.4	=	0.4	A	C	0.7	=	0.7	C
A	0.4	=	0.4	B	C	0.5	<	0.8	D
B	0.6	=	0.6	B	D	0.6	=	0.6	D

next subsection.

The LD game includes continuous variables. Thus, there are potentially innumerable varieties of dynamics in action, resource, and states. However, in the 'cooperative society' seen in this 2-person convex LD game, the number of game dynamics to be observed tends to be limited to few. In particular, only one identical game dynamics is often observed on all 60 hills. Such reduction of game dynamics is brought about by the articulated structure in the AGS diagrams caused by the strategies and the dynamical nature of this game. As will be shown in the followings, the limitation of the number of dynamics makes possible the above sufficient condition for stabilization of cooperation to be realized.

Mechanism of the stabilization of cooperative society and the formation of norms

Now let us study how the state of cooperation gets to be created and stabilized when the game dynamics are articulated.

Transition between cooperative societies from the viewpoint of evolutionary stability First, let us study how the transition from epoch A to epoch B takes place and how the type B dynamics is stabilized.

The left-hand side of Table 1 gives the approximate average score between the type A and B players during the transitional period from epoch A to epoch B (at the 1,250th generation), where the players of the two types are dominant but the population of the type A species is being replaced by the B species through the selection (Fig. 3). The game dynamics played between two lumberjacks A are type A, whose average score is 0.4 for both. Meanwhile, lumberjack B creates the type A dynamics for the game against lumberjack A, and thus the average score is again 0.4 for both, while for the game dynamics played between two lumberjacks B are type B, whose average score is 0.6. In other words, lumberjack B is able to deal with both type A and type B dynamics.[10]

Due to this dominance of the species B over A, the species will increase its population in the game-world,[11] to establish epoch B.

In general, evolutionary stability is discussed as follows. Suppose that $E(X, Y)$ is the score of Strategy X against Strategy Y. Then, the condition for Strategy I to be the *Evolutionary Stable Strategy (ESS)* is in general that for all Strategy J (except I), [1] $E(I, I) > E(J, I)$ or [2] $E(I, I) = E(J, I)$ and $E(I, J) > E(J, J)$. (13)

Following Table 1, it is clear that lumberjack B satisfies the condition [1]. As long as most existing strategies in this period are limited to either species A or species B,[12] the increase in the strategy B and the stability of the species-B society is thus explained. In other words, if game dynamics are articulated and many lumberjacks adopt dynamics of either type A or type B dynamics, then the transition from epoch A to epoch B can be explained by ESS. Similarly, we can explain the transition from epoch B to epoch C.

Decline of cooperation due to the continuous increase of generosity Next, let us take a look at the transition from epoch C to epoch D (Fig. 3). Since the fitness value drops at this transition, the mechanism described above cannot explain it.

Table 1 also gives the score between two lumberjacks from the species C and D during the transition period from epoch C to epoch D (at 1,685th generation). In the game between lumberjacks D and C, D exploits C. The average score of D is 0.8, which is larger than the average score between C Lumberjacks. Thus, if the distribution of strategies is concentrated on lumberjack C and lumberjack D, the lumberjack D can invade the society of species C.

The reason for the success of the invasion of the lumberjack D lies in *continuity* in the LD game. Among the strategies classified into type C, there are slight and continuous differences. Although the *type* of species stated above is based on the *attractor of the game dynamics*, the parameters in decision-making function is continuously distributed. In the early stage of epoch C immediately after the end of epoch B, the parameters are selected so that the species C do not lose the game against the species B (= type D) which are less cooperative. The species C at this stage have strategies 'strict' against the type-B(D) dynamics that are less cooperative. However, with the success of species C, almost all battles start to adopt the game dynamics type C. At this stage, it is better for a lumberjack to settle down to the cooperative type C game dynamics, as soon as possible. For example, Fig. 6 shows that the type C dynamics begin after 20 transient rounds. By decreasing the transient length, the average score is increased. As a result, with

---

[10] This statement only holds in the transitional period to epoch B and at the early period of epoch B, while at later generation, the lumberjack B may lose the ability to comply with A.

[11] This process is similar to the evolution in the imitation game (11), since B can imitate A but not otherwise.

[12] This is true when lumberjacks' strategies of the game-world are concentrated near the border between the two plateaus of type A and type B, in the AGS diagram.

generations, the generosity is increased among the type C dynamics. When this direction of evolution reaches some point, the lumberjack C is now so generous that he is no more strict against the lumberjack D who cuts the tree more frequently, and allows for the invasion of the relatively selfish strategy.[13]

Summary of the mechanisms to escape from the 'tragedy' Now let us summarize the formation, stabilization, and transitions of the cooperative society in the LD game, the DS game model of the *social dilemmas*:

[1] The premise for establishing cooperation lies in the articulation of the game dynamics as is seen in the AGS diagrams in Fig. 7. Whether articulation is possible or not depends on the dynamical law of the game-world in the first place. For some dynamical law, such articulation is difficult or impossible. For example, in the linear LD game adopting the piecewise linear map for the growth law of tree, articulation of the game dynamics is extremely difficult as stated in the following section. Furthermore, even in the convex LD game, the articulated structure can hardly be observed in a complete non-cooperative society. In the cooperative society of the convex LD game, some articulation is achieved through the interaction of the lumberjacks, as a digital structure created in the originally analog LD games.

[2] The cooperation norms of the society are formed as *stable game dynamics* generated by articulation. In spite of dilemma nature in the LD games, such dilemma is avoided with a dynamical structure created in the game. There are several such stable dynamics to form the cooperation, which corresponds to different epochs of stable society adopting a different norm. Succession of different societies is the transition among stable game dynamics, which mechanism can be analyzed by dynamics of payoff among the strategies articulated from continuous choice of parameters.

[3] After a cooperative society is achieved with articulated dynamics, it is sometimes taken over by a less cooperative society. This is because the norm adopted for cooperation becomes too generous with the continuous evolution of parameters, and the society allows for the invasion of relatively selfish strategies.

Bias to defection In this simulation, there are four occasions during which the evolution of the lumberjacks escalates to selfish actions. Fig. 8 shows the AGS diagram for the strategies of the two lumberjacks in a society when selfish actions are about to become dominant (the 140th generation). Although 1-person convex game has articulated structure in game dynamics (Fig. 1-(a)), the interaction of selfish cooperative two players destroys the structure in this case. Here, if a lumberjack acts a lit-

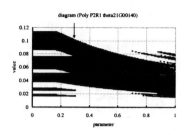

Figure 8: AGS diagrams of an uncooperative society (140th generation): The horizontal axis shows the parameter θ_{21} of the decision-making function of a lumberjack. To any θ_{21} values, the attractor of the game dynamics is the quasi-periodic motion.

tle more selfishly, the game dynamics will be less productive, and the total sum of the lumberjacks' benefits will be smaller, but the betrayer's benefit will be more. This, indeed, exemplifies the situation that we usually regard in the real world as the *tragedy* in the social dilemma. When the articulated structure is lost, the dynamics can change continuously toward the non-cooperative state with the competitive tree cutting. Hence the dilemma underlying in our problem cannot be resolved, and the cooperative society collapses. Now we can again see that the articulation based on the bifurcation plays a key role to keep away from the spread of selfish action.

Effect of the number of players and of the game dynamics

In this section, we survey the qualitative difference among the evolutionary phenomena caused by the change of 'the number of players' and 'game dynamics.' We will show that the effect can drastically change between the static and dynamic games.

Here, we study either 2- or 3- person games, while we choose either the linear map or the convex map for the dynamics of the tree growth (i.e., totally 4 cases). We have carried out three runs of evolutionary simulation (with different random numbers) for each setting. These three runs for each lead to the same qualitative behavior, from which we will show that the difficulty of cooperation for a larger number of players discussed in static game does not hold, at least in the convex map, and that the difference in tree dynamics is important in the evolution of cooperation.

In the **2-person linear LD game**, the lumberjacks start to compete in tree cutting without waiting, and the fitness value is lowered. They form a non-cooperative society around 500th generation, which lasts very long, up to approximately the 7,500th generation. Only after the generation, they start to wait for the tree growth, and the fitness value is stabilized at approximately 0.5. On the other hand, in the **3-person linear LD game**, the

---

[13] Similar mechanism of the collapse of cooperativity caused by the exceeding generosity is also seen in the collapse of the money in (16).

non-cooperative society formed around the 30th generation last up to the final generation of our simulation (the 20000th). Thus, in the linear LD game, a cooperative society cannot be achieved (or at least the achievement is more difficult), as the number of persons increases.

Next, we study the change by the number of players for the **2-person convex LD games**. For a two-person game, as stated in the previous sections, the cooperative society becomes almost completely stabilized from approximately the 4,500th generation with the fitness value fluctuating between 0.6 and 0.8. The average size of the tree in the game environment is approximately 0.25 in later generations, and a productive game environment is maintained. Meanwhile, in the **3-person convex LD game**, a cooperative society is formed at an earlier generation. Once a cooperative society is formed, it is never replaced by non-cooperative society. Here, the fitness value fluctuates around 0.5. In this 3-person convex LD game, period-15 dynamics[14] are often adopted to form the cooperation. The lumberjacks often adopt the period-15 cooperative dynamics, in which each lumberjack is required the patience to wait no less than 10 rounds of the other two players' tern. The average tree size is 0.35, and a more productive game environment is maintained than the 2-person game. In other words, in this dynamic game, the formation and maintenance of cooperative society is easier in a 3-person game than in a 2-person game.

To summarize, the cooperation is more difficult with the increase of the number of players in the linear LD game, as in the static game. In the convex game, however, a more cooperative society is created and is more easily stabilized for a three-person game. From the perspective of the static game, both games have same social dilemmas. Then why does a difference in dynamics in the tree growth result in such essential difference?

The answer lies in the bifurcation structure of the convex game. In the convex LD games, the productive period-15 game dynamics in the 3-person game has higher stability, with regards to the mechanism based on the articulation. In the 3-person game, the priod-15 attractor dynamics has a plateau in the AGS diagram, which has a higher stability against the invasion of other strategies. On the other hand, linear LD games can form only few articulated structures in the game dynamics, even in the case of a one-person game (see Fig. 1-(b)). As a result, few structure can be formed in linear LD games that could prevents the 'bias to the betrayal' that social dilemma inherently involves. Thus the effect of the number of players has the same effect as in the static game case.

---

[14] After each lumberjack does the period-5 action of "cut, wait, cut, wait, and cut," he waits for 10 rounds while the other two lumberjacks enter this period-5 action in turn. This rule is kept stable in this 3-person game.

Summary and discussion

Modeling of social dilemma by static games and DS games Examples of social dilemmas including the issues of garbage disposal, consuming pasture and drains on various resources were formulated as static games such as the n-person Prisoners' Dilemma by extracting some common dilemmatic characteristics. It is certainly important to direct attention to the common feature among certain phenomena, and extract essential factor. When we model a game-like interaction as an algebraic payoff matrix of static game models, radical abstraction is definitely needed. In the course of the abstraction, however, we may neglect what can be described only as dynamics, and it is possible that the parts omitted here may be essential to resolve social dilemma. In any static game which models a social dilemma, the increase of the number of players within any social dilemma always makes cooperation difficult. In a real community, however, we do not necessarily fall in tragedy, when we are faced with dilemmatic situations.

As a new approach to this issue, we introduced a dynamical systems game model. In the simulations of this paper, the above 'omitted parts' have an important effect on the avoidance of the tragedy. Surely, the formation of cooperative society gets more difficult in the linear game with the increase of the number of the players. Our study for the convex game shows, however, that the increase of the number of players does not bring about the less cooperative society. Rather, the tragedy is more easily avoided. In other words, dynamical characteristic of our game are more important than the effect of the number of players.

Here, the articulation of strategy, based on the bifurcation in the attractor dynamics plays a key role in forming the cooperative society. With such structure, social norms are created. Each dynamics corresponding to each cooperative society is separated far from the non-cooperative dynamics. Small change to the strategy or to the society cannot destroy the cooperative society. In the linear game, on the other hand, there is no (or few) articulation in the AGS diagram. Hence, the strategy can be mutated continuously to the non-cooperative one. Accordingly, the increase in the number of players has a same effect as the static game, in this case.

Note that the cooperation in the convex LD games does not require any external force, but is created spontaneously within the system. The social norms for cooperation are composed through the interactions among the lumberjacks and the game environment. Also, the sanctions to those who break the social standards are organized through the dynamical structure of the system, which is not implemented as *sanction strategies* at the stage of modeling. Such a social norm autonomously keeps its stability within the players' system as a certain rule for cooperation. Conversely, when we try to set a norm from the outside of the system, it is desirable to set it so that it meets with the stable rule given as an articulated structure in the DS game. Then it can be maintained steadily with the aid of the nature of the dynamical system.

Formation of dynamical cooperation in the real world and in a theoretical model An actual example of recovering from tragedy is seen in pasturing in North America (15). Besides this example, there are innumerable cases of social dilemma as to sharing resources within communities (14), where their solutions often use space-time structure such as the change of roles in turn to allocate resources.

When we actually try to avoid tragedy in consuming resources, we normally come up (consciously or unconsciously) to consider dynamical change of environment, such as the growth of the pasture, the degree of the restoration of the land, and the nutritional state of cattle. If we try to obtain certain amount of resources without fail, we need to manage the dynamics of the resources, and thus it will be necessary to have certain agreement for cooperation. As a result, we will begin to take actions such as "raising the resources jointly and then consuming them together" or "raising the resources and consuming them alternately." In many real cases, it is essential to consider the space-time structure in the environment, to avoid tragedy within social dilemmas. Such cooperation in reality is understood as a metastable solution of the corresponding DS game that reflects the nature of the dynamics in the resource. As is shown in our model, we often behave based on some cooperation rules, whether they are explicit or implicit.

In the traditional game theory, one cannot discuss such cooperation in the form of dynamics, let alone explain the stability of the cooperative state. In the first place, it neither can describe, in principle, the temporal change in resource nor can show the effect of dynamics of the game environment. Of course one could model such social dilemma with a static game by preparing strategies such as 'no grazing (cooperation),' 'grazing (defection),' and 'grazing n-cows (n-degree defection).' However, in order to handle most issues of social dilemma, it is important to consider the dynamics of action based on dynamics of resources, such as *timing* of grazing cows, depending on the state of the pasture, the nutritious state of cows and the economic state of cowherds.

In this paper, a novel solution is provided to the problem of cooperation within social dilemma, based on dynamical systems game(1). Norms for cooperation are organized spontaneously, based on the articulation structure in the strategy, which arises from the bifurcation of the attractor dynamics. As long as our world includes space-time structure by nature, the DS game gives a powerful theoretical framework to study various types of social and economic problems produced by multiple decision-makers.

Acknowledgments
The authors would like to thank T.Ikegami and S.Sasa for useful discussions. This work is partially supported by Grants-in-Aid for Scientific Research from the Ministry of Education, Science and Culture of Japan. The first author is supported by Japan Society for the Promotion of Science under the contract number RFTF96I00102.

References

1. Akiyama E. and K. Kaneko: Dynamical Systems Game Theory and Dynamics of games, to appear in Physica D.
2. Akiyama E. and K. Kaneko: Dynamical Systems Game Theory II — Application to Social Dilemma, in preparation.
3. Axelrod, R. , The evolution of cooperation, Basic Books, New York (1984)
4. Axelrod, R., An Evolutionary Approach to Norms, American Political Science Review, 80-4 (1986) 1095-1111
5. Axelrod, R. and D. Dion., The further evolution of cooperation, Science (Washington, D. C.), 242 (1988) 1385-1390
6. Boyd, R. and P. J. Richerson, The evolution of reciprocity in sizable groups. J Theor Biol. (1988) Jun 7;132(3):337-56.
7. Boyd, R. and P. J. Richerson, Culture and cooperation, 111-132 in J. J. Mansbridge, ed. Beyond self-interest, University of Chicago Press, Chicago.
8. Boyd, R. and P. J. Richerson, Punishment allows the evolution of cooperation (or anything else) in sizable groups, Ethology and Sociobiology, 13 (1992) 171-195
9. Hamilton, W.D., The genetical evolution of social behaviour, I, II. J Theor Biol. 7:1-52 1964
10. Hardin, G., The tragedy of the commons. Science 162 (1968) 1243-1248.
11. K. Kaneko and J. Suzuki, "Evolution to the Edge of Chaos in Imitation Game," Artificial Life III (1994) 43-54
12. N. V. Joshi, Evolution of cooperation by reciprocation within structured demes, Journal of Genetics, 66 (1987) 69-84
13. Maynard Smith, J. Evolution and the Theory of Games, Cambridge University Press, Cambridge (1982)
14. Ostrom, E., Gardner, R. and J. Walker, Rules, Games, & Common-Pool Resources, Univ. of Michigan Press (1994)
15. World resources, New York ; Oxford: Oxford University Press, (1994-95)
16. Yasutomi, A., The emergence and collapse of money, Physica-D 82 pp.180-94, 1995.

Interactions between Learning and Evolution: The Outstanding Strategy Generated by the Baldwin Effect

Takaya Arita and **Reiji Suzuki**
Graduate School of Human Informatics, Nagoya University
Chikusa-ku, Nagoya 464-8601, JAPAN

Abstract

The Baldwin effect is known as interactions between learning and evolution, which suggests that individual lifetime learning can influence the course of evolution without the Lamarckian mechanism. Our concern is to consider the Baldwin effect in dynamic environments, especially when there is no explicit optimal solution through generations and it depends only on interactions among agents. We adopted the iterated Prisoner's Dilemma as a dynamic environment, introduced phenotypic plasticity into strategies, and conducted the computational experiments, in which phenotypic plasticity is allowed to evolve. The Baldwin effect was observed in the experiments as follows: First, strategies with enough plasticity spread, which caused a shift from defect-oriented population to cooperative population. Second, these strategies were replaced by a strategy with a modest amount of plasticity generated by interactions between learning and evolution. By making three kinds of analysis, we have shown that this strategy provides the outstanding performance. Further experiments towards open-ended evolution have also been conducted so as to generalize our results.

Introduction

Baldwin proposed 100 years ago that individual lifetime learning (phenotypic plasticity) can influence the course of evolution without the Lamarckian mechanism (Baldwin, 1896). This "Baldwin effect" explains the interactions between learning and evolution by paying attention to balances between benefit and cost of learning. The Baldwin effect consists of the following two steps (Turney, Whitley and Anderson, 1996). In the first step, lifetime learning gives individual agents chances to change their phenotypes. If the learned traits are useful to agents and result in increased fitness, they will spread in the next population. This step means the synergy between learning and evolution. In the second step, if the environment is sufficiently stable, the evolutionary path finds innate traits that can replace learned traits, because of the cost of learning. This step is known as *genetic assimilation*. Through these steps, learning can accelerate the genetic acquisition of learned traits without the Lamarckian mechanism in general. Figure 1 roughly shows the concept of the Baldwin effect which consists of the two steps described above.

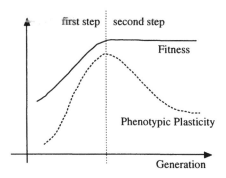

Figure 1: Two steps of the Baldwin effect.

Hinton and Nowlan constructed the first computational model of the Baldwin effect and conducted an evolutionary simulation (Hinton and Nowlan, 1987). Their pioneering work caused the Baldwin effect to come to the attention of the computer scientists, and many computational approaches concerning the Baldwin effect have been conducted since then (Arita, 2000). For example, Ackley and Littman successfully showed that learning and evolution together were more successful than either alone in producing adaptive populations in an artificial environment that survived to the end of their simulation (Ackley and Littman, 1991). Also, Bull recently examined the performance of the Baldwin effect under varying rates and amounts of learning using a version of the NK fitness landscapes (Bull, 1999).

Most of them including Hinton and Nowlan's work have assumed that environments are fixed and the optimal solution is unique, and have investigated the first step (synergy between learning and evolution). However, as we see in the real world, learning could be more effective and utilized in dynamic environments, because the flexibility of plasticity itself is advantageous to adapting ourselves to the changing world. Therefore, it is essential to examine how learning can affect the course of evolution in dynamic environments (Suzuki and Arita, 2000).

Our objective is to gain valuable insights into interactions between learning and evolution, especially into the Baldwin effect, by focusing on balances between benefit and cost of learning in dynamic environments: whether the Baldwin effect is observed or not, how it works, and what it brings over all in dynamic environments.

As one of the few studies that looks at both the benefits and costs of learning (in static environments), Menczer and Belew showed that interactions between learning and evolution are not beneficial if the task that learning is trying to optimize is not correlated with the task that evolution is working on (Menczer and Belew, 1991). Also, Mayley explored two criteria for the second step of the Baldwin effect by using NK fitness landscapes (Mayley, 1997). He concluded that two conditions, high relative evolutionary cost of learning and the existence of a neighborhood correlation relationship between genotypic space and phenotypic space, are the necessary conditions for the second step to occur.

In general, dynamic environments can be divided typically into the following two types: the environments in which the optimal solution is changed as the environment changes, and the ones in which each individual's fitness is decided by interactions with others. As the former type of environments, Anderson quantitatively analyzed how learning affects evolutionary process in the dynamic environment whose optimal solution changed through generations by incorporating the effects of learning into traditional quantitative genetics models (Anderson, 1995). It was shown that in changing environments, learning eases the process of genetic change in the population, while in fixed environments the individual advantage of learning is transient. Also, Sasaki and Tokoro studied the relationship between learning and evolution using a simple model where individuals learned to distinguish poison and food by modifying the connective weights of neural network (Sasaki and Tokoro, 1999). They have shown that the Darwinian mechanism is more stable than the Lamarckian mechanism while maintaining adaptability. Both studies emphasized the importance of learning in dynamic environments.

We adopted the iterated Prisoner's Dilemma (IPD) as the latter type of environments, where there is no explicit optimal solution through generations and fitness of individuals depends mainly on interactions among them. Phenotypic plasticity, which can be modified by lifetime learning, has been introduced into strategies in our model, and we conducted the computational experiments in which phenotypic plasticity is allowed to evolve.

The rest of the paper is organized as follows. Section 2 describes a model for investigating the interactions between learning and evolution by evolving the strategies for the IPD. The results of evolutionary experiments based on this model are described in Section 3. In Section 4, we analyze the strategy generated by the Baldwin effect in these experiments by three methods (ESS conditi-

tion, state transition analysis and qualitative analysis). Section 5 describes the extended experiments towards open-ended evolution in order to generalize the results in the previous sections. Section 6 summarizes the paper.

Model
Expression of Strategies for the Prisoner's Dilemma

We have adopted the iterated Prisoner's Dilemma (IPD) game as a dynamic environment, which represents an elegant abstraction of the situations causing social dilemma. IPD game is carried out as follows:

1) Two players independently choose actions from cooperate (C) or defect (D) without knowing the other's choice.

2) Each player gets the score according to the payoff matrix (Table 1). We term this procedure "round".

3) Players play the game repeatedly, retaining access at each round to the results of all previous rounds, and compete for higher average scores.

Table 1: A payoff matrix of Prisoner's Dilemma.

player \ opponent	cooperate	defect
cooperate	(R:3, R:3)	(S:0, T:5)
defect	(T:5, S:0)	(P:1, P:1)

(player's score, opponent's score)
$$T > R > P > S, 2R > T + S$$

In case of one round game, the payoff matrix makes defecting be the only dominant strategy regardless of opponent's action, and defect-defect action pair is the only Nash equilibrium. But this equilibrium is not Pareto optimal because the score of each player is higher when both of the players cooperate, which causes a dilemma. Furthermore, if the same couple play repeatedly, this allows each player to return the co-player's help or punish co-player's defection, and therefore cooperating with each other can be advantageous to both of them in the long run (Axelrod, 1984).

The strategies of agents are expressed by two types of genes: genes for representing strategies (GS) and genes for representing phenotypic plasticity (GP). GS describes deterministic strategies for IPD by the method adopted in Lindgren's model (Lindgren, 1991), which defines next action according to the history of actions. GP expresses whether each corresponding bit of GS is plastic or not.

A strategy of memory m has an action history h_m which is a m-length binary string as follows:

$$h_m = (a_{m-1}, \ldots, a_1, a_0)_2, \qquad (1)$$

where a_0 is the opponent's previous action ("0" represents defection and "1" represents cooperation), a_1 is the previous player's action, a_2 is the opponent's next to previous action, and so on.

GS for a strategy of memory m can be expressed by associating an action A_k (0 or 1) with each history k as follows:

$$GS = [A_0 A_1 \cdots A_{n-1}] \quad (n = 2^m). \quad (2)$$

In GP, P_x specifies whether each phenotype of A_x is plastic (1) or not (0). Thus, GP can be expressed as follows:

$$GP = [P_0 P_1 \cdots P_{n-1}]. \quad (3)$$

For example, the popular strategy "Tit-for-Tat" (cooperates on the first round, does whatever its opponent did on the previous round) (Axelrod, 1984) can be described by memory 2 as GS=[0101] and GP=[0000].

Meta-Pavolv Learning

A plastic phenotype can be changed by learning during game. We adopted a simple learning method termed "Meta-Pavlov". Each agent changes plastic phenotypes according to the result of each round by referring to the Meta-Pavlov learning matrix (Table 2). It doesn't express any strategy but expresses the way to change one's own strategy (phenotype) according to the result of the current round, though this matrix is the same as that of the Pavlov strategy which is famous because it was shown that it outperforms the popular strategy "Tit-for-Tat" (Nowak and Sigmund, 1993).

Table 2: The Meta-Pavlov learning matrix.

player \ opponent	cooperate	defect
cooperate	C	D
defect	D	C

The learning process is described as follows:

1) At the beginning of the game, each agent has the same phenotype as GS itself.

2) If the phenotype used in the last round was plastic, in other words, the bit of GP corresponding to the phenotype is 1, the phenotype is changed to the corresponding value in the Meta-Pavlov learning matrix based on the result of the last round.

3) The new strategy specified by the modified phenotype will be used by the player from next round on.

Take a strategy of memory 2 expressed by GS=[0001] and GP=[0011] for example of learning (Figure 2). Each phenotype represents the next action corresponding to

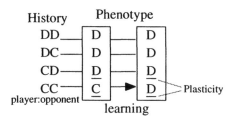

Figure 2: An example of Meta-Pavlov learning.

the history of the previous round, and the underlined phenotypes are plastic.

Let us suppose that the action pair of the previous round was "CC (player's action: cooperation, opponent's action: cooperation)" and the opponent defects at the present round. This strategy cooperates according to the phenotype and the result of the current round is "CD" (*Sucker's* payoff). The strategy changes own phenotype according to this failure based on the Meta-Pavlov learning matrix, because the phenotype applied at this round is plastic. The phenotype "C" corresponding to the history "CC" is changed to "D" in this example. Therefore, this strategy chooses defection when it has the history "CC" at the next time. Meta-Pavlov learning is intuitive and natural in the sense that it is a simple realization of reinforcement learning.

The values of GS that are plastic act merely as the initial values of phenotype. Thus we represent strategies by GS with plastic genes replaced by "x" (e.g. GS=[1000] and GP=[1001] $\rightarrow$ [x00x]).

Evolution

We shall consider a population of N individuals interacting according to the IPD. All genes are set randomly in the initial population. The round robin tournament is conducted between individuals with the strategies which are expressed in the above described way. Performed action can be changed by noise (mistake) with probability p_n. Each plastic phenotype is reset to the corresponding value of GS at the beginning of games. The game is played for several rounds. We shall assume that there is a constant probability p_d (*discount* parameter) for another round. The tournament is "ecological": The total score of each agent is regarded as a fitness value, new population is generated by the "roulette wheel selection" according to the scores, and mutation is performed on a bit-by-bit basis with probability p_m.

Average scores during the first 20 IPD games between new pair are stored, and will be used as the results of the games instead of repeating games actually, so as to reduce the amount of computation. Stored scores are cleared and computed again by doing games every 500 generation.

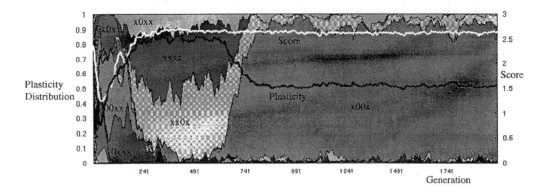

Figure 3: The experimental result (2000 generations).

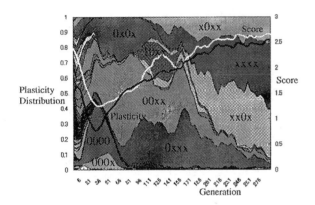

Figure 4: The experimental result (300 generations).

Evolutionary Experiments

Strategies of memory 2 were investigated in the evolutionary experiments described in this section. We conducted an experiment for 2000 generations using following parameters: $N = 1000$, $p_m = 1/1500$, $p_n = 1/25$ and $p_d = 99/100$.

The evolution of population for the first 2000 generations is shown in Figure 3 and that for the first 300 generations is shown in Figure 4. In each figure, the horizontal axis represents the generations. The vertical axis represents the distribution of strategies, and at the same time, it also represents both "plasticity of population" (in black line) and the average score (in white line). Plasticity of population is the ratio of "1" in all genes of GPs, and it corresponds to the "Phenotypic Plasticity" in Figure 1. The average score represents the degree of cooperation in the population, and it takes 3.0 as the maximum value when all rounds are "CC".

The evolutionary phenomena that were observed in experiments are summarized as follows. Defect-oriented strategies ([0000], [000x] and so on) spread and made the average score decrease until about the 60th generation, because these strategies can't cooperate with each other. Simultaneously, partially plastic strategies ([0x0x], [00xx] and so on) occupied most of the population. Next, around the 250th generation, more plastic strategies ([xxxx], [x0xx] and so on) established cooperative relationships quickly, which made the plasticity and average fitness increase sharply. This transition is regarded as the first step of the Baldwin effect.

Subsequently, the plasticity of population decreased and then converged to 0.5 while keeping the average score high. Finally, the strategy [x00x] occupied the population. The reason seems to be that the strategy has the necessary and sufficient amount of plasticity to maintain cooperative relationships and prevent other strategies from invading in the population. This transition is regarded as the second step of the Baldwin effect.

The evolutionary phenomena described above was observed in about 70% of the experiments, and the population converged to the strategy [x00x] in *all* experiments we conducted. Further analysis on this strategy will be conducted in the next Section. Another series of experiments has shown overall that the higher the mutation rate becomes, the faster the strategies tend to evolve. It has also been shown that the higher the noise probability becomes, the more All-D type strategies are selected, and the less the system becomes stable.

Figure 5 made us grasp the clear image of the evolutionary behavior of the system in the experiments. This figure shows the evolutionary trajectory of ten experiments drawn in the space of score and plasticity. We see the evolutionary process consists of 3 modes. The score decreases without increase of plasticity during an initial stage. The cause of this decrease is that defect-oriented strategies (e.g. [0000][000x]) spread in the initial randomly-created population. The score decreases nearly to 1.0 which is the score in the case of defect-defect action pair. When the score reaches this value, a "mode transition" happens and the first step of the Baldwin effect starts. In this stage, phenotypic plasticity gives chances to be adaptive. Therefore, score is correlated with plasticity, and approaches nearly 3.0, that is the

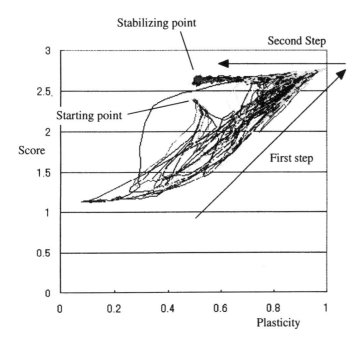

Figure 5: Two steps of the Baldwin effect.

score in the case of cooperate-cooperate action. Strategies with enough plasticity (e.g. [xxxx][x0xx][xx0x]) occupy the end of this stage. Then, another mode transition happens suddenly, and plasticity decreases gradually while keeping the score high. The plasticity decreases monotonously, and then, the population always converged to be homogenous, that is occupied with the strategy [x00x]. As is apparent from this figure, there were exceptions to which above description doesn't apply, however, it has been shown that the system always stabilized with [x00x] in the end.

Analysis of Meta-Pavlov [x00x]

ESS Condition

An ESS (Evolutionary Stable Strategy) is a strategy such that, if all the members of a population adopt it, no mutant strategy can invade (Maynard Smith, 1987). The necessary and sufficient condition for a strategy "a" to be ESS is:

$$E(a, a) > E(b, a) \quad \forall b, \tag{4}$$

or

$$E(a, a) = E(b, a) \quad and \quad E(a, b) > E(b, b) \quad \forall b, \tag{5}$$

where $E(a, b)$ is the score of strategy "a" when strategy "a" plays against strategy "b".

We conducted the iterated games between [x00x] (GS=[0000], GP=[1001]) and all 256 strategies with memory 2, and computed the average scores of them, so as to examine whether it satisfied the ESS condition

or not. The noise probability (p_n) was 1/25 and the discount parameter (p_d) was 99/100. The results are shown in Figure 6. The horizontal axis represents all strategies by interpreting the genotypic expression [$GSGP$] as an 8 bit binary number x (e.g. GS=[0000], GP=[1001] $\rightarrow$ 00001001_2=9). The vertical axis represents the relative scores of the strategy x, that is,

$$E([x00x], [x00x]) - E(x, [x00x]). \tag{6}$$

This graph shows that this value is always positive. Therefore, [x00x] is an ESS in the population of memory 2 strategies.

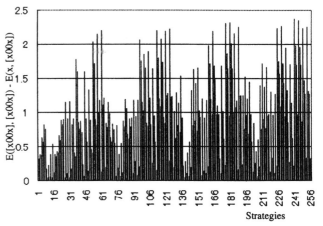

Figure 6: Relative scores of all strategies of memory 2 against [x00x].

State Transition Analysis

Figure 7 shows a state transition diagram of the Meta-Pavlov [x00x] strategy. Each state is represented by a box, in which the actions in the current round are described: the opponent's action on top and the [x00x]'s action on bottom (0: defect, 1: cooperate). The current values of plastic genes also discriminate the states, and they are described in the lower right corner (e.g. left "x"=0 and right "x"=1 $\rightarrow$ 01). Two arrows issue forth from each state, depending on whether the opponent plays C or D at the next round. Described actions of [x00x] in the destination box are identical, and it will be the next action of [x00x]. For example, the stabilized state of the game between [x00x] and All-D is expressed by a loop ("cycle 2" in this figure), which means that the game generates the periodic action pairs. The boxes without inputted arrows can be reached by noise.

Duration of the state "A" means that mutually cooperative relationship has been established. It is a remarkable point that if this relationship is abolished by the opponent's defection, a bit of *protocol* (cycle "1" in this figure) is needed to restore the damaged relationship as follows:

```
Opponent:  .. 1  1  0  0  1  0  1  1  1  ..
[x00x]:     .. 1  1  1  0  1  0  1  1  1  ..
       Mutual C |noise|fence-  | Mutual C
                       mending
```

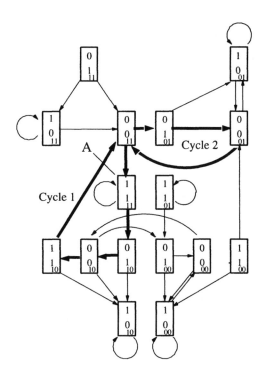

Figure 7: A state transition diagram of the Meta-Pavlov [x00x].

This minimal fence-mending is done exactly when an accidental opponent's defection by noise occurs after mutually cooperative relationship has been established in the game between [x00x] and itself. This property of [x00x] seems to play an important part in recovery from the broken relationship, and to make the strategy an ESS.

Qualitative Analysis on Phenotypic Plasticity

Many researchers in evolutionary computation or related fields have focused exclusively on the benefits on the phenotypic plasticity. Phenotypic plasticity enables the individuals to explore neighboring regions of phenotype space. The fitness of an individual is determined approximately by the maximum fitness in its local region. Therefore, if the genotype and the phenotype are correlated, plasticity has the effect of smoothing the fitness landscape, and makes it easier for evolution to climb to peaks in the landscape. This is the first step of the Baldwin effect.

However, there is the second step, because plasticity can be costly for an individual. Learning requires energy and time, for example, and sometimes brings about dangerous mistakes. In our computational experiments, the costs of learning are not explicitly embed in the system. The costs of learning are implicitly expressed by the behavior that is caused typically by noise. For example, when a noise happens to a game between [x00x] and [xxxx], plastic properties make the [xxxx] strategy play more C than [x00x] while they restore the damaged relationship, which generates [xxxx]'s loss. The optimum balance between plasticity and rigidity depends on the performance of the learning algorithm. In this context, the Meta-Pavlov learning algorithm gets along extremely well with [x00x], as will be shown in the extended experiments.

Here, we investigate why these two plastic genes in [x00x] remained in the second step of the Baldwin effect, that is, the significance of the two plastic genes. While the functions of these two genes are of course depend on the interactions among all genes, simple explanation could be possible based on the results of our qualitative analysis as follows:

- The left "x" (which describes the plasticity of the action immediately after D-D) is effective especially when [x00x] plays against defect-oriented strategies. For example, when [x00x] plays against All-D, [x00x] gets the *Sucker's* payoff once every three rounds (cycle 2 in Figure 7), and gets only 0.67 on average, caused by the plasticity of the left "x". However, All-D gets about 2.33, which supports the ESS property of [x00x] because [x00x] gets about 2.6 when it plays against itself, as follows:

```
[0000]:  .. 000000000000 ..  Average 2.33
[x00x]:  .. 010010010010 ..  Average 0.67
```

In contrast, for example, the game between the Pavlov strategy (Nowak and Sigmund, 1994) and All-D is as follows:

```
[0000]:  .. 000000000000 ..  Average 3
[1001]:  .. 010101010101 ..  Average 0.5
```

- The right "x" (which describes the plasticity of the action immediately after C-C) is effective especially when [x00x] plays against cooperate-oriented strategies. [x00x] can defect flexibly by taking advantage of the opponent's accidental defection by noise. The right "x" becomes 0, when C-C relationship is abolished by the opponent's defection as shown in Figure 7. Therefore, [x00x] exploits relatively cooperate-oriented strategies. Followings are the rounds between [x001] and [x00x]. The first 0 of [x001] represents an accidental defection by noise. Average scores are calculated only during the oscillation.

```
[x001]:  .. 1110011011011 ..  Average 1.33
[x00x]:  .. 1111010010010 ..  Average 3
```

On the other hand, for example, the game between the Pavlov strategy and [x001] is as follows:

```
[x001]:   .. 1110000000000 ..   Average 3
[1001]:   .. 1111010101010 ..   Average 0.5
```

These two properties of [x00x] are quite effective on the premise that it establishes strong relationship with itself. Actually, minimal fence-mending is realized by utilizing these two plastic genes (two times of learning each gene) which is represented by the "cycle 1" in Figure 7.

Extended Experiments towards Open-ended Evolution

Evolution of Learning Algorithms

So far, we have adopted the Meta-Pavlov learning method as an algorithm for modifying strategies by changing plastic phenotype. Here, we weaken this constraint, and shall focus on the evolution of not only strategies but also learning algorithms by defining the third type of genes.

In the experiments described in this section, each individual has genes for defining a learning method (GL), which decides how to modify the phenotype representing its strategies. GL is a four-length binary string composed of the elements of a learning matrix such like Table 2. The order of elements in the string is [(DD) (DC) (CD) (CC)]. For example, the Meta-Pavlov learning method described in the previous sections is expressed by [1001]. It could be said that the learning methods (GL) and the strategies (GS and GP) co-evolve, because the performance of learning methods depends on the strategies to which they will be applied.

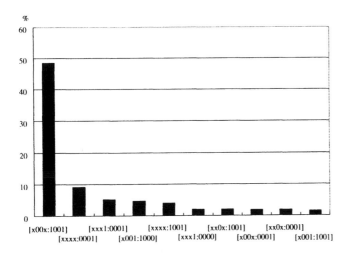

Figure 10: Average occupation of strategies.

Experiments were conducted under the same conditions as those in the previous experiments except for

GL. Initial population had 100 kinds of combinations of randomly generated GS, GP and GL, and each kind had ten identical individuals. Typical results are shown in Figure 8 and Figure 9. Each area in these figures expresses a (strategy, learning method) pair. For example, "x00x:1001" means the [x00x] strategy with the learning method [1001] (Meta-Pavlov). It is shown that Meta-Pavlov [x00x] and [x001:1000] occupied the populations and established a stable state in Figure 8 and Figure 9 respectively.

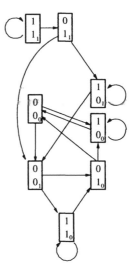

Figure 11: A state transition diagram of [x001:1000].

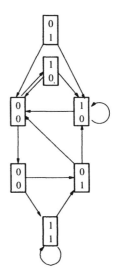

Figure 12: A state transition diagram of Prudent-Pavlov.

Figure 10 shows the average occupation of top ten (strategy, learning method) pairs in the 4000th generation over 60 trials. It is shown that Meta-Pavlov [x00x] occupied nearly half of the population in the 4000th generation on average. Meta-Pavlov [x00x] occupied the

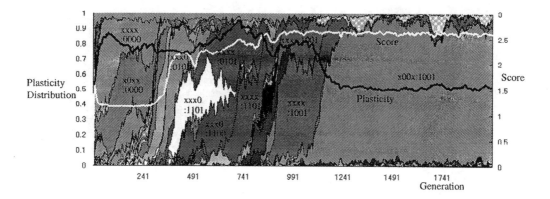

Figure 8: Evolution of learning algorithms and strategies (Case 1).

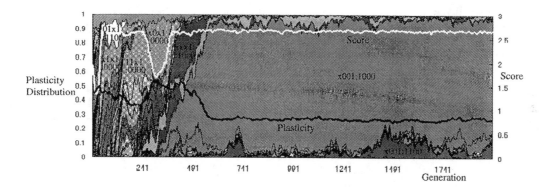

Figure 9: Evolution of learning algorithms and strategies (Case 2).

population and established a stable state (as shown in Figure 8) in 29 trials, [x001:1000] which is at the 4th in Figure 10 did so (as shown in Figure 9) in 3 trials, and no pairs occupied the population and established a stable state in the rest of trials. It follows from these facts that all but these two strategies in Figure 10 are invaded by mutants, though they can invade the population in certain conditions.

A state transition diagram of [x001:1000] is shown in Figure 11. We have found that this strategy has essentially the same property as that of "Prudent-Pavlov", whose state transition diagram is shown in Figure 12, though [x001:1000] has additional transient nodes, and there are subtle differences in expression of states and state transitions. Prudent-Pavlov can be interpreted as a sophisticated offspring of Pavlov (Boerljst, Nowak and Sigmund, 1997). Prudent-Pavlov follows in most cases the Pavlov strategy. However, after any defection it will only resume cooperation after two rounds of mutual defection. They are remarkable facts that in our experiments a derivative of such a sophisticated *human-made* strategy was generated automatically, and that the Meta-Pavlov [x00x] outperformed the other strategies including this strategy.

Evolution without limitation of memory length

We have conducted further experiments towards open-ended evolution. Two types of mutation, gene duplication and split mutation, were additionally adopted, which allows strategies to become complex or simple without restrictions. The gene duplication attaches a copy of the genome itself (e.g., [1101] → [11011101]). The split mutation randomly removes the first or second half of the genome (e.g., [1101] → [11] or [01]). Each mutation is operated on GS and GP at the same time. In this series of experiments, we adopted Meta-Pavlov learning without allowing the learning mechanisms to evolve for convenience of the analysis.

Initial population was composed of strategies of memory 1, each of which has randomly generated GS and GP which was set to [00] (no plasticity). The results are shown in Figure 13. In most trials, during the first hundreds of generations, the system oscillated ([01] → [11] → [10] → [00]) in the same manner as in the Lindgren's experiments (Lindgren, 1991). At the end of the period of oscillation, a group of memory 2 strategies was growing, and took over the population. After that, there were two major evolutionary pathways, both of which happened

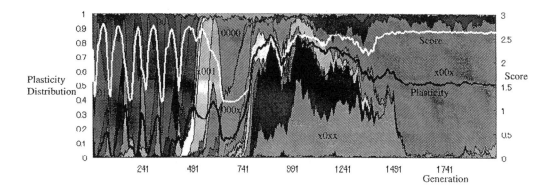

Figure 13: Evolution without limitation of memory length.

with nearly equal probabilities:

1) Strategies evolved showing the Baldwin effect as described in the previous sections. Later on, the system stabilized with [x00x] typically near the 1500th generation.

2) [x00x] entered the scene quickly, took over the generation, and the system stabilized with it.

It has been shown that which course the evolution takes depends on the state of the population while memory 2 strategies are growing. If the population is taken over by defect-oriented strategies before cooperate-oriented strategies emerge, the evolution tends to take the course 1). On the other hand, if the population is taken over by cooperate-oriented strategies without emergence of defect-oriented strategies of memory 2, then the evolution tends to take the course 2).

In most cases we observed, the system got stuck in the evolutionary stable state through either of the courses, though in rare cases the system didn't stabilize with [x00x] but stabilized with some mixture of various strategies of more than 2-length memory. The reason why strategies of more than 2-length memory rarely evolved is considered to relate to the mutation of learning mechanisms. The point here is that gene duplication changes the phenotype corresponding to the plastic genes because learning happens independently at two different points if a plastic gene is duplicated. Therefore, the evolution of phenotype could be discontinuous when gene duplication happens.

Discussion and Conclusions

The Baldwin effect has not always been well received by biologists, partly because they have suspected it of being Lamarckist, and partly because it was not obvious it would work (Maynard Smith 1996). Our results of the experiments inspire us to image realistically how learning can affect the course of evolution in dynamic environments. It is an important fact that a drastic

mode transition happens at the edge between the first step and second step of the Baldwin effect in the environments where the optimal solution is dynamically changed depending on the interactions between individuals as is clearly shown in Figure 5.

Furthermore, based on the results of our experiments, we could imagine biological adaptation as a *measuring worm* climbing around on the fitness landscape (Figure 14). The population of a species is represented by the worm. Its head is on the phenotypic plane and its tail is on the genotypic plane. These two planes are assumed to be correlated to each other to a high degree. The landscape is always changing corresponding to the state of the worm (interactions between individuals). The worm stretches its head to the local top (first step), and when it stretches itself out, it starts pulling it's tail (second step). In our experiments, the Baldwin effect was observed once every trial. We believe that the repetition of these two steps like the behavior of measuring worms will be observed in the experiments where the environment (e.g. payoff matrix) itself is also changing. Such view of the interactions between learning and evolution might simplify the explanation of *punctuated equilibria*. In fact, Baldwin noticed that the effect might explain that variations in fossil deposits seem often to be discontinuous (Baldwin, 1896).

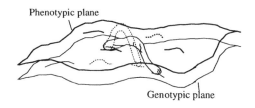

Figure 14: A "measuring worm" on the fitness landscape.

It has also been shown that the implications of the learning cost on the attribution of an individual's fitness score in dynamic environments is very different from those in static environments. High evolutionary cost of learning is one of the necessary conditions for the second step of the Baldwin effect to occur in general, as pointed out by Mayley (Mayley, 1997). However, in our model the learning costs are not explicitly embed in the system. In the experiments, the second step was dominated not by time-wasting costs, energy costs, unreliability costs and so on during the vulnerable learning period. Instead, it was dominated by the constraints of the performance of the learning algorithms themselves in the complex environment where it was impossible for any algorithm to predict opponents' behavior perfectly. These constraints are equivalent to the cost of learning and, therefore, they could cause the decrease in phenotypic plasticity.

The Baldwin effect generated the Meta-Pavlov [x00x] strategy, and the system stabilized with it. We have analyzed the property of the Meta-Pavlov [x00x] strategy, and have shown it's outstanding performance, which is rather a by-product to us. The excellent performance of the Meta-Pavlov [x00x] is also supported by the fact that in the extended experiments it outperformed a derivative of the Prudent-Pavlov which can be interpreted as a sophisticated offspring of the famous strategy Pavlov.

This model can be extended in several directions. It would be interesting to investigate the interactions between learning and evolution in the environments allowing competitive coevolution by using games where first-player and second player have different roles (Rosin and Belew, 1997). One obvious direction would be to attempt to reinterpret and evaluate our results concerning the interactions between learning and evolution in the context of pure biology.

Another direction would be to focus on the technical aspects of the evolutionary mechanism of varying phenotypic plasticity. It would be interesting to apply the automatic mechanism of adjusting the balance between evolution and learning in the fields of distributed AI or multi-agent systems.

References

Ackley and Littman, 1991. Interactions Between Learning and Evolution. *Proceedings of Artificial Life II*: 487–509.

Anderson R. W. 1995. Learning and Evolution: A Quantitative Genetics Approach. *Journal of Theoretical Biology* 175: 89–101.

Arita, T. 2000. *Artificial Life: A Constructive Approach to the Origin/Evolution of Life, Society, and Language* (in Japanese). Science Press.

Axelrod R. 1984. *Evolution of Cooperation*. Basic Books.

Baldwin J. M. 1896. A New Factor in Evolution. *American Naturalist* 30: 441–451.

Boerlijst M. C., Nowak M. A. and Sigmund K. 1997. The Logic of Contrition. *Journal of Theoretical Biology* 185: 281–293.

Bull L. 1999. On the Baldwin Effect. *Artificial Life* 5(3): 241–246.

Hinton G. E. and Nowlan S. J. 1987. How Learning Can Guide Evolution. *Complex Systems* 1: 495–502.

Lindgren K. 1991. Evolutionary Phenomena in Simple Dynamics. *Proceedings of Artificial Life II*: 295–311.

Mayley G. 1997. Landscapes, Learning Costs, and Genetic Assimilation. *Evolutionary Computation* 4(3): 213–234.

Maynard Smith J. 1982. *Evolution and the Theory of Games*. Cambridge University Press.

Maynard Smith J. 1996. Natural Selection: When Learning Guides Evolution. In *Adaptive Individuals in Evolving Populations: Models and Algorithms* edited by Belew R. K. and Mitchell M.: 455–457.

Menczer F. and Belew K. 1994. Evolving Sensors in Environments of Controlled Complexity. *Proceedings of Artificial Life IV*: 210–221.

Nowak M. A. and Sigmund K. 1993. A Strategy of Win-Stay, Lose-Shift that Outperforms Tit-for-Tat in the Prisoner's Dilemma Game. *Nature* 364(1): 56–58.

Rosin C. D. and Belew R. K. 1997. New Methods for Competitive Coevolution. *Evolutionary Computation* 5(1): 1–29.

Sasaki T. and Tokoro M. 1999. Evolving Learnable Neutral Networks Under Changing Environments with Various Rates of Inheritance of Acquired Characters: Comparison Between Darwinian and Lamarckian Evolution. *Artificial Life* 5(3): 203–223.

Suzuki R. and Arita T. 2000. How Learning Can Affect the Course of Evolution in Dynamic Environments. *Proceedings of Fifth International Symposium on Artificial Life and Robotics* to appear.

Turney P., Whitley D. and Anderson R.W. 1996. Evolution, Learning, and Instinct: 100 Years of the Baldwin Effect. *Evolutionary Computation* 4(3): 4–8.

Evolvability Analysis: Distribution of Hyperblobs in a Variable-Length Protein Genotype Space

Hideaki Suzuki

ATR Human Information Processing Research Laboratories
2-2-2 Hikaridai, Seika-cho, Soraku-gun, Kyoto 619-0288 Japan

Abstract

A variable-length protein genotype space (the whole amino-acid sequence space) is mathematically analyzed and a lower threshold density for adequate connectivity of functional (viable) genotypes is estimated. Functional genotypes are assumed to distribute as a 'hyperblob' which means a cluster or an island, and connectivity between hyperblobs is estimated using the theory of regular languages and the random graph theory. It is shown that the logarithmic value of the threshold density approximately decreases with an increase in the genotype length.

Introduction

Proteins are fundamental functional units of living organisms. The evolvability (the possibility of evolution of various kinds) of living things is greatly dependent upon the evolvability of proteins; hence, to enhance the evolvability of an evolutionary system, we have to augment the possible evolution of various proteins. The evolvability of protein molecules has been argued by several authors to date. About thirty years ago, Maynard-Smith (Maynard Smith, 1970) argued that 'if evolution by natural selection is to occur, functional proteins must form a continuous network which can be traversed by unit mutational steps without passing through nonfunctional intermediates.' In 1991, Lipman et al. (Lipman and Wilbur, 1991) conducted a numerical experiment using a two-dimensional conformation model of artificial proteins proposed by Lau et al. (Lau and Dill, 1989; Lau and Dill, 1990). Lipman et al. confirmed that the fictional proteins satisfy Maynard-Smith's condition. As these authors pointed out, the evolvability of proteins is largely determined by the connectivity of functional (viable) genotypes in the protein genotype space.

(Here and throughout the paper, we consider the protein genotype space as the amino-acid sequence space in which points represent the amino-acid sequences of the proteins.) Since non-functional proteins quickly die out and cannot be fixed in the population, mutations, namely, slight modifications of genotypes, cannot search for various functional genotypes if the functional genotypes are sparsely distributed in the genotype space. For high evolvability, the functional proteins have to be densely distributsteps so that the functional genotypes are interconnected by unit mutational steps.

Recently, by focusing on the secondary structure of RNA molecules, Schuster et al. (Reidys et al., 1997a; Reidys et al., 1997b; Reidys, 1997; Schuster, 1997; Reidys et al., 1998) studied the genotype space of the RNA base sequence and analyzed the connectivity of *neutral networks*. Their neutral network is a graph whose vertices represent genotypes with the same secondary structure and whose edges represent mutational changes between genotypes. They applied random graph theory to this network and derived a formula for the minimum occurrence probability of neutral mutants to ensure the high connectivity of a neutral network. However, as Schuster described in his paper (Schuster, 1997), their arguments are principally based upon the assumption that 'sequences folding into the same structure are (almost) randomly distributed in sequence space.' This prediction by inverse folding (Hofacker et al., 1994) is valid for the RNA secondary structure; and yet for the conformations of protein molecules, it is expected that the amino acid sequences that create the same three-dimensional structure are distributed in an *island-like* way in the sequence space (Nishikawa, 1993). Protein genotypes with the same phenotype are likely to be unevenly distributed, so that we cannot adopt the same assumption as Shuster's for the analysis of the protein genotype space.

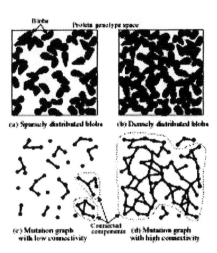

Figure 1: Basic concept of the genotype space analysis using blob distribution. (a) and (b) are symbolized figures of protein genotype space, and (c) and (d) are the corresponding mutation graphs. In (a) and (b), all genotypes are classified into functional ones (colored black) or nonfunctional ones (colored white). An undirected edge occurs between a node pair of the mutation graph if and only if corresponding blobs are interconnected in the genotype space. Evolvability is low for (a) and (c), and high for (b) and (d).

Based upon these notions, the author has recently analyzed the fixed-length genotype space of proteins and derived a quantitative condition for high evolvability (Suzuki, 2000a; Suzuki, 2000b). Figure 1 shows the basic concept of these papers (Suzuki, 2000a; Suzuki, 2000b). Protein genotypes with the same phenotype (function) are assumed to distribute as a *blob* which means a cluster or an island. A blob pair was regarded to be interconnected by mutation if they are adjacent or overlap in the genotype space, and, applying random graph theory (Bollobas, 1985; Palmer, 1985), the connectivity between blobs was quantitatively estimated. Although the studies reported in these papers first succeeded in formulating the minimum density of functional proteins to make a system evolvable, the studies had the following serious drawbacks. First, even if a pair of blobs are adjacent or overlap (have one common genotype at least), the size of a typical blob (here 'size' means the number of included genotypes) is so enormous

that the possibility of mutation bringing about the transition from one blob to another blob is extraordinarily small. This caused overestimation of the connection (transition) probability. Second, mutational modifications causing changes in the genotype length (deletion and insertion of amino acids) were neglected, and only the substitution of amino acid bases was considered as a modification of the genotype.

The present paper remedies these problems and studies the variable-length protein genotype space. A set of genotypes with the same phenotype is referred to as a *hyperblob* or *hblob* (which includes genotypes with different lengths), and the hblob is considered to be interconnected with another by mutation if the ratio of the size of the region common to another hblob compared to the size of the hblob is larger than some constant value. (The constant value is determined from the computational resource used in the system.) Like previous studies (Suzuki, 2000a; Suzuki, 2000b), the connectivity between hblobs is represented by a graph called a *mutation graph* whose nodes correspond to hblobs and whose *directed* edges represent mutational transitions between hblobs. (Note that unlike previous studies, the edges of the mutation graph for hyperblobs are directed ones.) It is known from the random graph theory (Bollobas, 1985; Palmer, 1985) that the connectivity of a random graph dramatically changes when the ratio of the edge number to the node number passes a particular threshold value. After conducting an experiment confirming this result for a directed random graph (Section II), two numerical estimations are made for the minimum hblob number needed for high connectivity of the mutation graph (Section III). The first one actually prepares a mutation graph of randomly created hblobs and studies the growth of its connectivity. From this experiment it is shown in Section IV that there is a threshold hblob number distinguishing between a region wherein the connectivity hardly increases with the hblob number and a region wherein the connectivity swiftly increases with the hblob number. The second calculation uses a prediction by the random graph theory. The occurrence probability of a directed edge in the mutation graph is measured using a Monte Carlo estimation method, and from this probability, the threshold hblob number is calculated (Section IV). The threshold values estimated by the two methods are compared.

Preliminary Experiment

This section's purpose is to conduct a numerical experiment for a *directed* random graph and derive an experimental formula relating the threshold node number to the occurrence probability of a directed edge. Before moving on to this experiment, it will be helpful to first describe the growth of connectivity of an *undirected* random graph briefly. Let N be the order (number of vertices) of an undirected random graph and M be the number of undirected edges of the random graph. According to the random graph theory (Bollobas, 1985; Palmer, 1985), it is known that when $M < N/2$, the orders of components (connected subgraphs) of a graph are much smaller than N, whereas when $M > N/2$, a giant component is likely to emerge in the graph and the order of the largest component is comparable to N. Hence, if we express the occurrence probability of an undirected edge between a pair of nodes by P, the threshold node number N_c of an undirected random graph is given by

$$\frac{N}{2} = M = \binom{N}{2}P = \frac{N(N-1)}{2}P$$
$$\therefore \; N_c = \frac{1}{P} + 1 \simeq \frac{1}{P}. \;\; \text{(undirected)} \quad (1)$$

In (Suzuki, 2000b), the author numerically studied the growth of connectivity of an undirected random graph and showed that when N passes N_c given by Eq. (1), the ratio of the average number of reachable nodes from one node compared to N suddenly begins to increase and swiftly approaches one.

A similar thing happens to a *directed* random graph. Let N be the total order of a directed random graph, N_c be the threshold number of N, P be the occurrence probability of a directed edge that connects a pair of nodes in a particular direction, and e be the ratio of the average number of reachable nodes from one node by directed edges compared to N. We start the experiment with $N = 0$, add nodes one by one, generate directed edges between a new node and older nodes using P, revise a reachable node list according to the connectivity, and calculate e from the reachable node list. Numerical trials are conducted ten times using a different random number sequence for each given P value. Figure 2 shows a part of the results given for different values of P. As is clearly shown in this figure, for a directed random graph as well, e suddenly

begins to increase after N passes a particular threshold value N_c. N_c is determined experimentally as

$$N_c \simeq \frac{1.15}{P}, \quad \text{(directed)} \quad (2)$$

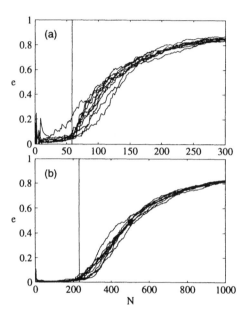

Figure 2: The growth of e as a function of N for a directed random graph with (a) $P = 0.02$ or (b) $P = 0.005$. The straight vertical lines represent N_c values given by Eq. (2).

Methods to Analyze Evolvability

Hyperblobs Represented by Regular Expressions

In this paper, the author represents a protein genotype by a sequence (string) of I characters chosen among K different characters (K is a fixed number and I is a variable). For a natural protein, an character can be compared to an amino acid base (I is about several hundred and $K = 20$), and for a subroutine of a machine-language programming system like Tierra (Ray, 1992; Ray, 1997; Ray and Hart, 1998), an character can be compared to an instruction (I is about several dozen and $K = 32$). In both example systems, the genotype of a protein/subroutine includes functionally important bases/instructions and functionally unimportant bases/instructions. The substitution or deletion of functionally important bases/instructions crucially changes the entire function of the molecule/subroutine, whereas the substitution, deletion, or even insertion of functionally unimportant bases/instructions usually has no influence on the functionality of the entire molecule/subroutine. Considering this characteristic, here we represent a hyperblob (a set of neighboring genotypes with the same function) by a sequence of I_u 'unsubstitutable' (functionally important) characters including insertions of arbitrary numbers of 'substitutable' (functionally unimportant) characters. I_s sequences are allowed to be inserted between some choice of positions between the unsubstitutable characters, and for each inserted sequence, characters are chosen out of K_s substitutable characters. Here the parameter ranges are $0 \leq I_u \leq I$, $0 \leq I_s \leq I_u + 1$, $0 \leq K_s \leq K$.

Such a set of sequences is simply represented by a kind of regular expression used in the theory of formal languages (Hopcroft and Ullman, 1969; Hopcroft and Ullman, 1979). Let, for example, an character set be represented by $\{\alpha, \beta, \gamma, \delta\}$ ($K = 4$). A regular expression for an hblob on this character set is a formula like

$$\alpha \cdot (\alpha + \gamma)^* \cdot \beta\alpha \cdot (\beta + \gamma + \delta)^*, \quad (3)$$

where '$\cdot$' represents *concatenation* ('$\cdot$' is often omitted), '$+$' represents *union* (or *or*), and

$$x^* = 1 + x + xx + xxx + \cdots$$

is a *closure* operation (1 represents a *null* string). Eq. (3) can be considered to represent an hblob by regarding the terms α and $\beta\alpha$ as unsubstitutable characters and the closure operations $(\alpha + \gamma)^*$ and $(\beta + \gamma + \delta)^*$ as the insertion of substitutable characters. In this example, the hblob parameter values are $I_u = 3$, $I_s = 2$, and $K_s = 2$ or 3.

The size of (the number of genotypes belonging to) hblob $S(I_u, I_s, K_s)$ is formulated as follows. As shown in Fig 3, the minimum length of the genotypes of an hblob is I_u, whereas the maximum length of the genotypes of an hblob is infinite. So, in this paper, we limit the length of genotypes to I_{max} and express $S(I_u, I_s, K_s)$ by the sum of section areas at length-I subspaces as

$$S(I_u, I_s, K_s) = \sum_{I=I_u}^{I_{max}} s(I; I_u, I_s, K_s) \quad (4)$$

To formulate $s(I; I_u, I_s, K_s)$, here the author adopts the assumption that the values of K_s

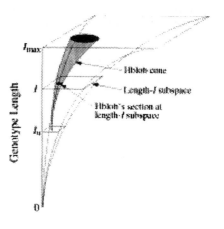

Figure 3: Symbolized picture of a hyperblob in the variable-length protein genotype space. The size of the length-I subspace increases exponentially with I. An hblob covers a cone-like region whose apex is located in the length-I_u subspace and whose base can extend freely toward large I values.

in a regular expression are the same (hereafter, this is assumed throughout the paper). Then $s(I; I_u, I_s, K_s)$ is approximately given by

$$s(I; I_u, I_s, K_s) \simeq \binom{I - I_u + I_s - 1}{I_s - 1} \cdot K_s^{I - I_u}. \quad (5)$$

See Appendix A for the detailed derivation.

Transition Probability between Hyperblobs

The analyses in the subsequent sections are based upon the following assumptions:

- Hblobs are uniformly distributed in the genotype space, allowing a pair of hblobs to overlap (have common genotypes). (Although no two hblobs can overlap actually, we here allow this possibility by assuming a random distribution of hblobs.)

- A population of protein genotypes in an hblob visits all inner genotypes uniformly, so that the transition probability from one hblob to another is calculated from the ratio of the size of the overlapped region compared to the size of the hblob.

- Although the transition from hblob-(a) to hblob-(b) actually happens if a set of mutant genotypes created from hblob-(a) overlaps with hblob-(b), we assumes that the transition happens only if hblob-(a) overlaps with hblob-(b).

Figure 4 symbolically shows the definition of the transition probability between a pair of hblobs. Using the last two assumptions, the transition probability from hblob-(a) to hblob-(b) is

$$r_{\text{a}\to\text{b}} \equiv \frac{S(\text{a}\wedge\text{b})}{S(\text{a})}. \quad (6)$$

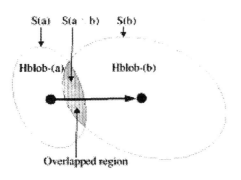

Figure 4: Symbolized figure of hblobs and corresponding mutation graph. $S(\text{a})$, $S(\text{b})$, and $S(\text{a}\wedge\text{b})$ are the size of hblob-(a), hblob-(b), and the overlapped region, respectively. In this figure, a directed edge from hblob-(a) to hblob-(b) occurs because the transition probability $r_{\text{a}\to\text{b}}$ satisfies $r_{\text{a}\to\text{b}} = S(\text{a}\wedge\text{b})/S(\text{a}) > r_0$, whereas a directed edge from hblob-(b) to hblob-(a) does not occur because $r_{\text{b}\to\text{a}} = S(\text{a}\wedge\text{b})/S(\text{b}) < r_0$.

To calculate $S(\text{a}\wedge\text{b})$, here we use the regular expression for an hblob and the non-deterministic finite automaton (NFA) which accepts the regular expression (Hopcroft and Ullman, 1969; Hopcroft and Ullman, 1979). If we are given two regular expressions for a pair of hblobs, we can make two NFAs that accept those expressions, combine those NFAs into one NFA, and derive a regular expression accepted by the combined NFA. The obtained regular expression represents genotypes common with the two original regular expressions. The author illustrates this procedure in the following.

Let the character set be $\{\alpha, \beta, \gamma, \delta\}$ ($K = 4$) and a pair of hblobs (hblob-(a) and hblob-(b)) be represented by regular expressions

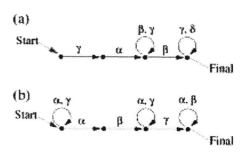

Figure 5: (a) NFA that accepts Expr. (7a), and (b) NFA that accepts Expr. (7b). Black dots represent states, straight arrows correspond to unsubstitutable characters, and loops correspond to closure operations (substitutable characters).

$$\gamma \cdot \alpha \cdot (\beta + \gamma)^* \cdot \beta \cdot (\gamma + \delta)^*, \quad (7a)$$
$$(\alpha + \gamma)^* \cdot \alpha \cdot \beta \cdot (\alpha + \gamma)^* \cdot \gamma \cdot (\alpha + \beta)^*. \quad (7b)$$

The parameter values for these hblobs are $I_u^{(\text{a})} = 3$, $I_s^{(\text{a})} = 2$, and $K_s^{(\text{a})} = 2$ for hblob-(a) and $I_u^{(\text{b})} = 3$, $I_s^{(\text{b})} = 3$, and $K_s^{(\text{b})} = 2$ for hblob-(b). NFAs that accept these expressions are shown in Fig. 5, and the schematic figures of these NFAs and the combined NFA are shown in Fig. 6. A string accepted by the NFA in Fig. 6(c) is accepted by both original NFAs, so that genotypes included in the overlapped (common) region of the two original hblobs is represented by regular expressions accepted by the NFA in Fig. 6(c), or in other words, the paths that begin at the starting state and end at the final state of Fig. 6(c). In the present example, there are two paths represented by the regular expression

$$\gamma \cdot \alpha \cdot \beta \cdot \gamma^* \cdot \gamma + \gamma \cdot \alpha \cdot \beta \cdot \gamma^* \cdot \gamma \cdot \beta^* \cdot \beta. \quad (8)$$

Here, the parameter values for the first term are $I_u = 4$, $I_s = 1$, and $K_s = 1$, and, for the second term, $I_u = 5$, $I_s = 2$, and $K_s = 1$. Generally speaking, the values of K_s are different for closure operations in a regular expression; and yet for simplicity, here we assume that K_s's in a common regular expression have a flat value equal to an integer nearest to $K_s^{(\text{a})} K_s^{(\text{b})}/K$ (which is one for the present example). (We also apply this assumption to all

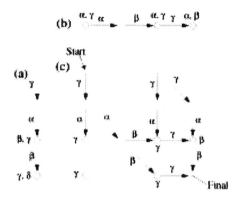

Figure 6: Schematic figures of NFAs that accept (a) Expr. (7a), (b) Expr. (7b), and (c) strings common with Exprs. (7a) and (7b). Black dots for states are omitted here, and loops are represented by white circles. (c) is created on the two dimensional grid whose vertical lines correspond to arrows of (a) and whose horizontal lines correspond to arrows of (b). In (c), a vertical arrow occurs if and only if the corresponding state of (b) has a loop whose character set includes the character of (a), a horizontal arrow occurs if and only if the corresponding state of (a) has a loop whose character set includes the character of (b), a loop occurs if and only if the corresponding states of (a) and (b) have loops whose character sets have common elements, and an oblique arrow occurs if and only if the corresponding arrows of (a) and (b) have the same character.

subsequent analyses.) Finally the transition probability $r_{a \to b}$ is derived as

$$r_{a \to b} = \frac{S(4,1,1) + S(5,2,1)}{S(3,2,2)}. \quad (9)$$

Strictly speaking, there is a possibility that regular expressions corresponding to different paths in the combined NFA have common strings; however we approximately neglected this possibility and calculated the numerator of Eq. (9) by the sum of the sizes of the common regular expressions. This approximation is also applied to all subsequent calculations. Eq. (9) can be evaluated for a fixed value I_{max} by substituting Eqs. (4) and (5).

Directed Mutation Graph

A directed mutation graph is a graph whose nodes correspond to hblobs and whose directed edges represent mutational transitions between hblobs (Fig. 4). A directed edge connecting a pair of hblobs occurs if and only

if the transition probability from the 'initial' hblob to the 'terminal' hblob r is larger than a particular constant value r_0. r_0 is determined so that if $r > r_0$, a population of genotypes distributed in an hblob might be able to find a common region with another hblob in a practical waiting time. (For natural proteins which have an enormous number of individuals in each generation, the value of r_0 is extremely small, but for an ALife system that is run in a computer, the r_0 value is taken to be rather large depending upon the computational resources used in the experiment.)

The evolvability of proteins is measured by the connectivity of the mutation graph. Let the 'evolvability ratio' e be defined as the ratio of the average number of reachable nodes from one node compared to the total node number of the mutation graph. When e is large and comparable to one ($e \sim 1$), there is a strong possibility that evolution starting with one genotype will explore through the whole genotype space by mutational transitions between hblobs and create a variety of genotypes in time. This makes proteins highly evolvable. When e is much smaller than one ($e \ll 1$), on the other hand, most hblobs are isolated and evolution starting at one genotype is very likely to be confined to the initial hblob or its neighborhood and mutation cannot explore through possible hblobs in the genotype space. This makes proteins less evolvable.

As was shown in the preliminary experiment, if a directed edge in the mutation graph occurs randomly according to the occurrence probability $P \equiv \text{Prob}(r > r_0)$, it is expected that e of the mutation graph will begin to increase drastically with the hblob number N after N exceeds some threshold value N_c. The author evaluates this threshold value with two different methods in the following section.

Direct Method

The first method is a direct one that creates a real example of the mutation graph of generated hblobs. In this method, regular expressions representing hblobs are created one by one using a random number sequence, the connectivity between a newly created hblob and previous hblobs is checked using the method of NFAs, the mutation graph is created according to the connectivity, and the evolvability ratio e is calculated from the mutation graph.

The hblob size S is determined by the hblob parameters I_u, I_s, K_s and the maximum genotype length I_{max} (Eqs. (4) and (5)). In the experiment, $I_{\max}$ is fixed to a particular constant number, and the values for I_u, I_s, and K_s are variously determined using Beta distributions. Beta distribution is a convenient probability distribution function on the range [0,1] whose mean and variance are freely chosen by adjusting parameters.

$$p_{\min}(I_u) \propto f_{\text{BET}} \left(\frac{I_u}{I_{\max}} ; \mu_{\min}, \sigma_{\min} \right) \quad (10a)$$

$$p_{\text{ins}}(I_s) \propto f_{\text{BET}} \left(\frac{I_s}{I_u + 1} ; \mu_{\text{ins}}, \sigma_{\text{ins}} \right) \quad (10b)$$

$$p_{\text{sub}}(K_s) \propto f_{\text{BET}} \left(\frac{K_s}{K} ; \mu_{\text{sub}}, \sigma_{\text{sub}} \right) . \quad (10c)$$

$p_{...}(X)$s are normalized so as to satisfy $\sum_X p_{...}(X) = 1$ and f_{BET} is a Beta distribution given by

$$f_{\text{BET}}(x; \mu, \sigma) = \frac{x^{v-1}(1-x)^{w-1}}{B(v, w)}, \quad (11a)$$

$$v = \frac{\mu^2 - \mu^3}{\sigma^2} - \mu, \quad (11b)$$

$$w = \left(\frac{1}{\mu} - 1 \right) \left(\frac{\mu^2 - \mu^3}{\sigma^2} - \mu \right) . \quad (11c)$$

v and w are determined so that μ and σ might be the average and the standard deviation of a Beta distribution, respectively. Throughout this paper, we use $\mu_{\min} = 0.4$, $\sigma_{\min} = 0.07$, $\mu_{\text{ins}} = 0.7$, $\sigma_{\text{ins}} = 0.05$, $\mu_{\text{sub}} = 0.7$, and $\sigma_{\text{sub}} = 0.05$ (Fig. 7). μ_{ins} and μ_{sub} are taken to be larger than $\mu_{\min}$ in order to consider a situation which allows the insertion of a fairly larger number of redundant characters than functionally important characters.

Numerical trials are conducted twenty times using different random number sequences, and for each trial, e is calculated as a function of N. The simulation is conducted on a Linux computer with Pentium II processor (333MHz) and 128MB main memory.

Monte Carlo Method

The second method calculates the edge's occurrence probability P by the Monte Carlo method and estimates N_c from theoretical formula Eq. (2) directly. Here we again assume that the hblob-size parameters I_u, I_s, and K_s obey the Beta distributions in Eqs. (10a)~(10c). Based upon this distribution, the expected value of P is a function of

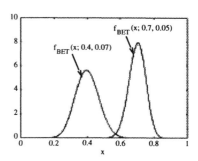

Figure 7: Beta distributions given by Eqs. (11a)~(11c) substituted with $\mu = 0.4$ and $\sigma = 0.07$ or $\mu = 0.7$ and $\sigma = 0.05$.

$I_{\max}$, K, $\mu_{\min}$, $\sigma_{\min}$, μ_{ins}, σ_{ins}, μ_{sub}, σ_{sub}, and r_0 and formulated as

$$\overline{P}(I_{\max}, K, \mu_{\min}, \sigma_{\min}, \mu_{\text{ins}}, \sigma_{\text{ins}}, \mu_{\text{sub}}, \sigma_{\text{sub}}, r_0)$$
$$= \sum_{I_u^{(a)}} \cdots \sum_{K_s^{(b)}} P(I_u^{(a)}, I_u^{(b)}, I_s^{(a)}, I_s^{(b)}, K_s^{(a)}$$
$$, K_s^{(b)}) \times p_{\min}(I_u^{(a)}) p_{\min}(I_u^{(b)}) p_{\text{ins}}(I_s^{(a)})$$
$$\times p_{\text{ins}}(I_s^{(b)}) p_{\text{sub}}(K_s^{(a)}) p_{\text{sub}}(K_s^{(b)}). \quad (12)$$

To calculate $P(I_u^{(a)}, \cdots, K_s^{(b)})$, we randomly generate a large number of pairs of hblob-(a) and hblob-(b), check if the transition probabilities between pairs ($r_{a \to b}$s) are larger than r_0 or not one by one, and calculate $P = \text{Prob}(r > r_0)$ statistically. The trial number of pairs is chosen so that at least five pairs of hblobs might satisfy $r > r_0$ or the number might not exceed one million. Since this calculation procedure is awfully time-consuming, the author adopts the following approximations to make an estimation within a practical waiting time. First, the author limits the number of paths in the combined NFA to 500,000 and discards paths exceeding this number. Second, the summation of Eq. (12) is carried out only for the terms satisfying

$$p_{\min}(I_u^{(a)}) \times \cdots \times p_{\text{sub}}(K_s^{(b)}) > p_{\text{th}}$$

using p_{th} determined by the condition that the partial sum of $p_{\min}(I_u^{(a)}) \times \cdots \times p_{\text{sub}}(K_s^{(b)})$ exceeds 0.95. In addition, the terms satisfying $0.8(I_u^{(a)} + I_u^{(b)}) \geq I_{\max}$ are also omitted from the summation because for these terms, the length (I_u values) of paths in the combined NFA is very likely to exceed $I_{\max}$. The calculated $\overline{P}$ is renormalized by the partial sum of $p_{\min}(I_{\mathrm{u}}^{(\mathrm{a})}) \times \cdots \times p_{\mathrm{sub}}(K_{\mathrm{s}}^{(\mathrm{b})})$. The evaluation program was written in C language and run on the same computer as that used in the experimental method.

Density of Functional Genotypes

In this subsection, the author describes the relation between the hblob number N and the density of functional protein genotypes ρ. If we neglect the overlapped regions between hblobs, a N value can be transformed to a ρ value using the average size of hblob $\overline{S}$ as

$$
\begin{aligned}
\rho &= \frac{N \times \overline{S}}{\text{Total number of genotypes}} \\
&= \frac{N \times \overline{S}}{\sum_{I=0}^{I_{\max}} K^I} \\
&= N \times \overline{S} \times \frac{K-1}{K^{I_{\max}+1}-1}.
\end{aligned} \quad (13)
$$

Like $\overline{P}$, $\overline{S}$ is calculated by

$$
\begin{aligned}
&\overline{S}(I_{\max}, K, \mu_{\min}, \sigma_{\min}, \mu_{\mathrm{ins}}, \sigma_{\mathrm{ins}}, \mu_{\mathrm{sub}}, \sigma_{\mathrm{sub}}) \\
&= \sum_{I_{\mathrm{u}}} \sum_{I_{\mathrm{s}}} \sum_{K_{\mathrm{s}}} S(I_{\mathrm{u}}, I_{\mathrm{s}}, K_{\mathrm{s}}) \cdot p_{\min}(I_{\mathrm{u}}) \\
&\quad \times p_{\mathrm{ins}}(I_{\mathrm{s}}) p_{\mathrm{sub}}(K_{\mathrm{s}}).
\end{aligned} \quad (14)
$$

Eq. (14) is evaluated numerically using Eqs. (4), (5), and (10a)~(10c). Eq. (13) is used for relating the threshold number N_{c} to the threshold density of functional proteins ρ_{c}.

Results

Figure 8 shows the results of the direct method given from a three-day simulation run. From this figure, we can conclude that the evolvability ratio e increases with the hblob number N. Compared to Fig. 2, it is harder to observe the existence of the threshold number N_{c} in these figures; and yet there certainly exists a positive number N_{c} below which e hardly increases with N and above which e increases fairly swiftly with N. These values are $N_{\mathrm{c}} \sim 200$ for (a) and $N_{\mathrm{c}} \sim 750$ for (b).

When the number of hblobs is smaller than these threshold values ($N < N_{\mathrm{c}}$), e is very likely to be near zero and mutation cannot easily cause changes from one hblob to another, resulting in limited protein evolvability. When the number of hblobs is larger than this value ($N > N_{\mathrm{c}}$), on the other hand, the number of hblobs reachable from one functional genotype swiftly increases with the number of hblobs.

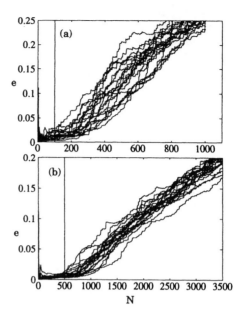

Figure 8: Evolvability ratio e as a function of hblob number N given by the direct method. Parameter values are $I_{\max} = 20$, $K = 10$, $\mu_{\min} = 0.4$, $\sigma_{\min} = 0.07$, $\mu_{\mathrm{ins}} = 0.7$, $\sigma_{\mathrm{ins}} = 0.05$, $\mu_{\mathrm{sub}} = 0.7$, $\sigma_{\mathrm{sub}} = 0.05$, and (a) $r_0 = 0.0001$ or (b) $r_0 = 0.0003$. The vertical straight lines show the theoretical estimation of N_{c} given from the extrapolation in Fig. 9(a).

When this happens, mutation can explore very widely through the genotype space and create a variety of functional protein genotypes, resulting in high protein evolvability. Accordingly, N_{c} can be regarded as the threshold value used for evaluating the extent of evolvability.

The reason why e does not increase drastically above the threshold values is inferred as follows. In the present experiment, the hblob parameters I_{u}, I_{s}, and K_{s} are chosen from the Beta distributions, and the sizes of created hblobs are diversely distributed. Because the occurrence probabilities of directed edges are strongly dependent upon the hblob sizes (Fig. 4), the difference in hblob sizes causes an uneven distribution of the occurrence probabilities of directed edges. This is considered to hinder e from increasing drastically in the region $N > N_{\mathrm{c}}$.

Figure 9 shows the results of the Monte Carlo method given by a five-day simulation run. Because the Monte Carlo trial number increases extraordinarily with an increase in

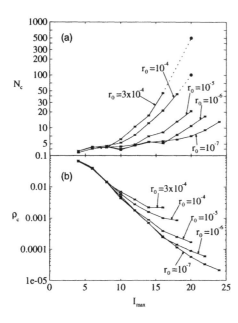

Figure 9: (a) Threshold hblob number N_c and (b) threshold density of functional genotypes ρ_c as a function of maximum genotype length I_{max} under different values of r_0. The other parameter values are $K = 10$, $\mu_{min} = 0.4$, $\sigma_{min} = 0.07$, $\mu_{ins} = 0.7$, $\sigma_{ins} = 0.05$, $\mu_{sub} = 0.7$, and $\sigma_{sub} = 0.05$. The dotted lines and black circles show the extrapolation for the values at $I_{max} = 20$. (b) was calculated from (a) using Eq. (13).

I_{max}, the results are given only for the limited ranges of I_{max}. We can say from Fig. 9(a) that $\log N_c$ approximately increases linearly with I_{max} in the regions of larger I_{max}. By extrapolating from this dependence, the author estimated the N_c values for $I_{max} = 20$, which are shown in Fig. 8 by vertical straight lines. Although the agreement between the two evaluation methods is not a precise one, the Monte Carlo method is considered to succeed in making a rough estimation of N_c in spite of a number of assumptions and approximations in the calculation.

Figure 9(b) says that for a fixed r_0, $\log \rho_c$ approximately decreases linearly with I_{max} although in the regions of larger I_{max}, the values deviate from this law depending upon r_0.

Discussion

To quantitatively estimate the evolvability of natural or artificial proteins, a hyperblob representing neighboring genotypes with the same function was introduced and its connectivity

was studied. A set of variable-length genotypes included in an hblob was expressed using the regular expression, and a method was established to estimate the connectivity between a pair of hblobs using the theory of the regular languages and automata. The connectivity of all hblobs was represented by a directed mutation graph, and by evaluating the connectivity, the minimum number of hyperblobs for high connectivity was calculated. It was concluded that the logarithmic value of the minimum density of functional genotypes for high evolvability approximately decreases with the maximum length of a protein genotype.

Evolvability has been one of the most widely studied topics in the study of artificial life. Bedau et al. (Bedau and Packard, 1992; Bedau et al., 1998; Rechtseiner and Bedau, 1999) have proposed using the neutral shadow model of a target system as a no-adaptation null hypothesis; Ray et al. (Standish, 1999; Ray and Xu, 2000) have studied several evolvability measures in Tierra; and the author and Ray (Suzuki and Ray, 2000) have proposed a set of design criteria to enhance the evolvability of an ALife system using a target system named SeMar (Suzuki, 1998; Suzuki, 1999; Suzuki, 2000c). Among them, the author's approach (Suzuki, 2000a; Suzuki, 2000b) has a unique characteristic in that it tries to study the distribution of functional genotypes in the *whole* protein genotype space. Although the results given in this paper (Fig. 9) cover only a limited region of parameter values, if more numerical experiments are conducted using the parameter values tailored for proteins (that is, functional units) of an ALife system, a derived ρ_c value might be used as a threshold value to discern whether or not the system has high evolvability. In order to make an ALife system highly evolvable, the system design must be optimized so that the ρ value (which can be estimated statistically) might be greater than ρ_c (Suzuki and Ray, 2000).

Appendix A: Derivation of Eq. (5)

$s(I; I_u, I_s, K_s)$ is the number of different strings with length I generated by the expansion of closure operations in a regular expression. That is approximately given by the product of the number of different index sets for the closure operations (represented by the repeated combination) and the number of different substitutable character sets in a string;

$$s(I; I_u, I_s, K_s) \simeq {}_{I_s}H_{I-I_u} \cdot K_s{}^{I-I_u}$$
$$= \binom{I - I_u + I_s - 1}{I_s - 1} \cdot K_s{}^{I-I_u}$$

Note that there is a possibility that the right-hand side of this equation might be an overestimation on account of the repeated counting of the same string. For example, although the expansion of $(\alpha + \beta)^* \cdot \alpha \cdot (\alpha + \beta)^*$ generates only 3 strings with length two ($\alpha\alpha$, $\alpha\beta$, and $\beta\alpha$), Eq. (5) gives $s(2; 1, 2, 2) = \binom{2}{1} \times 2 = 4$. This is because the string $\alpha\alpha$ is counted twice by distinguishing '$1 \cdot \alpha \cdot \alpha$' and '$\alpha \cdot \alpha \cdot 1$'.

References

Bedau, M.A., Packard, N.H.: Measurement of evolutionary activity, teleology, and life. In Proc. of *Artificial Life II* (1992) 431–461

Bedau, M.A., Snyder, E., Packard, N.H.: A Classification of Long-Term Evolutionary Dynamics. In Proc. of *Artificial Life VI* (1998) 228-237

Bollobás, B.: Random graphs. Academic Press, London (1985)

Hofacker, I.L., Fontana, W., Stadler, P.F., Bonhoeffer, L.S., Tacker, M., Schuster, P.: Fast folding and comparison of RNA secondary structures. *Monatshefte für Chemie* **125** (1994) 167-188

Hopcroft, J.E., Ullman, J.D.: Formal Languages and Their Relation to Automata. Addison-Wesley, Massachusetts (1969)

Hopcroft, J.E., Ullman, J.D.: Introduction to Automata Theory, Languages, and Computation. Addison-Wesley, Massachusetts (1979)

Lau, F.L., Dill, K.A.: A lattice statistical mechanics model of the conformational and sequence spaces of proteins. *Macromolecules* **22** (1989) 3986-3997

Lau, F.L., Dill, K.A.: Theory for protein mutability and biogenesis. *Proc. Natn. Acad. Sci. U.S.A.* **87** (1990) 638-642

Lipman, D.J., Wilbur, W.J.: Modelling neutral and selective evolution of protein folding. *Proc. R. Soc. Lond. B* **245** (1991) 7-11

Maynard Smith, J.: Natural selection and the concept of a protein space. *Nature, Lond.* **225** (1970) 563-564

Nishikawa, K.: Island hypothesis: Protein distribution in the sequence space. *Viva Origino* **21** (1993) 91-102

Palmer, E.M.: Graphical evolution. John Wiley & Sons, New York (1985)

Ray, T.S.: An approach to the synthesis of life. In Proc. of *Artificial Life II* (1992) 371-408

Ray, T.S.: Selecting Naturally for Differentiation. In Proc. of *Genetic Programming* (1997) 414-419

Ray, T.S., Hart, J.: Evolution of differentiated multi-threaded digital organisms. In Proc. of *Artificial Life VI* (1998) 295-304

Ray, T.S., Xu, C.: Measures of evolvability in tierra. In Proc. of *Artificial Life and Robotics* Vol. 1 (2000) I-12-I-15

Rechtseiner, A., Bedau, M.A.: A genetic neutral model for quantitative comparison of genotypic evolutionary activity. In Proc. of *European Conference on Artificial Life* (1999) 109-118

Reidys, C., Stadler, P.F., Schuster, P.: Genetic properties of combinatory maps - Neutral networks of RNA secondary structures. *Bull. Math. Biol.* **59** (1997) 339-397 or Santa Fe Working Paper #95-07-058 available at http://www.santafe.edu/sfi/publications/working-papers.html

Reidys, C., Kopp, S., Schuster, P.: Evolutionary optimization of biopolymers and sequence structure maps. In Proc. of *Artificial Life V* (1997) 379-386

Reidys, C.: Random induced subgraphs of generalized *n*-cubes. *Adv. in Appl. Math.* **19** (1997) 360-377

Reidys, C., Forst, C.V., Schuster, P.: Replication and mutation on neutral networks. Submitted to *Bull. of Math. Biol.*, or Santa Fe Working Paper #98-04-036 available at http://www.santafe.edu/sfi/publications/working-papers.html

Schuster, P.: Landscapes and molecular evolution. *Physica D* **107** (1997) 351-365 or Santa Fe Working Paper #96-07-047 available at http://www.santafe.edu/sfi/publications/working-papers.html

Standish, R.K.: Some techniques for the measurement of complexity in Tierra. In Proc. of *European Conference on Artificial Life* (1999) 104-108

Suzuki, H.: One-dimensional unicellular creatures evolved with genetic algorithms. In Proc. of *Joint Conference on Information Sciences* Vol. II (1998) 411-414

Suzuki, H.: A Simulation of Life Using a Dynamic Core Memory Partitioned by Membrane Data. In Proc. of *European Conference on Artificial Life* (1999) 412-416

Suzuki, H.: Minimum Density of Functional Proteins to Make a System Evolvable. In Proc. of *Artificial Life and Robotics* Vol. 1 (2000) 30-33

Suzuki, H., Ray, T.S.: Conditions to Facilitate the Evolvability of Digital Proteins. In Proc. of *Joint Conference on Information Sciences* Vol. I (2000) 1078-1082

Suzuki, H.: Evolvability Analysis Using Random Graph Theory. To be published in Proc. of *Asian Fuzzy Systems Symposium* (2000)

Suzuki, H.: An Approach to Biological Computation: Unicellular Core-Memory Creatures Evolved Using Genetic Algorithms. *Artificial Life* **5** (2000) 367-386

Influence of chance, history and adaptation on evolution in *Digitalia*

Daniel Wagenaar and Chris Adami

California Institute of Technology, Pasadena, CA 91125

Abstract

We evolved multiple clones of populations of *Digitalia*, a type of digital organism, to study the effects of chance, history, and adaptation in evolution. We show that clones adapted to a specific environment can adapt to new environments quickly and efficiently, although their history remains a significant factor in their fitness. Adaptation is most significant (and the effects of history less so) if the old and new environments are dissimilar. For more similar environments, adaptation is slower while history is more prominent. For both similar and dissimilar transfer environments, populations quickly lose the capability to perform computations (the analog of beneficial chemical reactions) that are no longer rewarded in the new environment. Populations that developed few computational "genes" in their original environment were unable to acquire them in the new environment.

Introduction

One of the central tenets of standard evolutionary theory is that characteristics of evolved populations can be explained by the process of adaptation, and thus that phenotypic differences, for the most part, have an adaptive value. That all of biological diversity is due to adaptation has been challenged in modern expositions of Darwinian theory. Kimura (1983), for example, proposes that there is a strong component of chance in evolution, while Gould and Lewontin (1979) stress the importance of history and contingency. Effects of chance are usually due to genetic drift and random mutations without value to the organism. History can become important if certain genetic changes (of adaptive value in the past) constrain or promote some evolutionary outcomes over others. To disentangle these effects, Gould (1989) has proposed to "replay the tape" of evolution to test its repeatability. Travisano, Mongold, Bennett and Lenski (1995) were the first to perform a rigorous experiment of this sort, albeit on a shorter timescale and with *E. coli* bacteria adapting to simple, artificial environments. The trait undergoing evolution in these experiments was fitness (measured as the Malthusian parameter). As a control, bacterial size (which in these environments is selectively neutral) was also monitored. This study found

that adaptation contributed most significantly to the evolutionary changes, often resulting in convergent evolution of fitness. The evolution of bacterial size, on the contrary, was influenced much more strongly by chance and history, as expected.

To date, this experiment has not been repeated with any other organism, nor could it be determined whether the relative effects of chance, history and adaptation are constant throughout evolutionary time. The advent of digital organisms opens the door to experiments much more alike to "replaying the tape of life", and also to test the validity of Travisano et al.'s results across organisms.

Digital organisms have been studied in a variety of experiments pertaining to evolution. They offer a tantalizing glimpse into the characteristics of living systems that do not share any ancestry with biochemical life on earth. These organisms are self-replicating strands of computer code competing for resources in a user-defined environment (the artificial "Petri" dish) within a computer's memory. Our *Digitalia* live in a world created and controlled through the AVIDA software developed at Caltech (Ofria 1999, Adami 1998, Lenski et al. 1999).

In the experiments reported here, we essentially follow the protocol of Travisano et al., measure similar phenotypic characteristics, and perform the same statistical analysis on the data obtained. However, due to the ease of these experiments with *Digitalia* (as compared to *E. coli* experiments), we are able to collect data much more frequently, allowing the observation of *changes* in the relative importance of chance, history, and adaptation as a function of time. While Travisano et al. used bacteria which had adapted to using glucose as their primary sugar, and studied their re-adaptation to a maltose environment, populations of *Digitalia* are transferred to new environments which award differing computational tasks. As the replication speed of *Digitalia* is mainly due to their computational prowess on random numbers in their environment (see Adami 1998, Ofria 1999) we can change landscapes simply by changing the set of computations which result in extra CPU time for the organism that achieves them. Thus, we adapt *Digitalia* to one landscape first, then transfer them to another and mon-

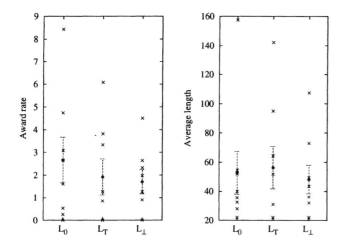

Figure 1: Award rates and average genome lengths attained by populations in their native landscape after 20,000 updates. Each cross represents a population. For the purpose of comparison across landscapes, in this plot (only), all $c_t^{(L)}$'s have been set to unity. Averages and their uncertainties are indicated by filled circles with errorbars.

itor their re-adaptation. "Immigrant" populations are expected to have a lower fitness in their new environment than populations that are native to that environment. We investigate how well the immigrants are able to recover this initial disadvantage, and whether in their new environment they carry with them long term effects from their evolutionary history preceding the transfer. In addition to fitness we also observe genome length, a phenotypic trait which we expect to be selectively neutral as far as the differences between our landscapes is concerned. This variable thus substitutes for the role played by bacterial size in Travisano et al.

Methods

We studied the effects of adaptation and history by evolving populations in one of three distinct environments. L_0 is the standard landscape used in most experiments to-date (Ofria 1999) rewarding a total of 76 different two- and three-input logical operations on random numbers. We split these logical operations into two orthogonal sets which are used to define the landscapes L_T and $L_\perp$[1]. We test the significance of history and adaptation in transfers $L_0{\to}L_T$ (similar landscapes) as well as $L_\perp{\to}L_T$ (transfer to a dissimilar one). In each experiment, we evolve eight separate populations in their environment of origin until they are well adapted, after which the entire populations are cloned five-fold and propagated in their transfer landscape. Genome length

[1]A definition of the landscapes in terms of their rewarded tasks can be found in the Appendix.

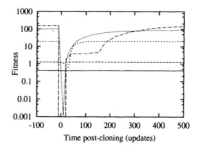

Figure 2: Impact on measured fitness of five cloned populations after re-insertion into their *original* landscape. In this re-transfer control, no adaptation occurs while the original fitness is recovered by AVIDA.

and parameters reflecting the average fitness of the populations are measured at various stages during evolution in the new environment, in order to be compared to the values at the time of the transfer.

To ascertain that the observed effects are not due to peculiarities of the landscapes we constructed, we check that L_T, $L_\perp$ and L_0 are equally challenging. The comparison of "award rates" (a measure reflecting fitness introduced below) for populations adapting to L_T, $L_\perp$ and L_0 depicted in Fig. 1 shows that this is so.

For *Digitalia* as for bacteria, average fitness can usually be measured directly and used as an indicator of the extent of adaptation. Here, we had to forgo this direct approach because the fitness of the transferred population is not well reflected in the measurements right after the transfer, simply because the AVIDA software cannot, at present, accurately monitor a genome's performance in a new environment until at least one replication cycle has been completed. As a consequence, the average fitness is incorrectly measured for a few hundred updates[2] (see Figure 2). This measurement error masks adaptive events occurring early after transfer and as a consequence would severely compromise the analysis.

Instead, we study the *award rate*, a variable closely related to average fitness. The award rate $\mathcal{A}^{(L)}(p)$ of a population p in a landscape L is defined as

$$\mathcal{A}^{(L)}(p) = \sum_t c_t^{(L)} \mathcal{F}_t(p) \, ,$$

where $\mathcal{F}_t(p)$ is the fraction of creatures in p that perform task t, and the coefficients $c_t^{(L)}$ are unity for tasks that are rewarded in landscape L, and zero for all other tasks. As opposed to average measured fitness, the award rate is not significantly affected by the cloning operation.

Following Travisano et al. (1995), we plot the *derived* (or adapted) value of the characteristic versus the *an-*

[2]Time is measured in arbitrary units called updates; every update represents the execution of an average of 30 instructions per individual in the population. A typical generation takes 5-10 updates.

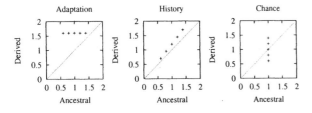

Figure 3: Derived versus ancestral values of hypothetical traits, the evolution of which is chiefly determined by adaptation (left panel), history (middle panel) and chance (right panel). Note that the effects of chance are demonstrated for a set of clones of a single ancestral genotype, whereas adaptation and history are illustrated for several independent ancestors. Adapted from Travisano et al. (1995).

cestral one at the time of transfer, in order to study the effects of chance, history and adaptation on the evolution of a population's characteristics (see Fig. 3).

Eight populations were evolved for 20,000 updates in each of the landscapes L_0, L_T and $L_\perp$, and cloned[3]. In both transfer experiments, cloned populations were propagated for an additional 10,000 updates of re-adaptation.

The relative contributions of adaptation, history, and chance to the evolution of traits such as fitness and genome length can be disentangled by studying the variance of the respective observable. A nested anova (see Sokal and Rohlf 1995) is used to determine what fraction of the variance (after evolution) should be attributed to the elements of history and chance. The contribution of adaptation is obtained from the average difference between derived values and ancestral values.

Award rate and genome length for each population are sampled over a range of 30 updates every 1,000 updates. The spread within these samples is used as the measurement error. This does not reflect the intra-population spread of the parameter under consideration, but only their natural short-time variability. Populations that had award rates less than 0.002 (as measured in their old landscapes) at the time of the transfer, were excluded from the analysis[4]. For the transfer $L_\perp \to L_T$, two out of eight sets of clones were excluded, while only one set of clones was excluded for the transfer $L_0 \to L_T$. None of

---

[3]We checked that propagation for an additional 10,000 updates in their native landscapes produced no significant further increase of award rate, so the populations can be said to be at or near equilibrium in their particular environments.

[4]Due to a special feature in the present physics of the AVIDA world, genomes can evolve that cannot alter their size, preventing any further adaptation. These cases can be considered anomalous as adaptation cannot be measured. We plan to alter the physics of replication in future versions of AVIDA to avoid such contaminations.

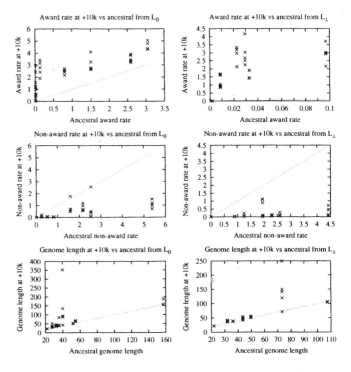

Figure 4: Award rates (top) and average genome length (bottom) for populations transferred from landscape L_0 and $L_\perp$ to landscape L_T: values at 10,000 updates after the transfer plotted against values at the time of the transfer. Middle panels show the 'non-award rate'.

the excluded populations attained a derived award rate significantly above 0.002 after transfer and propagation. The lowest recorded derived award rate for any of the other populations was 0.23 ± 0.05.

Results

We found that adaptation is the dominant component of evolutionary changes of fitness in *Digitalia*, mirroring the results of Travisano et al. obtained for *E. coli*. This can be seen even without statistical analysis by noting that the award rates in Fig. 4 are consistently higher after adaptation to a new landscape than upon transfer, independently from whether the transfer landscapes are similar to the ancestral one or not. Genome length did not change significantly as a result of re-adaptation, confirming that this trait is selectively neutral. Plotting the relative contributions of adaptation, chance, and history as a function of time (Figures 5, 6) reveals that adaptation of fitness is always more important than chance. It is dominated by history at first, but ultimately becomes the principal component of evolutionary change. Conversely, chance and history were dominant in the evolution of genome length. This indicates that genome length indeed does not discriminate between the different landscapes, i.e., that the amount of information in the environments is similar.

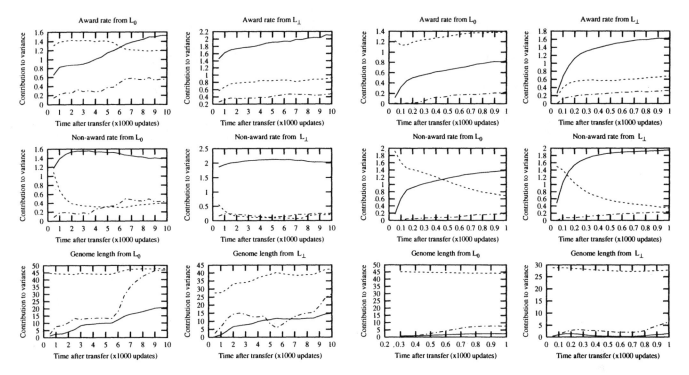

Figure 5: Contributions of adaptation (lines), history (dashes) and chance (dash-dotted) to variance in award rate (top), non-award rate (middle) and genome length (bottom) in populations transferred from landscape L_0 (left) and $L_\perp$ (right) to landscape L_T.

Figure 6: Expanded view of the early history (the first 1,000 updates) of adaptation shown in Figure 5. Award rates and genome lengths for updates earlier than 200 updates may be affected by the transfer (although no long term effects persist).

The adaptation effects on award rate and 'non-award' rate (the rate at which tasks are performed which are *not* rewarded in the landscape) mostly take place in the first 1,000 updates (see Figures 5, 6). After 10,000 updates the award rates of the transferred clones do not differ significantly from the award rates of populations that evolved *de novo* in landscape L_T for 20,000+10,000 updates, indicating convergent evolution.

Note that the dominance of adaptation and history over chance after 10,000 updates is statistically highly significant (see confidence limits in Figure 7), whereas the dominance of adaptation over history is not. The significance of this dominance should be ascertained in experiments in which the period of adaptation is extended.

Conclusions and Outlook

We have found that populations of *Digitalia* propagated within an artificial world created by the AVIDA software are able to adapt to a new environment even though they are already well adapted to another environment, and even if the new landscape is orthogonal to the old one in terms of rewarded behavior. The fitness of organisms in the new landscape continues to be strongly influenced by their history in the old landscape, more so

than in the bacterial experiments with *E. coli* referred to above, and this influence does not appear to decrease after 10,000 updates. However, the effect of adaptation was still increasing when each of the experiments was terminated. This suggests that follow-up experiments with longer adaptation in the transfer environment might be required to completely assess the influence of adaptation. As a significant part of the adaptation was found to take place in the first 1,000 updates after the transfer, it is imperative to repeat these experiments studying the actual average fitness of the population instead of the award rate, and verify the present results. Experiments of this nature with an updated version of AVIDA are planned for the near future (Smith and Wagenaar 2000).

The high speed of adaptation post transfer indicates that populations are able to change the tasks they perform, and so as it were rearrange their metabolism which provides the energy for their replication, by simple mutations only. This suggests that, although the organisms were highly adapted to their native landscape, they retained a significant amount of plasticity which allowed them to thrive in changed environments without much effort. The plasticity of genes in molecular biology is the subject of much discussion (see, e.g., Via et al, 1995) in particular with the advent of whole genomes in the age

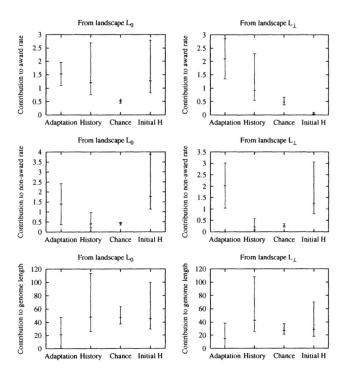

Figure 7: Confidence limits for the relative contributions of history, chance and adaptation at 10,000 updates post transfer. 'Initial H' is the initial spread of the variable due to history in the original landscape. This value is extremely low for the $L_\perp \to L_T$ award rate due to the very low initial fitness of these populations, as landscapes L_T and $L_\perp$ are orthogonal.

of bioinformatics. That it can also be observed in the evolution of digital organisms suggests that it may be a universal feature of evolution.

D. W. wishes to thank Justin Smith for useful discussions. This material is based upon work supported by the NSF under Award No. DEB-9981397.

Appendix – Avida parameters

Experiments reported here were performed with version 1.3.1 of the AVIDA software, which can be obtained from http://www.krl.caltech.edu/avida. We used populations of 60×60 organisms, with standard instruction set, and removal of the oldest as birth method. Genomes were subject to a copy mutation rate of 0.7% per instruction copied, and an insert/delete probability of 5% per generation. Rewards for performing logical functions were set as follows. Multiplicative merits for two input logic functions:

Operation	L_0		L_T		$L_\perp$	
	1st	2nd	1st	2nd	1st	2nd
$A \wedge B$	1.2	1.1	-	-	1.5	1.3
$A \vee B$	1.25	1.1	-	-	1.5	1.3
$A \vee \overline{B}$	1.2	1.1	1.5	1.3	-	-
$A \wedge \overline{B}$	1.25	1.1	1.5	1.3	-	-
$\overline{A \vee B}$	1.3	1.1	-	-	1.5	1.3
$\overline{A \wedge B}$	1.15	1.1	-	-	1.3	1.2
$A \neq B$	1.5	1.1	-	-	1.8	1.5
$A = B$	1.5	1.1	1.8	1.8	-	-

Multiplicative merits for three input logic functions:

Nos	L_0		L_T		$L_\perp$	
	1st	2nd	1st	2nd	1st	2nd
odd	1.5	1.1	-	-	2.2	1.5
even	1.5	1.1	2.0	1.4	-	-

Note that merits for landscape $L_\perp$ are consistently higher than for landscape L_T. This was done because a variant of landscape $L_\perp$ with lower merits turned out to be unlearnable. The merits for the trivial (one-input) operations were set to 1.05 for all landscapes. For calculating the award rates, however, only the two and three input logic functions are taken into account. Any non-unity merit caused that task to be counted with a weight $c_t = 1$.

References

Adami, C., 1998. *Introduction to Artificial Life*. Santa Clara: TELOS Springer-Verlag.

Gould, S.J., 1989. *Wonderful Life: The Burgess Shale and the Nature of History*. New York: Norton.

Gould, S.J. and Lewontin, R.C., 1979. The spandrels of San Marco and the panglossian paradigm: A critique of the adaptationist programme. *Proc. R. Soc. London Ser. B* **205**:581–598.

Kimura, M., 1983. *The Neutral Theory of Molecular Evolution*. Cambridge: Cambridge University Press.

Lenski, R.E., Ofria, C., Collier, T. C., and Adami, C., 1999. Genome complexity, robustness, and genetic interaction in digital organisms. *Nature* 400: 661–663.

Ofria, C. A., 1999. *Evolution of Genetic Codes*, Ph.D. thesis, Caltech, Pasadena, CA 91125.

Smith, J., and Wagenaar, D. A., 2000. Early stages of evolution of newly cloned digital organisms. Work in progress.

Sokal, R. R. and Rohlf, F. J., 1995. *Biometry, 3rd ed.* New York: W. H. Freeman.

Travisano, M., Mongold, J. A., Bennett, A. F., and Lenski, R. E., 1995. Experimental test of the roles of adaptation, chance and history in evolution. *Science* 267: 87–90.

Via, S. et al., 1995. Adaptive phenotypic plasticity—Consensus and controversy. *Trends Ecol. Evol.* **10**: 212-217.

Ecology and Extinction – Macroevolutionary Extinction Dynamics in a Simulated Ecosystem

Stevan Jay Anastasoff

Evolutionary and Emergent Behaviour Intelligence and Computation Group
School of Computer Science
University of Birmingham
Birmingham B15 2TT, United Kingdom

Abstract

A number of models of macroevolutionary extinction dynamics have been proposed during the past decade or so. Many of these models produce results that are in good accord with empirical data, as drawn from the fossil record. However, they all still suffer from significant shortcomings. A new simulation based model is presented here that attempts to address some of the weaknesses and limitations of these existing models. The simulation is driven by ecology level interactions amongst dynamically changing populations of species in a theoretical food web, coupled with the effects of external environmental stresses. The results of a number of extended runs of the simulation are presented and discussed. It is observed that it is the interactions between intrinsic ecological factors, and external environmental factors, that determine the specific extinction dynamics generated. Ecological factors appear key in defining the large-scale statistical trends of the system. Environmental factors appear to act as a sort of 'tempo keeper', determining the precise timing of extinction events within the large-scale framework. Overall, the results from the simulation suggest that macroevolutionary patterns of extinctions are primarily generated intrinsically by an ecosystem. Environmental factors are not so much a direct cause of extinction events, as a determinant of the precise timing of events that will, in any case, inevitably ensue.

Introduction

Over the past twenty years or so the dynamics, patterns, and underlying causes of extinction events throughout the history of life on this planet have been the focus of considerable scientific attention. The aim of this paper is to present a simulation of the dynamics of a set of interacting populations of different species over very large time scales. By analyzing the behaviour of this simulated ecosystem, and comparing it with empirical evidence derived from the fossil record, it is intended that further light can be shed on the problems associated with this study of extinction dynamics.

Empirical Extinction Data

In order to establish biological credibility for a computational simulation, it should ideally be possible to make direct quantitative comparisons between the results of the simulation and the actual available empirical data. In the case of the simulation to be presented here (and the other related computational and mathematical models which will be considered in the next section), the appropriate empirical data is derived primarily from the fossil record.

Trends and Patterns in the Fossil Record

There are several key statistical trends in the fossil record which are potentially of relevance to the models and simulations considered here. Of greatest importance is the distribution of extinction events of a particular size. One of the important features that almost all of the existing models of extinction dynamics share, is that they predict a power law distribution of extinction sizes. The fossil data (as drawn from compiled databases of the times of origination and termination of taxonomic groups such as those given by Sepkoski [33] and Benton [9]) can then be plotted to test this conjecture. It does indeed turn out that the available data is compatible with a power law distribution. Specifically, the probability of a given size of extinction event $p(s)$, in a given time step in the fossil record can be approximated by the formula $p(s) \propto s^{-\tau}$, with $\tau = 2.0 \pm 0.2$.

Other Empirical Data

Further empirical evidence for this power law distribution of extinction sizes can be found in the studies of Hawaiian avifauna by Keitt and Marquet [19]. In this work, the distributions and population sizes of various species of native and introduced birds across a range of Hawaiian islands were analysed. The introduction of new species in these cases resulted in increased levels of extinction, such that sufficient data could be collected on which to perform statistical analyses. Their conclusion is that in the case of the Hawaiian avifauna, extinction patterns do indeed appear to follow a power law form.

Recent Models of Extinction

Over the past ten years or so, a number of methods have been proposed to model the statistical macroevolutionary dynamics described above. In this section, some of these models will be briefly reviewed.

The Bak-Sneppen Model

The Bak-Sneppen model [7] is a simple model of large-scale coevolution. The system organises itself to a critical point at which 'avalanches' of evolutionary activity propagate throughout, giving a power law distribution of avalanche sizes. Although extinction is not addressed explicitly, these avalanches of activity can be interpreted as a possible biotic basis for extinction. The scale-free, power law behaviour seen is one characteristic of such self-organised critical systems. This suggests, therefore, that macroevolutionary extinction dynamics may themselves arise through a self-organised critical process, as indicated by the power law distribution of extinction sizes across the fossil record.

There are many problems with the Bak-Sneppen model as a simulation of extinction dynamics. Most obviously, there is no explicit notion of extinction at all in the system. The assumption is simply made that extinction patterns will be in some way correlated to general patterns of evolutionary activity. Additionally, only coevolutionary factors intrinsic to the system are considered, extrinsic factors are ignored. However, external environmental factors are widely held to be a primary cause of extinction events. This is largely due to the influential work of Alvarez[5] which examined evidence of an extra-terrestrial bolide impact at the time of the end-cretaceous mass extinction. Hallam and Wignall [15], in their summary of the proposed causes of the main Phanerozoic extinction events, list only the following, all extrinsic, possibilities: Bolide impact, volcanism, global cooling, global warming, marine regressions, and marine anoxia. More exotic theories, such as extinction through the ionising effects of cosmic ray jets as proposed by Dar, Laor, and Shaviv [12] are still being suggested.

Another possible distortion of the dynamics of the system is the lack of any speciation mechanism. The number of species is artificially maintained at a fixed level, whereas in biological systems species diversity is free to fluctuate. It seems reasonable to assume that the number and density of different species could have some affect on the probabilities and sizes of extinction events.

A good discussion of the Bak-Sneppen model, its weaknesses as a model of extinction dynamics, and its variants, can be found in Newman [26].

Environmental Stress Models

Extensions to the Bak-Sneppen model incorporating the effects of environmental stress have been proposed in Newman & Roberts [27], and Roberts & Newman [32]. These variants are also an improvement over the Bak-Sneppen model for describing extinction dynamics through the use of an explicit extinction mechanism.

Newman [25] has also investigated the same model, but using only environmental stress – no co-evolutionary mechanism is employed. This model produces near identical dynamics to the earlier version, suggesting that it is the dynamics caused by external stresses which are dominant. What is particularly interesting about this, is that Newman's model is demonstrably not self-organised critical. However, it still produces extinction dynamics that are power law in form. This suggests that the power law form of the fossil record data is not necessarily indicative of a self-organised critical extinction mechanism.

The models of both Newman and Roberts, and Newman, both still suffer from significant problems. For example, as with the Bak-Sneppen model, there is still no explicit speciation mechanism.

Speciation Models

Vandewalle and Ausloos [36] and Head and Rogers [16] have both proposed models which are derived from the Bak-Sneppen model but incorporate explicit speciation mechanisms. A further model, Wilke and Martinetz [37], incorporates speciation based on Newman's environmental stress model. Additionally, Amarel and Meyer [6] includes a speciation mechanism, which will be considered separately below. A brief mention should also be made here of the speciation mechanism used in the biogeographical model of Pelletier [28].

Vandewalle and Ausloos [36] and Head and Rogers [16] are primarily models of branching in phylogenetic trees. The models produce a number of interesting dynamics, but, like the Bak-Sneppen model, there are no explicit extinction mechanisms. This limits their direct relevance to the current study.

Wilke and Martinetz [37] take Newman's environmental stress model as a starting point. In each time step the number of expected new species is calculated, based on the current number of species and a proposed ceiling to the total number of species that the ecosystem can support. This number of new random species is introduced. Although the formula used does describe the growth in numbers of species in biological systems well, it is also somewhat arbitrarily imposed.

The biogeographical model of Pelletier [28] incorporates an interesting speciation mechanism. Pelletier bases his mechanism on the observation that genetic drift occurs at a faster rate in smaller populations. This leads him to adopt a speciation mechanism in which the rate of speciation of a species is inversely proportional to the population size of that species. However, this would be reasonable only if speciation rates could be directly correlated to rates of genetic drift, which seems at best a questionable assumption.

Ecology Based Models

Of most direct relevance to the model to be presented here are two food web based models, Amarel and Meyer

[6], and Abramson [1]. The belief is that this finer-grained ecological level is more appropriate for modeling the fundamentals of macroevolutionary dynamics than the more abstract representations used by the models described above. There is good supporting evidence for this importance of ecological factors in macroevolutionary dynamics. Maynard-Smith [24] gives a number of examples of ecological factors that are the direct cause of extinctions. Further discussion of the relevance of ecological factors, such as inter-species competition and trophic position, to macroevolutionary dynamics can be found in Jablonski [17], Sepkoski [34], and Lawton [20].

Abramson's model [1] is based around the population dynamics of a simple one-dimensional food chain, as described by a Lotka-Volterra type equation. The model does produce periodic cascades of extinction that propagate along the food chain in a manner that Abramson claims to be similar to that of mass extinction events.

The Amarel and Meyer model [6] involves more complex food webs, consisting of multiple trophic levels. The species are arranged on a two-dimensional lattice. One dimension represents trophic levels, the other represents ecological niches in that level. Although the Amarel-Meyer model lacks the population dynamic level of detail of the Abramson model, it does produce results more in line with the fossil data. Further analysis of this model can be found in Drossel [13].

Other Relevant Models

Brief mention should be made of two other relevant lines of research. The first is the work done by Adami [4, 3] based on Ray's *Tierra* system [31], in which evolutionary patterns in populations of self-replicating computer code were studied. Adami looked at the duration of different 'species' of code in extended runs of *Tierra*. He found a good fit to a power law distribution (apart from a fall-off towards high values) similar to that observed in the fossil record. Like Bak and Sneppen, Adami took this power law form as evidence of self-organised criticality in the system.

One further piece of work has had some influence in the development of the model presented here – the macroevolutionary algorithms of Marin and Solé [23]. Marin and Solé have developed an evolutionary optimisation algorithm similar in some ways to the genetic algorithm. However, instead of operating at the level of individual genomes, the macroevolutionary algorithm operates at the level of species. The traditional genetic algorithm operators of mutation and crossover are replaced by species level operators of diversification and extinction. Solé and Manrubia [35] have studied extinction dynamics in a model of species interactions closely similar to that used for macroevolutionary algorithms.

An Ecological Model of Macroevolutionary Extinction Dynamics

The model presented here is based around the evolutionary and ecological dynamics of populations of species in a simulated ecosystem. Predator-prey relationships, as represented by Lotka-Volterra type equations, form the basis for the ecological interactions of the system. In addition to the ecological interactions inherent in the ecosystem, the populations of each species are also affected by external environmental stresses, generated at random. The system is driven by low background levels of phyletic transformation and speciation that gradually change species over time and introduce new species, replacing those driven to extinction by ecological or environmental factors.

With the exception of Abramson [1], all the models discussed in the previous section considered species as the fundamental ecological unit. The Amarel-Meyer model [6], for example, defines the food web explicitly and only in terms of which species feed on which other species. However, the population based model used here, (and that used by Abramson) operates at a more appropriate level. As Eldredge [14] points out, there are two distinct sorts of biological hierarchy – genealogical and ecological. A species is a unit of genealogical organisation, an evolutionary lineage. It is a single layer in a hierarchy consisting of the various taxonomic ranks – order, family, genus etc. The fundamental organisational unit of an ecosystem (above the level of individual organisms) is, however, a population of conspecifics (Eldredge uses the term 'avatar' to describe such a population). In the ecological hierarchy it is these populations that constitute local ecosystems, which in turn build up regional ecosystems, and so on. It is, therefore, more appropriate to model ecological processes at this level of populations drawn from species, rather than treating a species as a discrete ecological unit on its own.

Structure of the Ecosystem

As in the Amarel-Meyer model, the ecosystem is structured around a two dimensional lattice. The vertical axis is representative of trophic levels within the system, while the horizontal axis is representative of individual ecological niches within each trophic level. Note that on the horizontal axis, physical proximity is not necessarily correlated with phyletic similarity – the arrangement of niches is quite arbitrary. Note also that although the word 'niche' is used to designate particular locations within the lattice, the true 'niche' of a species (in the usual sense of the word) is determined also by its trophic interactions. The word 'niche' will henceforth be used only in the former sense, unless specifically stated otherwise. Populations from each level feed on one or more populations from the next lowest trophic level. Species from the lowest trophic level are assumed to be autotrophic, deriving their energy directly from the environment.

The maximum number of trophic levels and niches per level are set at fixed values. The maximum number of trophic levels was set based on data from biological food webs taken from Cohen [11]. As will be seen later, the simulated food webs rarely develop to maximum possible trophic depth. Thus, the arbitrarily imposed limit is largely irrelevant – even if additional trophic levels were allowed, it is unlikely that food webs would ever evolve sufficient complexity to make use of these additional levels.

The justification for fixing the maximum number of ecological niches per level is based largely on studies of island biogeography (e.g. MacArthur [22], and Begon, Mortimer and Thompson [8]). Species-area curves can be plotted of numbers of species occurring on islands of varying sizes. From these curves, it can be seen that there is an apparent fixed maximum number of species that any given area can harbour. Although the number of available niches in the model is set arbitrarily, changing this value would simply result in the model being a simulation of the dynamics of species over varying geographical ranges.

Each species is parameterised in three ways: population density, predator/prey interactions, and tolerance to environmental stress. Population density is simply an integer value assigned on an arbitrary scale. Predation and prey interactions are modeled using two vectors, one each for interactions with predators and prey. The vectors map on to the trophic levels directly above and below that of the species, as appropriate. Each element in each vector contains a floating point value in the range [0, 1] representing the strength of the interaction. Values of 0 indicate that no interaction between the two species takes place. Note that the values in each predator vector are kept coherent with the values in the corresponding prey vector. If a predation value is reduced to 0 (through phyletic transformation), then the corresponding prey value is also automatically reduced to 0. Tolerance to environmental stress is represented by one or more floating point values in the range [0, 1], with a separate value being assigned for each environmental stress (see below for more details on environmental stress).

The ecosystem is initially seeded with a single randomly generated autotrophic species. It is then updated in discrete time steps by repeatedly applying the following operators, which are described in more detail in the following sub-sections. First, a reproduction operator updates the population density of each species. This is done using Lotka-Volterra type equations based on each population's interactions with its neighbouring trophic levels, modified by the effects of environmental stress. An extinction operator then removes any species whose population density has dropped below a critical thresh-

old. A phyletic transformation operator is then applied to each species with some fixed low probability, altering that species' environmental stress tolerances and trophic interactions. Finally, a speciation operator is applied to each species with some fixed low probability, resulting in the generation of new species.

The Reproduction Operator

The population density of each species is updated according to the following Lotka-Volterra type equation, derived from those used by Abramson [1], [2], and those given in Begon, Mortimer, and Thompson [8]:

$$\Delta n_z^i(t) = \sum_{j=1}^{N_{(z-1)}} k_z^{ij} n_{(z-1)}^j(t) - \sum_{j=1}^{N_{(z+1)}} g_z^{ij} n_{(z+1)}^j(t) - \sigma_z^i(t) n_z^i(t)$$

The term $n_z^j(t)$ gives the population density of species n in niche j of trophic level z at time t. N_z is the total number of niches in trophic level z. Predator and prey relationships are given by the variables k and g where k_z^{ij} is the jth element in the prey vector of the ith niche of trophic level z, and g_z^{ij} is the jth element in the predator vector of the ith niche of trophic level z. Environmental stress is given by the function $\sigma_z^i(t)$, the proportion of the population of niche i of trophic level z at time t being eliminated by environmental stress.

In essence this means that population growth of each species will be in proportion to the population density of its predators and prey times the ability of the species to exploit their prey and resist their predators respectively (modified by environmental stress).

For autotrophic species, it is assumed there is some fixed level of available environmental resources. The prey vector of an autotrophic species simply reflects its ability to exploit these resources. Population increases for autotrophic species are then calculated by multiplying the species' ability to exploit environmental resources by an assigned constant value.

The system is assumed to undergo some fixed number of different environmental stresses. In each time step, each stress is assigned a global value according to some probability distribution (linear, normal and bimodal distributions were used in various different runs). To calculate population decreases due to environmental stress, the tolerance levels of each species are compared in turn to the appropriate global stress levels. If tolerance exceeds stress for all stresses, then the population suffers no additional losses. If any stress exceeds the associated tolerance level, then a proportion of the population is killed off, based on the difference between stress and tolerance (as represented by the function σ in the equation above).

The Extinction Operator

After population densities have been updated by the reproduction operator, the extinction operator then removes any species whose population density has dropped below a critical threshold value. There are two reasons why this threshold is applied. The first is simply a matter of computational implementation. If population densities are allowed to become arbitrarily small, it makes it very difficult to reduce a population to a density of zero due to rounding of the integer values used. However, from a biological point of view there are minimum viable population sizes, as described for example in Raup [29]. A species dropping below its minimum viable population will almost inevitably be driven to extinction regardless of any other factors. For this reason, the use of an extinction threshold, as a representation of the minimum viable population size, seems justified.

A species whose population drops below the threshold value has its population density reduced immediately to zero. Additionally all trophic interactions involving the niche occupied by the newly extinct species are removed from the system.

The Phyletic Transformation Operator

In each time step, each species will be phyletically transformed with some given probability. Phyletic transformation potentially changes species in up to three ways. Firstly, the species' tolerance to environmental stress can be altered. Secondly, the strengths of a species' interactions with its existing predators and prey can change. Finally, the species can acquire new prey, randomly selected from the next lower trophic level, or lose existing prey. Note that any acquisition or loss of prey will also result in a corresponding acquisition/loss of the predator to the prey's predation vector.

The phyletic transformation operator uses a simple hill-climbing algorithm to determine whether phyletic transformation occurs in a selected species. Once a species has been selected for phyletic transformation, a variant on that species is generated. The two species are compared by looking at the expected population changes in the next time step assuming the rest of the ecosystem and environment remains constant. If the newly generated species has the higher expected population growth of the two, then it is assumed to be the 'fitter' and replaces the original. This is supposed to be broadly representative of microevolutionary selective forces acting to improve the overall fitness of the species.

Note, however, that this hill-climbing process will not necessarily act to maximise all of a species' parameters. During times of low environmental stress, there is no selective advantage to high stress tolerance levels. Selective forces on ecological interactions will therefore dominate, with phyletic drift pulling down stress tolerances. Likewise, during periods of intense environmental

stress, increases in stress tolerance will dominate, with drift weakening the ecological position of the species.

To generate a new phyletically transformed species, each of the parameters of the original species is examined in turn. Each is then assigned a new random value with some given low probability.

The Speciation Operator

The final stage in each time step is the application of the speciation operator. Each species is selected to undergo speciation with some given low probability. The speciation operator then works as follows. A new niche is selected at random from the same or a neighbouring trophic level. If this niche is empty then a speciation event will occur. If a speciation occurs, then a new species is generated to occupy the vacant niche. The new species will be a phyletically transformed version of the original. In the case of a speciation event causing a species to be formed on a different trophic level than its progenitor, it will be assigned a single random prey from the next lower trophic level. It will initially have no predators.

Some further justification of this speciation mechanism (which is essentially the same as that used by Amarel and Meyer [6]) should be given. In absence of any limiting factors, it would be expected that biodiversity (e.g. the total number of species in an ecosystem) would increase exponentially. Speciation mechanisms such as geographical isolation of sub-populations, or runaway sexual selection, are equally applicable no matter high many species are present. Therefore the more species an ecosystem contains, the faster would be the total rate of speciation across the ecosystem. However, there are of course limitations. In a finite ecosystem with only a limited amount of resources, competition for these resources will inevitably limit the number of species the ecosystem can support. If a speciation event were to occur such that the new species would occupy an ecological niche already occupied by an existing species, one of two things will occur. Most likely, the new species will simply not manage to establish itself, and the speciation will fail. The alternative is that the new species will usurp the extant species, driving it to extinction and taking its ecological position. The assumption is made in the model that this sort of competitive niche invasion happens only rarely, and for computational efficiency can be ignored altogether as being irrelevant to the macroevolutionary dynamics. Thus, speciation is only allowed to occur into vacant niches. Some justification for this is needed, however.

This mechanism of extinction through competitive replacement was of course the principal extinction mechanism envisioned by Darwin, and this influence remained strong for some considerable period. However, more modern studies suggest that it is not a significant factor in evolutionary dynamics, and the decision here to omit

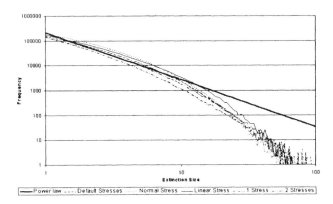

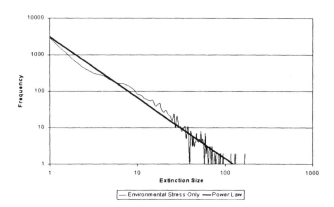

Figure 1: Graph of extinction rates for five simulation runs varying numbers and types of environmental stress. The power law here has an exponent of $\tau = 1.90$.

Figure 2: Graph of extinction rates for simulation run in which only environmental stress was used. The power law here has an exponent of $\tau = 1.67$.

it is well supported by palaeobiological evidence. See Raup [30] for some broad consideration of the issues involved, and Benton [10] for more detailed work. Benton performed an analysis of 840 families of tetrapods, concluding that "competitive replacement was apparently rare in the evolution of tetrapod families, and family originations were most often associated with expansion into new niches." ([10], pg. 204).

Results from a Number of Extended Runs of the Simulation

The simulation was run a number of times to test the effects of various parameter settings on the resultant dynamics. Except where noted, the following default principle parameter settings were used: 6 trophic levels with 500 niches per level; 4 environmental stresses using a bimodal probability distribution; an autotrophic level of 1000; extinction threshold of 50; a 0.02 probability of phyletic transformation per species per time step; and a 0.01 probability of speciation per species per time step. Each run consisted of 1,000,000 iterations. In each run data was collected on the number of extinction events per time step, the lifetime of species at the point of extinction, levels of environmental stress in each time step, mean tolerance to environmental stress across each trophic level and the ecosystem as a whole, number of species in each trophic level, mean number of prey species per predator species, and mean interaction levels between predators and prey. However, only selected results from the total collected will be presented and discussed here. Extinction rates were also calculated over periods longer than a single time step, and at very low resolutions. In the latter case, only partial

data was recorded, to reflect the low resolution of the fossil record. Both these data sets indicated potentially interesting avenues for further investigation, but are not directly relevant to the discussion below, and so will not be considered any further.

In addition to conducting runs varying the default parameter settings, several runs were also conducted in which only ecological or environmental factors were considered.

Extinction Rates

Figure 1 shows extinction rates (number of extinctions occurring in a single time step against rate of occurrence of an event of that size) using the default values given above, as well as four additional runs using various numbers and types of environmental stress. The results are plotted log-log, on which a power law form will appear as a straight line.

As can be seen, all five runs produced statistically very similar sets of results, suggesting that the dynamics of the system are robust to the type and number of environmental stresses. A power law with an exponent $\tau = 1.9$ has also been plotted, a value well within the range of available fossil data $\tau = 2.0 \pm 0.2$. Although the simulation curves initially follow this trend line very closely, higher extinction rates can be seen to be somewhat under-represented in the data according to the power law.

Figure 2 displays data from a run of the simulation in which only environmental stress was used. This data fits a power law form, with an exponent of $\tau = 1.67$, as has been plotted on the graph. This value is slightly lower than the range compatible with the fossil data.

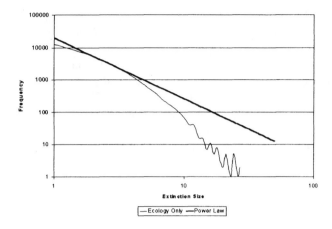

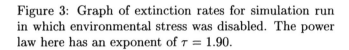

Figure 3: Graph of extinction rates for simulation run in which environmental stress was disabled. The power law here has an exponent of $\tau = 1.90$.

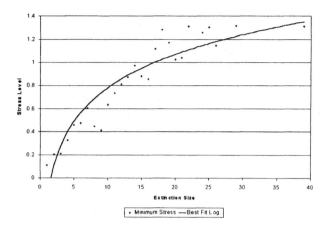

Figure 4: Graph of extinction size against lowest level of environmental stress at which an extinction event of that size occurred.

The data for figure 3 is drawn from a run in which the system was driven by ecological interactions alone, with environmental stress being held constant. Without the effects of environmental stress, the total number of species in the ecosystem was able to rise to considerably higher levels than in the other runs. Due to the increased computational demands that this entailed, the 'ecology-only' runs were limited to only 50000 generations. However, even on this shorter time-scale it is still clear that the statistical dynamics of the system remain almost constant when environmental stress is disabled. The same power law form as before, with $\tau = 1.9$, has been added to the graph. The simulation data again follows the pattern seen in figure 1, initially following a power law form closely with large extinction events being under-represented.

Figure 4 is derived from the default parameter settings described earlier. This figure shows the size of extinction events against the environmental stress level during the time step in which the extinctions took place. The graph has been generated from 10000 random data points taken from throughout the course of the run, by plotting the minimum level of stress during which an extinction event of that size occurred. A clear relationship can be seen here between the size of an extinction event and the level of environmental stress at the time at which that event occurred. A best-fit logarithmic curve has also been plotted, which can be seen to be a reasonable fit for the relationship between the two data sets. Note that the scale on the environmental stress axis is entirely arbitrary, and is here in the range $[0, 1.4]$.

Ecological Data

Additionally, data on the ecological interactions occurring during the default parameter run was taken. The primary purpose of recording this data was to assess the biological credibility of the assumptions underlying the simulation mechanisms. However, some ecological data will be also be relevant to the discussion of macroevolutionary dynamics in the next section.

So far, all results have been tested as robust to minor changes in rates of speciation and phyletic transformation. However, due to the very high computationl demands of the simulation, comprehensive parameter sweeps have yet to be completed.

Discussion

Methodological Issues

There are several methodological issues which the use of a simulation-based model, such as the one described here, raises. Given sufficient degrees of freedom (and the model here has plenty) almost any desired results may be achievable, not through the intrinsic dynamics of the simulation, but rather just through careful control of parameters. This is particularly evident when modelling extinction patterns, due to the low resolution and extent of the empirical data to be modelled. Almost any simulation resulting in a power law distribution of some result set could purport to be a simulation of extinction dynamics in good accord with the empirical evidence. Several precautions must therefore be considered.

In the first place, the mechanisms of the simulation must generally be justifiable in terms of the system being simulated. For example in this case, it would be very easy to justify the finite number of ecological niches by referring to the finite memory capacity of the computers on which the simulation is to be run. This would not, however, be satisfactory. This sort of problem is most evident in analytic models, which are intrinsically grounded by the requirements of solvability, rather than biological credibility. If the assumptions on which a model is based cannot be justified outside of that model, then it is difficult to see how a link of explanatory relevance can possibly be established between simulation and reality.

Secondly, the analysis of the model should extend throughout the domain being modelled, and not be limited to a single parameter. For example, as discussed below, measurements were taken of ecological factors in this simulation, such as lengths of food chains, in addition to the specific measurements of extinction events. A good simulation should be representative of the simulated domain in whichever way the data is to be considered.

Finally, a good simulation based model should be able to make predictions concerning the domain being simulated. This final point is perhaps the most important of all. The predictive power of a scientific theory has long been held as a cornerstone of its usefulness and validity. For a simulation based model to establish scientific credibility, it too should conform to this standard. Although no predictions will be presented here, there is certainly a large predictive scope for this simulation. For example, predictions could be made concerning the distribution of extinctions throughout different trophic levels, e.g. what proportion of autotrophic species are driven to extinction in a given size of extinction event?

Ecological Considerations in the Simulation

The data recorded included various measures of the ecological factors in the simulation. These can be compared with ecological data from biological systems in order to establish the credibility of the assumptions underlying the model. The data to be considered here include: maximum length of food chains; numbers of prey species per predator species; and relative abundance of species in different trophic levels. Some further discussion of issues regarding the comparison of simulated food webs with actual ecosystems can be found in Lindgren & Nordahl [21].

Begon, Mortimer and Thompson [8] point out that food chains rarely consist of more than four or five trophic levels. The data recorded here is in good accordance with this. For approximately 80% of the time, no food chains longer than five trophic levels occur in the ecosystem. Even at those times when food chains of six trophic levels do develop, they are few in number compared to shorter food chains (note that these values represent maximum chain lengths across the entire ecosystem).

The mean number of prey species per predator in a given time step was measured across the system at 1.64, with a standard deviation of 0.1, for the default parameter settings. This can be compared with the extensive data compiled in Cohen [11]. The value obtained from the simulation does appear to be very much on the low side when such a comparison is made – the mean value for all of Cohen's data is listed as 2.2. The simulation value is by no means implausible, however. The Georgia Salt Marsh food web, for example, contains only 1.75 prey species per predator, and that of the Florida Gastropods only 1.67.

Cohen also observes that the number of species in each trophic level is proportional to the number of species in the next lower trophic level. Cohen measures the constant of proportionality at 0.77 based on a limited sub-set of the available data. A constant of ≈ 0.5 was observed in the simulation – a not implausible value. The food webs of Costa Rican gastropods and lake-dwelling triclads both have values of exactly 0.5, for example, again as taken from Cohen's compilation.

Rates of Extinction – The Roles of Ecology and Environment

The distribution of sizes of extinction events is the single most important data set from the simulation. Although the data is highly suggestive of a power law form, it clearly does not take this form closely and consistently, with a notable under-representation of large extinction events. It should be noted that the finite bounds of the simulation might serve to distort the data, as pointed out for example in Kauffman [18]. The imposed limit on the maximum number of species the ecosystem can contain may result in a reduced number of large extinction events, since the system cannot expand to a point where the largest extinctions would occur. However, this under-representation of large extinctions is in fact in closer accord with the empirical fossil data than a straight power law form, also as observed in [18].

What is more interesting is a comparison of the curves for those runs in which environmental stresses were running in tandem with ecological factors, and the run in which environmental stress was the sole driving mechanism for the simulation. In all the runs in which the ecological factors were allowed to play a part, the resulting dynamics were near identical (see figure 1). In each case the curve follows the pattern described above, with exponents for the power law curve of $\tau \approx 1.9$ with under-representation of large extinction events. This same pattern was evident regardless of the number of environmental stresses, the probability distribution of those stresses, or even if there was any environmental stress at all. In part, this robustness of the dynamics to environmental stress is in line with the results from Newman [25]. In his model, the results were also highly robust to the particulars of stress generation. However, what is more surprising, and where these results differ from Newman's, is that the statistical properties of the system continue to remain the same even when there is no environmental stress at all. The only time at which the dynamic changes is when the simulation is run with no ecological factors. This strongly suggests that it is the ecological factors, and not the environmental stresses, which are the sole contributory force to the extinction dynamics of the complete system. (If x appears to remain the same regardless of y then it is a fair assumption that x is independent of y). However, matters are not that simple.

Figure 4 shows why. There is a direct correlation between the size of an extinction event and the minimum level of stress occurring in the same time step as an event of that size. Were it not for the robustness to the presence or absence of various types of stress described above, this would be taken as a strong indicator that it is environmental stresses that are the dominant factor in determining extinction patterns. How can these two apparently contradictory dynamics be reconciled?

One way to view the results would be to consider that the long-scale dynamics of the system are governed by ecological factors alone. Thus, when looked at in a long time frame, the extinction statistics of any run in which ecological interactions were enabled would appear the same regardless of environmental stress levels. Under this view, external stresses are responsible only for determining the exact timing of extinction events of a particular size, acting as a sort of 'tempo-keeper' over the effects of ecological interactions. Note that this is opposed to the traditional view of extinctions taken in modern palaeontology (as for example in Raup [29]). This opposing view holds that factors extrinsic to an ecosystem are the direct cause of large scale extinction events, and the dynamics of such events should be determined by the patterns of such extrinsic factors. Ecological factors are typically viewed as generating low levels of background extinction, independently from environmental stresses.

These ideas are, of course, in need of a more formal and rigorous explication. This could perhaps be done in terms of the density of trophic connections, or the saturation level of the ecological niches in the ecosystem. In particular, the dynamics of the simulation need to be examined at a much finer grained level to ascertain exactly what role different factors play in specific individual extinction events.

One advantage of the current simulation over any of the other models considered is that it is driven by sufficiently low level mechanisms that further detailed investigations are easily facilitated. Future work can focus in on extinction dynamics at any required level of resolution, right down to the dynamics of individual populations. This provides scope for developing a very thorough and detailed theory of extinction dynamics based around the dynamics of this simulation.

Conclusion

An ecology-based model of macroevolutionary dynamics has been presented. Results from several extended runs of the simulation have been considered, and discussed. The results suggest that patterns of macroevolutionary extinction dynamics are determined by the interactions between intrinsic ecological factors, and extrinsic environmental factors. Intrinsic factors appear to determine the large scale statistical dynamics of the system over long time scales. However, under finer grained analysis, specific extinction events can be seen to be highly correlated with specific levels of environmental stress.

Additionally, the finer-grained resolution of the simulation opens up considerable scope for further investigation. Predictions and analysis may be feasible that go beyond anything that is currently possible using existing models of extinction dynamics.

Acknowledgements

Thanks to Greg Werner and the members of the ALER-GIC group at the University of Sussex, and the EEBIC group at the University of Birmingham, for comments on various earlier versions of this paper. Thanks also to Brandon Anastasoff, Daniel Martin, and Greg Werner for the loan of additional computer resources. Supported by funding from the Engineering and Physical Sciences Research Council. Work carried out in part at the School of Cognitive and Computer Sciences, University of Sussex.

References

[1] Abramson, G., 1997. Ecological model of Extinctions. *Phys. Rev. E* 55.

[2] Abramson, G., & Zanette, D. H., 1998. Statistics of extinction and survival in Lotka-Volterra systems. *Phys. Rev. E* 57:4.

[3] Adami, C., 1995. Self-organized criticality in living systems. *Phys. Lett. A* 203:23.

[4] Adami, C., 1994. On Modeling Life. In Maes, M. & Brooks, R., eds., *Artificial Life IV, Proceedings of the fourth conference on artificial life.* Addison-Wesley.

[5] Alvarez, L. W., et al., 1980. Extraterrestrial Cause for the Cretaceous-Tertiary Extinction. *Science* 208:4448.

[6] Amarel, J. A. N. & Meyer, M., 1999. Environmental Changes, Coextinction, and Patterns in the Fossil Record. *Phys. Rev. Lett.* 82:3.

[7] Bak, P. & Sneppen, K., 1993. Punctuated equilibrium and criticality in a simple model of evolution. *Phys. Rev. Lett.* 71.

[8] Begon, M., Mortimer, M., & Thompson, D. J., 1996. *Population Ecology, third edition*. Blackwell Science, Oxford.

[9] Benton, M. J., 1993. *The Fossil Record 2*. Chapman and Hall, London.

[10] Benton, M. J., 1996. On the Nonprevalence of Competitive Replacement in the Evolution of Tetrapods. In Jablonski, D., Erwin, D. H., Lipps, J. H., eds., *Evolutionary Paleobiology*. University of Chicago Press, Chicago.

[11] Cohen, H. E., 1978. *Food Webs and Niche Space*. Princeton University Press, Princeton.

[12] Dar, A., Laor, A., & Shaviv, N. J., 1998. Life Extinctions by Cosmic Ray Jets. *Phys. Rev. Lett.* 80:26.

[13] Drossel, B., 1998. Extinction Events and Species Lifetimes in a Simple Ecological Model. *Phys. Rev. Lett.* 81:22.

[14] Eldredge, N., 1996. Hierarchies in Macroevolution. In Jablonski, D., Erwin, D. H., Lipps, J. H., eds., *Evolutionary Paleobiology*. University of Chicago Press, Chicago.

[15] Hallam, A. & Wignall, P. B., 1997. *Mass Extinctions and their Aftermath*. Oxford University Press, Oxford.

[16] Head, D. A. & Rodgers, G. J., 1997. Speciation and extinction in a simple model of evolution. *Phys. Rev. E* 55:3.

[17] Jablonski, D., 1996. Body Size and Macroevolution. In Jablonski, D., Erwin, D. H., Lipps, J. H., eds., *Evolutionary Paleobiology*. University of Chicago Press, Chicago.

[18] Kauffman, S., 1995. *At Home in the Universe: The Search for the Laws of Complexity*. Oxford University Press, Oxford.

[19] Keitt, T. H. & Marquet, P. A., 1996. The Introduced Hawaiian Avifauna Reconsidered: Evidence for Self-Organized Criticality? *J.theor. Biol.* 182.

[20] Lawton, J. H., 1994. Population dynamic principles. *Phil. Trans R. Soc. London B* 344.

[21] Lindgren, K. & Nordahl, M. G., 1994. Artificial Food Webs. In Langton, C. G., ed., *Artificial Life III, proceedings of the third workshop on artificial life*. Addison-Wesley, Mass.

[22] MacArthur, R. H., 1972. *Geographical Ecology*. Harper Row, New York.

[23] Marin, J. & Solè, R. V., 1999. Macroevolutionary Algorithms: a New Optimization Method on Fitness Landscapes. *IEEE Transactions on Evolutionary Computation*.

[24] Maynard Smith, J., 1989. The causes of extinction. *Phil Trans R. Soc. London B* 325.

[25] Newman, M. E. J., 1997. A model of mass extinction. *J. Theor. Biol.* 189:235.

[26] Newman, M. E. J. & Palmer, R. G., submitted. Models of Extinction: A Review. *Paleobiology*, submitted.

[27] Newman, M. E. J. & Roberts, B. W., 1995. Mass-extinction: Evolution and the effects of external influences on unfit species. *Proc. R. Soc. London B* 260:31.

[28] Pelletier, J. D., 1999. Species-Area Relation and Self-Similarity in a Biogeographical Model of Speciation and Extinction. *Physical Review Letters* 82:9.

[29] Raup, D. M., 1991. *Extinction: Bad Genes or Bad Luck*, Oxford University Press, Oxford.

[30] Raup, D. M., 1996. Extinction Models. In Jablonski, D., Erwin, D. H., Lipps, J. H., eds., *Evolutionary Paleobiology*. University of Chicago Press, Chicago.

[31] Ray, T. S., 1992. An approach to the synthesis of life. In Langton, C. G., Taylor, C., Farmer, J. D., & Rasmussen, S., eds., *Proceedings of Artificial Life IV*. Addison Wesley, Redwood City CA.

[32] Roberts, B. W. & Newman, M. E. J., 1995. A model for evolution and extinction. *Cornell University technical report CU-MSC-8-95-4*.

[33] Sepkoski, J. J., Jr., 1993. A compendium of fossil marine animal families, 2nd edition. *Milwaukee Public Museum Contributions in Biology and Geology* 83.

[34] Sepkoski, J. J. Jr., 1996. Competition in Macroevolution: The Double Wedge Revisited. In Jablonski, D., Erwin, D. H., Lipps, J. H., eds., *Evolutionary Paleobiology*. University of Chicago Press, Chicago.

[35] Solé, R. V. & Manrubia, S. C., 1996. Extinction and self-organized criticality in a model of large-scale evolution. *Phys. Rev. E* 54.

[36] Vandewalle, N. & Ausloos, M., 1995. Physical models of biological evolution. *Proceedings of the Symposium on the Evolution of Complexity*.

[37] Wilke, C. & Martinetz, T., 1997. Simple model of evolution with variable system size. *Phys. Rev. E* 56:6.

Quantification of Microscopic Events
in the Process of Long-Term Evolutionary Dynamics

Shinichiro Yoshii, Eiichi Miyamoto, and Yukinori Kakazu

Complex Systems Engineering, Hokkaido University
N-13, W-8, Kita-ku, Sapporo 060-8628, Japan
yoshii@complex.eng.hokudai.ac.jp

Abstract

This paper describes a new standpoint to further understanding the evolvability of A-life systems, that is, quantification of the dependency of system constituents. If a system
exhibits evolutionary activity, new events, including the development or extinction of systems constituents, may emerge one after the other. The occurrence of new events would
involve a changing complexion in the course of the future
direction of system behaviors. In this respect, there have
been a considerable number of studies on capturing long-
term dynamics in A-life systems in terms of, for example,
population dynamics. On the other hand, we focus on the
quantification of relational concepts between constituents to
clarify the occurrence of fundamental events affecting long-
term course of evolution. We apply a quantification method
to our A-life simulation model. Sequential analysis of the
obtained results enables us to visualize its long-term evolutionary dynamics. We discuss the effectiveness of our approach and the quantified evolutionary dynamics.

Introduction

Many A-life models have been proposed so far for the understanding of behaviors of living systems. Some of them
have given us new insights to divine the fundamentals of
living things, and some have been successfully applied to
the development of artifacts that can perform life-like activity.

A-life research usually consists of the conception of a
model, its realization through artificial media, and observation of simulated behavior in a collection of experiments.
One of the characteristics of A-life that attracts researchers
in this community is the emergent or evolutionary activity
of a system observed in simulation runs, though the meanings of these terms are still debatable.

While a number of simulation models have been proposed, one important future issue in A-life research would
be the development of tools to analyze and evaluate the
emergent activity of A-life systems. In particular, the most
striking and mystical term, emergence, should be reconsidered in respect of how to quantify evolutionary dynamics
so that we can compare emergent behaviors and specify an
event crucial for the future course of evolution. The terms
emergence and evolution are frequently used for behaviors
of A-life systems. However, excessive use of these terms

without the establishment of criteria for defining the activities of emergence and evolution is likely to devalue their
significance.

There seems to be two major streams of research related
to the above issue. One is a human diagnosing approach
similar to the Turing Test in the realm of artificial intelligence. Ronald et al. recently proposed an emergence test
consisting of the criteria of *design*, *observation* and *surprise* for diagnosing emergence (Ronald, Sipper, and Cascarrere 1999). They have tried to establish an emergence
test by assessing surprise on the basis of how easy (or how
difficult) it is for an observer to bridge the gap from design
to observation. In their study, they administered the test to
some major A-life models and assessed their emergence
levels.

The other major stream of research is a more statistical
approach. This approach is usually related to evolutionary
dynamics in an objective model. Bedau et al. devised a way
of measuring evolutionary activity from the viewpoint of
population genetics (Bedau and Packard 1992), and they
then applied their methodology to the fossil data obtained
from the real biosphere (Bedau et al. 1997). Their research
progressed to the classification of long-term evolutionary
dynamics (Bedau, Snyder, and Packard 1998), similar to
Wolfram's famous work on cellular automata (Wolfram
1984). Following on from Bedau's work, Taylor investigated the role of contingency by mutation in the evolutionary dynamics of his A-life model (Taylor and Hallam
1998). Furthermore, from the viewpoint of computational
complexity, we have described self-organized complexity
of our A-life model through an increase in the complexity
of interactions among constituents of the system (Yoshii,
Ohashi, and Kakazu 1998a).

The latter approaches described above are statistical in
the sense that those approaches analyze simulated behaviors by measuring some variables that appear at the stage of
design of a model or observation in terms of Ronald's test
criteria. However, we question whether the measurement
of such anticipatorily selected variables is appropriate for
understanding emergence. As for the concept of emergence,
there is a consensus that a holistic context may change depending on the local behavior of each system constituent
and, on the other hand, each local behavior is affected by
the holistic context. The holistic context of such systems as

a whole is dynamically organized through alteration in the relations among the constituents that interact with each other. In other words, the holistic context of a system is not determined before the system actually begins to function.

In this respect, we do not think that measurement based on anticipatory criterion variables is sufficient to aid in the capturing of essences of emergence. Rather, we think that a new standpoint is necessary. We therefore focus on alteration in the dependency relationship of system constituents following microscopic events involving their generation and extinction.

Since a dependency relationship is *qualitative* rather than *quantitative*, a method for transforming qualitative data into quantitative data is needed. In addition to this, criterion variables for statistical analysis of a holistic context are not available before the system functions. Thus, this paper introduces a method developed by Hayashi for quantifying a dependency relationship. Among the various methods developed by Hayashi, a method called Quantification Type IV (QTIV) can be applied to problems for which no criterion variables are available in advance (Hayashi 1964).

In this paper, we apply the QTIV to our A-life model for quantifying and clarifying microscopic events in the process of the long-term course of evolution. We represent the concept of a dependency relationship as a *preference* described with an ordinal scale. Preference data will be gathered from interactions between system constituents in the evolutionary process. Then, those data will be analyzed using the QTIV to generate a point that will be plotted on a *dependency projection space* (DPS). Based on the results of quantification, we show the process of changes in relations between the system constituents through long-term evolutionary dynamics.

Next, we briefly describe our A-life model, named PROTEAN, and then we describe the QTIV method. Lastly, we show some quantified evolutionary dynamics obtained by the QTIV and end up with discussion.

Our A-life Model: PROTEAN

Our test model for quantification of evolutionary dynamics, named PROTEAN (*Platform for Recursive Ontogenetic Turing-machine Ecosystem for Autopoietic Networks*), is an A-life system aimed at simulating autopoietic behaviors of computer programs described in the form of a Turing machine (TM) (Yoshii, Ohashi, and Kakazu 1998b). In the PROTEAN, survival games take place on a universal Turing machine (UTM), with dynamically changing ecological resources being competed for. The essential features of the PROTEAN are the implementation on a UTM, the interaction procedure between constituents, and the constitution of the system as a whole. Although some of its characteristics are similar to the models in former works by Ray (Ray 1991) and Adami (Adami 1994), the PROTEAN has the following unique characteristics.

• Universal Encoding of Algorithms by UTM

For the realization of a universal model for complex adaptive systems, the PROTEAN implements the notion of a UTM. Conceptually speaking, the PROTEAN has the capability of describing and producing any kind of algorithm computable by TM.

We call a system constituent of the PROTEAN a *scheme*. A scheme as a tape for the UTM is nothing more than the coded version of a TM that performs the task desired of the UTM. We can regard a scheme as a genotype that encodes the Turing machine's functionality, while a scheme decoded as a TM on the UTM is called a phenotype because it is dictated completely by that scheme.

• Interaction between Turing Machines

Figure 1 illustrates an overview of an interaction. First, a scheme, g_i, is interpreted by the UTM and it is then decoded into a phenotype (i.e., a specific TM, T_i). Next, the decoded TM, T_i, reads and operates on another scheme, g_j, as an input tape operated. If this tape is accepted, a new scheme is generated through the following template matching procedure. The newly generated genotype will dictate a new scheme, or in other words, a new TM. This means that such an interaction can realize a process whereby a program can directly operate on the description of another program to generate a new algorithm.

• Parallel Processing to Avoid the Halting Problem

There is no way to tell in advance whether a TM will accept its input tape or not. In order to avoid this insoluble problem, known as the halting problem, the PROTEAN uses parallel processing and compulsory termination: All of the schemes reside in a static amount of memory called an ecosystem, and each scheme occupies its own memory block dependent on its description length. Each scheme reads and operates, in parallel, the other schemes chosen at random. When a scheme accepts an input tape and some partition is vacant, the parent scheme allocates a memory block for its child there. On the other hand, if another scheme or schemes have already occupied that memory block, the parent can terminate their interaction process and rewrite the memory. Consequently, those schemes that are likely to face the halting problem are eliminated from the ecosystem.

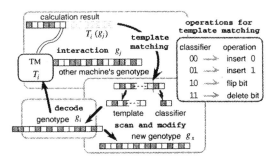

Figure 1: Interaction between schemes.

Furthermore, there are no arbitrary rules for self-organization or self-reproduction, as well as no special metaphors derived from living systems. Intrinsically, the PROTEAN modeling does not restrict the possibilities for algorithms: any kind of algorithm can be produced theoretically.

The Quantification Method

Originally, the QTIV was developed for the field of sociometry. When a preference table by paired comparison is given, the purpose of the QTIV is to quantify so that a pair of similitudes come close to each other and to arrange each element in the minimum dimension possible. The unique aspect of the QTIV is that it is applicable even when there is no advance information that is directly helpful for spatial arrangement.

Let e_{ij} $(i \neq j)$ be the degree of preference of an element i for j. In order to simplify the explanation of the QTIV, one-dimensional quantification is first described.

Now, giving one-dimensional coordinate value to each element, consider a Euclid square distance $(x_i - x_j)^2$ as a value denoting a distance between a pair of objective elements. In order to approximate a pair of similitudes, in other words, a pair whose preference for the other is great, and to sunder a pair whose preference for the other is small, determine $\{x_i\}$ that maximizes the following value Q:

$$Q = -\sum_{i=1}^{n}\sum_{j=1}^{n} e_{ij}(x_i - x_j)^2. \quad (1)$$

Q is the inner product of $-e_{ij}$ (negative preference) and a Euclid square distance that denotes the degree of difference between the objective elements. To maximize Q is equivalent to determining the quantity $\{x_i\}$ by which the non-preference $-e_{ij}$ and its Euclid square distance have the closest agreement. Here, for the normalization of data, let the average of $\{x_i\}$ be

$$\frac{1}{n}\sum_{i=1}^{n} x_i = 0, \quad (2)$$

and let the variance be

$$\frac{1}{n}\sum_{i=1}^{n}(x_i - \bar{x})^2. \quad (3)$$

In equation (1), the greater the preference of a pair of elements is for each other, the larger the value of e_{ij} becomes. As a result, the smaller a Euclid square distance $(x_i - x_j)^2$ is, the larger Q becomes. On the other hand, if a preference between a pair of elements is small, e_{ij} becomes small, and, thus, if a Euclid square distance $(x_i - x_j)^2$ is larger, the value of Q becomes larger.

Now, to determine the quantity $\{x_i\}$ results in solving an eigenvalue problem for a matrix $H = (h_{ij})$ that appears in the following equation:

$$\sum_{j=1}^{n} h_{ij}x_j - \beta x_i. \quad (i = 1, 2, \ldots, n). \quad (4)$$

The matrix $H(h_{ij})$ is defined by e_{ij}, which denotes preference as follows:

$$\left.\begin{array}{l} h_{ij} = h_{ji} = e_{ij} + e_{ji} \\ h_{ii} = -\sum_{j \neq i} h_{ij}(= -\sum_{j \neq i} h_{ji}) \end{array}\right\} \quad (5)$$

It follows that $\{x_i\}$ is equivalent to an element of an eigenvector for the largest eigenvalue β_1 in equation (4).

When multidimensional quantification is required, first, a q-dimensional quantity for i is defined as $\{x_i^{(1)}, x_i^{(2)}, \ldots, x_i^{(q)}\}$. In this context, consider the maximization of

$$Q = -\sum_{i \neq j}\sum e_{ij}\{(x_i^{(1)} - x_j^{(1)})^2 + (x_i^{(2)} - x_j^{(2)})^2 + \cdots \\ + (x_i^{(q)} - x_j^{(q)})^2\}. \quad (6)$$

By solving the above equation, elements of an eigenvector for eigenvalues, each of which is from the largest one to the q-th largest, become quantities of $\{x_i^{(1)}, x_i^{(2)}, \ldots, x_i^{(q)}\}$ respectively.

Quantified Evolutionary Dynamics

In this section, the quantification method QTIV is applied to simulation results obtained through runs of the PROTEAN.

Preference data gathered from local interactions are analyzed to generate points in a DPS. This paper uses two-dimensional analysis, because there is general agreement that up to the second largest eigenvector in the QTIV calculation is enough to quantify objective data. The position of each scheme in the 2-D DPS changes with time during their reproductive processes. Thus, quantified evolutionary dynamics are demonstrated.

We first begin with an example of a self-developmental process of a scheme that is deliberately chosen because of its interesting process in interaction, and then we demonstrate quantification of a more complicated case of evolution in the self-developmental process. The evolutionary dynamics of the population of schemes in the PROTEAN induces a set of orbits in the DPS. Different types of long-term evolutionary dynamics are discussed using the collection orbits across the sequential DPS.

Microscopic Snapshot of Dependency Relationship in Pure Self-Developmental Process

Here, a pure self-developmental process that starts from scheme 1001011000010101111011100101100100 is examined. The reason we call the self-developmental process discussed here "a pure self-developmental process" is that we check all of the possible interactions in advance and then use its step-by-step results for the demonstration of the effectiveness of the quantification method.

In the following, a 100-adic notation system is adopted for scheme descriptions, using 100 characters such as {A, ..., Y, a, ..., y, A, ..., Y, a, ..., y}, as the length of a bit-string does not allow full representation. The above scheme, for example, is described as BAWH1r under the notation system.

The characteristic of a series of interactions beginning from scheme BAWH1r is that the scheme first generates a new one operating on its own description, and then such an interaction process automatically progresses with the resources generated as follows:

$$\text{BAWH1r} \otimes \text{BAWH1r} \rightarrow \text{CBtPpb} \quad\dots\dots\dots (1)$$

$$\text{CBtPpb} \otimes \text{BAWH1r} \rightarrow \text{Ecngxt} \quad\dots\dots\dots (2)$$

$$\text{ECngxt} \otimes \text{BAWH1r} \rightarrow \text{IfbONf} \quad\dots\dots\dots (3)$$

$$\text{IFbONf} \otimes \text{BAWH1r} \rightarrow \text{QLDdtC} \quad\dots\dots\dots (4)$$

$$\text{QLDdtC} \otimes \text{BAWH1r} \rightarrow \text{hXGHFv} \quad\dots\dots\dots (5)$$

$$\text{hXGHFv} \otimes \text{BAWH1r} \rightarrow \text{QvNOdj} \quad\dots\dots\dots (6)$$

$$\text{QvNOdj} \otimes \text{BAWH1r} \rightarrow \text{BdrbdYK} \quad\dots\dots\dots (7)$$

$$\text{BdrbdYK} \otimes \text{BAWH1r} \rightarrow \text{CHjCHQM} \quad\dots\dots (8)$$

$$\text{CHjCHQM} \otimes \text{BAWH1r} \rightarrow \text{FPTFPyQ} \quad\dots\dots (9)$$

$$\text{FPTFPyQ} \otimes \text{BAWH1r} \rightarrow \text{KgnKgPY} \quad\dots\dots (10)$$

$$\text{KgnKgPY} \otimes \text{BAWH1r} \rightarrow \text{UmbUMwp} \quad\dots\dots (11)$$

$$\text{UMbUMwp} \otimes \text{BAWH1r} \rightarrow \text{qaCqaLW} \quad\dots\dots (12)$$

$$\text{qaCqaLW} \otimes \text{BAWH1r} \rightarrow \text{hBEhApl} \quad\dots\dots (13)$$

$$\text{hBEhApl} \otimes \text{BAWH1r} \rightarrow \text{BPCJPAwQ} \quad\dots\dots (14)$$

$$\text{BPCJPAwQ} \otimes \text{BAWH1r} \rightarrow \text{DfETfCMU} \quad\dots\dots (15)$$

where $g_i \otimes g_j \rightarrow g_x$ denotes an interaction, in which a scheme g_i that is decoded into T_i reads the other scheme g_j and then newly generates a scheme g_x.

These equations are a series of possible interactions that are likely to occur through a run of the PROTEAN with a seed of scheme BAWH1r. Although the interactions execut-

ed by the above schemes are actually quite trivial from a computational point of view, the interacting processes are sufficient for demonstrating the developmental dependency relationship.

Suppose there are n schemes that interact with each other. When an interaction $g_i \otimes g_j \rightarrow g_x$ occurs, giving one integer point to g_i and g_j for g_x, an $n \times n$ matrix is obtained. The way of scoring means g_x owes its generation to g_i and g_j. In this way, a preference table denoting the dependency relationship is defined.

For example, Table 1 shows a preference table of the interactions from (1) to (3) in the above series of interaction. Thus, by applying the quantification method QTIV to the preference table, a 2-D diagram of DPS is then obtained, as depicted in Figure 2. Note that the coordinate real value itself of each element in the DPS is not important. What has to be noted is the spatial arrangement of elements based on their degree of preference to other elements. A pair of elements that are close to each other has an affinitive relationship. A relational concept, which is qualitative data, must now be transformed into quantitative data with a scale described in the form of a Euclid distance.

	BAWH1r	CBtPpb	Ecngxt	IfbONf
BAWH1r	0	0	0	0
CBtPpb	2	0	0	0
Ecngxt	1	1	0	0
IfbONf	1	0	1	0

Table 1: Preference table obtained for interactions at the early stage.

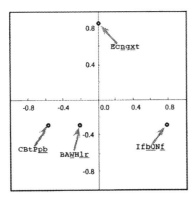

Figure 2: 2-D dependency projection space obtained by application of QTIV to the interactions in Table 1.

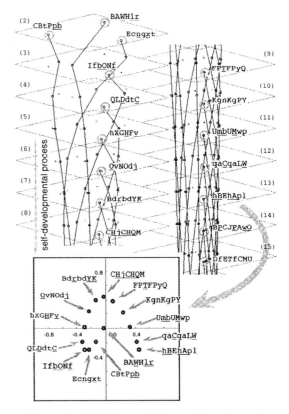

Figure 3: 3-D diagram that quantifies the self-developmental process and one close-up DPS snapshot selected among those sequential analyses.

235

Furthermore, a sequential collection of DPS by quantification through to interaction (15) is shown in Figure 3. The 3-D diagram that converges in a spiral form can be seen. One DPS snapshot among the sequential view is also depicted in Figure 3. It illustrates that affinitive relationships among the elements are represented as a curve consisting of a set of crowded dots.

Note that this figure is just drawn from the above data for a sample demonstration of a fact that the quantified method works well with our intuitive expectation. The actual evolutionary dynamics through a run of the PROTEAN is examined next.

Quantification of Evolutionary Dynamics through Computer Simulation

A significant difference in an evolutionary process from the above *pure* self-developmental process shown in Figure

3 is that the existing schemes have to now compete with the others for a static amount of resources, that is, memory space in the PROTEAN. Their fitness is dynamically dependent on their relations to others and the consequent environment thereof. At the same time, extinction of each scheme is now likely to occur.

Figure 4 illustrates evolutionary dynamics that starts from an interaction of scheme BAWH1r on the basis of the same quantification procedure as that used in the former experiment. As space is limited, abbreviations of the whole dynamics are shown in this figure. The reason the observed complexion is quite different in spite of beginning from the same seed as in the case of Figure 3 is because of the effect of survival games taking place under a dynamically changing selection pressure and because those local interactions and the environment thereof are, then, organizing a holistic context. The combination of a top-down selection pressure and a bottom-up local interactions makes the

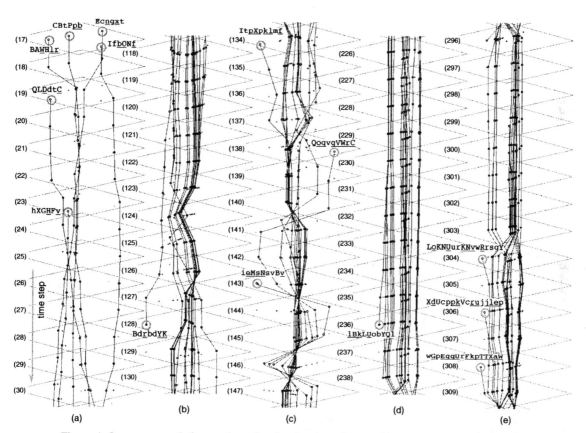

Figure 4: Long-term evolutionary dynamics that involve microscopic events obtained through the quantification of the dependency relationship of system constituents: (a) the generation of new system constituents at the early stage of evolution; (b) the extinction of a system constituent involving an alteration in the following evolutionary dynamics; (c) the generations of new constituents and drastic changes in the dependency relationship subsequent to those events; (d) a simple extinction; (e) transition from a stationary state to the generation rush of new constituents.

evolutionary dynamics more complicated than that of Figure 3.

Although the dynamics is complicated, Figure 4 shows some interesting patterns in the long-term evolution. Next, let us turn to the investigation of each case.

A Gradual Acceptance. Figure 4 (a) depicts dynamics at the early stage of evolution. Some schemes, such as QLDdtC and hXGHFv, are newly generated from the initial seed. Scheme IfbONf is once left out of the others and then takes a hand again in them by degrees after the generation of scheme hXGHFv. Remember that the closer to the center in each 2-D DPS the location of a constituent is, the more affinitive is the relationship that it has to the others. Scheme IfbONf is thought to have an important role in the subsequent dynamics.

Events Involving Drastic Changes. Figure 4 (b) and (c) show examples of the extinction and generations of schemes, respectively. Figure 4 (b) shows a process in which scheme BdrbdYK loses its adaptivity. That scheme eventually becomes extinct, and it follows from that event that the dependency relationship of the remaining schemes has been expanded. On the other hand, In Figure 4 (c), the schemes of ItpXpklmf, QogvgVWrC and ieMsNsvBv are generated. Drastic changes in the dependency relationship coincide with those events, implying that those generations are stimulatory in the evolutionary dynamics.

A Desolate Extinction. Unlike the case in Figure 4 (b), scheme lBkLUobYQl disappears in a long-term stationary state as depicted in Figure 4 (d). Very few changes occur, meaning that this scheme already has no importance for the current evolutionary dynamics at this moment.

Calm before Rush of Generation. While a stationary state is continuous, as shown in the upper part of Figure 4 (e), the generation of scheme LoKNUurKNvwRrsgY led to the birth of its offsprings. However, the evolutionary dynamics shows that the ecosystem manages to absorb those newcomers.

Discussion

We have described the quantification of evolutionary dynamics observed in computer simulations of an A-life model. In this section, the effectiveness of this quantification approach and future prospects are discussed.

The QTIV is a method for representing the relational concept between objective elements as their spatial arrangement through quantification of their relationships. The figures shown in the former section illustrate the dependency relationship of each element: An element that has affinities to the others is located close to the origin in the DPS because of the minimization of Euclid distances to those elements. Furthermore, the sequential analysis of such spatial arrangements has enabled us to visualize long-term evolutionary dynamics from the viewpoint of local interactions.

In this paper, a simple self-developmental process has been described for demonstration of the quantified evolutionary dynamics. The following results were obtained. When a system constituent is newly generated in the system, it appears far from the origin. If this constituent is subsequently capable of interacting with other constituents, it gradually comes close to those constituents. On the other hand, a constituent that is likely to become extinct goes away from the origin and disappears. At this moment, if a drastic alteration in the spatial arrangements follows, the extinct constituent is thought to have had an affinitive relationship with the other constituents. On the other hand, no distinctive change occurs when a constituent that has already lost its existence value becomes extinct. The extent of dynamics change following generation or extinction depends on the holistic constitution of the system at that time.

An important point is that the quantification of the dependency relationship indicates the possibility of clarifying microscopic events capable of influencing the future course of evolution. It could be said, for example, that the appearance of a new constituent in the spatial arrangement of DPS corresponds to emergence in a narrow sense. At the same time, the following changes in evolutionary dynamics are thought to show a complexion in which the holistic context is re-organized. In some previous works on evolutionary dynamics, a certain criterion for quantifying a holistic complexion has been introduced in advance. As described before, the anticipatory definition of such a criterion would not intrinsically agree with the essences of emergence and evolution. On the other hand, we have tried to represent ongoing dynamics of a system from the microscopic point of view, free from any pre-embodiment of criterion variables.

Motivated by such an intention, we dealt with a simple self-developmental process in this study. From the formal point of view, it is no more than a simple step-by-step developmental process. The results of simulation, however, revealed complex aspects ongoing in the system, unlike a pure step-by-step process. In the previous section, we described some interesting patterns among its long-term evolutionary dynamics. We are currently studying evolutionary dynamics consisting of schemes that execute more interesting computation in the theoretical sense, and we have confirmed that this quantification approach is effective even in more complex problems.

Lastly, let us now turn to another characteristic of the quantification approach, that is, the normalization of data. As shown in equations (2) and (3), the results obtained from this method are always normalized. The relative quantity through the normalization may be able to be applied in the future to calculation of the fitness value of each constituent. In the field of evolutionary computation, the fitness value is given absolutely from a certain pre-defined equation. On the other hand, some studies on experimental biology insist each system constituent defines its adaptivity as a dynamically changing relative value to the others (Matsuno 1989). Such a school argues in favor of an approach based on the normalized relational concept.

To summarize, an approach of quantification of local interactions has been presented as a possible tool for future research in A-life. Evolutionary dynamics of our A-life model have been represented in the form of the sequential transition of quantified local interactions. Evolutionary dynamics and the usefulness of our approach were discussed on the basis of results of computer simulations.

References

Adami, C. and Brown, C. T. 1994. Evolutionary Learning in the 2D Artificial Life System 'Avida'. Artificial Life IV, edited by R. A. Brooks and P. Maes. MIT Press. 377-381.

Bedau, M. A. and Packard, N. H. 1992. Measurement of Evolutionary Activity, Teleology, and Life. *Artificial Life II*, edited by C. G. Langton, C. E. Taylor, J. D. Farmer, S. Rasmussen. 431-461. Addison-Wesley.

Bedau, M. A., Snyder, E., Brown, T. and Packard, N. H. 1997. A Comparison of Evolutionary Activity in Artificial Evolving Systems and in the Biosphere. In Proceedings of Fourth European Conference on Artificial Life, edited by P. Husbands and I. Harvey. 125-134. MIT Press.

Bedau, M. A., Snyder, E., and Packard, N. H. 1998. A Classification of Long-Term Evolutionary Dynamics. *Artificial Life VI*, edited by C. Adami et al. 228-237. MIT Press.

Hayashi, C. 1964. Multidimensional Quantification of the Data Obtained by the Method of Paired Comparison. *Annals of the Institute of Statistical Mathematics*, Vol. XVI, pp. 231-245.

Matsuno, K. 1989. Protobiology: Physical Basis of Biology. CRC Press, Boca Raton, FL.

Ray, T. S. 1991. An Approach to the Synthesis of Life, *Artificial Life II*, edited by C. G. Langton, C. E. Taylor, J. D. Farmer, S. Rasmussen. 371-408. Addison-Wesley.

Ronald, E. M. A, Sipper, M., and Cascarrere, M. S. 1999. Testing for Emergence in Artificial Life. *Advances in Artificial Life*, 13-20. Springer.

Taylor, T., and Hallam, J. 1997. Studying Evolution with Self-Replicating Computer Programs. In Proceedings of Fourth European Conference on Artificial Life, edited by P. Husbands and I. Harvey. 550-559. MIT Press.

Taylor, T., and Hallam, J. 1998. Replaying the Tape: An Investigation into the Role of Contingency in Evolution. *Artificial Life IV*, edited by C. Adami et al. 256-265. MIT Press.

Wolfram, S. 1984. Cellular Automata as Models of Complexity. *Nature* 311: 419-424.

Yoshii, S., Ohashi, S., and Kakazu, Y. 1998a. Self-Organized Complexity in a Computer Program Ecosystem. *Artificial Life VI*, edited by C. Adami et al. 483-488. MIT Press.

Yoshii, S., Ohashi, S., and Kakazu, Y. 1998b. Modeling of Emergent Ecology for Simulating Adaptive Behavior of Universal Computer Programs, *From Animats to Animals 5*. 303-308. MIT Press.

An Ecolab Perspective on the Bedau Evolutionary Statistics

Russell K. Standish

High Performance Computing Support Unit
University of New South Wales
Sydney, 2052
Australia
R.Standish@unsw.edu.au
http://parallel.hpc.unsw.edu.au/rks

Abstract

At Alife VI, Mark Bedau proposed some evolutionary statistics as a means of classifying different evolutionary systems. Ecolab, whilst not an artificial life system, is a model of an evolving ecology that has advantages of mathematical tractability and computational simplicity. The Bedau statistics are well defined for Ecolab, and this paper reports statistics measured for typical Ecolab runs, as a function of mutation rate. The behaviour ranges from class 1 (when mutation is switched off), through class 3 at intermediate mutation rates (corresponding to scale free dynamics) to class 2 at high mutation rates. The class 3/class 2 transition corresponds to an error threshold. Class 4 behaviour, which is typified by the Biosphere, is characterised by unbounded growth in diversity. It turns out that Ecolab is governed by an inverse relationship between diversity and connectivity, which also seems likely of the Biosphere. In Ecolab, the mutation operator is conservative with respect to connectivity, which explains the boundedness of diversity. The only way to get class 4 behaviour in Ecolab is to develop an evolutionary dynamics that reduces connectivity of time.

Introduction

At Alife VI, Mark Bedau proposed some evolutionary statistics (Bedau *et al.*, 1998) as a means of classifying different evolutionary systems. The intent here is to find a general scheme analogous to Wolfram's (1984) classification scheme of cellular automata. Three statistics are proposed:

Diversity (D): The number of species or components in the system

Mean Cumulative Evolutionary Activity ($\bar{A}_{\mathrm{cum}}$):
Activity of a species is defined as the population count of that species, the vector n in Ecolab terms. Evolutionary activity subtracts from this the neutral or nonadaptive part. This is achieved by running a *neutral shadow model*, that is identical with the original model, except that natural selection must be "turned off". Finally, this activity is accumulated over the lifetime of the species, and then averaged over all species.

New Evolutionary Activity (A_{new}): This corresponds the the number of new species crossing a threshold, divided by the diversity.

Bedau describes four classes of evolutionary behaviour, as in the following table:

Class	$D(t)$	$A_{\mathrm{cum}}(t)$	$A_{\mathrm{new}}(t)$	Description
1	bounded	zero	zero	none
2	bounded	unbounded	none	unbounded, uncreative
3	bounded	bounded	positive	bounded, creative
4	unbounded	positive	positive	unbounded, creative

Note that in Bedau *et al.* (1998), only 3 classes are mentioned — class 2 was added later in his presentation at Alife VI. Bedau has applied his statistics to a number of artificial life models, including Echo (Holland, 1995) and Tierra (Ray,1991), none of which exhibit class 4 behaviour. By contrast, the same statistics applied to the fossil record (at least for the Phanerozoic — the period of time since the appearance of multicellular life in the Cambrian) — show a strong class 4 behaviour. Further, Bedau speculates that the global economy and internet traffic are also class 4, particularly as they show strong growth over a significant period of time. Since no artificial life systems to date appear to show class 4 behaviour, the gauntlet has been laid down to discover such a system to work out whether this classification difference is fundamental or not.

Ecolab (Standish, 1994), whilst not an artificial life system, is a model of an evolving ecology that has advantages of mathematical tractability and computational simplicity. It lies in between the extremely simplistic models of (for example) Bak and Sneppen (1993) or Newman (1997) and artificial life models of evolution such as *Tierra* or *Avida*. One of its key characteristics is that its dynamics are defined by the ecological interactions between the species, rather than ad hoc exogenous dynamics. The Bedau statistics are well defined for it, so it is interesting to see what class behaviour Ecolab has. Furthermore, an Ecolab-like model is possible for all artificial life systems (valid in a continuum limit). For example, the equations of motion for Tierra are given in Standish(Standish, 1997).

Ecolab

The Ecolab model (as opposed to the Ecolab simulation system) is based on an evolving Lotka-Volterra ecology. The defining equation is given by:

$$\dot{n} = r * n + n * \beta n + \gamma * \nabla^2 n + \mu(r * n) \quad (1)$$

where n is the species density, r the effective reproduction rate (difference between the intrinsic birth and death rates in the absence of competition), β the matrix of interaction terms between species, γ the migration rate and μ the mutation operator. All of these quantities (apart from β, which is a matrix) are vectors of length n_{sp}, the number of species in the ecology. The operator $*$ denotes elementwise multiplication. The mutation operator returns a vector of dimensionality greater than n_{sp}, with the first n_{sp} elements set to zero — in effect expanding the dimensionality of the space, a key feature of this system. For a more detailed exposition of the various properties of the model, in particular, the precise form of the mutation operator, the reader is referred to the previous published papers, as well as the Ecolab Technical Report, which are all available from the Ecolab Web Site[1].

For the purposes of this paper, it is worthwhile expounding a little on the properties of the mutation operator. It models point mutations in particular (other mutation types, such as recombination are simply not modeled within Ecolab). Point mutations in genotype space, which satisfy Poisson statistics, give rise random mutations, with locality, in phenotype space. Since the only phenotypic properties of interest to the model are the parameters r, β and γ the parameters are mutated according to a normal or lognormal distribution (according as the parameters are reals or positive (or negative) respectively), using a sample from the Poisson distribution for the width. The two parameters governing mutation (width of the Poisson distribution, and the rate at which mutations are attempted) are related via a simple proportional factor (called the "species radius (or separation)") that is kept constant throughout the simulations reported here. Each species has its own mutation rate — given as a vector μ.

Each of these phenotypic parameters are initialised from a uniform distribution. The relevant input parameters for a run are then maximum and minimum values for each of r, the diagonal of β, the offdiagonal of β, μ, γ and the species radiua ρ. The complete system may be scaled in the time dimension, fixed by what value is chosen for the timestep. In this case, $\max_i r_i = 0.1$, so one timestep corresponds to about a 14th of the doubling time of the fastest reproducing organism in the

ecology. This is a compromise between continuity of the simulation and computational expense. The ratio $\frac{\max_i r_i}{\max_i \beta_{ii}}$ roughly corresponds to the carrying capacity of the ecology. This is chosen to about 100 so that behaviour near the equilibrium is reasonably continuous rather than stochastic. The ratio of offdiagonal to diagonal terms relates to how negative definite β is. Since mutations tends to drive the matrix away from being negative definite (system stability), the maximum of the offdiagonal terms is chosen to make the initial system marginally unstable. The species radius $\rho = 0.1$ was chosen empirically to make new species *phenotypically* distinct from its parent species.

Having fixed the other parameters according to the above criteria, the remaining degrees of freedom are μ and γ. In this paper, we vary the maximum mutation rate in different simulations, but keep the distribution of migration rates fixed.

One other feature worth noting is that the mutation operator will also randomly add or drop connections between species, according to an exponential distribution. Thus, the mutation operator is in fact highly conservative — with the lognormally mutated parameters capped (in the case of μ and γ) or restricted by the requirements of boundedness (diagonal components of β)(Standish, 1998; Ecolab Technical Report).

Neutral Shadow Model

An important feature for improving the accuracy of the evolutionary statistics is the use of a *neutral shadow model*. This model should be as similar as possible to the original model, but with all selection turned off. In the case of Ecolab, this is accomplished by running a shadow population density vector n', and when n is updated, the shadow vector is updated by a random permutation of the updates. Thus each shadow species behaves in the long run like an average species. Activity is also tracked at the same time, with the activity vector being updated by the difference between the population density and the shadow population density, provided that difference is positive.

The new activity statistic A_{new} is computed by summing the number of species that have crossed a threshold. In (Bedau et al., 1998), this threshold is determined by plotting the activity distributions for both the original and the shadow model, and taking the cross-over point as the threshold. This turned out to be 50 individuals, rather than the arbitrary 10 individuals used in other Ecolab studies. In fact the two distributions are nearly equal over the range 10–50, but if an activity is above 50, then it is highly likely to be due to adaptive behaviour.

---

[1]http://parallel.hpc.unsw.edu.au/rks/ecolab.html

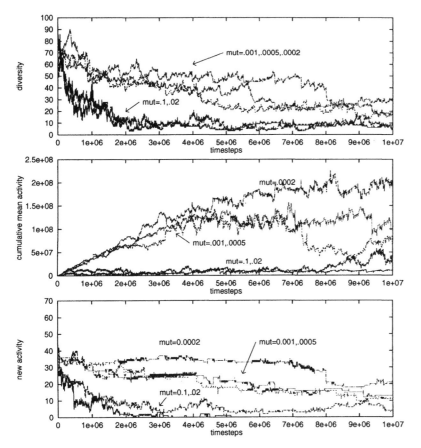

Figure 1: A typical run for panmictic Ecolab at varying mutation rates, showing the Bedau statistics: diversity, cumulative mean activity and new activity

Behaviour of Ecolab

Figure 1 shows the Bedau statistics for typical Ecolab runs (panmictic, or spatially independent case), as a function of mutation rate. When the mutation rate is too low, class two behaviour is seen. Diversity remains constant, and activity grows unbounded as the system rapidly sheds unviable organisms and tends to a stable ecology. Conversely, for high levels of mutation, class one behaviour is seen. There is a constant churn of organisms, that do not have any chance to generate activity. For intermediate levels of mutation, an interesting situation arises. Here, the number of mutant organisms that successfully invade the ecosystem roughly balances the number lost through extinction(Standish, 1998). Scale free behaviour is observed in a number of statistics, such as the distribution of species lifetime. These same 3 states of behaviour have been observed in Avida(Adami et al., 1998).

The code used for this simulation is available from the Ecolab web site as version 3.3 of the software. The model including the neutral shadow model is defined in shadow.cc, and a sample experimental script

given as bedau.tcl. The only parameters varied are the spatial dimensions and mutation(random,maxval).

The evolutionary statistics were also collected for a spatially dependent Ecolab, however due to some implementation difficulties, run lengths exceeding 1×10^6 timesteps have not been achieved prior to this paper's deadline. Broadly speaking, though, the same behaviour is seen as the panmictic case, although there is a period of diversity growth in the early period prior to settling on a higher level of diversity than the panmictic case.

This can be understood by considering two extremes of spatially dependent Ecolab models, namely zero migration and infinite migration. Infinite migration effectively corresponds to the panmictic case again, whereas zero migration corresponds to a number of cells, independent of each other, each running the panmictic model. So we would expect in the case of zero migration, the diversity (in the long run) should be proportional to the number of cells (or the total area). The in between case of finite nonzero migration should also show an increase in diversity with area, due to partial independence of each cell, but the increase should be sublinear, as migration causes some species to be identified between cells. Island Biogeography(MacArthur and Wilson, 1967) theory postulates that the relationship is $D \propto A^{-s}$ for some coefficient s, which presumably must depend in some fashion on the migration rates, but is generally in the range 0.2–0.35 for most empirical studies.

May's Stability Criterion

May(1972) proposed that random Lotka-Volterra webs would be unstable if

$$n_{\mathrm{sp}} < \frac{1}{s^2 C} \qquad (2)$$

where C is the connectivity, defined as the proportion of nonzero elements in β, and s is the interaction strength, defined as the standard deviation of the offdiagonal terms of β, divided by the average of the diagonal terms. Cohen and Newman(1985) showed that May's criteria does not hold for Lotka-Volterra systems in general, only a smaller class related to the models May studied. However, the inverse relationship between species number and connectivity does appear to hold(Pimm, 1982;Cohen and Newman, 1988;Cohen et al., 1990).

Stability is not a relevant property in Ecolab, as really the persistent state (which includes the stable state as a special case) is the attractor. However, the inverse relationship between diversity and connectivity does hold(Standish, 1998), for spatially dependent as well as panmictic cases. Therefore, in order for diversity to show an increasing trend, a corresponding decreasing trend must occur in connectivity. This ought to be true of the biosphere also, given the universality of this relationship.

As mentioned in section , the mutation operator is highly conservative with respect to connectivity. It assumes that a new species inherits the same connections as its parent, with random additions or deletions according to a symmetric distribution (just as likely to gain a connection as lose one). This has the effect of preserving the connectivity over time. In order for connectivity to decrease, different dynamics would need to be proposed, for example assuming that the mutant species did not compete with its parent.

One possibility for the cause of this growth in diversity is the mass extinctions, that have occurred a handful of times throughout the Phanerozoic. However, the only reasonable way of modeling this is to remove a random proportion of species from the ecology at a particular time. This operation does not alter the connectivity, as the links lost is exactly balanced by the reduced diversity. When implemented within Ecolab, one gets the characteristic rebound in diversity after the extinction event, however, the rebound is back to about the same diversity level as existed prior to the extinction.

Another possibility that actually would work in the right way is related to the fact that the Phanerozoic era corresponds to the breakup of the Pangaea supercontinent — firstly into Gondwana and Laurasia, then into the six continents we know today. Assuming that there is almost no migration between the continents (thus 6 equal-sized continents would support 6 times the diversity of one continent that size) and that the species-area law within a continent has $D \propto A^{.3}$, we would expect that a breakup of a single supercontinent into 6 equal sized pieces should produce $6^{1-.3} = 3.5$ times the diversity of the original supercontinent. This factor accounts for a significant fraction of the diversity growth since the Permian.[2](Benton, 1995)

Clearly this is a very rough "back of the envelope" calculation, but it is sufficient to show that continental breakup needs to be allowed for in determining if there is any intrinsic evolutionary processes driving diversity growth.

---

[2]In case anyone thinks that this result is an argument in favour of habitat fragmentation for promotion of diversity, this is a question of scale. Over short timescales habitat fragmentation is bad for diversity, as is any major environmental change. Only over evolutionary timescales will the diversity bounce back.

Acknowledgements

I wish to thank Mark Bedau for many illuminating discussions, and for assistance in developing the neutral shadow model for Ecolab. I also wish to thank the *New South Wales Centre for Parallel Computing* for computational resources required for this project.

References

Chris Adami, Ryoichi Seki, and Robel Yirdaw. Critical exponents of species-size distribution in evolution. In Chris Adami, Richard Belew, Hiroaki Kitano, and Charles Taylor, editors, *Artificial Life VI*, pages 221–227, Cambridge, Mass., 1998. MIT Press.

Per Bak and Kim Sneppen. Puntuated equilibrium and criticality in a simple model of evolution. *Phys. Rev. Lett.*, 71:4083, 1993.

Mark A. Bedau, Emile Snyder, and Norman H. Packard. A classification of long-term evolutionary dynamics. In Chris Adami, Richard Belew, Hiroaki Kitano, and Charles Taylor, editors, *Artificial Life VI*, pages 228–237, Cambridge, Mass., 1998. MIT Press.

M. J. Benton. Diversification and extinction in the history of life. *Science*, 268:52–58, 1995.

J. E. Cohen, T. Luczac, C. M. Newman, and Z.-M. Zhou. Stochastic structure and nonlinear dynamics of food webs. *Proc. R. Soc. Lond. B*, 240:607–627, 1990.

J. E. Cohen and C. M. Newman. When will a large complex system be stable? *J. Theo. Bio.*, 113:153–156, 1985.

J. E. Cohen and C. M. Newman. Dynamic basis of food web organisation. *Ecology*, 1988.

John H. Holland. *Hidden Order: How Adaption Builds Complexity*. Helix Books, 1995.

R. H. MacArthur and E. O. Wilson. *The Theory of Island Biogeography*. Princeton UP, Princeton, 1967.

R. M. May. Will a large complex system be stable. *Nature*, 238:413–414, 1972.

M. E. J. Newman. A model of mass extinction. *J. Theo. Bio.*, 189:235–252, 1997.

S. L. Pimm. *Food Webs*. Chapman and Hall, London, 1982.

Tom Ray. An approach to the synthesis of life. In C. G. Langton, C. Taylor, J. D. Farmer, and S. Rasmussen, editors, *Artificial Life II*, page 371. Addison-Wesley, New York, 1991.

R. K. Standish. Embryology in Tierra: A study of a genotype to phenotype map. *Complexity International*, 4, 1997. http://www.csu.edu.au/ci.

R. K. Standish. Statistics of certain models of evolution. *Phys. Rev. E*, 59:1545–1550, 1999.

R. K. Standish. The role of innovation within economics. In W. Barnett, C. Chiarella, S. Keen, R. Marks, and H. Schnabl, editors, *Commerce, Complexity and Evolution*, volume 11 of *International Symposia in Economic Theory and Econometrics*, pages 61–79. Cambridge UP, 2000.

Russell Standish. Cellular Ecolab. In Russell Standish, Bruce Henry, Simon Watt, Robert Marks, Robert Stocker, David Green, Steve Keen, and Terry Bossomaier, editors, *Complex Systems '98 — Complexity Between the Ecos: From Ecology to Economics*, page 80. Complexity Online, http://life.csu.edu.au/complex, 1998. also in *Complexity International*, **6** http://www.csu.edu.au/ci.

Russell K. Standish. Ecolab documentation. Available at http://parallel.acsu.unsw.edu.au/rks/ecolab.html.

Russell K. Standish. Population models with random embryologies as a paradigm for evolution. In *Complex Systems: Mechanism of Adaption*. IOS Press, Amsterdam, 1994. also *Complexity International*, **2**, http://www.csu.edu.au/ci.

S. Wolfram. Cellular automata as models of complexity. *Nature*, 311:419–424, 1984.

Mate Choice: Simple or Complex? *

Patricio Lerena
PAI Research Group, IIUF, University of Fribourg
Ch. du Musée 3, CH-1700 Fribourg, Switzerland
e-mail: lerena@mail.com

Abstract

Focusing on situations in which sexual preferences inspect several traits, this paper explores the question of how complex the process of assessing mate attractiveness might get, through self-organization, over the course of evolution: do sexual preferences evaluate traits in a simple, linear way or in a complex, nonlinear fashion? Arguments and simulation results (individual-based model in which preferences with different degrees of evaluation-complexity "compete" in a population) are presented, suggesting that a bias should exist, and that its direction should depend on levels of different types of noise (in perception and in the mate-evaluation process) and on mutation rates: while noise in perception and mutation favor simple preferences, noise in evaluation favors complex ones; for possibly the most plausible parameters, complex preferences are favored. Possible implications on the speciation process are mentioned. This novel way of looking at mate choice could be a rich source of new insights in this intriguing process.

Introduction

Perhaps the most fascinating split in sexual selection theory is that existing between the so-called "good genes" and "good taste" theories (or "schools") (Ridley, 1993). "Good genes" theories explain female mate-choice evolution on the basis of the viability "value" of genetic material contributed by males (e.g. (Hamilton and Zuk, 1982)). In contrast, Fisher's "good taste" school argues that female preferences for ecologically adapted males are not the rule and that conflict between natural and sexual selection forces can arise, leading to compromises. The arguments emphasize self-reinforcing aspects of mate choice and are based on the development of a correlation between consistently choosing preferences and their preferred traits (Fisher, 1930; Kirkpatrick, 1987). This correlation indirectly confers an advantage to such preferences, because they are likely to select individuals that carry genes for the very same preferences. The effect increases with choice consistency (Kirkpatrick, 1982), making discrimination capabilities relevant.

Within the context given by the "good taste" school, this paper explores the following questions. When several traits are being inspected in the evaluation of potential mates, are single traits contributing independently (additively) to the overall "sexiness" of the examined individual or are their contributions being combined in a nonlinear, complex way?[1] Does the answer to this question depend on mutation rate, noise in perception or noise in mate-evaluation? If it does, how?

Notice that (1) the picture of traits (trait-vectors) adapting to preferences by "adaptive steps" makes natural the use of the term *preference-landscape* to designate the underlying structure of sexual preferences, and (2) because more complexity in sexual preferences means more rugged preference-landscapes, the questions addressed in this paper can conveniently be discussed in terms of preference-ruggedness.

This paper states that the level of complexity in sexual preferences is not neutral to their evolution; different levels of interaction among single trait evaluations may differ in their evolutionary success. Furthermore, these differences should depend on the levels of noise in perception and in evaluation in fully different ways: as perception noise is increased from zero to highest lev-

---

Work supported by SNSF grant 20-05026.97. Thanks to P. Todd for discussions. Also thanks to the group PAI (www-iiuf.unifr.ch/pai), in particular to M. Courant and B. Hirsbrunner for support and advice, and to F. Chantemargue, S. LePeutrec, A. Pérez-Uribe and A. Robert for discussions and comments.

---

[1]A preference examining more than one trait is called here *complex*, if the contribution of single traits depends on the context given by the other traits under examination (i.e., there is an interaction among contributions of single traits to the overall mate value); in contrast, a preference is said to be *simple* if the contribution of every single trait to the overall mate value do not depend on the state of other traits (in a linear way).

els, the bias is expected to switch from high to low complexity; but for evaluation noise, only very low levels should favor low complexity, otherwise complex preferences should be favored. The effects of mutation should be qualitatively similar to those of perception noise. In order to discuss the causes for this expectations (detailed in the section presenting the results), it is convenient to distinguish two levels: (1) a functional level (ruggedness in preferences may affect their performance), and (2) an "evolutionary" level (shape and size of basins of attraction in preference-landscapes may affect their evolutionary future in a more indirect way).

The focussed situation is that in which trait-vectors in the population are concentrated (in terms of Hamming-distance) in a small, local region of trait-space. This is a plausible case for small populations, due in part to the effects of genetic drift.

To explore these questions, an individual-based simulation is used, in which (mutants of) two "competing" preferences with different complexity levels are initially present in a population that is then let evolve freely, and in the absence of natural selection, until one of them is fixed in the population. This allows to observe, over a big number of runs, fixation frequencies of preferences with different complexities.

The Simulation

This section describes the model used in this study, including variables, parameters, and details of the sets of runs performed.

Individuals

Individuals are composed of three parts: a binary *trait-vector*, a binary *preference-vector* and a *preference-landscape*. The trait-vector (denoted by $\vec{t}$) represents the list of all single traits taken into account in mate-choice by sexual preferences. The preference-vector (denoted by $\vec{p}$) is a list of single preferences for single traits, and allows to model the way in which traits are perceived. The preference-landscape (denoted by P) is a function defined over N-dimensional binary vectors and with values in $[0,1]$, which, as explained next, complements the preference-vector to determine the value of any potential mate.

Mate evaluation is depicted in Fig. 1. An individual i evaluating a potential mate j, compares first its preference-vector $\vec{p}_i$ with the traits $\vec{t}_j$ of individual j. The comparison is performed locus by locus using an exclusive or. This operation yields a (binary) *comparison vector*, which is then fed into i's preference-landscape P_i, finally yielding the value $p_i(j)$ (a real number in $[0,1]$). To sum up, $p_i(j) := P_i(\vec{p}_i \text{ xor } \vec{t}_j)$.

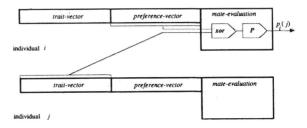

Figure 1: Evaluation by individual i of a potential mate j.

Phenotypes are genetically determined; two chromosomes code for an individual. The first one codes for the traits (trait-vector), the second one, for the preferences (preference-vector and the preference-landscape). Recombination takes place at this level but there is no intrachromosomal recombination. Mutation affects trait- and preference- vectors; these are directly encoded. Let us remark that the code used for preference-lanscapes is irrelevant, because these landscapes are inherited without modification; they are not changed by genetic operators. Nevertheless, a preference as a whole (composed of a preference-vector and a preference-landscape) *is* affected by mutation, because preference-vectors are mutated.

Notice that preference-vectors allow for a simple representation of single preferences, which determine *how* single traits are perceived. In addition, as clarified below, preference-vectors provide a simple means for modelling preference-mutation and perception-noise.

Preference-landscapes represent the core of the mate-evaluation system. This component models the interaction among the evaluations of single traits and is supposed to be stable compared to preference-vectors. A preference-landscape is represented by a table that codes an NK-landscape (Kauffman, 1993), with parameters N (dimension of preference-vectors), and K, which determines the preference-complexity level. The NK-model provides a simple, abstract, well known means of modelling different levels of interaction among components in their contribution to an overall value; results are available that characterize landscapes with different levels of interaction and many of them qualitatively hold across different parametrizations (number of interacting elements, distributions of values, etc.) (Kauffman, 1993)[2]; the nature of possible

---

[2] The results we are most interested in here are related to local maxima of NK-landscapes and their relationship to adaptation. With increasing values of K, the ruggedness of the landscape increases, the number of local-extrema grows very quickly, the number of adaptive steps to local maxima decreases in length, correlation among values of neighbors decreases and in particular, for neighbors of optima (its one-mutant variants), fitness drops become more and more

interactions in the present context is largely unknown, and asks for such a simple, robust way of modelling them[3].

As already mentioned, to make the model more realistic, two kinds of noise are introduced. Firstly, the level of *perception-noise* is given by the probability of perceiving, for each single trait in an examined trait-vector, a random value instead of its actual value. Secondly, *evaluation noise* perturbs computed mate-values; it is normally distributed around zero and its standard deviation will be called *evaluation noise level.* This noise is added to $p_i(j)$; the (noisy) result of this addition is a random variable, but its values will also be denoted by $p_i(j)$, in order to keep notation simple (disambiguated by context).

Competing Preferences Experiments

We turn now to the description of a single simulation-run, called here a "competing preferences experiment".

A small population of *70* individuals, with $N=16$, is evolved from an initial state in which (1) all trait-vectors are similar: an "original" trait-vector is randomly generated and mutated clones of it are produced to create the traits for the first generation (the mutation rate per basis in this initial phase is *0.02*); and (2) two different, randomly generated preferences with P-landscapes of different complexities (different values of K) are assigned, in equal proportions, to the individuals of the first generation[4]; preference vectors are mutated in the same way trait-vectors are mutated in this initial phase (identical mutation rate).

At each step, individuals choose a mate and reproduce. In the mate-choice phase of every single step, individuals are presented with a *choice-group* of *8* potential mates (randomly drawn), from which they select their mate. Mate-selection from this group by an individual i is performed as follows: for every individual j in the group, its value $p_i(j)$ is obtained (with traits previously perturbed if perception-noise is nonzero and with the value itself perturbed if evaluation-noise is nonzero); and the individual with the maximal value in the group is chosen for mating. Sexual reproduction follows; chosen individuals cannot refuse mating.

With the help of the effects of drift, one of the competing preference-landscapes becomes fixed after some steps. This event determines the halting point of a run. The K-value of the "winning" preference is

precipitous (Kauffman, 1993).

[3]Furthermore, alternative modelling approaches are being experimented; preliminary results obtained with them so far are qualitatively consistent with the ones presented in this text.

[4]The two "competing" P-landscapes are randomly chosen from a set of *1000* different, randomly generated ones.

recorded. Such runs are repeated *1000* times using different seeds for the pseudo-random number generators and using different preference-landscapes and different initial trait- and preference-vectors. The points plotted in the next section represent the average number of times one or the other K value is fixed (its fixation frequency) over these *1000* runs.

Such couples of K are compared for different levels of perception noise, evaluation noise and mutation rate.

Notice that choice is made on the basis of noisy mate-values; the individual with the highest noisy mate-value in the presented choice-group is chosen. This is a very simple and certainly plausible "algorithm"; there are many possible sources of noise, including fluctuations in the evaluation itself (two evaluations by and of the same individual may yield different values at different times or contexts) and limited cognitive capabilities (e.g. memory). Due to the many factors that may be involved, a normal distribution seems appropriate.

Sets of Runs

The default parametrization used in the sets of runs is the following: (1) mutation rate (per locus): *0.001*; (2) evaluation and perception noise: *0.05*. Differences in any particular setting will be mentioned.

In the first set of runs fixation frequencies of preferences with different complexity levels K_2 are compared to $K_1=0$ (*1000* runs for each pair).

In the next set of runs, mutation rate is varied for a fixed pair of K's: *0* vs. *4*. (*1000* runs are performed for each examined mutation rate.)

In the third set of runs, perception noise is varied for the same fixed pair of K's (also here *1000* runs are performed for each examined noise level).

The last set of runs is similar to the previous one (same number of runs too), but the parameter being varied for every fixed pair of K's is evaluation-noise.

In the next section, together with the presentation of the results for each of these set of runs, the theoretical arguments allowing to predict each of their key features are given (instead of introducing them before the results); this is just for brievity.

Results and Discussion

The results obtained with the described simulation have been tested for statistical significance (sign-test used due to the binomial nature of the data) and are presented with 95% CI of mean. Moreover, fixation frequencies greater (or equal) than *0.54* or smaller (or equal) than *0.46* highly significantly differ from the "null-value" *0.5*, which represents no frequency-differences (sign test, $p < 0.01$).

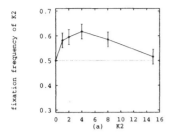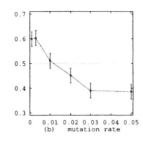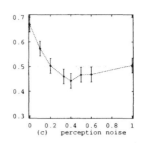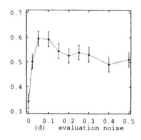

Figure 2: Fixation frequencies for different "competing-preferences-experiments". Each point represents the fixation frequency of K_2 when competing with $K_1 = 0$ (average over *1000* runs): (a) against different values of K_2 (mut. rate: *0.001*, perception and evaluation-noises: *0.05*; the plotted value of *0.5* for $K_2 = 0$ is the theoretical value), (b) against mutation rate ($K_2 = 4$, perception and evaluation-noises: *0.05*), (c) against perception noise ($K_2 = 4$, mut. rate: *0.001*, evaluation-noise: *0.05*), (d) against evaluation noise ($K_2 = 4$, mut. rate: *0.001*, perception-noise: *0.05*). Error bars reflect the 95% CI of Mean.

Varying K_2

The results for the first set of runs are presented in Fig. 2a, which shows an optimum for moderate complexity values. This is expected, because (1) some ruggedness in preference-landscapes allows for good discrimination of traits in any given small, local region of phenotype-space, such as the one covered by a small population (notice that in contrast, a very smooth landscape yields similar values in any such region); and (2) a too high K may lead to choice inconsistency at a functional level, due to perception noise, and at an evolutionary level, due to mutation (the explanations in the next two subsections should clarify this point).

Varying Mutation Rate

Fig. 2b shows how increasing mutation rate most affects complex preferences. For low levels of mutation, complex preferences evolve more frequently, but the situation is reversed for high levels of mutation, with simple preferences evolving more frequently. This is due to differences in the induced choice-inconsistencies through generations: for highly complex preferences, a single mutation usually produces important changes in the values assigned to traits; consequently, preference-trait correlations are strongly affected.

Varying Perception Noise

Fig. 2c shows how low levels of perception noise favor the evolution of complex preferences; in contrast, high levels of this noise are very perturbing for complex preferences and simple preferences are more likely to evolve. As already mentioned, this is due to the great choice inconsistency induced on complex preferences, for which a big change in the value assigned to a potential mate may result due to a single trait being wrongly perceived. Notice that the difference in frequencies is lost for highest levels of noise: perceived

traits are not correlated to real traits and consequently mate-values are randomly assigned, so that actually random mating takes place and the complexity level of the preference-landscape becomes unimportant.

Varying Evaluation Noise

In the last set of runs, evaluation noise is varied (Fig. 2d). For extreme low levels, simple preferences are favored, but as this noise is increased, more complex preferences become quickly favored. Let us first remark that in the total absence of any form of noise, simple and complex preferences are functionally similar: both are equally able to rank individuals in a group according to their traits. Nevertheless, they are not *evolutionary* equivalent: in particular, basins of attraction of local maxima get smaller as complexity increases, and the mean number of adaptive steps (available to traits when climbing preferences) decreases at the same time (from an average of $N/2$ steps for $K=0$ to an average of $ln(N)$ steps for $K=N-1$, approximately). By allowing only a small number of adaptive steps to traits, the positive effect preferences obtain when they correlate to traits they prefer can only take place a small number of times. Therefore, preferences with $K=0$ are favored in this case.

As evaluation noise increases, and the chances of less preferred traits being chosen increases in a way that is determined by the local shape of the preference-landscape (i.e., this shape determines such probability), landscapes with some ruggedness gain a functional advantage: they allow for better local discrimination.[5] At the same time, the evolutionary advantage that

---

[5]Notice that, because noise is normally distributed, slightly different mate-values (as for $K=0$) are easily swaped by this noise; in contrast, values with bigger differences (as for higher values of K) are swaped with a much lower probability (due to the bell-form of the distribution).

simplest landscapes had in the case without evaluation noise is now compromised. The reason is that although the number of possible adaptive steps remains the same, their *quality* is lowest for such landscapes: because only small differences exist in the mate-values assigned by preferences (in any small region of trait-space), choice consistency is reduced and this leads to weak preference-trait correlations. This problem affects rugged landscapes less, since they assign a wider spectrum of mate-values in a local region of trait-space and thus induce a more consistent choice. This, in turn, entails higher quality (stronger) preference-trait correlations. Fig. 2d can be seen as depicting the way in which the relative importance of differences in *quantity* of available adaptive steps decreases as their *quality* (magnitude) differs more and more in the compared landscapes. Finally, as in the case of perception noise, and for similar reasons, if levels of this noise are extremely increased, differences in fixation frequencies decrease.

Conclusion

The arguments and results presented strongly suggest that the level of "complexity" in sexual preferences influences their fixation frequencies. There appears to be a "most favored" level of complexity whose level depends on the levels of different forms of noise (perception and evaluation noise) and on mutation rate. In essence, while both high perception noise and high mutation rates favor simplest preferences (with the effect reversed for low levels of this noise), moderate and high evaluation noise favors more complex ones (with the effect reversed for very low levels of this noise).

The assumption of phenotypes concentrated in phenotype-space may of course restrict the applicability of these results. Nevertheless, the situation may be a plausible one for relatively small populations, due to the effects of genetic drift.

Another important point is the plausibility of the levels of noise. Although it is difficult to qualify some regions of "noise-space" as being more or less likely, it seems reasonable to exclude extremely low evaluation noise. Due to the many factors playing a role associated to mate evaluation as it is modelled here, including the reliability of memory (for mate-values) and the degree of consistency among repeated evaluations of identically perceived traits, it seems reasonable to assume evaluation-noises which are not very low.

Under such hypothesis, perception-noise and mutation rate can still influence the outcome (fixation frequencies). But let us just discuss the case in which mutation rate is low (the complementary discussion is trivial). It seems reasonable to think that in the case

of traits directly and clearly displayed during courtship behavior, noise in their perception should be low. This kind of traits should usually be well adapted to the targeted sensors, making them easy to examine. In such cases (low perception noise and moderate to high evaluation noise), the evolution of some level of complexity in sexual preferences is likely. Of course, the nature of some traits may make them difficult to examine; in cases in which most evaluated traits are of this kind, no matter how much evaluation noise is present, simplest sexual preferences have the highest chances to get fixed in a population.

With respect to testing the kind of hypotheses presented in this paper in real animals, it is clear that such a task would be difficult, but appears to be possible in principle. This should ideally be done for traits under low or no natural selective pressure.

Finally, speciation rates are possibly influenced by the degree of complexity in sexual preferences. Complex preferences should induce a high speciatiation probability: small changes in such preferences lead to big changes in the assigned mate values. A short time of reproductive isolation may suffice for such preferences to strongly diverge (in mate value assignation), becoming themselves a cause for reproductive isolation. This way of looking at mate choice is a potential source of new insights in this intriguing process.

References

Fisher, R. (1930). *The Genetical Theory of Natural Selection*. Clarendon Press, Oxford.

Hamilton, W. and Zuk, M. (1982). Heritable true fitness and bright birds: a role for parasites? *Science*, 218:384–387.

Kauffman, S. (1993). *Origins of Order: Self-organization and selection in evolution*. New York: Oxford University Press.

Kirkpatrick, M. (1982). Sexual selection and the evolution of female choice. *Evolution*, 36:1–12.

Kirkpatrick, M. (1987). Sexual selection by female choice in polygynous animals. *Ann. Rev. Ecol. Syst. 18 (43-70)*, 18:43–70.

Ridley, M. (1993). *The red queen: Sex and the evolution of human nature*. England: Penguin Books.

A Discussion of the Use of Artificial Life Models to Evaluate Gould's Hypothesis about Progress in Evolution

Paul Domjan

The Plan II Honors Program
The University of Texas at Austin
WCH 4.104
Austin, Texas 78712-1105
pdomjan@mail.utexas.edu

Abstract

Stephen Gould has introduced the hypothesis that progress in evolutionary history is due to contingency. Daniel Dennett has further suggested that this hypothesis could be confirmed or denied by artificial life models, in which multiple evolutionary histories can be produced and the role of contingency thus evaluated. While existing models do not allow for this evaluation, Calabretta *et al.* (1998) suggest a general form to be used in developing artificial life models to answer questions about biology. Using this form, a model is outlined which may be able to test this hypothesis.

Gould's System

In 1979, Stephen Gould and Richard Lewontin published their seminal article "The Spandrels of San Marco and the Panglossian Paradigm: A Critique of the Adaptationist Programme." This article argued that, although we have traditionally viewed all variation in traits over time as adaptive, many traits actually result from structural necessity or chance. In Gould and Lewontin's view, it is inappropriate to tell an adaptationist story for each individual trait of the organism, rather we should look at traits as largely the result of how the organism uses and changes what originally appears as a result of its form.

This simple critique is extended in Gould's 1989 discussion of the Burgess Shale, *Wonderful Life.* In *Wonderful Life,* Gould argues that, if we were to begin with the extreme diversity of multi-cellular life present at the time of the Burgess Shale and replay the tape over again and again, we would get a new collection of surviving phyla, and, thus, life would take a radically different course. Contingency plays a major role in the process of evolution, which selected the Burgess creatures that would survive. Nevertheless, one is still able to tell an adaptationist story about why certain types of creatures survived. Under this view of the Burgess shale, the appearance of intelligence and consciousness are the result of contingency in the history of life.

However, contingency and adaptationist selection are not incompatible. Rather, the issue at stake is to what degree each has contributed to the create the current collection of forms of life. The adaptationists argue that life appears in its current form entirely, or at least primarily, as a result of the specific adaptive advantage of each particular trait. Gould and Lewontin, on the other hand, argue that, while some traits may evolve for adaptive reasons, contingency is the most important factor in determining survival and many traits appear because of contingency.

In Gould's 1996 work, *Full House,* he argues that the maximum complexity, which is closely related to maximum organism size, of organisms has increased not as a result of some essential direction in evolution. This claim has two parts: (1) If size has no adaptive value, but there is a minimum viable size, then a tail of high complexity organisms will develop, and (2) in the biosphere size has no adaptive value. Organism size has a minimum, namely the smallest a prokaryote can be, but no maximum. Thus, without any particular tendency towards increased complexity, random variation will tend to produce larger, more complex organisms. However, the mode of organism complexity remains the prokaryote, who lie very close to the minimum organism size. Furthermore, there is no direct line of increase in complexity. Rather, largely unrelated organisms occasionally stumble onto the large end of the complexity scale. As such, any replay of the tape would lead to a vastly different collection of complex organisms, though not necessarily a different distribution of sizes. Thus, thus "progress," that is continual increase in organism complexity, is not a result of the adaptive value of complexity, but of the neutral value of complexity.

Dennett's Critique of Gould

Daniel Dennett (1995) has criticized Gould for not using a computer, and, as a result of his technological incompetence, ignoring the possibility that computer models of evolution produced by the field of Artificial Life might actually allow one to replay the tape of life again and again, seeing which direction evolution will take. Interestingly, Dennett does not do this either. However, were one to do the experiment, either each run will produce vastly different results, suggesting that Gould is correct, or, as Dennett theorizes, each run will find the same "Good Moves in Design Space" each time it is replayed, producing similar results each time. Thus, Dennett believes that every time the tape is replayed, certain useful traits, like consciousness, will be arrived at in some way by some species. This simple criticism opens the possibility that Artificial Life modeling could develop real instantiations of evolution, which could then be studied to evaluate Gould's criticism.

Modeling Evolution: Replaying the Tape

Artificial Life is the study of synthetic, computer based, systems which are actual instantiations of some properties of real living organisms. Thus, while 'Boids' (Reynolds 1987), an artificial model of flocking behavior, does not include any real birds, the 'boids' in it are actually flocking. That is, each 'boid,' based on its surroundings, is making decisions about what direction to move, and, as such, the 'boids' move together as a flock. This sort of model could be used to analyze and categorize flocking behavior.

Unfortunately, open ended evolution is much more complicated than flocking. The time scale of flocking is very short: either the organisms move in a flock or they do not. Furthermore, flocking is a property of small collections of organisms, not a property of a larger, time-indexed system formed by the totality of all the organisms. Evolution, however, takes place across time and is a property of the evolving system, not any smaller collection of individuals in the system. There are three steps which must be completed in order to evaluate Gould's hypothesis. First: a model of open ended evolution must be developed. Second: the model must be tested to determine the role of contingency in its evolution. Third: the model must be tested to determine whether it approximates evolution in the biosphere. We will find that, while artificial life models exhibit one sort of evolution, this is not the same sort of evolution as in the biosphere. Thus, while they tell us something about the nature of evolution, they do not help us evaluate Gould's hypothesis about the natural of evolution in the biosphere.

Modeling Evolution

Tierra (Ray 1992) is probably the most famous artificial evolving system ever developed. The system has an ancestor organism, a simple computer program, which has a start sequence, a reproductive loop, and a copy sequence. The system evolves for a period (a couple of hours at most on a relatively fast computer), develops a relatively complicated ecosystem, which includes parasitism and, occasionally, social behavior, and then stops changing significantly. While there is still some genomic change in the system, this change can all be attributed to genetic drift rather than adaptation.

Ray and Hart (1998) have developed an extended, multi-threaded (multi-cellular) version of Tierra, Network Tierra, which incorporates two different types of cells. Their initial data seems to suggest that Network Tierra suffers from the same problem as Tierra: the environment presents a problem, which the organisms then proceed to solve. Once the organisms have developed an efficient solution to the problem, adaptive evolution stops (Bedau et al. 1997a).

Although there are many artificial life models, some of the major ones being Echo (Holland 1994, 1995), Bugs (Bedau and Packard 1992), and Tierra derivatives, like COSMOS (Taylor 1997) and Avida (Adami and Brown 1994), all of these models are similar in character to Tierra. They all continue evolving until they solve the problem presented by their environment. Thus, in order use these models to evaluate Gould's hypothesis, we must determine, among other things, whether a model which only evolves to a point can tell us about evolution in the biosphere.

Determining the Role of Contingency in Evolution

In 1992, Ray formulated the following agenda for artificial life:

> Because biology is based on a sample size of one, we cannot know what features of life are peculiar to earth, and what features are general, characteristic of all life.... A practical alternative to an inter-planetary or mythical biology is to create synthetic life in a computer. "Evolution in a bottle" provides a valuable tool for the experimental study of evolution and ecology. (371-2)

While evolution on earth cannot be replayed, computer evolution can be replayed many times. However, artificial life has only recently been used to address this question. Taylor and Hallam (1998) attempt to further illuminate this issue by looking at multiple runs of COSMOS (Taylor 1997) to determine what role contingency plays in COSMOS. The authors find that, regardless of the seed, the simulation produces similar results, thus, they conclude that, in artificial life models, while contingency effects the specifics of the outcome, the general result is determined by the problems presented by the environment. In Dennett's (1995) formulation, while Stephen Gould's existence may be due to contingency, the fact that someone would inhabit his office at Harvard is not. Taylor and Hallam's results point in this general direction. However, Taylor and Hallam realize that these initial results do not shed much light on Gould's hypothesis.

Nevertheless, it may be that there are central differences between the current generation of artificial life systems and the biosphere which cause multiple runs of an artificial life model to evolve in basically the same direction, making such systems unfit to evaluate Gould's claim. In Gould's (1996) view, increased body size is the result of drift away from a left wall with little or no pressure to evolve in either direction. Thus, given time, complexity will continue to drift away from the simplicity of the modal bacter. Artificial life models, however, usually force organisms to compete for clock cycles on the computer. As such, larger organisms are significantly less fit than their better optimized cousins, because they require more clock cycles to reproduce, and, thus, these models are not a system in with size has no adaptive value, but there is a minimum size. Because these models do not fulfill the conditions of the disjunctive first part of Gould's claim, they cannot be used to evaluate this claim, and they certainly cannot be used to evaluate the empirical hypothesis of the second part of his claim. Therefore, they cannot be used to evaluate the claim. It may be that, while contingency essentially plays a major role in the development of a tail of complex organisms, leading away from the modally dominant bacterium is the biosphere, such a tail is unable to develop in an artificial model, because large organisms are significantly maladapted.

Evaluating the Models

One should be hesitant to use artificial life models to make claims about the nature of the biosphere, although, as Calabretta *et al.* (1998) show, evolutionary models can be very useful in investigating a simple property of a system. Firstly, one needs to isolate the feature. Secondly, with a sufficient understanding of the feature to begin with, one must build an artificial life model which incorporates the aspects of the feature pertinent to the problem in question. Finally, this artificial life model can be used to directly manipulate the feature and evaluate the consequences.

In Calabretta *et al.*'s case, modularity of structures and recursive design was investigated by developing neural networks, some with modules and some with the possibility of developing modules. Both sets of networks were presented with a problem, and those with modules where able to solve it faster. However, upon examining the networks which began without modules, it was found that they developed much more specific modularity, thus allowing the researchers to see how modules could be seen as functional units. This research was successful, in large part, because the structure the researchers were investigating was simple and they were able to easily isolate the aspects of it which were pertinent to their problem.

Bedau *et al.* (1997a, b) has done extensive work in classifying evolutionary system, with the aim of showing that artificial life systems are qualitatively different than the biosphere. Bedau's work is based on using a neutral variant of a simulation, in which genotype does not affect survival, to determine how long genotypes persist in the absence of selection. This data is then used to determine whether genotypes, in the presence of natural selection, persist longer than they would in the neutral model, indicating that they are persisting because they are better adapted. Using this data, Bedau *et al.* (1998) find that artificial life models either never engage in adaptive evolution, or stop changing after a point.

It is assumed that:

[T]he mere fact that a [taxonomic] family appears in the fossil record is good evidence that its persistence reflects its adaptive significance. Significantly

maladaptive taxonomic families would likely go extinct before leaving a trace in the fossil record. (Bedau *et al.* 1998: 229)

This begs the question by assuming that genomes which appear in the fossil record are evolutionarily significant. Because it is impossible to create a neutral model for the biosphere, it may be that the biosphere, like the artificial life models, also has solved the problem presented by its environment, and, now, is simply drifting through evolutionary space relative to the basic structure of its environment, meaning that, at an early stage in evolution, organisms developed solutions which allowed them to survive in the biosphere (Domjan 1999). This is not to say that there are no further evolutionary pressures, on the contrary constant threats, being eaten by tigers for example, may appear to threaten organisms. However, being eaten by tigers is not a problem of the structural features of one's environment, but a problem created by the other organisms in one's environment. Both the tiger and the animal it is trying to eat have solved the basic problem of being alive in the biosphere.

This would even be consistent with Bedau *et al.*'s assumption about taxonomic families. That is, if the biosphere has solved the problem presented by its environment, none of the surviving taxonomic families will be "significantly maladaptive." In Tierra, for example, once adaptive evolution stops, random genomic change persists, and the new genotypes which survive are not significantly maladaptive. Rather, they are not significantly better adapted than the original ancestor organism. Furthermore, the fact that the majority of organisms are bacteria further supports this possibility. If the environment of the biosphere presented a problem which required an increase in complexity to solve, bacteria would have been superseded by larger organisms. However, because this has not happened, it stands to reason that a bacteria is capable of solving the problems presented by our biosphere. We cannot dismiss Gould's claim on the basis of artificial life evidence produced by existing models because they may not be sufficiently analogous to the biosphere, either because a structural feature of their environments (i.e. competition for clock cycles) imposes significant downward pressure on organism size, or because, unlike the biosphere, they do not engage in long run adaptation. These problems show us that, using our current generation of artificial life models, we are not able, as Calabretta *et al.* (1998) did, to isolate and implement those features of evolution (i.e. the first half of the conditional part of Gould's claim) which are pertinent to the question at hand.

Towards Testing Gould's Hypothesis

As Calabretta *et al.* (1998) show, models can be useful to investigate problems in biology. In order to do this, one must isolate the problem in question. I believe that this could be done in the case of Gould's hypothesis about progress in evolution. The question to be answered is as follows:

In a system where there is no evolutionary pressure in terms of size, a system where size does not significantly affect survival, will organisms continue to grow in size, developing the tail that Gould (1996) describes, while the mode of life remains next to the wall of minimum complexity?

Tierra is not such a system, because large size is maladaptive. However, if such a model were developed, and the answer to the above question was yes, the conditional first portion of Gould's claim would be largely confirmed. The only remaining problem would be to determine whether size has a positive adaptive value in the biosphere. However, if the species at the end of the tail do not fall into a direct lineage (Gould 1996: 171-2), it would seem that this model does approximate our biosphere. Furthermore, once such a model was developed, its taxonomic lineage could be compared to that of the biosphere through a taxonomic analysis. McShea (1994, 1996) provides such an analysis of portions of the fossil record based on large-scale evolutionary trends. If the taxonomic lineage of produced by this next generation artificial life model was statically similar to that of the biosphere, as evaluated by McShea, this would add significant support to the later portion of Gould's claim.

Conclusion

Although there is no existing model which allows us to evaluate Gould's hypothesis, it might be possible to develop one. If Gould is correct, this model would also provide us with the first model of evolution which evolves complexity, providing a basis for a great deal of other research in artificial life and evolutionary biology. However, because of the time the biosphere required to evolve the complexity it currently exhibits, it is unlikely that even a correct model will provide the complexity necessary to study other problems in artificial life and evolutionary biology in the near future. Nevertheless, this appears to be a fruitful direction for both artificial life and evolutionary biology to begin to move in.

Acknowledgments

My research is supported by grants from the Dean of Liberal Arts, Plan II Honors Program, and Department of Philosophy at the University of Texas at Austin.

References

Adami, C., Belew, R. K., Kitano, H., and Taylor, C. E. (1998) (eds.), *Artificial Life VI* (UCLA, Proceedings; Cambridge, Mass.: MIT Press).

———, Brown, C. T. (1994), 'Evolutionary learning in the 2D artificial life system "avida"', in Brooks, R. and Maes, P. (1994), 377-381.

Bedau, M. A., Synder, E., Packard, N. H. (1998), 'A Classification of Long-Term Evolutionary Dynamics', in Adami *et al.* (1998), 228-237.

——— Synder, E., Brown, C. T., and Packard, N. H. (1997a), 'A Comparison of Evolutionary Activity in Artificial Evolving Systems and in the Biosphere', in Husbands and Harvey (1997), 293-301.

——— and Packard, N. H. (1992), 'Measurement of Evolutionary Activity, Teleology, and Life', in Langton *et al.* (1992), 431-61.

——— and Brown, C. T. (1997b), 'Visualizing Evolutionary Activity of Genotypes', *Adaptive Behavior*. Forthcoming.

Brooks, R. and Maes, P. (1994), *Artificial Life IV* (Cambridge, Mass.: MIT Press).

Calabretta, R., Nolfi, S., Parisi, D., Wagner, G. P. (1998), 'A Case Study of the Evolution of Modularity: Towards a Bridge between Evolutionary Biology, Artificial Life, Neuro- and Cognitive Science', in Adami *et al.* (1998), 275-284.

Cowen, G. A., Pines, D., Meltzer, D. (1994) (eds.), *Complexity: Metaphors, Models and Reality* (Redwood City, Calif.: Addison-Wesley).

Dennett, D. C. (1995), 'Tinker to Evers to Chance: The Burgess Shale Double-Play Mystery', *Darwin's Dangerous Idea* (New York: Simon & Schuster), 299-312.

Domjan, P (1999), 'Are Romance Novels Really Alive? A Discussion of the Supple Adaptation View of Life', in Floreano *et al.* (1999), 21-25.

Floreano, D. *et al.* (1999) (eds.), *Advances in Artificial Life* (Berlin: Springer-Verlag).

Gould, S. J. (1992), *Full House: The spread of excellence from Plato to Darwin* (New York: Harmony Books).

——— and Lewontin, R. C. (1979), 'The Spandrels of San Marco and the Panglossian Paradigm: A Critique of the Adaptationist Programme', *Proceedings of the Royal Society* B, 205, 581-598.

———. (1989), *Wonderful Life: The Burgess Shale and the Nature of History* (New York: W.W. Norton).

Holland, J. H. (1994), 'Echoing emergence: objectives, rough definitions, and speculations for echo-class models', in Cowen, G. A. *et al.* (1994).

————. (1995), *Hidden Order: How Adaptation Builds Complexity* (Helix Books).

Husbands, P., and Harvey, I. (1997) (eds.), *Proceedings of the Fourth European Conference on Artificial Life* (Cambridge, Mass.: MIT Press).

Langton, C., Taylor, C., Farmer, J. D., and Rasmussen, S. (1992) (eds.), *Artificial Life II* (Santa Fe Institute Studies in the Science of Complexity, Proceedings, 10; Redwood City, Calif.: Addison-Wesley).

McShea, D. W. (1994), 'Mechanisms of large-scale evolutionary trends', *Evolution* 48: 1747-1763.

————. (1996), 'Metazoan complexity and evolution: is there a trend?', *Evolution* 50: 477-492.

Ray, T. S. (1992), 'An Approach to the Synthesis of Life', in Langton *et al.* (1992), 371 408.

———— and Hart, J. (1998), 'Evolution of Differentiated Multi-threaded Digital Organisms', in Adami *et al.* (1998), 295-304.

Reynolds, C. W. (1987), 'Flocks, Herds, and Schools: A Distributed Behavioral Model', Proceedings of SIGGRAPH '87, *Computer Graphics* V 21/4: 25-34.

Taylor, T. (1997), "The COSMOS artificial life system', Working Paper 263, Department of Artificial Intelligence, University of Edinburgh. Available from http://www.dai.ed.ac.uk/daidb/people/homes/timt/papers/.

———— and Hallam, J. (1998), 'Replaying the Tape: An Investigation into the Role of Contingency in Evolution', in Adami *et al.* (1998), 256-265.

Should seeds fly or not?

Tatsuo Unemi*

Department of Information Systems Science
Soka University, Hachiōji, Tokyo, 192-8577 Japan

Abstract

This paper describes some results of our computer simulation concerning ecological competition on the target area of seed dispersal. It is better for any kinds of plants to disperse their seeds as far as possible because it might spread in new frontier earlier than the others. But, it would be better to put the seeds down just at the neighbor position when the environmental condition is stable. From drawing a fitness landscape for distance and area of dispersal through a computer simulation, it was revealed that both of these strategies are locally optimal to gain more reproductive success, and neighboring strategy is the best when the environment is unchanged and uniform. We examined a type of evolutionary process to investigate the effects of three kinds of environmental parameters, scale of disturbance, death-sprout ratio, and geographical granularity of fertility. The population initialized by random parameters converged into either or both of two types of species, far and broad dispersal and neighboring reproduction. For all of three parameters, the experimental results showed that the probability to converge into dispersal of longer distance becomes greater corresponding to the degree of environmental change in time and space.

Introduction

Known as *seed dispersal*, some types of plants are facilitated to distribute their seeds efficiently using animals, birds, wind, stream of river, and so on (Howe 82; Ueda 99). Burrs cling to animals' fur. Birds and monkeys eat fruits but the seeds are excreted at the other place. Seeds of dandelion fly in the wind. Walnuts and coconuts drift with the stream and the current. A lot of biologists have investigated many types of dispersal strategies from view points of seed morphology, symbiosis between plants and animals, and evolutionary ecology.

As summarized in (Howe 82), it seems reasonable that the further and broader area seeds can reach the more adaptive against environmental changes, because it provides advantage to be able to occupy new frontier faster. On the other hand, it is also the fact that the other types

* Currently staying at AI Lab., IFI, University of Zürich, Winterthurerstr. 190, CH-8057 Zürich, Switzerland.

of strategies for plant propagation use bulbs, stolons, rootstocks, and so on. These are to propagate new plants near around the old one rather than far away. It is also explainable by the probability of good environmental condition for sprout and growth at the position where a seed reaches. It is unreliable in the other place far away from the ancestor. The fact that both strategies exist in plants *as-we-know* suggests both of them are the candidates of the optimal solutions for efficient propagation. If we accept this hypothesis, it would be an interesting issue to illustrate the fitness landscape of seed dispersal region under some conditions. Some models of seed dispersal have already been proposed in such as (Chambers 94). They are based on the field observation to build a diagram of the effects among seed production, dispersal, germination, and death from the view point of ecology, but are not used to draw the fitness landscape from the view point of evolutionary theory. This paper provides a hint to consider the selective forces that produces sophisticated morphology for seed dispersal.

The following part of this paper describes our design of the model of plant propagation, drawing of the fitness landscape via competitions between two species that disperse seeds into the areas of different distance and width, and then results of our simulation of evolutionary process concerning the effects of three kinds of environmental parameters, scale of disturbance, death-sprout ratio, and geographical granularity of fertility.

Model of propagation

Our design of the model of propagation is as follows. A plant occupies a circle of constant diameter d on the ground, two dimensional plane of continuous Euclidean space. In the initial state, a number of plants are placed in random positions so as not to collide each other. In each simulation step, each of plants produces one seed with probability P_s and dies out with probability P_d. The position of seed is determined from two parameters associated with the ancestor plant. The seed sprouts and grows in diameter d if there is no other plant within the distance d. The order of execution among plants is randomly shuffled in each step. The region of seed

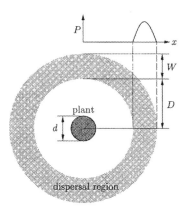

Figure 1: Dispersal region of seeds from a plant in our model.

dispersal for a plant is shaped as shown in Figure 1. It is represented by shortest distance D and the width W. The probability $P(x)$ to select a distance x is defined by the following equation.

$$P(x) = \begin{cases} \dfrac{\pi}{2W} \sin \dfrac{x-D}{W}\pi & \text{if } D < x < D+W \\ 0 & \text{otherwise.} \end{cases} \quad (1)$$

The above probability is practically realized in our simulator by calculating the value of x using following expression.

$$x := \frac{\arccos(2u-1)}{\pi} \quad (2)$$

where u is a random number of uniform distribution within $[0,1)$. The orientation is determined using a random number of uniform distribution.

Both of D and W are inherited from a mother plant to daughter seeds. Each of our experiments starts with 3,000 plants at random positions. We assume the field of propagation is a square space of which length of the edge is 100 times of plant's diameter d, and is formed as a torus to prohibit the effects of the boundaries, that is, the upper and lower edges and the left and right edges are connected respectively.

Drawing a fitness landscape

To investigate the shape of the fitness landscape on dispersal region, we examined competitions between two species of different D and W exhaustively for $D = 1, 2, \ldots, 7$[1] and $W = 0, 1, \ldots, 6$, totally $(7 \times 7) \times (7 \times 7 - 1)/2 = 1176$ matches. The other parameters are set up as $P_s = 1$ and $P_d = 0.1$. We stop each of matches when either species is extinguished or 1,000 steps passed. Figure 2 summarizes the result of matches, which indicates

[1] Practically, we used 1.01 instead of 1 for the value of D to avoid erroneous collision among plants caused by error of numerical computation.

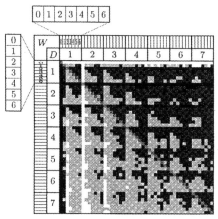

☐ The left species extinguished the upper one within 1,000 steps.
▨ The left species dominated the upper one at the 1,000th step.
▩ The upper species dominated the left one at the 1,000th step.
■ The upper species extinguished the left one within 1,000 steps.

Figure 2: Result of matches between species of different parameters for dispersal region.

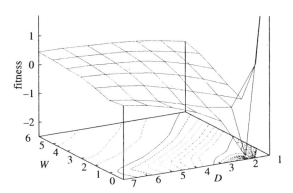

Figure 3: Fitness landscape drawn from the result of ten times of exhaustive matches.

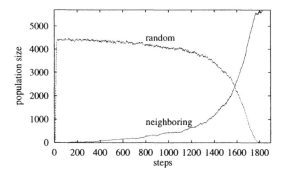

Figure 4: Changes of population sizes in the competition between neighboring and random strategies. One plant for each species is placed in the field at the initial state.

$D\backslash W$	0	1	2	3	4	5	6
1	5.54	0.43	−1.05	−0.80	−0.49	−0.21	−0.04
2	−2.37	−0.95	−0.76	−0.46	−0.19	−0.01	0.13
3	−1.03	−0.81	−0.50	−0.17	0.05	0.16	0.25
4	−0.59	−0.55	−0.22	0.04	0.20	0.28	0.33
5	−0.27	−0.30	−0.01	0.18	0.27	0.34	0.38
6	−0.06	−0.12	0.12	0.28	0.34	0.39	0.43
7	0.08	0.01	0.19	0.33	0.37	0.42	0.44

Table 1: Fitness of species calculated by exhaustive matches.

larger values of both D and W tend to make it tougher except the cases $(D, W) = (1, 0)$ and $(1, 1)$.

To draw the fitness landscape for D and W, we gives h points to the winner defined by the following equation.

$$
h = \begin{cases} \dfrac{1000}{M} & \text{if } M < 1000 \\[2mm] \dfrac{N_w - N_l}{N_w + N_l} & \text{otherwise,} \end{cases} \tag{3}
$$

where M is the number of steps until the match stops, and N_w and N_l are the number of plants at the final step of the winner and the loser respectively. The loser loses the same amount of points after the match. Table 1 shows the average points of each species after ten times of exhaustive combinations by separated random number sequences, and Figure 3 shows its shape for intuitive understanding. It would be reasonable that further and broader region of seed dispersal provides more reproductive success, because it increases a chance to put the seed at an appropriate position.

However, it might seem strange that *neighboring* strategy that puts the seeds at just adjoining side, that is $(D, W) = (1, 0)$, is the best, even though all of the environmental conditions are uniform around the field. There is no evidence of unreliability to sprout and grow anywhere. The fact is that neighboring strategy has beaten all of other species as shown in the upper and left edges of Figure 2. To investigate the process that the neighboring strategy extinguishes others, we examined a match with random strategy that puts seeds at random positions. Figure 4 shows the changes of the number of plants starting from *one-by-one* to the opponent's extinction. The random strategy propagates fastly on the early stage, but the neighboring strategy gradually broadens its territory as shown in Figure 5. The population size of the latter one increases exponentially since the probability of interference by the former one decreases proportionally to that population size itself. One of the reasons why the latter one beats the former one is because placement of a child at neighboring side realizes the minimum distance between plants and leads to higher density of occupation. Higher density brings more reproductive success. The data from

our simulation to draw Figure 4 supports this explanation as the maximum size of random and neighboring populations were respectively 4483 and 5713.

The result that the neighboring strategy is the best does not coincide with the phenomena in the nature that we know. Sophisticated mechanisms for seed dispersal could never appear through the evolutionary process if the similar phenomena to our simulation had occurred generally on the earth. The following section gives consideration on some environmental parameters that affects the fitness landscape and provides more reproductive success for long distance dispersal over the neighboring strategy.

Effects of environmental parameters

We designed an evolutionary process to investigate the effects of some environmental parameters, to reduce the CPU time relatively to the above exhaustive method. Each of D and W is represented in a 16 bits unsigned integer that is copied from mother to daughter erroneously under a mutation rate μ, the probability to flip a bit for each. We set $\mu = 0.001$ in our experiments. Each integer is proportionally transformed into a floating point number in $[0, 5]$ from the integer in $[0, 2^{16} - 1]$. We examined 20 trials of separated random number sequences for each parameter settings. Each trial starts with 3,000 plants with random positions and random integers for D and W using uniformly distributed random numbers produced by `drand48` Unix library function.

Scale of disturbance

The advantage of seed dispersal is ability to reach and occupy any new frontier faster, that is, larger D and W would gives more reproductive success than neighboring strategy when large-scale disasters occur frequently. As simulating disastrous disturbance, we introduce a procedure to kill all of the plants within a size of circular area placed randomly with a constant frequency. We examined a variety of diameter A of disturbance area that occurs once per ten steps to investigate the effects of the scale of disturbance. Figure 6 shows the result of the simulation for $A = 10, 15, \ldots, 40$. It indicates the

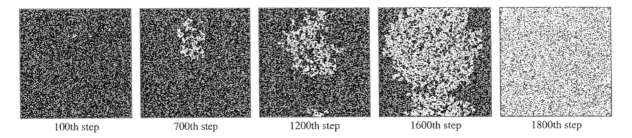

| 100th step | 700th step | 1200th step | 1600th step | 1800th step |

Figure 5: Changes of spatial distribution of plants in the competition between neighboring and random strategies. A black circle indicates a plant of random strategy, and a white circle indicates a plant of neighboring strategy.

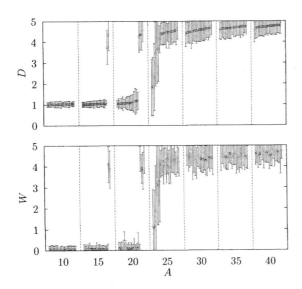

Figure 6: Average values and standard deviation of D and W among plants in the population at 5,000th step with 20 cases for each value of the size of disturbance area A.

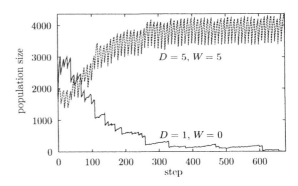

Figure 7: Changes of population sizes in the competition starting from 1,500 plants for each. The diameter of disturbance area $A = 30$ and it occurs once per ten steps.

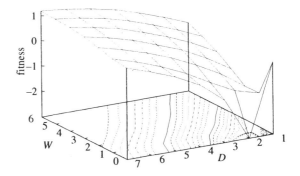

Figure 8: Fitness landscape drawn from the result of ten times of exhaustive matches in $A = 30$.

effect of small scale ($A = 10$) of disturbance is negligible, but large scale ($A \geq 30$) of disturbance exchanges the positions of two strategies. This result supports our prediction.

To see the process of population changes, we examined a competition without mutation between two strategies, $(D, W) = (1, 0)$ and $(5, 5)$. The match starts from 1,500 plants at random positions for each. Figure 7 shows the changes of population sizes when $A = 30$. The result indicates that the sizes of both species shrink at every occurrence of disturbance, and the wider strategy rapidly recovers the diminution, but it is difficult for neighboring strategy. Figure 8 illustrates the shape of fitness landscape drawn by same method described in the previous section, that is, ten times of exhaustive matches. Comparing with Figure 3, it is clear that the fitness of neighboring strategy decreases and further and broader

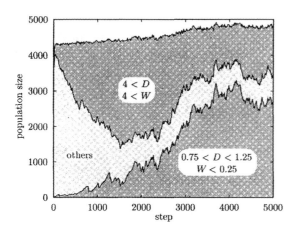

Figure 9: Changes of population sizes in a case of $A = 25$ on three species, neighboring strategy ($0.75 \leq D < 1.25$ and $W < 0.25$), wider dispersal ($4 \leq D$ and $4 \leq W$), and others.

dispersal has more fitness than that.

When $A = 15$ and 20, the population alternatively converges into either strategy. More than half number of cases in $A = 25$ fell into wider dispersal, but the other cases proceeded to mixture of the two optimal species as shown in Figure 9 that illustrates population sizes of three species, $0.75 \leq D < 1.25$ and $W < 0.25$, $4 \leq D$ and $4 \leq W$, and others.

Some readers might be interested in seeing what happen when the frequency is changed. It is obvious that low frequency has small effect but high frequency makes the same effect of large-scale disturbance. The expected damage by disturbance would be proportional to both size and frequency. It would be the similar effect for same value of A/T where T is the number of interval steps. But if both A and T are large, the pattern of population changes would be largely fluctuated. The sudden extinction of neighboring strategy tends to occur, for example.

In the real world, size and frequency of disturbance is various depending on the type such as a flood, storm, landslide, earthquake, volcano's explosion, and so on. It causes fluctuation of the fitness and has affected the evolutionary process of the vegetable kingdom on the earth.

Death-sprout ratio

The reason why the neighboring strategy wins is high probability of a chance that plants touches each other. This probability becomes low if mother plants disappear sooner relatively to the sprout of daughter seeds. We examined evolutionary process for various values of probability of death P_d to see the effect of rapid death of plants. Figure 10 shows the results for $P_d =$

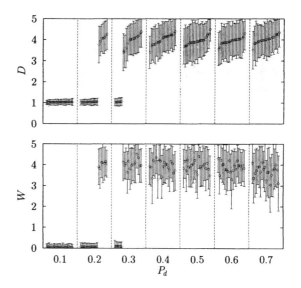

Figure 10: Average values and standard deviation of D and W among plants in the population at 5,000th step for a variety of the probability of death P_d.

$0.1, 0.2, \ldots, 0.7$.

As we expected, high probability of death ($P_d > 0.4$) causes convergence to long distance dispersal. This tendency would accelerate the ability to move to a more fertile field from a barren place. It might have been a factor for acceleration of evolutionary changes. The reason why it is difficult to converge into coexistence of two separated optimal solutions for any value of P_d is that the environmental condition is quite uniform around the field in contrast with the case of disturbance examined in the previous section.

Geographical granularity of fertility

In the real seed, both inner and environmental condition determines the possibility of sprout and growth. The environmental condition includes not only the distance to the other plants but also fertility around the seed. It is difficult to measure the degree of real fertility, but at least it is obvious that the distribution is uneven over the field in any granularity, depending on soil, rocks, water, slopes, and so on. Saying with other words on the reason why the neighboring strategy wins, it is because daughter seeds always grows at the neighbor side of the mother plant where the condition should be good. If the granularity of fertility is too fine relatively to the size of plant, this condition would not be satisfied since it loses a guarantee of fertility around mother plants. To certify this hypothesis, we introduce uneven distribution of probability of seed production P_s on the field, by placing a same number of random points p for $P_s = 1$ and 0, and

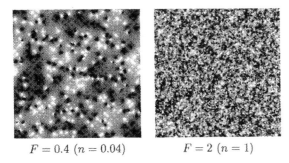

$F = 0.4 \ (n = 0.04)$ $F = 2 \ (n = 1)$

Figure 11: Example distribution of the seed production rate P_s to test the effect of geographical granularity of fertility.

using the following equations for smooth interpolation.

$$P_s(a) = \begin{cases} P_s(p) & \text{if } a = p \\ \dfrac{\sum_p P_s(p) \cdot w(a,p)}{\sum_p w(a,p)} & \text{otherwise} \end{cases} \quad (4)$$

$$w(a,p) = \frac{\prod_q \text{dist}(a,q)^2}{\text{dist}(a,p)^2} \quad (5)$$

where $P_s(a)$ is the seed production rate at position a, and $\text{dist}(a,p)$ is the Euclidean distance between a and p.

Here we denote the fineness of granularity by $F = 2\sqrt{n}$ where n the number of points p per unit area. Unit area is a square of which edge has same length with the diameter of plant. $1/F$ is theoretically the average value of distances from each point p to the nearest other point[2]. To reduce the computation cost, we use only the points within $\text{dist}(a,p) < 8/F$ for each a. Figure 11 shows examples of the distribution in the field when $F = 0.4$ and 2.

Figure 12 shows the results of evolutionary process for $F = 0.2, 0.4, \ldots, 2.4$. As we predicted, further and broader dispersal surely gains advantage when the granularity is fine ($F \geq 2$). However, the different phenomenon is observed for more even fertility in comparison with the previous two cases. Figure 13 shows the distribution of average values of D and W at 5,000th step, which indicates that combination of small D and large W is also good but combination of large D and small W is not. For more precise analysis, we examined the exhaustive matches again for $F = 0.8$ and 1.6. Figure 14 shows the fitness landscapes drawn from the results. In contrast with the cases described above, large value of D is not good though large value of W is still good when $F = 0.8$ except near values of neighboring strategy. This means that dispersal to wide region provides more reproductive success but it needs to fall seeds also near the

---

[2] The proof is omitted because it is too long. It is trivial that the average value of distances between the nearest points is proportional to $1/\sqrt{n}$ because of the dimension.

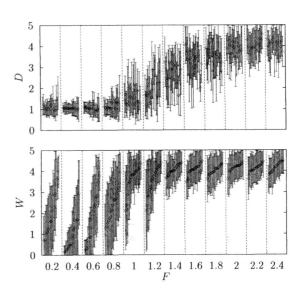

Figure 12: Average values and standard deviation of D and W among plants in the population at 5,000th step for a variety of the geographical granularity of fertility F.

mother plant under relatively sparse granularity. The similar effect can be observed even when $F = 1.6$, but large value of D still has enough advantage.

On the real plants, size of plant is different among species. The above result suggests that it is better for a large plant to disperse the seeds in a further and broader region, because granularity of any type of geographical distribution is relatively fine for it. The opposite tendency might be concluded for a small plant, but it would be hard because a small size of community of small plant is easily extinguished by disturbance in the real environment.

Conclusion

We drew a fitness landscape for distance and width of seed dispersal through a computer simulation. It revealed that both of neighboring and wide range strategies are locally optimal to gain more reproductive success. Contrary to the phenomena in the real world, the neighboring strategy is the best in our first result. We examined a type of evolutionary process to investigate the effects of three kinds of environmental parameters, scale of disturbance, death-sprout ratio, and geographical granularity of fertility. For all of three parameters, the experimental results showed that the probability to converge into dispersal of longer distance becomes greater corresponding to the degree of environmental changes in time and space. Sophisticated mechanisms for seed dispersal are thought to be produced through evolutionary process in changing environment as

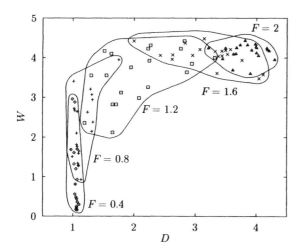

Figure 13: Average values of D and W among each population at 5,000th step for a variety of F.

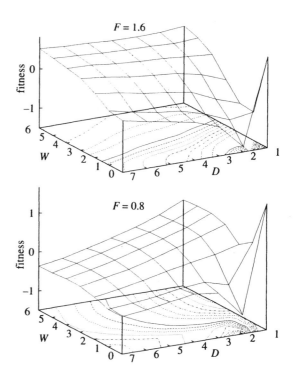

Figure 14: Fitness landscape drawn from the result of ten times of exhaustive matches in $F = 1.6$ and 0.8.

explained since more than one hundred years ago. The results supports this explanation and shows quantitative characteristics on the balance between these two opposite strategies. Some types of plants have both abilities to realize neighboring and wide range dispersal, such as a lawn, field horsetail, bamboo and so on. This type of function is thought to have evolved under the fitness landscape with two peaks.

Migration of organisms affects evolutionary process as an explanation of large scale evolution, as suggested in (Eldredge 89). Though plants don't move by itself differently from animals, seed dispersal realizes migration of plants for long distance, sometimes over the sea. From this point of view, understanding on the characteristics of seed dispersal could help our understanding on the evolutionary process of organisms.

Considering the real community of plant, there are various size of plants, various number and size of seeds, the cost of long distance dispersal, vertical structures, pollination, geographical boundaries, symbiosis with animals and insects, and other complex relations among all entities in the nature, that we ignored in the simulation. Phenotipic diversity of plants and coexistence of a variety would be caused by these and other factors. It would depend on the objective to decide which factors we should consider in the next work. Specially, local dispersal has disadvantage caused by parent-offspring and inner offsprings conflict. To consider this type of effect, we should introduce the variable size of occupation area determined through competition among plants and seedlings.

We hope this study could be a milestone for Artificial Life approaches to understand some side of evolutionary and ecological characteristics of life.

Acknowledgement

The author thanks the reviewers for providing a lot of helpful and valuable comments from various points of view. He also would like to thank the organizers and participants of the domestic workshop of Artificial Life held at Nagoya University in February of this year for their useful comments on the previous work.

References

Chambers, J. C. 1994. A day in the life of a seed: movements and fates of seeds and their implications for natural and managed systems. *Ann. Rev. Ecol. Syst.* 25: 263–292.

Eldredge, N. 1989. *Macroevolutionary Dynamics – Species, Niches, and Adaptive Peaks.* McGraw-Hill, Inc.

Howe, H. F. and J. Smallwood. 1982. Ecology of seed dispersal. *Ann. Rev. Ecol. Syst.* 13: 201–228.

Ueda, K. (ed). 1999. *Seed dispersal,* Tsukiji Shokan (in Japanese).

IV Robots and Autonomous Agents

Connecting Brains to Robots: The Development of a Hybrid System for the Study of Learning in Neural Tissues

Bernard D. Reger[1], Karen M. Fleming[1], Vittorio Sanguineti[2], Simon Alford[3]

and Ferdinando A. Mussa-Ivaldi[1]

[1]Department of Physiology, Northwestern University Medical School, Chicago, IL 60611
[2]Dipartimento di Informatica Sistemistica e Telematica, Universit di Genova, Italy
[3]Department of Biological Sciences, University of Illinois at Chicago, Chicago, IL 60607

Abstract

We have developed a hybrid neuro-robotic system based on a two-way communication between the brain of a lamprey and a small mobile robot. The purpose of this system is to offer a new paradigm for investigating the behavioral, computational and neurobiological mechanisms of sensory motor learning in a unified context. The mobile robot acts as an artificial body that delivers sensory information to the neural tissue and receives command signals from it. The sensory information encodes the intensity of light generated by a fixed source. The closed-loop interaction between brain and robot generates autonomous behaviors whose features are strictly related to the structure and operation of the neural preparation. In this paper we provide a detailed description of the hybrid system and we present experimental findings on its performance. In particular, we found (a) that the hybrid system generates stable behaviors; (b) that different preparation display different but systematic responses to the presentation of an optical stimulus and (c) that alteration of the sensory input lead to short and long term adaptive changes in the robot responses. The comparison of the behaviors generated by the lamprey's brainstem with the behaviors generated by network models of the same neural system provides us with a new tool for investigating the computational properties of synaptic plasticity.

Introduction

Since its inception, robotic science has given great contributions to the study of motor learning and control in humans and other biological systems (Hildreth and Hollerbach 1987). The most notable contribution has been the determination of what interesting computational problems must be solved by the brain as well as by an intelligent machine when either one must control the mechanical interaction between limbs and environment. Theories concerning what computational problems must be solved by an intelligent system have been called "competence" theories (Marr 1982) to distinguish them from "performance" theories, concerning the physical processes that are actually chosen to solve a problem. In this paper, we present a first attempt to utilize a robotic system for investigating the neural processes underlying sensory motor adaptation, that is for understanding a distinctive feature in the performance of biological systems. Our goal is to develop a computational and experimental framework for relating the neurobiological study of neural plasticity-the modification of neuronal excitability following past experience of input and output patterns- to the behavioral functions that are supported by neural plasticity.

The framework that we have developed is a hybrid system, which establishes a two-way signal interaction between a mobile robot and brain tissue maintained alive *in vitro* from the reticular formation of the lamprey- a primitive eel-like vertebrate. In this experimental arrangement, the brain and the robot are interconnected in a closed loop. They communicate through an interface that transforms (a) light information from the robot's optical sensors into electrical stimulation to the lamprey's brainstem, and (b) recorded neural activity from two brainstem nuclei into motor commands to the robot's wheels (Figure 1). We have chosen the lamprey for this first study because of the easy access in this preparation to a system of very large neurons- the Muller cells in the reticular formation- that integrate command and sensory signals directed to the spinal motor centers.

From the standpoint of a neurobiologist, this neuro-robotic system can be regarded as a system for complementing the electrophysiological study of neuronal properties with an artificial behavioral context. We must stress the adjective "artificial", because the signals that normally would travel along the circuits that we are stimulating are signals of vestibular rather than

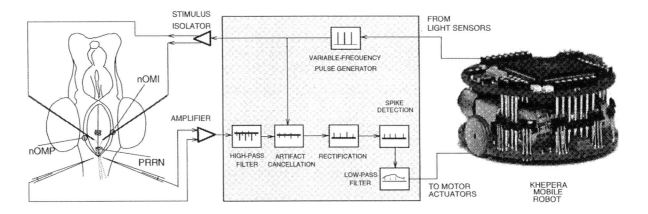

Figure 1: Robot Signal Flow Block Diagram. The neuro-robotic interface (shaded center region) translates the light sensor data from the robot (right) into a stimulation pattern for the lamprey preparation (left). The neural response is converted into motor signals by the interface.

visual origin. In this brain stem preparation we have selected a portion of neural circuitry that in normal circumstances combines vestibular signals and motor commands to stabilize the orientation of the body during swimming (Rovainen 1979; Deliagina 1992a; Deliagina 1992b; Orlovsky, Deliagina, and Wallén 1992). This system has been shown to be adaptive, as unilateral lesions of the vestibular capsules are followed by a slow reconfiguration of neural activities until the correct postural control is recovered (Deliagina 1995; Deliagina 1997). In our hybrid system, vestibular signals are replaced by light intensity signals. As the vestibular signals have a right and left source- the two vestibular capsules - so do the two light intensity signals originating from sensors on the right and left side of the robot. Therefore, the natural stabilizing behavior, in which the lamprey would track the vertical axis, corresponds, in the hybrid system, to a positive phototaxis, that is a tendency of the robot to track a source of light. We are convinced that the properties of the information processing associated with natural behaviors may be explored by observing the information processing associated with the artificial behavior. This, in a way, is a consequence of the abstract and generalized nature of information. An obvious advantage of our hybrid system, always from the point of view of experimental neurobiologists, is that, unlike natural motor behaviors, artificial behaviors do not interfere mechanically with the electrophysiological setup. In any study involving intra- or extracellular recording, even the slightest motion of the tissue tends to cause unwanted displacements of the electrodes.

From the perspective of neural computation, the hybrid system provides a means to test models of information processing by direct interaction with a biological neural network. As we detail in the methods section, the behavior of the robotic system is described by a relatively simple- and yet nonlinear- system of differential equations. To the extent that the brain properties may be considered stationary (over the time scale of robot movements), these equations describe an autonomous system whose properties are modulated by the structure of the neural pathways and connections intervening between stimulating and recording electrodes. Conversely, the observation of the sensory-motor behaviors that emerge from this system offer an insight into the computational structure of the neural system. The search for such an insight is what drives our study.

Here, we report three initial findings of this study. First, we have succeeded in obtaining stable behaviors over extended periods of time, characterized by repeatable motor responses to a light source. Second, in different preparations, we have observed different responses ranging from light tracking to light avoidance. Through simple simulations, we show how these different responses may be readily accounted for by different patterns of connectivity between stimulation and recorded signals. Finally, we have observed plastic adaptive changes following the unilateral alterations of the sensory inputs. These findings provide supporting evidence for the use of neuro-robotic systems in the study of the neurobiological mechanisms of sensory motor learning.

Methods

In this section we describe the components of our hybrid neuro-robotic system, the experimental setup used to assess its performance, and the basic computational model that characterizes the system's behavior.

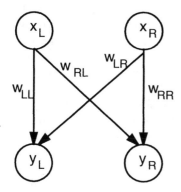

Figure 2: Simple two layer neural network with two inputs, two outputs, and four weights.

The neural preparation

The neural component of the hybrid system is a portion of the brainstem of the Sea Lamprey in its larval state. In larvae of Sea Lamprey Petromyzon marinus, anesthetized with tricane methanesulphonate (MS222, 100-200 mg/l), the whole brain was dissected and maintained in continuously superfused, oxygenated and refrigerated (9-11°C) Ringer's Solution (NaCl, 100.0 mM; KCl, 2.1 mM; CaCl2, 2.6 mM; MgCl2, 1.8 mM; glucose, 4.0 mM; NaHCO3, 25.0 mM); details in Alford et. al. (Alford, Zompa, and Dubuc 1995).

We recorded extracellularly the activity of neurons in the a region of the reticular formation, a relay that connects different sensory systems (visual, vestibular, tactile) and central commands to the motor centers of the spinal cord. We placed two recording electrodes in the right and left Posterior Rhombencephalic Reticular Nuclei (PRRN). We also placed two unipolar tungsten stimulation electrodes among the axons of the Intermediate and Posterior Octavomotor nuclei (nOMI and nOMP). These nuclei receive inputs from the vestibular capsule and their axons form synapses with the Rhombencephalic neurons on both sides. The impedance of the stimulation electrodes ranged between 1 and 2 MΩ. Recording electrodes were glass micropipettes filled with 1M NaCl (1.5-10 MΩ impedance). The recorded signals were acquired at 10kHz by a data acquisition board (National Instruments PCI-MIO-16E4) on a Pentium II 200MHz computer (Dell Computer Corp.).

Electrode placement

While the axons of the nOMI remain ipsilateral, those of the nOMP cross the midline. As a result, the activity of one vestibular capsule affects both the ipsi- and contralateral reticulo spinal (RS) nuclei. We placed each stimulating electrode near the region in which

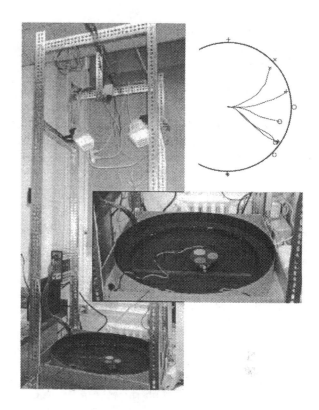

Figure 3: Robot setup. Using a pattern of colored circles (lower right), the overhead camera tracks the robot. Trajectories are plotted, each symbol representing a target light. (upper right).

the axons of the nOMI and nOMP cross (Figure 1). This placement of the electrodes also induced predominantly excitatory responses in the downstream neurons. The recording electrodes were placed on either side of the midline, near the visually identified neurons of the PRRN. To verify the placement of the stimulating electrodes we delivered brief single stimulus pulses (200μs) and observed the response in both the ipsi- and contralateral PRRN neurons. Once it was determined that the stimulation electrodes were properly placed, the recording electrodes were moved caudally in order to pick up population spikes.

The robot

The robot system is the base Khepera module (K-Team). Its small size allows us to use a small workspace (Figure 3). A circular wall was constructed with a 2 foot diameter and then painted black to reduce the amount of reflected light. Placed along the circumference of the robot are eight sensors each providing proximity and light intensity information. The sensors are located on opposite sides of the robot's midline at 10°, 45°, 85°, and 165° from the front position. Two wheels provide a means of locomotion for the small robot. Our computer system communicates with the robot through the serial port and a custom designed LabVIEW© application. Eight lights are mounted at the edge of the robot workspace at 45° intervals. The lights were numbered one through eight moving counter clockwise with light number one located at the right most position (0°). The lights are computer controlled using the digital outputs of our acquisition card. These lights generate the stimulus that elicit a phototactic response.

The interface

The interface acts as an interpreter between the neural signals and the robot control system (Figure 1). It is responsible for the transformation of the robot's light sensor information into vestibular inputs and then processing in real time the neural activity of the reticulospinal nuclei and translating it into motor commands for the robot.

Stimulus The light intensities detected by the robot sensors determine the frequencies at which the right and left vestibular pathways are stimulated. As stated above, there are eight light sensors on the robot. We weighted the sensors to give the greatest strength to sources of light that come at 45° and to ignore the rear sensors. The weighted sum of the sensors on each side is multiplied by a gain factor which determines the maximum stimulation frequency. The final result is the frequency at which we stimulate each side. We use the digital counter on the acquisition board to generate a pulse train. This pulse train is delivered to the neural preparation by the tungsten electrodes after passing through ISO-Flex stimulus isolators.

Neural Response The spiking activity of the PRRN as recorded near the axons is analyzed through a five step process. The signal picked up by the recording electrodes contains a combination of spikes, stimulus artifacts, excitatory and inhibitory postsynaptic potentials (PSP) and noise. To suppress the slow PSP components, this signal is first put through a high pass filter (cutoff at 200 Hz). The output of this filter contains high frequency noise, stimulus artifacts, and the spikes generated by multiple neurons in the vicinity of the electrode. Stimulus artifacts are canceled by zeroing the recorded signals over temporal windows of 4 ms following the delivery of each 200 μs stimulation pulse. The remaining signal is rectified, and a threshold is applied to separate the spikes from the background noise- under the assumption that the spike amplitude is much larger than the noise amplitude. The resulting train of spikes, is put through a low pass filter (5 Hz) which effectively generates a rate coded signal. The mean of this signal is used as a control signal for each of the robot's wheels.

The interface is calibrated so as to account for random differences between the recorded responses from the left and right side of the brainstem. Indeed, the net intensity of the signal picked up by each electrode depends on a number of uncontrollable factors, such as the actual distance from signal sources. To compensate for these random factors, we make the working assumption that when both left and right sides are stimulated at the same frequency, the same motor response should be obtained on each side of the robot. This corresponds to stating that all initial asymmetries between right and left side are artifacts. Accordingly, all initial difference between right and left responses to the same right and left signals are balanced by regulating two output gains.

In most cases, the right and left sides of the neural preparation were connected both in input and in output with the corresponding sides of the robot (direct mode). However, as discussed below, in some cases it was necessary to implement a reverse mode option. When connected in reverse mode, the right recording electrode is connected through the interface to the controller of the left wheel and vice versa.

Movement acquisition

The robot position and orientation is sampled and acquired using an overhead color camera (Ultrak STC-630A). The image frames are analyzed using a Newton Research Labs Cognachrome 2000 Vision System. The Cognachrome vision system is capable of simultaneously tracking up to three different colors. We have chosen a blue, red, and pink colored circle arranged in an equilateral triangle (Figure 3). The Cognachrome system captures video frames at 60 Hz, and then each frame is analyzed to determine the center and area of each colored centroid. If all three centroids are visible, the orientation of the robot is calculated and the mean of all centers is calculated to determine the center of the robot. If the area of a centroid drops below a specified amount, the remaining two centroids are used to determine both the position and orientation. This reduces the probability that the position and orientation are lost due to partial occlusion of the set of centroids.

Trajectories induced by the same light stimulus were quantitatively compared using a figural distance measure (Conditt, Gandolfo, and Mussa-Ivaldi 1997). The figural distance between two trajectories, A and B, is based on the repeated measure of the Euclidean distance between each point in one trajectory and all the points in the other. If the trajectory A has n points, {A(1), A(2), . . . , A(n)}, and the trajectory B has m points, {B(1), B(2), . . . , B(m)} , then one derives the n-dimensional vector

$$dist_{A-B}(i) = \min_{j} (\| A(i) - B(j) \|)$$

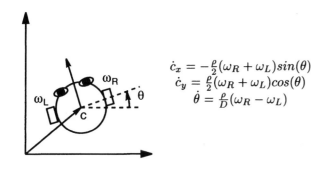

$$\dot{c}_x = -\frac{\rho}{2}(\omega_R + \omega_L)sin(\theta)$$
$$\dot{c}_y = \frac{\rho}{2}(\omega_R + \omega_L)cos(\theta)$$
$$\dot{\theta} = \frac{\rho}{D}(\omega_R - \omega_L)$$

Figure 4: The dynamics of the robot are described by three nonlinear first order differential equations.

and the m-dimensional vector

$$dist_{B-A}(j) = \min_i (\| A(i) - B(j) \|)$$

Then, the figural distance between A and B is defined as

$$\varepsilon(A, B) = \sum_i (dist_{A-B}(i) + dist_{B-A}(j))/(m + n)$$

The figural distance between two trajectories is a symmetric measure of the difference between the shapes of the respective paths. In each experimental set we considered movements to five different targets. Then, we constructed a net figural distance between two sets by summing the figural distances between trajectories to the same lights.

Simulation

To simulate the artificial behaviors generated by the cyborg we consider the interaction of three systems: a) the robot's motor system b) the robot sensory system and c) the lamprey's brain.

Robot motor system. The dynamics of the mobile robot are described by a system of three nonlinear first-order differential equations (Figure 4a). Here, (c_x, c_y) are the coordinates of the Khepera's center with respect to a fixed laboratory frame, θ is the angle of the line passing through the wheels (the axle) with respect to the x-axis of the same frame, ρ is the wheel radius (0.3cm) and D is the axle length (5.3cm). The state of this system is described by the 3D vector (c_x, c_y, θ). The input is the 2D vector, (ω_L, ω_R), of angular velocities of the left and right wheel.

Light sensors. The intensity signal generated by each sensor (i_L, i_R) is inversely proportional to the square distance to the light source (Figure 5b).

The angle ϕ is the "preferred direction of the sensor", that is the direction of maximum response. The source is fixed in the environment and has an emission

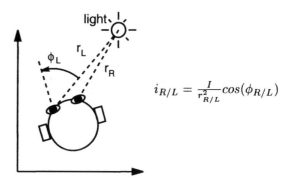

$$i_{R/L} = \frac{I}{r_{R/L}^2}cos(\phi_{R/L})$$

Figure 5: The response of the robot's sensors are a function of distance and angle from the light source.

intensity, I. Under these assumptions, the intensity signals, (i_L, i_R), are both functions of the robot's state: $i_{R/L} = i_{R/L}(c_x, c_y, \theta)$.

Lamprey's brain. Unlike the robot, the operation of the brain is essentially unknown. The purpose of the hybrid system is actually to investigate the computational properties of this neural tissue. In our simulations, we considered an extremely simplified linear model of this neural system (Figure 2). There are two inputs- the light intensity signals used as stimuli - and two outputs - the angular velocities of the wheels. These signals are connected by a "weight matrix", W, whose elements may be taken to represent the strengths of the connections

between inputs and outputs. Positive weights represents excitatory connections and negative weights inhibitory connections.

The whole system. When all the above components are assembled into a single system, one obtains three differential equations in which the rate of change of the state vector depends only upon the state vector itself, and not on time:

$$\dot{c}_x = f_1(c_x, c_y, \theta \mid W)$$
$$\dot{c}_y = f_2(c_x, c_y, \theta \mid W) \quad (1)$$
$$\dot{\theta} = f_3(c_x, c_y, \theta \mid W)$$

This is called an autonomous system. Here, we have emphasized that the particular behaviors emerging from this autonomous system are determined by the parameters that describe the behavior of the neural system and that are assumed to be time-invariant (or, at least, to be varying on a time scale that is much longer than the scale of each behavior). In this first simulated approximation, the neural parameters are fully expressed by a 2x2 matrix W. But, of course, to capture with high accuracy the behavior of the real system it will be necessary to utilize more complex models.

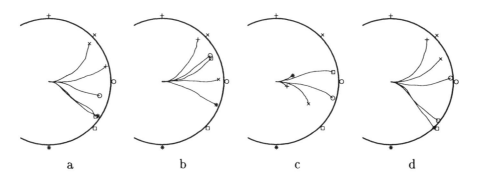

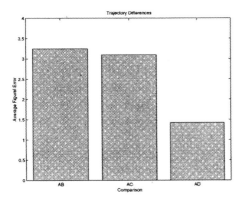

Figure 6: Unstable robot trajectories. Panels a and b were separated by 10 min, b and c by 5 min, and c and d by 0 min. The trajectories generated by the light marked with a star (*), circle (o), and square (□) vary greatly from one trial to the next.

Figure 7: Figural error of the unstable trajectories. The first bar is the error between panels a and b of Figure 6, second bar for panels a and c, and third bar for a and d.

Results

Stability

In these experiments, the lamprey's brain was maintained *in vitro* for periods ranging from 4 to 8 hours. In most cases, the preparation maintained its full responsiveness across the entire experiment. In addition to the overall health of the preparation, other factors affecting the persistence and stability of behaviors are (a) the displacement of the electrodes within the neural tissue and (b) local damage to neural tissue caused by repeated stimuli. Figure 6 shows the behaviors generated by what we considered to be an unstable preparation. The four panels display four consecutive experimental sets separated by intervals of 10min, 5min and 1min. During each experimental set, the lights indicated by different symbols along the workspace boundary were turned on in sequence. We collected trajectory data from the moment the light turned on until the robot either stopped moving or reached the edge of the workspace. A single trial set contains the trajectories collected as the robot

reacted to each light. It is evident that the trajectories in the four panels of Figure 6 change rather drastically from trial to trial. This variability is quantified by the figural error plots in Figure 7. One may see that there is a particularly strong variation between trial 1 and trials 2 and 3. Whenever we observed this kind of unstable behavior - quantified by a net figural error larger than 2.5cm - we moved the stimulation and/or recording electrodes to different sites. If these adjustments did not result in some improvement, we discarded the preparation. Figure 8 shows a set of stable behaviors. The panels are arranged as in Figure 6. The variability of the trajectories in Figure 8 is expressed by the figural errors shown in Figure 9. Although some amount of variability between different trial sets is still visible, the predominant positive phototaxis is evident in all panels and the overall trajectory shapes are similar for trajectories elicited by the same lights. We considered for further analysis only preparations with stability comparable or better than shown in this example, as determined by a figural error of less than 2.5cm.

Behavioral responses

The features of the trajectories generated by the neurorobotic system depend upon the pattern of neural connections between stimulating and recording electrodes. In a first approximation, one may represent the operation of these connections by the linear two-layer network of Figure 2. We have combined this simple network model with a simulator of the Khepera dynamics. The response of the combined system to a source of light is described by a set of three nonlinear first-order autonomous equations (Equation 1). By simulating these equations we could predict the general features of trajectories corresponding to different patterns of stimulation/recording connectivity. The structure of the connection matrix, W, establishes the sign of the ensuing phototactic behavior. In case of pure ipsilateral excitatory connectiv-

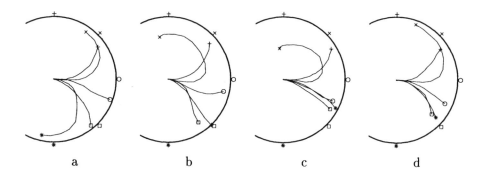

Figure 8: Stable robot trajectories. Panels a and b were separated by 10min, b and c by 5min, and c and d by 0min.

ity (right-to-right and left-to-left), the off diagonal terms are both zero and the diagonal terms are positive. When the diagonal terms are equal (that is when the connectivity matrix is proportional to the unit matrix,) then the resulting behavior is a negative phototaxis- i.e. movement away from the light source - as shown in Figure 10b. In contrast, if there is purely contralateral excitatory connectivity (right-to-left and left-to-right), the diagonal terms are zero and the off diagonal terms are positive. The resulting trajectories (Figure 10a) correspond to positive phototaxis- i.e. movements toward the light. A broad spectrum of intermediate behaviors (an example is in Figure 10c) is obtained by matrices with both diagonal and non-diagonal terms and with different degrees of asymmetry.

the rapid drop in light intensity as the Khepera moved away from the light source. Because of scattering and other phenomena not included in the model, the actual drop in light intensity was more pronounced than the simulated drop. This effect is compound by the presence of friction, which is also not included in the model. To obtain, in cases like this one, a higher sensitivity of the observed behaviors in response to the different light sources, we biased our preparation toward positive phototaxis by selecting the reverse mode option that is by connecting the right electrode to the left wheel controller and vice versa. This operation is equivalent to exchanging off-diagonal with diagonal weights in the connectivity matrix

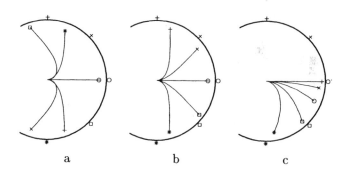

Figure 10: Different matrices W are used with the two layer network model and the robot simulation to generate a) negative, b) positive, and c) mixed phototaxis.

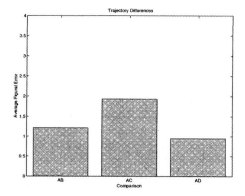

Figure 9: Figural error of the stable trajectories. The first bar is the error between panels a and b of Figure 8, second bar for panels a and c, and third bar for a and d.

Depending on the placement of the electrodes in the actual neural tissue, we were able to observe both positive and negative phototaxis, as well as intermediate behaviors (Figure 11). It is worth observing that negative phototaxis with the actual system tended to result in shortened trajectories compared with negative phototaxis in the simulator (Figure 11c). This is likely due to

Adaptive modification of artificial behavior

The neural component of the hybrid system is a portion of the brainstem, the reticular formation, that normally combines vestibular information with other sensory inputs and descending commands. The outcome is a neural signal that modulates the ongoing activity of the spinal cord for the control of swimming movements (Grillner et al. 1993; Grillner and Mastuishima 1991). A significant feature of this circuitry is its ability to modify the

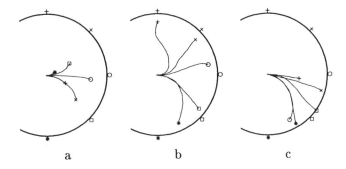

Figure 11: In these trials, the robot displays behavioral patterns that can be classified as: a) negative, b) positive, and c) mixed phototaxis.

efficacy of its own synaptic connections in response to sustained patterns of stimulation. Both long-term potentiation (LTP) and long-term depression (LTD) have been documented (Schwartz et al. 1998). We wished to explore the possibility of using our system for observing the effects of plastic changes on artificial behaviors and for separating, on the basis of this observation, the effects of long-term changes from those of short-term changes.

To generate adaptive changes in the neural preparation, we doubled the sensitivity of the light sensors on the left side of the robot while leaving unchanged the sensitivity on the right. This alteration induced the change in behavior shown in Figure 12a and 12b. The trajectories with the initial setting of gains are displayed in Figure 12a whereas the trajectories in Figure 12b were obtained immediately after the change in the left light sensors. It should be noted that the initial behavior of system was of mixed phototaxis exhibiting a counterclockwise curl. It is possible to observe a predominant clockwise rotation of the trajectories. Immediately following the acquisition of the trajectories, the robot was placed in its home position (center of workspace with the "nose" facing the light number 1, at 0°) and fixed in place so that it could not move. Light number one was turned on for a period of five minutes. Although both sides of the robot were exposed to approximately the same amount of light, the increased sensitivity on the left side doubled the corresponding frequency on the left side of the lamprey. Following this extended period of stimulation, a second set of trajectories was recorded (Figure 12c). These trajectories were highly distorted, compared to those obtained in the initial phase of the experiment. There is a strong clockwise rotation together with the formation of circular patterns. Such circular patterns are a typical sign of strong imbalance between right and left channels.

This particular experiment was conducted in reverse mode. Therefore, the clockwise rotation of the trajecto-

ries, which reflects an increase in speed of the left wheel (and/or a decrease of the right), is due either to an increase in response of the right reticulo-spinal neurons (and/or to a decrease in responsiveness of the left neurons).

Considering that the preparation had a predominant ipsilateral response- because the reversed response was predominantly a positive phototaxis- these changes are likely to reflect a depression of the synapses in the left reticular nucleus rather than a potentiation of the right neurons.

A final, third set of trajectories was recorded after a 5 minutes resting period (Figure 12d). Here, the lamprey's brain appears to have over compensated for the change in synaptic efficacy induced by the prolonged stimulation at rest. Comparing the trajectories in Figure 12c to those in Figure 12d, it appears that in the last stage of the experiment the trajectories have a large counter clockwise shift. Although these are preliminary results, it is possible to speculate that this rotation reflects not only the end of the short-term change seen after the prolonged stimulation, but also the onset of a trend toward the adaptive compensation of the initial response to the change in sensor balance. Such a long-term compensation could be accounted for by an unsupervised Hebbian regulation of synaptic plasticity elicited during the trials in which the robot moved in response to the light stimulus.

Discussion

The work described in this paper is a first step toward the realization of a hybrid neuro-robotic system for the investigation of the neurobiological basis of sensorimotor learning and behavior. We have created a system in which a portion of neural tissue from the lamprey's reticular formation is connected through a computer interface with a small mobile robot. The optical sensors on the robot determine the parameters of the electrical stimuli delivered to the vestibular axons of the lamprey. The signals recorded from the neural populations with which these axons form synapses are used as control signals for the robot's movement.

Why did we use an actual physical robot instead of a simulated virtual body? This is an important question, and we must stress that physical robots are not necessarily a better choice than simulated bodies. Indeed, simulated bodies offer the possibility to probe the neural tissue with an infinite range of possible transformations. In this work, we have chosen to use a physical robot for two reasons. First, we see this study as an initial step toward the development of fully-integrated neuro-robotic systems, where machines and biological tissue coexist in mutual interaction. The ability to create such hybrid systems is of obvious relevance to the development of ad-

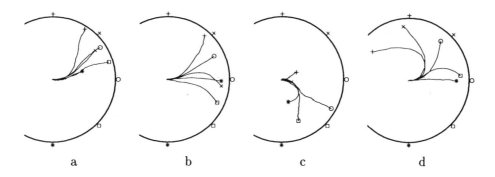

Figure 12: Sensor sensitivity was unilaterally doubled. Trajectories were recorded a) before any changes in sensitivity were made, b) immediately following this change, c) after five minutes of steady stimulation, and d) after a five minute resting period.

vanced prostheic devices. Furthermore, we expect that in some future time, the creation of brain/machine interfaces will allow us to tap into the extraordinary computational power of living tissues. The second, and more opportunistic reason, for using a physical robot is that, maybe contrary to intuition, it is in practice much easier to interface a Lamprey's brain with our mobile robot than with a simulated model of it. Using a simulated system, we would have to deal with an increased overhead introduced by the need to synchronize the input/output from the interface with the simulation cycles. In addition, while physical systems have some inherent stability associated with the presence of dissipative elements, simulated system may easily display "unphysical" instabilities induced, for example, by the presence of discretization errors.

The idea of using neural signals for driving mechanical apparatus is certainly not new. Research in prosthetic devices has long been pursuing the use of myoelectrical signals for controlling artificial replicas of the limbs (Abul-Haj and Hogan 1990). More recently, Chapin and coworkers (Chapin et al. 1999) have developed an experimental paradigm in which the signals recorded from a population of neurons in the motor cortex of the rat were used to drive a mechanical lever which controlled the the release of a food reward. The study of Chapin and coworkers has provided us with new evidence that motor cortical activity may be dissociated from the activity of limb muscles. The same cortical activity observed when the reward was obtained by a movement of the paw could also be maintained when the same reward was obtained by a movement of the mechanism and with the paw at rest. A distinctive feature of our hybrid system is that it exploits a closed-loop relation between the neural tissue and robot, which operates as an artificial body. In this closed-loop arrangement, the movements determined by activities in the reticular neurons cause changes in the robot's exposure to the light generated by a fixed source.

These changes, in turn, cause a variation in the electrical stimulus that is responsible for the activities in the same reticular neurons. This paradigm is well suited for investigating the operation of Hebbian learning mechanisms (Edeline 1996; Pennartz 1997; Shors and Matzel 1997; Grzywacz and Burgi 1998) by which the strength of a given synapse is modified based on the correlation between pre and postsynaptic activities.

We have found that, with some exceptions, our neuro-robotic system generated stable behaviors over extended periods of time. The lamprey's brain can indeed be maintained alive *in vitro* for entire days. In these experiments- which lasted only a few hours- we have assessed stability by observing the repeatability of the trajectories triggered by light sources placed at different locations. We do not need to stress that the stability of the behaviors generated by our preparation is a necessary condition for proceeding with further analysis and, in particular, with investigations that assume that the neural connectivity remains invariant over the time scale of individual sensory-motor responses.

The second finding of our study is the observation of different type of phototaxis in different preparations. We observed light-seeking behaviors (positive phototaxis), light aversion behaviors (negative phototaxis) and linear combinations of light seeking and light aversion (mixed phototaxis). A simple linear model is sufficient to account for these different types of behavior on the basis of the amount of ipsi- and contralateral connections between stimulating and recording electrodes.

We must acknowledge, at this point, that our work has been profoundly inspired by some ideas that Valentino Braitenberg expressed almost 20 years ago, at the beginning of the "connectionist revolution" (Braitenberg 1984). In a delightfully entertaining book, Braitenberg described how relatively simple connections between sensors and motors could endow some imaginary mechanical vehicles with life-like behaviors. These are behaviors that could easily be interpreted as intelligent or emotional responses to environmental stimuli. While the sensory-motor responses generated by our neuro-robotic system are not as remarkable as some of the behaviors described in that book, this system may be regarded as an implementation of Braitenberg's ideas and, in particular, of the idea of connecting the study of cellular brain structures with the observable responses that may be supported by these structures. As a parallel to Braitenberg's "experiments in synthetic psychology", one could call the studies with the neuro-robotic system an experiment in synthetic neurobiology.

Finally, we have observed systematic adaptive responses induced by the selective alteration of the sensor signals on one side of the robot. In particular, we have observed a strong change of behavior followed by gradual return toward the initial responses. The possibility to generate adaptive changes in the robot's behavior opens the way to using the neuro-robotic interface for studying the transformations induced in the brain tissue by long and short- term modifications of synaptic properties. This system offers the possibility of substituting the actual brain tissue with a computational model of its neurons and its connections. The comparison of biological adaptive changes with their simulated counterparts may provide us with new means to directly investigate the computational properties of synaptic plasticity.

Acknowledgments

This work has been sponsored by ONR grant #N00014-99-1-0881 and AASERT grant #N00014-97-1-0714.

References

Abul-Haj, C. J. and Hogan, N. 1990. Functional assessment of control systems for cybernetic elbow prostheses – Part II: Application of the technique. *IEEE Transactions on Biomedical Engineering* 37(11): 1037-1047.

Alford, S., Zompa, I. and Dubuc, R. 1995. Long-Term Potentiation of Glutamatergic Pathways in the Lamprey Brainstem. *Journal of Neuroscience* 15(11): 7528-7538.

Braitenberg, V., 1984. *Vehicles*. Cambridge, Massachusetts : MIT Press.

Chapin, J. K., et al. 1999. Real-time control of a robot arm using simultaneously recorded neurons in the motor cortex. *Nature Neuroscience* 2(7): 664-670.

Conditt, M. A., Gandolfo, F. and Mussa-Ivaldi, F. A. 1997. The Motor System Does Not Learn the Dynamics of the Arm by Rote Memorization of Past Experience. *Journal of Neurophysiology* 78(554-560).

Deliagina, T. G. 1995. Vestibular compensation in the lamprey. *NeuroReport* 6: 2599-2603.

Deliagina, T. G. 1997. Vestibular Compensation in Lampreys: Impairment and Recovery of Equilibrium Control During Locomotion. *Journal of Experimental Biology* 200: 1459-1471.

[Deliagina 1992a]Deliagina, T. G., et al. 1992a. Vestibular control of swimming in lamprey II - Characteristics of spatial sensitivity of reticulospinal neurons. *Experimental Brain Research* 90: 489-498.

Deliagina, T. G., et al. 1992b. Vestibular control of swimming in lamprey III - Activity of vestibular afferents: convergence of vestibular inputs on reticulospinal neurons. *Experimental Brain Research* 90: 499-507.

Edeline, J. M. 1996. Does Hebbian synaptic plasticity explain learning-induced sensory plasticity in adult mammals? *Journal of Physiology* 90(3-4): 271-276.

Grillner, S. and Mastushima, T. 1991. The Neural Network Underlying Locomotion in Lamprey - Synaptic and Cellular Mechanisms. *Neuron* 7: 1-15.

Grillner, S., et al., "The Neural Generation of Locomtion in the Lamprey: An Incomplete Account," in: *Neural Origin of Rhythmic Movements*, Cambridge University Press, Cambridge, 1983,

Grzywacz, N. M. and Burgi, P. Y. 1998. Toward a biophysically plausible bidirectional Hebbian rule. *Neural Computation* 10(3): 499-520.

Hildreth, E. C. and Hollerbach, J. M., "Artificial Intelligence: computational approach to vision and motor control," in: *Handbook of Physiology, Section 1: The Nervous System, Volume V: Higher Functions of the Brain, Part II*, F. Plum, ed., American Physiological Society, Bethesda, Maryland, 1987, 605-642.

Marr, D., 1982. *Vision*. San Francisco, California : WH Freeman and Company.

Orlovsky, G. N., Deliagina, T. G. and Wallén, P. 1992. Vestibular control of swimming in lamprey I - Responses of reticulospinal neurons to roll and pitch. *Experimental Brain Research* 90: 479-488.

Pennartz,C. M. 1997. Reinforcement learning by Hebbian synapses with adaptive thresholds. *Neuroscience* 81(2): 303-319.

Rovainen, C. M. 1979. Electrophysiology of Vestibulospinal and Vestibuloreticulospinal Systems in Lampreys. *Journal of Neurophysiology* 42(3): 745-766.

Schwartz, N. E., et al. 1998. Synaptic plasticity and sensorimotor learning in Lamprey's reticulospinal neurons. *Society for Neuroscience Abstracts*.

Shors, T. J. and Matzel, L. D. 1997. Long term potentiation: what's learning got to do with it? *Behavioral & Brain Sciences* 20(4): 597-614.

A 'Fitness Landscaping' Comparison of Evolved Robot Control Systems

Stevan Jay Anastasoff

Evolutionary and Emergent Behaviour Intelligence and Computation Group
School of Computer Science
University of Birmingham
Birmingham B15 2TT, United Kingdom

Abstract

Most work in evolutionary robotics has focussed on evolving neural network based control systems, rather than high level control programs using genetic programming. One reason for this is that the fitness landscapes generated by a genetic programming framework may be poorly suited to the evolutionary techniques employed. The aim of this paper is to investigate this claim. The first part of the paper demonstrates two simulations of evolving populations of simple robots, one based on genetic programming, the other based on evolved neural nets. The latter part of the paper then introduces a technique for 'fitness landscaping', that is for constructing 3-D visualisations of the sorts of complex, high dimensionality fitness landscapes involved. This technique is applied to the results from the two simulations, and the resulting representations are discussed.

Introduction

Traditional wisdom in the field of evolutionary robotics holds that neural network based control systems are a more suitable evolutionary architecture than genetic programming [5], [10]. Compared to neural nets, the primitives of a genetic programming framework are very high level. The result of this is potentially a "...coarse-grained fitness landscape with steeper precipices." ([5], p366). However, a significant amount of work in evolutionary robotics has still been carried out using genetic programming, with some good levels of success [7], [8], [12], [9].

In the simulations to be presented here these two evolutionary robotics paradigms will be compared, in order to evaluate the statement quoted above. The two approaches are used to develop control systems for basic navigational behaviour (wall following) using a 2-D representation of a simple autonomous robot. Samples from each generation of the two evolutionary processes are then stored in order to be used to create visualisations of the sorts of fitness landscapes generated. These can then be compared with the sorts of landscapes expected from the earlier claim.

The 'fitness landscaping' process involves mapping random samples from the evolving populations on to a 2-D lattice according to the level of similarity between the contents of the cells of the lattice. A simulated annealing algorithm determines the distribution of the samples in the lattice. This optimises the distribution such that the most cells have the most in common with their neighbours. The fitness level of the contents of each cell is then mapped in order to generate the final 3-D visualisation. This whole process will be described in more detail in the later parts of the paper.

Two Evolutionary Simulations

The following sections describe in detail the simulation environment, together with the genetic programming and evolved neural net algorithms used.

The Robot Simulation

The simulation models a 2-D robot together with its environment, consisting of a close-ended corridor containing several bends (two right-hand bends, followed by two left-hand bends). This scenario is derived from [12]. The robot's behaviour is then governed by an evolved control program or neural network fed in by the appropriate evolutionary framework. The robot itself is a somewhat simplified version of that described in [5]. It consists simply of two motors, one either side, capable of rotating backwards and forwards independently, and two sensors. The sensors are 'antenna' or 'whisker' like proximity switches. These independently return either a value of true, if they are in contact with an object, or false otherwise. They are located at the front of the robot, protruding at approximately 30 degrees to either side. They are of approximately the same length as the body of the robot.

The model is based on an actual simple Lego robot. The parameters used have been selected to make the simulation correspond as closely as possible to this real robot. A small amount of noise, as has been demonstrated to be beneficial (in [12] for example), was also introduced. The noise levels were again intended to be consistent with comparative runs with the real robot. The entire simulation has been developed using the programming language POP-11, running the PopBugs simulation software.

Both the robot being simulated, and the environment in which it is situated, were both intentionally kept as simple as possible. The idea here was that by having a simpler simulation, simpler (and thus easier to model) fitness landscapes would be generated.

Evolving the Control Programs

A very much standard genetic programming framework was used to supply the control programs for the simulation. The genetic operators used, as well as the particular functions and terminals, were based closely on those found in [7]. An initial population of 50 control programs was randomly generated using the 'ramped half-and-half' method, to create a wide diversity of different sized and shaped starting tree structures. The population size was somewhat smaller than those generally used in evolving robot control programs, but this can be justified by the simplicity of the robot and problem being simulated. The fitness of each of these programs was then assessed by running the simulation and measuring the distance the robot managed to travel along the length of the corridor. The simulation was stopped and distance measurements were taken once the robot's wheels had rotated a fixed number of times. Simulations in which the robot's wheels stopped moving for two consecutive time steps were finished automatically at that point. Parents for each successive generation were chosen using a method of tournament selection. Two candidates were chosen, and the fitter was allocated as the parent with a probability of 0.9. For each pair of parents so chosen, crossover occurred along a random branch in the program tree with a probability of 0.8. Otherwise the two parents were reproduced without crossover. This was repeated enough times to fill the next generation. Fifty consecutive generations were run in total.

The function and terminal sets used were quite small, again as stated above with the intention of producing a simpler and more easily represented fitness landscape. Three functions were used: the two conditionals *ifRight* and *ifLeft* each take two arguments. If the appropriate side sensor is activated the first argument is evaluated, otherwise the second argument is evaluated. The third function used is the program connective *prog*. This again takes two arguments, and evaluates each in turn. The terminal set consists of four operators: *forwards*, *backwards*, *turnLeft* and *turnRight*. Each of these moves the robot in the stated manner; forwards and backwards each move the robot one full body length, turnLeft and turnRight turn the robot approximately 20 degrees in the appropriate direction (all values averaged over noise).

Evolving the Neural Networks

The neural net controllers and the evolutionary framework used to develop them were based in part on those used in [5]. Each controller consisted of a 3-layer feed-forward neural net containing two input nodes (one con-

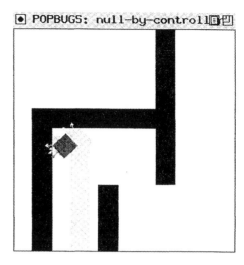

Figure 1: Evolved robot control program, population best from the first generation of the genetic programming framework.

nected to each sensor), two 'hidden' nodes, and four output nodes used to control the two motors. Each node was a noisy linear threshold device, allowing for a much broader range of possible interesting behaviours. This principally differs from those networks used in [5] through the reduced number of input nodes, and the fixed number of 'hidden' nodes.

Each network was encoded genetically as a fixed length chromosome consisting of 22 floating point numbers. Since the network topology was fixed, each weight and threshold value in any net could be assigned by the value at a specified location in a given chromosome. A standard 'vanilla' GA, using single point crossover and tournament selection was then used to evolve a population of network chromosomes. As in the genetic programming framework, a population size of 50 was used, again being evolved for a total of 50 generations.

The Resulting Control Systems

A variety of different behaviours evolved over the course of the simulations.

Screen shots for four of these can be seen in the accompanying figures 1 through 4. These same basic classes of behaviour, and the progression through them, were present for both evolved neural networks and control programs. Note that in these figures the sides of the boxes designate the edges of the environment and are impassable.

Those solutions that produced movement in a straight line achieved the most basic level of success, traveling as far as the first bend (figure 1). These programs dominated the first few generations. The next level of sophistication, the 'right-turners' (figure 2), rapidly usurped these, however. These travel forward in a straight line until bumping into a wall, and then turn right. This be-

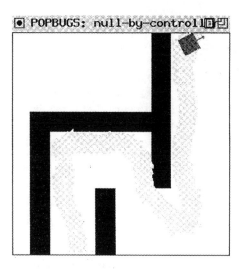

Figure 4: Evolved neural net, population best from the fiftieth (final) generation of the evolved neural network framework.

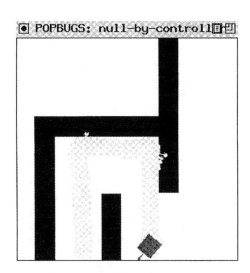

Figure 2: Evolved robot control program, population best from the third generation of the genetic programming framework.

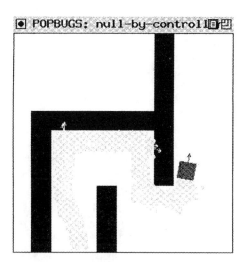

Figure 3: Evolved neural net, population best from the seventeenth generation of the evolved neural network framework.

haviour manages to get them as far as the third bend. Within ten to twenty generations the first of the more sophisticated strategies could be seen to be emerging (figure 3). These cause the robot to veer over to the left until bumping into a wall, and then turn right. This behaviour can be effectively described as 'find the wall on your left hand side and follow it'. This strategy is sufficient to drive the robot the full length of the corridor. However, in the early versions, the optimal turning ratios were yet to evolve, with far too much time spent turning too sharply. Throughout the remaining generations this final behaviour gradually came to dominate both types of control system (although only very late on in the run did it eventually completely supercede the right-turning behaviour), slowly becoming more refined. By the final generation something like the solution demonstrated in fig. 4 had evolved in both evolutionary systems. The turning ratios have been optimised, and the robot is capable of traversing the full length of the corridor.

It is perhaps worth noting that despite the extremely limited number of test cases (one!), perfectly generalised corridor following behaviours have evolved. Although the early strategies exploited features peculiar to the single test case (e.g. right-turning behaviours), the final best scorers are perfectly capable of following any corridor (employing the simple strategy of finding a wall and then sticking to it).

Quantitative Comparison of the two Algorithms

Figure 5 shows a comparison of the performance over time of the evolved neural networks and control programs. Up to the tenth generation, the genetic programming framework can be seen to perform slightly better than the evolved neural networks. However, from the tenth generation onwards, the neural networks have a slight, though clear and consistent, advantage over the evolved control programs. This suggests that the genetic programming algorithm may be slightly better at finding the simpler strategies, such as moving straight ahead or right-turning, while the neural network algorithm may have a slight advantage in moving from these simpler strategies on to the more sophisticated strategies needed to traverse the full length of the corridor.

This conclusion would fit in well with the earlier claim concerning the structure of the fitness landscapes generated by the two approaches. This will be discussed further later, in conjunction with the relevant fitness landscape visualisations.

The 'Fitness Landscaping' Algorithm

Having looked at the simulation environments and the two evolutionary frameworks, the next sections will present the 'fitness landscaping' algorithm employed. The idea behind this landscaping algorithm is to generate a 3-D visualisation of the fitness landscapes produced by particular evolutionary approaches to a given problem. The evolved populations are randomly sampled, and simulated annealing is used to map these samples on to a 'similarity lattice'. The idea is that syntactically similar structures should be close by each other in the lattice. Once fitness values have been overlaid, it can then be seen whether or not (and how) semantic similarity (demonstrated by fitness) corresponds to syntactic similarity (demonstrated by proximity in the lattice).

According to the initial claim quoted at the beginning of the paper, the neural network landscape should be significantly smoother, with gradual changes in fitness evident in neighbouring localities in the landscape. The genetic programming landscape should illustrate sharper precipices, a result of the higher level primitives being employed.

Other Visualisation Techniques

A number of multidimensional, and other, visualisation techniques have been employed for studying evolutionary fitness landscapes. Summaries of some of these can be found in [11]. The technique to be employed here is most similar to Sammon mapping, initially presented in [13] and described in application to evolutionary algorithms in [4]. It also bears much in common with the Sammon mapping derivative Genotypic-Space mapping [3].

Sammon mapping is a technique for detecting structure in a set of N vectors of dimensionality L, and mapping this structure on to a set of N 2- or 3-dimensional vectors. This is done by minimising the dissimilarity between the distance between vectors in the higher dimensional and lower dimensional vector sets. Any domain appropriate measure of distance can be employed, although most typically Euclidean distance is used. The level of dissimilarity is given by the equation:

$$E = \frac{1}{\sum_{i<j} \delta_{ij}} \sum_{i<j} \frac{(d_{ij} - \delta_{ij})^2}{\delta_{ij}}$$

where δ_{ij} denotes the distance between the ith point and the jth point of the higher dimensionality vector set, and d_{ij} denotes the distance between the ith point and the jth point of the lower dimensionality vector set. A steepest gradient descent method is then used to minimise this value.

There are a number of significant problems that arise when trying to apply Sammon mapping to the domain under consideration here. Firstly, suitable distance metrics must be found. This is particularly important for the evolved control programs, where even the dimensionality of the solutions does not remain constant. More seriously, the computational cost of comparing large numbers of points becomes rapidly prohibits any but the smallest landscapes to be generated in a reasonable amount of time. In fact, the computational complexity increases exponentially with the size of the mapping [3]. Additionally, the gradient descent method used has a tendency to get stuck on sub-optimal minima, as observed in [11].

The proposed solutions to these problems are discussed in more detail below. In summary, computational overheads are reduced by generating dissimilarity measurements only from local neighbourhoods (as suggested by [3]), and stochastically sampling the candidate solutions, rather than comparing all possible candidates. Further, simulated annealing has been used to generate overall dissimilarity measurements significantly lower than those obtained using the normal gradient descent method.

Developing the Algorithm

The landscaping algorithm was initially developed using a simpler evolutionary model than the robot-navigation models described above. A simple bit-string genetic algorithm was employed. The exact algorithm used was the standard GA used in [2]. This was applied across a number of optimisation problems taken from the test suites employed in [2] and [1].

In each case the genetic algorithm was run, and a total of 400 (enough to fill a 20x20 lattice) bit-strings were randomly selected from throughout the run, chosen evenly from throughout each generation. These were initially distributed across the lattice at random.

In existing work applying Sammon mapping to evolutionary algorithms (eg [4]), the highest fitness member of each generation is generally used. However, the solutions generated by an evolutionary approach to a problem emerge from the interactions of all members of all populations throughout the course of a run. It is likely therefore that there will be characteristic features

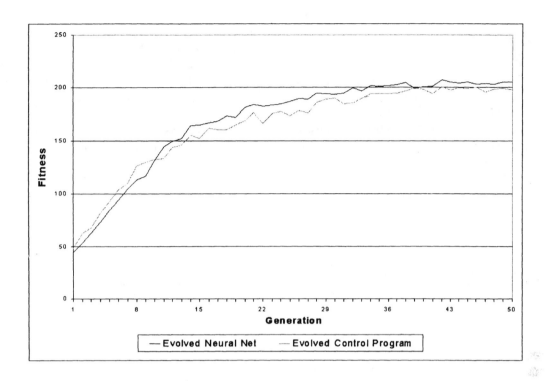

Figure 5: Comparison of fitness over time for the evolved neural networks and evolved control programs. Values averaged over 10 simulation runs.

of a particular evolutionary path that will not be captured simply by considering the fittest member of each generation. However, generating a mapping from all solutions generated throughout the course of a run is an extrememly computationally demanding task. For this reason, stochastic sampling of each generation was used, selecting enough members from each population to generate a reasonable landscape, while keeping computational overheads manageable.

The similarity of two bit-strings can be defined as the number of loci that are identical in each. The total similarity value of the lattice is then calculated by taking each cell in turn, and summing its similarity with each of its neighbours. The sum of all such neighbourhood similarities is the total similarity value for that particular lattice arrangement. Dissimilarity can be calculated as a simple inverse of this similarity value.

The simulated annealing algorithm was then utilised in order to minimise this dissimilarity value. A straight gradient-descent algorithm was also tested, but generally resulted in arrangements with lower final dissimilarity values. The simulated annealing algorithm worked as follows: two cells in the lattice were swapped at random, and the total dissimilarity of the new arrangement was calculated. The new arrangement was then accepted or rejected based on the change in total dissimilarity and according to a Boltzmann probability function. The fol-

lowing annealing schedule was used: the temperature was initially set to 5000, and reduced geometrically by a factor of 0.9 to a minimum of 1. The algorithm was iterated 1000 times at each temperature level, with a total of 150 iterations of this process. These parameter settings were determined largely by trial and error, based on an initial algorithm taken from [1].

Once the simulated annealing was complete, the fitness values for each cell in the lattice were plotted to generate the final 3-D visualisation.

Landscaping the Evolved Control Programs

The next stage was to adapt the process as used in the simpler GA cases to the more complex evolution of the robot control systems. The only essential difference would be the similarity functions used. The functions used would have to reflect the ease with which one program or neural net could be transformed into another, something akin to an inverse of Hamming distance [6]. After some experimentation the following procedure was found to give reasonable results for the genetic programming population: for every sub-tree from the first program that is also a sub-tree of the second program, increase the similarity by one. As it stands, this would produce an asymmetric result, such that the similarity between x and y is not necessarily the same as the similarity between y and x. To rectify this, the same process is then applied in the opposite direction. The procedure then becomes: for every sub-tree in the first program that is also in the second, and for every sub-tree in the second that is also in the first, increase similarity by one. This still has the slight problem that large programs will, through this definition, tend to produce higher similarity values than small programs. This is settled by normalising according to the combined lengths of the two programs and scaling up to give a value from 0 to 20 (this range chosen somewhat arbitrarily as matching the range produced by the earlier genetic algorithm test cases).

For the neural network chromosome, similarity could be defined much more simply. A sum was taken of the differences between the values of each gene in the two chromosomes being compared. This was then inverted and scaled to again give a value in the range 0 through 20.

Further, the additional complexity of applying the annealing algorithm while using the more involved genetic programming similarity function resulted in enormously increased computational demands. For this reason, and given the limitations of available resources, the annealing schedule was reduced somewhat. Each temperature setting was iterated only 500 times, and a total of only 100 iterations of this process occurred. For consistency this same schedule was used also for the neural network landscaping process. The final resulting fitness landscape visualisations can be seen in figures 6 and 7.

For comparative purposes, the evolved neural network landscape before application of simulated annealing to minimise the dissimilarity is also shown in figure 8. This landscape has effectively been generated by distributing evolved solutions across the lattice at random, and then plotting their fitness values.

Discussion

The Landscaping Algorithm

In the initial development of the landscapiong algorithm, characteristics of each of the fitness functions being tested were evident to some degree in the generated visualisations. In 'plateau' type landscapes, wide, flat areas, gave way to narrow ridges, leading to sharp peaks. When fitness changes in these landscapes it does so sharply, but with clear paths still progressing from plain to ridge to peak. Other 'linear' type landscapes produced visualisations that undulated up and down smoothly, but with peaks rising progressively across the course of the visualisation.

These test cases are certainly very suggestive that key features of the underlying structure of the fitness landscape are being captured by the visualisation. However, a lot more work is really needed on a wider variety of problems to really determine the extent to which the mapping algorithm is capable of capturing the structure of any sort of complex landscape.

Comparing the neural network visualisation with the randomly generated landscape using the same data is also suggestive that the algorithm is capturing at least some relevant features. Although any interpretation of the landscapes is a highly subjective matter, there certainly does appear to be a tighter grouping of peaks, rising steadily from lower regions at the extreme corners of the final visualisation, than in its random counterpart.

The Robot Control System Landscapes

Two important differences can be noted in the landscape visualisations. In the first instance, as noted above, it can be seen that the peaks for the neural network approach are far more tightly bunched than those on the genetic programming landscape. Secondly, it can be seen that the genetic programming landscape is criss-crossed with numerous deep crevasses and canyons. Overall, there can certainly be seen to be more structure to the neural network landscape, suggesting a less problematical search space for a given evolutionary algorithm to navigate. These observations lend credence to the conventional evolutionary robotics wisdom.

Some further analysis in conjunction with figure 5 can be speculated on. Initially, increases in fitness in both systems are fairly level. In both landscapes very low regions are adjacent to mid level regions. However, the evolutionary paths needed to reach peaks can be seen to be somewhat different in each case. Many of the mid level plateaus in the genetic programming landscape are separated by crevasses from peaks, where as the neural network landscape rises more consistently towards the highest peaks. This may help explain the slightly higher mean fitness levels achieved by the neural nets after mid-level fitnesses have been achieved.

However, it certainly seems plausible (if not likely) that key characteristics of the fitness landscapes in this particular situation are a product at least as much of the particular fitness function as of the evolutionary control strategy. For a proper comparison, a much wider range of control problems should ideally be addressed. The particular corridor following problem addressed here will tend to produce fitness plateaus, with large jumps in fitness occurring as each bend in the corridor is reached and passed. Other types of problem may produce different structured fitness dynamics that will interact in different ways with opposing evolutionary strategies. Only if the two approaches produce similar landscapes on a variety of problems can it really be said that the original claim is supported.

It should perhaps be emphasised that any analysis of generated landscapes at this stage is really very much a subjective and speculative process. The process used, however, has still demonstrated some value in suggesting possible lines of research into understanding the behaviour of the specific domain under consideration. Fur-

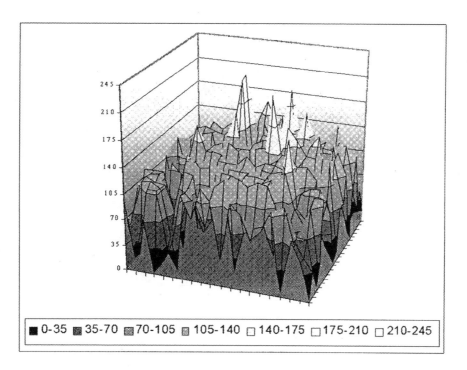

Figure 6: Fitness landscape visualisation of evolved neural net robot controllers

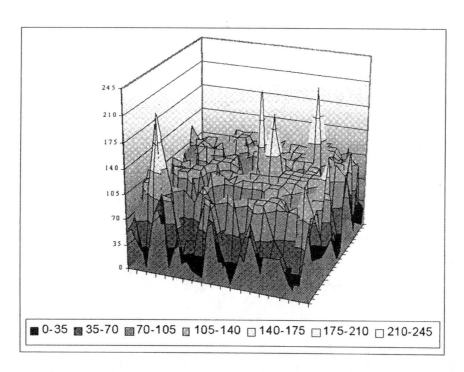

Figure 7: Fitness landscape visualisation of evolved high level robot control programs

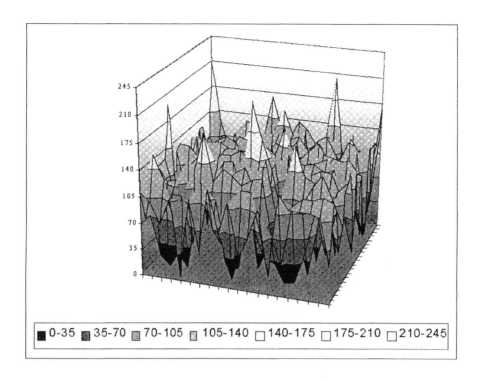

Figure 8: Random landscape visualisation of evolved neural net robot controllers

ther work in analysing the resultant visualisations is intended to quantify more specific properties.

Conclusions

The two approaches taken to evolving corridor following behaviour in a simple autonomous robot resulted in a variety of different solutions, varying in their level of success. However, the desired behaviour was evolved, using both genetic programming and evolved neural networks. A method for visualising evolutionary fitness landscapes was then demonstrated, and applied both to some sample genetic algorithm problems, and to the evolution of the two differing types of robot control system. The resulting landscapes provide some (although very inconclusive) support for the claim that genetic programming techniques result in fitness landscapes poorly suited to evolutionary robotics in comparison to evolved neural nets. In particular, the visualisations suggested one possible (though again highly speculative) reason why neural nets had improved performance over control programs only for moderate to high fitness levels. Further investigation, directly involving a wider selection of problems, would absolutely be necessary to establish anything more conclusive. However, the methodology employed here does still hold promise. The techniques developed could be used as one possible tool in this sort of a more thorough investigation.

Acknowledgements

Thanks to James Neil for comments on the fitness landscaping algorithm. Supported by funding from the Engineering and Physical Sciences Research Council. Work carried out in part at the School of Cognitive and Computer Sciences, University of Sussex.

References

[1] Ackley, D. H.: Bit Vector Function Optimization. In L. Davis, ed., *Genetic Algorithms and Simulated Annealing*. Morgan Kaufmann, 1987.

[2] Anastasoff, S. J.: Evolving Mutation Rates for the Self-Optimisation of Genetic Algorithms. In Floreano, D., Nicoud, J. D. & Mondada, F., eds. *Proceedings of the 5th European Conference on Artificial Life*. Springer-Verlag, 1999.

[3] Collins, T.: Genotypic-Space Mapping: Population Visualization for Genetic Algorithms. Technical Report, KMI-TR-39, Knowledge Media Institute, the Open University, Milton Keynes, UK, 1997.

[4] Dybowski, R., Collins, T. D., & Weller, P.D.: Visualization of Binary String Convergence by Sammon Mapping. In Fogel, D. B., Angeline, P. J., & Back, T., eds. Evolutionary Programming V, Proceedings of the Fifth Annual Conference on Evolutionary Programming. MIT Press, 1996.

[5] Harvey, I., Husbands, P., Cliff, D.: Issues in Evolutionary Robotics. In Meyer, J., Roitblat, H. L., Wilson, S. W., eds. *From Animals to Animats 2, Proceedings of the Second International Conference on Simulation of Adaptive Behaviour*. MIT Press, 1993.

[6] Kauffman, S. A.: *The Origins of Order: Self-Organization and Selection in Evolution*. Oxford University Press, 1993.

[7] Koza, J. R.: *Genetic Programming: On the Programming of Computers by Means of Natural Selection*. MIT Press, 1992.

[8] Koza, J. R.: *Genetic Programming II: Automatic Discovery of Reusable Programs*. MIT Press, 1994.

[9] Lee, W. P., Hallam, J., Lund, H. H.: A Hybrid GP/GA Approach for Co-evolving Controllers and Robot Bodies to Achieve Fitness Specified Tasks. In *Proceedings of IEEE 3rd International Conference on Evolutionary Computation*. IEEE Press, 1996.

[10] Nolfi, S., Floreano, D., Miglino, O., Mondada, F.: How to Evolve Autonomous Robots: Different Approaches in Evolutionary Robotics. In Brooks, R. A., & Maes, P., eds. *Artificial Life IV. Proceedings of the Fourth International Workshop on the Synthesis and Simulation of Living Systems*. MIT Press, 1994.

[11] Pohlheim, H.: Visualization of Evolutionary Algorithms – Set of Standard Techniques and Multidimensional Visualization. In W. Banzhaf, et al., eds. *Proceedings of the 1999 Genetic and Evolutionary Computation Conference*. Morgan Kaufmann, 1999.

[12] Reynolds, C. W.: Evolution of Corridor Following Behavior in a Noisy World. In Cliff, D., Husbands, P., Meyer, J. A., Wilson, S. W., eds. *From Animals to Animats 3. Proceedings of the Third International Conference on Simulation of Adaptive Behavior*. MIT Press, 1994.

[13] Sammon, J. W. jr.: A Nonlinear Mapping for Data Structure Analysis. In *IEEE Transactions on Computers*, vol. C-18, pp. 401-409, 1969.

Evolving Physical Creatures

Hod Lipson and Jordan B. Pollack

Computer Science Department, Brandeis University
415 South Street, Waltham, MA 02454, USA
lipson@cs.brandeis.edu

Abstract

One of the prevailing characteristics of natural life is *autonomy*. The field of Artificial Life has so far addressed the notion of autonomy mostly in terms of power and behavior. In this paper we attempt to extend the notion of autonomy to include also *design and fabrication*: We claim that not only should artificial creatures be able to operate untethered and without external guidance, but they should also be, like living systems, *autonomously designed and fabricated* without external intervention. Only then can we expect synthetic creatures to bootstrap and sustain their own evolution. In this work we demonstrate for the first time a path that allows transfer of virtual diversity into reality, and so reduce this key principle of Artificial Life into practice. Our approach is the use of only elementary building blocks in both the design and embodiment. We describe a set of preliminary experiments evolving electromechanical systems composed of thermoplastic, linear actuators and neurons for the task of locomotion, first in simulation then in reality. Using 3D solid printing, these creatures then replicate automatically into reality where they faithfully reproduce the performance of their virtual ancestors.

Introduction

One of the prevailing characteristics of natural life is *autonomy*. The field of Artificial Life has so far addressed the notion of autonomy mostly in terms of power and behavior. In this paper we attempt to extend the notion of autonomy to include also *design and fabrication*: Artificial creatures should not only be able to operate untethered and without external guidance, but they should also be, like living systems, *autonomously designed and fabricated* without external intervention. We hypothesize that only then can we expect synthetic creatures to bootstrap and sustain their own evolution.

The lack of full autonomy has caused a dichotomy in Artificial Life research: Objects of study in this field are either digital creatures that are diverse and dynamic but remain virtual, or hand designed and constructed robots that are physical but have a predominantly fixed architecture. Indeed, studies in the field of evolutionary robotics reported to date involve either entirely virtual worlds (Sims 1994; Komosinski 1999), or, when applied in reality, adaptation of only the control level of manually designed and constructed robots (Husbands and Meyer, 1998) with a predominantly fixed architecture. Other works involving

real robots make use of high-level building blocks, like wheeled platforms and C-subroutines comprising significant pre-programmed knowledge (Leger, 1999)

We thus seek automatically designed and constructed physical artifacts that are functional in the real world, diverse in architecture (possibly each slightly different), and producible at low cost and large quantities. So far these requisites have not been met.

Structure of this paper

This paper is structured as follows: First, we outline our approach for autonomy, based on the use of elementary building blocks, and contrast it with current approaches in evolutionary robotics. We then describe a set of preliminary experiments examining the ability to achieve fully automated design and transfer into reality, both with as little human intervention as possible. We make no claims as to the evolutionary computation itself: we use a simple evolutionary algorithm and hence do not elaborate on it beyond providing sufficient details for replication. We then present results of both virtual and real machines evolved for the task of locomotion, and compare their performance. In performing this comparison we essentially complete a physical synthetic evolution cycle. We then conclude with some remarks on the significance of these results.

Elementary Building Blocks

The approach we propose is based on the use of only elementary constituents in both the design and fabrication process. As building blocks become more elementary, external knowledge associated with them is minimized, and at the same time architectural flexibility is maximized. Similarly, use of elementary building blocks in the fabrication process allows it to be more simple and systematic. The use of elementary building blocks also minimizes potential inductive bias that might be introduced inadvertently into the evolutionary substrate. In theory, if we could use only atoms as building blocks, only laws of physics as constraints and only atom-manipulation as a fabrication process, then this principle would be maximized. Earlier reported work on evolution complete-creatures used higher level knowledge and consequently limited architectures (like only tree structures, Sims 1994; Komosinski 1999) and resulted in expedited convergence to

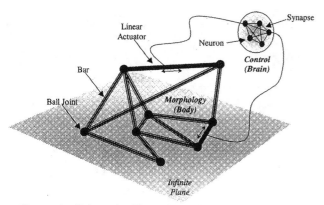

Figure 1. Schematic illustration of an evolvable robot, containing only linear bar/actuators and control neurons. Some bar architectures form rigid substructures

working solutions, but at the expense of exploration power and reduced claims of innovation that might be attributed to the algorithm.

Experiments

In a set of experiments we used bars as building blocks of structure, neurons as building blocks of control, and additive fabrication (Demos *et al*, 1998) as a production process. Bars connected with free joints can form trusses that represent arbitrary rigid, flexible and articulated structures as well as multiple detached structures, with revolute, linear and planar joints at various levels of hierarchy. Similarly, sigmoidal neurons can connect to create arbitrary control architectures such as feed-forward and recurrent nets, state machines and multiple independent brains. Additive fabrication allows automatic generation of arbitrarily complex physical structures and series of physically different bodies. A schematic illustration of a possible architecture is shown in Figure 1. The bars connect to each other through ball-and-socket joints, neurons can connect to other neurons through synaptic connections, and

neurons can connect to bars. In the latter case, the length of the bar becomes governed by the output of the neuron, essentially making it a linear actuator. No sensors were used at this stage.

Starting with a population of 200-1000 machines that were initially comprised of zero bars and neurons, we conducted evolution in simulation. The fitness of a machine was determined by its locomotion ability: the net distance its center of mass moved on an infinite plane in a fixed duration. The process iteratively selected fitter machines, created offspring by adding, modifying and removing building blocks, and replaced them into the population. Figure 2 shows a typical progress of fitness of creatures as function of generations. (See Appendix A for details of the experiments).

We use only elementary operators of mutation that introduce least external knowledge. This process continued for several hundred generations. Both body (morphology) and brain (control) were co-evolved simultaneously. Although it is common practice in the field to separate and evolve a control for a fixed morphology and vice versa, in nature there is no such distinction – like a chicken and egg – neither came first. Coevolution has been successfully used to solve problems such as sorting networks (Hillis, 1990), and cellular automata (Juillé and Pollack, 1998).

The simulator we used for evaluating fitness supported

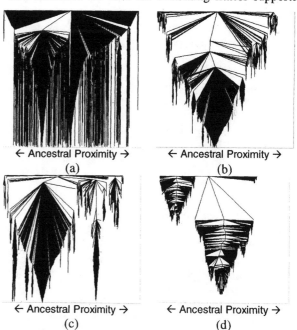

Figure 3. Phylogenic trees of several different evolutionary runs. Vertical axis represents generations and horizontal axis represents ancestral proximity. Trees exhibit various degrees of divergence and speciation: (a) extreme divergence, (b) extreme convergence, (c) intermediate level, and (d) massive extinction. The trees were thinned and depict several hundred generations each.

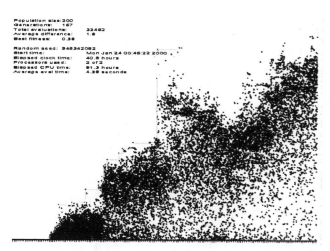

Figure 2. A typical plot of solution fitness (0-0.38) as function of generation (167).

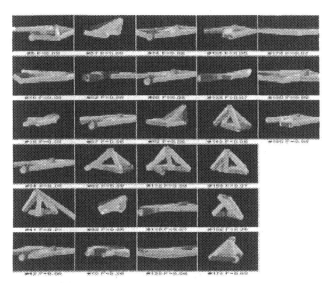

Figure 4. A sample instance of an entire generation, thinned down to show only different individuals. Note the two prevailing species.

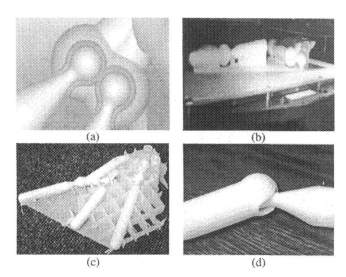

Figure 5. (a) Automatically fleshed joints, (b) replication process, (c) pre-assembled robot in mid print, (d) joint close-up

quasi-static motion in which each frame is statically stable[1]. This kind of motion is simpler to transfer reliably into reality, yet is rich enough to support low momentum locomotion (See Appendix B for details of the simulation). Typically, several tens of generations passed before the first movement occurred. For example, at a minimum, a neural network generating varying output must assemble and connect to an actuator for any motion. Various patterns of evolution dynamics emerged, some of which are reminiscent of natural phylogenic trees. Figure 3 presents examples of extreme cases of convergence, speciation, and massive extinction. A sample instance of an entire generation, thinned down to only unique individuals is shown in Figure 4.

Fabrication

Selected robots out of those with winning performance were then *automatically* replicated into reality: their bodies, which exist only as points and lines, are first converted into a solid model with ball-joints and accommodations for linear motors according to the evolved design (Figure 5a). The solidifying stage was automatic but used a hand-coded procedure describing a generic bar, joint, and actuator. The virtual solid bodies were then materialized using commercial rapid prototyping technology (Figure 5b). This machine uses a temperature-controlled head to extrude thermoplastic material layer by layer, so that the arbitrarily evolved morphology emerges as a solid three-dimensional structure (Figure 5c) without tooling or human intervention. The entire pre-assembled machine is printed as a single unit, with fine plastic supports connecting between moving

---

[1] This kind of motion is similar to popular plasteline animation, also known as *claymation*.

parts; these supports break away at first motion (Figure 5d).

The resulting structures contained complex joints that would be difficult to design or manufacture using traditional methods (see Figure 6b). Standard motors are then snapped in, and the evolved neural network is executed to activate the motors.

Results

In spite of the relatively simple task and environment (locomotion over an infinite horizontal plane), surprisingly different and elaborate solutions were evolved. Machines typically contained around 20 building blocks, sometimes with significant redundancy (perhaps as means against mutation, Lenski *et al*, 1999). Not less surprising is the fact that some exhibited symmetry, which was neither specified nor rewarded for anywhere in the code; a possible explanation is that symmetric machines are more likely to move in a straight line, consequently covering a greater net distance and acquiring more fitness. Similarly, successful designs appear to be robust in the sense that changes to bar lengths would not significantly hamper their mobility. The corresponding physical machines (3 to date) then faithfully reproduced their virtual ancestors' behavior in reality.

Table I: Comparison of performance of physical creatures versus their virtual origin. Values are net distance [cm] center of mass traveled over 12 cycles of neural network. Distances in physical column are compensated for scale reduction (actual distance in parentheses).

Distance traveled [cm]	Virtual	Physical
Tetrahedron (Figure 6a)	38.5	38.4 (35)
Arrow (Figure 6b)	59.6	22.5 (18)
Pusher (Figure 6c)	85.1	23.4 (15)

Table 1 compares the performance of the physical creatures versus their virtual origin. While the distances traveled in differ significantly between simulation and reality, it should be noted that the actual mode of locomotion was faithfully replicated. The difference stems from slipping of the limbs on the floor surface, implying that the simple Newtonian friction model we used was insufficiently accurate. Three samples are shown and described in Figure 6, exploiting principles of ratcheting (6a), anti-phase synchronization (6b) and dragging (6c). Others (not shown here) used a sort of bi-pedalism, where left and right "limbs" are advanced in alternating thrusts. Some mechanisms moved articulated components to produce crab-like sideways motion. Other machines used a balancing mechanism to shift friction point from side to side and advance by oscillatory motion.

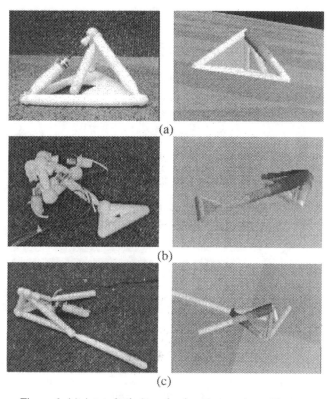

(a)

(b)

(c)

Figure 6. (a) A tetrahedral mechanism that produces hinge-like motion and advances by pushing the central bar against the floor. (b) This surprisingly symmetric machine uses a 7-neuron network to drive the center actuator in perfect anti-phase with the two synchronized side limb actuators. While the upper two limbs push, the central body is retracted, and vice versa. (c) This mechanism has an elevated body, from which it pushes an actuator down directly onto the floor to create ratcheting motion. It has a few redundant bars dragged on the floor, which might be contributing to its stability. These machines perform in reality is the same way they perform in simulation. Motion videos of these robots and others can be viewed at http://www.demo.cs.brandeis.edu/golem

Conclusions

In summary, while both the machines and task we describe in this work are fairly simple from the perspective of what teams of human engineers can produce, and what biological evolution has produced, we have demonstrated for the first time a robotic bootstrap, where robotically designed electromechanical systems can be robotically manufactured. We have carefully minimized human intervention both in the design and in the fabrication stages. Besides snapping in the motors, the only human work was in informing the simulation about the universe it could expect to be manufacturable.

Without reference to specific organic chemistry, life is an autonomous design process that is in control of a complex set of chemical factories allowing the generation and testing of physical entities which exploit the properties of the medium of their own construction. Using a different medium, namely off-the-shelf rapid manufacturing, and evolutionary design in simulation, we have replicated this autonomy of design and manufacturing. Our claimed advance, namely the ability to move artificial evolution from simulation to the real world is not a mere curiosity; rather, some claim that if indeed artificial systems are to ultimately interact and integrate with reality, they must learn, evolve and be studied in it (Beer, 1990). Technological advances in multi-material rapid prototyping, MEMS and nano-fabrication, higher fidelity of physical simulation and increased understanding of evolutionary computational processes thus open exciting opportunities in more fully automating this path towards artificial life.

Acknowledgements

This research was supported in part by the Defense Advanced Research Projects Administration (DARPA) Grant No. DASG60-99-1-0004. Hod Lipson acknowledges the generous support of the Fischbach Postdoctoral Fellowship.

Appendix A

Details of the evolutionary simulation

Experiments were performed using version 1.2 of GOLEM (Genetically Organized Lifelike Electro Mechanics), which can be obtained from http://www.demo.cs.brandeis.edu/golem. We carried out a simulated evolutionary process (Holland, 1975; Koza, 1992): The fitness function was defined as the net Euclidean distance that the center-of-mass of an individual has moved over a fixed number (12-24) of cycles of its neural control. We started with a population of 200-1000 null (empty) individuals. Random seed was randomized. Individuals were then selected, mutated, and replaced into the population in steady-state as follows: The selection function can be either random, fitness proportionate or rank proportionate. The mutation operators used to generate an offspring can be any of the following (with probability): Small mutation in length of bar or neuron synaptic weight (0.1), removal/addition of a small dangling bar or unconnected neuron (0.01), split vertex into two and add a small bar or split bar into two and add vertex (0.03), attach/detach neuron to bar (0.03). The dice is rolled until at least one mutation is applied. After mutation, a new fitness is assigned to the individual by means of a simulation of the mechanics and the control (see details below). The offspring is inserted into the population by replacing an existing individual either chosen at random, chosen in inverse-proportion to its fitness, or using similarity proportionate elitism (deterministic crowding, Mafoud, 1995). Various permutation of selection-replacement methods are possible; we typically used fitness-proportionate or rank selection with random replacement, or random selection with similarity proportionate replacement where similarity was approximated by the distance in the ancestral tree. The process continued for 500-5000 generations (approx 10^5 to 10^6 evaluations overall). The process was carried both serially and in parallel (on a 16-processor computer). On parallel computers we noticed an inherent bias towards simplicity: Simpler machines could complete their evaluation sooner and consequently reproduce more quickly than complex machines.

Appendix B

Details of the simulation

Both the mechanics and the neural control of a machine were simulated concurrently. The mechanics were simulated using quasi-static motion, where each frame of the motion is assumed to be statically stable. This kind of motion is simple to simulate and easy to induce in reality, yet is rich enough to support various kinds of low-momentum motion like crawling and walking (but not jumping). The model consisted of ball-joined cylindrical bars with true diameters. Each frame was solved by relaxation: An energy term was defined, taking into account elasticity of the bars, potential gravitational energy, and penetration energy of collision and contact. The degrees of freedom of the model (vertex coordinates) were then iteratively adjusted according to their derivatives to minimize the energy term, and the energy was recalculated. Static friction was also modeled. The use of relaxation permitted handling singularities (e.g. snap-through buckling) and under-constrained cases (like dangling bar). Noise was added to ensure the system does not converge to unstable equilibrium points, and to cover simulation-reality gap. Material properties modeled correspond to the properties of the rapid prototyping material (E=0.896GPa, ρ=1000Kg/m^3 σ_{yield}=19MPa). The neural network was simulated in discrete cycles. In each cycle, actuator lengths were modified in small increments not larger than 1 cm.

Simulator physics

Static solution of each frame is achieved by defining a global system energy term H and then slightly modifying each of the systems degrees of freedom so as to lower H according to its partial derivatives. This process continues until relaxation is reached or instability is determined. The terms included in H define the richness of the simulation. For example, a basic model including bar flexion and gravitational energy would be:

$$H = \sum_1^n k\delta^2 + \sum_1^n mgh$$

where n is the number of bars, and for each bar i, m represents the mass, h represents the average height, and the term k represents the stiffness,

$$k_i = \frac{E_i A_i}{l_i}$$

where E is the material's module of elasticity, A is the cross section, and l is the length of the bar, and δ represents the difference between the bar's current length and its original length, given by

$$\delta_i = \sqrt{(v_x - w_x)^2 + (v_y - w_y)^2 + (v_z - w_z)^2} - l_i$$

and v,w represents the two current endpoints of the bar. Differentiating the total energy H with respect to each of the degrees of freedom (coordinates of the endpoints, in this case), produces direction of adjustment (second derivatives would produce a more accurate adjustment, etc.).

$$\Delta w_x = \frac{\partial H}{\partial w_x} \cdot \Delta s$$

where Δs (the relaxation factor) represents the adjustment magnitude for each iteration. Each degree of freedom is then updated by its adjustment value, and H is recalculated. Small Δs produce a stable but slower convergence, whereas large Δs can solve static frames faster but may run into stability problems. This process is repeated until adjustments go below a certain threshold.

Although this solution method is slower that simultaneous solutions, (e.g. finite elements), it is capable of solving correct highly-nonlinear cases (e.g. contact collision and snap-through buckling) and, more importantly, under-constrained states: for example, a truss in mid fall which is not yet fully supported (by six rigid-body equations), as it falls into position. Indeed as machines move and function within the simulator they often pass through such stages.

As additional real-world energy terms are added to H, it becomes more accurate. We have included terms for friction, collision between bars, external forces and noise.

References

Beer R. D., (1990) *Intelligence as Adaptive Behavior*, Academic Press

Dimos. D, Danforth S.C., Cima M.J., (1998), *Solid Freeform and Additive Fabrication*, MRS Symposium, Boston Massachusetts

Hillis, W. D. (1990). Co-evolving parasites improve simulated evolution as an optimizing procedure. *Physica.*

Holland J., (1975) *Adaptation in natural and artificial systems*, University of Michigan Press

Husbands P., Meyer J. A., (1998), *Evolutionary Robotics*, Springer Verlag

Juillé, H. and Pollack, J. B. (1998). Coevolving the "Ideal" Trainer: Application to the Discovery of Cellular Automata Rules. *Proceedings of the Third Annual Genetic Programming Conference* , Madison, Wisconsin, July 22 - 25, 1998

Komosinski M., Ulatowski S., (1999) Framstics: Towrds a simulation of a nature-like world, creatures and evolution, ECAL '99, pp. 261-265

Koza J., (1992) *Genetic Programming,* MIT Press

Leger C., (1999), "Automated Synthesis and Optimization of Robot Configurations: An Evolutionary Approach", *Ph.D. Thesis*, Carnegie Mellon University

Lenski R. E., Charles O., Collier T., Adami C., (1999) "Genome Complexity, robustness and genetic interactions in digital organisms", *Nature* 400, pp. 661-664

Mahfoud S. W., (1995), "Niching methods for genetic algorithms", *Ph.D. Thesis,* University of Illinois at Urbana-Champaign

Sims K., (1994) Evolving Virtual Creatures, Computer Graphics, SIGGRAPH '94, pp. 15-22

Evolving Insect Locomotion using Non-uniform Cellular Automata

Edgar E. Vallejo
Computer Science Department
Tecnológico de Monterrey
Campus Estado de México, México
evallejo@campus.cem.itesm.mx

Fernando Ramos
Computer Science Department
Tecnológico de Monterrey
Campus Morelos, México
framos@campus.mor.itesm.mx

Abstract

This article presents a model for the evolution of loco-motion behavior in a simulated insect. In our model, locomotion is defined over a discrete state space using non-uniform cellular automata. The architecture of the model is inspired from the distributed model for leg coordination proposed by Cruse. We apply a genetic algorithm to a population of non-uniform cellular automata to evolve locomotion behaviors. We demostrate that this model can be used to evolve several commonly observed gaits of insects. Additionally, we show that the evolutionary process yielded periodic attractors which are invariant from the initial conditions.

Introduction

The problem of insect locomotion has been approached with several methods. Brooks (1989) has used a subsumption architecture in building a robot that walks. Beer (1990) has used a recurrent neural network to control locomotion in a simulated insect. Spencer (1994) has demonstrated that genetic programming can be used to evolve locomotion with a minimum of assumptions about how walking should occur.

Some insect inspired locomotion models suggest that locomotion may be viewed as an emergent property of local interactions between the mechanisms responsible for the control of individual legs (Ferrell 1995). These models share several properties with cellular automata: parallelism, locality of interactions and simplicity of components. We use this observation in building a model based on non-uniform cellular automata. The model gives a simulated insect the ability of walking.

Previously, cellular automata have been applied to the study of many different emergent collective phenomena (Gutowitz 1991). More recently, cellular automata have been used as the foundation of artificial life models (Sipper 1995). We propose the use of cellular automata for the synthesis of agent behaviors.

Research in autonomous agents has focused for many years on the design of computational models capable of synthesizing agent behaviors. Pfeifer and Scheier (1999) argue in favor of parsimony when modeling agents: if there are a number of competing models, the more parsimonious ones are to be preferred. Cellular automata are general and simple (Sipper 1995). Additionally, they provide a formal framework for understanding the emergent and dynamical properties of agent behaviors. However, the design of the cellular automata local interaction rules to perform global computation can be extremely difficult to accomplish.

Das, Crutchfield, Mitchell and Hanson (1996) have demonstrated that an evolutionary process can be used to produce globally coordinated behavior on a distributed system. Using a genetic algorithm to evolve cellular automata, they showed the evolution of spontaneous synchronization and emergent coordination.

In our model, locomotion is defined over a discrete state space using non-uniform cellular automata (Sipper 1997). In non-uniform cellular automata, the cellular rules not need to be identical for all cells. The architecture of the model is inspired from the distributed model for leg coordination proposed by Cruse (1990). We apply a genetic algorithm to a population of non-uniform cellular automata to evolve locomotion behaviors. Our results indicate that this model can be used to evolve several commonly observed gaits of insects, including those resulting from amputation.

We demonstrate that a simple, parsimonious model based on non-uniform cellular automata is capable of explaining the essence of coordination in locomotion behavior. Additionally, we show that the evolutionary process yielded periodic attractors which are invariant from the initial conditions. These periodic behaviors represent general solutions to the insect locomotion problem.

Insect Locomotion

Insect locomotion is notably robust and flexible. Insects can walk over a variety of terrains. In addition, they can also adapt their gait to the loss of up to two legs without serious degradation of performance. To provide support and progression, the movement of the legs must be coordinated by the locomotion system (Beer & Chiel 1995).

Wilson (1966) defined several terms associated with insect locomotion:

Protraction: The leg moves towards the front of the body.

Retraction: The leg moves towards the rear of the body.

Power stroke: The leg is on the ground where it supports and then propels the body. In forward walking, the leg retracts during this phase.

Return stroke: The leg lifts and then swings to the starting position of the the next power stroke. In forward walking, the leg protracts during this phase.

Anterior extreme position: In forward walking, this is the target position of the return stroke.

Posterior extreme position: In forward walking, this is the target position of the power stroke.

Cruse (1990) developed a model that employs stimulus and reflexes to generate leg motion and gait coordination. There are three basic excitatory and inhibitory influences responsible for the coordination between the legs. These influences affect the threshold for beginning the return stroke by adjusting the posterior extreme position of the receiving leg. It has been shown that these three influences provide the necessary and sufficient conditions for sustained movement.

The distributed model for leg coordination proposed by Cruse may be viewed as a *dynamical system* in which the set of the state variables consists of the positions of the legs and their corresponding extreme positions at every time step. The *dynamical law* of the system can be formulated in terms of the excitatory and inhibitory influences received by each leg.

This perspective of the locomotion model shares several properties with cellular automata: parallelism, locality of interactions and simplicity of components. Also, cellular automata are dynamical systems. We use this observation in building a locomotion model based on non-uniform cellular automata.

The Model

The artificial insect we use is based on the model proposed by Beer (1990). The design of the body of the artificial insect is shown in figure 1.

The insect can raise and lower its legs, and is assumed to be constantly moving. That is, when the

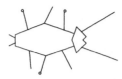

Figure 1: Artificial insect (Beer, 1990)

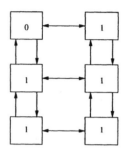

Figure 2: Cellular automata locomotion model

insect raises one of its legs, a protraction movement is automatically activated. Similarly, when the insect lowers one of its legs, a retraction movement is automatically activated. Finally, both the return stroke and the power stroke are assumed to terminate in one time step. The return stroke has no effect when the leg is positioned at its anterior extreme position. Similarly, the power stroke has no effect when the leg is positioned at its posterior extreme position.

In this work, we conceive locomotion as an emergent behavior. Locomotion emerges as a result of the interaction of two non mutually exclusive behaviors: support and progression.

The artificial insect exhibits the support behavior at any time step if the polygon formed by the supported legs contains the center of mass of the body (Beer 1995). The artificial insect exhibits the progression behavior if the legs are coordinated over a number of time steps in order to move the body of the insect.

The distributed model for leg coordination is modeled using a non-uniform cellular automaton consisting of six cells: L_1, L_2, L_3, R_1, R_2 and R_3. Each cell is a possible different finite state machine that transits between power stroke and return stroke depending on the state of the adjacent cells. Each cell assumes a "0" state during its return stroke and a "1" state during its power stroke. An example configuration of a cellular automaton is shown in figure 2. In this configuration, cell L_1 is in its return stroke and the other cells are in their power stroke.

The neighborhood state function g of each cell α is determined by the direction of the influences indicated

α	$g(\alpha)$
L_1	$\{L_2, R_1\}$
L_2	$\{L_1, L_3, R_2\}$
L_3	$\{L_2, R_3\}$
R_1	$\{R_2, L_1\}$
R_2	$\{R_1, R_3, L_2\}$
R_3	$\{R_2, L_3\}$

Table 1: Neighborhood state function

L_1	L_2	R_1	L_1
0	0	0	1
0	0	1	1
0	1	0	0
0	1	1	1
1	0	0	0
1	0	1	0
1	1	0	0
1	1	1	1

Table 2: Transition function for cell L_1

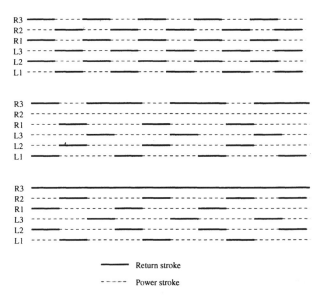

Figure 3: Experiment 1: gaits

in the distributed model for leg coordination proposed by Cruse. Table 1 shows the fuction g.

We define 28 configurations in which the insect is assumed to be supported. The insect is stable when at least 4 legs are in the "1" state. Additionally, we define six stable configurations with 3 legs.

A gait is defined in terms of a propagation of the cellular automaton. Most of the propagations are inherently periodic (limit cycles).

Experiments

Experiment 1

In this experiment we applied a genetic algorithm to a population of non-uniform cellular automata. We consider fixed initial conditions in which all the legs are in their power stroke.

We use a generational genetic algorithm with linear scaling but without elitism (Goldberg 1989) . These considerations provide the opportunity to obtain several different solutions to the problem in the same run.

Genome representation The representation of the genome consists of the concatenation of the values of the state transition function of each cell using the order imposed by the state transition table. For example, consider the transition function shown in table 2.

The right column of the state transition function is codified in the genome using the order imposed by the table. In this way, values are indexed in the genome and can be accessed directly. Concatenating the values

for all cells yields a genome of length 64.

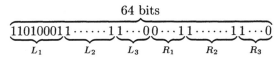

Fitness function Fitness is defined as the distance traveled by the insect as long as it remains stable. We assume that the insect takes one step if it is stable and if at least one return stroke is immediately followed by a power stroke.

Locomotion behavior is inherently periodic. This observation indicates that the evaluation of individuals can reach an infinite loop. To avoid this problem, we define a maximum distance traveled by the insect.

Parameters of the runs We use a population of 256 individuals, and a maximum of 100 generations. We set the crossover and mutation probabilities to $p_c = 0.6$ and $p_m = 0.001$, respectively. Also, we set the maximum distance traveled by the insect to 64 steps.

Results We performed several runs with this model. In each run, the evolutionary process yielded individuals capable of sustained movement. The gaits obtained are similar to gaits observed in insects, including those resulting from amputation. Figure 3 shows some of the gaits obtained from this experiment.

The first gait corresponds to the tripod gait. In the second gait, the cell R_2 converges to the power stroke. This is interpreted as a lesion in which the leg drags. In the third gait, the cell R_3 converges to the return stroke. This is interpreted as an amputation of the leg.

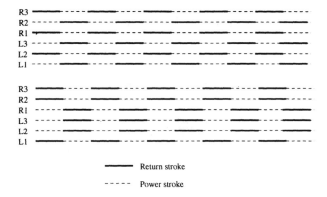

Return stroke

Power stroke

Figure 4: Experiment 2: periodic attractors

State	Phase
1	power stroke
2	power stroke
3	power stroke
0	return stroke

Table 3: States of the leg

State	Protraction	Retraction
1	0	2
2	0	3
3	0	3
0	0	1

Table 4: State transitions

Experiment 2

In this experiment we explore the generality of the solutions found in our experiments. The genetic algorithm searched for a non-uniform cellular automaton capable of producing a sustained movement from all valid initial conditions. We considered all the 28 configurations in which the insect is assumed to be supported.

We used the genetic algorithm of experiment 1.

Results. We performed several runs with this model. In each run, the evolutionary process yielded individuals capable of sustained movement. The gaits obtained behave as periodic attractors. From every initial conditions, the cellular automata converge to a particular gait. Two of the periodic attractor obtained are shown in figure 4.

In most cases, the periodic attractors obtained correspond to different versions of the tripod gait and all of the initial conditions converge to the same gait. In other cases, some of the initial conditions converge to a tripod gait and the rest converge to a different tripod gait. These results show the existence of several periodic attractors that separates the state space into different basins of attraction sets.

Experiment 3

The model considered in previous experiments is incapable of producing several gaits observed in insects. This problem arises because the model captures the approximation to the extreme positions in a limited way. A solution to this problem is to extend the neighborhood of each cell. However, with this modification the distributed nature of the model will be affected.

Another solution to this problem is to extend the state space of each cell. We present a modification of the model in which the power stroke is split into 3 states. These states represent different advance degrees of the leg. Invariably, the return stroke is as-

sumed to reach the anterior extreme position in one time step. With this modification, each leg may assume four different states as shown in table 3.

Our model is restricted to forward walking. This consideration constrains the transitions between states of each leg. These constrains, in turn, provide the possibility of reducing the range of the state transition function of each cell. In consequence, we consider 2 possible values "0" and "1" corresponding to the actions of protraction and retraction. The indicated action returned by the state transition function is used to update the state of each leg as indicated in table 4.

Genome representation The genome representation for this model is essentially the same as in previous experiments. However, due to the extension of the state space of the cellular automata, the length of the genome increases to 768.

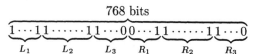

Fitness function. Fitness is defined as the distance traveled by the insect as long as it remains stable. We assume that the insect takes one step if it is stable and if at least one protraction movement is immediately followed by a retraction movement or if at least three retraction movements are executed by legs that are in states 1 or 2.

Parameters of the runs. We use a population of 2048 individuals, and a maximum of 100 generations. We set the crossover and mutation probabilities to $p_c = 0.6$ and $p_m = 0.001$, respectively. Also, we set the maximum distance traveled by the artificial insect to 64 steps.

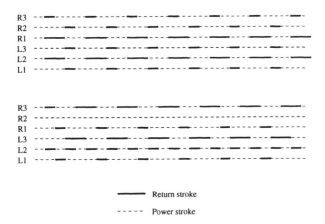

——	Return stroke
- - - -	Power stroke

Figure 5: Experiment 3: gaits

Results. We performed several runs with this model. In each run, the evolutionary process yielded individuals capable of sustained movement. The gaits obtained in this experiment exhibit more complex periodic behaviors. Two of the gaits obtained with this experiment are shown in figure 5.

Conclusions and Future Work

Cellular automata are recognized as simple, general and explanatory models for a wide variety of emergent collective phenomena. We propose the application of cellular automata to the synthesis of agent behaviors.

Depending on the research goals, cellular automata can be very useful models when modeling agents. Agent behaviors can be coded in a simple, parsimonious form. Additionally, the behavior produced by cellular automata can be analysed using the formal framework of the theory of dynamical systems.

In this work, we demostrate the application of cellular automata to the synthesis of locomotion behavior on a simulated insect. We use an evolutionary process to shape the dynamical law of the system. The model was capable of evolving several commonly observed gaits in insects.

When we explore the generality of the solutions, the evolutionary process yielded periodic attractors which are invariant from the initial conditions. The indentification of attractors and their corresponding basins of attraction sets can be very important when modeling agents for which the conditions of the environment are uncertain. These behaviors are invariant to perturbations.

The focus of this study has been the evolution of locomotion behavior. An immediate extension of this work is the consideration of other agent behaviors. Other extensions include a more formal analysis of the behavior of the system using tools provided by the theory of dynamical systems, such as the identification of separatrices of the state space in the presence of several periodic attractors. These studies will provide a framework for understanding the emergent and dynamical properties of agent behaviors.

References

Beer, R. 1990. *Intelligence as Adaptive Behavior. An Experiment in Computational Neuroethology*, Academic Press.

Beer, R. 1995 A dynamical systems perspective on autonomous agents. *Artificial Intelligence* 72: 173-215. Elsevier.

Beer, R. and H. Chiel. 1995. Locomotion Invertebrate. In M. Arbib ed., *The Handbook of Brain Theory and Neural Networks.* The MIT Press.

Brooks, R.A. 1989. A Robot that Walks: Emergent Behaviors from a Carefully Evolved Network. *Neural Computation.* **1:2**, pp 365-382.

Cruse, H. 1990. What Mechanisms Coordinate Leg Movement in Walking Arthropods. *Trends in Neurosciences.* **13**, pp 15-21.

Das, R., J. Crutchfield, M. Mitchell, M. and J. Hanson. 1995. Evolving globally synchronized cellular automata. En Eshelman, L. J., ed., *Proceedings of the Sixth International Conference on Genetic Algorithms.* Morgan Kaufmann.

Ferrell, C. 1995. A Comparision of Three Insect-Inspired Locomotion Controllers. *Robotics and Autonomous Systems.* **16:2-4**, pp 135-159. Elsevier.

Goldberg, D. E. 1989. *Genetic Algorithms in Search, Optimization and Machine Languages.* Addison Wesley.

Gutowitz, H. ed. 1991. *Cellular Automata. Theory and Experiment.* The MIT Press.

Pfeifer, R. and C. Scheier (1999) *Understanding Intelligence.* The MIT Press.

Sipper, M. 1995. Studying Artificial Life using a Simple, General Cellular Model. *Artificial Life* **2:1**. pp 1-35. The MIT Press.

Sipper, M. 1997. *Evolution of Parallel Cellular Machines. The Cellular Programming Approach,* Springer-Verlag.

Spencer, G. 1994. Automatic Generation of Programs for Crawling and Walking. In K. Kinnear (ed.) *Advances in Genetic Programming.* pp 335-353, The MIT Press.

Wilson, D. 1966. Insect Walking. *Annual Review of Entomology* **11**, pp 103-122.

From Directed to Open-Ended Evolution
in a Complex Simulation Model

Maciej Komosiński[1] and Ádám Rotaru-Varga[2]

[1]Institute of Computing Science,
Poznan University of Technology
Piotrowo 3A, 60-965 Poznan, Poland

[2]School of Computing Science
Simon Fraser University
Vancouver, Canada

Maciej.Komosinski@cs.put.poznan.pl

http://www.frams.poznan.pl/

Abstract

The problem of achieving open-ended evolution in complex systems is studied in this paper. We propose various techniques that support it, and present a gradual approach. These techniques are used and tested in the *Framsticks* system, which is a realistic, three-dimensional artificial life simulator with rich capabilities. Specifically, as there are no strong constraints imposed on the structure (body) and control (neural network brain), this system is suitable for testing open-endedness. In Framsticks, energetic requirements form the basis for competition, while interactions occurring on various levels act as a source of complexity and variation. The way towards open-ended evolution is discussed, developmental genotype encodings are proposed, and the results of so-far experiments are presented.

Introduction

Those of us who have ever run an artificial evolution experiment probably know the excitement of observing evolution unfolding under our guidance, and arriving at the solutions we imagined. Even bigger is our excitement when evolution surpasses our initial expectations, producing results never imagined. More exciting as it is, spontaneous, open-ended evolution rarely occurs. There are several theoretical works investigating reasons for this phenomenon and providing insights how to approach it. There are also several practical works focusing on *open-ended* evolution using relatively simple models (Adami and Brown 1994, Balkenius 1995, Channon and Damper 1998). In this work, we analyze such evolution in a complex, realistic system. We present theoretical arguments, discuss why open-endedness is important to study, why it is hard to accomplish, and what are some of its prerequisites. The gradual approach to achieving open-endedness is also described.

Our system, *Framsticks*, was implemented with several hierarchical objectives in mind: a fast simulation of a complex, realistic 3D world; autonomous creatures with free-form body and brain; ability to design and test creatures interactively; possibility of evolution of creatures to solve predefined problems (locomotion, swimming, food-finding, etc.); and ultimately, *spontaneous evolution of complex creatures* and behaviors. The development of Framsticks (Komosiński and Ulatowski 1999, Komosiński 2000) has satisfied these objectives, except for spontaneous evolution, which is our current focus. The property of open-endedness in such an environment may bring meaningful results, with potential references to natural evolution and the real world.

The creatures in Framsticks are composed of connected 'sticks', equipped with muscles, sensors and a neural network. They are capable of interacting with the environment (walking, swimming, ingesting), and each other (locating, pushing, hurting, killing, eating, etc.). The virtual world can contain a combination of flat land, hills, water, and various objects. Despite the simplified simulation (made possible by the sticks), this setup allows for complex behaviors.

Evolution operates on a population of genotypes (describing brains and bodies), using mutation and crossover. Several predefined criteria can be used to direct evolution (fitness). A host of interesting behaviors (Komosiński and Ulatowski 1997) is a result of this technique (some of them are reported here). Using a general criterion for selection, such as life span, favors reproduction of organisms which can survive efficiently, but does not restrict them to specific behaviors. This is one of our techniques for approaching open-ended evolution. Other proposed techniques include coevolution and natural selection.

Potentially, as there is no limitation imposed on the complexity of neural networks, creatures can display sophisticated behaviors. They can sense (and possibly remember) their environment, find out their orientation, locate food sources and other creatures. With the proper use of sensors and complex brains, creatures could discriminate between various morphologies of other habitants, sense their movement, exhibit preferences, and various group and social behaviors. The simulation of the virtual environment allows for *embodiment of intelligence*, which is argued to be an important factor when studying intelligence (Hofstadter 1979, Bajcsy 1998).

The following section describes our ideas on approaching open-endedness in a complex system. Section III presents the Framsticks system in more detail, and section IV characterizes the proposed genetic representations. Finally, we describe some of our evolutionary experiments, and outline future work.

Towards Open-Ended Evolution

As evolution in artificial systems had materialized in numerous works in the past decade, the distinction between *directed vs. open-ended* evolution was identified by several authors. This theme, referred to as "spontaneous evolution", "incremental adaptation" (Cliff, Harvey and Husbands 1993), or "perpetuating evolution" (Channon and Damper 1998), has been described from several perspectives, often independently. As the concept of open-ended evolution does not have well-defined boundaries, we are prompted to present our interpretation.

Intuitively, open-ended evolution happens if a system continues evolving. A more rigorous formulation is as follows: a system displays open-ended evolution when some internal complexity of the system can (and does) continually increase, with minimal outside involvement by the designer of the system. A more practical question is how to design and build such a system. Based on various suggestions and results, we compiled a list of necessary conditions (which does not have to be sufficient):

- the simulation model is not limited in complexity,
- interactions occurring on different levels act as a source of complexity,
- evolution works on highly evolvable genetic representations.

Since Framsticks satisfies mostly the first two requirements, our current focus is on more evolvable genotype representations. The internal complexity of creatures and inter-creature interactions are potentially unlimited. There are several types of other interactions: between parts of a creature, between creatures and environment, between groups of creatures (species), etc.

As we achieved interesting results with Framsticks with less and less human intervention, we learned that there are *degrees* to open-endedness. At one extreme, the designer's involvement can be total, when every detail is designed directly (this case is more interesting from an engineering standpoint than from artificial life). Evolution is given some role when it is used to improve hand-built creatures. When evolution is used to create creatures that satisfy some predefined criteria, human involvement is still present, but on a different level. Finally, in an open-ended scenario, the designer's involvement is minimal, and potential goals of evolution are not known *a priori*.

Based on this succession of scenarios, we feel that there is a gradual approach to open-endedness: it is not an all-or-nothing property. True open-endedness is difficult to achieve, but we believe Framsticks advances in the right direction.

Framsticks Ecosystem

System Architecture

Framsticks creatures live in a virtual world. However, a distinction is introduced between the gene pool and the living phenotypes. The capacity of the world and the capacity of the gene pool can be adjusted, thus constituting a simulation ratio. When the interactions between creatures can be ignored, only one phenotype may be simulated at a time. On the other hand, when the interaction is important and coevolution is to take place, the number of concurrently living organisms may be high, possibly equal to the capacity of the gene pool. The more individuals are simulated, the stronger is the competition between species (groups of individuals which exhibit similar methods of achieving fitness, for example, by acquiring energy).

Such separation of the pool of genes and the world of phenotypes results in greater flexibility of the system architecture, where it is possible to adjust the degree of coexistence. It also allows for easy distribution of parallel evolutionary processes. The state of simulation is considered as the state of the gene pool only. When saving the evolutionary snapshot, the world is disregarded (the environment and all living entities represent an enormous amount of data), and only the gene pool is saved.

The creatures are revived based on the genotypes from the gene pool. Selection may be random or proportional to the weighted sum of the predefined criteria (if directed evolution is to be used). The directed model means that evolution is considered as the process of optimization, as it is in genetic algorithms (Goldberg 1989) and many artificial life simulations (Sims 1994). It may be assumed in simplified reality, that reproduction is generally proportional to the life span of individuals (and living requires energy acquisition). Then it is possible to mimic some aspects of open-ended evolution with the 'directed' model of evolution, when the fitness criterion (and selection) is directly connected with the survival and reproduction abilities: the life span criterion can be used as the estimate of reproduction and fitness.

Physical Simulation

Framsticks simulates a three-dimensional world and creatures. We decided to use such a sophisticated environment for evolution, expecting that a range of complex, various stimuli affecting organisms will be the origin of dynamic development and emergence of interesting behaviors. All kinds of interaction between physical objects are considered: static and dynamic friction, damping, action and reaction forces, energy losses after deformations, gravitation, and uplift pressure – buoyancy (in a water environment).

One should note that there is always a tradeoff between simulation accuracy and simulation time. We need a fast simulation to perform evolution, on the other hand the system should be as realistic (detailed) as possible to produce realistic (complex) behaviors. As we expect emergence of more and more sophisticated phenomena, the evolution has to be longer and the simulation must be less accurate, but faster.

Currently, in order to make the simulation fast and due to the computational complexity, some aspects were discarded: collisions between parts of an organism itself and the movement of a water medium were both ignored. Including these in the simulation would make it more realistic, but would not introduce significant, qualitatively new phenomena. Meaningful interactions were considered more important than very realistic, but too slow simulation (Maes 1995).

The basic element is a stick made of two flexibly joined particles (finite element method is used for simulation). Sticks have specific properties: biological (muscle strength, stamina, energetic: assimilation, ingestion, and initial energy level), physical (length, weight, friction), and concerning stick joints (rotation, twist, curvedness). Biological properties are mutually exclusive: an increase in one property of a stick results in a decrease of the rest, which is the price of specialization.

Muscles are placed on stick joints. There are two kinds of muscles: bending and rotating. Positive and negative changes of muscle control signal make the sticks move in either direction – it is analogous to the natural systems of muscles, with flexors and extensors. The strength of a muscle determines its effective ability of movement and speed (acceleration). A stronger muscle consumes more energy during its work.

Framsticks have currently three kinds of *receptors* (*senses*): those for orientation in space (equilibrium sense, gyroscope), detection of energy (smell) and detection of physical contact (touch). A sample framstick equipped with these elements is shown on figure 1.

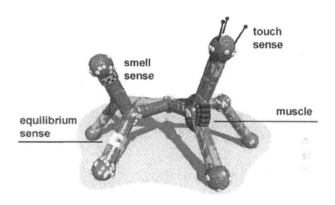

Fig. 1. Receptors (equilibrium, touch, smell) and effectors (muscles) in Framsticks

Neural Network

In order to expand the expression abilities of neural networks, the standard artificial neurons were enriched with a few additional parameters which could alter the behavior of each neuron independently. Since Nature was our inspiration, we did not introduce sophisticated and unnatural processing units – as in (Sims 1994) – because it is possible to construct complex modules (integrating, differentiating, summing, subtracting, and generators with different shapes) from simple neurons.

The additional neuron parameters (*force*, *inertia* and *sigmoid*) are under genetic control. They affect the way neurons work: *Force* and *inertia* influence changes of the inner neuron state (speed and tendency of changes, respectively). The *sigmoid* coefficient modifies the output function. Details and sample neuronal runs can be found at (Komosiński and Ulatowski 1997).

The neural network can have any topology and complexity. Neurons can be connected with each other in any way (some may be unconnected). Inputs can be connected to outputs of another neurons, constant values, or senses, while outputs can be connected to inputs of another neurons, or to effectors (muscles).

Genotype Representations

Various genotype encodings are proposed in order to study the processes of evolution in different setups, and to test their suitability for open-ended evolution. All of the representations combine "body" and "brain" in the same genotype, so that both morphology (body made of sticks) and control (brain made of neurons) evolve simultaneously. The simplest one is the basic direct encoding, which is least restrictive and fully exploits the capabilities of the world simulator (the genotype simply lists all the parts and details of an organism). Other encodings are described below.

Recurrent Direct Encoding

In this representation, the genotype describes all the parts of the corresponding phenotype. Small changes in the genotype cause small changes in the resulting creature. Stick phenotypic properties (see "Physical Simulation" in section III) are represented locally, but propagate through a creature's structure (with decreasing power). That means that most of the properties (and neural network connections) are maintained when a part of a genotype is moved to another place. Control elements (neurons, receptors) are associated with the elements under their control (muscles, sticks). Only tree-like structures can be represented (no cycles allowed).

While body is made of sticks, brain is made of neurons with their connections described *relatively*. Such a way of describing connections lets sub-networks come through the crossover operation: the whole set of neurons (a module) can be moved to another place in the genotype (and in the creature), possibly with limbs, and can still be operational.

The operations of mutation, crossing over and repair (to validate minor representation errors) are introduced.

Developmental Encoding

In nature, the genetic code of complex organisms does not encode their body layout directly, but rather their *process of development*. There are several hypothesized benefits of a developmental encoding in an evolutionary system (natural or artificial): higher evolvability, higher adaptation to environmental effects, compactness (Rotaru-Varga 1999). Based on these theoretical assumptions, we are developing a genotype encoding which is development-oriented, similar to encoding applied for evolving neural networks (Gruau 1996). An interesting merit of developmental encoding is that it can incorporate *symmetry* and *modularity*, features commonly found in natural systems, yet difficult to formalize.

The developmental encoding is similar to the recurrent direct one, but codes are interpreted as commands by cells (sticks, neurons, etc.). Cells can change their parameters,

and divide. Each cell maintains its own pointer to the current command in the genetic code. After division, cells can execute different codes, and thus differentiate themselves. The final body (phenotype) is the result of a development process: it starts with an undifferentiated ancestor cell, and ends with a collection of interconnected differentiated cells (sticks and neurons). Codes affecting stick and neuron parameters are identical to the representation in the direct encoding. Currently the developmental process is not a part of the simulated world, but it is implemented as an external process during genotype decoding. However, a developmental process inside the simulation would add a new set of interactions between the environment and genes, through the environment affecting the development of a creature.

Other Encodings

To test various representations of creatures (morphologies and brains), and the influence of the representations on the evolutionary process, other encodings are also proposed.

In the "direct similarity development" encoding, all the parts of a creature are described in the genotype, but each part is a separate object. Each object has some multi-dimensional "links", which are connected together during the embryogeny process in the way that maximizes their similarity. Thus the final creature is developed.

In the "implicit embryogeny development" encoding, the genotype encodes a set of rules which are used during embryogeny to develop a creature. The rules concern spatial emission of some "chemicals", which affect growth of a creature and future activation of rules themselves.

Further experiments with all the encodings and their comparison are yet to be done. Another suggestions and ideas of representations can also be relatively easily incorporated in Framsticks.

Discussion of Results and Future Work

Our first experiments concerned the study of locomotion and orientation, so the fitness was defined as speed (on the ground or in water). Recurrent direct encoding was used. Many walking and swimming species evolved during evolutionary runs, and we were able to see the evolution of ideas of "how to move" (Komosiński and Ulatowski 1997).

In one evolutionary run, a limb of an efficiently-moving creature was doubled while crossing over, and after some further evolution the organism was able to move with two limbs – one for pushing back and one for pulling. We also noticed a case when a limb was simultaneously bent and rotated, which was a more effective method of pushing against the ground. In one case a neuron used its saturation to produce delayed signals, which is a kind of a simple "short-term memory". More sophisticated creatures which

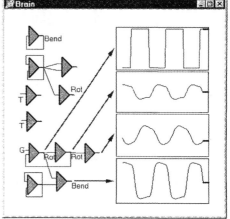

Fig. 2. A simple walking framstick. Two touch receptors (letter T), one gyroscope (G). Note the redundancy and random character of brain (six top-most neurons, including all touch sensors, do not influence the movement: touch neurons are unconnected, and the two muscle-controlling neurons are passive). Three (out of five) muscles are working; oscillations are produced only by the equilibrium sensor (G).

were evolved could not be easily examined because of their high complexity. A simple walking example is shown on figure 2. Even though most of body and brain designs seem to be redundant and not optimal, manual removing of some parts often makes individuals less fit. This happens due to various implicit dependencies (like feedbacks, etc.), which are overlooked or their influence is too complex to predict.

The experiments concerning evolutionary improvement of human predesigned structures have also been conducted. An example is the successful evolution of control (the neural network) for a hand-designed morphology in order to obtain a creature that lives long. Therefore, creatures were required to find energy sources and ingest them.

Some of the creatures display realistic behaviors even though they were not intended to imitate the real animals, as it was investigated in (Ijspeert 1999, Cruse et al. 1998). An example is a "salamander" (14 sticks, 16 neurons), which, after directed evolution of the neural network with fitness defined as speed, walks in a realistic way (see fig. 3). Another such example is swimming creatures; a swimming individual is studied in (Komosinski 2000).

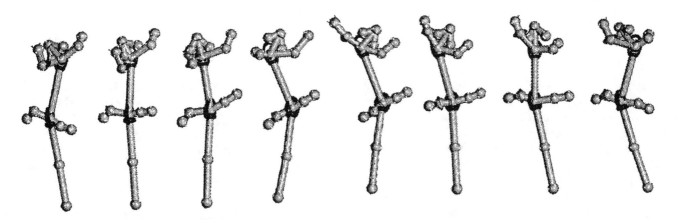

Fig. 3. A salamander-like, predesigned framstick. After evolution of brain walks in a realistic way.

A serious problem preventing open-endedness is when evolution is trapped in a local fitness maximum region. We attacked this problem from several sides: more evolvable genetic representations can induce a fitness landscape with less local peaks, and coevolution can dynamically change the fitness landscape.

When the survival of a creature does not depend on itself and the environment only, but also on interaction with other creatures (possibly from other species), the fitness landscape becomes dynamic (by coevolution). Effectively, a local maximum is 'lowered' by species trying to exploit it. Coevolutionary dynamics had been analyzed extensively in prey-predator (2 species) setups. The importance of coevolution is captured, among others, by (Steels 1997):

> Whereas evolution in itself causes an equilibrium to be reached, co-evolution causes a self-enforcing spiral towards greater complexity.

Framsticks has the potential for multi-species coevolution. Creatures can interact in various ways (push, eat, kill, etc.), and species can form and evolve simultaneously. However, a condition is that the number of simultaneously evolved creatures is high, which results in slow simulation.

We plan to improve the system architecture, and make the selection process more natural. A process of selection and reproduction which is external to the simulation does not resemble the natural situation. In a natural (or 'implicit') selection setup, creatures can trigger reproduction themselves when certain conditions are met (regarding energy, age). Thus the actual fact of survival is used instead of the probability of survival – numerical fitness (Cliff, Harvey and Husbands 1993). While natural selection makes it more difficult to 'breed' creatures to solve specific subtasks, it has better potential for maintaining perpetual evolution.

Our future work will also concern defining a similarity function on the phenotype space (to allow automatic analysis of populations and species, and to allow artificial speciation in directed evolution setup). We will compare various genotypic encodings and their influence on efficiency of directed and open-ended evolution. More receptors may be introduced. If the increase in computing power permits, we will use a more accurate (but slower) physical simulator, so that the evolved creatures will behave in a more realistic way.

Creating complex evolutionary systems, like Framsticks, raises the problem of analysis of the achieved results. After long efforts to *synthesize* artificial, realistic, open-ended ecosystems, we face the problem of *analyzing* something that was meant to become complex. Analysis of such results may rather be qualitative and behavioral, as simple quantitative measures are insufficient to capture the *way* and *method* of evolved successful individuals.

Investigation of sophisticated creatures requires support of human-friendly tools and intelligent automatic systems, or human experts.

Thus we decided to develop a possibility of worldwide participation (Komosiński and Ulatowski 1997) in the process of setting up experiments, their distributed execution, and analysis of results. We encourage the ALife community to take part in this collective enterprise.

The achieved results are promising, and we hope to contribute to a large-scale, open-ended evolutionary experiments and analysis. We are looking forward to exploit full functionality of the simulator and learn about the results of such experiments.

Acknowledgements

Maciej Komosiński wishes to thank KBN, Polish State Committee for Scientific Research, for the research grant supporting this work.

References

Adami, C., and Brown, C. T. 1994. Evolutionary learning in the 2D artificial life system "Avida". In Rodney A. Brooks and Pattie Maes (eds.), Artificial Life IV: MIT Press/Bradford Books.

Balkenius, Ch. 1995. Natural Intelligence in Artificial Creatures, Ph. D. Thesis, Lund University Cognitive Studies.

Bajcsy, R. 1998. Active Perception. In Proceedings of the IEEE (76) 8, 996-1005.

Channon, A. D., and Damper, R. I. 1998. Perpetuating evolutionary emergence. In From Animals to Animats 5, Proceedings of the Fifth International Conference on Simulation of Adaptive Behavior. Rolf Pfeifer, Bruce Blumberg, Jean-Arcady Meyer, and Stewart W. Wilson (eds): MIT Press.

Channon, A. D., and Damper, R. I. 1998. Evolving Novel Behaviors via Natural Selection. In Artificial Life VI, Proceedings of the Sixth International Conference on Artificial Life. Christoph Adami, Richard K. Belew, Hiroaki Kitano, and Charles E. Taylor (eds): MIT Press.

Cliff, D., Harvey, I., and Husbands, P. 1993. Incremental evolution of neural network architectures for adaptive behaviour. In M. Verleysen (ed.), Proceedings of the First European Symposium on Artificial Neural Networks, 39-44. Brussels, Belgium: D-facto/Editions Quorum.

Cruse, H., Kindermann, T., Schumm, M., Dean, J., and Schmitz, J. 1998. Walknet – a biologically inspired network to control six-legged walking. In *Neural Networks* 11 1435-1447: Elsevier Science Ltd.

Goldberg, D. E. 1989. *Genetic Algorithms in Search, Optimization, and Machine Learning*: Addison-Wesley Publishing Co.

Gruau, F. 1995. Modular Genetic Neural Networks for 6-Legged Locomotion. In Artificial Evolution European Conference, Lecture Notes in Computing Science, 201-219. Alliot, J.-M., Lutton, E., Ronald, E., Schoenauer, M., Snyers, D., eds: Springer, 1996.

Hofstadter, D. R. 1979. *Goedel, Escher, Bach: an Eternal Golden Braid.* NY: Basic Books.

Ijspeert, A. J. 1999. Synthetic Approaches to Neurobiology: Review and Case Study in the Control of Anguiliform Locomotion. In Proceedings of 5th European Conference on Artificial Life, 195-204. September 13-17, 1999, Lausanne, Switzerland: Springer-Verlag.

Komosiński, M., and Ulatowski, Sz. 1997. Framsticks Internet site, documentation, and sample experiments, http://www.frams.poznan.pl/

Komosiński, M., and Ulatowski, Sz. 1999. Framsticks: towards a simulation of a nature-like world, creatures and evolution. In Proceedings of 5th European Conference on Artificial Life, 261-265. September 13-17, 1999, Lausanne, Switzerland: Springer-Verlag.

Komosiński, M. 2000. The World of Framsticks: Simulation, Evolution, Interaction. In Proceedings of 2nd International Conference on Virtual Worlds, Paris: Springer.

Maes, P. 1995. Artificial Life meets Entertainment: Interacting with Lifelike Autonomous Agents. In Special Issue on New Horizons of Commercial and Industrial AI 38, 11 108-114. *Communications of the ACM*: ACM Press.

Rotaru-Varga, A. 1999. Modularity in evolved artificial neural networks. In Proceedings of the 5th European Conference on Artificial Life, Lausanne, 1999. D. Floreano, J.-D. Nicoud, F. Mondada, eds: Springer.

Sims, K. 1994. Evolving 3D Morphology and Behaviour by Competition. In R. Brooks, P. Maes (eds.), Artificial Life IV Proceedings, 28-39: MIT Press.

Steels, L. 1997. Synthesising the origins of language and meaning using co-evolution, self-organisation and level formation. In: Hurford, J., C. Knight, and M. Studdert-Kennedy (ed.), Evolution of Human Language: Edinburgh Univ. Press.

Artificial Neural Development for Pulsed Neural Network Design

- Generating Place Recognition Circuits of Animats -

Masayasu Atsumi

Department of Information Systems Science, Faculty of Engineering, Soka University
1-236 Tangi-cho, Hachioji-shi, Tokyo 192-8577, JAPAN, matsumi@t.soka.ac.jp

Abstract

We propose the artificial neural development method that generates the three-dimensional multi-regional pulsed neural network arranged in three layers of the nerve area layer, the nerve sub-area layer, and the cell layer. In this method, the neural development process consists of the first genome-controlled spatiotemporal generation of a neural network structure and the latter activity-dependent regulation of it. In the first process, by decoding a genome, 1)a nerve sub-area is generated in each nerve area and neurons are produced in it, 2)axonal outgrowth target sub-areas are recognized according to the attraction and repulsion rule, and 3)synapse formation is controlled under the topology preservation projection rule between origin cells and target cells. In the latter process, 4)programmed cell death occurs under control of spiking activity and a neurotrophic factor, then 5)synaptic efficacy is regulated according to the spike-based hebbian rule and weakened synapses are eliminated as a result of competition of spiking activity. For design of genomes, the steady state genetic algorithm is introduced and it is applied to initial genomes partially designed manually. To evaluate our artificial neural development method, simulation experiments are conducted to generate a pulsed neural network of an animal-like robot (animat) which moves in an environment. We evolve and develop an animat's place recognition circuit that contains the place cell area. The place recognition performance is evaluated in an environment where an animat comes into existence and in another environment where the animat enters after development. Through these experiments, we show our artificial neural development method is useful for generating a biologically realistic pulsed neural network of the animat.

Introduction

The design of artificial neural networks can result in the design of artificial genomes with lowered dimensions by mimicking the neurogenesis process of living things, automatically generating artificial neural networks based on genomes' decoding and neurergic regulation. The neurogenesis process of living things is roughly divided into the first genome-controlled process and the latter activity-dependent process. The first process forms a nerve structure under control of a genome and the latter process regulates it dependent on spiking activity caused by interaction with environments.

Various artificial neural development mechanisms can be formulated according to which part of the neurogenesis process is modeled. For example, (Vaario, Onitsuka and Shimohara 1997), (Rust *et al.* 1997a), (de Garis 1994) have treated mainly axonal outgrowth and synapse formation. (Cangelosi, Parisi and Nolfi 1994), (Eggenberger 1996), (Kitano 1995), (Dellaert and Beer 1996) have modeled cell division, cell differentiation and cell migration in addition to those processes. (Rust *et al.* 1997b), (Vos, Heijst and Greuters 1997) have focused on programmed cell death and synapse elimination especially.

In this paper, we propose the artificial neural development method that generates three-dimensional multi-regional pulsed neural networks arranged in three layers of the nerve area layer, the nerve sub-area layer, and the cell layer. The reason the pulsed neural network is adopted is that it is suitable for synchronous spiking expression within the scale of milliseconds, which is considered to be one of information expression in the brain. The three-layered structure is introduced to model neural structure at a level above that of the individual neuron but below that of an overall neural system, for example modules such as brain regions and their columnar organization of the nerve system of living things. The nerve system is a functional network of those modules, and, in general, modularity and functional connection of modules are indispensable for the design of a complex system. The multi-regional neural structure is designed by providing arrangement, properties, and connections of nerve areas. In our method, the design process is composed of the first genome-controlled process and the latter activity-dependent process by modeling the following features in the neurogenesis process of living things. The features of attention in the first process are as follows. 1) A genome predetermines the spatiotemporal structure of neural development, that is it is predetermined according to a genome when and where neurons are arranged and what properties they possess. 2) As for connection formation by the axon outgrowth among areas, axonal outgrowth targets are recognized according to attraction and repulsion between receptor factors on axons of presynaptic cells and guide factors on postsynaptic cells, and synapses are formed under the topology preservation projection between origin cells and target cells. The features of attention in the latter process are as follows. 3) For connection regulation among areas, the size of origin cells and target cells is adjusted based on the programmed cell death, and the target government among presynapses is controlled based on the synapse elimination.

By the way, since complex neural networks need complicated genomes, it becomes difficult to design all genomes by hand. So it is useful to apply evolutionary computation to the design of genomes, however it is also difficult to design all the complex genome structure by evolutionary computation because the computation cost is expensive and it often fails due to premature convergence. Accordingly a method for partially applying an evolutionary algorithm to manual coded genomes is introduced in our method, and we use the steady state genetic algorithm as an evolutionary algorithm.

To evaluate our artificial neural development method, simulation experiments are conducted to generate the place recognition circuit of an animal-like robot (animat) which moves in an environment. It is known in rats that assembly of place cells that fires at the same time express a place in an environment. It is also examined that the place recognition circuit is a multi-regional circuit which contains the place cell area, the head-direction cell area, the outer-world sensory area, and so forth. For these reasons, we evolve and develop an animat's place recognition circuit that contains the place cell area. The Bayesian place reconstruction mapping (Zhang *et al.* 1998) from a spiking pattern of place cells to a place in an environment is used for evaluation of animat's place recognition, and the accuracy of this reconstruction is used as fitness for evolution. The place recognition performance is evaluated in an environment where an animat comes into existence and in another environment where the animat enters after development. Through these experiments, we discuss utility of our artificial neural development method and some properties of generated place recognition circuits.

This paper is organized as follows. In section 2, we present our artificial neural development method. Section 3 describes the formation of the animat's place recognition circuit and the genome to generate it genetically. Section 4 describes experimental results on the animat's place recognition circuit genesis and evaluation of place recognition performance. Finally in section 5, we present conclusions and ongoing research.

Artificial Neural Development NEUROGEN'2000

Overview of NEUROGEN'2000

Our artificial neural development method, that is named NEUROGEN'2000, generates pulsed neural networks (Gerstner 1999) arranged in three layers of the nerve area layer, the nerve sub-area layer, and the cell layer, expressed in the three-dimensional coordinate system of the anterior-posterior axis, the dorsal-ventral axis and the right-left axis. As a result, each cell has a three-dimensional coordinate triplet $((X_a,Y_a,Z_a), (X_s,Y_s,Z_s), (X_c,Y_c,Z_c))$, where (X_a,Y_a,Z_a) is a nerve area coordinate, (X_s,Y_s,Z_s) is a nerve sub-area coordinate in the nerve area, and (X_c,Y_c,Z_c) is a cell coordinate in the nerve sub-area.

Our genome represents a schedule of what kind of nerve sub-area is developed in each nerve area at each development stage. Therefore a genome consists of a matrix of genes whose order is defined by the number of nerve areas multiplied by the length of development stages.

The artificial neural development process in NEUROGEN'2000 is outlined as follows. A population of initial genomes is set by manually coding a part of genomes and automatically the rest. For each genome, by decoding a gene for each nerve area at each development stage,

Step 1: a nerve sub-area is generated and neurons are produced in it,

Step 2: axonal outgrowth target sub-areas are recognized according to the attraction and repulsion rule,

Step 3: synapse formation is controlled under the topology preservation projection rule between origin cells and target cells.

Then for a pulsed neural network generated,

Step 4: programmed cell death occurs under control of spiking activity and a neurotrophic factor,

Step 5: synaptic efficacy is regulated according to the spike-based hebbian rule and weakened synapses are eliminated as a result of competition of spiking activity.

Fitness of each neural network is evaluated in an environment, and the steady state genetic algorithm evolves the specified parts in genomes based on fitness values.

Target Recognition and Synapse Formation

The genome is a matrix of genes whose order is defined by the number of nerve areas multiplied by the length of development stages and it represents a schedule of what kind of nerve sub-area is developed in each nerve area at each development stage. Each gene expresses properties of a nerve sub-area and neurons in it, that is developed in a nerve area at a certain development stage. These properties include a direction of nerve sub-area generation in a nerve area, the number of neurons, the neuron's response type of excitation or inhibition, target nerve areas, the guide factor and the receptor factor for target sub-area recognition, the number of axon branches, and the amount of neurotrophic factor.

The artificial neural development is executed by decoding these genes in specified nerve areas at specified development stages. First, a sub-area is generated in the specified direction in a nerve area and neurons of specified number with specified response type are produced. Each sub-area is assigned a three-dimensional coordinate in the nerve area and each neuron is assigned a three-dimensional coordinate in the nerve sub-area. Second, target nerve sub-areas are selected according to the target recognition rule from among nerve sub-areas in the specified target area. The target nerve sub-area recognition is performed based on attraction and repulsion between the specified receptor factor on axons of presynaptic neurons and the specified guide factor on postsynaptic neurons. The target recognition rule is defined as follows.

[**Target recognition rule**] Let R and G be binary vectors of length n by which the receptor factor and the guide factor are expressed. For $R=(r_1,...,r_n)$ and $G=(g_1,...g_n)$, let define $(R \cap G)_i=1 \Leftrightarrow r_i=1 \wedge g_i=1$, and $R \subseteq G \Leftrightarrow R \cap G=R$. Then the target recognition rule is defined as follows: if $R \subseteq G$ then attraction between R and G occurs else repulsion of R against G occurs. □

Thirdly, axons grow toward the target nerve sub-area and synapses are formed between neurons in the origin nerve sub-area and the target nerve sub-area. The topology among coordinates of neurons in the origin nerve sub-area and the target nerve sub-area is preserved according to the topology preservation projection rule described below.

[**Topology preservation projection rule**] The topology preservation projection from a certain nerve sub-area (S_O) to another nerve sub-area (S_T) is a mapping $M: S_O \rightarrow 2^{ST}$, from a cell in S_O to a set of cells in S_T, that satisfies the following condition:

for any $c_{O1}(x_{O1},y_{O1},z_{O1})$, $c_{O2}(x_{O2},y_{O2},z_{O2}) \in S_O$, there exist $c_{T1}(x_{T1},y_{T1},z_{T1}) \in M(c_{O1})$ and $c_{T2}(x_{T2},y_{T2},z_{T2}) \in M(c_{O2})$ s.t. $(x_{O1}-x_{O2})(x_{T1}-x_{T2}) \geqq 0$, $(y_{O1}-y_{O2})(y_{T1}-y_{T2}) \geqq 0$, $(z_{O1}-z_{O2})(z_{T1}-z_{T2}) \geqq 0$,

where $c_{ij}(x_{ij},y_{ij},z_{ij})$ represents a cell c_{ij} and its coordinate (x_{ij},y_{ij},z_{ij}) in a nerve subarea. □

Synapse formation is performed by the following procedure under control of this topology preservation projection rule, the number of axon branches, axon concentration parameters, and the amount of a neurotrophic factor on postsynaptic neurons. First of all, a certain neuron is selected as a pioneer neuron, and a pioneer axon's main branch selects a target cell, then its sub-branches select target cells in its neighborhood under the limit in the number of axon branches. Next, neurons around the pioneer neuron grow axons sequentially in the manner that a main branch selects a target cell under the topology preservation projection between main branches, then sub-branches select target cells in its neighborhood under the limit in the number of axon branches. The amount of a neurotrophic factor of target cells limits the number of presynapse formation and axon concentration parameters for a main branch and sub-branches also give a concentration constraint of a main branch and sub-branches on a target nerve sub-area respectively. The axon outgrowth and synapse formation continues for a certain development stages after neuron production.

Activity-dependent Synapse Regulation

We formulate a model of a pulse neuron based on the Spike Response Model proposed by (Gerstner 1999). In our model, activity of a neuron is formulated as follows. Let T be a set of discrete times whose time unit is defined in the scale of one or a few milliseconds. The membrane potential $u_i(t)$ of neuron i at time $t(\in T)$ is formulated by

$$u_i(t) = \begin{cases} \eta(t-t_i^f)+\sum_{j\in \Gamma_i} p_{ij}(t)\cdots\cdots t-t_i^f > \delta_{abs} \\ undefined\cdots\cdots\cdots\cdots\cdots otherwise \end{cases} \quad (1)$$

where t_i^f: recent firing time before time t, $\eta(t-t_i^f)$: negative contribution to membrane potential that is called a refractory function, and represents negative effect for the relative refractory period, δ_{abs}: time length of the absolute refractory period, Γ_i: a set of presynaptic neurons of neuron i, $p_{ij}(t)$: synaptic potential from presynaptic neuron $j(\in \Gamma_i)$ to postsynaptic neuron i.

The refractory function η is given by

$$\eta(s) = -\eta_0 \exp(-\frac{s-\delta_{abs}}{\tau_{rel}}) \quad (2)$$

where τ_{rel}: time constant of relative refractoriness, η_0: a positive constant that represents the magnitude of relative refractoriness.

The synaptic potential $p_{ij}(t)$ from presynaptic neuron j to postsynaptic neuron i at time $t(\in T)$ is formulated by

$$p_{ij}(t) = (p_{ij}(t_{ij}^a)+w_{ij}(t_{ij}^a))\exp(-\frac{t-t_{ij}^a}{\tau_{syn}}) \quad (3)$$

where τ_{syn}: time constant of synaptic potential, $w_{ij}(t)$: synaptic efficacy from presynaptic neuron j to postsynaptic neuron i at time t, t_{ij}^a: recent time of impulse arrival before time t from neuron j, that is computed as $t_{ij}^a=t_j^f+\Delta_{ij}^{ax}$ using the conduction delay Δ_{ij}^{ax} of impulse from neuron j to neuron i.

Then, for a threshold θ, a neuron i fires when $u_i(t) \geqq \theta$ and $u_i(t-1) < \theta$ hold. As a result, a set of firing times of neuron i is given by the following expression

$$F_i = \{t+\Delta^{ir} \in T \mid u_i(t) \geq \theta \wedge u_i(t-1) < \theta\} \quad (4)$$

where Δ^{ir} is an impulse rising period.

The transmission of impulse takes the conduction delay Δ_{ij}^{ax} proportional to a distance between neurons, and is calculated from three-dimensional coordinate triplets of them.

Next we formulate synaptic efficacy regulation based on the spike-based hebbian rule. Hebbian rule generally says that when both presynaptic neuron and postsynaptic neuron fire synchronously, the synaptic efficacy increases, otherwise decreases. That is, change in synaptic efficacy is determined according to whether postsynaptic neuron fires for an impulse arrival from presynaptic neuron. Let t_i^f be recent fire time of neuron i, and t_{ij}^a be recent impulse arrival time from neuron j to neuron i. Then

(1) if postsynaptic neuron i fires when an impulse arrives from presynaptic neuron j, all synaptic efficacy w_{ik} for all presynaptic neurons $k(\in \Gamma_i)$ are regulated according to

$$\Delta w_{ik}(t_i^f) = \lambda(|w_{ik}(t_i^f)|) \times |p_{ik}(t_i^f)| \times W(t_i^f - t_{ik}^a), \qquad (5)$$

(2) if postsynaptic neuron i don't fire when an impulse arrives from presynaptic neuron j, synaptic efficacy w_{ij} for presynaptic neuron $j(\in \Gamma_i)$ is regulated according to

$$\Delta w_{ij}(t_{ij}^a) = \lambda(|w_{ij}(t_{ij}^a)|) \times |p_{ij}(t_{ij}^a)| \times W(t_i^f - t_{ij}^a). \qquad (6)$$

In above formulas, λ is a learning rate function dependent on the magnitude of synaptic efficiency and is given by the following expression

$$\lambda(w) = \lambda_0 \exp(-\frac{(w - w_0)^2}{2\lambda_1^2}) \qquad (7)$$

where w_0: an initial value of synaptic efficacy, $\lambda_0(0 \leq \lambda_0 \leq 1)$ and $\lambda_1(\lambda_1 \geq 0)$: constants.

An initial value of synaptic efficiency can take a different value in excitatory synapse and inhibitory synapse. $W(s)$ is a learning window as a function of the delay s between postsynaptic firing and presynaptic impulse arrival, and expressed as (Gerstner *et al.* 1996)

$$W(s) = \begin{cases} (A_+ - A_-) \exp(-\dfrac{s - \Delta^{ir}}{\tau_0}) \cdots\cdots\cdots s \geq \Delta^{ir} \geq 0 \\ A_+ \exp(-\dfrac{\Delta^{ir} - s}{\tau_+}) - A_- \exp(-\dfrac{\Delta^{ir} - s}{\tau_-}) \cdots s < \Delta^{ir} \end{cases} \qquad (8)$$

where $A_+ > A_- > 0$, $\tau_0 > 0$, and $\tau_- > \tau_+ > 0$.

Parameters A_+, A_-, τ_+, τ_-, and τ_0 can take different values for excitatory synapses and inhibitory synapses respectively, and are set to values in Table 3. As a result, the excitatory synaptic efficiency $w_{ij}(t)(\geq 0)$ is regulated according to

$$w_{ij}(t) \leftarrow \max(0, w_{ij}(t) + \Delta w_{ij}(t)), \qquad (9)$$

and the inhibitory synaptic efficacy $w_{ij}(t)(\leq 0)$ is regulated according to

$$w_{ij}(t) \leftarrow \min(0, w_{ij}(t) + \Delta w_{ij}(t)). \qquad (10)$$

We can see the following properties as for synaptic efficacy regulation from these formulas. (a) For an excitatory synapse $(p_{ij}(t) \geq 0)$, synaptic efficacy is reinforced when the postsynaptic neuron fires for a presynaptic impulse arrival and weakened when it do not fire. For an inhibitory synapse $(p_{ij}(t) \leq 0)$, synaptic efficacy is reinforced when the

postsynaptic neuron do not fire against a presynaptic impulse arrival and weakened when it fires. (b) Since change in synaptic efficiency is proportional to $|p_{ij}(t)|$, a continuous stimulation like the tetanus stimulation causes large regulation. (c) Since change in synaptic efficiency is large for the first stimulation and gradually becomes small as indicated in $\lambda(w)$, synaptic efficiency becomes steady for repeated reinforcement.

In our artificial neural development method, spiking activity affects the neural network structure through the programmed cell death and the synapse elimination. The programmed cell death adjusts the size of an origin neuron group and a target neuron group in case of living things. It is known that the interception of presynaptic inputs and the loss of connection with target nerve areas cause the programmed cell death (Hall 1992). In our method, the programmed cell death is controlled according to the following rule.

[**Programmed cell death rule**] (1) If a neuron forms no synapses with target cells, it dies because it is supplied no neurotrophic factor. (2) If a neuron receives no impulses from presynaptic neurons throughout development, it dies because of stimulation interception. □

As for the synapse elimination, it is considered that eliminated synapses are those defeated at the target government competition among presynapses through spiking activity. The synapse elimination can be considered to play the role of selective stabilization of connections between nerve areas. According to this speculation, weakened synapses are considered to be doomed to elimination and the following synapse elimination rule is applied to a set of excitatory presynapses and a set of inhibitory presynapses for each neuron respectively.

[**Synapse elimination rule**] Let m be a mean of synaptic efficacy of presynapses for each neuron, and σ be their standard deviation. Then synapses with lower efficiency than $(m - k\sigma)$ are eliminated where $k \geq 0$. □

This rule is an extension of the tree-wide rule of (Rust *et al.* 1997b).

Animat's Neural Structure and Genome for Place Recognition

Animat's Place Recognition Circuit

Activity of neurons that occurs in response to an animal's position is especially observed from neurons of dorsal hippocampus. These neurons are called place cells because it is considered that groups of these neurons code information on places where an animal is situated. From experiments of rats, correlation between a rat's position and activity of an individual place cell is weak but there is a strong correlation between a set of place cell firing and a rat's position (Delcomyn 1998). This means that a rat recognizes his/her location with a place cell assembly. A lot of models of the place recognition circuit that contains

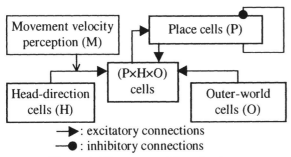

Figure 1: Place recognition circuit

place cells for navigation of animals and robots have been proposed (Trullier *et al.* 1997). In this paper, a model of the artificial place recognition circuit shown in Figure 1 is supposed based on some of these works. As for receptive cells, we suppose outer-world cells (O) and head-direction cells (H). The movement velocity modulates receptor potential of head-direction cells. Central circuits for place recognition are supposed to be the interconnected circuit of place cells with (P×H×O) cells and the lateral inhibitory circuit for place cells. Lateral inhibition is introduced to support place recognition based on the competition principle.

Animat's Neural Structure and Genome

Our animat has a body of the diameter 40 cm and has four sonar sensors and four infrared sensors for outer-world sensing on the frontal right and left body respectively. The infrared sensor measures distance in the range from 0 to 15 inches. The sonar sensor measures distance in the range from 0 to 255 inches. It has also a head-direction sensor as an internal sensor. As an effector, it has two differential-drive wheels. Figure 2 shows the composition of our animat.

One receptive cell is assigned to each infrared sensor and six receptive cells are assigned to each sonar sensor. Infrared receptive cells and sonar receptive cells have the selective reactivity to distance and they selectively generate receptor potential for the specific range of distance measured by sensors. An infrared receptive cell has the reactivity to short distance. Six sonar receptive cells for each sonar sensor have the different reactivity, that is each reacts to one of six ranges for distance from 0 to 255 inches. Sixteen

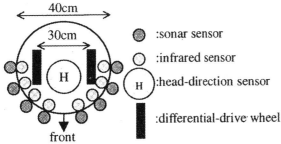

Figure 2: Sensors and actuators of the animat

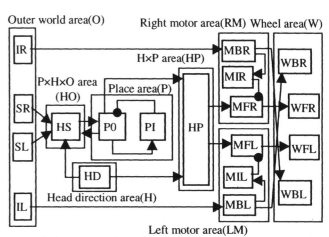

IR,IL: infrared receptive sub-areas, SR,SL: sonar receptive sub-areas, HD: head-direction receptive sub-area, WFR,WFL: wheel effector sub-areas (forward), WBR,WBL: wheel effector sub-areas (backward), P0,PI: place sub-areas, HS: (P×H×O) sub-area, MFR,MBR,MIR,MFL,MBL,MIL,HP: motor sub-areas

Figure 3: The animat's neural structure

head-direction cells are assigned to the head-direction sensor, and each selectively reacts in one direction of every 22.5° and generates the receptor potential. When membrane potential as the temporal sum of receptor potential exceeds a threshold, a spike is generated. As for the effector, plural effector cells are assigned to right and left wheels respectively. Effector cells compute the velocity of the wheel in proportion to the number of spikes received in a certain interval.

Figure 3 shows a skeleton of the hypothetical neural network of the animat that contains the place recognition circuit shown in Figure 1. The neural network roughly consists of the forward movement controller and the backward movement controller. The forward movement is performed based on place and head-direction recognition by signals from sonar and head-direction sensors. The backward movement is performed based on signals from infrared sensors. Table 1 shows the composition of the

NA	NSA	NOC	RF	NAB	GF	ANF
O	SL	24	1010	4		
	SR	24	1010	4		
	IL	4	0110	6		
	IR	4	0101	6		
H	HD	16	1010	4		
W	WFL	2			0101	5
	WFR	2			0110	5
	WBL	2			1010	5
	WBR	2			1001	5

NA: Nerve Area, NSA: Nerve Sub-Area, NOC: the Number Of Cells, RF: Receptor Factor, NAB: the Number of Axon Branches, GF: Guide Factor, ANF: the Amount of Neurotrophic Factor

Table 1: Composition of receptor and effector areas

NA	DS	DD	NOC	TNA	RT	GF	RF	NAB	ANF	NSA
HO	1	0,0,0	VAR	S,H,P	+	1010	0011	VAR	VAR	HS
P	2	0,0,0	VAR	HO,HP,P	+	0011	1010	VAR	VAR	P0
	3	-1,0,0	VAR	P	-	1010	0011	VAR	VAR	PI
HP	1	0,0,0	20	H,LM,RM	+	1010	0011	2	8	HP
LM	2	0,-,-	10	S,W,LM	+	0110	1001	2	8	MBL
	3	0,+,-	10	W	+	0011	0101	2	16	MFL
	4	0,0,-	10	LM	-	1001	0011	6	8	MIL
RM	2	0,-,+	10	S,W,RM	+	0101	1010	2	8	MBR
	3	0,+,+	10	W	+	0011	0110	2	16	MFR
	4	0,0,+	10	RM	-	1010	0011	6	8	MIR

The left table shows an animat's genome. Each row is a gene indexed by a nerve area(NA) and a development stage(DS). The right table shows the correspondence of genes with nerve sub-areas in Figure 3.

Table2: An animat's genome

DS: Development Stage, DD: Development Direction, TNA: Target Nerve Areas, RT: Response Type, NA,NOC,GF,RF,NAB,ANF,NSA: see table 1

receptor area for sensors and the effector area for wheels.

When a specification of receptor areas and effector areas like Table 1 is given by a user, the artificial neural development of NEUROGEN'2000 can generate a neural network that connects these receptor areas and effector areas by developing genomes evolved automatically. However, it takes enormous computation cost to evolve genomes that generate such a neural network we suppose to construct in this paper, even if it can be done. On the other hand, it requires a serious load to design all the genome for generating a neural network like Figure 3 by hand. In order to design a place recognition circuit of good place recognition performance, there are several difficult problems, for example, how many neurons are necessary in each nerve area, what connections are necessary between nerve areas, and so forth. In NEUROGEN'2000, hybrid manual and evolutionary genome design method resolve these problems.

The genome of the animat is shown in Table 2 by a table form. The number of nerve areas is 5 and the number of development stages is 4, so the gene matrix of (5, 4)-type represents the genome. In the table, the "VAR" sign expresses a part that is determined by evolutionary computation, and these nine values are evolutionally determined in the experiments of the next section. In general, an arbitrary part of genomes can be determined by evolutionary computation.

Experiments

Testbed

As an environment to evaluate the artificial neural development and evolution of the animat's artificial place recognition circuit, we use a simple T-maze in Figure 4. The T-maze has three arms of 120cm in width and 160cm in length. The maze is sectioned in grids of 10cm.

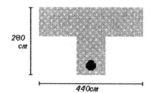

Figure 4: T-maze with 3 arms

The animat can move to an arbitrary direction at a speed within a certain range, and is located on a lattice point that is nearest to the moved point.

Decoding a genome through four development stages develops a neural network. The period of the target recognition for axonal outgrowth is set to 3 development stages. Activity of a neural network is computed at every time unit. The interval of outer-world sensing by sonar and infrared sensors is set to 5 and the interval of movement computation is set to 50. The time length of spiking before synapse elimination is set to 10000. The σ value for synapse elimination is set to 3. Main parameters which characterizes a pulse neuron are shown in Table 3.

Parameters for the steady state genetic algorithm are as follows. The selection method is set to the ranking selection, the replacement method is set to the conditional ranked replacement, and the non-selection age is set to 2. As for genetic operators, the creeping mutation is used for mutations of the number of cells, the number of axon branches, and the amount of a neurotrophic factor. The mutation rate is set to 0.5. These operators change the number of cells, the number of axon branches, and the amount of a neurotrophic factor between 10 and 100, 1 and 10, and 1 to 20 respectively.

The Bayesian place reconstruction mapping from a spiking pattern of place cells to a place as a physical variable in an environment (Zhang et al. 1998) is used for evaluation of animat's place recognition. The Bayesian place reconstruction consists of the sampling phase and the

	excitatory	inhibitory	δ_{abs}	4
W_0	1.0	3.0	η_0	1.5
A_+	0.5	0.35	τ_{rel}	3
A_-	0.2	0.3	τ_{syn}	10
τ_0	10	10		
τ_+	5	5	θ	3.0
τ_-	20	20	Δ_{ir}	2
			λ_0	0.01
			λ_1	0.5

Table 3: Parameters of pulse neurons

reconstruction phase. In sampling, the spatial occupancy $P(X)$ of the animat and the firing rate map $f_i(X)$ are computed. The spatial occupancy is the probability for the animat to visit each spatial position $X=(x, y)$. The firing rate map is the average firing rate of a place cell i while the animat is at each position X. In reconstruction, the one-step reconstruction method and the two-step reconstruction method are used to compute a reconstructed position of the animat from the spatial occupancy and the firing rate map. Let $n_t=(n_1,...,n_N)$ be the numbers of spikes fired within the t-th time window, where n_i is the number of spikes of cell i. The one-step reconstruction method computes the conditional probability $P(X_t|n_t)$ for the animat to be at each position X_t. The two-step reconstruction method introduces a continuity constraint of positions and computes the conditional probability $P(X_t|n_t, X_{t-1})$ for the animat to be at each position X_t, given the preceding position X_{t-1}. In these probability distribution, the most probable position is taken as the reconstructed position of the animat at the t-th time window. In experiments, the time window is set to 50. Since length of the time window for reconstruction is equal to the interval of movement, the pulse assembly of place cells counted in the time window can be considered to code a place where the animat is situated.

As a fitness function for evolution, the following function is used to evolve an animat who can recognize places accurately and move widely.

$$fitness = \frac{(V-1)^\beta}{\alpha + E} \qquad (11)$$

where V: the number of visit positions, E: the reconstruction error that is computed as the sum of distance between true positions and reconstructed positions, α and β: constants.

In experiments, one-step reconstructed positions are used for the fitness computation, where the sampling time length and the reconstruction time length are set to 10000. The constants α and β are set to 250 and 1 respectively.

Results: Place Recognition Circuit Genesis

We could generate animats that achieve high fitness values by repeated simulations of population size 10 and generation length 15. Figure 5 shows true positions, the one-step reconstructed positions, and the two-step reconstructed positions of an animat with a high fitness value, where the sampling time length is set to 30000 and the reconstruction time length is set to 10000. We can observe several erratic jumps in the trajectory of the one-step reconstruction. These erratic jumps are also observed when positions are reconstructed from firing data of rat's place cells in (Zhang *et al.* 1998), and our reconstructed trajectory is similar to those results. In case of the two-step reconstruction, reconstruction error is less in comparison with the case of the one-step reconstruction. This is also similar to the result in (Zhang *et al.* 1998). As a type of place recognition error, there is so-called perceptual aliasing in which different places are judged to be the same place due to incomplete perception. It is considered that integration of view information and dead reckoning information is necessary to avoid this type of error. Since erratic jumps are resolved by a continuity constraint using information from two consecutive times, it is suggested that erratic jumps are induced by the perceptual aliasing. We guess that place recognition is not complete at hippocampal place cells of rats since erratic jumps are observed in their one-step reconstructed trajectories. It is remarkable that our reconstructed trajectories are similar to those reconstructed from firing data of rat's place cells, including feature of

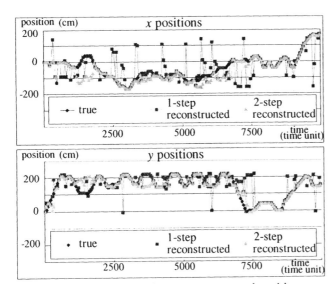

Figure 5: True position, 1-step reconstructed position, and 2-step reconstructed position of an animat with a high fitness value

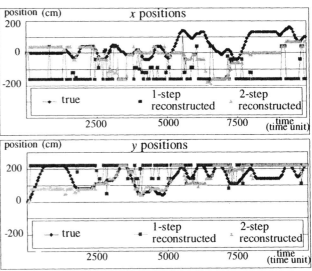

Figure 6: True position, 1-step reconstructed position, and 2-step reconstructed position of an animat with a low fitness value

erratic jumps. Figure 6 shows a place reconstruction result of an animat with a low fitness value, who can not recognize places well. By these experiments, it was confirmed that there were place recognition circuits that were able to reconstruct places well and that were not able to reconstruct places at all. It was also confirmed that the well-cognitive circuits could be designed by evolutionary computation. That is, it was found that the evolutionary design could be achieved as evolutionary search of finding a suitable combination of nine parameters in the genome of Table 2 in combination with the artificial neural development.

Table 4 shows a part of the genome and the neural network composition of the above animat with a high fitness value. Many synapse connections are generated among place recognition nerve sub-areas as assumed in Figure 1. This suggests that the assumed circuit satisfies a necessary condition for place recognition. However, it seemed that it was difficult to distinguish high-fit animats with low-fit animats only by these connections.

It was also observed that the place reconstruction performance of high-fit animats was superior for a short sampling time length but was inferior for a long sampling time length. In fact, the place reconstruction performance was far better for 10000 sampling time length than for 30000 sampling time length and was slightly worse for 50000 sampling time length than for 30000 sampling time length, that is the case in Figure 5. We presume that one cause for decrease in reconstruction accuracy is the increase of visits to the same place from different directions. It is one of typical place recognition errors in which a place is judged to be different places for visiting the place from different directions. Since it is considered that integration of view information and head-direction information is necessary to

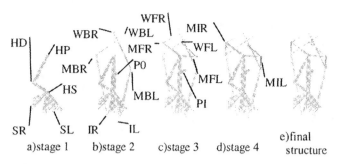

Figure 7: The development of the artificial neural network

avoid this type of error, the interconnected circuit of place cells with (P×H×O) cells is supposed to realize the integration to set the equivalence between different views from the same place. That is, it is supposed that a spiking pattern specific to a place occurs in the place cell area as a result of integrating view information and head-direction information in this interconnected circuit. However, it became clear this circuit was insufficient to achieve this information integration. To resolve this problem, it is necessary to evolve more parts of the place recognition circuit structure to find a better one or to assume another place recognition circuit scheme.

The development process of the artificial neural network of the above high-fit animat is shown in Figure 7 and numbers of cells and synapses in the place recognition circuit before and after the programmed cell death and the synapse elimination are shown in Figure 8. It is observed sizes of origin neuron groups and target neuron groups are adjusted by the programmed cell death to meet roughly the size of the nerve sub-area HS.

NA	DS	NOC	NAB	ANF		NSA
HO	1	25	8	10		HS
P	2	72	3	9		PO
	3	63	4	8		PI

	NOC	the number of synapses(row→column)						
		SR	SL	HD	HS	HP	PI	PO
SR	24	0	0	0	96	0	0	0
SL	24	0	0	0	96	0	0	0
HD	13	0	0	0	32	46	0	0
HS	25	0	0	0	0	0	0	200
HP	17	0	0	0	0	0	0	0
PI	28	0	0	0	0	0	0	107
PO	55	0	0	0	26	85	155	0

The top tables show the animat's genes and their correspondence with nerve sub-areas in Figure 3. Each row is a gene indexed by a nerve area and a development stage. The bottom table shows the number of cells and synapses of the animat.

NA, DS, NOC, NAB, ANF, NSA: see table 1 and 2.

Table 4: A part of the genome and the neural structure of the animat with a high fitness value

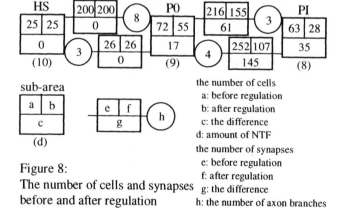

Figure 8:
The number of cells and synapses before and after regulation

the number of cells
a: before regulation
b: after regulation
c: the difference
d: amount of NTF
the number of synapses
e: before regulation
f: after regulation
g: the difference
h: the number of axon branches

Results: Place Recognition Performance

We evaluate the place recognition performance of animats in an environment where they first enter after development, in comparison with the place recognition performance in an environment where they came into existence. It is reported that place cells shows specific spatial firing immediately after an animal is introduced in a new environment, that is place recognition do not depend on learning. This experiment examines whether our place recognition circuit shows such a characteristic.

For that purpose, we use animats of good place recognition performance and animats of bad place recognition performance in the T-maze with three arms of Figure 5, and measure place recognition performance of these animats in the T-maze with seven arms of Figure 9. In this experiment, animats start from each of initial positions A to E in the T-maze of Figure 9, and moves freely in it. The Bayesian place reconstruction mapping is used for the evaluation of place recognition performance. That is, animats start from each initial position and move freely in the T-maze firstly for the sampling time length 30000 and then for the reconstruction time length 10000, and errors between true positions and reconstructed positions are computed.

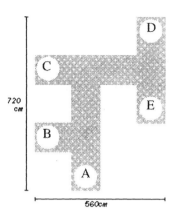

720 cm

560cm

A,B,C,D,E: Initial positions

Figure 9: T-maze with 7 arms

	Initial positions					
	A	**B**	**C**	**D**	**E**	
G1	61.0	43.6	45.0	71.8	44.8	53.2
G2	59.5	54.2	37.6	92.0	48.0	58.3
G3	44.3	58.7	75.0	77.4	71.9	65.5
G4	67.4	48.0	52.0	44.5	51.9	52.8
B1	170.8	187.9	180.9	113.1	111.1	152.8
B2	111.0	120.7	106.4	109.0	291.4	147.7
B3	126.5	209.1	108.1	148.2	152.1	148.8
B4	187.3	90.3	136.9	89.3	127.7	126.3

Average reconstruction error in the case of the 1-step reconstruction. The sampling time length is 30000 and the reconstruction time length is 10000. The unit is the centimeter (cm).

Table 6: Place recognition performance in T-maze with 7 arms

Table 5 shows place recognition performance of animats in the T-maze of Figure 5. Four animats in Table 5(a) are high-fit ones and have good place recognition performance. On the other hand, four animats in Table 5(b) are low-fit ones and have bad place recognition performance. Table 6 shows what place recognition performance these animats display in the T-maze of Figure 9. As for animats that had good place recognition performance in the T-maze with three arms, they displayed good place recognition performance in another T-maze with seven arms. As for animats that had bad place recognition performance in the T-maze with three arms, there was a tendency that they displayed worse place recognition performance in another T-maze with seven arms.

These results suggest that the place recognition ability is an innate ability dependent on the composition of the place recognition circuit.

Conclusion and Future Work

We have proposed the artificial neural development method NEUROGEN'2000 that generates the three-dimensional multi-regional pulsed neural network arranged in three layers of the nerve area layer, the nerve sub-area layer, and the cell layer. Then we have applied it to develop an animat's place recognition circuit that focuses on the place cell area. The Bayesian place reconstruction mapping from a spiking pattern of place cells to a place in an environment was used for evaluation of animat's place recognition performance. As a result of simulation experiments, it was found that performance of place recognition depended on the composition of the place recognition circuit and the well-cognitive circuits could be designed by our artificial neural development method in combination with evolutionary computation. That is, we could generate a multi-regional place recognition circuit with the place cell area that showed place recognition features similar to rat's place cells. Also it was suggested that the place recognition ability was an

	S10000-R10000		S30000-R10000	
	1-step	**2-step**	**1-step**	**2-step**
G1	26.4	11.6	52.9	41.3
G2	15.9	7.3	46.6	26.4
G3	19.3	5.4	47.4	24.1
G4	19.0	11.2	50.5	17.1
Mean	20.2	8.9	49.4	27.2

(a)results of animats with high fitness values

	S10000-R10000		S30000-R10000	
	1-step	**2-step**	**1-step**	**2-step**
B1	96.7	96.9	92.7	75.8
B2	84.4	79.1	183.3	98.1
B3	82.8	98.2	90.5	90.7
B4	89.4	115.4	83.4	98.6
Mean	88.3	97.4	112.5	90.8

(b)results of animats with low fitness values

Average reconstruction error that is the average distance between true positions and reconstructed positions. The unit is the centimeter (cm). The label "S10000-R10000" means that the sampling time length is 10000 and the reconstruction time length is 10000. The label "S30000-R10000" means that the sampling time length is 30000 and the reconstruction time length is 10000. The label "1-step" represents 1-step reconstruction and the label "2-step" represents 2-step reconstruction.

Table 5: Place recognition performance in T-maze with 3 arms where animats came into existence

innate ability dependent on the composition of the place recognition circuit. By these results, it was confirmed that NEUROGEN'2000 was useful for designing biologically realistic multi-regional pulsed neural networks.

However, there are a lot of future works and the following researches are ongoing at present. The first is to find genetic and/or nerve-structural factors of good place recognition performance by analyzing genomes and nerve circuits of good place recognition performance and bad place recognition performance. This may suggest how the development process contributes to generation of good place recognition circuits. The second is genetically to design the place recognition-triggered navigation circuit, that is the goal-seeking action circuit in combination with the place recognition circuit. As for the goal-seeking behavior, learning after growth plays an important role in addition to development and evolution.

References

Cangelosi, A., Parisi, D. and Nolfi, S. 1994. Cell Division and Migration in a 'genotype' for Neural Networks. *Network* 5: 497-515.

Delcomyn, F. 1998. *Foundations of Neurobiology*. W. H. Freeman and Company, New York.

Dellaert, F. and Beer, R. D. 1996. A Developmental Model for the Evolution of Complete Autonomous Agents. In *Proceedings of the Fourth International Conference on Simulation of Adaptive Behavior*, 393-401.

de Garis, H. 1994. An Artificial Brain - ATR's CAM-Brain Project Aims to Build/Evolve an Artificial Brain with a Million Neural Net Modules Inside a Trillion Cell Cellular Automata Machine. *New Generation Computing* 12: 215-221.

Eggenberger, P. 1996. Cell Interactions as a Control Tool of Developmental Processes for Evolutionary Robotics. In *Proceedings of the Fourth International Conference on Simulation of Adaptive Behavior*, 440-448.

Gerstner, W., Kempter, R., van Hemmen,J.L. and Wagner,H. 1996. A Neuronal Learning Rule for Sub-millisecond Temporal Coding. *Nature* 383: 76-78.

Gerstner, W. 1999. Spiking Neurons. In *Pulsed Neural Networks*, The MIT Press.

Hall, W. H. 1992. *An Introduction to Molecular Neurobiology*, Sinauer Associates, Inc.

Kitano,H. 1995. A Simple Model of Neurogenesis and Cell Differentiation based on Evolutionary Large-Scale Chaos. *Artificial Life* 2: 79-99.

Rust, A. G., Adams, R., George, S. and Bolouri, H. 1997a. Designing Development Rules for Artificial Evolution. In *Proceedings of the 3rd International Conference on Artificial Neural Nets and Genetic Algorithms*, 509-513.

Rust, A. G., Adams, R., George, S. and Bolouri,H. 1997b. Activity-based Pruning in Developmental Artificial Neural Network. In *Fourth European Conference on Artificial Life*, 224-233.

Trullier, O., Wiener, S. I., Berthoz, A. and Meyer,J.A. 1997. Biologically based Artificial Navigation System: Review and Prospects. *Progress in Neurobiology* 51: 483-544.

Vaario, J., Onitsuka, A. and Shimohara, K. 1997. Formation of Neural Structures. In *Fourth European Conference on Artificial Life*, 214-223.

Vos, J. E., Heijst, J. J., Greuters, S. 1997. Programmed Cell Death during Early Development of the Nervous Systems, Modelled by Pruning in a Neural Network. In *Spatiotemporal Models in Biological and Artificial Systems*, IOS Press, 192-199.

Zhang, K., Ginsburg, I., McNaughton, B. L. and Sejnowski, T. J. 1998. Interpreting Neuronal Population Activity by Reconstruction: Unified Framework with Application to Hippocampal Place Cells. *Journal of Neurophysiology* 79: 1017-1044.

Emergent SMA-Net Robot Control by Coupled Oscillator System

Yuichi Sato[1], Takashi Nagai[1], Hiroshi Yokoi[1], Takafumi Mizuno[2] and **Yukinori Kakazu[1]**

[1]Laboratory of Autonomous Systems Engineering, Graduate school of Hokkaido University, Hokkaido, Japan
[2]National Institute of Bioscience and Human Technology, Tsukuba, Ibaraki, Japan

Abstract

This paper proposes a method for controlling a flexible SMA (Shape Memory Alloy) - Net robot by using a coupled oscillator system. The control method is inspired by the emergent behavior control mechanism of slime mold amoebae and using highly distributed information processing system. The control method enables basic behaviors of movement to attractive stimulus and escape from unattractive stimulus.

Introduction

Flexible structure robot and it's control system are important domains in recent robotics research activities. This paper proposes emergent control method of flexible SMA-Net robot by using a coupled oscillator system. This control method is inspired by the emergent behavior control mechanism of slime mold amoebae.

Recently, many kinds of flexible structure robot have been proposed by Dittirch (Dittrich, Andreas and Wolfgan, 1998), Paap (Paap, Dehlwisch, and Klaassen, 1996) and Jammes (Jammes, Hiraki, and Ozono, 1996) et al. The random morphology ("RM-") robot, proposed by Dittrich, consists of six servos coupled arbitrarily by thin metal joints. Genetic Programming is used for the control system of the RM-robot (Dittrich, Andreas and Wolfgan, 1998). The GMD-Snake, developed by Paap consists of many joints and is actuated by motors and strings.

Most of the flexible structure robots are composed of many actuators and mechanical components. The first difficulty in construction of a flexible structure robot is to make actuators and mechanical components simple and compact. The second difficulty is the control system of actuators, which in responsible for the coordination of actuators and enables objective behaviors.

To solve the first difficulty, we propose a flexible structure robot that is actuated by shape memory alloy actuator (SMA-Net robot). Because this robot use a SMA spring as an actuator and skeleton, fewer mechanical components are needed.

To solve the second difficulty, we propose a control method that uses a coupled oscillator system. It is well known that a coupled oscillator system creates the emergent coordinate behavior of slime mold. Cellular slime mold (Dictyostelium discoideum) is famous for its inter-cellular communication by means of chemical substances (cAMP, etc.).

There is no special difference among cells, each of which works as a unit, and there is no complex design in the interaction among cells. However, in the case of starvation or a dry situation, the cells use chemical substances as a coupled oscillator, enabling emergent information processing. Then, the cells make slug shape pseudoplasmodium to move to a suitable place. Slime mold has the ability of basic information processing such as aggregation to foods or escape from aversion stimulus. Slime mold can do a cooperative behavior in whole cells for these motions. In such behavior, global organization emerges spontaneously due to coupled oscillator systems.

This paper is composed as follows. First, we show the effectiveness to use SMA actuator for flexible structure robot. Then, the SMA-Net robot is introduced. A control method for the robot using a coupled oscillator system is proposed in next section. The effectiveness of the proposed control method is discussed on the basis of simulation results. Finally, our conclusions are presented in last section.

Flexible Structure Robot and Shape Memory Alloy Actuators

The first SMA appeared about 50 years ago, and titanium-nickel SMA that is the most famous SMA was discovered in the 1960s. SMA undergo a crystalline phase change when heated or cooled above ambient. This phase change is accompanied by a change in elastic modulus (Johnson and Martynov, 1997). SMA wire actuator has an ability to produce large forces and displacement by direct electoric drive. This cause the decrease of electoric and mechanical parts of the robot

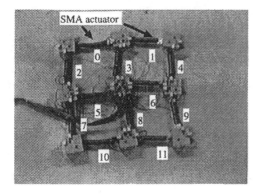

Figure 1: SMA-Net Robot

and increase robustness. These simplifications of the parts is an important factor to build the flexible structure robot that consists of simple elements. And the we can apply oscillator system that is observed in pulse-beat of muscle to the direct electoric drive of SMA actuator. This is the reason for selecting SMA wire as the locomotion system of our flexible structure robot.

SMA-Net Robot

We develop SMA-Net robot that is shown in Fig.1.

The SMA-Net robot consists of twelve SMA spring actuators and micro-computers (BASIC-Stamp). The SMA actuators are connected to nine nodes. There have been many studies on the modeling and control of SMA actuators (Shu, Lagoudas, and Hughes, 1997). SMA actuator have been used in heat engines, flange-bolts for space mission, etc. Recent technology has enabled the formation of an micro SMA actuator on a silicon substrate. In these applications the SMA actuator is merely used as an actuator. However, in our SMA-Net robot, the SMA actuators are used not only as actuators but also as a skeleton of the robot. Therefore, the SMA actuators work like muscle fiber.

Nodes with one SMA actuator can not move themselves. But, cooperation or synchronization with other SMA actuators generates movement of the nodes and more power than one actuator. To generate such cooperation among actuators, appropriate periodic on-off timing, that is, rhythmic contraction of the actuator, is needed. We therefore decided to use a coupled oscillator system. In a coupled oscillator system, oscillators that generate periodic on-off timing are locally connected and affected by connected oscillators. This connection enables both synchronization and phase difference of actuators' on-off timing.

In the next section, we propose a control system for the SMA-Net robot using a coupled oscillator system.

Coupled Oscillator System

The rhythmic contraction of slime mold has been analyzed by means of a locally coupled oscillator system. The rhythmic contraction can be explained as entrainment synchronization phenomena in spatiotemporal pattern formation or various kinds of nonlinear oscillatory phenomena. Recently, chemotaxis and other cellular behaviors of slime mold have been studied using a coupled oscillator model (Miyake, Tanabe, Murakami, Yano, and Shimizu, 1996).

In this section, we propose a method for generating control signals by using a coupled oscillator system. First, the concept of a nonlinear oscillator is introduced. Then, we describe a local coupled oscillator model for making synchronous oscillation. Finally, we describe how oscillator amplitude is converted to control signals.

van der Pol Oscillator

Oscillators can be categorized into linear and nonlinear oscillators. The nonlinear oscillator has the feature of synchronization by entrainment. Oscillation of a linear oscillator depends on the initial condition, and it is difficult to use it as control signal in the SMA-Net robot. We therefore selected a nonlinear oscillator that follows equation (1). Each actuator i contains this oscillator to generate a control signal.

$$\alpha_i \ddot{x}_i(t) + \beta_i(1 - x_i(t)^2)\dot{x}_i(t) + x_i(t) = 0 \qquad (1)$$

This oscillator is called a van der Pol oscillator, and it has stable oscillation attractor. Although the oscillator starts from an arbitrary initial condition, it generates the same oscillation except for phase.

Parameter α_i determines the frequency of oscillation and β_i determines nonlinearly. When $\beta_i = 0$, equation (1) generates a simple sine wave.

Local Coupled Oscillator System

Each oscillation x_i, which is generated by the van der Pol oscillator equation (1), is the same wave. However, there are phase difference among each oscillator. Therefore, synchronization of each oscillation is needed to generate cooperation among the actuators and intelligent behavior of the SMA-Net Robot. In this section, we propose a coupled oscillator system for generating such synchronization of an oscillator.

Coupled Map Lattice (CML) (Kaneko 1989) and Globally Coupled Map (GCM) (Kaneko 1990) are well-known models of a coupled oscillator system. These models were proposed for analysis of spatiotemporal chaotic phenomena such as rhythmic contraction of a cell group and heat turbulent flow etc.

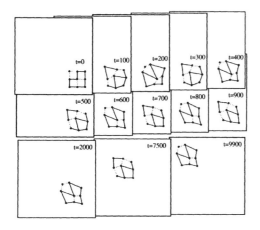

Figure 2: Diagonal Movement of SMA-Net Robot

In this paper, we introduce a simplified form of the CML. Equation (2) shows the local coupled oscillator system that is used for the control of the SMA-Net robot.

$$x_i(t+1) = (1 - \epsilon)x_i(t) + \frac{\epsilon_i}{n(N_i)} \sum_{j \in N_i} x_j(t) \qquad (2)$$

$$(0.0 \leq \epsilon_i \leq 1.0)$$

$n(N_i)$ means the size of the neighborhood oscillator set N_i, and parameter ϵ_i determines the strength of connection among oscillators.

Control Signal by Coupled Oscillator

To generate a control signal for the SMA actuator, equation (3) is introduced. The control signal is affected not only by the actuator's oscillator but also by the connected oscillators.

$$s_i(t) = x_i(t) - \frac{\xi_i}{n(N_i)} \sum_{j \in N_i} x_j(t) \qquad (3)$$

$$(0.0 \leq \xi_i)$$

Parameter ξ_i determines the strength of connection among oscillators. There is no effect of other oscillators, when ξ_i equals 0.0.

The power of the SMA actuator is turned on and the SMA actuator expands in the case of $s_i(t) > 0$. Conversely, the power of the actuator is turned off and the SMA actuator shrinks in the case of $s_i(t) \leq 0$.

Experiments and Results

To examine the ability of the proposed coupled oscillator control method, we built the simulator of SMA-Net robot and executed three experiments. The first one

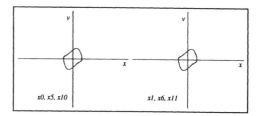

Figure 3: Trajectory of Oscillator Amplitude (synchronous state)

is SMA-Net robot control by simple sine wave. The second one and third one are control method by using local coupled oscillator system. In the second experiment, the proposed control method enables movement to attractive stimulus. On the other hand, the third experiment is examines escape behavior from unattractive stimulus.

Robot Control by Sine Waves

First, to examine our simulator, we gave the SMA-Net robot the sine wave control signals $s_i(t)$ generated by the following parameters and initial conditions.

$$\alpha_i = (100.0/\pi)^2, \beta_i = 0.0$$

$$\epsilon_i = 0.0, \xi_i = 0.0,$$

$$N_i = \emptyset,$$

$$x_i(0) = 0.0,$$

$$\dot{x}_i(0) = \begin{cases} +100.0/\pi & (i = 0, 2, 6, 8) \\ -100.0/\pi & (i = 1, 3, 5, 7) \\ 0.0 & (\text{otherwise}) \end{cases}$$

The values of oscillators $x_i(t), (i = 0, 1, 2, 4, 5, 6, 7, 8)$ are determined by equations (1)-(3). Other oscillators $x_i(t), (i = 4, 9, 10, 11)$, were set to 0.0.

These control signals generate periodical contraction of the actuator and enables diagonal movement of the SMA-Net robot (see Fig.2). Because the phase difference between $s_i(t), (i = 0, 2, 6, 8)$ and $s_i(t), (i = 1, 3, 5, 7)$ is π, two sets of actuators expand alternately.

Robot Control by Coupled Oscillator System

In the next two experiments, we examined SMA-Net robot control by using a local coupled oscillator system. The following parameters and initial conditions were selected for these experiments.

$$\alpha_i = 1.0, \beta_i = 1.0$$

$$\epsilon_i = 0.05, \xi_i = 1.0,$$

$$N_0 = \{1\}, N_5 = \{6\}, N_{10} = \{11\},$$

$$N_1 = \{0\}, N_6 = \{5\}, N_{11} = \{10\},$$

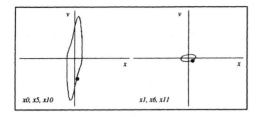

Figure 4: Trajectory of Oscillator Amplitude (attractive)

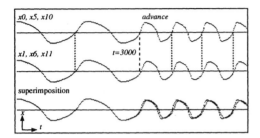

Figure 5: Amplitude of Oscillator (attractive)

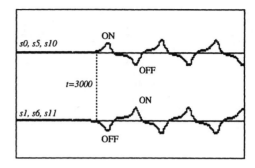

Figure 6: Control Signal of SMA Actuator (attractive)

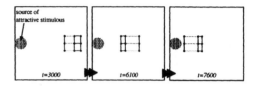

Figure 7: Motion of SMA-Net Robot for attractive stimulus

$$1.5 \leq x_i(0) \leq 2.5,$$
$$1.5 \leq \dot{x}_i(0), \leq 2.5$$

The value of oscillators $x_i(t)$, $(i = 0, 5, 10, 1, 6, 11)$ follow equation (1)-(3). Other oscillators $x_i(t)$, $(i = 2, 3, 4, 7, 8, 9)$ are set to 0.0.

Although initial conditions $x_i(0)$ and $\dot{x}_i(0)$ are selected randomly, after several hundreds time steps, two sets of oscillators $\{x_0, x_5, x_{10}\}$ and $\{x_1, x_6, x_{11}\}$ synchronize by using the entrainment. Fig.3 shows stable trajectory of oscillator. In this figure, the values of $x_i(t)$ is plotted in horizontal axis and the values of $v_i(t) = \dot{x}_i(t)$ are plotted in vertical axis.

Movement in response to an Attractive Stimulus In this section, we examine the parameters of a coupled oscillator that enable movement in response to an attractive stimulus. After 3,000 time steps, parameters of oscillators x_i, $(i = 0, 5, 10, 1, 6, 11)$ change as follows.

$$\alpha_i = \begin{cases} 0.15 & (i = 0, 5, 10) \\ 1.0 & (i = 1, 6, 11) \end{cases},$$
$$\beta_i = \begin{cases} 0.5 & (i = 0, 5, 10) \\ 1.0 & (i = 1, 6, 11) \end{cases}$$

These parameters mean that oscillators on the left side of the horizontal SMA-Net actuators (0, 5, 10) are stimulated by an attractive stimulus and the frequency becomes high. In the van der Pol equation (1), the parameter α_i becomes smaller, and the frequency

becomes higher. The stable oscillation of two sets of oscillators changes, and these trajectories are shown in Fig.4.

These oscillations becomes asynchronous, and the phase of oscillators $\{x_0, x_5, x_{10}\}$ is more advanced than that of $\{x_1, x_6, x_{11}\}$. Fig.5 shows a time sequence of $x_i(t)$. In this figure, the superposition of two time sequences of amplitude shows the transition from a synchronous state to an asynchronous state after 3,000 time steps.

Fig.6 shows a time sequence of the control signals of an actuator. Until 3,000 time steps, because two sets of oscillators are synchronous, $s_i(t)$ is 0.0 and the power of the actuators is turned off. After 3,000 time steps, the power of actuator (0, 5, 10) is turned on first, and then the power of the actuator (1, 6, 11) is turned on.

Fig.7 denotes movement in response to an attractive stimulus. The movement is generated by control signals shown in Fig.6.

Escape from an Unattractive Stimulus In this section, we examine parameters of a coupled oscillator that enable escape from an unattractive stimulus. After 3,000 time steps, parameters of oscillators x_i, $(i = 0, 5, 10, 1, 6, 11)$ change as follows.

$$\alpha_i = \begin{cases} 1.5 & (i = 0, 5, 10) \\ 0.15 & (i = 1, 6, 11) \end{cases},$$
$$\beta_i = \begin{cases} 0.8 & (i = 0, 5, 10) \\ 0.5 & (i = 1, 6, 11) \end{cases}$$

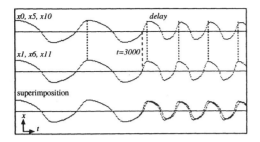

Figure 8: Amplitude of Oscillator (unattractive)

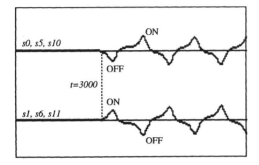

Figure 9: Control Signal of SMA Actuator (unattractive)

These parameters mean that oscillators on the left side of horizontal SMA-Net actuators (0, 5, 10) are inactivated by an unattractive stimulus and the frequency becomes low. On the other hand, oscillators on the right side of horizontal SMA-Net actuators (1, 6, 11) are activated to escape from an unattractive stimulus. In the van der Pol equation (1), the parameter α_i becomes bigger, and the frequency becomes lower.

These oscillations become asynchronous, and the phase of oscillators $\{x_0, x_5, x_{10}\}$ is more delayed than that of $\{x_1, x_6, x_{11}\}$. Fig.8 shows a time sequence of $x_i(t)$. In this figure, as in Fig.8, the superposition of two time sequences of amplitude shows the transition from a synchronous state to an asynchronous state after 3,000 time steps.

Fig.9 shows the time sequence of control signals of the actuator. Up until 3,000 time steps, since two sets of oscillators are synchronous, $s_i(t)$ is 0.0 and the power of actuators is turned off. After 3,000 time steps, the power of actuator (1, 6, 11) is turned on first, then the power of actuator (0, 5, 10) is turned on.

Conclusions

In this paper, we propose an emergent control method for an SMA-Net robot using a coupled oscillator system. The results of computational simulations showed that our proposed method enables basic behaviors, that is, movement in response to an attractive stimulus and escape from an unattractive stimulus, by locally interacted information processing.

In the proposed control system, the sensor - action mapping is translated in the change of parameters of coupled oscillator. In the action decision system of real slime mold, chemical substances (e.g. cAMP) works as coupled oscillators. The result of simulation shows our proposed method implements the action decision method of amoebae of slime mold.

In the experiments, the parameters were manually selected for the objective behavior. However, in future works, the most suitable parameters for objective behaviors will be selected by Genetic Algorithm. We intend to apply the proposed method to a real SMA-Net robot and to increase the number of actuators and extend the SMA-Net robot to a three-dimensional composition.

References

Dittrich, P., B. Andreas, and B. Wolfgan. 1998. Learning to move a robot with random morphology. *Evolutionary Robotics*. Lecture Notes in Computer Science 1468: Springer-Verlag, p. 165–178.

Paap, K. L., M. Dehlwisch, and B. Klaassen. 1996. GMD-snake: A semi-autonomous snake-like robot. *Distributed Autonomous Systems 2*. Springer-Verlag, pp. 71–77.

Jammes, L., M. Hiraki, and S. Ozono. 1996. Development of snake-like robot - designed concept and simulation of planar movement mode -. *Distributed Autonomous Systems 2*. Springer-Verlag, pp. 78–88.

Shu, S.G., D. C. Lagoudas, and D. Hughes. 1997. Modeling of a flexible beam actuated by shape memory alloy. *Smart Material and Structure*: 265–277.

Miyake, Y., S. Tanabe, H. Murakami, M. Yano, and H. Shimizu. 1996. Environment-dependent self organization of positional infomation field in chemotaxis of physarum plasmodium. *Journal of Theoretical Biology* 178: 341–359.

Kaneko, K. 1989. Spatiotemporal chaos in one- and two- dimensional coupled map lattice. *Physica D* 37: 60–82.

Kaneko, K. 1990. Clustering, coding, switching, hierarchical ordering, and control in network of chaotic elements. *Physica D* 41: 137–172.

Johnson, A. D., V. V. Martynov. 1997. Application of Shape-Memory Alloy Thin Film. International Organization on Shape Memory and Superelastic Technologies.

V Communication, Cooperation, and Collective Behavior

Developmental Insights into Evolving Systems:
Roles of Diversity, Non-Selection, Self-Organization, Symbiosis

Norman L. Johnson

Theoretical Division MS-B216
Los Alamos National Laboratory
Los Alamos, NM 87545
nlj@lanl.gov

Abstract

A developmental view of evolving systems (ecological, social, economical, organizational) is examined to clarify 1) the role of selection processes versus collective, non-selective processes, 2) the origins of diversity and its role in system performance and robustness 3) the origin of explicit subsystem interactions (cooperation/symbiosis) that enhance individual and system performance, 4) the preconditions necessary for further evolutionary development, and 5) the effect of environmental timescales with adaptation timescales. Three sequential stages of evolving systems (based on the work of Salthe) are proposed: a *Immature stage* dominated by highly decentralized, selective processes with chaotic local and global dynamics, a *Mature stage* dominated by non-selective, self-organizing processes with global robustness but locally chaotic dynamics, and a *Senescent stage* dominated by rigid interactions with global fragility. A simple model problem with many optimal and non-optimal solutions - an agent solution to a maze - illustrates the entire developmental history. Within the model, the agents evolve their capability from a random approach to an optimized performance by natural selection. As the agents develop improved capability, natural selection becomes rare, and an emergent collective solution is observed that is better than the performance of an average agent. As the collective, self-organizing structures are incorporated into individual capability within a stable environment, constraints arise in the agent's interactions, and the system loses diversity. The resulting Senescent system exhibits reduced randomness due to the rigid structures and ultimately becomes fragile. Depending on the degree environmental change, the Senescent system will either "die," or collapse under environmental stress to the Mature or Immature stage, or incorporate the constraints system-wide into a new hierarchical system. The current study adds to the literature on developmental systems by finding: Transitions between stages are dependent on the degree of sustained environmental stability and how exclusive cooperation (e.g., symbiosis) in a subsystem can originate, and how it results in a decline in diversity.

Introduction

One challenge for researches in evolving systems, whether for economic, ecological, political, social or organizational systems, is a common viewpoint for both understanding the dominant processes in their models and a basis by which to compare their models to others. The current work grew out of research on how self-organizing groups can solve problems better than experts and the consequent need to understand non-selective, self-organization processes[1] (Johnson et al. 1998) and how it relates to natural selection (Johnson 2000). Both selective and non-selective processes are observed in real systems and have validity. Both processes increase global fitness, but by distinctly different mechanisms. But, because each originates from very different viewpoints (competitive versus non-competitive agents), can they be reconciled into a single understanding? It was from this starting point that the current study began. A detailed examination of the dynamics and properties of non-selective self-organization was done (Johnson 1998), but offered no real insight into how this approach to global function related to selective processes. What was finally discovered was that these seemingly different approaches are representative of different stages of an evolving system (Johnson 2000). And the apparent conflict is resolved by a developmental perspective of evolution, based on the sustained work by Salthe

---

[1] An example of non-selective self-organization is foraging in social insects (Bonabeau et al. 1999), expressing what is called "swarm" intelligence. In the absence of selection (no ants die), a foraging group can be observed to perform greater than the capability of the individual (e.g., the path to the food is shorter for the average group than the best individual). These final paths represent the collective action of many individuals solving their own path problem, in a manner that is ultimately useful to the entire population but which is never expressed as a goal at the level of the individual. Another example is path formation in humans and the method for book referral used by Amazon.com (Johnson 1998).

(1989, 1993, 1999) for ecosystems. Within this developmental view, an extension of the model problem developed for non-selective self-organization is used to illustrate the full developmental history of evolving systems. This study begins to address many questions of interest in the fields of evolutionary systems: How do groups achieve higher global functionality? How can the robustness of the system be improved? What global properties are prerequisites for greater functionality? How does the system "boot-strap" itself to a higher functionality? The approach taken is review the developmental view of evolving systems, and then propose simple model problem to illustrate the stages of the theory and the mechanisms for transitions from one stage to another. While the simple model is not rich as real systems or highly developed simulation models, its simplicity enables clear discussion of the processes characterized by the general theory.

Developmental View of Systems

In this section a developmental view of evolving systems is presented, a summary of the work of Salthe (1989, 1993, 1999) and its extension to a other systems, economic, political, social, and organizational (Johnson 2000). The focus, here, is on the interplay of natural selection with non-selective processes, on the role of diversity, and on the transference of global emergent properties to subsystems. While three stages are presented below, it is understood that these are just points in a continuous development. Furthermore, to simplify the presentation, the systems are assumed to be homogeneous in the progress of maturation (sub-components are not mixed across states). More likely, systems will have multiple stages of development simultaneously, particularly as a system increases in complexity and partially independent subsystems undergo cycles of maturation and failure and become out of phase.

Three stages for the development of any evolutionary system are: *Immature systems* - high selection pressure, rapid component and global variations, highly decentralized, low "complexity", low symbiosis (minimal interactions and dependencies), high entropy production; *Mature systems* - low selection pressure, high diversity, multiply interconnected and robust; *Senescent or aging systems* - similar to Mature systems, but interactions become restricted and rigid (lower variation in interactions - low entropy production), and the resulting system is fragile.

Table 1 presents a variety of properties for the different stages. The following definitions are used in the table. *Diversity* is the uniqueness of the attributes of the individuals. *Interconnectivity* is the degree of interaction between individuals/subsystems. *Chaotic dynamics* is the sensitivity of states (local or global) to small changes. *Decentralization* is the degree of autonomy of an individual's actions. *Individual flexibility* is the degree that the individual can survive changes in the environment or other sub-components; *system flexibility* (robustness) is the degree of survivability of the global system to environmental change or sub-component failure. *Entropy* is the measure per unit of the randomness expressed within the constraints of the system.

Immature Stage Immature systems are characterized by "hard" selection (Fisher 1930; Wallace 1970), with the corresponding aspects of relatively high competition and high mutation rates. The interdependence of individuals within Immature systems is initially low (highly decentralized), but increases as interdependencies and global structures form. Because of the role of selection for improving group fitness, diversity of the populations is essential for adaptation of the system to changes in environment. Because diversity is consumed by selection, diversity's role is as an investment for future adaptation and does not contribute to current system performance (the average state of the individual fitness) (Johnson 2000). In fact, the presence of diversity in an Immature system at any one time lowers system fitness because of high degree of individual failure.

Mature Stage The evolution of a Immature system to a Mature system is the creation of interdependency from the increasing diversity of individuals or subsystems (Kauffman 1993). As a consequence, overall system fitness shifts from a consequence of selection to improved fitness resulting from non-selective, self-organizing processes (Johnson 2000). Diversity in Mature systems is essential to the *current* fitness of the system and occurs, not as a result of new niche formation (a mechanism requiring selection), but just from the random processes of mutation without hard selection: *survival of the fittest becomes survival of the adequate* - also called soft selection (Salthe 1972; Wallace 1970). The development of global structures results in less chaotic dynamics on a global level, but the responsiveness of the system is retained by having chaotic dynamics at the local level (Johnson 1998). Similarly, the global system becomes more robust, due to redundancy and contingencies in the subsystems, a consequence of randomly generated diversity and the flexibility of the interdependencies between subsystems. Given sufficient environmental variability, the system will remain at a Mature state, because interdependencies are sufficiently dynamic to prevent rigid interdependencies from forming.

Property/Process	Immature	Mature	Senescent
Diversity	Increasing	High	Declining
Interconnectivity	Low and increasing	High and redundant	Declining and rigid
Chaotic dynamics:			
locally	High	High	Low
globally	High	Low	Low
Selection -	High	Low - preserves status quo	Low
Competition -			
Individual turnover			
Source of new diversity	Niche creation	Random generation	None
Group improvement	By individual selection	By collective processes	Same
Decentralization	High	Medium due to high interconnectivity	Low
Flexibility:			
individual	Low	High due to elastic interactions	Low
global	Varied	High due to redundancy	Low
Entropy production	High	Moderate	Low
Rate of environment change for stability	Varied	Slowly varying	None or little

Table 1: Stages of development in evolving systems.

Senescent Stage The transition from a Mature to Senescent system occurs as the consequence of a relatively stable environment. The self-organizing, flexible structures that are advantageous in the Mature state become exclusive, reinforced and rigid; entropy production is reduced. One form of these rigid structures is symbiotic (mutualistic or parasitic) relationships. Expressed another way, the emergent properties, and consequent advantages, in the Mature system are replaced by explicit (non-emergent) properties at the level of the individual. This is advantageous in a stable environment, because it eliminates the chaotic and unpredictable nature of a flexible, dynamics of the Mature stage. This transference is argued to be the origin of explicit cooperative behavior between individuals in many systems.

Death or Hierarchical Resolution The final outcome of the evolutionary cycle is dependent on many factors. One possibility is system-wide failure, resulting in the loss (death) of all constituents. While rigid structures in Senescent systems offer advantages to the subsystems, they are detrimental to the robustness of the global system. Senescent systems can fail if environmental stress is sufficient to break critical interdependencies. Then, due to the rigidity of the system, a global collapse of the system can result. From a global perspective, the collapse can return a Senescent system to a Mature or Immature stage. Alternatively, the advantageous rigid structures can be subsumed system-

wide, and the evolutionary process can begin again on top of the structure. The adoption of DNA encoding or formation of a cell nucleus in life is an example of useful structures being incorporated system-wide, allowing variation then to occur on top of these global structures. The possibility that rigid structures can be subsumed system-wide is a path to developing hierarchical systems.

Examples of Each Stage

A prime example of a Immature systems is the field of Evolutionary programming and Genetic algorithms, summarized by Fogel (1999) and characterized by the use of algorithms using population-based variation and selection. Methods based on non-selective processes, such as the simulations presented in the Mature stage below, are absent.

An example of a Mature stage is a mature ecosystem, composed of diverse species, where each individual living to fulfil their own needs, resulting in a stable system that benefits all. While competition and selection occurs in Mature ecosystems, the global fitness (e.g., robustness) is due to the non-competitive interactions of a diverse community (Johnson 2000). The creation of new diversity is continual, not because of selection, but from lack of selection. The ecosystem is locally chaotic (species and individual interactions are unpredictable), but the global system is robust and insensitive to details of the chaotic nature. For social systems (organizational or political), most of the above observations for ecosystems can be also made. In particular, the unappreciated aspects of social networks in organizations provide problem solving capability and contingencies that directly result from diverse individuals or groups (Linstone 1999).

Examples of fully Senescent systems are rare due to their fragility. Very old ecosystems, such as the Australian rainforest that drains into the Great Barrier Reef, are good examples. Interactions are either highly specialized, such as a single species pollinating another single species, or highly restricted, such as the limited predators of the extremely poisonous tree frogs. The American automotive industry a decade ago reflected a system that was highly evolved but had limited flexibility, with few and fixed interdependencies.

A Model Problem of Evolution

A model problem is presented to illustrate the developmental perspective presented in the last section. The model problem as a Mature system has been studied in detail (Johnson 1998) and is extended here to the other stages. The model problem is the solution of a sequential problem (e.g., as in Fig. 1), which has many optimal and non-optimal solutions, solved by agents. While this maze problem in Fig. 1 is quite simple from a global perspective, it serves as a representation of more complex processes: the solution of a problem that has many decisions points and many possible solutions and that has difficulty greater than that solvable optimally by one individual. A more realistic landscape would not change the underlying processes that are observed in this simple model. It is argued (Johnson 1998) that all evolutionary systems are sequential in nature (every action of an individual has a prior, different action leading to the present state), and that the current model is an abstracted representation of real systems.

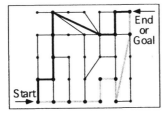

Figure 1: The example maze. Two of the 14 minimum length paths are highlighted.

The solution process for a single agent is divided into a *Learning phase* where simple rules of movement are used to explore and learn about the problem domain. Because the agents have no global sense of the problem, they initially explore the problem until the goal is found. The learning process can be thought of as an agent exploring the maze randomly and leaving "breadcrumbs" behind to aid in their search for the goal, thereby avoiding fruitless paths. Then in an *Application phase*, this "learned" information (bread crumbs or path preferences) is then used by the agent to solve the problem again, typically with a shorter path[2] as a consequence of eliminating unnecessary loops. Essentially, the agent in the Application phase follows the path with the most breadcrumbs.

The following assumptions are made and are discussed in detail elsewhere (Johnson 1998; Johnson 2000). These assumptions were analyzed and found to not be critical to the conclusions of this study. 1) The information available to an individual at a decision point (node) is independent of the path that they took to get there, i.e., the solution is path independent. Said alternatively, only local information is used in the decision - there is no global perspective by the individual. 2) Individuals "solve" the same problem, both in goals and in a common world view.[3] 3) Finally, individuals have identical assessment of the value of information.

Simulations of the Model Problem

A variety of strategies for an individual agent was used to solve the model problem (see Table 2). For example, the *random walk method* starts the agent at the beginning node, and then the next node is randomly selected. The process is repeated, each time selecting from all possible nodes connected to the present node, until the goal is reached. The *no-backstep random walk method* is the random selection of a new node excluding the node that was just vacated. And the *non-repeating random walk method* is the selection, if possible, of only untried links. The *Learning Rules* are a set of rules that mimic the idea of laying bread crumbs (or pheromones) down to aid the search - and giving more bread crumbs to the last link taken, so if the node is returned to and all nodes have been tried, then the last link used will be preferred.

These different Learning methods are differentiated by the degree the learned information used - from the extreme of being ignored in random walk to being optimized in the Learning Rules.[4] In the Application phase, the agents use the bread crumbs of the Learning phase to solve the maze again (see Table 2). The Application Rules are the same for all methods, and basically pick the path with the greatest "breadcrumbs."

---

[2]Note that "path length" is the number of segments in the path, not the actual path length.

[3]The common world view is taken to mean that the possible options that an agent has are identical. This does not mean that the preferred options are the same, only that the possible options are the same.

[4]Note that although multiple agents exist, *the agents solve the problem independent of the other agents*; this restriction is removed in Senescent version of the simulations.

Why is the performance in the Application phase better than in the Learning phase? Fig. 2 illustrates how an extraneous loop is eliminated by the Application phase for an individual. This mechanism is comparable to those argued for collective improvement in ant simulations (Bonabeau et al. 1999).

Learning method	Average	Standard deviation
Random walk (RW)	48.8 *(123)*	55 *(103)*
No-backstep RW	38.6 *(64)*	40 *(66)*
Non-repeating RW	33.7 *(51)*	36 *(51)*
Learning Rules	12.8 *(34)*	3.1 *(24)*

Table 2: Path Lengths for the Application phase for a population of 100 agents. The quantities in brackets are for the Learning phase.

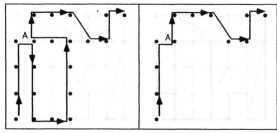

Figure 2: Plots of the paths in the two phases for a representative individual. The Application phase, at right, removes the extra loop at point A from the Learning phase, at left.

A Immature Stage of the Model Problem

The methods in Table 2 represent different strategies in solving the problem by an individual agent. While not implemented within the current simulations, a selective process can be developed where the agents with different strategies compete with each other, while repeatedly solving the maze, redoing the Learning and Application phases fresh each time. If the agents above a certain path length (number of links in a path) are eliminated and replaced with a new agent that uses a strategy sampled from the surviving agents (a genetic approach), then the population would become dominated by the most successful strategy - a typical result of natural selection. In the absence of mutations and with sufficient selection pressure, all diversity of capability would be lost.

Because the initial search is random, a collection of individuals using the same method shows a *diversity of experience* (knowledge of different regions of the maze), *diversity of preferences* (different preferred paths at any one location in the maze), and *diversity of performance* (different numbers of steps). The differences

in "experiential" and "preferential" diversity, as opposed to capability diversity, are a direct consequence of the redundancy in the solution space of the problem: because there are many paths of equal length, there is no selection pressure for one path over another. If a selection pressure is added to encourage the agents to occupy vacant portions of the maze (say food sources) and if some "genetic" memory of the experience of the maze is passed onto new offspring ("turn left at node 5"), then the different regions of the maze would become equally populated. This is argued to be comparable to the "filling" of niches as a form for diversification in the process of natural selection. If a load-carrying capacity were defined to be the global fitness of a group of agents, then this experiential diversification would result in a maximum for global fitness.

The above description of a potential simulation scenario illustrates that the model problem can capture the attributes in Table 1 associated with Immature systems. Because of the predominance of natural selection in evolutionary approaches, comparable examples of Immature systems are abundant in the literature (Fogel 1999).

Mature Stage of the Model Problem

In this section the collective effects of having multiple agents solve a common problem is examined. The central question is how contributions from diverse agents can be organized in a manner that is useful to both the individuals and to the global system. The approach taken below is to present a collective solution to the model problem, but one which does not require individual interaction (competition or cooperation) or selection in any form. Therefore, the model problem is specifically constructed to isolate the emergent collective effects.

Forming a collective solution Suppose a group of individuals is heading to a cafe and all have prior experience with finding the cafe. At each corner, they combine their own experiences without discussion, and then chose a preferred path based on this collective information, using the same rules they each used as individuals. Said another way, suppose a group could see the bread crumbs of all of the members, and they pick the path with the greatest amount of bread crumbs. The realization of this metaphor in the current model problem is to use a linear combination of the each individual's experiences at each node in the maze for all the individuals in a group (Johnson 1998). Then the same Application rules as used for the individual are used on this group information to find a group solution. In some sense, the collective is a "super-informed" individual in that it has access to more information, but has the same capability as an individual. Figure 3 shows the simulation results for groups of increasing size for the different learning methods listed in Table 2. In Fig. 4, the simulations using the Learning Rules are shown (the Novice and Established concepts are discussed shortly). Because of the variation in performance for small groups, the simulations results in Figures 3 and 4 are ensemble averages of many simulations. To easily identify the improvement of the collective over the average performance of the individuals in the group (*the collective advantage*), the collective path lengths are normalized by the average of the performance of the individuals making up the ensemble (around 12.8 on average).

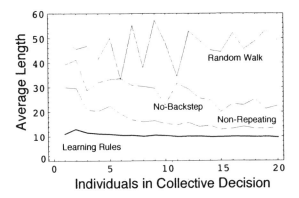

Figure 3: The path length for different numbers of groups of individuals using the various rules in Table 2. Each point on a curve is an average of 50 simulations. The two curves for the novice are for a selection of different individuals in the group.

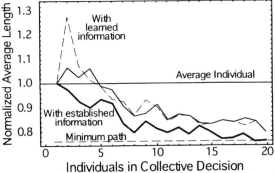

Figure 4: The normalized path length for different numbers of groups of individuals using the Learning Rules. The two curves for the "learned information" groups are for two different random selections of individuals.

In the repeated solution to an unchanging prob-

lem, we tend to remember only the information that is needed and forget extraneous information associated with unused paths. Here, the equivalent effect is for the agent to remember only "established" or reinforced information along paths, thereby "forgetting" unused paths. The process of forgetting unused information does not change the performance of an individual agent, because both the learned and established information produces the same path in the Application phase. An established individual experience is created from the learned experience by retaining information used in an individual solution, and forgetting unused information (e.g., the information used on the right in Fig. 2). In Fig. 4, the collective performance of the learned and established individual information is shown. The significance of the difference in performance is discussed in the section below on diversity.

The Collective Advantage as Emergence The model problem is developed with three requirements: 1) rules of the agents do not use global information, 2) the agents do not include logic for finding a shorter path and 3) the agents learn and apply information independently (do not interact or cooperate). These three requirements assure that any observed global property or functionality cannot be predicted from the properties of the individual. In Figs. 3 and 4 we observe that, with the exception of the random walk simulation, large collectives perform better than the average individual making up the collective and for collectives of sufficient size using the Learning Rules, the collective solution converges to one of minimum paths (the other methods converge to a non-optimal solution, as noted below). Therefore, the occurrence of a shorter path length for large collectives over the average individual is an emergent property of the system.

One mechanism of the collective improvement is the collective equivalent of the elimination of extraneous loops, similar to process of improvement of the individual in Fig. 2. Because the collective bread crumbs are a superposition of the information of many individuals, collective bread crumbs can contain extraneous loops as observed in Fig. 2, but which are only partially closed in the individual contributions. Fig. 5 illustrates how this can occur. By combining the information of multiple individuals, collective bread crumbs contain complete information and the extraneous loops can be removed from the group path. This emergent global property was found to be a robust property of the model problem and insensitive to many alterations or removal of assumptions in the model (Johnson 1998). The only exception was for groups composed of individuals learning by the random walk method (discussed next).

What is the significance of an emergent global property in this model? Before explicit forms of symbiosis and interdependencies between individuals can evolve, there first needs to be some emergent expression of the advantages of dependencies (Hemelrijk 1997). This is the classic "boot-strapping" dilemma of all evolving systems (Kauffman 1993).

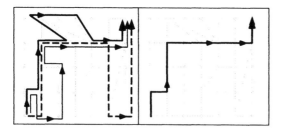

Figure 5: Paths of three individuals in the Learning phase, at left, combine to form the Collective Path, at right. Extraneous loops in the collective solution are eliminated from incomplete loops in the individual contributions.

The above discussion leads to the question of how the emergent capability in the model problem could be used by the individual or collective in a way which is not a consequence of explicit cooperation. The following variation is proposed. Suppose many individuals solve the maze simultaneously. At any given time, if there is more than one individual at a node, then a collective decision is made and the group continues to the next node together. Because the likelihood of the formation of a group is random and no selection of information from a subset of the group is made, this is an example of a combination of information that is both random and non-selective. The consequence is that the randomly formed group would follow the previously identified collective solution from that point on, assuming no other individuals are added to the group. If other individuals or groups are added during their progress to the goal, the additional members would only improve the collective solution.

Coupling of individual and group performance
The results in Fig. 3 are essentially a study in which the problem difficulty (maze) is held constant and the individual's capability is varied. A comparable study was done in which the mazes were made more complex while the individual's capability was held constant (Johnson 1998). The following conclusions were drawn. 1) A simple maze to a good individual solver is trivial, and no collective advantage occurs. 2) Mazes of greater difficulty than can be trivially solved by an individual are solved optimally by large groups. 3) An extremely difficult maze to an individual with fixed capability

leads to a random individual solution and no collective advantage is observed. These conclusions indicate that harder and harder problems cannot be solved by larger and larger groups of individuals with constant ability. Or, equivalently, the individual must have some capability (i.e., not random) which can be amplified in groups. These statements indicate that the collective advantage is coupled to the individual problem solving ability and the global problem difficulty. Hence, an essential aspect of the development of an evolving system can be identified: *the process of natural selection is needed to increase the performance of the individual to the point which the emergent global structures can amplify the weak individual signals.* This, then, explains why the groups formed from individuals that learn by the random walk method fail to show a collective advantage: there is no coherent signal that can be amplified by the collective solution.

Learning Method	Collective Application Phase	
	Novice	*Established*
Random Walk (RW)	32 (0.31)	21 (0.46)
No-Backstep RW	13 (0.32)	9 (0.42)
Non-Repeating RW	10 (0.30)	10 (0.44)
Learning Rules	9 (0.38)	9 (0.60)

Table 3: Path Lengths for the Application phase for different Learning phases for a population of 100. In the parentheses are the diversity measures.

Diversity and collective advantage What property of the group can be used to predict the occurrence of the collective advantage? The best correlation found was a measure of diversity which gives more weight to the contributions of the individuals that have the least commonality with the group - see (Johnson 1998) for a formal definition. In other words, individuals with potential experiences that are not shared by others are the most important. As shared potential experiences increase at a node, the weighting is less, until it is finally zero if all members of the group share the same potential experiences. The words "potential experience" are used, because no consideration is given to the magnitude of the preferences at a node.

In Table 3, a summary is given for collectives of 100 members for the various individual learning methods for a single simulation. The values of the experiential diversity are given - where the diversity is normalized such that it is between zero and one. Groups using established information are found to have a higher diversity measure and perform better than groups using novice information. Because the correlation for the results using the random walk learning method does not follow the same trend as the other simulations, this suggests that the effect captured by the experiential diversity measure is not the only reason for higher performance of the collective, and some accounting of quality of information, such as the relative performance of the individual, is also required.

Chaos and robustness in the simulations The model problem for the Mature stage expresses both chaos (locally) and robustness (globally). A detailed study of the local chaotic nature of the simulations (Johnson 1998) indicates, for example, that the specific path (sequence of nodes) of the collective solution is sensitive to the addition of one individual, even for arbitrarily large groups. This is a consequence of the problem domain containing multiple paths of equal fitness. But, the ability of the collective to find a minimum path is not chaotic, but is stable to small changes.

The robustness of the global solution can be demonstrated by also evaluating the sensitivity of the model problem to noise. Noise in this context is the random replacement of valid information in the individual's contribution to the collective, thereby creating *false* information. Fig. 6 shows the effect of the addition of noise in simulations with different frequencies: 0.0, 0.3, 0.7 and 0.9, where 0.0 represents the simulation using the Learning Rules in Fig. 2. These results are insensitive to the magnitude of the noise, as long as it is less than the maximum weighting of a path. The collective solution is observed to be remarkably insensitive to the addition of noise at low frequencies. Even at higher frequencies, the noise only delays the collective advantage to larger groups.[5] This is a clear demonstration how diversity makes the collective decision robust. The above results support the conclusion that the diversity measure is also a measure for the robustness of the collective solution.

Senescent Stage of the Model Problem

The essential difference between the Mature and Senescent systems is the formation of rigid interactions in the place of flexible ones. In the model problem the Mature-Senescent transition is captured by the introduction of feedback of the collective experience to either individuals or groups during the Learning or Application phases. This approach is comparable to the feedback of the combined pheromone trails to the indi-

---

[5]Note that an individual performance is much more sensitive to noise than the collective. This occurs because noise leads an individual to parts of the maze for which they have no experience from the Learning phase. In unexplored regions, the Application Rules degenerate to a random walk approach. For collectives, particularly large collectives, experience is available throughout the entire maze and, therefore, the collective cannot be misdirected by false information to unknown parts of the maze.

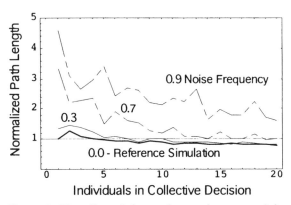

Figure 6: The effect of the random replacement of the individual's contribution to the collective for different frequencies of replacement.

viduals paths in ant-motivated simulations (Bonabeau et al. 1999). While this added feature to the model problem has not been examined in detail, sufficient evidence suggests that the feedback effect is significant and has the expected consequences.

Suppose in the Application phase of the collective, described in the prior section, as the groups increase in size they make use of the collective experience of the previous group. This was implemented in the simulations presented in Fig. 4 by letting the bread crumbs for the Application phase of the current group size be a linear combination of the previous group's bread crumbs and the current group's bread crumbs. This combined collective experience was then passed onto the next size group. When only 10 percent of the prior group's information is used, the convergence to a minimum path occurred by a group size of 8, instead of 20. Even more importantly, the local chaotic nature of the collective solution was lost: the same minimum path is selected as a consequence of the positive reinforcement of a single path in repeated collective solutions. Although not examined, the robustness to noise of this collective solution is expected to decrease with the loss of group diversity. These simulation results suggested the following analysis of the model problem.

Suppose in the following discussion that a collective experience has been generated, as describe in the section for the Mature stage. If this collective experience is used by a new individual during its Learning phase, then the individual will identically follow the choices made by the collective in the collective Application phase. A unique individual experience will not be created by the individual, because there is no random exploration that leads to diversity. This is because the individual begins the Learning phase with no experience (zero bread crumbs), and the presence of the collective experience will dominate the learning of the individual. The resulting individual experience from this collectively-enhanced learning process will cause the individual in its Application phase to duplicate (to within paths of equal preference) the path of the collective - a minimum path for large collectives. This process of individual learning based on the experience of the collective will filter out the information of alternative paths that were originally part of the collective experience, which was the basis for the system robustness. Hence, the individual achieves optimal performance, but at the cost of loss of individual robustness. If a new group is created from these "collectively-enhanced" individuals, they will express zero diversity and exhibit none of the robustness of the prior diverse collective. Furthermore, this "collectively-enhanced" group will perform optimally, but never better than the collective of identical individuals: the collective advantage will not occur, even if the "collectively-enhanced" individuals did not find a minimum path.

The above understanding is all within the context of a "stable" environment for the model problem: the beginning node and end node (goal) are unchanged in the above process. Suppose that the above process is repeated, but the end node is changed between the formation of the collective experience and the Learning phase of the "collectively-enhanced" individual. The consequence is that the collective experience would not correspond to the current goal, and the individual would resort to a random search, resulting in a return to a diverse population of individuals and a model performance similar to the Mature state describe earlier. If the goal change is not significantly different, then the system will come to a new equilibrium based on the old experience. If the goal is changed significantly, then the earlier structures (dominant paths) will be eliminated and replaced by new structures. If the simulations also included the "genetic selection" described in the Immature stage of the problem, the performance of the individual may be sufficiently poor to trigger a return to the Immature stage and selection of new capabilities.

The above process is argued to be the mechanism by which a diverse and robust population, as described in the Mature phase section become a Senescent system. In a stable environment, the emergent properties of the interactions (e.g., the collective path) can be incorporated by the individual to optimize their performance, but only at the expense of the loss of robustness of the collective. Depending on the rate of change in the environment, the system will either remain flexible in its interactions (slow rate of change for the Mature stage) or create rigid structures (no change for the transi-

tion from the Mature to Senescent stage) or lose all rigid structures (rapid change from the Senescent to Immature stage). In all of the above discussion the extent of the problem domain was unchanged. Supposed that the present maze is only one of many overlapping mazes, each with its own agents. Within this context, then, the formation of rigid structures on one level represents the incorporation of a subsystem structure, upon which further variation can occur on top of or around it. This models the formation of a hierarchical system (Mayer and Rasmussen 1998).

Summary

The main *theme* of this work is the presentation of a developmental theory of evolving systems (ecological, social, political, organizational). The main *purpose* is to understand of the roles of diversity and mechanisms of higher functionality by processes of non-selection. It is only within a developmental perspective that non-selective processes can be compared to the predominant explanation of natural selection as the source of functionality in evolving systems.

The developmental view of evolution identifies three sequential stages: a *Immature* stage dominated by highly decentralized, selective processes and chaotic dynamics (local and global), a *Mature* stage dominated by non-selective, self-organizing processes and global robustness, and a *Senescent* stage dominated by rigid interactions and global fragility. The allowable transitions of an evolving system through these developmental stages depend both on the rate of environmental change (external pressure) and the occurrence of prerequisite processes of the stage before. These prerequisites are 1) high diversity of individuals, 2) sufficient local interconnectivity between individuals or subsystems, 3) sufficient capability of the individual relative to the global challenges that can be amplified by self-organizing processes, and 4) mechanisms for the capturing of emergent properties at the global level in the individual properties.

A model problem was proposed and used to illustrate the processes and transitions of a developmental view. It is significant that one simple model yields insights into abstract concepts (self-organization, selection, diversity, robustness, origin of cooperation, etc.) and lends credence to the view that a sequential problem with multiple solutions solved by an diverse agents is an appropriate general model for evolving systems. Comparable model problems of evolution of the iterated prisoner's dilemma and cellular automata are not as general.

The following insights into developmental systems are captured by the simple model. 1) Diversity is best defined as the measure of the uniqueness of the members in a group. This measure correlates consistently with group or system robustness for all the stages and for the non-selective, self-organizing collective advantage. 2) Diversity is expressed in rich systems in many ways: capability, experience, preferences and performance. Selection in a system may reduce one type of diversity, but may not affect other types. 3) Diversity can arise at random in groups of agents *of identical capability* when a system has little or no selection pressure (survival of the adequate, instead of survival of the fittest). Diversity does not have to arise by selection or from agents of different capability. 4) Random creation of diversity can contribute directly to both global performance (collective self-organization) and robustness, above that of an individual and in the absence of any selection from the population. 5) The process of collective, non-selective self-organization can duplicate the system-wide advantages of explicit cooperation. 6) The performance of the collective self-organization is coupled to the individual performance and global problem difficulty. 7) These emergent processes are the precursors of cooperative advantages that are often attributed to individuals and are the key transitional mechanism from the Mature to Senescent stage. 8) Mature systems can express both local chaos (entropy) and global stability (robustness) simultaneously. Both are a direct consequence of the diversity of the system. 9) Exclusive and explicit interdependencies, characteristic of the Senescent stage, reduce entropy/diversity and consequently the robustness of a system. These interdependencies are a transfer of the collective self-organization properties observed in the Mature phase into the properties of the individuals. These explicit interdependencies form only in stable environments. 10) Hierarchical systems can form by the incorporation of these exclusive interactions into global structures, thereby, creating a new landscape for variation and return to an Immature or Mature stage of development.

Acknowledgements

The author gratefully acknowledges insightful conversations with Stanley Salthe and many other colleagues that a shared common world view. This research is supported by the Department of Energy under contract W-7405-ENG-36.

References

Bonabeau, E., M. Dorigo, and G. Theraulaz. 1999. *Swarm Intelligence: From Natural to Artificial Systems.* New York: Oxford University Press.

Fisher, R.A. 1930. *The Genetic Theory of Natural Selection.* New York: Oxford Univ. Press.

Fogel, L.J. 1999. *Intelligence through Simulated Evolution: Forty years of evolutionary programming.* Edited by A.M.J. Albus and L.A. Zadeh. New York: John Wiley.

Hemelrijk, C.K. 1997. Cooperation without Genes, Games or Cognition. In *Fourth European Conference on Artificial Life,* edited by P.H.a.I. Harvey. Cambridge: MIT Press.

Johnson, N.L., S. Rasmussen, C. Joslyn, L. Rocha, S. Smith, and M. Kantor. 1998. Symbiotic Intelligence: Self-organizing knowledge on distributed networks driven by human interactions. In *Artificial Life VI,* edited by C. Adami, R.K. Belew, H. Kitano and C.E. Taylor. Cambridge, MA: MIT Press.

Johnson, N.L. 1998. *Collective Problem Solving: Functionality Beyond the Individual:* http://ishi.lanl.gov/Documents1.html.

Johnson, N.L. 2000. Importance of Diversity: Reconciling Natural Selection and Noncompetitive. Processes. In *Closure: Emergent Organizations and Their Dynamics,* edited by J.L.R. Chandler and G.V.d. Vijer. New York: New York Academy of Sciences.

Kauffman, S. 1993. *The Origins of Order: Self Organization and Selection in Evolution.* New York: Oxford University Press.

Linstone, H.A., 1999. *Decision Making for Technology Executives: Using multiple perspectives to improve performance.* Boston: Artech House.

Mayer, B. and S. Rasmussen. 1998. Self-Reproduction of Dynamical Hierarchies in Chemical Systems. In *Artificial Life VI,* edited by C. Adami, R.K. Belew, H. Kitano and C.E. Taylor. Cambridge, Mass.: MIT Press.

Salthe, S.N. 1972. *Evolutionary Biology.* New York: Holt, Rinehart and Wilson.

Salthe, S.N. 1989. Self-organization of/in Hierarchically Structured Systems. *Systems Research* 6:199-208.

Salthe, S.N. 1993. Development and Evolution: Complexity and change in biology. Cambridge: MIT Press.

Salthe, S.N. 1999. Energy, development and semiosis. In E. Taborsky (ed.) *Semiosis, Evolution, Energy: Towards a Reconceptualization of the Sign.* Shaker Verlag: 245-261.

Smith, J.B. 1994. *Collective Intelligence in Computer-Based Collaboration.* New York: Erlabum.

Wallace, B. 1970. *Genetic Load: Its Biological and Conceptual Aspects.* Upper Saddle River: Prentice-Hall.

Reducing Collective Behavioural Complexity through Heterogeneity

Josh C. Bongard[1]

[1]Artificial Intelligence Laboratory, University of Zurich, Switzerland
bongard@ifi.unizh.ch

Abstract

In this paper, the correlation between behavioural heterogeneity and behavioural complexity within groups of cooperating agents is investigated. This investigation is accomplished using the Legion system, a type of evolutionary algorithm for evolving group behaviours in which behavioural differences among agents in the group is subject to selection pressure. Two collective task domains are studied, and two types of control architecture for the agents are used. From the experiments reported here it is concluded that increased behavioural heterogeneity within a group leads to reduced average control complexity, and also that reducing the maximum size of control architectures results in the evolution of increased behavioural heterogeneity. It is argued that this correlation helps to clarify the relationship between robustness, division of labour and variation within cooperating agent populations, and also that heterogeneity can be a useful tool for robot group design.

Introduction

In the literature on cooperative robotics and agent-based systems, most studies investigate groups of homogeneous robots or agents with identical morphologies and sensor to motor mappings (Dorigo 1999; Reynolds 1987). At most, agents have access to the same behavioural repertoire, but exhibit differing propensities to perform certain tasks (Theraulaz *et al* 1991; Bonabeau 1998). Agents in (Sims 1994) exhibit large morphological and behavioural variation, but the implications of this variation for collective problem solving was not investigated.

Exceptions include work by Mataric *et al*, who have implemented groups of robots in which heterogeneity is realized through spatial differentiation within the task space in order to minimize physical interference (Fontan & Mataric 1996; Goldberg & Mataric 1997), or by implementing a dominance hierarchy: inferior robots can only perform a subset of the basis behaviours available to more dominant robots (Mataric 1995). In (Parker 1994), morphological heterogeneity

was studied: physical robots have non-overlapping sets of sensors and effectors. In (Balch 1998), physical and simulated robots with distinct motor schemata are referred to as behaviourally heterogeneous populations.

Arkin and Hobbs (1992) delineate the advantages of heterogeneity within robot groups, such as redundancy and divison of labour. Similarly, the focus of this paper is to shed light on the correlation between behavioural heterogeneity and divison of labour by investigating how the average behavioural complexity of agents in a group decreases as heterogeneity increases.

In order to study this relationship, an evolutionary algorithm for evolving group behaviours, the Legion system, is employed (Bongard 2000). Evolutionary approaches to heterogeneity and collective problem solving include work by Bull (1996), who presents an island-model genetic algorithm for encoding classifier systems used to control a quadruped robot; in (Potter 1995), cascade neural networks (Fahlman 1990) are evolved for parity computation using an incremental genetic algorithm. In both investigations, however, the behavioural niches of the population are predetermined. The Legion system improves on a genetic programming model for evolving heterogeneity (Luke & Spector 1996). In the Legion system, evolutionary changes in group heterogeneity are influenced by the selection pressure of the given task domain, and evolutionary increases in heterogeneity occur through the biologically-inspired dynamic of gene duplication and differentiation (Ohno 1970; Ohta 1988).

By using an evolutionary approach to intra-group heterogeneity, a measure of heterogeneity independent of the specific behaviours of the task domain is possible. This stands in contrast to another, domain-specific measure of heterogeneity developed for learning robot groups (Balch 2000).

Armed with a system for evolving groups of agents to perform some collective task, a measure of heterogeneity and a clear definition of collective behavioural complexity, it is possible to measure correlations be-

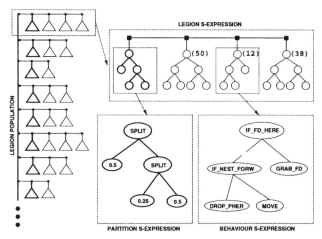

Figure 1: **The genetic programming representation of the Legion system** The bracketed numbers next to the three behaviour s-expressions denote the percentage of agents from a population that would be assigned to that behavioural class. The percentages, and the number of behavioural classes, are determined by the partition s-expression.

tween behavioural heterogeneity and specialization. In the next section, a more detailed description of the Legion system is given, along with definitions of the heterogeneity measure and collective behavioural complexity.

In the results section, the Legion system is applied to two task domains. Within both domains, two sets of simulations are peformed, in which control of individual agents is accomplished using a genetic programming and a neural network architecture, respectively. In the discussion session, the correlation between behavioural heterogeneity and complexity in these four sets of simulations is analyzed. We conclude with the implications of this work for robotics research, and avenues for future research.

Methodology

The Legion System

The Legion system improves on the paradigm introduced in (Luke & Spector 1996), which in turn extends the concept of automatically defined functions in genetic programming (Koza 1994). The Legion system evolves s-expressions that encode a suite of behaviours to be used by agents cooperating on a collective task: the s-expression partitions the agent population into separate behavioural classes, and then assigns distinct behaviours to the agents of each class. A graphic representation of the Legion system is given in Fig. 1.

Depending on the task domain, agent groups will begin to exhibit increased behavioural heterogeneity over

evolutionary time if the task domain favours heterogeneous groups: distinct behavioural classes will form which progressively differentiate over subsequent generations. If the task domain does not require group heterogeneity, behavioural classes fail to form (Bongard 2000).

Crossover in the Legion system is accomplished by *restricted breeding* (Luke & Spector 1996): given two Legion s-expressions s_1 and s_2 with partition and behaviour s-expressions $\{p_1, b_{1,1}, b_{2,1} \ldots b_{i,1}\}$ and $\{p_2, b_{1,2}, b_{2,2} \ldots b_{j,2}\}$, the partition s-expressions of the two children are created by sub-tree crossover of p_1 and p_2, and the behaviour s-expressions are created by the pairwise crossings of $\{(b_{1,1}, b_{1,2}), (b_{2,1}, b_{2,2}), \ldots (b_{i,1}, b_{i,2})\}$, where $i \leq j$.

From the mechanics of the Legion system, it becomes clear that selection pressure can alter the number of behavioural classes and the number of agents assigned to them (via alterations to the partition s-expression), and can also alter the amount of differentiation between behavioural classes (via alteration of the behaviour s-expressions through restricted breeding). This process occurs by duplication and subsequent differentiation of behaviour s-expressions during a run of the Legion system. This process was modelled on the evolutionary concept of multigene families, which are produced by gene duplication and differentiation (Ohta 1988; Ohno 1970). It has been argued that the amount of heterogeneity in an agent population is dependent on the number, membership and differentiation of the behavioural classes in an agent group (Balch 1998). Because the behaviours of the agents in each behavioural class in the Legion system is under evolutionary control, it follows that the amount of heterogeneity in the agent populations evolved by the Legion system is determined by the selection pressure of the task domain.

Although the Legion system uses s-expressions to evolve the group behaviours, other types of control architectures are evolved for individual agents. Fig. 2 shows the set-up used for evolving the weights of neural networks, which are then used to control agents within a group.

The Travelling Mailman Problem

In this paper the Legion system is applied to two collective tasks. The first task is synthetic, and is referred to as the Travelling Mailman Problem, or the TMP. The TMP was designed in order to test the Legion system on a task domain in which both homogeneous and heterogeneous populations can optimally solve the given task (Bongard 2000).

Consider a city with s streets that produces

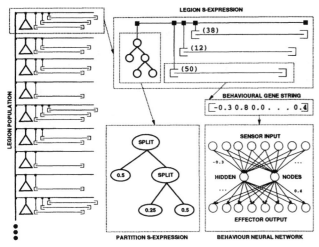

Figure 2: **A hybrid representation of the Legion system** In this model, individual agents generate behaviour using a neural network. All agents within a group have the same neural net architecture, but connection weights may differ between the behavioural classes within a group. The paritioning of the agent group is determined by the partition s-expression; the connection weights for each behavioural class are encoded in a real-valued vector.

$\{l_1, l_2, \ldots, l_s\}$ letters each day, which are collected by a group of mailmen. Each mailman can collect one letter each day. The goal of the mailman group is to arrange themselves across the streets in the city so as to minimize the amount of uncollected mail. At the beginning of each simulation, each mailman indicates the street number which will be his mail route for the duration of the simulation. The total amount of uncollected mail at the end of the simulation is given by

$$\sum_{i=1}^{n} \sum_{j=1}^{s} \left\{ \begin{array}{lll} u_j - m_j & : & u_j > m_j \\ 0 & : & u_j \leq m_j \end{array} \right. , \qquad (1)$$

where s is the number of streets, n is the number of iterations in the simulation, u_j is the amount of uncollected mail at street j, and m_j is the number of mailmen servicing street j.

In Table 1, the information necessary for applying the Legion system to the TMP is given.

In the hybrid Legion system, the action of each mailman is controlled by the neural network specified by the behavioural class to which the mailman belongs. The architecture of the neural network associated with each mailman is fixed: the number of input nodes is set to s, the number of streets in the city; the number of hidden nodes is specified at run time, and is denoted by h; and the number of output nodes is set to $\lceil(\log_2(s))\rceil$. Now consider a mailman m assigned to a

Fitness Function	Equivalent to equation 1
Termination Criteria	250 generations completed
Non-terminal Nodes	Description
IF_ST_CAP (Arity = 2)	j = evaluated left branch k = evaluated right branch if $u_j > m_j$, move to street j else move forward k streets
PLUS (Arity = 2)	left branch + right branch
Terminal Nodes	Constants 0, 1
Population Size	500
Generations	250
Selection Method	Tournament selection Tournament size = 2
Max Tree Depth	7
Mutation Rate	1%

Table 1: **Legion System Parameters for the Travelling Mailman Problem** The fitness function is a decreasing function; lower fitness values imply a more fit solution.

behavioural class b. The first s floating-point values of the behaviour gene string b are used to label the s input nodes: if one of the values is negative, the corresponding input node is disabled; if the value is positive, it is multiplied by 10, and thus indicates a specific street in the city. The remaining $h(s + \lceil(\log_2(s))\rceil)$ floating-point values are used to label the connections in the network.

The values of the binary input nodes are determined as follows: each input node i with label l_i is checked to see whether i is active. If i is active, i is set to 1 if $u_{l_i} > m_{l_i}$, where u is the amount of uncollected mail on street l_i, and m is the number of mailmen currently servicing street l_i.

The values of the hidden and output nodes are computed using the standard sigmoid function $\frac{1}{1+e^{-x}}$, where x is the summed input to the node. The output nodes are rounded to binary values. Finally, the binary array of output nodes is treated as a binary value, and converted to a decimal value. This value determines the street number to which mailman m is moved.

From the above description it follows that the number of different streets on which a mailman within some behavioural class i bases its action is equivalent to the number of unique positive labellings of the input nodes of the neural network associated with i. This value is denoted by s_i. Thus for some mailman group with c behavioural classes, the average number of streets influencing the action of each mailman in the group is given by

$$e_s = \frac{m}{\sum_{i=1}^c m_i s_i},\qquad(2)$$

where m_i is the number of mailmen assigned to behavioural class i, and m is the total number of mailmen. The value e_s is henceforth referred to as *environmental specialization*: as e_s increases, the mailmen assigned to each behavioural class rely, on average, on a smaller fraction of the total environmental information available to them.

Food Foraging

The second task studied is food foraging in simulated ant colonies (Arkin & Ali 1994; Bennett 1996). Twenty ants must forage within a 32 by 32 toroidal grid for food placed at two food sources, and return as much food as possible to a single nest. Ants may lay and sense pheromones. At each time step of the simulation, each ant performs one action, based on the state of its local environment, and its own state.

The fitness function used to evaluate the performance of an ant colony is given by

$$f + r + \sum_{i=1}^n t_i.\qquad(3)$$

In the fitness function, f stands for functionality. Given an ant colony $(a_1, a_2, \ldots, a_n)$, f is set to 0 if no ant attempts any behaviour; 1 if at least one ant attempts one of the three actions *grab food*, *drop pheromone* or *move*; 2 if at least two ants a_i and a_j attempt one of these three actions, and the actions of a_i and a_j are distinct; and 3 if at least three ants a_i, a_j and a_k attempt one of the three actions, and the actions of a_i, a_j and a_k are distinct. The functionality term f is used to motivate initial Legion populations to evolve ant colonies with high functionality.[1]

In later generations, ants removing food from the food piles are rewarded by r, the number of food pellets removed by the colony from the food piles. The final term of the fitness function rewards colonies for returning food to the nest quickly: n is the number of food pellets returned to the nest, and t_i is the number of time steps remaining in the simulation when food pellet i was returned to the nest. Table 2 provides the

---

[1]In (Bennett 1996), a similar fitness function to that of equation 3 was employed, but the functionality term f was not used. Because of this, evolved behaviours reported in (Bennett 1996) were produced with a population size of 64000 over 80 generations. These solutions were roughly as fit as the evolved solutions reported in this work, which were generated using a population size of 400 over 250 generations.

Fitness Function	See equation 3
Termination Criteria	250 generations elapsed, or all food returned to nest
Non-terminal Nodes	Description
IF_FD_HERE	Ant standing on food pellet
IF_FD_FORW	Food in front of the ant
IF_CARRYING_FD	Ant is carrying food pellet
IF_NEST_HERE	Ant is standing on the nest
IF_FACING_NEST	The ant is facing the nest
IF_SMELL_FOOD	Food pellet next to ant
IF_SMELL_PHER	Pheromone next to ant
IF_PHER_FORW	Pheromone in front of ant
Terminal Nodes	Description
MOVE_FORW	Move one cell forward
TURN_RT	Turn 90° clockwise
TURN_LT	Turn 90° counterclockwise
MOVE_RAND	Choose random direction; Move forward two cells
GRAB_FD	Pick up food pellet if here
DROP_PHER	Drop pheromone
NO_ACT	Do not perform any action
MOVE_DROP	Move forward one cell; drop pheromone
Population Size	400
Generations	250
Selection Method	Tournament selection tournament size = 2
Max Tree Depth	7
Mutation Rate	1%

Table 2: **Legion System Parameters for the Food Foraging Problem** The fitness function is an increasing function; higher fitness values indicate a more fit solution.

information necessary for applying the Legion system to the food foraging problem.

In applying the hybrid Legion system to the food foraging problem, the actions of each ant are determined by the fully connected feed forward neural network associated with its behavioural class. The neural network of each ant is constructed of eight input and eight output nodes, and a number of hidden nodes specified at run time. Specifically, each input node is associated with one of the conditional non-terminal nodes listed in table 2.

The values of the eight floating-point output nodes are then calculated as for the TMP, but are not rounded to binary values. The eight output nodes represent the eight actions that are available to the ants, and are listed in Table 2. The output node with the maximum value is found, and the action corresponding to this node is performed by the ant.

The Heterogeneity Measure

Evolution within the Legion system proceeds based on a fitness function, which indicates the relative performance of an agent group on a given task. As pointed out in (Balch 1998), the heterogeneity of an agent group is a function of both the number of behavioural classes within the group, and the differences across the behavioural classes. Consider then a group of n agents partitioned by the Legion system (or some other evolutionary algorithm) into a set of behavioural classes $B = \{b_1, b_2, \ldots, b_c\}$. Let f be the fitness of this agent group. Let $P = \{p_1, p_2, \ldots, p_{2^c-1}\} - \emptyset$ be the power set of B. We can then iteratively assign agents in the group to the behavioural classes of p_i and compute the fitness $f(p_i)$ of the group. Each behavioural class in p_i is assigned $\frac{n}{|p_i|}$ agents. We can now define the heterogeneity measure as

$$H = 1 - \frac{\sum_{p_i \in P}(\sum_{j=1}^{|p_i|} |a_j|)f(p_i)}{(\sum_{p_i \in P}\sum_{j=1}^{|p_i|} |a_j|)f} \quad (4)$$

It follows from this that the heterogeneity measure H for a completely homogeneous agent group—one which contains a single behavioural class—is zero. If agents from a group containing several behavioural classes are constrained to a subset of those classes, and perform the overall task poorly, H will approach unity.[2] In this way, H indicates not only the heterogeneity of an agent group (the number of, and differences between behavioural classes), but also the division of labour within the group: agents within each behavioural class perform some partial task which contributes positively to the overall task. This is formalized as

$$H = \begin{cases} 0.0 & : \quad \text{if} \quad \forall p \in P, f(p) = f \\ > 0.0 & : \quad \text{if} \quad \exists p' \in P, f(p') < f, \\ & : \quad \text{and} \quad \forall \bar{p} \in P - p', f(\bar{p}) = f \\ 1.0 & : \quad \text{if} \quad \forall p \in P, f(p) = 0 \end{cases} \quad (5)$$

Behavioural Complexity

The concept of *collective behavioural complexity* refers to the average amount of computation performed by an individual agent in determining which action to perform, based on its sensory input. In Fig. 1, agents are controlled by s-expressions, the nodes of which indicate sensor information, sensor to action mappings and a series of action primitives. The total number of s-expression nodes encoded in an agent population is $\sum_{i=1}^{n} s_i$, where n is the number of agents within

the group, and s_i is the number of nodes encoded in the s-expression controlling the behaviour of agent i. If the agents in a group are partitioned into a set of behavioural classes,

$$t = \frac{\sum_{i=1}^{c} n_i s_i}{n} \quad (6)$$

gives the average amount of control structure used to generate behaviour for an agent group, where c is the number of behavioural classes in the group, n_i is the number of agents in class i, and s_i is the total number of nodes encoded in the s-expression generating behaviour for agents in class i. Similarly,

$$v = \frac{\sum_{i=1}^{c} n_i v_i}{n} \quad (7)$$

gives the average amount of control structure actually used by agents within this group[3], where v_i represents the average number of s-expression nodes evaluated by agents within behavioural class i during the length of the group simulation.

Behavioural complexity in the case of neural network controlled agents is characterized as the number of nodes and connections within the neural network. The complexity of control architectures can then be reduced by reducing the number of hidden nodes in fully-connected neural networks. This in turn has the effect of reducing the total number of connections in the network, and forcing the network to perform dimensionality reduction on the input space (Bishop 1997). For the two problem domains here, this results in limiting the amount of sensory information that can influence an agent's action.

Results

Two sets of 30 runs of the Legion system applied to the TMP were executed. In the first set of runs, the maximum possible behavioural classes was limited to three. Since mailman groups tend to evolve heterogeneous behaviours over evolutionary time (Bongard 2000), in the second set of runs, only one behavioural class was allowed, forcing groups to evolve completely homogeneous solutions. In this way, it was possible to evolve mailman groups exhibiting a wide range of heterogeneity and fitness. The H, t and v values, given by Eqns. 4, 6 and 7 respectively, were recorded for the most fit group after each generation. Figs. 3 and 4 show the correlations between H and t, and H and v, for mailman groups with similar fitness values.

---

[2] For task domains in which decreasing fitness values indicate increased fitness, the numerator and denominator in Eqn. 4 are swapped.

[3] Because internal nodes in the s-expression can be conditional statements, some of the encoded nodes are not evaluated, based on the current state of the agent's local environment.

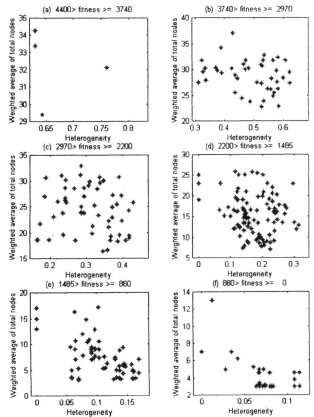

Figure 3: Correlations between heterogeneity and the weighted average of s-expression nodes encoded in the most fit mailman group after each generation. Mailman groups were partitioned into classes of similar fitness.

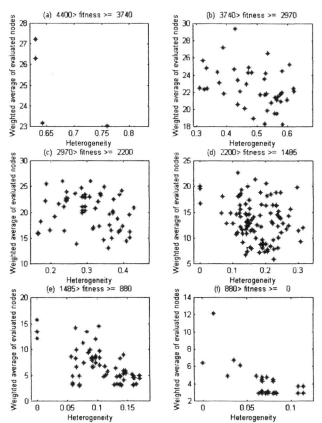

Figure 4: Correlations between heterogeneity and the weighted average of s-expression nodes evaluated by the most fit mailman group after each generation.

In applying the hybrid Legion system to the TMP, three sets of 30 runs were performed in which individual mailmen contained neural networks with two, four and six hidden nodes, respectively. The most fit mailman group was extracted from the end of each generation of the three sets of 30 runs, and the environmental specialization of the group was computed using Eqn. 2. The groups were partitioned into network type and fitness category: the e_s values of the groups within each partioned were averaged. Fig. 5 plots the differences between the average e_s values for these partitions.

Similarly, three sets of 30 runs of the Legion system applied to the food foraging problem were executed. Since ant groups tend to evolve homogeneous behaviours over evolutionary time (Bongard 2000), in order to evolve ant groups with both high fitness values and large H values, the second set of 30 runs used the fitness function Hf, where f is defined in Eqn. 3. Figs. 6 and 7 show the correlations between H and t, and H and v, for ant groups with similar fitness.

In applying the hybrid Legion system to the food

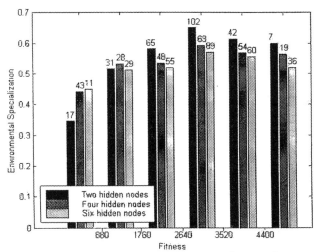

Figure 5: Differences in environmental specialization for mailman groups evolved using the hybrid Legion system. Figures above the columns indicate the number of groups included in the partition.

foraging problem, three sets of 30 runs were executed: each set of runs contained ants controlled by a neural network with two, four and six nodes, respectively.

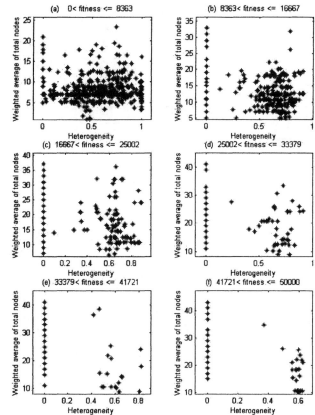

Figure 6: Correlations between heterogeneity and the weighted average of s-expression nodes encoded in the most fit ant group after each generation.

Discussion

The agent groups reported in Figs. 3, 4, 6 and 7 were partitioned into similar fitness categories in order to minimize the positive correlation between s-expression size and fitness witnessed in genetic programming simulations (Langdon & Poli 1997). It is hypothesized that this positive correlation tends to weaken any negative correlation detected between group heterogeneity and s-expression size within that group. On the other hand, by partitioning the groups into finer fitness categories, the average number of groups falling within any one category drops, and reduces the statistical significance of any negative correlation detected within a category. It was found that six fitness categories sufficed for both the TMP and the food foraging problems: a sufficient number of agent groups filled each category, and negative correlation was detected in several categories, despite the mitigating positive correlation between solution size and fitness.

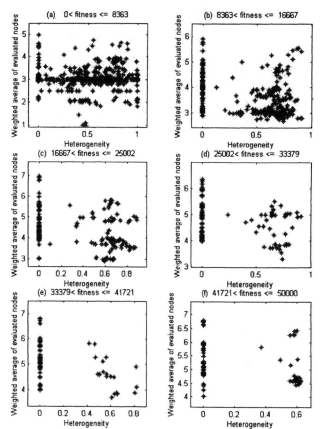

Figure 7: Correlations between heterogeneity and the weighted average of s-expression nodes evaluated by the most fit ant group after each generation.

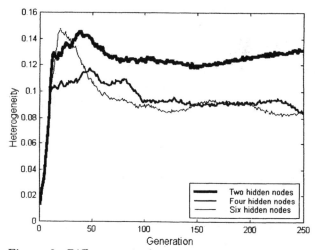

Figure 8: Differences in heterogeneity for ant groups evolved using the hybrid Legion system.

tween solution size and fitness.

As the heterogeneity, fitness and behavioural complexity distributions of agent groups from both task domains did not derive from any underlying parametric distribution, the non-parametric Spearman rank cor-

Fitness Range	Data Points	Rank Correlation	Correlation Significance
[4400,3740)	4	-0.737	0.262135
[3740,2970)	43	-0.210	0.175451
[2970,2200)	57	-0.182	0.174957
[2200,1485)	98	-0.160	0.114399
[1485,880)	64	-0.504	0.000021
[880,0]	30	-0.528	0.002707

Table 3: Correlation between heterogeneity and average number of s-expression nodes encoded in mailman groups for the TMP. Refer to Fig. 3.

Fitness Range	Data Points	Rank Correlation	Correlation Significance
[4400,3740)	4	-0.948	0.051316
[3740,2970)	43	-0.388	0.010065
[2970,2200)	57	-0.228	0.087731
[2200,1485)	98	-0.200	0.048173
[1485,880)	64	-0.504	0.000021
[880,0]	30	-0.561	0.001257

Table 4: Correlation between heterogeneity and average number of s-expression nodes evaluated in mailman groups for the TMP. Refer to Fig. 4.

Fitness Range	Data Points	Rank Correlation	Correlation Significance
[0,8363)	572	0.015	0.706355
[8363,16667)	376	-0.103	0.045612
[16667,25002)	227	-0.261	0.000064
[25002,33379)	125	-0.348	0.000067
[33379,41721)	93	-0.464	0.000002
[41721,50000]	78	-0.525	8.1×10^{-11}

Table 5: Correlation between heterogeneity and average number of s-expression nodes encoded in ant groups for the food foraging problem. Refer to Fig. 6.

Fitness Range	Data Points	Rank Correlation	Correlation Significance
[0,8363)	572	0.016	0.690497
[8363,16667)	376	-0.159	0.001939
[16667,25002)	227	-0.261	0.000066
[25002,33379)	125	-0.253	0.004399
[33379,41721)	93	-0.314	0.002110
[41721,50000]	78	-0.318	0.004441

Table 6: Correlation between heterogeneity and average number of s-expression nodes evaluated in ant groups for the food foraging problem. Refer to Fig. 7.

relation test (Noether 1991) was employed to test for correlation within the fitness categories.

The results of the tests for correlation for the TMP are given in Tables 3 and 4. Correlation significance is given as the two-sided significance level of the rank correlation's deviation from the null hypothesis (Press *et al* 1992): the null hypothesis is that there is no correlation between group heterogeneity, measured by H as defined in Eqn. 4, and that group's behavioural complexity, measured by either t or v as in Eqns. 6 and 7, respectively.

Table 3 indicates a negative correlation between heterogeneity and the average number of s-expression nodes encoded in the mailman groups for the two last fitness categories: the less fit mailman groups with fitness values between 4400 and 1485 did not show a significant drop in control architecture size in response to increased heterogeneity.

In the case of the average s-expression nodes evaluated by the group, as shown in Table 4, there is a negative correlation between H and v for all six fitness categories. Again though, the two final categories containing the most fit mailman groups have much better statistical significance that the previous four categories.

Tables 5 and 6 report the correlations between H and t, and H and v for the simulated ant groups shown in Figs. 6 and 7, respectively. In the case of the food foraging problem, significant negative correlation is found for ant groups in the last four fitness categories. In the case of group heterogeneity versus the average number of encoded s-expression nodes in the group shown in Table 5 and Fig. 6, the negative correlation grows for increasingly fit agent groups. For H versus v shown in Table 6 and Fig. 7, the negative correlation stays relatively constant for the last four fitness categories. This implies that the proportion of *unused* control code in the ant groups, $t - v$, decreases with increasingly heterogeneous groups: this suggests that heterogeneity may be useful for counteracting the phenomenon of bloat encountered in many genetic programming models (Langdon & Poli 1997).

These results suggest that for increasingly fit agent groups, increased heterogeneity signals an increase in the division of labour within these groups: on average, relatively less control structure is used by individual agents within heterogeneous groups to perform as well on a collective task as a corresponding homogeneous group.

Figs. 5 and 8 support this argument from another approach: by constraining the neural network size of individual agents, heterogeneity within agent groups tends to increase. Fig. 8 shows that groups containing agents controlled by neural networks with four and six hidden nodes tend to evolve more homogeneous behaviours, whereas groups with agents controlled by neural networks with only two hidden nodes tend to evolve more heterogeneous behaviours. It is hypothesized that this dynamic occurs because agents forced to rely on a reduced amount of sensory information tend to specialize to subtasks requiring only a subset of the sensory information available to them. In this way, behavioural classes emerge, composed of agents which use non-overlapping subsets of information in the neural network input space, and thus the heterogeneity measure of the group is non-zero.

The application of the hybrid Legion system to the TMP further supports this claim: the environmental specialization measure given in Eqn. 2 explicitly guages the amount of environmental information accessed by any one agent. Fig. 5 shows that for agent groups in the last four fitness categories, groups with more constrained control architectures tend to access the state of less streets than mailmen from groups with correspondingly more control structure. This relationship demonstrates that constrained control architectures explicitly increases specialization, at least for the TMP.

Conclusions

In this report, the relationship between the size and the amount of heterogeneity in evolved control architectures for agent groups has been studied. A biologically-inspired type of evolutionary algorithm, the Legion system, was used for this purpose.

Specifically, it was determined that for evolved group behaviours, heterogeneous groups tend to contain less control structure than similarly fit homogeneous groups. This correlation was found in the two collective task domains investigated here, and for agents with two types of control structure. The hybrid Legion system experiments show that the correlation between solution size and heterogeneity is not simply an artefact of the genetic programming paradigm, but is rather a result of a deeper relationship between reduced control structure, heterogeneity and specialization. The hybrid Legion system also shows that the heterogeneity measure used here not only measures variation within an agent group, but also the specialization among different behavioural classes within that group.

The implications of this work suggest that heterogeneity in evolved control structure for robotics research is a useful avenue of study. If we assume that behaviour is a result of the mapping between sensory information and effector commands in a robot, then us-

ing the Legion system to evolve behaviours for groups of robots may lead to a reduction in the average number of sensor/effector pairings required for each robot. It may also lead to a reduction in the intricacy of the transformations from sensory information to effector commands. This could be helpful in the domain of evolved circuit design of robot control architectures, the workings of which are often difficult to analyze (Thompson 1999).

Also, because heterogeneity depends on selection pressure in the Legion system, this paradigm may prove to be a useful tool in theoretical evolutionary biology, such as for investigating the origins of eusociality in insects (Drogoul 1995).

The concept of robustness is closely allied with that of heterogeneity: it has been pointed out (Arkin & Hobbs 1992; Wilson 1990), that the benefits of diversification must be balanced against reduced redundancy in heterogeneous agent systems. A promising future area of study would be to introduce individual agent failure into the Legion system, and investigate how this affects heterogeneity in the evolved group behaviours.

It is hoped that the work presented herein serves as sufficient motivation for study into how heterogeneity may be used not only to address some of the technical and engineering challenges faced in the evolutionary design of robot control architectures, but also more general, theoretical implications of evolved heterogeneity for biologically inspired collective problem solving.

Acknowledgements

The author would kindly like to thank faculty and students of the School of Cognitive and Computing Sciences at the University of Sussex, without whom this work would not have been possible. Special thanks to Inman Harvey, for generous contributions of time, thought, criticism and encouragement.

References

Arkin, R. C. & J. D. Hobbs. "Dimensions of Communication and Social Organization in Multi-agent Robotic Systems". In Meyer, J.-A., H. L. Roitblat & S. W. Wilson (eds.), *Procs. of the Second Intl. Conf. on the Simulation of Adaptive Behaviour*. MIT Press, Cambridge, MA, pp. 486–493. (1992)

Arkin, R. C. & K. S. Ali. "Integration of Reactive and Telerobotic Control in Multi-agent Robotic Systems". In Cliff, D., P. Husbands, J.-A. Meyer & S. W. Wilson (eds.), *Procs. of the Third Intl. Conf. on the Simulation of Adaptive Behaviour*. MIT Press. (1994)

Balch, T. *Behavioral Diversity in Learning Robot Teams*. PhD thesis, Georgia Institute of Technology. (1998)

Balch, T. "Hierarchic Social Entropy: An Information Theoretic Measure of Robot Group Diversity". *Autonomous Robots*, 8:3, July, to appear. (2000)

Bennett, F. H. "Automatic Creation of an Efficient Multi-Agent Architecture Using Genetic Programming with Architecture-Altering Operations". In Koza, J. R., D. E. Goldberg & D. B Fogel (eds.), *Procs. of the First Annual Conference on Genetic Programming*. MIT Press, pp. 30–38. (1996)

Bishop, C. M. *Neural Networks for Pattern Recognition*, Oxford University Press, Oxford, UK, pp. 316–17. (1997)

Bonabeau, E., A. Sobkowski, G. Theraulaz & J.-L. Deneubourg. "Adaptive Task Allocation Inspired by a Model of Division of Labour in Social Insects". *Sante Fe Institute Tech. Rep. 98-01-004.* (1998)

Bongard, J. C. "The Legion System: A Novel Approach to Evolving Heterogneity for Collective Problem Solving". To appear in *Procs. of the Third European Conf. on Genetic Programming*. Edinburgh, UK, April 14-15. (2000)

Bull, L. & C. Fogarty. "Evolutionary Computing in Multi-Agent Environments: Speciation and Symbiogenesis". In Voigt, H.-M., W. Ebeling & I. Rechenberg (eds.), *Parallel Problem Solving from Nature IV*. Springer-Verlag, Berlin, pp. 12–21. (1996)

Dorigo, M., G. Di Caro & L. M. Gambardella. "Ant Algorithms for Discrete Optimization". *Artificial Life*, 5(2):137–172. (1999)

"MANTA: New Experimental Results on the Emergence of (Artificial) Ant Societies". In Gilbert, N. & R. Conte (eds.), *Artificial Societies: the Computer Simulation of Social Life*. UCL Press, London, UK. (1995)

Fahlman, S. & C. Lebiere. "The Cascade-Correlation Learning Architecture". *Carnegie Mellon University tech. rep. CMU-CS-90-100.* (1990)

Fontan, M. S. & M. J. Mataric. "A Study of Territoriality: The Role of Critical Mass in Adaptive Task Division". In Maes, P., M. Mataric, J.-A. Meyer, J. Pollack & S. W. Wilson (eds.), *Fourth Intl. Conf. on Simulation of Adaptive Behavior*. MIT Press. (1996)

Goldberg, D. & M. J. Mataric. "Interference as a Tool for Designing and Evaluating Multi-Robot Controllers". In *Procs. of the Fourteenth Natl. Conf. on Artificial Intelligence*. MIT Press, Cambridge, MA. (1997)

Koza, J. R. *Genetic Programming*. MIT Press. (1992)

Koza, J. R. *Genetic Programming II: Automatic Discovery of Reusable Programs.* MIT Press. (1994)

Langdon, W. B. & R. Poli. "Fitness Causes Bloat". *Second On-Line World Conference on Soft Computing in Engineering Design and Manufacturing.* Springer-Verlag, London, pp. 13–22. (1997)

Luke, S. & L. Spector. "Evolving Teamwork and Coordination with Genetic Programming". In Koza, J. R., D. E. Goldberg, D. B. Fogel & R. L. Riolo (eds.), *Procs. of the First Annual Conference on Genetic Programming*. MIT Press. (1996)

M. J. Mataric. "Designing and Understanding Adaptive Group Behavior". *Adaptive Behavior* 4(1):51–80. (1995)

Noether, G. E. *Introduction to Statistics: The Nonparametric Way.* Springer-Verlag, New York, NY, pp. 236–37. (1991)

Ohno, S. *Evolution by Gene Duplication.* Springer-Verlag, New York, NY. (1970)

Ohta, T. "Multigene and Supergene Families". *Oxford Surv. Evol. Biol.*, vol. 5, pp. 41–65. (1988)

Parker, L. *Heterogeneous Multi-Robot Cooperation.* PhD thesis, Massachussets Institute of Technology. (1994)

Potter, M. & K. De Jong. "Evolving neural networks with collaborative species". In *Procs. of the 1995 Summer Computer Simulation Conference* Ottawa, Canada. (1995)

"Press, W. H., S. A. Teukolsky, W. T. Vetterling & B. P. Flannery". *Numerical Recipes in C.* Cambridge University Press, Cambridge, UK, pp. 640–42. (1992)

Reynolds, C. "Flocks, Herds and Schools: A Distributed Behavioural Model". *Computer Graphics*, vol. 21, pp. 25–34. (1987)

Sims, K. "Evolving 3D Morphology and Behaviour by Competition". In Brooks, R. and P. Maes (eds.), *Artificial Life VI*. MIT Press, pp. 28–39. (1994)

Theraulaz, G., S. Goss, J. Gervet & J.-L. Deneubourg. "Task Differentiation in *Polistes* Wasp Colonies: a Model for Self-organizing Groups of Robots". In Meyer, J. A. & S. W. Wilson (eds.), *Procs. of the First Intl. Conf. on the Simulation of Adaptive Behaviour*. MIT Press, pp. 346–355. (1991)

Thompson, A. "Analysis of Unconventional Evolved Electronics". *Communications of the ACM*, 42:4, pp. 71–79. (1999)

Wilson, E. O. *The Ants.* Belknap Press, Cambridge, MA. (1990)

Studying Attention Dynamics of a Predator in a Prey-Predator System

Shin I. Nishimura*

Intelligent Systems Division, Electrotechnical Laboratory (ETL), AIST, MITI, Japan.
1-1-4 Umezono, Tsukuba, Ibaraki 305-8568.

Abstract

Mathematical treatment of attention dynamics of a predator chasing prey is studied. In our model, prey and a predator move around a two-dimensional surface. Real numbers that correspond to individuals of prey are used. A predator selects prey of the highest number. If prey's motion is ordered, the predator selects the nearest individual. If prey's motion is disordered, the predator tends to select the individual which has split away from a collective, ie. alone. It is emphasized that partially disordered motion, is rather difficult for the predator to select an individual.

Introduction

Many animal species show group motion. In the ocean, millions of fish make a school in which almost all individuals have the same heading direction. On the other hand, in the sky a lot of birds fly collectively, as if they dance. It is no doubt that flocking or schooling reduces dangerous accidents for prey species. Many researchers believe that collective behavior of prey functions as a protection against predators(Breder, 1959; Partridge, 1982; Cusing and Jones, 1968). One effective method of protection is to break predators' decision making. Some reports say that predators invest a lot of energy into deciding their target. For example, Kruuk studied spotted hyenas over years(Kruuk, 1972). When hunting, hyenas spend a lot of time in deciding a target. Either an old, or a weak individual often becomes the prey of hyenas. But old and weak features are not the reason they become prey. When Kruuk puts an arbitrary marker on an individual prey (e.g. eland), the individual is almost always taken as a target for hyenas. Thus it is emphasized that easily recognizable prey can easily become a target for predators. This is why prey all of the same size (especially fish) gather, since predators lose criteria of targeting(Peuhkuri, 1997). They have to look for small differences among their prey.

In this paper, we study one predator's selection mechanism of prey gathering. What mechanism of targeting is effective for a predator? How does it depend on the

*fishin@etl.go.jp

prey's grouping behavior? We tend to try to find mechanisms independent of a predator's internal structure, for instance, structure of eyes, a brain, and so on. This paper does not address the issues of how to catch prey and how to escape from predators. All problems of a predator's selection mechanism will be thought of as problems of selecting "priority functions" introduced in the next section.

A Predator's Selection Mechanism

For mathematical treatment of a predator's selection mechanism of prey, we use real numbers: each number corresponds to each individual of prey. Such a number gives "priority" for a predator attacking prey. As the most natural definition, a predator attracts attention to the prey individual corresponding to the highest number. We called the individual the "candidate" and the action "attention". If a candidate changes one after another, a predator might hardly catch any prey. We define that a predator decides to chase an individual, (which is merely called the "target") only if it attracts attention to the same candidate in a certain interval, I_a without a break. This action is called "locking on".

This definition reflects a problem of how to calculate priority. Before discussing this, we have to clarify treatment of prey. Roughly speaking, there are two categories of properties that prey have. One is static properties, namely, shape, color, smell, and so on. The other is dynamical properties, for example, position, velocity, acceleration and so on. Only considering the latter, makes our investigation rather simplified. In this paper, all individuals of prey and a predator are thought of as "material points", which have been widely discussed in many papers, especially fish schooling studies (Breder, 1954; Shimoyama et al., 1996; Sannomiya et al., 1996; Sannomiya and Dousrari, 1996; Niwa, 1994; Aoki, 1980; Nishimura and Ikegami, 1998). Along the above line, priority of prey becomes computed from position, velocity and acceleration of prey. More details will be discussed in the next section.

Assume many individuals of prey are chased by one predator. In general, the priority for the ith individual,

p_i can be written as follows:

$$p_i = F(w_i : w_1, w2, ..., w_{i-1}, w_{i+1}, ..., w_n), \qquad (1)$$

where w_j is a function of the i individual and the predator. F is called a "priority function".

Note that the function $F(\cdot : \cdot)$ is invariable against permutation of variables after a colon, that is,

$$F(w_i : ..., w_j, ..., w_k, ..., w_n)$$
$$= F(w_i : ..., w_k, ..., w_j, ..., w_n) \quad , \qquad (2)$$

with regard to any pair of numbers, j and k. This means that the function is independent of correspondence of prey individuals to natural numbers. Destruction of the invariability causes unnatural results, for example, results depending on programming techniques (cf. a usage of memories).

In this paper, we assume that the priority function F is linear because of its simplicity. The linear function has terms about current position of prey as well as delayed data, given by the following:

$$p_i = \sum_{l=0} s_l |\vec{r}_i(t_l) - \vec{r}_a(t_l)|, \qquad (3)$$

where $\vec{r}_i$ indicates the position of the ith individual of prey, $\vec{r}_a$ does that of the predator and t_l does the time step labeled l. t_0 means the current step, whereas $t_l : l > 1$ does the delayed steps. s_l are constants. Comparing Equation (3) with (1), there is no variable after a colon in Equation (3), that is the ith priority depends on only the ith position. This definition automatically preserves the independence of permutation as written above. In this paper, the max of l is two, which means total terms equal three.

Note that non-linear functions, such as neural networks, will be discussed in **Discussion** briefly.

Equation of Motion

All individuals in the model move around a two-dimensional, unbounded surface. First we discuss motion of prey. We desire a model in which many behaviors emerge with only a few parameters. Fortunately, we have a good model studied by (Shimoyama et al., 1996), who have studied collective motion of many animal species. We use the model with a small modification:

$$\frac{d\vec{r}_i}{dt} = b \sum \frac{1 + e \cdot \cos \psi_{ij}}{1 + e} \vec{f}_{ij} + d_p \vec{n}_i \qquad (4)$$

$$\frac{d\theta_i}{dt} = h_p \cdot \sin(\phi_i - \theta_i), \qquad (5)$$

where $\vec{n}_i = (\cos \theta_i, \sin \theta_i)$, n_i indicates the heading direction of the i individual, θ_i does the angle notion of the heading direction, ϕ_i does the angle notation of the velocity direction and ψ_{ij} does the relative angle between

the ith heading direction and the direction from the ith to jth individuals. b, e, d_p and h_p are constant parameters. The term $\frac{1 + e \cdot \cos \psi_{ij}}{1 + e}$ indicates a visual field of the ith individual. The larger the parameter e, the smaller the visual field is. When e equals zero, the individual's visual field becomes a disc. We omit an inertial term from the original equation by Shimoyama et al., since the term seems to be ineffective, which is also reported in their paper. It is noted that by choosing appropriate parameters, only searching the parameter e leads to almost all patterns in this equation. $\vec{f}_{ij}$ indicates mutual interaction given by the following:

$$\vec{f}_{ij} = (\frac{1}{r_{ij}} - \frac{1}{r_{ij}^2}) \frac{\vec{r}_{ij}}{r_{ij}}, \qquad (6)$$

where $\vec{r}_{ij} = \vec{r}_j - \vec{r}_i$ and $r_{ij} = |\vec{r}_{ij}|$. This interaction is called Lenard Jones potential, introduced by Breder(Breder, 1954) for fish schooling, and has an effect of maintaining a distance between the ith and jth individuals.

As has been written, the predator chases a target after it locks on to the target. We can express the behavior of the predator by the following equation:

$$\frac{d\vec{r}_a}{dt} = g \cdot (\vec{r}_t - \vec{r}_a)/|\vec{r}_t - \vec{r}_a| + d_a \vec{n} \qquad (7)$$

$$\frac{d\theta_a}{dt} = h_a \cdot \sin(\phi_a - \theta_a), \qquad (8)$$

where $\vec{r}_t$ is the position of the target, g, d_a and h_a are constants. Roughly, the value of d_a and g mean the velocity before and after locking on to a target, respectively. Before locking on, g is set to zero. If distance between a target locked on and the predator becomes smaller than a certain value, the score of the predator rises by one, and the target is removed from the surface. The predator becomes the state of attention and finds a target.

Collective Motion of Prey

Except the parameter e, all parameters of prey's motion defined in the previous section are fixed, since only searching the parameter e is enough to find all patterns of motion. The parameters are set as follows: $b = 0.9$, $d_p = 7.0$ and $h_p = 100$.

The following four distinct collective behavior are found by the dynamics which were reported by (Shimoyama et al., 1996).

- **Marching**
  The elements form a regular triangular crystal, moving at a constant velocity. The formation is stable against disturbance. (Figure 1-(1))

- **Oscillation**
  Several group motions exhibit regular oscillations: (i)

Wavy motion of the cluster along a linear trajectory. (ii) A cluster circling a center outside the cluster. (iii) A cluster circling a center inside the cluster. (Figure 1-(2))

- **Wandering**
  Although the lattice-like order inside the cluster persists, the center of the cluster can wander quite irregularly. Chaotic intermittency of motion is found. (Figure 1-(3))

- **Swarming**
  The most irregular motions are found in swarming. Although the cluster persists, lattice-like order is broken completely. The velocity of the elements has a large distribution, and the mobility of the cluster is small. (Figure 1-(4))

In this paper, we call marching and oscillation ordered motion, wandering partially disordered motion, and swarming disordered motion.

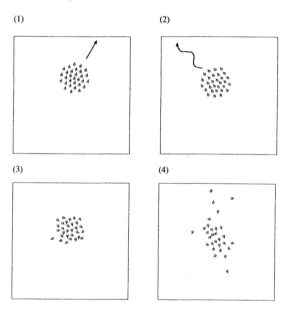

Figure 1: An individual of prey is marked by a circle with a line that is the heading direction. (1)Marching: $e = 0$ (2)Oscillation: $e = 0.2$ (3)Wandering: $e = 0.5$ (4) Swarming: $e = 0.9$. Details are written in the body of the paper.

Results of Prey with a Predator

We investigate the characteristics of this model, by searching parameters of a priority function of a predator, and the parameter e of its prey. Although there are three parameters of a priority function, we can reduce one degree of freedom. Instead of any three real numbers, s_0, s_1 and s_2, we can use s_0/S, s_1/S and s_2/S,

where $S = \sqrt{s_0^2 + s_1^2 + s_2^2}$ since a priority function is linear. It is convenient to adopt the following notation:

$$
\begin{aligned}
s_0 &= \sin \alpha \\
s_1 &= \cos \alpha \sin \beta \\
s_2 &= \cos \alpha \cos \beta
\end{aligned}
\tag{9}
$$

since $\sqrt{s_0^2 + s_1^2 + s_2^2}$ equals 1 automatically. Note that the pair, (α, β) indicates a point of a sphere with the radius of 1. We run a simulation for 1000 steps. We run 100 simulations for one set of parameters, and average the results to avoid dependence on initial conditions.

Parameters are set as follows: $I_a = 0.08$, $d_a = d_p = 7.0$ and $g = 10$. It is important that g is greater than d_p since the predator always catches a target.

The relation between averaged scores of a predator and e is shown in Figure 2. Averaging was the over all (discrete) values of α and β. At the point $e = 0.5$, the averaged score was the smallest. At both ends, the scores are the maxima.

We pick three values for e: $e = 0$, $e = 0.5$ and $e = 0.9$. At the those three points, we calculate Lyapunov spectrums of prey's (not including predator's) motion shown in Figure 3. [1]

At the value of $e = 0$, all Lyapunov exponents are less than zero, since the prey's motion is ordered. At the value of $e = 0.5$, some larger exponents are greater than zero. This is relatively a low dimensional chaotic motion. Finally, at the value of $e = 0.9$, more exponents are larger than zero, that is, high dimensional chaos. From Figures 2 and 3, it is asserted that when prey's motion becomes low dimensional chaos, the predator gets the smallest score rather than during high dimensional chaos. This result does not necessarily depend on the magnitude of the largest Lyapunov exponent.

In relation to the above three points, we also show the relation between parameters of the priority function and scores of the predator. Figure 4 consists of three contour plots. Corresponding to the value of $e = 0$, one peak ridge of scores can be found in Figure 4-(1). Figure 4-(2) corresponds to the value of $e = 0.5$, it is comparatively flat. Figure 4-(3) with regard to the value of $e = 0.9$ has two peak ridges, one of which is nearly the same as Figure 4-(1), whereas the other is new.

Let us look at snapshots of prey and a predator for Figure 4. Figure 5 and 6 correspond to Figure 4-(1) and 4-(2), respectively. In Figure 5, we set parameters from the point $(0, \pi)$ in the ridge of Figure 4-(1) marked by a cross. It is clear that a predator locks on to the relatively nearest individual of prey. On the other hand, we use parameters from the point $(0, 2\pi)$ in the second ridge of Figure 4-(3) (also marked by a cross) for Figure 6. With contrast to Figure 5, the predator chases the farthest individual.

---

[1]How to calculate Lyapunov exponents was written in (Bennetin et al., 1980; Shimada and Nagashima, 1979).

340

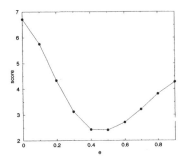

Figure 2: The horizontal axis is e and the vertical axis is the scores of a predator.

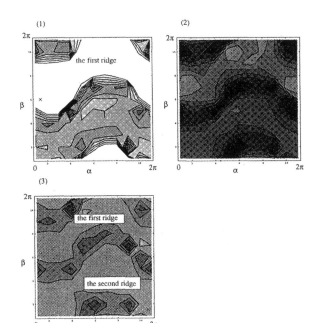

Figure 4: Contour plots of predator's scores. Horizontal and vertical axis are α and β. $0 < \alpha, \beta < 2\pi$. The lighter area is the higher score. (1)$e = 0$. One ridge can been seen that is called the first ridge. (2)$e = 0.5$. The contour is rather flat compared to the others. (3) $e = 0.9$. There are two ridges, one of which is nearly the same as the first ridge of (1), and the other is new, called the second ridge.

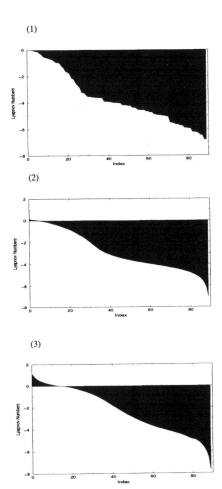

Figure 3: Lyapunov spectrums of prey. Horizontal axis is indices and vertical axis is Lyapunov exponents arranged in order of decreasing size. The graphs are filled in black from a horizontal line of zero to Lyapunov exponents. Sub-figures (1), (2) and (3) correspond to the values of $e = 0$, $e = 0.5$ and $e = 0.9$, respectively.

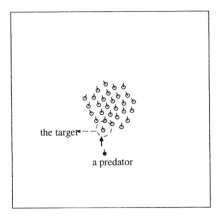

Figure 5: A snapshot of prey and a predator. White circles are the prey and the black circle is a predator. A dashed circle indicates the target of the predator. $\alpha = 0$ and $\beta = \pi$.

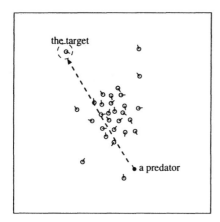

Figure 6: A snapshot. $\alpha = 0$ and $\beta = 2\pi$.

Discussion

The summary of all the results are as follows. If prey's motion is ordered, in other words, if all Lyapunov exponents are negative, the predator chases the nearest individual. However, if its motion is disordered, that is, a lot of Lyapunov exponents are positive, the predator locks on to the farthest one as well. The predator gets the smallest score when prey's motion is partially disordered, that is, when a few Lyapunov exponents are positive.

It is trivial that the predator chases the nearest individual. Rather we are interested in chasing the farthest individual. In the most disordered pattern, one individual often rushes out from a cluster. The class of priority functions given in this paper cannot directly detect such an individual. However, in many cases, the individual goes often far away from a cluster of prey. Consequently, the predator catches the individual rushing out from a cluster. It is very dangerous for an individual to rush out or split away from a cluster in the natural world. For instance, when there were two prey fish as well as a predator fish in a tank, the predator fish caught neither fish. However, if the researcher removed one prey fish, the predator instantly ate the remaining prey(Partridge, 1982).

Difficulty of partially disordered patterns for the predator, is an interesting topic. A cluster of those patterns has few crevices. In addition, this cluster moves in a disordered manner. The order of the priority easily changes, since the priority is a function of position of prey. Thus, the predator has to have a sufficiently precise priority function. In the natural world, similar flocking behavior can be found. One good example is starlings(Edmunds, 1974; Wilson, 1995). When they are attacked by a falcon, they gather into one flock and fly in zigzags. The falcon often runs against the flock haphazardly. This motion of the starlings may be partially

disordered.

We address the feature works. In this paper, priority functions are linear. However, nonlinear functions are also interesting. What priority function is required corresponding to partially disordered patterns? What about chaotic functions? Possibly, chaotic functions have an affinity to partially disordered motion. Observations of flocks or schools attacked by a predator must be important. Of cause, it is difficult to calculate Lyapunov exponents from only data. In addition, two dimensional data in a video film transformed into tree dimensional data is also complicated. It is expected that similar calculations to Lyapunov exponents, which is somewhat easy to be computed from two dimensional data in a video film.

One effective strategy of prey is "dispersing". Herds of many prey species radially disperse when a predator approaches. After that, prey make a herd again. It is believed that this behavior confuses a predator's selection mechanism(Driver and Humphries, 1988; Edmunds, 1974). I speculate that dispersing is an easier task than zigzagging since many species, birds, insects, fish, and so on, show dispersing behavior, whereas only some bird species fly in zigzags. The reason is that, since a nervous system has time delay effects, it seems to be difficult for birds to avoid splitting away from a herd without prediction of other individuals' motion. In order to disperse however, individuals do not need prediction but to detect a predator approaching. My future paper will investigate effects of dispersing.

Conclusion

We studied attention dynamics of a predator through priority functions. Although the results of this paper were not novel but predictable, we succeeded in mathematically formulating those results. In order to obtain generic results, all we have to do is to investigate more classes of priority functions, since we studied one special class of function in this paper. Nevertheless, future works will be based on the results of the paper.

Acknowledgments

I thank Dr. Gordon Cheng, Mrs. Katrina Cheng and Dr. Yasuo Kuniyoshi. The present work is partly supported by the COE program by STA, and ETL.

References

Aoki, I. (1980). An analysis of the schooling behavior of fish: internal organization and communication process. *Bull. Ocean Res. Inst. Univ. of Tokyo*, 12:1–62.

Bennetin, G., Giorgilli, A., and Strelcyn, J. M. (1980). Lyapunov characteristic exponents for smooth dynamical systems and for Hamiltonian systems: a method for computing all of them. *Meccanica*, 15:9.

Breder, C. M. (1954). Equations ascriptive of fish school

and other animal aggregations. *Ecology*, 35(3):211–129.

Breder, C. M. (1959). Studies on social groupings in fishes. *Bulletin of the American Museum of Natural History*, 117.

Cusing, D. H. and Jones, F. R. H. (1968). Why do fish school? *Nature*, 218:918–920.

Edmunds, M. (1974). *Defense in Animals*. Longman Group Limited, London.

Driver, P. M. and Humphries, D. A. (1988) *Protean Behavior*. Oxford University Press.

Kruuk, H. (1972). *The spotted hyena: a study of predation and social behavior*. University Chicago Press.

Nishimura, S. I. and Ikegami, T. (1998). Emergence of collective strategies in a prey-predator game model. *Artificial Life*, 3:243–260.

Niwa, H. (1994). Self-organizing dynamic model of fish schooling. *J. of Theor. Biol.*, 171:123–136.

Partridge, B. L. (1982). The structure and function of fish schools. *Sci. Am.*, 246:90–99.

Peuhkuri, N. (1997). Size-associative shoaling in fish: the effect of oddity on foraging behavior. *Anim. Behav.*, 54:271–278.

Sannomiya, N. and Dousrari, M. A. (1996). A simulation study on autonomous decentralized mechanism in fish behavior model. *International Journal of Systems Science*, 27:1001–1007.

Sannomiya, N., Nagano, K., and kuramitsu, M. (1996). A behavior model of fish school with three individuals. *SICE*, 32:948–956.

Shimada, I. and Nagashima, T. (1979). A numerical approach to ergodic problems of dissipative dynamical systems. *Prog. Theor. Phys.*, 61:1605.

Shimoyama, N., Sugawara, K., Mizuguchi, T., Hayakawa, Y., and Sano, M. (1996). Collective motion in a system of motile elements. *Physical Review Letters*, 76(20):3870–3873.

Wilson, E. O. (1995). *Sociobiology*. Harvard University Press.

On the Emergence of Possession Norms in Agent Societies

Felix Flentge[1] and **Daniel Polani**[1,2] and **Thomas Uthmann**[1]

[1] Institut für Informatik, Johannes Gutenberg-Universität, D-55099 Mainz, Germany
[2]Institut für Neuro- und Bioinformatik, Medizinische Universität Lübeck, D-23569 Lübeck, Germany

Abstract

Our paper studies the emergence of social norms and their subsequent influence in a simulated society of artificial agents. These norms and their propagation in a society is put in close relation with the meme concept of Dawkins. In particular, the norms studied in this paper are concerned with the possession of goods. Here a global norm regarding possession of goods may result from the collective dynamics of the society and arise from purely local agent interactions. The role of sanctions and costs of enforcing sanctions and their relationship to the establishment of a possession norm are studied.

Introduction

An important part of the research in the field of Artificial Life is directed towards the modeling and simulation of fundamental properties of living beings. Because of their high complexity, human societies which are the topic central to social sciences pose a particular challenge to this kind of modeling. A simulation of qualitative properties of such systems can help to test and to improve models and theories from social sciences. In turn, these theories can be useful in constructing multi-agent systems, which poses difficulties typically found in human societies.

A prominent aspect of human social systems are social norms. They have various functions, like the control of cooperation and the reduction of aggression. Perhaps the most important aspect of norms is that they make the individuals' behavior more predictable. By determining the behavior of an individual in a given situation, norms give other individuals the possibility to adapt themselves to the situation more easily and to act with respect to the expected behavior.

Thus norms are extremely important factors in reducing the complexity of social situations, which is of particular relevance for individuals that have typically limited resources. Furthermore, since the validity of norms extends beyond the concrete situation, they allow a coordinated behavior without tedious agreement or coordination protocols.

The goal of our paper is on the one hand the simulation of the emergence of norms and, on the other hand, the study of the effects of a norm on a society. The norms result from the dynamics of an agent society where the agents have the possibility to follow or not to follow a certain norm and to sanction or not to sanction behavior that deviates from the norm.

In existing approaches for the simulation of norms two separate aspects are studied: either the possibility that a norm emerges or not (Axelrod, 1986; Coleman, 1986/87) or the effects of an explicitly given norm (Conte/Castelfranchi, 1995; Saam/Harrer, 1999). Our model combines both approaches: it remains open whether a norm emerges or not and, in addition, the effects of the norm on the agent society are studied.

First, we will create a connection between norms and the concept of memes as put forward by Dawkins (Dawkins, 1996) that shall play an important role in our agent society model. The model itself is an extension of the sugarscape model of Epstein and Axtell (Epstein/Axtell, 1996) and will be described in the subsequent section. We will present our results and give some conclusions.

Social Norms and Memes

A *norm* can be considered as a general behavior code that is more or less compulsory and whose nonobservance is punished by sanctions. Norms often result from regularities in behavior and are enforced by institutions only after some time. We shall see that such an "unscheduled" emergence of norms can be smoothly combined with Dawkins' concept of memes. Dawkins suggests to apply the theory of evolution to cultural tradition.

Dawkins calls the "building blocks" of cultural tradition *memes* (which correspond to the *genes*, the building blocks of genetic information transfer). Memes thus encode cultural traits. They propagate from brain to brain by imitation. Behaviors are imitated and ideas are adopted. Some memes can propagate better than others and establish themselves with time. Many memes are passed on by education. Children assume many of their parents' behavioral patterns. Genetic and memetic inheritance is tightly coupled. The main difference between them, however, is that the memes can change during life while the genes can not. Norms in this sense can also be regarded as memes. They can propagate from individual to individual and are connected with certain behaviors. An individual follows a certain norm if it carries the corresponding meme.

The Model

The norm which we model in our studies is a norm concerned with the possession of goods, in short a *possession norm*. We view possession norms as regular behaviors that relate to the act of acquisition, to the state of possession itself and particularly to the respect of the possessions of others. Our model allows the development of possession norms for the individual agents. Our concept of possession relates to the possession of a good (a plot of land) producing a good important for survival (food).

We will say that a possession norm has developed in the agent society if a large number of agents has a behavior code respecting possessions claimed by others and if non-observance of this code is sanctioned. Possession does not exist *a priori*, and can only persist if enough individuals behave appropriately and respect it. We chose the *sugarscape* model from Epstein/Axtell, 1996 as a basis for our model since the concept includes a landscape that produces food (sugar) and it allows an extension towards the simulation of norm development in a natural way.

The Sugarscape Model

The sugarscape model by Epstein and Axtell consists of a landscape that provides sugar and agents which feed upon this sugar and can move around the landscape. In our simulations the landscape consists of a 50×50 grid of individual cells with periodic boundary conditions (torus).

For the agents and the cells of the landscape certain rules are defined which describe the interaction between the agents and between agents and cells. Each such rule can be switched on and off individually, being valid for all agents or cells. Each agent is activated once per simulation step and performs the rules valid for agents. The agents are activated in a random order each step. First the agent and then the cells are updated, according to the corresponding rules.

Each cell has a maximum sugar capacity and a current sugar level. The sugar capacity is chosen in such a way that one obtains a landscape with two central regions with the maximum capacity, around which the sugar capacity slowly drops. A cell can be *occupied* by a single agent at most.

For the landscape cells the *sugarscape growback rule* G_α is valid: in each simulation step the sugar content of

a cell grows by α units until it reaches the sugar capacity of the cell (α is an integer).

The agents have a certain metabolism. It is given by an integer value and determines the number of sugar units used up from an agent's internal sugar level per simulation step. Each agent has a certain *vision range* which determines how far an agent can see in the four main directions of the landscape grid (n, s, e, w). Agents cannot see in diagonal directions. The vision range is fixed at the creation of an agent and chosen from an externally given interval.

At its turn, an agent can move to an arbitrary unoccupied cell inside its vision range and collect all the sugar on this cell. It then uses up the amount of sugar given by its metabolism. Agents die if they have no more sugar or when reaching a certain maximum age determined at their creation (a random value from a given interval). Agents perform the *Agent Movement Rule M* every simulation step.

- *Agent Movement Rule M*

  - search the unoccupied cells in vision range with a maximum current sugar level
  - if several such cells are present, choose the closest
  - go to this cell
  - collect all sugar in the cell

Interactions between agents can only take place between direct neighbors in the four main grid directions. Agents can reproduce. There are two types of agents, male and female. The agent sex is determined randomly with the probability of 1/2 at the creation (birth) of an agent. An agent is fertile if its age is in a certain age interval and if the agent has at least as much sugar as it had at its birth. Fertile agents perform the *Agent Sex Rule S*.

- *Agent Sex Rule S*

  - select a neighboring agent randomly
  - if this neighbor is fertile, if the neighbor belongs to the opposite sex, and if at least one of the agents has an unoccupied neighboring cell, a new agent is created and placed on this cell
  - if the sugar level of the current agent is high enough to reproduce again, and if there are other neighbors that were not selected this update, repeat rule S for another neighbor

The new agent inherits metabolism and vision range with a probability each of 1/2 from one of both parents. From each parent the agent obtains half of the amount of sugar that the respective parent possessed at its own birth. If an agent is "wealthy" enough, it can reproduce several times per simulation step.

Cultural Tags and Memes

Each agent has several *cultural tags* which can assume one of two different values each. The number of tags is the same for all agents. The tags represent the cultural attributes of agents. After moving, each agent performs the *Cultural Transmission Rule K*[1].

- *Cultural Transmission Rule K*

  - for each neighbor choose randomly one of its tags
  - set the neighbor's tag to the same value as the current agent's tag.

In our extended model, these tags play an extremely important role and will determine agent behavior in a very direct way. Therefore from now on we will denote these tags as *memes*. Rule K then describes the dynamics of meme propagation via imitation. Memes are inherited by children from their parents at birth and thereby simulate the influence of the parents on their children via education. Later the parents may adopt some memes from their children via rule K, though.

Model Extension towards Norm Simulation

To study the emergence and the impact of possession norms, we extend and partly modify the sugarscape model. In our model, we give the agents the option to acquire resources, namely a "plot of land" (i.e. cells). For this purpose, agents can mark cells. Only unmarked cells can be marked and only if occupied by the agent that wants to mark it. Cell marks can be seen by an agent if the cell is in its vision range. On death of an agent, all its cell marks are deleted.

An important aspect of the model is that marking a cell does not automatically mean that this cell becomes the property of the agent. For this to happen, it is necessary that also other agents respect the cell as being in possession of the first agent and behave accordingly. Here the meme model plays a central role. The first two memes of our agents determine their behavior[2].

The first meme (the *possession meme*) determines the behavior of an agent with respect to marked cells. Agents with an active possession meme consider marked cells as zero sugar cells and do not collect sugar from these cells. Under rule M they will not move to these cells unless all other cells in their vision range are zero sugar cells. Cells are only marked by agents with an active possession meme.

The second (*sanction*) meme determines the behavior when norm violations are observed. Norm violations are only registered by agents with an active possession meme

and inside their range of vision. If an agent has an active sanction meme, it will sanction all norm violations observed by it. On being sanctioned, the "wrongdoer", i.e. the norm violating agent, loses a certain amount of sugar from its internal sugar level. The "avenger", i.e. the sanctioning agent, also incurs a cost and has to pay a certain amount of sugar.

These model extensions introduce a necessity to change the movement rule M since otherwise the agents' rate of reproduction is too low. The reason for this is that agents often have no incentive to move to a cell neighboring a possible mating partner because these are often already marked by some other agent. Therefore we introduce the movement rule MF. With this rule, agents still try to maximize the amount of sugar they collect, but their first priority is reproduction.

- *Movement Rule MF*

  - if current agent is not fertile, perform movement rule M
  - else search all unoccupied cells in vision range neighboring a fertile agent of the opposite sex
  - if no such cell is found, perform rule M, else go to a cell with a maximum amount of sugar
  - if several such cells are present, choose the closest
  - collect the sugar of the cell

In our extended model, the landscape of the original model is modified in such a way that the sugar capacity of the original landscape (ranging from 0 to 4) is multiplied by the factor of 10. This allows larger and finer variations of the values for metabolism in combination with the growback rule G_α. Among other aspects this enables us to prevent that agents with a (minimal) metabolism of 1 can survive by just sticking to a cell where in every simulation step at least a single unit of sugar grows back.

An important change w.r.t. the original model is to start the simulation with a random agent age between zero and the maximum age. Otherwise one has a very unnatural age distribution at the beginning of the run and has to wait for a large number of generations until this transient behavior normalizes.

Furthermore, in the original work metabolism and vision range evolve by time. Instead, we vary these values systematically from run to run, but keep them fixed during each individual run. This enables us to filter out the additional effects introduced by the variations of these parameters during the runs and to interpret our results more easily.

Results

In our simulations we keep one part of the parameters fixed and the other one varies systematically within certain limits. Our choice of the fixed parameters is based

---

[1]In the original work (Epstein/Axtell, 1996), K denotes a combination of two rules. Here we only focus on the Cultural Transmission Rule which we denote by K for simplicity.

[2]Other memes exist, following the original sugarscape model, but they are not used here and only influence the cultural transmission probability (see Results).

on the values Epstein and Axtell have chosen for their simulations. So the sugar level of an agent at the time of birth lies between 50 and 100 sugar units and the maximum age between 60 and 100. As mentioned above the birth age of the first agent generation (at the beginning of a simulation) is distributed uniformly within this interval. The fertility for *males* starts between the age of 12 and 15 and ends between the age of 50 and 60. For *females*, fertility also begins between the age of 12 and 15, it ends between the age of 40 and 50.

The agents have 11 memes of whom only the first two are explicitly used in our scenario. Since only one meme at a time is affected by the interactions between agents, the probability of involving one of the two memes is given by 1/11. The first meme (the possession meme) is responsible for the possession norm, the second meme (the sanction meme) for carrying out sanctions against norm violations. The rules MF, F, G_1, and K are active. At the beginning of a run, 400 agents were placed in the landscape.

Epstein and Axtell used their model only to demonstrate the occurrence of certain phenomena and therefore performed just a single run for each parameter set. In contrast to this, to attain a broad statistical basis, our experiments consist of 100 runs for each parameter set, each with 2000 steps (updates). Metabolism rate and range of vision are varied systematically in double steps each from two to ten. So it is possible to analyze the enforcement and the effects of norms under conditions differently favorable for the agents. The chosen landscape is the original landscape with the tenfold sugar capacity for each cell.

Effects of Possession Norms

First we intended to clarify the effects of the possession norm "do not collect sugar on someone else's cells". For this purpose, we compare two runs: in one run the whole population has an active possession meme, in the other run the possession meme is deactivated for the whole population. The respective meme state does not change throughout the entire run, as at birth the agents adopt the memes of their parents and memes are always transmitted without defect.

The Figs. 1 and 2 show the number (frequency) of runs for which the agent populations survived the 2000 simulation steps. In all other runs the entire population became extinct. Agents respecting the possession norm can survive much better under unfavorable conditions (high rate of metabolism and small range of vision) as agents not aware of the norm. Without possession norm the agents in fact only can survive with a metabolism rate of two. The range of vision has almost no effect on the survival frequency. In the case a possession norm exists the range of vision has a great impact on the survival frequency because it is of high importance to have a far-reaching range of vision, since no sugar can be col-

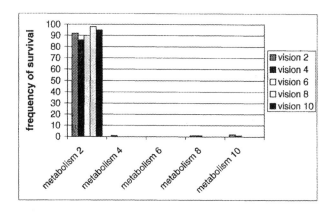

Figure 1: Frequency of survival without possession norm. The figure shows the number of runs in which the agent population survived 2000 steps versus various metabolism rates and vision ranges

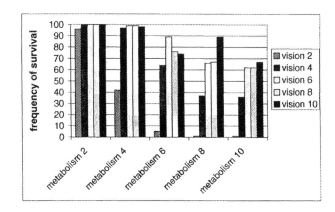

Figure 2: Frequency of survival with possession norm

lected in many fields as they belong to other agents. The reason for the higher survival probability is that agents can usually collect more sugar per step in this case. If they possess several fields, they can wait until it is really profitable to collect sugar from the corresponding fields. In such a way, for an agent with a metabolism rate of two a possession of two fields is sufficient. It can shuttle back and forth between these two fields and collect two pieces of sugar each time.

Establishment of Possession Norms

Evolution Dynamics without Sanctions An established possession norm is a definitive advantage for the survival of an entire agent population. Thus the question arises whether a possession norm can establish itself with only genetic and memetic evolution as its driving force. For this purpose, we perform runs where only half of the agents have an active possession meme in the initial population.

The result is: the survival frequency in this case is just

as bad as if no possession norm exists. A closer analysis of the individual runs shows that the meme always disappears first and then the population becomes extinct. It turns out that the agents without possession norm can reproduce significantly better on the short term, though ultimately leading the population to extinction.

Introduction of Sanctions In order to get the agents to obey norms, sanctions for norm violations have to be introduced. Fig. 3 shows the survival frequency for a punishment of four and sanction costs of zero. At the beginning, the possession meme was active in half of the population and the sanction meme in the entire population. Therefore, all agents respecting the norm sanction its violation.

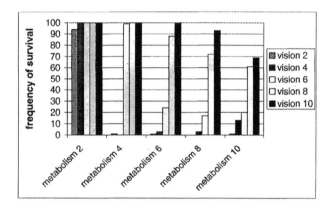

Figure 3: Frequency of survival with 50% possession meme, 100% sanction meme and a punishment of four

In this case the survival frequency is considerably better than without sanctions. Particularly for large values of the vision range very good values for the survival frequency are achieved. Because at high ranges of vision more agents can observe a norm violation and every agent knowing the norm is sanctioning a violation, overall punishment is considerably higher than at small ranges of vision. All in all it can be stated that with an increasing level of punishment the survival frequency of the population increases as well, even for small ranges of vision.

Therefore, the norm can assert itself quite well with sanctions. However, this is only valid as long as sanctions are not combined with costs. Further simulations show that, if costs exist, the norm cannot be asserted that easily because the sanction meme now is combined with disadvantages for its owner. The sanctioning agent now has to bear the costs for the sanctions, without obtaining an immediate advantage. Thus for increasing costs the percentage of the sanction meme decreases and, in turn, the probability of the possession meme to assert itself is reduced.

Conclusion

The simulation runs showed the emergence of possession norms under certain conditions. We speak of an established norm if a large number of agents behaves correspondingly. Thus, it represents an emergent quality of the agent society. Furthermore, the possession norm has been proven extremely useful for the survival of the agent population. Without the possession norm the agents become extinct much earlier in the case of unfavorable conditions than with possession norm.

However, the dilemma consists in the fact that there is a short-term advantage to agents ignoring the possession norm. The short-term interest of the agents to collect as much sugar as possible reduces the chances for the society to survive. An advantage for the individual results in a disadvantage for the society. Sanctions can offer a way out from this dilemma by reducing the short-term advantages. However, a new problem arises if sanctions are combined with costs and an individual may therefore avoid to set up sanctions. In real societies this problem is often solved by *institutions* which guarantee the obedience to certain norms and which are responsible for imposing sanctions.

In this case, costs have still to be paid. But institutions can exert a pressure on all individuals to share these costs. Thus, those may be kept lower on average and it reduces the probability that an individual can profit from the norm without contributing the cost necessary for its support. However, there is no guarantee that an institution will indeed be able to gather the relevant support from the individuals. For the future, it is intended to simulate the evolutionary emergence of institutions.

References

R. Axelrod: An Evolutionary Approach To Norms, in: American Political Science Review, Vol. 80 No. 4, December 1986, pp. 1095-1111.

J. S. Coleman: The Emergence of Norms in Varying Social Structures, in: Angewandte Sozialforschung, Jg. 14,1 , 1986/87, pp. 17-30.

R. Conte, C. Castelfranchi: Understanding the functions of norms in social groups through simulation, in: N. Gilbert, R. Conte (eds.): Artificial Societies. The computer simulation of social life, London, 1995, pp. 252-267.

R. Dawkins: The Selfish Gene, Oxford University Press, 1989.

J. M. Epstein, R. Axtell, Growing Artificial Societies, Cambridge, MA, 1996.

N. J. Saam, A. Harrer: Simulating Norms, Social Inequality, and Functional Change in Artificial Societies, in: Journal of Artificial Societies and Social Simulation vol. 2, no. 1, www.soc.surrey.ac.uk/JASSS/2/1/2.html, 1999.

Investigating the Mechanisms Underlying Cooperation in Viscous Population Multi-Agent Systems

James A. R. Marshall and Jonathan E. Rowe

Dept. of Computer and Information Sciences
De Montfort University
Milton Keynes MK7 6HP
UNITED KINGDOM
{jmarshall | jrowe}@dmu.ac.uk

Abstract

This paper presents a viscous population multi-agent system, which is claimed to provide scope for the emergence of cooperation both through iterated interaction and through kin selection. Theoretical examinations of iterated interaction and kin selection within the model are conducted and compared with empirical results. It is concluded that the model does allow for the operation both of iterated interaction and kin selection. The methods presented in the paper allow the operation of the two mechanisms to be distinguished in any instance of the model.

Introduction

The Prisoner's Dilemma (PD), first formalised by Tucker, is a well known metaphor for social interactions between individuals in which there is a dilemma over whether to act cooperatively or selfishly. In the single shot PD selfish behaviour is the only rational outcome (Nash 1950,1951). However (Axelrod 1984) proposed iterated interaction in the PD as a mechanism for promoting cooperation.

An alternative explanation for the evolution of cooperative behaviour in populations exists in (Hamilton 1964)'s theory of kin selection. Kin selection theory expects cooperative behaviour to be favoured in situations in which an individual's behaviour has an effect on the reproductive success of its relatives. In particular, so-called "viscous populations", in which individuals have limited dispersal and so tend to interact with relatives, provide such a situation. However, in these same populations, there is a potential force opposing the operation of kin selection, namely competition among relatives for finite resources.

Kin selection has previously been examined through computer simulation by (Oliphant 1994). Oliphant's model used genetic algorithms in a one-dimensional spatial environment to model individuals playing the non-iterated PD with their neighbours, and producing offspring within the same neighbourhoods. The results from the model showed that cooperation could emerge in such an environment through kin selection.

This paper presents a multi-agent evolutionary model with the PD as the model of social interaction. The model allows for the evolution of cooperation through kin selection as in (Oliphant, 1994) as well as through iterated interaction. A mathematical investigation of the influence of repeated interaction and kin selection in the model is presented, and conclusions are drawn on empirical results in the light of these investigations.

The Model

The model described in this paper was implemented using the Swarm software developed at the Santa Fe Institute (http://www.santafe.edu/projects/swarm). Source code for the model is available from one of the author's websites (http://www.mk.dmu.ac.uk/~jmarshall).

Model Overview

The model used comprises a population of agents, and an environment in which they are situated. The environment is simply a grid of cells, with each edge of the grid wrapped around to meet its opposite edge, thus forming a torus. This type of toroidal environment has become widespread in artificial life models, such as that presented in (Nowak and May 1993). Each cell is capable of housing any number of agents from zero upwards. Cells are used as the local area of interaction, i.e. agents can only play the PD with, and mate with, agents in the cell they currently inhabit. At each time step, agents are able to move to any of the eight adjacent cells with a certain probability. In addition to this physical environment, there are eight environmental variables, namely the mutation rate, the parameterised crossover rate, the maximum population the environment can support, the initial population, the death probability, the movement probability, the initial agent energy, and the energy cost for living. These variables are described in the following sections.

Agent Description

Each agent is defined as having a chromosome, an energy level, and a memory of PD interactions with other agents. For every other "opponent" that an agent has interacted with during its lifetime the agent remembers both the last actions of itself and its "opponent" (cooperate (C) or defect (D) in each case). This memory is used to determine the action an agent will take next time it meets the same "opponent". The mapping of this interaction history to an action is achieved by the agent's strategy chromosome.

The agents' chromosomes specify characteristics of the agents, and are used during interaction and mating. These chromosomes are based on (Holland 1975)'s pioneering work using genetic algorithms in adaptive artificial systems.

Following (Mar and St. Denis 1994), a five loci chromosome is used, each locus having two alleles, a cooperation allele and a defection allele. This chromosome describes a "two-dimensional" strategy, which specifies a cooperate or defect action based on the previous action of each interacting individual. Four loci are used to specify actions based on this previous pair of interactions. The remaining locus is used to specify an initial action when the agent has never interacted with its opponent before.

Lastly, the agents also have an energy level, initialised at their "birth" and decreased by a certain amount every time step. Agents "die" when their energy level reaches zero, and the only way to replenish this energy and thus survive longer is by receiving payoffs from PD interactions with other agents. The PD payoffs used were temptation $T = 5$, reward $R = 3$, punishment $P = 1$ and sucker's payoff $S = 0$.

Model Operation

The operation of the model is now described. In the first stage an initial population is created, of a size specified by an environmental variable, and randomly distributed over the environment. Each agent's initial energy level is set to a specified level. In the second stage, each agent's energy level is decreased by the "living cost" specified. Any agent whose energy level reaches zero is removed from the population. The next stage terminates agents randomly according to the probability of death in the environment. These two methods of termination correspond to "death by starvation" and by "natural causes". In the fourth stage all the agents move to an adjacent cell with a certain probability. This probability can be changed to vary the population viscosity, the implications of which are discussed in the next section. Next, agents are randomly paired up within each cell, and each pair plays one round of the Prisoner's Dilemma. The action each agent chooses will be determined by their interaction history together, and their individual strategy chromosomes. As a result of this, agents' energy levels are increased in the next stage by the payoff received from the PD interaction. Finally the agents in a cell are again randomly paired, and produce offspring with a probability $1 - P$, where P is the proportion of the maximum possible population that the current population represents, and $0 <= P <= 1$. On completion of this stage the model re-enters the second stage and executes as before.

A pair of agents reproduce to combine their genetic material into one offspring agent, whose energy level is initialised appropriately. The genetic operators used in the reproduction process are (Spears and DeJong 1991)'s parameterised crossover operator, and a standard mutation operator.

Kin Selection

A Necessary Condition for Kin Selection

There is one fundamental necessary condition for the operation of kin selection, that the behaviour of an individual has an effect on the reproductive success of its relatives (Hamilton 1964). Our model addresses this condition in two ways. First, the local random movement of agents leads to PD interactions among relatives, as a result of the agents' mating neighbourhoods and interaction neighbourhoods being identical. Second, agents' behaviour has an effect on relatives' reproductive success, because payoffs from PD interactions are used as "energy" which is necessary for an agent's continued survival. Therefore exploitation of a relative (or indeed any other agent) will directly harm them by depriving them of vital "energy". This can directly contribute to their "death". The method of local interaction and mating described above creates what is termed a viscous population, which provides a suitable environment for kin selection to operate. As (Hamilton 1964) said, "we would expect to find giving traits commonest and most highly developed in the species with the most viscous populations whereas uninhibited competition should characterise species with the most freely mixing populations."

The Opposing Forces of Kin Selection

When considering kin selection as a potential explanation for cooperative behaviour, it is important to note the possible existence of mechanisms counter-acting its operation. Such a counteracting mechanism could exist in the form of competition between relatives for finite resources. (Wilson, Pollock and Dugatkin 1992) studied a viscous population model and concluded that the effects of these opposing mechanisms exactly cancelled each other, reducing evolutionary fitness to the simplest form of individual fitness. (Taylor 1992a) concluded the same for homogenous environments, as well as extending their results to patch-structured populations (Taylor 1992b). However others, such as (Kelly 1992,1994) and (Queller 1994), argued against these claims. (Kelly 1994) in particular suggested that regulation of a population at the global rather than local level would still allow kin selection to be effective. The model studied in this paper implements such global population regulation.

Examination of the Implicit Shadow of the Future

Based on the model described above, it is possible to

make some simple calculations about the probability of repeated interaction between two agents. Due to the undecidability inherent in such models (Grim 1994), even those without stochastic elements, these are limited to calculations from one time step to the next. However, we can still use these calculations to get an impression of the implicit "shadow of the future" (Axelrod 1984). The shadow of the future is defined as the probability that two agents will interact again at some point in the future. If the shadow of the future is large then iterated interaction is likely, and so cooperation is favoured. If, however, the shadow of the future is small, then single iteration encounters are common, and so defection is favoured.

The following equation calculates the probability of two consecutive interactions between the same agents, where m represents the agents' movement probability, d the environmental death probability, and p the size of the local population in an agent's cell. It should be noted that it is difficult to take into account the probability of an agent's death through running out of energy.

$$\frac{(1-d)^2}{(p-1)}\left((1-m)^2+\frac{m^2}{8}\right)$$

Repeated interaction in the model is a geometric process, ignoring death by "starvation". Given one interaction, the probability there will be another is $1/(1-q)$, where q is the repeated interaction probability from the previous formula.

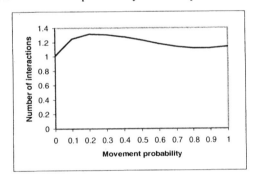

Figure 1. Expected number of interactions between agent pairs, under varying movement probability

The results of this calculation are presented in figure 1 above. The parameters used to draw this graph are: $d = 0.05$. As local population level p varies according to movement probability m in the model, the value for p at each value of m was taken from empirical data gathered from the model.

Examination of Inclusive Fitness of Cooperation and Defection Alleles

Not all the loci on an agent's chromosome are subject to kin selection. Some loci are used only in interactions where both agents will see the same interaction history and therefore use the same locus to determine their actions. Hence kin selection cannot operate on these loci. However interactions using the remaining loci will always result in each interacting agent using a different locus to determine their action. At these loci alleles can benefit copies of themselves in an "opponent" agent through increasing the fitness of the "opponent" agent, possibly at the expense of agent in which they are carried. For these loci it is interesting to calculate the inclusive fitness of cooperation and defection alleles, to gain a greater understanding of the potential for the operation of kin selection within the model. This calculation is presented below. In the following equations, T, R, S and P represent the payoffs in the Prisoner's Dilemma, c represents the frequency of cooperators in the environment, r represents the relatedness of the agents at the locus of the allele whose fitness we are considering.

Inclusive fitness of cooperation allele:
$$(Rc + S(1 - c)) + ((Rc + T(1 - c))r)$$
Inclusive fitness of defection allele:
$$(Tc + P(1 - c)) + ((Sc + P(1 - c))r)$$

In each of these equations, the first term represents the fitness of the allele in one interacting agent, the second term represents the fitness of a copy of that allele in its "opponent" agent, multiplied by a coefficient of relatedness. This is the standard inclusive fitness calculation (Hamilton 1964), interpreting payoff from a Prisoner's Dilemma interaction as reproductive potential. Each fitness term is calculated as the average payoff received by the agent weighted by the frequency of cooperators in the environment (i.e. the probability that an agent's "opponent" will cooperate in an interaction).

These two equations were used to draw the graph shown below in figure 2, given the following parameters: $T = 5$, $R = 3$, $S = 0$, $P = 1$.

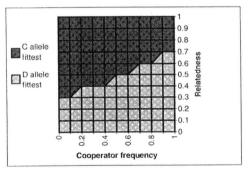

Figure 2. Relative fitness of cooperation and defection alleles at loci subject to kin selection

The inclusive fitness results in figure 2 above clearly show that the model supports kin selection. This can be seen by observing that, for the loci we are considering, the C allele is fittest at high relatedness levels.

Presentation and Discussion of Results

Experiments were run to observe the effect of varying population viscosity in the model, through changing the movement probability of agents, as follows: the experiments were conducted in 11 parameter configurations, sweeping the movement probability from 0 to 1 inclusive in increments of 0.1. For each parameter configuration 20 runs of 1000 time steps each were conducted. For each run, the average cooperation over the entire run was calculated, and this was used to calculate the average and standard deviation of the cooperation level for each value of the movement probability. The average number of interactions between any two interacting agents was also calculated, as well as the average relatedness of interacting agent pairs. Relatedness was calculated using the following formula from (Collins and Jefferson 1991), where f_i = frequency of allele 0 (for example) at locus i, l = number of loci being compared and $0 \leq D \leq 1$. The relatedness of two interacting agents was therefore calculated as 1-D, with $l = 5$.

$$D = \frac{\sum_{i=1}^{l}(1 - 4(0.5 - f_i)^2)}{l}$$

The other model parameters were as follows: environment width and height = 10, maximum agent population = 200, initial agent population = 75, mutation rate = 0.01, parameterised crossover rate = 0.1, death probability = 0.05, initial agent energy = 9, living cost = 3.

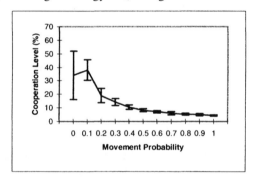

Figure 3. Global cooperation level among all agents in model

In figure 3 the cooperation level, averaged globally over all types of interaction, behaves as predicted both by iterated interaction and kin selection theory, increasing as population viscosity increases (agent movement probability decreases). It is interesting to note, however, that cooperation levels remain low across all movement probabilities, only approaching the mean at the lowest movement probabilities.

Figure 4 below presents the average number of interactions within agent pairs in the experiment, while figure 5 presents the average relatedness of pairs of interacting agents.

In figure 4, average number of interactions could be considered to be well correlated with average cooperation level in the model. Both values exhibit the same increase under decreasing movement probability, peaking at movement probability 0.1, subsequently falling at movement probability 0. In fact the only obvious difference between the curves of average cooperation level and average interaction length is the magnitude of fall at movement probability 0.

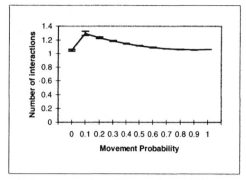

Figure 4. Average number of interactions within agent pairs

Comparing the theoretical predictions shown in figure 1 with the empirical results presented in figure 4, it can be seen that there is a close match, apart from some discrepancies at movement probabilities 0.1 and 0.2. It is suggested that the reason for these discrepancies is that the theoretical calculations cannot take account of death by "starvation", which may become increasingly important at the lower movement probabilities.

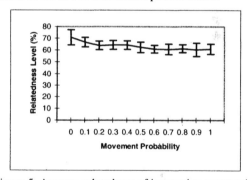

Figure 5. Average relatedness of interacting agent pairs

Figure 5 shows a high degree of relatedness within interacting agent pairs. However this relatedness level is comparatively constant over movement probability.

In addition to these results, further cooperation and relatedness data was collected from the experiments, focussing on a subset of the initial data. Figure 6 and figure 7 below present the average cooperation level and relatedness of agent pairs respectively, for interactions determined by the two loci subject to kin selection. The motivation for studying this subset of the data was presented in the section entitled "Examination of Inclusive Fitness of Cooperation and Defection Alleles".

Figure 6 shows a noisy cooperation level that does not vary with movement probability, but remains around the mean cooperation level found in the absence of selection.

Figure 7 shows the same high relatedness level observed in figure 5. However the data presented in figure 7 is less noisy, with a smaller standard deviation on the average, than that presented in figure 5. It is interesting to note that while relatedness does increase as population viscosity increases (movement probability decreases), this increase is not as pronounced as (Hamilton 1964) suggested it might be. Possible explanations for this may include the existence of forces counteracting the operation of kin selection, as outlined under the heading "The Opposing Forces of Kin Selection". Alternatively, the particular parameters used for the experiments, more particularly the PD parameters, may not allow kin selection to operate effectively. This is suggested by the cooperation levels presented in figure 6, which are based around the mean cooperation level found in the absence of selection, and do not show any trend over varying movement probability.

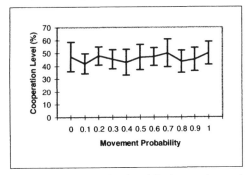

Figure 6. Average cooperation level for interactions using loci subject to kin selection

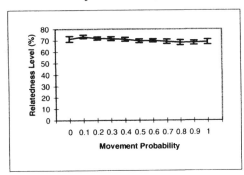

Figure 7. Average relatedness of interacting agent pairs for interactions using loci subject to kin selection

Returning to the theoretical result presented in figure 2, and taking the empirical relatedness level from figure 7, we can suggest an explanation for the observed mean cooperation level by examining the parameter space of relative C and D allele fitness. Looking at figure 2 with a relatedness level of between 0.6 and 0.7 (taken from figure 7), we find at cooperator frequencies 0.7 to 1 the threshold of relative C allele/D allele fitness. There is little difference between the fitness of the C and D alleles at this point in the parameter space, thus there is an absence of strong selection on these loci, resulting in the mean cooperation level observed.

Conclusions

This paper has presented a theoretical examination of the possible mechanisms underlying emergent cooperation in a viscous population multi-agent system. These examinations, in conjunction with empirical data from the model, conclude that the model presented allows for cooperation to emerge both through iterated interaction and through kin selection. The methods presented allow for the differentiation of the two mechanisms underlying cooperation in different cases of the presented model.

References

Axelrod, R. 1984. *The Evolution of Cooperation*. Penguin Books, London.

Collins, R. J. and Jefferson, D. R. 1991. Selection in Massively Parallel Genetic Algorithms. In Proceedings of the Fourth International Conference on Genetic Algorithms , 249-256.

Grim, P. 1997. The Undecidability of the Spatialized Prisoner's Dilemma. *Theory and Decision* 42, 1:53-80.

Hamilton, W. D. 1964. The Genetical Evolution of Social Behaviour (I and II). *Journal of Theoretical Biology* 78:1-16,17-52.

Holland, J. H. 1975. *Adaptation in Natural and Artificial Systems*. University of Michigan Press.

Kelly, J. K. 1992. Restricted Migration and the Evolution of Altruism. *Evolution* 46, 5:1492-1495.

Kelly, J. K. 1994. The Effect of Scale Dependent Processes on Kin Selection: Mating and Density Regulation. *Theoretical Population Biology* 46:32-57

Mar, G. and St. Denis, P. 1994. Chaos in Cooperation: Continuous-Valued Prisoner's Dilemmas in Infinite-Valued Logic. *International Journal of Bifurcation and Chaos* 4,4 :943-958.

Nash, J. F. 1950. Equilibrium Points in N-Person Games. *Proceedings of the National Academy of Sciences of the United States of America* 36:48-49.

Nash, J. F. 1951. Non-Cooperative Games. *Annals of Mathematics* 54:286-295.

Nowak, M.A. and May, R.M. 1993. The Spatial Dilemmas of Evolution. *International Journal of Bifurcation and Chaos* 3, 1:35-78.

Oliphant, M. 1994. Evolving Cooperation in the Non-Iterated Prisoner's Dilemma: The Importance of Spatial Organisation. In Proceedings of the Fourth Artificial Life Workshop, 349-352.

Queller, D.C. 1994. Genetic Relatedness in Viscous Populations. *Evolutionary Ecology* 8:70-73.

Spears, W. M., DeJong, K. A. 1991. On the Virtues of Parameterized Uniform Crossover. In Proceedings of the Fourth International Conference on Genetic Algorithms, 230-236.

Taylor, P. D. 1992a. Inclusive Fitness in a Homogenous Environment. *Proceedings of the Royal Society of London B* 249:299-302.

Taylor, P. D. 1992b. Altruism in Viscous Populations – An Inclusive Fitness Model. *Evolutionary Ecology* 6:352-356.

Wilson, D. S., Pollock, G. B. and Dugatkin, L. A. 1992. Can Altruism Evolve in Purely Viscous Populations? *Evolutionary Ecology* 6:331-341.

Aintz: A study of emergent properties in a model of ant foraging

Lars Kroll Kristensen

Dept. of Computer science, Aarhus University, Aarhus Denmark
kroll@daimi.au.dk

Abstract

In this paper I discuss the notion of multiple levels of emergence, and motivate its usefulness as a conceptual tool. I argue that a multi agent system containing emergent properties can be divided into i) Axiomatic behaviour ii) Emergent behaviour and iii) World rules. I show how to use this division as an informal design tool, when designing and analysing emergent properties. Specifically I discuss where and why to place the axiomatic level, given the goal of the model.
Further I present an ant model, called Aintz. I show that the very simple basic behavior of the individual agents cause complicated emergent behaviors of multiple levels. I use the Aintz model to exemplify the notion of multiple levels of emergence.

Introduction

The field of Artificial Life investigates the fundamental properties of living systems and attemps to capture these in artificial media, such as computers. One of the most important concepts that have been identified is emergent behavior. This concept, and the study of it, is important because it provides an elegant way to create complex systems that are:

(i) Robust
(ii) Distributed
(iii) Extendable.

These qualities are all desirable in most types of computer systems. Emergent properties, in the context of computer programs, are often defined as being properties that are a consequence of the interactions of the behaviors programmed into the simpler parts of the system. If these properties sum up to "more" than the sum of their parts, they are said to be *synergetic*. The terminology is useful for identifying fuzzy properties of complex systems that might be hard to describe formally. This paper will address the issue of emergent behavior, introduce a simple model with a number of emergent properties, and provide some conceptual tools for the design of emergent properties.

Emergent Behavior

A formal definition of emergent properties was proposed by (Baas 1993):

P is an emergent property of S^2

iff

$P \in Obs^2(S^2)$, but $P \notin Obs^2(S^1_{i_1}) \; \forall i_1$

In Baas' definition, S^2, the second order structure, is the result R of applying interactions Int^1 to the primitives, S^1, and the observable properties of the primitives $Obs^1(S^1)$:

$S^2 = R(S^1, Obs^1(S^1), Int^1)$

This means, that a Property P of S^2 is emergent iff it is observable on S^2 but not below. For instance, while observing the flightpath of a group of birds, one might conclude that the group forms a flock. If this property isn't observable by looking at individual flightpaths, the flocking property is said to be an emergent behaviour of the group.

I will adopt a less strict approach to my definition of emergent properties, and define them in terms of systems that contain a number of agents, each of which behave according to a traditional sequential program. These agents all inhabit the same world, and are able to modify it, and each other, using a set of *atomic actions.* The different agents must run concurrently, either on a distributed system, or parrallelized on a single CPU. In a system like this the *axiomatic behavior* is the code according to which the individual agent behaves. This behavior is predictable and formally describable, using normal computer science formalisms.

The *emergent behavior* of the system is the behavior arising as a consequence of the axiomatic behavior. This includes the axiomatic behavior, but much more interestingly, it includes the interactions of the individual agents, the *history of changes* that the agents leave in the environment and the interactions of the agents with this history. This behavior is sometimes so complex and difficult to predict that the easiest way to observe it is to simulate it.

The history of changes at time t is the sum of all the interactions between the agents and the environment, and between individual agents from time 0 until t. This history defines the *state* of the system at time *t*. The state itself doesn't need to record the entire history of changes, indeed whenever it does, this part of the history becomes a part of the state. Interestingly the state of the system at time *t* quite often has dramatic impact on the behavior of the system at time *t+1*.

Some emergent behaviors have higher order consequences themselves.

I will name these consequences N-ary emergent effects. Primary emergence would thus be the direct consequences of the axiomatic behavior, secondary emergence would be the consequences of primary emergent behavior etc. It is interesting to identify these different levels since this allows us to reason informally, yet still structurally, about emergent effects.

For instance, a property P of the system that is not explicitly coded, but is deduced to be a direct consequence of the axiomatic behavior, can be classified as a primary emergent effect. Other emergent effects must be higher level emergent effects.

This definition should be used to structure the thoughts of the implementor in the design and evaluation phases of the programming process.

Thus, what "direct consequence" means exactly, is dependant on the implementor's ability to predict and deduce emergent effects.

Introducing the Aintz

To exemplify some of the topics that I have introduced in the previous sections, I will now introduce an ant model that displays many of these qualities. The model, Aintz, is inspired by biological ants. It is not a study of Ant systems, such as the one presented by (Dorigo et al. 1997). It resembles the MANTA system, as described in (Drogoul et al. 1992) and even more the system of "dockworker" robots described in (Drogoul et al. 1993) The Aintz system is a simulation of foraging by artificial pseudoants, here called "Aintz"[1]. There are different types of aintz, all of which employ different strategies in this foraging.
The model is based on some basic objectives:

1. Aintz shouldn't individually be able to do things biological ants can't do

2. Aintz should operate according to local rules, without global control.

[1]Named so to illustrate that they are not ants

3. Only the search for food is modelled.

Aintz can leave pheromone trails in the environment, detect them, and react to them. They always know the direction to their home. Further, pheromone trails evaporate linearly, and the infinitesimal amount of food an Aint eats is ignored. Instead of focusing on expenditure of energy, the model is focused on transformation of biomatter, specifically from food items to Aintz, and back. The total biomatter of the system is kept constant, some of the biomatter existing as food, and some of it as Aintz.

The World

The world of the Aintz is a two dimensional lattice, with a 4 neighbourhood topology. A grid cell, (here called a place) doesn't have any topological dimensionality. Anything at the same place is completely colocated. The edges of the lattice wrap around, thus forming a torus. A place can hold any number of Aintz of the same color, or race, but not two of different color. Any place can also hold exactly one scent or pheromomone. Finally a place can hold any non-negative number of food items. Some places are special places, called hills. These places serve as the home for any Aintz produced there.

Whenever a hill holds more than F_a fooditems, it transforms F_a of its fooditems into a new Aint of its own color. This is done to simulate that eggs have been laid, larvae fed etc. All of this is compacted to the single action of transforming biomatter from food to an Aint. The scent of a place may be either neutral or of a certain color. Neutral scents are ignored and handled as no scent. Colored scents have a certain strength that decreases linearly in time. Whenever a scent is added to a place it either completely replaces, combines with, or is overpowered by the scent that is already present. A combination occurs when the two scents are of the same color. Replacement/overpowering occurs when they are different, in which case the stronger scent overpowers the weaker. Scents do not dissipate in the environment.

Whenever an Aint dies, it is transformed back into F_a fooditems. These can be reintroduced into the world in different ways: They may be dropped at the spot where the Aint died, to simulate a carcass that Aintz might eat (cannibalistic reentry) or it may be dropped randomly into the world, either uniformly or normally around the center (Random reentry).

The Aintz

There are four different types of Aintz, identified with red, blue, cyan and white color. The generalized Aint implements the common behavior of all Aintz which mainly is bookkeeping, graphics rendering, etc.

All of the Aintz have access to a number of *atomic actions*:

- **Move**
  Moves the Aint to a neighbouring place.

- **Drop Pheromone**
  Drops a pheromone item of a certain strength and color.

- **Pick up food**
  If any food is at the same place as the Aint, the Aint picks up one food item.

- **Drop food**
  If the Aint is carrying a food item then it drops it.

In addition, the Aintz can sense the pheromone of the place on which they are standing, and of the neighbouring places. Further they can sense the direction to home.

The generalized Aintz' axiomatic behavior can be stated as a subsumption architecture. (Brooks 1987) This means that the Aint has an ordered list of core behaviors, higher level behaviors *subsuming*, taking over, in special situations.

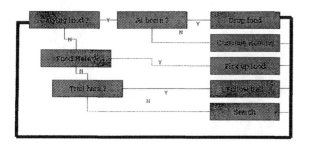

Figure 1: subsumption architecture of the general Aint

Thus an Aint will first check whether it is carrying food or not. If it is, it will check whether it is at home or not. If it is at home, and carrying food, it will drop the food. If it is not at home, but carrying food, it will "Go Home", which here means that it moves one step in the direction of home. While an Aint is homing (while carrying food) it will leave a pheromone trail pointing in the opposite direction of its current travel direction. This trail can be seen as a small arrow pointing in one of the four directions, N, S, E or W. Thus a homing Aint will leave these small arrows pointing towards wherever it found the food it is bringing back.
The "Pick up food" behavior simply removes the food from the world, and adds it to the food the Aint is carrying. The Aintz energy is then set to 250, to represent

that the Aint eats a bite of the food. This energy measure should not to be confused with the biomatter, neither in the Aint, nor in the food items. It is a mechanism to ensure that Aintz die if they don't find at least one food item every 250 timesteps.
While performing the "Follow trail" behavior, the aint reads the direction of the trail it is standing on, and takes a step in this direction. The trails of different types of Aintz are identical, thus a blue Aint can sense and follow a trail left by a red Aint.
While performing the "Search" behavior, in the case of the general Aint, it simply chooses a random compas direction and moves a step in this direction. Whenever a timestep has passed, the Aint loses a unit of energy. If the Aintz energy drops to zero, it dies, to simulate starvation. If an Aint moves into a place where an Aint of a different color already stands it also dies. This rule simulates combat between Aintz. All Aintz share this architecture and basic behavior, whereas the "search" behavior differs between Aintz of different colors.
The red Aintz simply use the default behavior of choosing a random direction. (Random search).
The blue Aintz (see fig. 2) leave a pink pheromone trail while searching, and when they need to choose a direction in which to move, they choose randomly among those neighbouring places containg the least pink pheromone. Thus, they avoid the pink pheromone. (Spreading search)

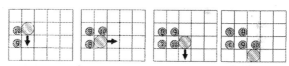

Figure 2: movement pattern of the blue Aint, avoiding the pheromone markers

The cyan Aintz (see fig. 3) have a gene-string of directions. At each timestep the genetic Aint chooses the next gene in its gene-string as the direction it will move. Whenever a genetic Aint drops food at home, it also leaves its gene-string. Whenever the anthill produces a new Aint, it performs a simple one point cross over combination of the last two gene-strings for the new Aint. There is also a certain probability[2] that the new Aint will be completely random. (Mutation) Thus Aintz that bring more food home have a better chance of spawning offspring. (Genetic search)

The white Aintz are a mixed population of half random searchers, and half genetic searchers. Whenever a

---

[2]1% in the current implementation

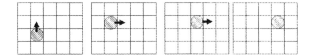

Figure 3: movement pattern of the cyan Aint, with gene-string North,East,East (NEE)

new white Aint is produced, there is a 50% chance it will be a new random Aint, and a 50% chance it will be a genetic Aint. (Hybrid)

Results

The system described above has a large number of interesting dynamic properties, arising from the local interactions of the different types of agents and objects in the world. In the following section I will focus on the efficiency of the different colors of Aintz, which is equal to their ability to proliferate in competition with the other Aintz. It is important to note that this efficiency is dependant on many factors. The most important goals are summarized below.

1. Avoid starvation

2. Gather food

3. Avoid getting killed by other Aintz

4. Kill other Aintz

Points 1 and 2 are related, since whenever food is found, starvation is avoided and gathering is in progress. The reason why I list them as different goals, is that 1) is concerned with the survival of the individual Aint, whereas 2) is concerned with the proliferation of the colony. Number 3 and 4 are quite difficult to accomplish, since the Aintz cannot sense each others presence.

Example runs

Figure 4 shows a screenshot of a typical run, in which food is reintroduced cannibalistically, i.e. the dead aint is replaced by a number of food items on the place it dies. Upper left is red aintz, upper right is cyan, lower left is blue and lower right is white aintz. Note how trails radiate from each hill. Note also that blue is much more efficient than the others.

Figure 5 shows a run where food is reintroduced randomly, i.e. whenever an aint dies, F_a food items are reintroduced randomly into the world. Note the circular area around blue (bootom left) which is continually patrolled and therefore empty of food. Note also that cyan is significantly more efficient than the others under these circumstances.

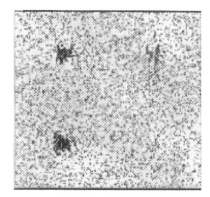

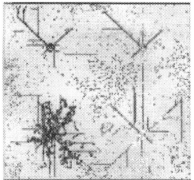

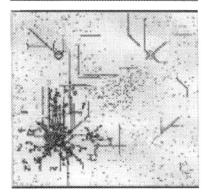

Figure 4: situation 1 after 61, 3006 and 10012 steps

The Red Aintz

The red Aintz, the random searchers, try to accomplish the goals in the simplest way compared to the other types. Whenever a red Aint is carrying food and is at home, it drops the food. Whenever a red Aint is carrying food and is not at home, it will move a step in the direction of home and leave a pheromone marker in the place it reaches, which will point towards the place it left. This is the axiomatic behavior of the red Aint carrying food. It clearly has the primary emergent effect that the Aint eventually will reach home and drop the food there. The effect that the Aint will leave a trail of

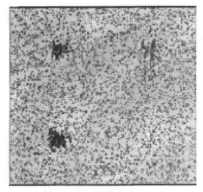

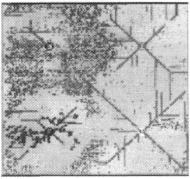

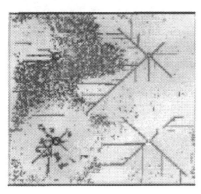

Figure 5: situation 2 after 73, 5009 and 10012 steps

pheromone markers, leading from home to wherever the fooditem was found, is another primary emergent effect. Whenever a red Aint is not carrying food and is standing on a pheromone trail, it will take a step in the direction of the trail. Whenever it is not carrying food, and is not standing on a trail, it will move a step in a random direction. This axiomatic behavior has the direct consequence that red Aintz seem to "seek" trails, albeit inefficiently, and follow them when they bump into them.

The Blue Aintz

The blue Aintz have the exact same properties as the red Aintz described above, except when they are searching for food or trails: Whenever a blue Aint is not carrying food and is not standing on a pheromone trail, it will examine the scent of all neighbouring places, and move to the place in its neighbourhood with the least amount of pink pheromomone. If several of the neighbouring places contain the same (least) amount, it will choose randomly among them. This axiomatic behavior has an interesting primary emergent effect: The individual blue Aint will tend to avoid its own footsteps. It will drive itself away from the places it has recently visited, thus tending to cover more ground than the red Aintz. It will also drive itself away from other blue Aintz, thus being less likely to examine areas that have already been examined by other blue Aintz.

The Cyan Aintz

The cyan Aintz move according to a sequence of directions in which they move one step in each timestep, and restart the sequence when they reach the end. This results in a cyclic movement pattern, for instance NNESN. This movement pattern can be reduced to the resulting movement NNE, since SN cancel each other (except of course if it runs into food or a trail). Because of this cyclic movement pattern, the cyan Aintz will tend to move in a line directly away from home, except for those that have an empty resulting movement (e.g. NS). Since all four directions initially have an equal chance of appearing in the gene-string of the initial Aintz, the cyan Aintz will initially fan out in a star-pattern.

The White Aintz

The white Aintz are composed of 50% random searchers, and 50% genetic searchers. Unsurprisingly, their behaviour is a mixture of the behaviours of the red and cyan Aintz.

Discussion of the Aintz model

The Aintz described so far have a number of emergent properties. Properties that are consequences of the axiomatic rules coded into the agents themselves. Some of these emergent properties have second order emergent properties. These properties are described in the following section.

Primary emergence

Primary emergent effects are caused indirectly by the hardcoded behavior of the single agent. It is not the behavior itself, for instance the "Homing" behavior of the general Aint that moves it one step closer to home at each timestep, is not emergent. It is axiomatic. However, the fact that this behavior leads the Aint to its home is an emergent behavior, though somewhat trivial. A less trivial primary emergent property, is the tendency of a single blue Aint to choose a certain direction, and stick to it. This property emerges from the behavior that the blue Aint will avoid its own tracks. The axiomatic behavior changes the environment and thus the basis for its own later decisions.

Secondary emergence

Secondary emergent effects are effects that emerge from (simpler, first order) emergent behaviors. An example is the tendency of cyan Aints to evolve towards higher efficiency: Cyan Aintz leave their genes at the hill whenever they bring food back home. This means that these Aintz have a higher probability of getting "offspring", in the sense that their genes are perpetuated. This in turn makes the colony as such evolve towards behaving like the most efficient food gatherers. Note that the fitness of a cyan Aint is an emergent property. It is a property that isn't recorded anywhere in the system, but it is still observable. We can see that certain directions are favoured, while others disappear.

Another example of secondary emergence is the tendency of blue Aintz to spread out to a certain radius, and then continually sweep this area around their own home (territorialism). This effect is partly a derivation of the blue Aintz tendency to avoid not only their own tracks but also those of their peers, and partly due to the fact that when a blue Aint returns home, and starts from home, it will choose the path with the least pink pheromone, i.e. the path left unused for the longest time, from home to "unexplored country". This path then becomes the path used most recently, and it won't be revisited until it becomes the oldest path again.

This primary emergent property has the secondary emergent property of territoriality, since it leads to the blue Aintz continually patrolling the areas that haven't been visited for a while.

Tertiary emergence

Tertiary effects are rarer, and less readily identified. However, the fact that blue Aintz are much more suited to competition involving mass combat than the other Aintz is primarily because the blue Aintz are territorial. If Aintz of different colors meet each other, one will die, and the other will live. This is an advantage for the colony that manages to "harvest" the cadaver and obviously a disadvantage for the colony that lost an Aint. Therefore it is an advantage to stay together, and relatively close to the hill, since the probability of harvesting the cadavers of a confrontation is then higher. Thus, the advantage in warlike scenarios for the blue Aintz emerges from their territorialism, which emerges from their tendency to avoid each other, which finally emerges from their individual tendency to avoid their own tracks. Therefore it is a tertiary emergent property.

Discussion of emergence

The classical Alife approach to the design of emergent behavior is to identify the task which needs to be solved, and then to look at nature for inspiration. Some aspect of a natural system that solves a similar task is then modelled in sufficient detail, and simulated. Hopefully, the model exhibits the same emergent properties as nature. If not, the natural system is reinvestigated, the model is refined and the cycle is repeated. This is the approach taken by people such as Dorigo et al. in solving graph-problems by ant colony optimisation (Dorigo et al. 1997). It is also the approach of the MANTA system (Drogoul et al. 1992), although in the MANTA system, the domain modelled isn't the natural world, but rather the laboratory conditions under which biological ants are studied.

This approach has the nice property that since the natural system solves the interesting task, all we need to do is to simulate nature sufficiently accurately to make the model solve the same task.

The disadvantage is that it is quite difficult to model nature "sufficiently accurately", especially since it is also a goal to model the minimum necessary to obtain the desired effects.

The "natural" system described above, does not need to be found in nature. Important lessons for real-world applications can be learned from observing artificial toy-worlds, such as for instance the Aintz model.

Assume someone wants to program a number of micro-robots to kill weeds on a golfcourse. Which of the Aintz should be used as inspiration?

The blue Aintz are territorial and will constantly patrol the area, keeping the weeds down. Therefore they appear to be a natural choice. If we wish to solve the task of unbounded exploration/mapping then territoriality becomes a problem. In that case, the simpler red Aintz seem more relevant.

Choosing atomic actions

The choice of atomic actions is closely related to the choice of axiomatic behaviour. To illustrate the differences between the two, consider the case of robotics: The atomic actions correspond closely to the physical capabilities of the robot: If the robot has wheels, one of the atomic actions of the robot could be MoveWheelForward, another might be TurnXDegrees etc. The atomic actions are the building blocks of the axiomatic behaviour. Therefore, the atomic actions are a level lower, in terms of abstraction, than the axiomatic behaviour.

The set of atomic actions is the set of interactions (i) between agents and the world and (ii) between agents and each other. Therefore, the atomic actions define the state-space of the system. For this reason, it is important to know which state-space the model should explore. I for instance, was interested in the foraging techniques of ants. To forage, it is necessary to be able to move, pick up, and drop items. I also wanted to see how much cooperation was possible by using simple non-dissipative trails. Therefore the Aintz were given the ability to lay out, and sense these trails.

Choosing axioms

In order to minimize the complexity of the model, while still being sufficiently accurate, it is crucial to identify the necessary axiomatic behaviors correctly. If a property P is wanted in the system, it should be carefully decided whether P is an Axiomatic behavior or an emergent behavior. This decision should be based on the level of modelling detail. If for instance the model is about the flocking behavior of birds then changing velocity should be an axiomatic behavior for the single bird (Reynolds 1987). If the focus is on the actual locomotion then the aerodynamics of the agent and the moving parts of the agent (it's limbs) could be modelled as axiomatic behavior (Sims 1994). This choice should be based on the goal of the research. It is useful to keep in mind that a lower level of axiomatic behavior also allows higher level emergent effects, as demonstrated in the Aintz system. For instance, it should be possible to obtain flocking results similar to the ones obtained in the boids model (Reynolds 1987), based on boids that propel themselves not by axiomatically accelerating, but by moving their limbs, thus causing acceleration, like in Karl Sims model. (Sims 1994).

However, due to the extra amount of computation, this approach will be more time consuming. It is important to know where and why to implement actions as axiomatic. It is a good idea to disregard aerodynamics and simply implement acceleration axiomatically if the focus is on unordered flocks, like swallows or gnats. However, if the focus is (evolving) ordered flocks, like for instance geese that fly in a V formation, then aerodynamic effects may indeed be crucial, and the axiomatic level should be on locomotion.

The pioneering paper "The motility of Microrobots" (Solem 1994) deals with the physical realities faced by a designer of microrobots. The interesting part here, is that the paper describes the world rules, and atomic actions possible for such microrobots. This defines the axiomatic level, and therefore the level at which the behaviour of these machines should be modeled. For instance, this investigation shows that jumping seems to be the most efficient method of transport for earthbound robots. A programmer interested in having a large number of microrobots perform some task involving movement should therefore choose jumping as an atomic action, and proceed from there.

More work of this kind is very much needed, since it allows work on the behaviour of the machines to start,

even though the machines themselves are not available yet. In the context of nano and microrobotics, the concept of emergent behaviour is very interesting, since micromachines can be assumed to have limited computing ability. Therefore their control programs will probably be rather simple. However, if these simple programs are designed carefully, interesting and useful emergent effects of a large number of these robots can be obtained.

Combining Axioms

Having decided at which level to model the domain, the next step is to implement the aximatic behaviors, and to combine them in such a way that the emergent behaviors arise. Here again there are several approaches. Brooks (Brooks 1987) introduced the concept of the subsumption architecture. The idea is to decompose the behaviour of the agent (in Brooks case a robot) into functional units. Thus for instance a robot could primarily be concerned with obstacle avoidance, secondarily with exploration and tertiarly with mapbuilding.

This concept is quite useful for the designer of emergent properties, since it provides a way to structure the individual agent, and cause it to choose among a number of axiomatic behaviors based on it's current state. Even though Brooks uses subsumption architectures for reactive systems, this approach is quite valid for other types of agents as well. There is no a priori reason to restrict the use of subsumption architectures to simple Stimulus-response agents. In fact, actions may be designed to change some state of the agent and stimuli may come from the agent itself. In such a way, there may be several processes going on inside the agent, which result in a cognitive agent as an emergent effect.

The World

The agents invariably live in some sort of world. The world, in this context, is the sum of all the rules that apply to the agents, as well as the medium in which they interact. It is the "laws of nature" of the model.
In the case of the Aintz model, the world is the lattice which represents the spatial properties of the model, the rules for transformation of biomatter ("birth" in the homes and reentry of biomatter when an Aint dies), the rules of combat, and the rule of pheromone evaporation. Obviously these laws have dramatic impact on the systems behavior.

For instance, the rule of biomatter reentry: Whenever an Aint dies, F_a food items will be reintroduced into the system. This can happen either by depositing the food items as a pile of food where the Aint died (cannibalistic reentry) or by reintroducing the fooditems somewhere random (Random reentry).

The Blue Aintz are significantly less efficient in random reentry compared to cannibalistic reentry, but interestingly the cyan Aintz (genetic search) are *more* efficient with random reentry. Assuming the Cyan home is W and N of the center of the normal distribution of food, the cyan genepool just needs to discover that E and S are "good genes", while N and W are "bad genes". This becomes much harder to discover if the Aintz going N and W leave their carcasses N and W of the home. Then those that follow will find them, producing even more Aintz going N and W. If on the other hand, the Aintz going N and W die, disappear, and reappear as food *in the center of the world*, the "good genes" are much easier to evolve.

This is an example of how the global rules can have dramatic impact on the behavior of the system. In the context of designing emergent behavior, the interaction with the world can be used to affect highlevel emergent effects, without nescessarily altering the lowlevel effects. We can use the global rules to affect those things that are difficult to affect by altering the local rules.

Continuous vs. Discrete worlds

In the context of emergent behavior, only few people seem to be interested in continuous worlds. Most models in this area are based on two or three dimensional lattices. Notable exceptions are (Reynolds 1987) and (Sims 1994). The reason for this is obvious: It is much easier to code, and less computationally expensive.

However, there are important differences between nature and lattice worlds

I have worked with both types of worlds (continuous and discrete) and the only thing that seems clear to me is that even small changes in the "world rules" can have enormous consequences in any emergent system. When modeling nature as a discrete lattice world, important aspects of the domain might be lost. For instance, when going from a discrete lattice based world to a continuous one, the concept of extensionality is introduced into the model. In this context, extensionality is the physical body of the agent. Specifically its size and shape. The work of (Sims 1994) shows that morphology plays a role in locomotion. Morphology and extensionality in general might play a very important role in interaction between agents.

The concept of extensionality might very well lead to more natural emergent effects. The boids model (Reynolds 1987) may work in a lattice based world, but not nearly as smoothly and naturally[3]. The work of (Sims 1994) certainly becomes much more artificial, when transferred to a lattice based world. The Aintz model would have to be changed dramatically to translate the axiomatic behaviors to continuity. In fact, the very choice of discrete vs. continuous modelling poses some severe restrictions on other choices of the model

---

[3] At the EVAlife workshop, Mr. Reynolds said that his primary reason for working with continuous rather than discrete models were aestetic. "It looks nicer".

later on.

The euclidian vs. non-euclidian distance problem is nicely exemplified in the Aintz model. If an Aint is at grid point (0,0) and moves 5 steps north, and then 5 steps east, it will find itself at gridpoint (5,5). If it moves in a straight line[4], from (0,0) to (5,5) it *also* moves 10 steps. In euclidian space, the shortest distance between two points is a line. In the Aintz lattice world, a line, an angle, indeed any string of steps, containing 5 N and 5 E will get the Aint to the same place.

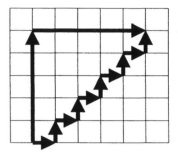

Figure 6: The geographical distance problem

If I had chosen an 8 neighbourhood topology instead of the 4 neighbourhood topology, I would have encountered similar problems.

This global rule is an artefact of the choice of a lattice world. Any lattice world will suffer these problems. Nature is not a lattice world.

Conclusions

I have addressed the notion of multiple levels of emergence, as introduced by Baas (Baas 1993), and have argued for the usefulness of this term. I have shown some ways to use the notion as a conceptual tool, when designing and analysing emergent properties. Specifically I have argued for the importance of choosing where to put the axiomatic level, depending on the goal of the model.

Further, I have discussed the differences between discrete and continuous worlds and have argued that discrete lattice worlds present serious problems with regard to geographical distances.

I have presented an ant model, called Aintz. I have shown that the very simple basic behavior of the individual agents cause quite complicated emergent behaviors,

---

[4] Or rather, the closest approximation of a straight line in this world

of multiple levels. I have used the Aintz model to exemplify the notion of multiple levels of emergence.

Future work: Evolving emergence

In the context of the Aintz model, it would be interesting to try to evolve the current axiomatic behaviours, along with others. The atomic actions would thus form the core of this new implementation of the Aintz model, along with a genetic programming algorithm to evolve the behavior of the individual Aint. In fact this amounts to choosing the basic actions of the Aintz, together with the evolutionary algorithm as the axiomatic behavior of the system, thus making the behaviors of the Aintz (the genetic code) the primary emergent effects. Another aspect to examine is the issue of discrete vs. continuous worlds. A viable approach would be to evolve the emergent behaviours as described above, first in a lattice-world, then in a continuous, and see if different strategies emerge. It is possible that the introduction of extensionality of actions (in continuous time) and agents (in continuous space) will prove to be important.

Acknowledgements

I thank the other members of the EVAlife group of the Computer science departement of Aarhus University, for their invaluable inspiration and useful comments. Particularly I thank Dr. Thiemo Krink for support, insight, trust and inspiration.

References

Aintz
The Aintz system is available on the WWW as an applet.

www.daimi.au.dk/~kroll/alife/

Baas, N.A. "Emergence, Hierarchies and Hyperstructures". *Artificial Life III, ed. Christopher G. Langton.Addison-Wesley Publishing 1993.*

Brooks, Rodney A. "A Robust Layered Control System for a Mobile Robot" , *MIT AI Lab Memo 864, September 1985.*

Drogoul et. al. 1992 "Multi-Agent Simulation as a Tool for Modeling Societies: Application to Social Differentiation in Ant Colonies" in *Proceedings of MAAMAW'92*

Drogoul A. & Ferber J "From Tom-Thumb to the Dockers: Some Experiments with Foraging Robots" in "*From Animals to Animats II*", *MIT Press, Cambridge, pp. 451-459, 1993.*

Dorigo M. & L.M. Gambardella (1997). "Ant Colonies for the Traveling Salesman Problem." *BioSystems, 43:73-81. (Also technical Report TR/IRIDIA/1996-3, IRIDIA, Universite Libre de Bruxelles.)*

Solem, J.C. (1994). "The motility of Microrobots", in: *Artificial Life III, ed. Christopher G. Langton.Addison-Wesley Publishing 1993.*

Reynolds, C. (1987) "Flocks, Herds & Schools: a distributed behavioral model" *SIGGRAPH 87*

Sims, Karl (1994) "Evolving virtual creatures" *SIGGRAPH 94 proceedings.*

A Co-evolution model of Scores and Strategies in IPD games: toward the understanding of the emergence of the social morals

Yoshiki Yamaguchi, Tsutomu Maruyama and **Tsutomu Hoshino**

Institute of Engineering Mechanics and Systems, University of Tsukuba

1-1-1 Ten-ou-dai Tsukuba Ibaraki, 305-8573 JAPAN

E-mail: {yoshiki, maruyama, hoshino}@darwin.esys.tsukuba.ac.jp

Abstract

It was shown that the evolution of a world consisting of agents that play Iterated Prisoner Dilemma (IPD) games each other is open-ended by Lindgren. The behavior of the world is very sensitive to the values of the payoff matrix used in IPD games, because the values have great influence on the population dynamics of the world. In general, the values are fixed throughout the simulations. In the real world, however, morals and the behaviors of individuals that follow the morals have been evolved influencing mutually.

In this paper, we propose a co-evolution model of agents and scores of IPD games toward the understanding of emergence of social morals. The co-evolution model consists of two layers. In the first layer, scores for IPD games are evolved using a genetic algorithm. Scores vary within the range of dilemma games, and scores that attract more agents in the second layer will gradually increase. In the second layer, agents play IPD games with all other agents following the scores that they believe and are evolved using Lindgren's model.

Simulation results showed that the values of the scores evolve toward the score which gives more payoffs for cooperative strategies and less payoffs for defective strategies as the strategies of the IPD agents are evolved. The results also showed that small colonies of defective strategies repeatedly appear and disappear throughout the simulations.

Introduction

It was shown that the evolution of a world consisting of agents that play Iterated Prisoner Dilemma (IPD) games each other is open-ended by Lindgren (1). In the model, strategies of the agents are evolved by mutations, and the agents that obtained more payoffs (using a given payoff matrix) will gradually increase its ratio in the total population. The behavior of the evolution of the world is very sensitive to the values of the payoff matrix, because the values have great influence on the population dynamics of the world. For instance, when a defective strategy is created by mutations, the balance of the values of the payoff matrix decides whether the strategy can increase in co-operative worlds or not. In general, the values are fixed throughout the simulation. In the real world, however, it seems that morals (which have strong influences on the behaviors of individuals like the values of the payoff matrix in IPD games) are gradually refined as the behaviors of individuals in the world are evolved and vice versa.

In this paper, we propose a co-evolution model of agents and scores of IPD games toward the understanding of emergence of social morals. The co-evolution model consists of two layers. In the first layer, scores for IPD games are evolved using genetic algorithms. Scores vary within the range of dilemma games, and scores that attract more agents in the second layer and obtain more payoffs (total payoffs obtained by the agents that follow the scores) will gradually increase. In the second layer, agents play IPD games with all other agents following the scores that they believe and are evolved using Lindgren's model. With this co-evolution model, we can observe how the scores are evolved as the strategies of the IPD agents are evolved, and how the scores lead the evolution of the behaviors (strategies) of the IPD agents.

Co-evolution Model

In this section, we describe the details of the co-evolution model. Our model has two layers; agents' layer and scores' layer. Figure 1 shows the overview of the model. In each layer, the size of the circle shows the ratio of each score and agent in the total population.

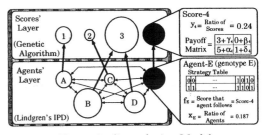

Figure 1: Co-evolution Model

The Agents' Layer

In the agents' layer, each agent plays IPD games with all other agents following Lindgren's IPD model(1), and agents that obtained more payoffs will increase its ratio in the next generation. Each agent follows one of the payoff matrix (score) in the score's layer, and their payoffs are given based on the payoff matrix (score). Therefore, agents with same strategies may get different payoffs even if they play games against same agents.

The Scores' Layer

Table 1 shows the payoff matrix used in our model.

Table 1: Payoff Matrix for Co-evolution Model

		Agent 2	
		Cooperate	Defect
Agent 1	Cooperate (C)	Reward $3+\gamma_i$	Sucker $0+\beta_i$
	Defect (D)	Temptation $5+\alpha_i$	Punishment $1+\delta_i$

Our matrix has four parameters; α_i, β_i, γ_i and δ_i. These parameters are evolved using a simple genetic algorithm, and the scores (matrices) that attracted more agents and obtained more payoffs (total payoffs obtained by agents that follow the scores) will increase its ratio in the next generation.

By changing the values of these parameters, categories of games become the one of the followings.

$$T>P>R>S \quad \text{Deadlock}$$
$$T>R>P>S \quad \text{IPD}$$
$$T>R>S>P \quad \text{Chicken}$$
$$R>T>P>S \quad \text{Stag Hunt}$$

Figure 2 shows the relationship between a score (matrix) and categories of the games.

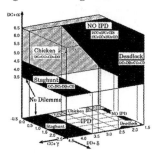

Figure 2: Scores and the Category of Games ($\beta_i = 0$)

In order to simplify the figure (a figure by four parameters is four dimensional), we fix the value of β_i to zero. In Figure 2, the projected figure on the floor (γ_i and δ_i plain) shows the relationship between the two parameter when α_i and β_i are fixed to -1.5 and

0. The black parts show areas out of dilemma games (CC>DC>CD>DD).

We limited the range of parameters as shown in the equation below. By this limitation, scores move only in the range of dilemma games.

$$-1.5 \leq \alpha_i, \beta_i, \gamma_i, \delta_i \leq 1.5$$

Evolutionary Operations for Scores

Mutation At the initial state, there is only one score (matrix) in the scores' layer, and its parameters are all zero. Therefore, all agents in the agents' layer follow the score.

Figure 3 shows how a new score is created by mutations. When a mutation happens to a score, its ratio (y_1) in the total population is divided to $(1-\tau)y_1$ and $\tau \times y_1$, and $\tau \times y_1$ is given to the new score created by the mutation. The ratio of the agents that follows the score is also divided to $(1-\xi)x_a$ and $\xi \times x_a$, and $\xi \times x_a$ follows the new score.

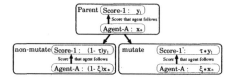

Figure 3: Mutation of Scores

In the mutation of the score, the parameters are changed using a gaussian function $f(s,t,u,v)$ below.

$$f = \frac{1}{4\pi^2} \frac{1}{\sigma_s} e^{-\frac{(s-\alpha_i)^2}{2\sigma_s^2}} \frac{1}{\sigma_t} e^{-\frac{(t-\beta_i)^2}{2\sigma_t^2}} \frac{1}{\sigma_u} e^{-\frac{(u-\gamma_i)^2}{2\sigma_u^2}} \frac{1}{\sigma_v} e^{-\frac{(v-\delta_i)^2}{2\sigma_v^2}}$$

We fixed the value of σ_s, σ_t, σ_u and σ_v to $\frac{1}{3}$.

Crossover Figure 4 shows how new scores are created by the crossover operation.

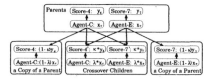

Figure 4: Crossover of Scores

First, two scores are selected at random, and then parameters in the payoff matrices are crossovered. The values of parameters are changed using the function $g(s,t,u,v)$ below.

$$g(s,t,u,v) = \frac{1}{4\pi^2} \frac{1}{\sigma_s} \frac{1}{\sigma_t} \frac{1}{\sigma_u} \frac{1}{\sigma_v} e^{-p_g}$$

$$p_g = \frac{\left(s-\frac{\alpha_i+\alpha_j}{2}\right)^2}{2\sigma_s^2} + \frac{\left(t-\frac{\beta_i+\beta_j}{2}\right)^2}{2\sigma_t^2} + \frac{\left(u-\frac{\gamma_i+\gamma_j}{2}\right)^2}{2\sigma_u^2} + \frac{\left(v-\frac{\delta_i+\delta_j}{2}\right)^2}{2\sigma_v^2}$$

We fixed the value of $(\sigma_s, \sigma_t, \sigma_u, \sigma_v)$ as shown below.

$$\sigma_s = \frac{1}{3} \cdot \left|\frac{\alpha_j-\alpha_i}{2}\right|, \quad \sigma_t = \frac{1}{3} \cdot \left|\frac{\beta_j-\beta_i}{2}\right|, \quad \sigma_u = \frac{1}{3} \cdot \left|\frac{\gamma_j-\gamma_i}{2}\right|, \quad \sigma_v = \frac{1}{3} \cdot \left|\frac{\delta_j-\delta_i}{2}\right|$$

Agents that follow the scores which are crossovered are divided into two groups ($(1 - \lambda)x_c$ and $\lambda \times x_c$, $(1-\lambda)x_e$ and $\lambda \times x_e$, respectively), and $\lambda \times x_c$ and $\lambda \times x_e$ follow new scores created by the crossover operation as shown in the figure 4.

Disappearance of Scores The total number of scores (number of scores with difference values of the parameters because scores with same values are grouped) in the scores' layer is limited , and when the number of scores exceeds the limit, scores that have less followers are deleted. When a score is deleted, the agents that follow the score select one of the scores in the scores' layer at random.

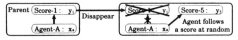

Figure 5: Disappearance of Scores

Population Dynamics

The ratio of score k (matrix k) in the total population is expressed as y_k. The population dynamics of scores are decided using the total payoffs of agents that follow the scores. Therefore, the point for score k in one generation is

$$p_k = \sum_i s_i \ (added\ only\ if\ agent_i\ follows\ the\ score)$$

The ratio of the score in generation t+1 is decided by the following equation, where $y_i(t)$ is the ratio of the agent in generation t and d_{score} is a constant.

$$y_i(t+1) = y_i(t) + d_{score} \cdot y_i(t)\left(\frac{p_i - p_{ave}}{p_{ave}}\right)$$

Simulation Results

We simulated the behavior of our co-evolution model changing the range of parameters in the payoff matrix (table 1) as follows.

1. changing the parameters without any limitation within the range below.

$$-1.5 \leq \alpha_i, \beta_i, \gamma_i, \delta_i \leq 1.5$$

2. adding the limitations as follows.

$$\alpha_i + \delta_i = 0, \quad \beta_i + \gamma_i = 0$$

3. narrowing the range as shown below and with the limitation above.

$$-1.0 \leq \alpha_i, \beta_i, \gamma_i, \delta_i \leq 1.0$$

By changing the range and the limitations, we can observe several kinds of behaviors of the evolution, though most of them proceed toward cooperative worlds.

Other Parameters

Table 2 shows the parameters used for the IPD agents, and table 3 shows the parameters for the scores. In the tables, number of agents/scores means the number of agents/scores with different strategies/values. In the simulations, agents/scores with same strategies/values are grouped and treated as one strategy/score.

Table 2: Parameters for Agents

$Generation$	generation	90000
$Iteration$	iteration times of games	1024
N_{agent}	number of agents	1024

Table 3: Parameters for Scores

N_{score}	number of scores	256
p_{p_score}	score mutation rate	1×10^{-4}
τ	score distribution rate	0.1
ξ	agent distribution rate	0.1
p_{c_score}	score crossover rate	1×10^{-4}
κ	score distribution rate	0.1
λ	agent distribution rate	0.1
d_{score}	population dynamics	1×10^{-4}

Simulation Results with no Limitation

In this case, parameters in the matrix table can move within the range below without any limitation.

$$-1.5 \leq \alpha_i, \beta_i, \gamma_i, \delta_i \leq 1.5$$

Figure 6 shows the average point of agents. In this simulation, agents following scores with larger payoffs (namely larger values of parameters) can get more payoffs in any situations of the IPD games. Therefore, all parameters of scores go to +1.5 in despite of strategies of agents that follow the scores. A table in figure 6 (right-side) shows this payoff matrix. With this table, a defective strategy created in a cooperative world by mutation can obtain more payoffs compared with the original matrix (all parameters are zero). However, cooperative strategies can also get more payoffs even if they are defected, and the whole world always proceeds to cooperative world.

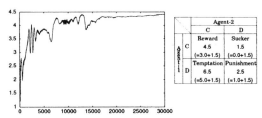

Figure 6: Average Points of Agents (No Limitation)

Simulation Results with the Limitation

By adding the following limitations, all agents are faced with more strict dilemma situations. With the first equation, scores that give more payoffs to cooperative strategies also give smaller payoffs when defected, and with the second equation, scores that give more payoffs to defective strategies also give smaller payoffs when defected.

$$\alpha_i + \delta_i = 0, \quad \beta_i + \gamma_i = 0$$

Figure 7 shows the relation ship between scores and categories of games (left-side) and a payoff matrix with this limitation (right-side). In the figure 7, a point in the center shows the original matrix used in Lindgren's model. The two vertical axes show the payoff obtained when the players' moves are CC and CD, while the two horizontal axes show the payoff obtained when the players' moves are DC and DD.

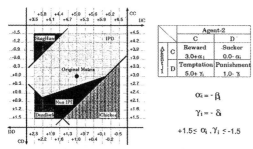

Figure 7: Scores and the Categories of Games

With the limitation above, simulations show that the world will fall into one of the followings and become stable.

Cooperative world

The left graph in figure 8 shows the average point obtained by agents in this case. At the earlier generations, the world is not stable yet, and many strategies appear and disappear. As the strategies are evolved, the scores are also evolved and gather to the score shown in a table of the figure 9 (right-side). Figure 9 (left-side) shows the distribution of scores at generation 90,000. All of the scores are in Staghunt area, because it gives more point (4.5) for cooperative strategies, and smaller point (3.5) for a defective strategy even if it defects cooperative strategies.

Defective world

With much smaller probability, defective strategies dominate the world. The right graph in figure 8 shows the average point obtained by agents in this case. Even in the defective world, the scores also gather to Staghunt area shown in the figure 9. When defective strategies dominate the world after the parameters in scores are fully evolved toward the values shown in a table in figure 9 (right-side), the defective strategies can get much payoff (2.5) by defecting each other, and cooperative strategies can not invade into the defective world unless many cooperative strategies are created at the same time by

mutations, which are almost impossible with the parameters shown in the table 3.

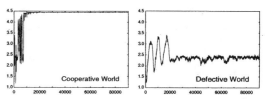

Figure 8: Average point of Agents (from 0 to 90000)

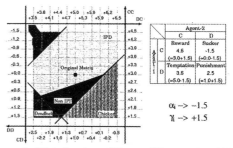

Figure 9: Distribution of Scores (Generation 90000)

Simulation Results with Narrower Range

When moving the parameters in the scores as shown in the previous subsection, all scores gather to Staghunt area and they do not move any more. Therefore, we simulated the co-evolution model with narrower range of the parameters as shown below.

$$\alpha_i + \delta_i = 0, \quad \beta_i + \gamma_i = 0, \quad -1.0 \le \alpha_i, \beta_i, \gamma_i, \delta_i \le 1.0$$

Figure 10 shows the new relationship between scores and categories of games, which is a port of the figure 7.

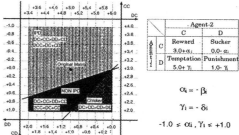

Figure 10: Scores and the Categories of Games

Figure 11 shows the average point and the number of scores, and the detail of the change from generation 25000 to 35000. In these figures, the left vertical axis shows the average point and the right vertical axis shows the number of scores. As shown in these figures, the average of the average point throughout the simulation is almost three, and this means that the world is almost dominated by cooperative strategies. As shown in the figures, the average point becomes worse when the number of scores increase.

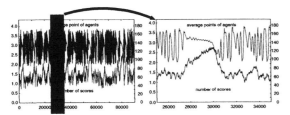

Figure 11: Average Points and the number of Scores

Figure 12 shows the ratio of the scores with in the range below which is close to the border with the Staghunt area. As shown in the figure, scores in the range always occupy the most of the population of scores.

$$-1.0 \leq \alpha_i, \beta_i \leq -0.8, \quad +0.8 \leq \gamma_i, \delta_i \leq +1.0$$

Figure 13 shows the distribution of scores. As shown in these figures, the scores repeat the cycles that the scores once gather to the area shown above, and then disperse. When the scores gather to the area, most of the strategies are cooperative, and the value of the parameter α_i in the score has no influence on their payoffs. Therefore, the value of α_i can be mutated freely (neutrality of mutation). However, this mutation makes it easier for defective strategies (created by mutation of strategies) to invade the cooperative world. Then, the number of the defective strategies increases and the average point becomes worse at this phase. By the invasion of the defective strategies, the scores that allows the invasion die out, and the world is dominated by the scores which are severe against defective strategies again.

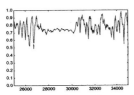

Figure 12: Ratio of Dominant Scores (25000-35000)

Conclusions

Simulations of the co-evolution model showed that scores of IPD games are gradually evolved toward cooperative scores, which give more payoffs for cooperative scores and less payoffs for defective strategies as the strategies of the agents are evolved, under all three situations we have tested. With these cooperative scores, cooperative strategies dominate the world with very high probability.

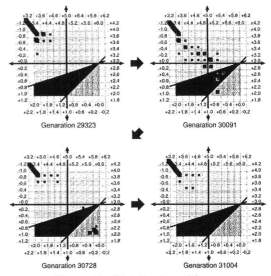

Figure 13: Distribution of Scores

When the range of the values of the scores are limited to more strict dilemma game areas, we can observe cycles as follows. First, all values of all scores move to the cooperative scores which give less payoffs for defective strategies as described above. Under the cooperative circumstances, values of scores which have no influence on cooperative moves (namely values which decide the payoffs for defective strategies) are mutated freely (neutrality of mutation). Then, this mutation allows invasion by defective strategies created by mutations. Then the cooperative scores which allow the invasion die out, and the world is dominated by the cooperative scores which are severe against defective strategies again.

We believe that cycles we observed in our co-evolution model are also observed in the real world, though there are many future works left. The behavior of the model is very sensitive to all parameters used in the simulations. We need to analyze the behaviors of the model varying the parameters fixed throughout the simulations in this paper. Furthermore, we need to investigate how the scores are evolved when the agents are placed in a world with a concept of distance, and costs for maintaining the scores obtained by the co-evolution.

References

[1] K. Lindgren, Evolutionary Phenomena in Simple Dynamics, *Artificial Life II*, (1991) pp.295–312.

[2] David B. Fogel, Evolving Behaviors in the Iterated Prisoner's Dilemma, *Evolutionaly Computation Volume 1(1)*, (1993) pp.77–97.

[3] Y. Yamaguchi, T. Maruyama and T.Hoshino, High Speed Hardware Computation of co-evolution Models, *5th European Conference on Artificial Life*, (1999) pp.566–574.

On the Effect of "Stock Alerts" in an Agent-Based Model of a Financial Market

Norberto Eiji Nawa[1,2], **Katsunori Shimohara**[1,2] and **Osamu Katai**[2]

[1]Dept. 6, ATR Human Information Processing Research Labs.
2-2 2-chome Hikari-dai, Seika-cho, Soraku-gun, Kyoto 619-0288, Japan
[2]Graduate School of Informatics, Kyoto University
Yoshida-honmachi, Sakyo-ku, Kyoto 606-8501, Japan

Abstract

Nowadays, an increasing number of web sites allow people to directly trade stocks and several other types of investments in financial markets through the Internet. These sites offer many conveniences to the customers using the network infrastructure, such as "stock alerts", which are mechanisms that promptly notify the customer when a certain stock reaches a threshold price value, set in advance by himself. The expected action from the stock alert user, upon receiving this notification, is to place an order to buy or sell stock shares, as the threshold price is supposedly associated with the profit the user is aiming for. The question addressed in this paper is how features like "stock alerts" affect the overall behavior of a financial market, as the number of individual stockholders grows due to the increased access to the market allowed by the Internet. Initial simulation results with an agent-based model of a financial market are presented.

Introduction

Trading stocks has never been so easy. Nowadays, an increasing number of web sites allow people to directly trade stocks and monitor market indexes and stock prices through the Internet. While financial newspapers and magazines may still be the investors' first source of information, many web sites offer services that deliver up-to-date news to their customers, making information broadly accessible, and what is more important to be noticed, practically instant. Despite their recency, financial web sites may strongly reinforce the popularization of the stock trading as an individual activity, a phenomenon that is already witnessed in the USA (New York Stock Exchange 1998), where the number of individuals owing corporate stocks, through mutual funds or other types of accounts, has been increasing continuously.

Taking this into account, it seems natural, as much as inevitable, that in the future the behavior of individual stockholders will play an increasingly major role in defining the market aggregate behavior. Interestingly, up to now mainstream economics has largely overlooked issues concerning the rationality and cognitive power of the individual agents, building models that regard economic agents as perfect decision makers, empowered with unlimited error-free decision making mechanisms, disregarding any factor of psychological or social nature that could possibly influence their reasoning processes.

Specifically in the context of financial markets, experiments conducted by Fischhoff and Slovic (Plous 1993) showed that the majority of the subjects were sure they could predict future trend changes in the price of a stock, if given information about stock prices and trends, though less than half of the subjects could actually do so. The work concluded that the forecasting errors were mainly due to the overconfidence of the subjects in their predictions, which illustrates well the fact that to assume perfect decision making by the economic agents is too restrictive a condition, since other factors might affect the reasoning process conducted by them. These issues have been long discussed in the field of economics (Simon 1982). The question that arises now, regardless of which economic theory explains better the reality of markets, is how the popularization of stock trading brought by the development of the Internet infrastructure, as an activity held by individuals investors, will affect the behavior of financial markets? The Internet introduces a new *tempo* to the system, as information and action are almost immediate, thus it is important to clarify its influence in financial and other types of markets.

This increase in the visibility of the market may also reinforce social movements in where investors have their behaviors influenced by other investors' behaviors. As Shiller points out (1989), the idea that social movements such as fashions and fads have deep implications in the sometimes idiosyncratic behavior of an asset's price is not consensual among the different schools of thought in economics. Indeed, the new reality may turn out to be even more distinct from the theory.

In this paper, the influence of a feature found in some of the major financial sites, which we will generally call "stock alert", is investigated by means of simulations with an agent-based model of a stock market, and some preliminary results are presented.

Stock Alerts

Conceptually, a stock alert is a very simple mechanism offered by financial sites like E*Trade[1], where it is called *smart alert*, and by Yahoo! Finance[2], where it is called *instant stock alert*. It works as a "watchdog" on a given stock's price; the price series is constantly monitored, and as soon as it achieves a certain threshold price value, the stock alert sends a notification to the customer, by email or through a ticker displayed in her computer. The stock and the threshold price value are freely customized by the user.

It is fair to assume that, in the large majority of the cases, an action from the customer will immediately be triggered by the stock alert's notification; if there are no time requirements involved in the action of buying or selling stocks, i.e. if there is no need to the customer react quickly, much of the significance of using such a feature vanishes away. In a sense, the stock alert user may even feel compelled to correspond fast; it may strengthen the ludic aspect of trading stocks, as in lottery games or other wagering markets.

Moving to the point-of-view of a stock alert user, there are some aspects to be considered. Online traders may have several analytic tools at hand to help them make their investment decisions, and in fact, some web sites offer such packages for the subscribers. Nevertheless, the population of individual traders can still be considered highly heterogeneous, a mix of sophisticated traders and more simplistic ones. Conjecturing on the characteristics of ordinary individual stockholders, it is difficult to envisage a scenario where the majority of the investors make complex calculations to find out a price value that will possibly maximize the profit within the desired period time. Ordinary investors may rather set their strategies and objectives according to subjective perceptions of the market dynamics, on what they "feel" would be most likely to happen to the stock price according to their past experience. They may also adopt more timid strategies and be more susceptible to behavioral changes on the arrival of economic and political news that put their investments at stake.

Another point that could be conjectured concerning stock alert users is that they will have expectations and beliefs transcripted in the form of price values, which in the ultimate level synthesize several factors: what the investor thinks it is a *feasible* value to be reached, which is a belief that may vary with time, as the traders observe the market develop and their needs change; the period of time the investor is willing to wait until he can accomplish the trade, which is an individual demand. What is most important to notice is that *prices* in monetary currencies are values that

are usually dealt with in several moments of a person's daily life. It is relatively easy to give a concrete value in terms of price for subjective and personal notions such as "expensive" or "cheap", "worthy" or "good profit", though people may not be aware or able to explain why and how they arrived to such values. Setting up price values as conditions for taking actions, as in the framework prepared by stock alerts, may facilitate traders to work with genuine and intuitive evaluations of worthinesses.

Stock alerts (indeed, online brokering mechanisms in general), thus, can be considered to be fast two-way connections linking a heterogeneous population of investors to the market. On the one hand, it quickly delivers information to the investors, and due to the very situation in which they are immersed (following the rationale that those who need to react immediately would be the most eager to make use of such convenience), it demands a quick reaction. The effect this feature may bring to financial markets, in the form of an increased influence by individual investors, is still to be confirmed. Though this may sound mere speculation wandering in the gray area of popular economic models, it should be recalled that the impact of the Internet in financial markets is still unclear, as it brings a complete new set of characteristics and dynamics to the markets (see (Varian 1998) for a short review of works in the field).

Experiments

The aim of this set of experiments was to identify eventual differences in the dynamics of the stock price due to the introduction of investors that behave as if they were using a stock alert, i.e. buying shares when the price gets below a lower limit and selling when it gets above an upper limit.

Experiments were performed using our implementation of the Santa Fe Artificial Stock Market (Arthur et al. 1997), using the Swarm Simulation System libraries (Minar et al. 1996). Our main interest in using the Santa Fe Artificial Stock Market is to have a basic, but nevertheless realistic, working framework of a stock market model. The market is comprised by a population of traders and a market specialist, whose function is to set up the price of the stock, based on the traders' orders to buy or sell stocks. The traders have to determine in each time step the composition of his investment portfolio, i.e. the distribution of money between a risk-free asset that pays a constant interest rate and stock shares. The dividend returned by the stock shares varies with time, following an autoregressive process unknown to the traders.

Several economic indicators reflect the market's state and are made available to the traders to guide their decisions; these indicators convey information on the fundamental value of the stock, i.e. the relation between the current stock price and the dividend, and also information on trends, which is usually used by

[1]http://www.etrade.com
[2]http://finance.yahoo.com

brokers who perform technical analysis. The traders in the model make use each one of a classifier system (Holland *et al.* 1986), whose IF <condition> THEN <prediction> rules determine the composition of the portfolio. A rule will fire if the current market status matches its <condition> part. The <prediction> part calculates the rule's forecast of the stock price in the next time step (p_{t+1}) plus the dividend (d_{t+1}), $E[p_{t+1} + d_{t+1}]$. This estimation is obtained through a linear function of the type $a_i(p_t + d_t) + b_i$, that take as arguments the current values of stock price and dividend. This estimation is fed into a CARA utility function (Constant Absolute Risk Aversion) to calculate the amount of shares the trader will buy or sell. The set of rules is periodically optimized by means of a genetic algorithm (Holland 1992).

It should be recalled that the main point clarified by Arthur *et al.* is the existence of two market dynamic situations, which are realized depending on how frequently the traders are allowed to learn, i.e. update their forecasting rules according to the genetic operations. Briefly explaining, they found that under low learning rates, a market regime characterized mainly by low trading volume, relatively low price variations, and the absence of technical trading emerged. This was named *rational-expectations regime*. As the learning frequency was raised, a more realistic market regime emerged, with the occurrence of temporary price bubbles, crashes and technical trading. This was called the *complex or rich psychological regime*. We attempted to reproduce the rich psychological regime in our own implementation by setting up a suitable set of values to the model parameters.

Two distinct simulation sessions were conducted. In the first session, a market with $N = 50$ traders, each one equipped with its own classifier system, was let run for a 300,000 days. These traders were named "complex traders". As we wanted to investigate eventual effects brought by the use of stock alerts, in the second session the market had $N = 50$ traders, from which 8% of the traders were models of stock alert users and the rest constituted by "complex traders". This model was first let run for 150,000 days with the stock alert users kept inactive, reproducing an identical situation to the first session (with the difference in the total number of traders). In the second period of the simulation, the stock alert users were allowed to participate in the market. This scheme supposedly allows the "complex traders" to learn about the market before the stock alert users get in to action.

The "complex traders" optimized the rules using a simple genetic algorithm. The frequency of update (learning ratio) was in average 1/1000. The total number of rules was set to 100. Whenever the learning algorithm was activated, the rules were ranked according to their forecast accuracies and 5% of the low ranked rules were optimized. To carry out a rule optimization, first a rank-selection scheme was used to choose a sec-

ond rule from the rule base. The information contained in the second rule was then "mixed" with the original rule; subsequently, the reformed rule would suffer mutation in its contents. The initial values of the parameters a, b of each rule were randomly selected from a uniform distribution in the interval $[0.7 \cdots 1.2]$ and $[-10 \cdots 19.002]$, respectively. (The rule update procedures of the genetic algorithm and the values of other relevant parameters of the model were the same as in (Arthur *et al.* 1997).)

Modeling the Stock Alert Users

Compared to the "complex traders", the models of the stock alerts users have a much simpler structure. They were basically built on two threshold prices: MBuy and mSell. Whenever the market price of the stock share was below the MBuy, with half chance the stock alert users put an order to buy one unit of a share. In the same way, whenever the price was above the mSell, with half chance the stock alert users put an order to sell whatever amount they had at that moment of stock shares. This is in accordance to the hypothesis that stock alert users in real markets may make offers or bids to the market as soon as the stock price reaches a given threshold.

MBuy and mSell were initialized as follows. A random number would be first sampled from a normal distribution with mean 100 and variance 15. This range of initial values was arbitrarily chosen from what it was thought reasonable, after some inspection of the prices practiced in the market model. Then, a second number would be generated by adding or subtracting, with equal chances, a random number from the interval $[-5 \cdots 5]$ to the first number. The lowest of the two numbers would be the initial MBuy and the other the initial mSell.

Stock alert users were also equipped with a simple parameter adjusting algorithm, that worked as follows. In case of a successful transaction, the stock alert user would count the number of successful orders, and when that number reached a threshold, S_i, he would increase his margin, aiming larger profits, and reset the number of successful orders. If the market can not clear, the orders are unsuccessful and each stock alert user would increment the number of deceptive orders in a row. When the number of unsuccessful orders reached a threshold, U_i, the user would decrease his margin, demanding more modest profits. The number of unsuccessful orders would be reset whenever a successful order was accomplished, or when it reached the threshold U_i.

The value of MBuy was decreased when the last order to buy shares was successful and the number of successful orders $= S_i$ according to the following rule:

$$\text{MBuy}_{\text{new}} = \text{MBuy}_{\text{old}} - |\text{MBuy}_{\text{old}} - p_t| \times \delta \qquad (1)$$

Correspondingly, if the last order to buy was unsuccessful and the number of unsuccessful orders $= U_i$,

MBuy would be increased as follows:

$$\texttt{MBuy}_{\texttt{new}} = \texttt{MBuy}_{\texttt{old}} + |\texttt{MBuy}_{\texttt{old}} - p_t| \times \delta \qquad (2)$$

For the upper limit, if the last order was a successful order to sell and the number of successful orders $= S_i$, mSell would be incremented as follows:

$$\texttt{mSell}_{\texttt{new}} = \texttt{mSell}_{\texttt{old}} + |\texttt{mSell}_{\texttt{old}} - p_t| \times \delta \qquad (3)$$

Correspondingly, if the last order to sell was unsuccessful and the number of unsuccessful orders $= U_i$, mSell would be decreased as follows:

$$\texttt{mSell}_{\texttt{new}} = \texttt{mSell}_{\texttt{old}} - |\texttt{mSell}_{\texttt{old}} - p_t| \times \delta \qquad (4)$$

The parameter adjusting rules described above are very simple, based on the rationale that an unsuccessful order is a signal that the trader is being greedy, leading her to decrease the margin, in order to increase the chances of having a successful order in the future. On the other hand, too many successful orders makes the trader increase her margin, trying to obtain higher profits. They do not perform any technical analysis nor directly try to detect price trends.

Values of S_i, U_i were randomly sampled in the interval $[10 \cdots 20]$, and δ was initialized for each stock alert user by sampling random numbers in the interval $[0.0005 \cdots 0.001]$.

Results and Discussion

Experiments were realized with the model described above and were divided in two sessions; in the first session, the population of traders was composed only by "complex traders" (identified by only CT), and in the second session, "complex traders" and "stock alert users" (identified by CT and SAU) were present in the market. In each session, 20 instances of the model were run. A set of 20 different random seeds was prepared to initialize the models in both sessions, so that the dividend series d_t would be pairwise identical. The results obtained are summarized in Tables 1, 2, 3, and 4[3].

Table 1: Summary statistics for the stock price, after 300,000 days (average of 20 runs)

	only CT	CT and SAU
mean	93.35	93.63
standard deviation	6.83	6.94
kurtosis	0.88	3.94
skewness	-0.11	-0.44

The first point that calls attention is related to the higher kurtosis presented by the model with stock alert users, in Table 1 and 3, relative respectively to the

[3]Kurtosis is defined as $\mu_4/\mu_2^2 - 3$, and skewness is defined as $\mu_3/\mu_2^{3/2}$, where μ_i denotes the ith moment $\langle x^i \rangle$.

Table 2: Summary statistics for the stock price, first 150,000 days (average of 20 runs)

	only CT	CT and SAU
mean	93.07	93.75
standard deviation	7.09	7.19
kurtosis	1.13	1.16
skewness	-0.10	-0.11

Table 3: Summary statistics for the stock price, last 150,000 days (average of 20 runs)

	only CT	CT and SAU
mean	93.63	93.50
standard deviation	6.54	6.65
kurtosis	0.48	5.76
skewness	-0.09	-0.72

total period of simulation and the last 150,000 days, i.e. the situation where the complex traders cohabitate the market with the stock alert users after being alone for 150,000 days. Apparently, this is due to a simulation artifact, generated by the sudden appearance of the stock alert users in the market. Looking to the results in Table 4, which takes in consideration only the last 100,000 days, i.e. leaving out the singular period when the stock alerts suddenly become active, indeed both kurtosis and skewness values are less than the values presented in the situation where only "complex traders" are present. For the agent-based model of the market, the market becomes more silent when the stock alert users are around.

Table 5 shows the mean and standard deviation values for MBuy and mSell of the stock alert users at the end of the simulation, averaged through the 20 runs.

It is important to notice that the complex traders and the stock alert users are inherently distinct in what relates to the reasoning process that leads to an action. The complex traders make use of a CARA utility function to determine the composition of the portfolios; on the other hand, the stock alert users adopt a simplistic (intuitive) strategy of acting on the market when the price is higher/lower than their individual reservation prices. As mentioned before, it is possible that in the

Table 4: Summary statistics for the stock price, last 100,000 days (average of 20 runs)

	only CT	CT and SAU
mean	93.59	93.44
standard deviation	6.48	6.35
kurtosis	0.49	0.33
skewness	-0.09	-0.08

Table 5: Initial and final mean and standard deviation for MBuy and mSell (average of 20 runs, population composed of 4 stock alert users and 46 complex traders in each run, after 300,000 days)

	$\overline{x}_{t=0}$	$\sigma_{t=0}$	$\overline{x}_{t=3e5}$	$\sigma_{t=3e5}$
MBuy	98.23	4.49	80.94	0.87
mSell	100.60	4.28	101.40	4.07

real world, stock alert users perform complex calculations to set the values of MBuy, mSell. However, we could conjecture that the majority of the investors are "modest" players; this does not imply that they are not willing to make big profits, but that the ordinary investor is prudent and aim for conservative gains. Recall that in (Arthur et al. 1997), it was found that the "complex psychological regime" was a product of the high learning rate of the traders, put in another way, the trader's eagerness to incorporate knowledge on the market in their reasoning processes. If the arrival of conveniences such as stock alerts leads to the creation of a class of stockholders that are well informed but at the same time circumspect when dealing with their investments, less prone to perform risky actions but nevertheless quick in acting on the market, the Internet could eventually induce a "calming down" effect in the stock prices.

Conclusions

In this paper, the influence of "stock alerts" in an agent-based model of a stock market, which is based to a great extent on the Santa Fe Artificial Stock Market, was investigated and preliminary results were presented.

There are some weaknesses in this study that should be pointed out. First, the assumptions made on the stock alert users' behaviors are not based on observations, nor are they derived from market indications, neither have anything close to a strong theoretical reasoning to support them. Rather, they are just speculative conjectures, that nevertheless are sound with the scenario pictured by common sense reasoning. Whether these assumptions are consistent or not is still an open question, for which to answer properly would require the collection and analysis of behavioral data from online stock traders. Following to that, it would be interesting to try similar experiments in a larger model to detect eventual issues related to the scale.

The popularization of stock trading as an individual activity may increase the amount of "noise trading" in the market, as defined in (Long et al. 1993), due to the arrival of a large population of non-experts into play. How the usage of stock alerts will influence the amount of "noise" in the market should also be an interesting topic of study.

While still an open issue, the question of how the Internet will change the behavior of financial markets should include also a discussion on the impact of computational autonomous agents. As advocated in (Kephart et al. 1998), it is likely that in a close future the Internet will be populated by software agents, who will interact with other software agents and with human users in order to facilitate transactions and disseminate information. Such a scenario is as interesting as it is unknown, and deserves further investigation.

References

W. Brian Arthur, John H. Holland, Blake LeBaron, Richard Palmer, and Paul Tayler. Asset pricing under endogenous expectations in an artificial stock market. In W. Brian Arthur, Steven N. Durlauf, and David A. Lane, editors, *The Economy as an Evolving Complex System II*, pages 15–44. Addison-Wesley, 1997.

John H. Holland, Keith J. Holyoak, Richard E. Nisbett, and Paul R. Thagard. *Induction - Process of Inference, Learning, and Discovery*. MIT Press, Cambridge, MA, 1986.

John H. Holland. *Adaptation in Natural and Artificial Systems: An Introductory Analysis with Applications to Biology, Control, and Artificial Intelligence*. MIT Press, Cambridge, MA, 1992. First Edition, The University of Michigan Press, 1975.

Jeffrey O. Kephart, James E. Hanson, and Jakka Sairamesh. Price and niche wars in a free-market economy of software agents. *Artificial Life*, 4(1):1–23, 1998.

J. Bradford De Long, Andrei Shleifer, Lawrence H. Summers, and Robert J. Waldmann. Noise trader risk in financial markets. In Richard H. Thaler, editor, *Advances in Behavioral Finance*, pages 23–58. Russel Sage Foundation, 1993.

Nelson Minar, Roger Burkhart, Chris Langton, and Manor Askenazi. The swarm simulation system: A toolkit for building multi-agent simulations. Santa Fe Institute Working Paper #96-06-042, at http://www.santafe.edu/projects/swarm/overview.ps, 1996.

New York Stock Exchange Fact Book. At http://www.nyse.com/pdfs/factbook/factbk98.pdf, 1998.

Scott Plous. *The Psychology of Judgment and Decision Making*. McGraw-Hill, 1993.

Robert J. Shiller. *Market Volatility*. MIT Press, 1989.

Herbert A. Simon. *Models of Bounded Rationality: Behavioral Economics and Business Organization*, volume 2. MIT Press, 1982.

Hal R. Varian. Effect of the Internet on financial markets. At http://www.sims.berkeley.edu/~hal/Papers/brookings-paper.pdf, September 1998.

The first author receives support from CNPq by grant number 200050/99-0.

Semiotic schemata:
Selection units for linguistic cultural evolution

Frédéric Kaplan

Sony CSL - Paris - 6 Rue Amyot, 75005 Paris
LIP6 - UPMC - 4, Place Jussieu F-75252 Paris
E-mail: kaplan@csl.sony.fr

Abstract

Words, like genes, are replicators in competition to colonize our brains. Some, by luck or thanks to their intrinsic qualities, manage to spread in entire populations. In this paper we take the approach of cultural selectionism to study the emergence of communication systems in a population of agents. By studying simple models of word competition in noisy environments, we define the basic dynamics of such systems. We then argue for their generality and introduce the notion of semiotic schemata, generic replicators that account for the different competitions that are going on during lexicon formation. Eventually, we present a synthesis of the dynamics using this new formalism.

Introduction

Genetic and cultural systems can both be seen as complex evolving dynamic architecture. In this paper we will discuss a particular paradigm for understanding the dynamics of cultural systems: *selectionism*. The comparison between genetic evolution and cultural evolution , popularized by Dawkins's *memes*, has proved to be fruitful (Dawkins, 1976; Dennett, 1995; Blackmore, 1999). By analogy with genes, Dawkins defines *memes*, as cultural replicators. Dawkins defines a replicator as "any entity in the universe which interacts with its world, including other replicators, in such way that copies of itself are made" (Dawkins, 1984). In genetics, replicators are single genes or fragments of genetic material. Evolutionary genetics study the *competition* between genetic replicators, how some of them are *selected*, how some of them *disappear*. Metaphorically, we can talk about the survival of some replicators and the death of others. From a similar perspective we could say that cultural replicators are in competition to colonize our brains. Like for genetic replicators, criteria such as *fecondity* or *fidelity* in the copying process, are useful notions to understand the victories or defeats of some memes against others.

The idea of *cultural selectionism* has been applied in the study of a particular kind of cultural evolution: the evolution of communication systems and languages. Indeed languages, like organisms, could be seen in competition with one another. They try to "survive" by being used by speakers. This particular perspective, different from contemporary linguistic approaches, is not really new. In 1937, Arsene Darmesteter wrote "the life of words" already announcing this paradigm (Darmesteter, 1937). Today, several researchers in Artificial Life talk of their work in similar terms (Batali, 1998; Hurford, 1998; Kirby, 1999a; Kirby, 1999b; Steels, 1997).

In our previous work, we have explore the emergence of complex communication systems, in particular the coupling between creation of grounded categories and lexicon formation (Steels and Kaplan, 1999c; Steels and Kaplan, 1999b; Steels and Kaplan, 1999a), and the effect of noise on the evolution of such systems (Steels and Kaplan, 1998b; Steels and Kaplan, 1998a). This paper is a synthesis, using the cultural selectionism paradigm, of the results we have obtained with complex architectures. We study basic model of linguistic evolution, simple enough to account for most experimental results in the field. In identifying these basic dynamics, we try to point out the different competitions that are going on during lexicon self-organisation. The main difficulty is to define the *right selection unit*. Like for memes, it is uneasy to precisely define what are the replicators in the cultural linguistic evolution. Should we consider competition between words, meanings or larger parts of languages?

The paper is organised as follows. In the next section, we study very simple models showing the competition between words for naming the same referent when there is noise in the environment. Then, we move to more complex architectures and analyse how these dynamics evolve when a lexicon is emerging to name a set of objects under the presence of noise during word transmission. We then argue that the dynamics identified are general and apply to other kinds of competition that are present during the self-organisation of a communication system. To account for all these different competitions, we introduce the notion of semiotic schemata, which are general replicators for linguistic evolution. We present a synthesis of the dynamics using this new formalism.

Competition between words

In this section, westudy the basic dynamics that enable one linguistic convention to be collectively chosen by a population of agents.

Positive feedback loop

Model 1.1 Each agent a in a population of N agents is defined by a single *preference vector* $(x_a^1, x_a^2, .., x_a^N)$. x_a^i represents the score that agent a gives to the word convention i. Agents interact through a very simple protocol. Two agents are picked at random in the population. One agent is speaker and the other one is hearer. When an agent is speaker, it uses the convention associated to the highest score in its *preference vector*. The convention is transmited to the hearer, and the latter simply increases by 1 the score of the convention used by the speaker in its own *preference vector*. Initially the population starts with N agents, each agent a having a single bias for one preferred convention which is modelised by a vector of size N $(0, 0, .., 0, 1, 0, .., 0)$.

Exp 1.1 (N=50, 1 run). Figure 1 shows the competition of the different word conventions for 50 agents trying to impose their word. The positive feedback loop introduced in this simple model creates a winner-take-all situation where one word dominates. The word that finally wins has no special properties and any new run of the simulation would lead to the selection of a different word.

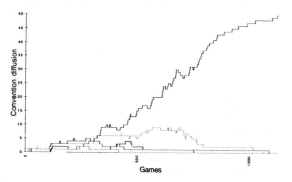

Figure 1: Word competition in a population of 50 agents. Each curve represents the diffusion of a word convention in the population. Eventually, one convention dominates being used by the 50 agents (Exp 1.1)

Implicit evaluation

Model 1.2. We now consider a set of words of unequal qualities. For instance, some are more resistant to noise. We model that in a very crude way by associating to each convention a mutation probability $P_m(W_i)$ between 0 % and 100%. For each game, a random test is done to check whether the word has been transmitted successfully or not. In case of failure, the word is transformed to another word randomly picked among all the possible ones.

For this experiment we choose a mutation probability that grows linearly with the word number. Thus for word W_i, the formula is:

$$P_m(C_i) = \frac{i}{\text{Number of words}} \qquad (1)$$

Exp 1.2.a (N=50, 1000 runs) Figure 2 shows, for 1000 simulations, the distribution of the winning words for a population of 50 agents. Words with low mutation rates have been selected. An external observer could say that the agents are doing a *collective optimisation*. They are naturally converging towards the best words. The phenomena is based of a *implicit evaluation* of the solution similar to the one described for forraging behavior in ant colony (Dorigo *et al.*, 1997). It means that the agent are not evaluating individually the quality of each word for choosing the more robust ones. Illadapted words simply mutate more often and cannot propagate as easily as the others.

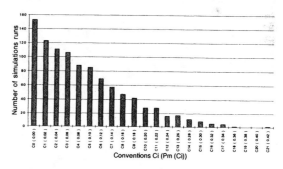

Figure 2: Distribution for 1000 simulation runs of the winning words for a population of 50 agents (Exp.1.2.a)

Reorganisation in the presence of an agent flux

In some cases though, the population might not converge towards the most robust words. The most important danger is *premature convergence*. If, for instance, a very good word appears in the population more lately during the experiment, it is probable that it will not be picked up because the positive feedback loop would have already caused the agents to converge towards a suboptimal one.

To consider an open population where agents are entering and leaving the population can correct this effect. Indeed, new agents entering the population have no special preference for the dominant word. They can discover the best solution and maybe, if it is really more robust than the one currently dominating, the outsider might eventually win.

Exp 1.2.b (N=50, different runs for different P_r). The following experiment is the same as the previous one, excepted that an agent flux, defined by the probability P_r of replacing an old agent by a new one, is applied. We want to see if this flux leads to a better selection of the words. For several values of P_r, we measure the proportion of simulation runs that end up with one of the three best word dominating. Beyond a certain value of P_r, the flux is too high to achieve convergence. The reorganisation can only be active near the edge of this threshold. Figure 3 shows this effect. We can draw an analogy between this effect and the role of temperature in optimisation techniques such as simulated annealing.

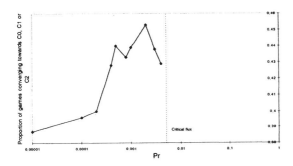

Figure 3: Proportion of simulation runs converging toward one of the three best solutions for different agent flux P_r (Exp 1.2.b)

In more complex models, an agent flux can also have a regularising effect. Because it increases the chance of picking up "good" linguistic conventions, conventions easier to learn will tend to be selected. Simon Kirby has illustrated this feedback loop on regularity in his simulation on the emergence of compositionality (Kirby, 1999b).

Conclusions

In this section we have identified the basic dynamics in the competition between different word conventions for naming one object.

- **Positive feedback loop**. If each agent is trying to induce the diffusion of each convention in the population, in order to use the one the most widely spread from its own point of view, then a positive feedback loop is created, leading to the domination of one convention. This dynamic is "blind" and does not prefer any convention *per se*.

- **Implicit evaluation of solutions**. But if some conventions are less easy to transmit, they will implicitly be left aside. Thus, best conventions tend to be chosen by the population.

- **Reorganisation and regularisation with an agent flux**. The presence of a flux of agent in the system avoids premature convergence. Better conventions (more robust, easier to learn) tend to be selected. The arrival of new agents enables a continuous parallel search for solutions that can replace the ones currently dominating. If needed it can cause a reorganisation in the communication system.

Competition during lexicon formation

In this section, we consider the case of the emergence of a lexicon: a mapping between a set of words and a set of objects.

Model 2.1. In this new model, the agents have to agree on names for a set of M objects. Each agent has an associated memory where are stored associations between words and objects. They use this memory to code an object into a word and to decode a word into an object. When several solutions are possible the agents choose the association with the highest score. Their associatiove memory is initially empty. Associations are progressively created as the agent interacts with other agents. As in the model of the previous section, a positive feedback loop enables lexicon self-organisation. Several experiments have shown that with such an architecture, a coherent lexicon emerges. Each word becomes associated to a single object and each object to a single name (Arita and Koyama, 1998; Steels, 1996; Steels and Kaplan, 1998b; Ferrer Cancho and Sole, 1998; Hutchins and Hazlehurst, 1995; Oliphant, 1997).

We consider a noisy environment where word transmission is difficult. Each word is an integer value between 0 and 1000. Each time a word is transmitted, a random number betwen $-B/2$ and $+B/2$ is added to the word. B is a measure of the global noise level. Each agent is equipped with a *filter* enabling him to select all the words in his associative memory which are at a distance D less than $D = B$. The structure of an interaction is the following:

1. The speaker randomly chooses an object o_1 between the different objects available and uses a word w_1 to name this object. If he doesn't have words associated with this object, the agent creates a new one (a random integer between 0 and 1000).

2. The word w_1 is transmitted to the hearer with an alternation between $-B/2$ and $+B/2$. The word heard is w_1'.

3. The hearer selects all the possible associations with a word close to w_1' (at distance less than B). If no association is available, the speaker indicates what was the subject and the hearer creates a new association between w_1' and the object o_1. If several associations are possible, the hearer chooses the one with the highest score: (w_2, o_2) .

4. if $o_1 = o_2$ the game is a success.

In case of success, the hearer increases the score of the association (w_2, o_2) with $+\delta$ and diminishes the score of competing associations (synonyms and homonyms) with $-\delta$. In case of failure, the hearer decreases the score of (w_2, o_2) with $-\delta$, the speaker indicates what was the subject and the hearer increases the score of the association (w_2, o_2) with $+\delta$, otherwise it creates it. Associations are initialy created with a 0 score. In the following experiment we take $\delta = 1$.

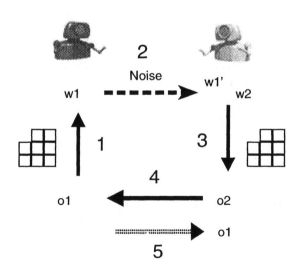

Figure 4: Interaction between agents using an associative memory (Model 2.1.).

Distinctivity

In the simple model of the previous section we have shown that collective dynamics lead to choose the "best" words to name an object. What are the best words in the current model ? A good word is a word than an agent will not confuse with another one that has a different meaning. A "good" lexicon should have set of words clearly distinct from one another depending on the object they name.

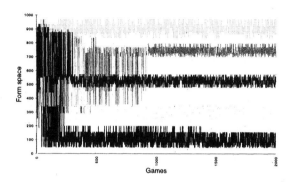

Figure 5: Evolution of the forms in the form space. After a first period of ambiguity, five well separated bands are forming to name each object (Exp. 2.1.a)

Exp 2.1.a $(N = 10, M = 5, B = 100)$ Figure 5 shows the evolution in the word space of the word associated with 5 objects in an experiment involving 10 agents. After an initial ambiguity period, five well separated bands in the word space are clearly indentifiable. Agents do not converge on a unique word form for each object. Each agent uses a different word. But as lexicon self-organisation is going on, these words tend to be very similar. For each object, they form a band in the word space which is cleary distinct form bands associated with other objects. No confusion is possible.

Figure 6 plots the same data as figure 5 showing the "average" word of each band. On this graph, it is easier to see the collective optimisation of distinctivity leading to a solution compatible with the level of noise present in the environment.

These results are somehow similar to the ones obtained by Bart de Boer (de Boer, 1997; De Boer, 1999). De Boer shows how the collective dynamics and noise lead a population of agents to converge towards a set of vowels optimally distributed in the phonological space in order to favor distinctiveness between them. Such emerging phonetic systems have high similarity with real ones as observed in natural languages.

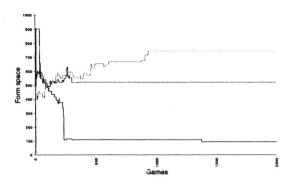

Figure 6: Evolution of the "average" forms in the form space (Exp. 2.1.a)

Compromise between distinctivity and robustness

Model 2.2 In the previously described experiments our model of a word - an integer - was very crude. In this section, each word is now a numeric chain of variable length. Each character of the chain is a number between 1 and 9. Noise is modelised by a probability of alteration P_m equal for each character. When a character mutates, it is simply replaced by a random character between 1 and 9.

As in the previous model, the hearer can look up in its lexicon for the chains that are "close" to the transmitted words. We define a distance D_c between word chains, similar to the traditional Hamming distance.

Let w_1 and w_2 be two words, the length of w_1 being either smaller or equal to the length of w_2. Let $w_1(i)$ and $w_2(i)$ be the character in position i in each of the chain. We define D_c as being the sum of the distance between the character of both chains to which is added 10 times their length difference, $l_2 - l_1$:

$$D_c(w_1, w_2) = \sum_i \|w_1(i) - w_2(i)\| + 10.(l_2 - l_1) \quad (2)$$

For instance the chains 1-4-5-2 and 1-4-5-7-3 are at a distance $5 + 10 = 15$. In the interaction the hearer selects the chains which are at a distance less than the threshold D.

We see that, with such a mechanism, too long or too short chains are naturally less adapted. Indeed, the longer a chain is, the more risk it has to be altered during transmission. In the first model, we have seen that such words generally loose the competition. But on the other side, if the lexicon is only composed of very short words, a single mutation might very often lead to confusion. A *compromise between word robustness and distinctivity* must be found: Short words are robust but easy to confuse, long words are easy to distinguish but difficult to transmit correctly.

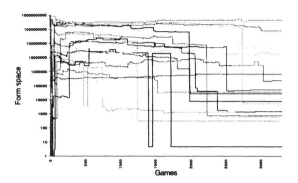

Figure 7: Example of evolution of the "average" words for 10 agents naming 20 objects with $D = 20$ and $P_m = 0.1$

Exp 2.2.a ($N = 10$, $M = 20$, $D = 20$, $P_m = 0.1$, **1 run**) Figure 7 shows the evolution of average words for 10 agents naming 20 objects. As no word contains the "0" character, we can still visualize the situation in a one dimensional word space. In this representation, the values between 1 and 9 represent 1 word character, between 11 and 99, two word characters, etc. Considering this formalism, a logarithmic scale is appropriate. Each new division shows a new class of words. On the graph we observe that for this single run, as expected, the words of intermediary length constitute the majority of the final lexicon.

Exp 2.2.b ($N = 10$, $M = 20$, $D = 20$, $P_m = 0.1$, **100 runs**) We have repeated experiment 2.2.a. a hundred times and analysed the distribution of all the words used by the agents after 5000 games (at this point, we have observed experimentally that the lexicon reaches a stable state). The results of the distribution of the word length are shown on figure 8. The distribution has a peak around words of length 3. Words too long or too short are less present in the final vocabularies.

Exp 2.2.c ($N = 10$, $M = 20$, $D = 5$, $P_m = 0.1$, **100 runs**) The result of another series of experiments with a reduced noise tolerance level ($D = 5$) are shown on figure 9. The peak is now for words of length 2. As the noise level tolerance is reduced, a larger set of shorter words can be used as long as the tolerance is sufficient to cope with the noise level. It is the case in the conditions of these experiments.

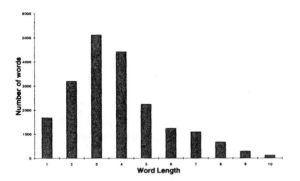

Figure 8: Distribution of word length on 100 simulation runs for 10 agents naming 20 objects with $D = 20$

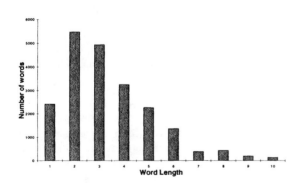

Figure 9: Distribution of word length on 100 simulation runs for 10 agents naming 20 objects with $D = 5$

Natural lexicon drift

We have seen with the model 2.1. that in a noisy environment, agents can converge on a stable system in which distinct bands in the word space are associated with distinct meanings. As we see in figures 5 and 6 this repartition in separated band does not evolve anymore once a stable solution has been found.

Exp 2.1.b ($N = 20, M = 2, B = 400, P_r = 0.01$) Graph 10 shows the evolution of the average form in the presence of an agent flux defined by a probability of replacing an old agent by a new one $P_r = 0.01$, for a population of 20 agents naming 2 objects. We see on the graph that the center of the bands are spontaneously evolving as new agents are entering the system. We will call this effect: the *natural lexicon drift*.

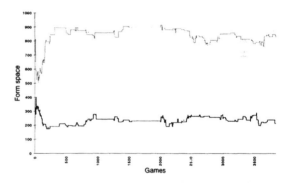

Figure 10: The natural lexicon drift. Spontaneous evolution of the "average" forms in presence of an agent flux (Exp 2.1.b)

This effect is easily understandable. A new agent tends to converge on words belonging to the existing bands for each meaning to express. But within this band, it has no reason to converge towards the exact center of the band. Thus the center is moving as the flux of new agents enters the system. The higher the agent tolerance on noise, the higher the amplitude of this drift.

These form bands are evolving spontaneously without any functional reasons. But this does not exclude that external pressures can direct these dynamics in a direction or another. The natural lexicon drift provides novelty and thus can lead to a more efficient reorganisation if needed.

We have shown in (Steels and Kaplan, 1998a) that this effect was active for more complex agent architectures including, like the simple agents of our model, a tolerance mechanims to cope with noise in the environment. Scott C. Stoness and Christopher Dircks have reproduced these results using another architecture based on neural networks (Dircks and Stoness, 1999). This shows that for such systems, collective dynamics are much more important than actual architecture and implementation details.

Conclusions

In this section we have explored the basic dynamics identified in the previous section with more complex models. Our conclusions are the following:

- **Distinctivity**. Noise during word transmission favors sets of words that are cleary distinct from one another when they mean different things. We experimentally observed the emergence of well separated bands in the word space. Each band is associated with a different word meaning.

- **Compromise between distinctivity and robustness**. When words can have different lengths, a compromise must be found between distinctivity and resilience to noise. Short words are easy to transmit but easy to confuse, long words are difficult to transmit correctly but are easily distinguishable from one another. We experimentally observe the convergence towards words of intermediary sizes.

- **Natural lexicon drift**. In the presence of noise and agent flux, we experimentally observe a spontaneous not functional lexicon evolution. This continuous exploration of the form space can lead to a more efficient reorganisation of the lexicon if needed.

Synthesis: the semiotic schemata
Other kind of competitions

We have considered in the previous sections very simple models in which meanings are simply discrete symbols without any particular properties. In some more complex architectures (Steels and Kaplan, 1999c; Steels and Kaplan, 1999b; Steels and Kaplan, 1999a) we have shown that when meanings are categories discriminating properties of the objects of the world, additional competitions can be observed. Some categories might be general and other specific. For instance, one might be used for describing a very particular shade of green, and another one for describing green objects in general. Depending on the environment and on the objects that need to be discriminated, sometimes general categories will be sufficient, sometimes specific ones will be needed.

Categories, like words, are competing with one another. Considering ompetition between isolated categories is not sufficient. The quality of a category needs to be evaluated regarding the category set to which

it belongs. A specific category might survive if other categories are present to "back it up".

Associations linking words and meanings of different qualities can also be seen as competing units. A widely spread association has a real advantage, even on an association linking a very solid and easily distinguishable word with a very often used category.

In Simon Kirby's work, more complex system are competing with one another (Kirby, 1999b). He discusses the victory of a compositional system on a idiosyncratic one because the first one, being more regular, is easier to learn.

Eventually, several competitions can be observed at the same time, each of them involving part of languages of different sizes and types.

Semiotic schemata

Our point in this paper is to suggest that even though the kind of replicators involved in language emergence can be very different, the dynamics are merely always the same ones. These dynamics are the same as the one we have identified with the simple models of word competition that we have studied in the previous sections.

In (Steels and Kaplan, 1999b; Steels and Kaplan, 1999a), we have introduced the notion of *semiotic landscape* to analyse the complex dynamics involved in the emergence of word meanings. A semiotic landscape is a complex network linking objects, categories and words with associations of "different weight". If, for instance, the weight between a word and a meaning is strong in a semantic landscape its means that this associations is frequently observed in the agents behavior. All the cultural replicators, that we have identified in our experiment, can be seen as partial specification of a semiotic landscape. Words and meanings are simple nodes. Associations are couple of nodes and their links. Lexicons are more complex configurations.

Analysing what should be the right selection unit for studying artificial genetic dynamics, John Holland has introduced the notion of *schema* (Holland, 1995). A schema is a partial specification of the genome. By analogy, we introduce the notion of *semiotic schema* as a partial specification of a semiotic landscape. A word is a semiotic schema, a meaning is a semiotic schema, associations, sets of distinct words and even set of associations are semiotic schemata.

Order of a semiotic schema

A genetic schema, in artificial systems like genetic algorithms, can be modeled as a chain using only three characters 0, 1 and * (Holland, 1995). Each character of the chain can be seen as a particular gene which can take two values 0 or 1 or be undetermined, in which case it takes the value *. For instance, if the length of the genome is limited to $L = 10$ characters, $s = (*, *, *, 1, *, *, 0, *, *, *)$ is a possible schema, specifying only the genes number 4 and 7.

Schemata can be compared by studying their spreading in the population. For instance, if the schema s is present in all the population members, it means that all the genomes have the value 1 for the fourth gene and the value 0 for the seventh one. *Landscapes* are representations of the distribution of schemata at a given state of the evolution. For each possible schema, the number of agents in which this schema is present can be plotted. In practice, such landscapes are difficult to draw as the number of possible schemata can be very high. For a genome of length L, 2^L different schemata are possible. Yet, the landscape metaphor is a good starting point to visualize the competition between schemata.

Unfortunately, the situation for semiotic schemata is a bit more complex. It seems that several kinds of competitions are going on in parallel involving schemata of different complexity. We can introduice the notion of *order* of a schema. Words, groups of words, meanings and groups of meanings are first order semiotic schemata. Associations and lexicons are second order semiotic schemata. For a given order, semiotic schema can be modeled exactly like genetic schema using a chain composed with the characters 0, 1 and *.

For instance, the results of Exp 1.2.a, where 50 words of decreasing quality were in competition, can be seen as the competition between schema of Length $L = 50$ where each character codes for a word in the word space. Figure 2 shows that schemas starting with "1" in the first positions have a higher fitness than others. Results of Exp 2.1.a can be seen as the competition between schema of length $L = 1000$ and show that schemata including equidistributed "1" in the word space have a higher fitness than schemata where "1" are close from one another. Results of Exp 2.2.b show that schemata including words of intermediary length have a higher fitness than others.

The same analysis can be done for schemata of order 2. In these schemata, each character is a possible association of the lexicon. The length L of these schemata is equal to the product of the number W of possible words and the number M of objects to name. Once a stable mapping has been found, the semiotic landscape of such systems is defined by M distinct peaks corresponding the the M objects to name.

Selection dynamics

Semiotic schemata are replicators. The more complex they are, the more difficult it is for them to replicate.

Their competition progressively structures the semiotic landscapes defining the common lexicon which is emerging. In particular kinds of word competition that we have studied in this paper, we observed three kinds of selection dynamics:

- **Individual choices select good schemata.** An agent has a way of evaluating semiotic schemata. The agent will use the ones that have proved to be efficient for communicating in past interactions. In the models we presented, a score was monitoring the success and the failure of each association and thus indirectly measuring their diffusion in the population. This dynamics create a positive feedback loop leading some semiotic schemata to be used more and more often. For this individual selection, schemata that are widely spread, resilient to noise and easy to learn have a selective advantage.

- **Agent flux ensures regularisation and reorganisation.** Individual selection is responsible of the lock-in effect on particular schemata. Because of premature convergence, the schemata chosen might not be the most efficient to communicate. The presence of a constant agent flux in the system puts additional pressure on the system for selecting really good schemata. An efficient schema might have been constructed by some individuals but appeared, later on, to be too difficult to transmit culturally to each new generation of agents. This agent flux creates a positive feedback loop on simplicity and therefore on regularity. The more regular and easy to learn a schema is, the more likely it is to pass the "generation bridge". The agent flux is also responsible for a continuous exploration of new possible schemata. Newborn agents might find simpler and more efficient solutions. If they are really good schemata they might replace the existing dominant ones.

- **Neutral dynamics ensures spontaneous novelty.** We have also observed some neutral dynamics. Schemata might be victims of neutral drifts similar by some aspects to process described in neutralist evolutionary theories (Kimura, 1983). Neutral dynamics are observed when a small level of noise causes inter-individual variations for schemata and new agents are regularly entering the population.

We guess that these dynamics are the most important for a large kind of semiotic schemata. But this remains to be tested in future works.

The role of noise

Semiotic schemata are always used in a particular context of a given environment. In the model we presented the effect of environments was limited to the addition of noise during the transmission phase. Noise has a double effect on semiotic schemata:

- **Noise as a diversity generator.** In artificial genetic evolution, noise could be assimilated to different mutations and errors that can appear during the copying phase of the genetic schemata. Noise is a diversity generator. In our dynamics also, noise could be a source of novelty for the creation of new semiotic schemata.

- **Noise as a pressure for selecting good schemata.** But its most important role is in the destabilisation of ill-adapted schemata. Words not distinct enough from one another could not survive in the presence of noise. Too long words are avoided. Only robust categories that are efficient in noisy environments are selected, etc.

Adaptation not optimisation

At the beginning of the experiments, pool of semiotic schemata are unstructured. Then, the dynamics select sets of schemata that are well adapted to the environment in which the agents are communicating. During this process, we might be tempted to say that the "quality" of the schemata increases. But, like for species natural evolution, optimisation stops once adaptation is reached. We have seen that in the presence of noise, well separated bands of words were emerging. Though, once a stable solution was found, this optimisation of distinctivity stops. This effect has been also observed in more complex architectures where residual polysemy was observed (Kaplan, 2000). In all these situations, there is no absolute optimisation, only the search for stable solutions adapted to the environment. Once a set of stable schemata emerges, it can be considered as a higher level schema that might enter in competition with other higher level schemata.

Conclusion

The understanding of cultural dynamics involved in the emergence of communication systems is only at its beginning. In this paper, we have shown on simple models some basic mechanisms that organise the selection of words in the emergence of a lexicon. We believe that these mechanisms apply to a larger set of replicators that we call semiotic schemata. Adapted semiotic schemata are culturally selected through individual choices in a population continuously renewed. As this process goes on, shared communication systems emerge.

Acknowledgment

This research was carried out at in the Language Group of the Sony Computer Science Laboratory in Paris. I thank Luc Steels, Angus McIntyre, Jelle Zuidema and Sébastien Picault for precious comments on this work.

References

T. Arita and Y. Koyama. Evolution of linguistic diverstity in a simple communication system. In C. Adami, R. Belew, H. Kitano, and C. Taylor, editors, *Proceedings of Artificial Life VI*, pages 9–17. MIT Press, June 1998.

J. Batali. Computational simulations of the emergence of grammar. In J. Hurford, C. Knight, and M. Studdert-Kennedy, editors, *Approaches to the Evolution of Language: Social and Cognitive bases*, pages 405–426. Cambridge University Press, Cambridge, 1998.

S. Blackmore. *The meme machine.* Oxford Univesity Press, 1999.

A. Darmesteter. *La vie des mots.* Delagrave, Paris, 1937.

R. Dawkins. *The selfish gene.* Oxford University Press, 1976.

R. Dawkins. Replicator selection and the extended phenotype. In E. Sober, editor, *Conceptual issues in evolutionary biology.* The MIT Press, Cambridge, Ma., 1984.

B. de Boer. Generating vowel systems in a population of agents. In P. Husbands and I. Harvey, editors, *Proceedings of the Fourth European Conference on Artificial Life*, Cambridge, MA, 1997. MIT Press.

Bart De Boer. *Self-organizing phonological systems.* PhD thesis, VUB University, Brussels, 1999.

Daniel Dennett. *Darwin's Dangerous Idea. Evolution and the Meaning of Life.* Simon and Schuster, New York, 1995.

C. Dircks and S. Stoness. Effective lexicon change in the absence of population flux. In D. Floreano, J-D Nicoud, and F. Mondada, editors, *Advances in Artificial Life (ECAL 99)*, Subserie of Lecture Notes in Computer Science, pages 720–724, Cambridge, MA, 1999. Springer-verglag.

M Dorigo, V. Maniezzo, and A. Colorni. The ant system: optimisation by a colony of cooperating agents. *Transactions on Systems, Man and Cybernetics Part B*, 26(1):29–41, 1997.

R. Ferrer Cancho and R. Sole. Naming games through distributed reinforcement. manuscript, 1998.

J. Holland. *Hidden order: How adaptation builds complexity.* Addison-Wesley Publishing Company, 1995.

J. Hurford. Social transmission favours linguistic generalisation. In C. Knight, J. Hurford, and M. Studdert-Kennedy, editors, *The emergence of Language.* To appear, 1998.

E. Hutchins and B. Hazlehurst. How to invent a lexicon: the development of shared symbols in interaction. In N. Gilbert and R. Conte, editors, *Artificial Societies: The Computer Simulation of Social Life.* UCL Press, 1995.

F. Kaplan. *L'émergence d'un lexique dans une population d'agents autonomes.* PhD thesis, LIP6 - Université Paris VI, 2000.

M. Kimura. *The neutral theory of molecular evolution.* Cambridge University Press, 1983.

S. Kirby. *Function, selection and innatensess: The emergence of language universals.* Oxford University Press, 1999.

S. Kirby. Syntax out of learning: the cultural evolution of structured communication in a population of induction algorithms. In D. Floreano, J-D Nicoud, and F. Mondada, editors, *Advances in Artificial Life (ECAL 99)*, Subserie of Lecture Notes in Computer Science, pages 694–703, Cambridge, MA, 1999. Springer-verglag.

M. Oliphant. *Formal approaches to innate and learned communicaton: laying the foundation for language.* PhD thesis, University of California, San Diego, 1997.

L. Steels and F. Kaplan. Spontaneous lexicon change. In *Proceedings of COLING-ACL 1998*, pages 1243–1249, Montreal, August 1998. ACL.

L. Steels and F. Kaplan. Stochasticity as a source of innovation in language games. In C. Adami, R. Belew, H. Kitano, and C. Taylor, editors, *Proceedings of Artificial Life VI*, Los Angeles, June 1998. MIT Press.

L. Steels and F. Kaplan. Bootstrapping grounded word semantics. In T. Briscoe, editor, *Linguistic evolution through language acquisition: formal and computational models.* Cambridge University Press, 1999.

L. Steels and F. Kaplan. Collective learning and semiotic dynamics. In D. Floreano, J-D Nicoud, and F. Mondada, editors, *Advances in Artificial Life (ECAL 99)*, Subserie of Lecture Notes in Computer Science, pages 679–688, Cambridge, MA, 1999. Springer-verglag.

L. Steels and F. Kaplan. Situated grounded word semantics. In *Proceedings of IJCAI 99*, pages 862–867, Stockholm, 1999.

L. Steels. Self-organizing vocabularies. In Chris Langton, editor, *Proceeding of Alife V*, Nara, Japan, 1996.

L. Steels. The synthetic modeling of language origins. *Evolution of Communication Journal*, 1(1):1–34, 1997.

The cultural evolution of syntactic constraints in phonology.

Luc Steels(1,2) and Pierre-Yves Oudeyer (1)
(1) Sony Computer Science Laboratory - Paris
(2) VUB Artificial Intelligence Laboratory - Brussels
steels@arti.vub.ac.be

Abstract

The paper reports on an experiment in which a group of autonomous agents self-organises through cultural evolution constraints on the combination of the individual sounds (phonemes) in their repertoires. We use a selectionist approach whereby a repertoire evolves by mutations of patterns, constrained by functional pressures from perception and production and the need to conform to the group.

Introduction

Language was commonly viewed in the 19th century, including by Charles Darwin, as a living system which evolves in a cultural fashion. This changed with the structuralist movement in linguistics that dominated research in the 20th century. Structuralism emphasises the formal description of language as an idealised system at a specific moment in time, which is largely innate. This approach has therefore not produced significant explanatory formal models on how language has emerged or how it evolves. Principles and modeling techniques from artificial life research can make a major contribution, although they need to be applied to cultural rather than genetic evolution. This paper reports on a case study in the cultural evolution of a particular nontrivial aspect of language, namely phonology.

The sound system of a natural language like English is constrained in two ways: The repertoire of individual sounds (phonemes) that speakers of a particular language are able to produce, recognise, and reproduce are a subset of all the possible sounds that the human vocal apparatus can in principle produce (Ladefoged and Maddieson,1996). For example, English does not use the vowel [y] (pronounced as in French "rue") whereas French does. Second, a language *constrains* the set of possible sound combinations. For example in English [mb] can occur at the end of a word as in "lamb" but not in the beginning, whereas in some African languages this is possible (as in Swahili "mbali" (far)).

Sounds fall into classes and the classes form a combinatorial system. It follows that the emergence of the sound system of a specific language has two aspects: (1) the emergence of a repertoire of individual sounds and (2) the emergence of an additional level of syntactic complexity constraining their combination. This paper is concerned with the problem how phonological classes and combinatorial constraints may emerge and continue to evolve. It is a concrete case study on how a level of (syntactic) complexity and systematicity might self-organise from independent units through cultural evolution.

The emergence of constraints on sound combinations requires that (1) sounds become grouped into classes, and (2) that combinatorial constraints among members of these classes become conventionalised in the population. Three theories have been put forward to explain the origin and acquisition of such phonological constraints. The most widely accepted theory at the moment, which developed from structuralist research in the line of Jakobson, and Chomsky and Halle (1968), is that the categorisation of sounds is based on a set of innate distinctive features (like voiced, fricative, etc.) and that their combinatorial constraints are a subset of universal combination principles chosen by setting some parameters. This suggests a genetic origin of phonology and a maturational approach to language acquisition (see e.g. Dresher, 1992). The second theory takes an empiricist track and proposes that the sound system of a language is acquired through an inductive statistical learning process which delineates classes based on the distribution of sounds in the inputs given to the learner. Most research within a connectionist framework follows this line (see e.g. Plaut and Kello, 1999).

This paper adopts a third alternative which is based on selectionist principles. It proposes that there are mechanisms in each agent that generate in a basically random fashion possible sound systems and variations on sound systems, but that the set of possibilities is constrained by two selectionist forces: There are functional constraints coming from production and recognition, for example, the sound "wrljts" is much more difficult, if not impossible, to reproduce easily and to recognise reliably compared to a sound like "baba". Self-organisation due to a positive feedback loop between use and success acts as a secondary selectionist force to ensure that speakers of the same language share the same conventions. This selectionist hypothesis has been put forward by a number of authors (Lindblom, MacNeilage, and Studdert-Kennedy (1984), Steels (1997a)) and has been suggested for the evolution of grammatical complexity as well (Hashimoto and Ikegami (1996), Kirby (1999), Steels (1997b)).

Our research group has developed a general framework for exploring this selectionist approach, not only for speech (De Boer, 1999) but also for the origins of lexicons (Steels and Kaplan, 1999) and grammar (Steels, 1997b). The framework assumes a population of distributed autonomous agents that take turns playing a consecutive series of games. Each game exercises some aspect of language (sound production and sound recognition in the case of experiments in phonetics) and is followed by adaptation based on feedback from the outcome of the game. For investigating speech, we have been employing imitation games in which the speaker produces a random sound from his repertoire, the hearer recognises the sound and attempts to reproduce it. Feedback is based on the speaker's judgement whether the hearer's sound is indeed the one the speaker produced. Adaptation includes the adoption of a new sound, shifting of a sound (in perceptual or production space), or elimination of a sound from the repertoire. Speakers may occasionally create new sounds by adopting a new randomly chosen configuration of the articulators. So far it has not only been shown that a repertoire of sounds (albeit only vowels) can emerge from such games but also that the possible repertoires satisfy the tendencies observed universally in human vowel systems as long as the speech apparatus and the hearing system are reasonably realistic with respect to human speech (De Boer,1997). Some preliminary work has been done on syllables (Redford, et.al. 1998) but not yet through multi-agent simulations.

This paper adopts the same framework for studying the emergence of sound combinations, more specifically syllables, like "pa", "bri", "art", etc. A typical language has about 250 to 300 possible syllables which are then combined into higher order units like words. At this point we have only studied the problem formally, i.e. by assuming an abstract articulatory space, and

an abstract perceptual space. This way we can study more generally how complex units may form from simple ones in a collective self-organising process. But the abstractions do not take away the major problems that need to be dealt with:

- *The inverse mapping problem.* The key difficulty in acquiring a sound repertoire is to learn how to move the articulators to reproduce a particular sound, based only on acoustic information about the sound. Because the articulatory space has many degrees of freedom which do not map directly onto the dimensions of the perceptual space this problem cannot be solved analytically, even if a good physical model would be available to the language learner. The inverse mapping problem is exacerbated by the fact that a smooth transition between positions in articulatory space may lead to non-smooth transitions in perceptual space.

- *Combinatorial explosion.* The number of possible syllables exponentially increases with the size of a repertoire (a 20 phoneme repertoire gives rise to 160,000 possible combinations of size 4 for example), so an exhaustive search for viable combinations is excluded.

- *Contextual influence.* Sounds are influenced by the context due to coarticulatory side-effects (Hardcastle and Hewlett, 1999). For example, a vowel before a nasal consonant (as in "on") is already slightly nasalised. Even the [k] in the word "cow" already shows the first signs of the lip-rounding associated with [w]. Consequently sounds in isolation are acoustically different from sounds in combination. In fact, some consonants cannot even be pronounced without context.

- *Continuous parameters* The distinctive features traditionally used in abstract phonology cannot be assumed as given. The parameters controlling the articulators are continuous. For example, the horizontal position of the tongue can go from high to low. Different languages carve up this continuum in different ways. The data from perception is continuous as well, unsegmented and uncategorised. A realistic simulation should therefore include a mechanism for mapping features onto the speech signal, rather than assuming that phonological features are given.

- *Memory limitation.* It is unrealistic to assume that agents store large sets of examples to which they can repeatedly return, as is done in many statistical learning approaches. We need case by case online learning without memory of past cases. Storing sounds or gestures in full detail is to be avoided as well as it would require enormous memory resources.

We have found it useful to decompose the evolution of phonological complexity into three transitions. For each transition we discuss (1) what is needed in terms of cognitive architecture to enable the increased complexity, (2) what results we obtained in simulating the transition, and (3) what selectionist pressure justifies the additional complexity. The three steps are: from individual (static) sounds to complex (dynamic) sounds, from complex undifferentiated sounds to sound patterns, and from sound patterns to categorial constraints.

From static to dynamic sounds

Individual phonemes

Our starting point are earlier simulations, in particular those by De Boer (1997), which clearly demonstrate the viability of the selectionist approach for individual isolated static sounds. The simulations assume a population of agents, which can be changing if we want to research questions of language transmission or language contact, that play a consecutive series of imitation games. Each agent is capable to store a repertoire of sounds in an associative memory. A sound has two components: (1) A target in the articulatory space, for example, for the sound [u] as in "boot", the horizontal tongue position is towards the back, vertical tongue position is up towards the roof, and lips are rounded. (2) A region (with a prototypical midpoint) in acoustic space, made up by the sound's energy level within certain frequency bands, known as formants. For example, the [u] sound is typically found in a region around F1=276 Hz, F2=740 Hz and F3=2177 Hz (Vallee, 1994). Agents must be able to control their articulatory apparatus to reproduce a sound in their repertoire and they can recognise which sound was produced based on a similarity match between the signal heard and the sound's corresponding region in acoustic space.

For each imitation game, two members are selected randomly from the population. The first agent acts as speaker, the second as hearer. The speaker chooses one sound from his repertoire and puts his articulators in the prescribed position, thus generating an acoustic signal. The hearer perceives the signal in terms of his perceptual space and retrieves the sound that is closest to the signal. Then the hearer reproduces his version of the sound which is perceived and categorised again by the speaker. If the speaker agrees that this is the same sound, he gives a positive feedback otherwise a negative one.

Based on this feedback the agents update their associative memory. When the sound could be correctly recognised and reproduced, their respective scores go up. Otherwise, the hearer may either move the perceived sound closer to the one heard by hill-climbing (both in acoustic and articulatory space) or add a new one when the sound heard was too far from the ones existing so far in his repertoire. A score is kept of the use and success of sounds. Those sounds that consistently fail or do not occur frequently enough are discarded. The computer simulations carried out by de Boer (1997) have abundantly shown that a collective repertoire of individual vowels indeed self-organises through these mechanisms and that the emerging repertoires exhibit the same characteristics as those found in natural phonologies, specifically they occupy preferentially the extrema of the vowel space and then start filling up the spaces in between.

A key property of a selectionist approach is that perceptual analysis only has to be able to *differentiate* the sounds that are effectively in the repertoire because this is enough to retrieve the motor program producing the sound. This requires less fine-grained feature extraction than if the position of the articulators has to be recovered for an inverse-mapping. For example, in Japanese no firm distinction exists between [l] and [r] so that Japanese speakers can be (and are) less sensitive than English speakers for this distinction. At the same time, imitation has to be only as precise as required to distinguish the different sounds. An imitation may sound inadequate for the ears of the speakers of one language, whereas it is perfectly fine from the perspective of another language.

Complex sounds

The major limitation of the simulations of de Boer is that only individual static sounds are employed, so that the problem of co-articulatory effects and non-linearity due to sequencing of articulatory targets do not appear. However, human languages make obviously use of more complex, connected sounds as opposed to individual sounds. Hence the speaker must be able to produce an articulatory gesture (Browman and Goldstein, 1992) and thus generate a continuous sound signal in time. The signal appears as a trajectory in the acoustic space of the hearer (usually called a signature). The main reason why languages use complex dynamic sounds is because the set of individual sounds reliably producable by the human vocal apparatus is limited and so this would severely restrict the semantic potential of a language (individual speakers have vocabularies of at least 50,000 words). Many sounds (particularly the consonants such as [t]) are almost impossible to produce and reliably perceive in isolation. So the pressure to develop a broader repertoire of phonetic building blocks pushes the emergence of more complex sounds, and we believe that this has inevitably given rise to the development of a complex phonology as we show in the paper.

But let us start first from a situation where the agents use complex sounds but no phonology yet. This means that a complex dynamic sound is taken as an individual unit *in toto*. A sound in the associative memory of the agents then consists on the one hand of a complete articulatory gesture, and on the other hand of a complete signature in acoustic space. Selectionist forces coming from reproducability and perception are again at work to restrict the set of possible sound combinations (as discussed by Lindblom and Maddieson, 1988). Agents now need a more complex matching function to find the sound that is similar to the ones in their repertoire. The matching function needs to be complex because it includes a time aspect, and because due to the difficulty of articulation, speaker variance, influence of environmental noise, and other factors such as the variable speed of speech, two acoustic signatures for the same sound will never be exactly the same.

Oudeyer (1999) has performed a series of imitation game experiments from this perspective, using exactly the same population dynamics as in the de Boer experiments. A realistic synthetic articulator built by Eduardo Miranda based on the Cook synthesiser (Cook, 1989) and real signal processing was used. It was shown that imitation was not harder than with static sounds and that a repertoire of shared sounds could emerge. This may seem a paradoxical result, because there are now many more degrees of freedom (14 for articulation). But in some sense the task is easier because when the articulatory space contains many more dimensions the agents have a less hard time to find regions that can be reliably be distinguished. Hence two complex sounds which humans perceive as being very different might nevertheless be considered as successful imitations by the agents.

These simulations showed at the same time the strong limitations of this approach:

- *Memory usage*: A new pair (articulatory gesture, acoustic gesture) needs to be stored for every complex sound in the language and it needs to be stored with full articulatory or acoustic precision. This is obviously not tenable from the viewpoint of memory storage and retrieval.

- *Repertoire size*: When the sound repertoire becomes larger, the acoustic space and the articulatory space become more crowded, making a more precise recog-

nition and more accurate reproduction necessary. The inverse-mapping problem then becomes more prominent and agents have a much harder time to collectively self-organise a shared repertoire.

- *Language acquisition*: The language learner is required to learn every sound separately, which puts a strong limit on the speed of language acquisition. It leads to difficulty in language acquisition which can be observed in the simulations because new sounds have a hard time to propagate in the rest of the population.

Oudeyer (1999) has shown that these difficulties impose strong limitations on the repertoire size that can be generated and maintained by a population. Specifically, the repertoire could never get beyond 20 complex sounds. So when lexicon and grammar pressure for a wider repertoire, the system will not be able to deliver. Each of the three dimensions above needs to be improved: Some form of drastic compression is needed to keep memory usage within bounds. Reproduction and recognition somehow need to handle much tighter regions in articulatory and acoustic space, and the spreading of new sounds must become faster. We believe that these three forces provide the positive selection pressure for the emergence of a phonological system.

From complex sounds to sound patterns

This brings us to the second transition. It consists in breaking up the elements of a complex sound into components which each have the properties of individual sounds discussed earlier: a target in articulatory space and a target in acoustic space. Such a break-up results in an enormous compression. The trajectory between the articulatory targets can be filled in by the motor system and need no longer be stored. The perceptual system gets the main targets but can ignore what happens in between. Once a complex sound is broken up in individual sounds, the same individual sound can be re-used in other complex sounds, giving an additional compression. Such an approach is also beneficial from the viewpoint of language acquisition because once an individual sound has been learned, it can be used to recognise an unknown complex sound.

Handling Patterns

To achieve this transition, the speech memory of the agents must be pattern-based, a realistic assumption born out by various types of psychological evidence (MacNeilage,1998). A pattern consists of a sequence of slots and possible fillers of each slot (figure 1). Patterns are widely used throughout the brain for various

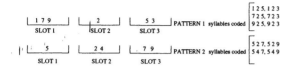

Figure 1: Two example patterns are shown together with the syllables they cover. Individual sounds are denoted by numbers.

tasks and could therefore easily have been recruited for speech. A specific realisation of a sound pattern corresponds to a syllable, such as "bla". Other syllables that could be governed by the same pattern are: "pla", "blo", "bli", "pli", etc. The possible fillers of a slot are individual sounds (phonemes) which are stored in a sound-memory similar to the one used in the imitation games discussed in the previous section. Each sound is an association between a point in articulatory space (an articulatory target) and a region with prototypical midpoint in acoustic space. Each slot-filler pair in a pattern has a score, decomposed into the use and success of the filler in the pattern. An agent's phonetic repertoire consists of a set of sounds, a set of sound patterns, and a specification what sounds can fill which slot in each pattern.

When an agent produces a complex sound (a syllable), he first selects a pattern and then selects a sound for each slot in the pattern. The sounds provide a series of consecutive articulatory targets (goals), starting from the rest state. The way the agent tries to achieve these targets is similar to other motor control tasks, like skiing down a slope with targets set on the way, and so we have used similar behavior-based techniques as those used for example in mobile robotics (Steels and Brooks (1995)) to plan and execute articulatory gestures. Each target acts as an attractor pulling the articulators involved towards it. At the same time an articulator exhibits inertia which slows down the approach to the target (for example the tongue can only move at a certain speed) and takes into account feedback towards the goal which repels the movement towards the target when there is a risk of overshooting. These different dynamical forces acting on a trajectory are captured in the following equations

$$pos(t + 1) = pos(t) + f(c_1 * attract(t) +$$
$$c_2 * feedback(t) + c3 * inertia(t)$$
$$attract(t) = \sum_{i/t_i > t} (-c_4(t_i - t) + c_5) \frac{pos(t)g_i}{norm(pos(t)g_i)}$$
$$feedback(t) = pos(t)g_i - \frac{t_i - t}{TIMESTEP} pos(t-1)pos(t)$$
$$inertia(t) = pos(t-1)pos(t)$$
$$f(v) = if(norm(v) > MAXSPEED)$$
$$then (MAXSPEED \frac{v}{norm(v)}) else v$$

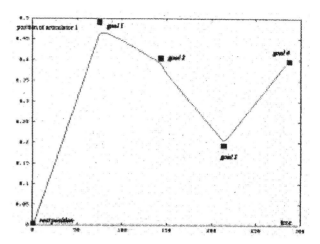

Figure 2: This figure shows the trajectory of one articulator for a pattern consisting of four slots.

These equations determine the position (pos) of an articulator at time t (in milliseconds), given $(g_0, t_0)(g_1, t_1)...$ where each g_i is an articulatory target and t_i the timing of the target. $c_1...c_5$ and $MAXSPEED$ are parameters of the model and $norm(v)$ is the norm of vector v.

An example of a trajectory (for only one single articulatory dimension) is shown in figure 2. It has been produced using the same realistic articulatory synthesiser as for the Oudeyer experiment discussed in the previous section. There are 14 trajectories for each of the articulatory dimensions.

Due to the inertia factor there is influence of the previous target on the next and due to the fact that the next target already starts pulling, there is anticipation just as in human speech. The articulatory synthesiser introduces additional co-articulatory effects that show up in the final signal. Realising a trajectory is not a matter of simply touching each target one after the other. Often there is no time to reach a target or a target can only be approached slightly before moving on to the next one. For instance, we see that the trajectory in figure 3 does not completely touch goal1. Sometimes two targets are so far apart that a trajectory is not possible. This puts strong constraints on the set of viable sound combinations that can be used by the agents. Agents can be seen as experimentally determining which sound combinations 'work' from an articulatory point of view.

The articulatory trajectories enacted by the speaker generate a trajectory in the perceptual space of the hearer. Figure 3 shows the outcome of the trajec-

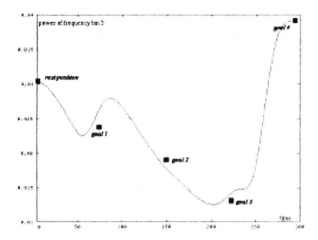

Figure 3: Trajectory of one of the acoustic dimensions produced by the articulatory trajectory partially represented in Figure 2.

tory partially described in figure 2, after considerable smoothing. A phonetic event is defined as a significant change of the first derivative in an acoustic trajectory. For each phonetic event a possible corresponding sound is retrieved by comparing its point in acoustic space to those stored with the individual phonemes in memory. Because the articulatory gesture seldom creates a perfect path, the acoustic signature can never be expected to yield a perfect match. Occasionally the match will be so bad that it is not possible to recognise the individual sounds.

The trajectory in figure 3 illustrates this quite clearly. The different acoustic goals of the individual sounds have been superimposed. Goal1 is more or less reached although a bit earlier than expected. Goal2 does not show up as a significant phonetic event, i.e. significant change in the direction of the trajectory, at all, instead there is a phantom event (due to non-linearity) between goal1 and goal2 which does not correspond to any articulatory target and hence cannot be recognised as a sound. Goal3 can be recognised although there is another confusing phonetic event close to it. Goal4 has been reached completely. The difficulty of aligning articulatory targets with acoustic targets puts a second strong constraint on the set of possible sound combinations: The individual targets in the pattern must be recognisable despite the distortion caused by combining them. The particular sequence of goals in figure 2 and 3 is an example of a non-viable pattern, because of the non-linear phenomena between goal1 and goal2.

Once individual phonemes have been recognised, the hearer needs to find the syllable in his pattern memory that matches the series of phonemes. Assuming that the hearer has found a pattern (or possibly more than one), he then reconstructs the articulatory gesture by setting the relevant articulatory targets and re-synthesising the complex sound. When the pattern did not exist yet in the hearer's repertoire, the hearer signals ignorance but nevertheless attempts to reproduce the syllable based on recognition of the individual sounds in the perceived syllable. The speaker then in turn attempts to recognise the syllable produced by the hearer based on his own repertoire and gives positive or negative feedback depending on whether a matching pattern could be found or not.

The agents adapt their memory based on the outcome of the imitation game. The scores of the slot-fillers that were used go up in the case of full success. In case of imitation success but cultural failure (the hearer did not possess the syllable although he could imitate it based on stringing together individual phonemes), the speaker decreases the scores of the slot-fillers in the pattern that was used and the hearer tries to incorporate the pattern in his own repertoire. There are two conditions: (1) the existing pattern must match sufficiently close (which means that there is only a difference in one slot), and (2) the new slot-filler must be compatible with the other slot-fillers already in the pattern. If this is not the case a totally new pattern is constructed if there is memory available. Finally, if imitation was not successful at all, only the speaker updates his memory by lowering the scores of the slot-fillers involved.

Occasionally the speaker generates a new pattern by a random combination of sounds from his sound repertoire, or mutates a pattern by trying out another slot-filler than those used so far. As only a subset of all possible combinations is viable (either from the viewpoint of perception or production) there is no guarantee that a randomly selected sound may be fittable in a pattern and so 'natural selection' from perception or production weeds out these patterns. Moreover because speaker and hearer locally exchange information whether a pattern is shared, the population pressure acts as secondary selectionist force on what repertoire will form. The population can be seen as performing a collective search for a shared set of patterns through a process that is similar to spin-glass style relaxation. Agents have a strong limitation on the set of possible patterns they can store in memory, so that they are forced to prune patterns to make way for new ones. The pruning criteria are based on how many different slot-fillers a pattern has and on the score of the slot-fillers.

As in a genetic system or in the immune system, mutation rates need to be regulated. Mutation gives rise to innovation and therefore to expansion of the repertoire. But when mutations are too rapid, the syllables cannot spread in the population and so imitation success starts to drop. The mutation rate has therefore been coupled to success. An agent stops mutating patterns when the success rate is not sufficiently high (below 85 %), so as to give a chance to absorb new syllables or have new syllables be absorbed by the rest of the population. Of course, if the agents have to lower their mutation rate the increase in the repertoire goes less fast. So this is a way to measure the efficiency of the language acquisition process.

Simulation Results

The mechanisms skeched above have been integrated in a computer simulation. An example of the results obtained with this simulation is shown in the following figures. These results are for a population of 30 agents which does not change. Agents start with a repertoire of 10 individual sounds which are derived using isolated sound imitation games, as in the de Boer experiment, and their objective is to construct a shared repertoire of sound patterns with these sounds. The parameters (e.g. maximum speed of articulators or timing between two goals) are set such that about 50 percent of syllables of length between three and five are not viable from an articulatory or perceptual point of view. Figure 4 shows the success percentage over time every 50 games and figure 5 the increase in the size of the repertoire.

There are four phases. In the first phase, success rapidly rises to almost 100 percent success. This is a phase where the patterns have basically one slot-filler so there is little ambiguity how they have to fit together. The situation is similar to learning individual syllables. However, in phase 2, the success rate drops because of incompatibilities between the different agents. There are two cases. Either an agent cannot incorporate a pattern and therefore creates a new one, or there is more than one possibility in which case patterns proliferate.

Here is an example of the first case. Suppose agent-1 and most of the other agents have the pattern $[\{1,2\},\{3\},\{5\}]$ but agent-2 the pattern $[\{1,6\},\{3,4\},5]]$. Suppose [2,4,5] is not possible due to articulatory constraints. Then if agent-1 produces the pattern [2,3,5], agent-2 cannot integrate it because it clashes with the other slot-fillers in his pattern. Agent-2 therefore has to construct a new pattern (when there is space in his memory) and it

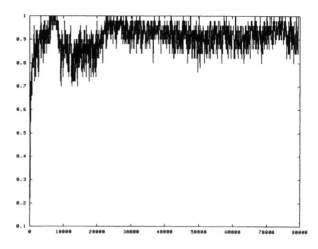

Figure 4: Each point represents the percentage of success for 50 games with 30 agents

will take a while before one of the two patterns is pruned to allow pattern coherence with the rest of the group. Here is an example of the second case. Suppose that two agents have both the following patterns $[\{1,2\},3,\{5,7\}]$ and $[8,\{3,2\},\{5,6\}]$. Now agent-1 produces a variation: $[9,3,5]$. Agent-2 has two ways to incorporate it. Either by extending pattern-1 which yields $[\{1,2,9\},3,\{5,7\}]$ or by extending pattern-2 which yields $[\{9,8\},\{3,2\},\{5,6\}]$. Each extension gives many more additional patterns (which is in se a good thing because it speeds up the growth of the repertoire) but it makes it also more difficult to establish coherence.

Because mutation stops when agents fail in the imitation game, the search process is given time to recover and eventually agents settle on a repertoire of patterns. This has happened in phase 3. Systematicity is now present and success as well as repertoire size move up. In phase 4 a plateau is reached. Due to memory limitations, agents can not store more patterns and given that patterns are already quite complex, it becomes harder and harder to extend them. So the repertoire does not change much anymore and imitation success reaches a stable state.

These simulation results show that a shared repertoire of patterns can indeed emerge in a population with a pattern-based memory. The advantages compared to the earlier solution, where complex sounds were viewed as undifferentiated units, are obvious: (1) To store a complex sound we only need to store the number of slots, the relative timing over these slots, and which individual sound is a possible filler in each

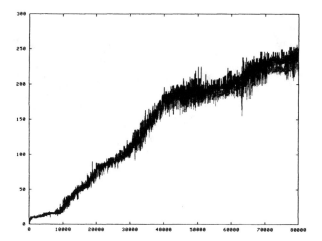

Figure 5: Evolution of the size of the syllable set over time.

slot. So there is an enormous compression of information compared to earlier on. (2) The size of the repertoire now approaches easily (after about 80,000 games) the typical repertoire of human languages, whereas only a dozen stable sound combinations could be handled in the same time frame when no patterns are used. (3) Propagation of new syllables (meaning speed of language acquisition) goes faster, particularly in phase 3 when patterns are already in place. But note that agents must slow down their mutation rate in order to give patterns a chance to stabilise before new ones are introduced.

From sound patterns to categorial constraints

We now turn to a third transition, enabling a full-fledged phonology. The human brain categorises almost anything it is confronted with and then exploits this categorisation to gain in efficiency and reliability. This is also what we postulate as having happened for speech.

Introducing phonological categories

There are many possible sources of categories, given the system described above. For example, certain regions in acoustic space (say high presence of nasal resonance) are systematically associated with certain regions in articulatory space (opening or closing of the nasal cavity), or, the sounds filling a particular slot in a pattern form a natural class, because all of them are viable as fillers of a particular slot and hence satisfy certain articulatory and perceptual constraints. So

categories can emerge naturally from the agents' efforts to find variations on patterns and the success they have with particular variations, they do not have to be innate as assumed in abstract generative phonology. We now demonstrate that these natural phonological categories can lead to an additional optimisation in memory usage and in language acquisition.

Let us assume that agents use the natural classes that form in one pattern as a basis for exploring variations in other patterns, both as speaker for producing a new syllable and as hearer for guessing in which pattern they need to incorporate a new syllable. This way the speaker is more likely to produce a new syllable that is viable and the hearer is more likely to integrate the new syllable within existing patterns. This should improve the speed of language acquisition. We call this mechanism *analogy exploitation* because agents construct extensions of one pattern by analogy with another one.

More technically assume that there is a set of patterns $P = \{p_1, ..., pn\}$ where each pattern has a set of slots $p = [s_1, s_2, ...]$ and a set of phonemes associated with each slot. The class associated with slot s_1 in pattern p_1 is denoted as C_{p_1,s_1}. We define a distance metric $\delta(C_1, C_2)$ on the set of phonological classes simply based on the set of common elements.

Assume now that the speaker has an existing pattern p_j but wants to construct a new variation. Rather than selecting a random phoneme from the repertoire of phonemes as the new filler of one of the slots s_k, the speaker searches for the class of sounds C_i that has the shortest distance δ to C_{p_j,s_k}. Then he picks a random member from C_i which was not already in C_{p_j,s_k} and produces the syllable coded by the mutated pattern. For example, the speaker with the patterns given in figure 1 might decide to employ phoneme 1 for constructing a variation of slot3 in pattern2 because $\{1,7,9\}$ have all occurred together as possible slot fillers of slot1 in pattern1. There is no guarantee that the mutation is viable because phonological categories are context-dependent. The possible fillers of a given slot depend on the fillers of the adjacents slots because of co-articulation, and thus the same set of fillers may be unsuited in another context. Nevertheless, the chance that it will be viable is much higher than with a random mutation.

On the side of the hearer, the same mechanism can be used to guess better how a new syllable should be incorporated in the existing repertoire. The hearer computes which patterns cover the new syllable with minimum variation and then use the same distance computation to decide what pattern to change. Most often only one possibility remains and incorporation

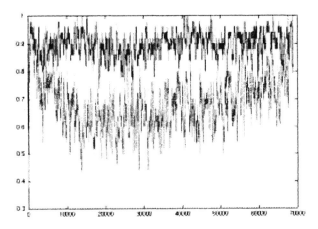

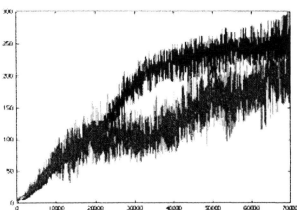

Figure 6: Evolution the percentage of success for 30 agents with analogy exploitation (top) and without (bottom).

becomes a lot less problematic.

Simulation Results

The above mechanisms have again been implemented and subjected to extensive testing. The effect of analogy exploitation can be seen by increasing the mutation rate (from 1 in 100 to 1 in 10 games). Because of this high mutation rate, agents which learn less efficiently will have trouble to acquire the phonology. Recall that mutation stops when success is below 85 % so that agents without analogy exploitation will still catch up but it will take much more time. This is confirmed by the results of experiments displayed in figure 6, which shows the imitation success, and figure 7 which shows the repertoire size. This experiment involves 30 agents and patterns of size 3. We see in figure 6 that the agents which exploit analogy have a consistently higher success rate than those without. Phase 1 is very short. As soon as agents start to make variations on patterns in phase 2, the group without analogy falls below the threshold at which new mutations would occur. A closer inspection of the simulation traces shows that successive incorporation errors lead to incompatibilities that were very difficult to resolve.

Slower learning can be seen in the evolution of the repertoire size. The group which exploits categorial analogy builds a repertoire much more quickly. The graph shows for each data point (i.e. every 50 games) the highest and the lowest repertoire size in the group. We therefore see that the agents which exploit analogy are also much more coherent throughout.

Figure 7: Evolution of number of syllables over time for a group of agents with (top curve) and without (bottom curve) analogy explotation.

An example of one set of patterns formed with analogy exploitation is shown in figure 8. Five natural classes have formed. C2 is reused five times. When classes (as opposed to individual phonemenes) are stored we get increased compression although the compression is less dramatic than for the second transition.

Conclusions

The paper has explored a selectionist approach towards the problem how complex syntactic conventions could arise in a population of agents through cultural evolution. New variations are generated and then tested to see whether they are viable from the viewpoint of articulation, distinguishable from the others from the viewpoint of perception, and culturally present in the group. We have introduced three major transitions: from single static sounds to complex undifferentiated sounds, from complex sounds to patterned sounds, and from patterned sounds to categorial constraints. Each transition is caused by recruiting a cognitive device: pattern-based memory as opposed to unit-based memory, and analogical exploitation of the naturally emerging classes, more specifically to restrict the search for patterns that are viable and culturally shared by the speaker or the hearer.

The simulation results obtained so far are extremely encouraging. Nevertheless a lot remains to be done. Most of our future work will be targeted towards adding progressive realism to the simulation. We need to apply the principles discussed in the paper to much more elaborate articulatory synthesisers in which realistic coarticulation effects occur. The recent work of

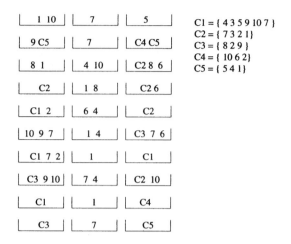

$$C1 = \{ 4\ 3\ 5\ 9\ 10\ 7 \}$$
$$C2 = \{ 7\ 3\ 2\ 1 \}$$
$$C3 = \{ 8\ 2\ 9 \}$$
$$C4 = \{ 10\ 6\ 2 \}$$
$$C5 = \{ 5\ 4\ 1 \}$$

Figure 8: Some example patterns in one agent and their slot fillers. Five classes have been formed.

Redford (1999) is an important source of insights on this matter. We will then be able to subject the solutions proposed in this paper to much greater difficulties in acoustic analysis - and we expect to show that categorial constraints can play a major role. Finally we need to couple the mechanism that creates phonological forms to the users of these form, namely lexicon and grammar.

From a broader perspective, we believe that the paper illustrates well how fundamental principles discussed in the context of biological systems are equally present in culturally evolved systems such as language. These principles include self-organisation through a positive feedback loop between use and success and selectionism which combines a process of variation with a process of pruning under natural pressure.

Acknowledgement

We thank Bart de Boer from the VUB AI Laboratory in Brussels for valuable comments on this paper.

References

Browman, C.P. and L., Goldstein (1992) Articulatory Phonology: An Overview. Phonetica, 49, 155-180.

Chomsky, N. and M. Halle (1968) The Sound Pattern of English. Harper Row, New york.

Cook, P.R. (1989) Synthesis of the singing voice using a physically parameterized model of the human vocal tract. Proceedings of the International Computer Music Conference. The MIT Press, Cambridge. pp. 69-72.

De Boer, B. (1999) Investigating the Emergence of Speech Sounds. In: Dean, T. (ed.) Proceedings of IJCAI 99. Morgan Kauffman, San Francisco. pp. 364-369.

Dresher, E. (1992) A learning model for a parametric theory in phonology. In: Levine, R. (ed.) (1992) Formal Grammar: Theory and Implementation. Oxford University Press.

Hardcastle, W.J. and N. Hewlett (eds.) (1999) Coarticulation. Theory, Data and Techniques. Cambridge University Press, Cambridge.

Ladefoged, P. and I. Maddison (1996) The Sounds of the World's Languages. Blackwell Publishers, Oxford.

Lindblom, B., P. MacNeilage, and M. Studdert-Kennedy (1984) Self-organizing processes and the explanation of phonological universals. In: Butterworth, G., B. Comrie and O. Dahl (eds.) (1984) Explanations for Language Universals. Walter de Gruyter, Berlin. pp. 181-203.

Lindblom, B., and I Maddieson (1988) Phonetic Universals in Consonant Systems. In: Hyman, L. and C.Li (eds.) Language, Speech and Mind. Routledge,London, pp. 62-79.

MacNeilage, P.F. (1998) The Frame/Content theory of evolution of speech production. Behavioral and Brain Sciences, 21, 499-548.

Plaut, D. and C. Kello (1999) The Emergence of Phonology from the Interplay of Speech Comprehension and Production: A distributed Connectionist Approach. In: MacWhinney, B. (ed.) The Emergence of Language. Lawrence Erlbaum, Mahweh, NJ.

Redford, M.A., C. Chen, and R. Miikkulainen (1998) Modeling the Emergence of Syllable Systems. In: Proceedings of the Twentieth Annual Conference of the Cognitive Science Society. Erlabum Ass. Hillsdale.

Redford, M. A. (1999) An Articulatory Basis for the Syllable. Ph.d. thesis. The University of Texas, Austin.

Steels, L. (1997a) The synthetic modeling of language origins. Evolution of Communication, 1(1):1-35.

Steels, L. (1997b) The origin of syntax in visually grounded robotic agents. In: Proceedings of IJCAI-97, Morgan Kauffman Pub. Los Angeles.

Steels, L. and R. Brooks (1995) The Artificial Life Route to Artificial Intelligence. Building Embodied Situated Agents. Lawrence Erlbaum, New Haven.

Oudeyer (1999) Experiments in emergent phonetics, Rapport de Stage de 2eme annee de magistere informatique et modelisation, Ecole Normale Superieure de Lyon, Submitted to COGSCI'2000.

Vallee, N. (1994) Systemes vocaliques: de la typologie aux predictions. These. ICP, Grenoble.

VI Methodological and Technological Applications

Modeling the Role of Neutral and Selective Mutations in Cancer

C. C. Maley[1] and S. Forrest[1,2]

[1]Department of Computer Science, UNM, Albuquerque, NM 87131
[2]Santa Fe Institute, 1399 Hyde Park Road, Santa Fe, NM 87501

Abstract

The transformation of normal cells into cancerous cells is an evolutionary process. Populations of precancerous cells reproduce, mutate, and compete for resources. Some of these mutations eventually lead to cancer. We calculate the probability of developing cancer under a set of simplifying assumptions and then elaborate these calculations, culminating in a simple simulation of the cell dynamics. The agent-based model allows us to examine the interactions of neutral and selective mutations, as well as mutations that raise the mutation rate for the entire cell. The simulations suggest that there must be at least two selectively neutral mutations necessary for the development of cancer and that preventive treatments will be most effective when they increase this number.

Cancer

Cancer is an evolutionary problem. This is the basis for both its virulence and our difficulties in treating it. The dynamics of cancer cells demonstrate the sufficient conditions for natural selection: heritable variation in the population and differential reproduction based on that variation. The variation in the population of precancerous cells (Fujii et al., 1996; Barrett et al., 1999) arises from the normal process of somatic mutations as well as the dramatic rise in mutation rates that is characteristic of the progression to cancer (Paulovich et al., 1997). Differential reproduction of the mutants is accomplished through phenomena such as the subversion of check points in the cell cycles of the mutants (Sherr, 1996). Nowell, 1976, argued for the importance of evolution in cancer more than two decades ago. Any mutations that redirect more of the body's resources to the cancer cells will be selected. This includes the invasion of new tissues and metastasis. The fact that the population of cells includes significant heterogeneity means that most treatments will not eradicate all the cells, leaving some resistant cells. Furthermore, since each patient's cells evolve through an independent set of mutations and selective environments, the resulting population of cancer cells in each patient is likely to be unique. This suggests that general treatments that will work for all, or even most, patients will be difficult to find. The fact that evolution within a tumor works against us in cancer means that not only is cancer an evolutionary problem, but that it will only be solved as an evolutionary problem.

Artificial life provides approaches that are ideal for addressing such evolutionary problems. The field of artificial life has grown up around evolutionary theory (Collins and Jefferson, 1992; Maley, 1998; Levin et al., 1997), and for good reason. When we try to represent heterogeneous populations of individuals interacting in a spatially structured environment, it is difficult to represent and analyze such systems with tractable mathematics. Computational models can help to extend analytical theory to the dynamics of systems with heterogeneous populations that are interacting and evolving. Computational models can help to test the simplifications necessary to reduce the biological system to a mathematically tractable formulation. At its best, artificial life models applied to theoretical biology lead to testable hypotheses.

This paper extends an analytical model of the risk of developing cancer and derives testable hypotheses about the genetic nature of the development of cancer from these models. We focus on a type of esophageal cancer known as esophageal adenocarcinoma, and its precancerous state, which is known as Barrett's esophagus (Reid, 1991; Neshat et al., 1994; Barrett et al., 1999).

Estimating Cancer Risks

Two dominant characteristics of cancer cells are their genetic instability (Lengauer et al., 1998) and uncontrolled proliferation (Kastan, 1997). The most commonly mutated tumor suppressor gene across all cancers is p53 (Smith and Fornace, 1995). The loss of this gene results in genetic instabilities (increased mutation rate), often with the loss or duplication of entire chromosomes (Smith and Fornace, 1995). The appearance of such aneuploid cells in Barrett's Esophagus is one of our most reliable indicators of a poor prognosis (Neshat et al., 1994). In contrast, p16 (a.k.a. CDKN2A and INK4a) is a gene thought to be responsible for shifting a cell from a proliferative state to a quiescent state (G0) (Sherr, 1996). Loss of a p16 allele is associated with the spread of cells with that mutation throughout the Barrett's region. But, at least in Barrett's Esophagus, mutations in both p53 and p16 are not sufficient to cause cancer (Barrett et al., 1999). How many other genes are involved and what are their roles?

There is a body of mathematical modeling work which argues that the development of cancer is best understood as a sequence of two or more stages (Moolgavkar and Luebeck, 1990; Moolgavkar, 1999; Little, 1995; Luebeck and Moolgavkar, 1994; Sherman and Portier, 1996). The two stages might be called "precancerous" and "malignant." The two-stage model involves at least 6 rate parameters: the rate of cells changing from a normal state to the precancerous state, the rate of reproduction of precancerous cells, the rate of loss of precancerous cells, the rate of cells changing from the precancerous to the malignant state, and the rates of reproduction and loss of the malignant cells. These parameters appear to be sufficient to fit the model to most epidemiological data on the incidence of cancer. Moolgavkar, 1999 argues "without ancillary biological information there is little point to fitting models postulating more than two stages to tumor incidence data." It has been shown that models which fail to include the stochastic birth and death dynamics of cells in the stages give different results than those models which do include those dynamics (Luebeck and Moolgavkar, 1994). These stage models, also promoted by experimentalists (Fearon and Vogelstein, 1990), abstract away the evolutionary dynamics of cancer. Progression to cancer is seen as a progression through a linear sequence of stages, rather than a diversification into a phylogeny of cell lines. There are no interactions between cells in these models, such as competition for resources.

Theoretical work could potentially help guide research into this fundamental area of cancer genetics. For example, we could ask, if cancer requires 2 (or more) selective mutations in genes such as p16, what is the chance of developing cancer? Or, if a mutation in a gene such as p53 boosts the mutation rate, how would this affect the probability of getting cancer? Since we have good epidemiological data on the probability of getting cancer, we can then make guesses as to the number and kind of mutations that are necessary for its development. We will begin with some simple analytical calculations and incrementally elaborate them until we are forced to move to a simulation-based model of the evolution of cancer.

Loeb's Paradox

In 1991 Loeb formulated the following paradoxically calculation for the incidence of cancer. From the literature on human cell cultures he takes a per base pair, per cell division mutation rate of 10^{-10} (Oller et al., 1989; Monnat Jr., 1989; Fukuchi et al., 1989;

Seshadri et al., 1987). He estimates that there are approximately 10^{16} cell divisions in a human lifetime. Finally, there are on the order of 10^9 base pairs in the human genome. Putting this together, we should expect $10^{-10} \times 10^{16} \times 10^9 = 10^{15}$ mutations in our cells during a human lifetime. If we are interested in the incidence of cells with two mutations at any loci, then this should occur $10^{-10} \times 10^{-10} \times 10^{16} \times 10^9 = 10^5$ times in a human lifetime. However, if a genetic disease requires 3 mutations to occur in the same cell, this should happen only once in 10^5 people. The chance of incurring 4 mutations is astronomically small. If these mutations must occur in specific loci, such as the coding regions of tumor suppressor genes and oncogenes, then the probability of developing cancer would be even smaller. Yet we believe that cancer requires a whole series of mutations (Armitage and Doll, 1954; Renan, 1993; Stein, 1991; Renan, 1993), and cancer is a frequent event during human lifespans.

Mutator Phenotype

One explanation for this paradox, offered in Loeb, 1998, is the idea of a "mutator" phenotype. Loeb's calculation changes if an early mutation, perhaps in p53, increases the mutation rate in the rest of the cell. Let us assume that the first event in this progression is a mutation that raises the mutation rate by c_m. Let μ be the mutation rate per locus per cell generation, k_m the number of critical genes necessary and sufficient to cause cancer, l_c the number of loci in a critical gene vulnerable to a cancer causing mutation, and let n_b be the number of cells in a human lifetime. To be generous, we will estimate that there are 100 different genes which, if they mutated, might raise the mutation rate. The expected number of cells that will independently develop cancer should be:

$$E[\text{Tumors}] = n_b \left[1 - (1 - \mu)^{l_c 100}\right] \left[1 - (1 - c_m \mu)^{l_c}\right]^{k_m} \tag{1}$$

Where $(1 - \mu)^{l_c 100}$ is the chance that a cell avoids a mutation in all $l_c 100$ loci that would produce the mutator phenotype. Thus $1 - (1 - \mu)^{l_c 100}$ is the probability that a cell has a mutation in at least one of the 100 genes that lead to the mutator phenotype. Here $c_m \mu$ is the increased mutation rate. Loeb estimated $n_b = 10^{16}$ and $\mu = 10^{-10}$. There are approximately 10^3 loci in a human gene at which point a deletion, insertion, or substitution is likely to affect the polypeptide which that gene encodes. So we will consider $l_c = 10^3$. Comparison of normal and malignant cell cultures has estimated a change in mutation rate due to malignancy of 1 to 3 orders of magnitude (Seshadri et al., 1987). If we assume that cancer requires the initial mutation in the mutator gene and then 3 more mutations, a total number of mutations that was astronomically unlikely in

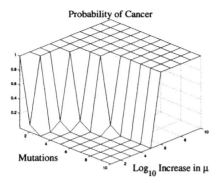

Figure 1: The expected number of cancerous cells that will develop during a person's lifetime. Two parameters are examined. The first parameter c_m is the increase in the mutation rate μ due to an initial mutation creating a mutator phenotype. This was calculated over the range of 10^1 to 10^{10}. The second parameter k_m is the number of mutations that are necessary and sufficient to cause cancer once the mutator phenotype has appeared, from 1 to 10. The expected number of cancerous cells has been truncated at 1.

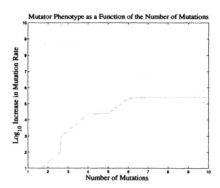

Figure 2: The predicted relationship between the increase in mutation rate of a "mutator phenotype" versus the number of mutations necessary to cause cancer after the appearance of the mutator phenotype. This an isocline calculated from Figure 1. This figure assumes a 0.4 probability of developing cancer during a lifetime. If the development of cancer requires many mutations, then the mutator phenotype would have to raise the mutation rate by at least 5 orders of magnitude.

Loeb's original estimation, and we assume that the mutator phenotype increases the mutation rate by 3 orders of magnitude, $c_m = 10^3$, then cancer should develop in $10^{16}[1 - (1 - \mu)^{10^3 10^2}](1 - (1 - 10^{-10} 10^3)^{10^3})^3 \approx 0.1$ cells in a human's lifetime. Figure 1 shows the log_{10} expected number of cancer cells dependent on k_m the number of mutations required and c_m the increase in the mutation rate due to the mutator phenotype. We have truncated the data at an expected single tumor because we are interested in the probability of developing cancer at least once.

Figure 1 shows that there is only a narrow window of mutation rate and number of sufficient mutations to develop cancer that result in realistic probabilities for developing cancer. In the United States, the chance of developing cancer during one's entire lifetime is approximately 40% (Ries et al., 1998). Figure 2 shows a view of the isocline where the probability of developing cancer is 40%. From this we can predict the relationship between the change in the mutation rate due to the emergence of the mutator phenotype and the number of mutations that are sufficient to cause cancer. For example, Figure 2 suggests that if the development of cancer requires 6 or more mutations after the initial rise in the mutation rate, then that initial increase must raise the mutation rate by at least 5 orders of magnitude.

Clonal Expansion

Loeb notes that Nowell, 1976, proposes another solution to his paradox. Some mutations can have selective effects and so increase the population of cells with that muta-

tion (Nowell, 1976). We can elaborate Loeb's calculations with the assumption that the necessary mutations along the progression to cancer all have selective effects. Thus, if a cell incurs such a mutation, it will increase in frequency to some number n_t which is approximately equal to the number of cells in a tumor. Again μ is the mutation rate, k_m the number of critical genes, l_c the number of loci in a critical gene vulnerable to a cancer causing mutation, and n_b is the number of cells in a human lifetime. We will assume that the mutations can occur in any order.

The chance of the first mutation occurring is 1 minus the chance that it doesn't occur:

$$Pr[\text{first mutation}] = 1 - (1 - \mu)^{l_c k_m n_b} \qquad (2)$$

This will cause the cell with that mutation to expand to n_t cells. ¿From then on, each new mutation has n_t chances of occurring in a background of cells carrying all the previous mutations. The probability that the remaining $k_m - 1$ mutations occur is then:

$$Pr[\text{other mutations}] = \left[1 - (1 - \mu)^{l_c n_t}\right]^{k_m - 1} \qquad (3)$$

Let us make some reasonable assumptions for the values of l_c, n_t and k_m. To estimate n_t we will consider Barrett's Esophagus, a precancerous condition of the esophagus studied by the Reid lab at the Fred Hutchinson Cancer Research Center. Biopsies collected from the neoplastic tissue of the patients typically include 10^6 cells in a 2mm by 5mm section of epithelium. The entire Barrett's region averages approximately a surface area of 50mm by 60mm, or 10 biopsies by 30 biopsies. So

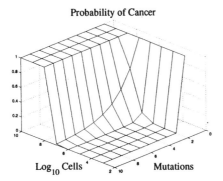

Probability of Cancer

Log₁₀ Cells   Mutations

Figure 3: The probability of developing cancer during a person's lifetime. Two parameters are examined. The cell population size to which a selected mutant grows has been calculated over the range of 10^3 to 10^{10}. The second parameter is the number of selected mutations that are necessary and sufficient to cause cancer, from 1 to 10. These calculations estimate that if the selected population size is below 10^6 there is little chance of developing cancer. If it is 10^8 or above, a person is guaranteed to develop cancer during their lifetime.

the entire surface area can be sectioned into 300 biopsies of 10^6 cells, for a total of 3×10^8 cells. Since mutant clones are often observed to have expanded over the entire Barrett's region of a patient, it seems reasonable to set $n_t = 10^8$. Let us consider the case where $k_m = 4$ mutations are necessary to cause cancer. Recall that Loeb calculates the chance of 4 mutations occurring in the same cell to be astronomically small. Then,

$$Pr[\text{first mutation}] = 1 - (1 - 10^{-10})^{4 \times 10^{19}} \approx 1 \quad (4)$$

This number is so close to 1 that most computers cannot represent it as anything other than 1. So many cells are generated in a human lifetime that there are probably many cells that carry a mutation at any given locus. The interesting dynamics lie in the sequence of mutations that follow the first one:

$$Pr[\text{other mutations}] = \left[1 - (1 - 10^{-10})^{10^{11}}\right]^3 = 0.99986 \quad (5)$$

Given our assumptions, we estimate that 4 specific selected mutations are almost certain to occur in the lifetime of an individual. Of course, our estimates may be off. Figure 3 shows the probability of suffering cancer as a function of the number of cells to which selected mutant expands (n_t) and the number of selective mutations necessary and sufficient to cause cancer (k_m).

Figure 3 shows a precipitous drop in the probability of experiencing cancer as we reduce our estimate of the number of cells in a tumor from 10^8 to 10^6. The SEER

report from the National Cancer Institute (Ries et al., 1998) estimates the lifetime probability of being diagnosed with cancer in the US is 45% for men and 38% for women (for all races and cancer sites combined). To match this estimate, our rough calculations suggest that in general cancers would require 3 selected mutations and those mutant clones would tend to spread to populations of 10^7 cells. Of course, this is an extremely simplified model of the incidence of cancer. We have not accounted for any environmental effects, genetic predispositions, or indeed any mutations that are necessary but do not spread through selection. Nevertheless, the elaboration to Loeb's calculations shows that Nowell's insight resolves the paradox. We develop cancer because the cells in the neoplastic tissues are evolving.

Both Mutator and Selective Effects

The two elaborations of Loeb's calculations consider selective and mutator mutations separately. A more realistic view of the development of cancer would likely consider both selective mutations and mutations that raise the mutation rate, and their interactions. In addition, there may be "neutral" mutations which have no effect on cell proliferation rates or mutation rates.

Consider the case in which a mutator or neutral mutation arises in a cell of the tumor. There is no reason to believe that this mutation would spread rapidly in the tumor. Without a selective advantage, such a mutation would be unlikely to grow to dominate the entire tumor. Meanwhile, if a selective mutation occurred in a cell which lacked the mutator or neutral mutation, the selective mutation would tend to expand throughout the tumor and thereby displace the mutant population with the mutator or neutral mutation. Thus, it is important to keep track of both the cells with the mutator or neutral mutations, as well as the cells that are free of those mutations but may yet suffer selective mutations. Each subpopulation can be characterized by the number of selective, mutator, and neutral mutations it has suffered, along with its population size. A set of difference equations can describe the growth dynamics of these subpopulations, as well as mutations that move cells from one subpopulation to another. But what growth dynamics should we use? The fundamental dynamic of biological reproduction is exponential. Is this a reasonable representation of tumor dynamics in humans?

In the esophagus, as in most of the digestive tract, cells along the lining (epithelium) are constantly being sloughed off and destroyed. These losses are replenished by the division of cells in the lining. In the case of Barrett's Esophagus, these cells are precancerous and hyperproliferative. The estimated turnover time is about once a week (Madara, 1995). As the cells are spatially structured as a two-dimensional layer (lining of a cylinder), there are severe spatial constraints restricting exponential growth. Further, cell division (mitosis) is a local process, and so most new cells must compete for space with their immediate ancestors. The easiest way to represent a heterogeneous population of cells growing in a two-dimensional environment is with a two-dimensional model resembling a cellular automaton.

The Model We represented the the states of all the precancerous cells in the lining of the Barrett's region of an esophagus. We instantiated this as a two-dimensional discrete-event simulation in the shape of a column with "wrap-around" boundaries on the left and right sides, but not on the top and bottom. The state of a cell in this grid has four components: the number of selective mutations it has suffered (0-4), the number of neutral mutations it has suffered (0-4), whether or not it has suffered a mutation that increases its mutation rate (a "mutator" mutation), and its age (0-16). The population of cells is updated serially in a time step which represents approximately half a day. The time until the next reproduction (mitotic) event for each cell is drawn from a normal probability distribution with a mean of 8 times steps and a standard deviation of 2 time steps. Each selective mutation has the effect of doubling the replication rate of the cell. Thus a cell that has incurred 2 selective mutations reproduces 4 times as fast as a normal cell. When a cell divides, the new cell has a 50% chance of displacing one of the 9 cells, selected with uniform probability, in the 3 by 3 cell neighborhood centered on the parental cell. A run of the model began with all cells at age 0 with no mutations. With each time step representing 12 hours, we ran the model for 54,000 time steps (approximately 74 years), or a human lifetime. This put practical limitations on the number of cells we could model, with a maximum of 256 by 256 (65,536) cells. In the future we hope to model more realistic tumor sizes with approximately 10^8 cells.

We model the mutation rate as a Bernoulli process. The probability of a cell changing state is

$$Pr[\text{mutation}] = 1 - (1 - \mu)^{(S+N+M)l_c n_p} = P \quad (6)$$

Where $\mu = 10^{-10}$ is the mutation rate per base pair per cell generation, $S, N,$ and M are the numbers of selective, neutral, and mutator genes sufficient and necessary to cause cancer if mutated, $l_c = 10^3$ is the number of critical base pairs (loci) in each gene at which a mutation could have a carcinogenic effect. In most cases, we assume that these mutations "knock out" the gene by either turning it off or destroying the functional effects of the normal protein produced by the unmutated gene. The last parameter, $n_p = 2$, is the number of independent pathways to cancer. This is an estimate of the number of genes in which a mutation will have the same carcinogenic effect. If a cell had at least one mutator gene mutated then μ increased by 10^3. This parameter for the increase in the mutation rate was called c_m in our earlier calculations. We primarily experimented

with parameters $S, N,$ and M, with some exploration of μ, l_c and the degree of increase in μ due to the mutator phenotype. A cell was called malignant if it had S selective mutations and N neutral mutations. We assumed that the mutator phenotype was not necessary for malignancy but only played a facilitating role through the increase in the mutation rate of the selective and neutral genes.

A Bernoulli process can be simulated by calculating the interarrival time for the next success. That is, instead of flipping a biased coin with probability of success P for each trial of the Bernoulli process, we can ask when the next success will happen. The probability mass function for the interarrival time k, the number of trials up to and including the next success, of a Bernoulli process is the geometric distribution:

$$Pr[k] = P(1-P)^{k-1} \quad (7)$$

for $k = 1, 2, \ldots$. The expected value of k is $E[k] = 1/P$. When P is very small, as it is for most mutation rates, this function very gradually drops off. In this case, for the purposes of efficiency, it is reasonable to approximate $Pr[k]$ as a uniform distribution from 0 to $2/P$, which has the same expected value $E[k] = 1/P$ although a smaller variance. We calculated this with a single call to the pseudorandom number generator, using a version of Knuth's subtractive method (Knuth, 1981, pp. 171–172) to generate the pseudorandom numbers. We assume that the processes of DNA synthesis and cell division are the primary causes of mutations. Thus, in our model mutations only occur at cell division (Paulovich et al., 1997; Zheng et al., 1993). Mutations have an equal probability of occurring in the new or parental cell.

At the end of a run we measured the proportion of cells that suffered enough mutations to cause cancer (S and N). We ran the model at least 50 times for each parameter setting. A grey-scale picture of the model in the midst of a run is shown in Figure 4.

Results A run of the model was considered to have led to cancer if the final population had at least 1 cell with number of mutations required for malignancy (S and N). Figures 5 and 6 show the resulting probability of developing cancer as a function of the number of selective mutations S and neutral mutations N necessary and sufficient for developing cancer. Figure 5 shows the probabilities when there is no mutator gene to raise the background mutation rate. Figure 6 shows the results of the same parameter configurations when there is a mutator gene that may also mutate and thereby raise the mutation rate by 3 orders of magnitude.

Figure 7 is an extraction of a single curve from Figures 5 and 6 where $N = 1$. Figure 7 also shows the 90% confidence intervals around these curves calculated by treating the probability of developing cancer as a Bernoulli process. When there is no mutator gene in the

Figure 4: A view of the model running. The cells are color coded by lineage. The lighter grey lineages share an ancestor that suffered a selective mutation. This mutant clone is in the process of sweeping through the entire tissue.

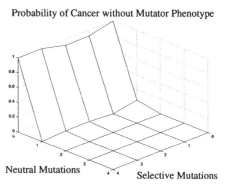

Figure 5: A plot of the probability of developing cancer as a function of the number of selective and neutral mutations necessary and sufficient to cause the disease. These probabilities have been calculated in the absences of a mutator gene. The probabilities are dominated by the number of neutral mutations that are necessary. The probabilities were calculated by at least 50 runs of the agent-based model with only 4096 cells.

system, the probability of developing cancer decreases with the number of selective mutations that are required. This seems reasonable in light of our earlier calculations. However, in the presence of a mutator gene that can raise the mutation rate at any time, the probability of developing cancer actually increases with the number of necessary selective mutations. This is because a selective mutation generates a large number of new cells, each cell representing a potential new mutation. When the mutation rate is high enough, this forms a positive feedback system in which one selective mutation generates the next selective mutation and so on until the system reaches malignancy.

Our explorations of other parameters in the system all show a relationship between the parameters and the probability of developing cancer that is either linear or sub-linear. In all of these cases we assume that 1 neutral and 2 selective mutations is necessary and sufficient for the development of cancer. .The exponents for these relationships were derived from the slope of the line that was fit to the log transformation of the data. It should be noted that in all cases the line was fit with only 3 or 4 data points, and so the results should be taken only as a qualitative indication of the dynamics of the system. Figures 8 and 10 are log-log plots of the relationship of the parameter to the probability of developing cancer. Figures 9 and 11 had to be plotted as a log-linear plots due to the calculated 0 probability of developing cancer in some instances. Figure 8 shows that the probability of developing cancer increases as a square root (in the presence of a mutator gene) or linear function (in the absence of a mutator gene) of the number of cells produced by a selective mutation. Figure 9 shows that the

probability of developing cancer increases in proportion to the square root of the mutation rate. Figure 10 shows that this probability also increases in proportion to cube root of the change in mutation rate caused by a mutation in the mutator gene, i.e., the difference between the normal and the mutator phenotype. Finally, Figure 11 shows that the probability of developing cancer increases roughly in proportion to the number of base pairs in the genes at which a mutation can have a carcinogenic effect. Of course, since probabilities are bounded at 0 and 1, these relationships may break down as they near those boundaries.

Discussion

Other researchers have studied the relative merits of the two solutions to Loeb's paradox (Tomlinson and Bodmer, 1995; Tomlinson et al., 1996). Tomlinson et al., 1996 concluded that selective mutations alone are sufficient to explain the mutations observed in cancer. In their investigation of the mutator phenotype, they investigated the case where either 2 or 6 neutral mutations were necessary to cause cancer (Tomlinson et al., 1996). They assumed the mutator phenotype raised the mutation rate from 10^{-8} to 10^{-4}. They found that in the case of requiring 2 neutral mutations, cancer often developed before the mutator phenotype appeared, but with 6 required mutations, the mutator phenotype would appear before cancer. They argue that the importance of a mutator cell will be wiped out if any of the other mutations have selective effects. Our results do not support this. The presence of a few selective mutations amongst many neutral mutations has little effect. However, the

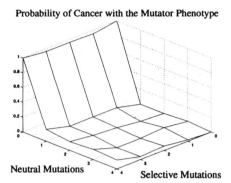

Probability of Cancer with the Mutator Phenotype

Neutral Mutations   Selective Mutations

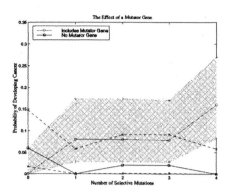

Figure 6: A plot of the probability of developing cancer as a function of the number of selective and neutral mutations necessary and sufficient to cause the disease. These probabilities have been calculated in the presence of a mutator gene that raises the background mutation rate from 10^{-10} to 10^{-7} when it is mutated. The mutator gene has the same probability of mutation as the other genes, and thus the background mutation rate may change at any time during the run of the model. This contrasts with Figure 1 in which we assumed that the mutator gene had been mutated before the other genes. Note that the probability of developing cancer rises with the number of selective mutations involved irrespective of the number of necessary neutral mutations.

Figure 7: The interaction of mutator and selective genes. The solid lines show the probability of developing cancer as a function of the number of necessary selective mutations. In all cases 1 neutral mutation was required. Both solid lines are surrounded by their 90% confidence intervals shown in dotted lines for the mutator case and dashed lines for the case without a mutator gene. The confidence interval for the mutator case is shaded in grey. There is a synergy between large numbers of necessary selective genes and the mutator gene. In the presence of a mutator gene, the probability of developing cancer actually increases with the number of necessary selective genes. In this case the process of developing cancer has a sort of positive feedback effect that quickly generates malignant cells. In the absence of a mutator gene the probability of developing cancer goes down with the number of necessary selective genes.

combination of selective and mutator mutations dramatically increases the probability of developing cancer, as is shown in Figure 7.

An important aspect of both the analysis of selective and mutator mutations in cancer is that the parameters of the predictions are observable and thus the predictions are experimentally testable. Data is becoming available on the population sizes of cells with selective mutations, and it is becoming feasible to measure the mutation rate in cells with mutator phenotypes, perhaps through the loss of p53. Similarly, it should be possible to derive accurate measurements of the number of critical loci in any given gene relevant to the development of cancer. In the model we assumed this number was about 10^3 for all genes, an estimate that could be improved significantly. In the foreseeable future we will be able to reduce the ranges of the significant parameters in the model when information about the number and kinds of mutations that are sufficient for the development of cancer is determined.

Our simulation of the development of cancer is only a toy model and as such it avoids many of the known complexities of the biological system. We have implicitly assumed that each mutation is independent of the others, and so can occur in any order. Further, we have not explicitly represented the phenomenon of dominance

in which a recessive phenotype might require two mutations before it appeared. However, this could be represented by the combination of a neutral mutation, which occurs first, and a selective mutation, which would follow the neutral mutation. We have also ignored the effects of cell senescence. Most cells stop dividing after some number of divisions have shortened the telomeres to the point where they no longer protect the ends of the chromosomes.

Only one type of selective effect has been modeled. However, mutations can have strong selective effects without changing the generation time of a cell. Mutants that tend to compete successfully for space, either by displacing their neighbors or by resisting displacement by future competitors, would also spread in the population. There are probably a variety of other genetic innovations that would have beneficial phenotypic effects. Most of these could be represented and explored in an elaborated model.

Our model of the mutator phenotype is probably inappropriate. We have modeled the mutator phenotype as a dramatic boost in the background mutation rate. This assumes that mutations occur independently throughout

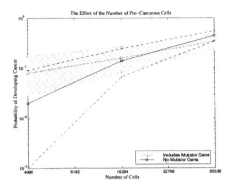

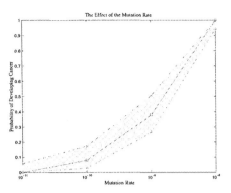

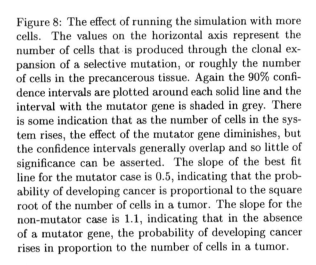

Figure 8: The effect of running the simulation with more cells. The values on the horizontal axis represent the number of cells that is produced through the clonal expansion of a selective mutation, or roughly the number of cells in the precancerous tissue. Again the 90% confidence intervals are plotted around each solid line and the interval with the mutator gene is shaded in grey. There is some indication that as the number of cells in the system rises, the effect of the mutator gene diminishes, but the confidence intervals generally overlap and so little of significance can be asserted. The slope of the best fit line for the mutator case is 0.5, indicating that the probability of developing cancer is proportional to the square root of the number of cells in a tumor. The slope for the non-mutator case is 1.1, indicating that in the absence of a mutator gene, the probability of developing cancer rises in proportion to the number of cells in a tumor.

Figure 9: A log-linear plot with %90 confidence intervals of the effect of changing the background mutation rate μ. The increase in the chance of developing cancer is roughly proportional to the increase in the mutation rate. If we fit a line to the log-transform of the axes, ignoring the 0 value, the probability of developing cancer is proportional to the square root of the mutation rate (the slope = 0.5).

Prediction 1 *The development of cancer requires at least 2 selectively neutral mutations.*

Our model of 2^{16} cells with 1 neutral and 2 selective mutations sufficient for developing cancer, in the presence of a mutator gene, led to a cancer incidence of 35%. With a more realistic number of cells in a tumor, perhaps 10^8, the simulated incidence of cancer would be unrealistically high. Requiring more selective mutations only makes the incidence of cancer higher. Thus, cancer must require more selectively neutral mutations.

Prediction 2 *The development of cancer involves a number of neutral mutations that is within the same order of magnitude as the number of selective mutations.*

the genome. However, our archetypal candidate for a mutator gene, p53, seems to cause the loss (and gain) of whole chromosomes as well as prevent the repair of damaged DNA. In the case of chromosome loss, mutations in genes are not independent and tend to occur in massive clusters. Furthermore, we have not modeled the effects of deleterious mutations. We would expect an increase in the background mutation rate to also increase the frequency of deleterious mutations, which would result in a selective disadvantage, and sometimes fatal, effect on the host cell.

Finally, we have completely ignored the immune response. We know that the human immune system sometimes attacks precancerous and cancerous cells (Jantscheff et al., 1999), but the details of these dynamics are still unknown. The immune system would clearly have selective effects on the populations of cells. The immune system could lower the probability of developing cancer relative to our estimates.

The simplifications of our models and our ignorance of realistic parameter values prevent us from making highly focused experimental predictions. However, the qualitative behaviors of the models do lead to two predictions:

The clonal expansion of a selective mutant produces a large population of mutant cells and involves a large number of cell divisions in which new mutations may arise. The chance of a neutral mutation is greatly enhanced if it follows a selective mutation. However, if more neutral than selective mutations are required for the development of cancer, then the neutral mutations form bottlenecks in path to cancer and make malignancy more unlikely. This was at the heart of Loeb's paradox. On the other hand, if few neutral mutations but many selective mutations are required, and there exist mutator genes in the system, then the mutator genes and the selective genes form a positive feedback system that accelerates the system towards cancer. Since mutations in p53 are common across most forms of cancer (Smith and Fornace, 1995), it is reasonable to suppose that there is a mutator gene in the system. In this case, with few necessary neutral mutations, the probability of developing

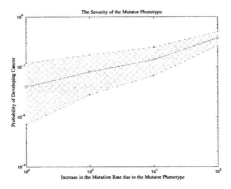

Figure 10: A log-log plot with %90 confidence intervals of the results from adjusting the effect of the mutator phenotype. The horizontal axis shows the change in the background mutation rate caused by a mutation in the mutator gene. The probability of developing cancer increases in proportion to the cube root (exponent = 0.3) of the change in the mutation rate due to the mutator phenotype.

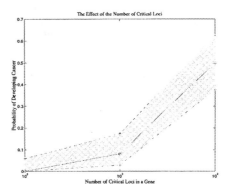

Figure 11: A log-linear plot with %90 confidence intervals of the effect of varying the assumed number of base-pairs or loci in a gene at which a mutation could have a carcinogenic effect. In the case of a tumor suppressor gene, this would correspond to the number of different mutations that could knock out the gene. It is difficult to fit a line to the log transform of the data since 1 of the 3 data points is 0. If we guess that the probability of getting cancer when each gene has 10^2 critical loci is between 0.01 and 0 (we only ran the model 50 times so we lack the resolution to distinguish probabilities this low), and replace that 0 value with 0.005, the slope of the line is 1. Thus, the probability of developing cancer is roughly proportional to this number of critical loci in a gene.

cancer is too high to be realistic. Thus, the number of necessary neutral mutations must be close to the number of selective mutations. Our guess for the meaning of "close" is the same order of magnitude.

What insights might we derive from these results for the treatment or prevention of cancer? All of the analyses suggest that neutral mutations are the bottleneck in the development of cancer. This implies that an effective prevention program would be one in which would add at least one additional neutral mutation to the set of necessary mutations for the development of cancer. In other words, we should try to add bottlenecks to the development of cancer. This might, for example, be achieved by treatments for which the precancerous cells would have to generate recessive mutations in order to escape the treatment and to progress on towards cancer. If the susceptible phenotype is completely dominant, then a mutation in one of the two alleles of a homozygous dominant cell will have no phenotypic effect and will thus be selectively neutral. Similarly, cocktails of multiple drugs (Hughes and Frenkel, 1997) that require mutations at multiple sites in order to develop resistance to all of the drugs in the cocktail might be particularly effective.

The authors gratefully acknowledge the support of the National Science Foundation (grants IRI-9711199 and CDA-9503064), the Office of Naval Research (grant N00014-99-1-0417), and the Intel Corporation. Thanks to Terry Jones for coding Knuth's pseudorandom number generator, Keith Wiley for writing the GUI (and capturing it in the act), the UNM adaptive group and two anonymous referees for their comments. A special thanks to Miller Maley for his advice on modeling probabilistic processes. We would also like to thank the Brian Reid lab at the Fred Hutchinson Cancer Research Center for their welcome, enthusiasm, and generous collaboration. Without their support, this paper could not have been written.

References

Armitage, P. and Doll, R. (1954). The age distribution of cancer and a multi-stage theory of carcinogenesis. *British Journal of Cancer*, 8:1–12.

Barrett, M. T., Sanchez, C. A., Prevo, L. J., Wong, D. J., Galipeau, P. C., Paulson, T. G., Rabinovitch, P. S., and Reid, B. J. (1999). Evolution of neoplastic cell lineages in Barrett oesophagus. *Nature Genetics*, 22:106–109.

Collins, R. and Jefferson, D. (1992). The evolution of sexual selection and female choice. In Varela, F. J. and Bourgine, P., editors, *Toward a Practice of Autonomous Systems: Proceedings of the First European Conference on Artificial Life*, pages 327–336, Cambridge, MA. MIT Press.

Fearon, E. R. and Vogelstein, B. (1990). A genetic model for colorectal tumorigenesis. *Cell*, 61:759–767.

Fujii, H., Marsh, C., Cairns, P., Sidransky, D., and

Gabrielson, E. (1996). Genetic divergence in the clonal evolution of breast cancer. *Cancer Research*, 56:1493–1497.

Fukuchi, K., Martin, G. M., and Monnat Jr., R. J. (1989). Mutator phenotype of werner syndrome is characterized by extensive deletions. *Proceedings of the National Academy of Sciences USA*, 86:5893–5897.

Hughes, R. S. and Frenkel, E. P. (1997). The role of chemotherapy in head and neck cancer. *American Journal of Clinical Oncology*, 20:449–461.

Jantscheff, P., Herrmann, R., and Rochlitz, C. (1999). Cancer gene and immunotherapy: Recent developments. *Medical Oncology*, 16:78–85.

Kastan, M. B. (1997). Molecular biology of cancer: The cell cycle. In DeVita, Jr., V. T., Hellman, S., and Rosenberg, S. A., editors, *Cancer: Principles and Practice of Oncology, Fifth Edition*, pages 121–134. Lippincott-Raven Publishers, Philadelphia, PA.

Knuth, D. E. (1981). *The Art of Computer Programming, vol. 2*. Addison-Wesley, Reading, MA.

Lengauer, C., Kinzler, K. W., and Vogelstein, B. (1998). Genetic instabilities in human cancers. *Nature*, 396:643–649.

Levin, S. A., Grenfell, B., Hastings, A., and Perelson, A. S. (1997). Mathematical and computational challenges in population biology and ecosystems science. *Science*, 275:334–343.

Little, M. P. (1995). Are two mutations sufficient to cause cancer? *Biometrics*, 51:1278–1291.

Loeb, L. A. (1998). Cancer cells exhibit a mutator phenotype. *Advances in Cancer Research*, 72:25–56.

Luebeck, E. G. and Moolgavkar, S. H. (1994). Simulating the process of malignant transformation. *Mathematical Biosciences*, 123:127–146.

Madara, J. L. (1995). Epithelia: Biologic principles of organization. In Yamada, T., editor, *Textbook of Gastroenterology, second edition*, pages 237–257. J. B. Lippincott Co., Philadelphia.

Maley, C. C. (1998). *The Evolution of Biodiversity: A Simulation Approach*. PhD thesis, Massachusetts Institute of Technology, Cambridge, MA.

Monnat Jr., R. J. (1989). Molecular analysis of spontaneous hypoxanthine phosphoribosyltransferase, mutations in thioguanine-resistant HL-60 human leukemia cells. *Cancer Research*, 49:81–87.

Moolgavkar, S. H. (1999). Stochastic models for estimation and prediction of cancer risk. In Barnett, V., Stein, A., and Turkman, K. F., editors, *Statistics for the Environment 4: Pollution Assessment and Control*, pages 237–257. John Wiley and Sons, Ltd., New York, NY.

Moolgavkar, S. H. and Luebeck, E. G. (1990). Two-event model for carcinogenesis: Biological, mathematical, and statistical considerations. *Risk Analysis*, 10:323–341.

Neshat, K., Sanchez, C. A., Galipeau, P. C., Cowan, D. S., Ramel, S., Levine, D. S., and Reid, B. J. (1994). Barrett's Esophagus: A model of human neoplastic progression. *Cold Spring Harbor Symposia on Quantitative Biology*, 59:577–583.

Nowell, P. C. (1976). The clonal evolution of tumor cell populations. *Science*, 194:23–28.

Oller, A. R., Rastogi, P., Morgenthaler, S., and Thilly, W. G. (1989). A statistical model to estimate variance in long term-low dose mutation assays: Testing of the model in a human lymphoblastoid mutation assay. *Mutation Research*, 216:149–161.

Paulovich, A. G., Toczyski, D. P., and Hartwell, L. H. (1997). When checkpoints fail. *Cell*, 88:315–321.

Reid, B. J. (1991). Barrett's esophagus and esophageal adenocarcinoma. *Gastroenterology Clinics of North America*, 20:817–834.

Renan, M. J. (1993). How many mutations are required for tumorigenesis? Implications from human cancer data. *Molecular Carcinogenetics*, 7:139–146.

Ries, L. A. G., Kosary, C. L., Hankey, B. F., Miller, B. A., and Edwards, B. K., editors (1998). *SEER Cancer Statistics Review, 1973-1995*. National Cancer Institute, Bethesda, MD.

Seshadri, R., Kutlaca, R. J., Trainor, K., Matthews, C., and Morley, A. A. (1987). Mutation rate of normal and malignant human lymphocytes. *Cancer Research*, 47:407–409.

Sherman, C. D. and Portier, C. J. (1996). Stochastic simulation of a multistage model of carcinogenesis. *Mathematical Biosciences*, 134:35–50.

Sherr, C. J. (1996). Cancer cell cycles. *Science*, 274:1672–1677.

Smith, M. L. and Fornace, A. J. (1995). Genomic instability and the role of p53 mutations in cancer cells. *Current Opinion in Oncology*, 7:69–75.

Stein, W. D. (1991). Analysis of cancer incidence data on the basis of multistage and clonal growth models. *Advances in Cancer Research*, 56:161–213.

Tomlinson, I. P. M. and Bodmer, W. F. (1995). Failure of programmed cell death and differentiation as causes of tumors: Some simple mathematical models. *Proceedings of the National Academy of Sciences of the United States of America*, 92:11130–11134.

Tomlinson, I. P. M., Novelli, M. R., and Bodmer, W. F. (1996). The mutation rate and cancer. *Proceedings of the National Academy of Sciences of the United States of America*, 93:14800–14803.

Zheng, C.-J., Byers, B., and Moolgavkar, S. H. (1993). Allelic instability in mitosis: A unified model for dominant disorders. *Proceedings of the National Academy of Sciences USA*, 90:10178–10182.

From Individuals to Populations, Approaches to the Study of Biological Emergent Phenomena

Anthony Liekens

Vrije Universiteit Brussel (Brussels, Belgium)

<mooby@alife.org>

Abstract

An assorted range of approaches have contributed to our understanding of the oscillatory behavior of population sizes in predation models. Among these are Mathematical Biology, Statistics and Artificial Life (ALife). In this paper, I will give a review of these different approaches. In addition, another approach, based on Evolutionary Game Theory, is proposed and discussed.

This paper also suggests that a complementary study of both the Mathematical, Artificial Life and Game Theory approach is needed to explain some of the mysticism surrounding the global emergent behavior of local predator-prey relationships.

Introduction

Emergent behavior

The first time you see a magician perform a trick, you're probably not able to explain the trick that just *emerged* before you. If you analyze and see the trick over and over again, you learn which ingredients are needed to recreate or *synthesize* the illusion yourself. If you still can't give a proper explanation of the inner workings of the trick, you could have missed a crucial mechanism in your analysis. Artificial Life researchers often look at phenomena in living systems and societies of agents in a similar manner. They try to synthesize biological phenomena using the simplest possible set of building blocks.

Even with some of the analytical components explaining the phenomenon partially, the analysis can still reach a dead end. At such point the illusion surrounding it tends to be called *emergent*:

"Emergence is often invoked in an almost mystical sense regarding the capabilities of behavior-based systems. It is true that what occurs in a behavior based system is often a surprise to the system's designer, but does the surprise come because of a shortcoming of the analysis of the constituent behavioral building blocks and their coordination, or because something else?" Arkin [Ark98], page 105

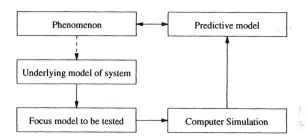

Figure 1: A model of modeling, showing the different phases of modeling a system and testing of its behavior.

Different approaches have contributed to our understanding of the oscillatory behavior of predator and prey population sizes in predation models. While we expect a stable configuration of these sizes to emerge from the system, they persist in having oscillations over time. Mathematical Biologists have made a couple of attempts to explain this phenomenon, but seem to ignore the difference between local inter-individual interactions and the global behavior of the system. Artificial Life, on the other hand, specifically studies this emergent behavior, but often neglects a sufficient and thorough analysis of the model itself. Evolutionary Game Theory has the power to explain *why* a particular phenomenon exists, where Mathematical Biology rather describes *what* happens. Finally, Artificial Life can shed light on *how* it happens. This paper combines these three approaches to find an explanation for the emergent phenomenon of oscillating predator and prey populations, whose magic disappears due to the analysis.

For an in-depth discussion of emergence, I refer to [Ron99].

Modeling modeling

In [Dac90, Gun90, Gie94, Mor99], the authors propose a model of modeling itself. A simplified version of the model is shown in Fig. 1.

A first step in the study of a model of a certain phenomenon is the definition of a hypothesis for the system, which is the underlying model of the system to be tested.

From this hypothesis, a set of rules is derived, to which the agents in the model should behave. This is the focus model to be tested for the phenomenon at hand. A computer simulation is built according to this set of rules. The output data from this computer simulation is then moulded into a predictive model, which then could be matched against the real life data of the phenomenon at hand.

Emergence arises when the set of rules - the focus model to be tested - is implemented, analyzed and moulded into the predictive model. When the set of local rules is matched against unexpected global behavior, emergence is discovered, due to the lack of explanatory capacities of the computer simulation used to study the system and its emerging behavior. This paper argues that using a bigger range of computational models, instead of a single computer simulation experiment, should be used to find a more profound match between the set of local rules and the global behavior of the system. Only by thorough analysis of the emerging behavior the unexpected predictive model *can* become the "obvious" model.

Complexity is not the art of discovering emergent phenomena, but the art of revealing the systems being at the basis of the emergent behavior.

Emergence in predator-prey systems

One of the first rules in any ecology handbook is that "an ecosystem urges towards a constant working level". With this rule, a student in ecology would expect that in a dynamical system of interacting predator and prey individuals, the population sizes would converge to an equilibrium over time. In this equilibrium, there are enough prey individuals to feed the predator population, and precisely enough predator agents to predate the prey. The system could then remain in this equilibrium. The surprise that the population sizes oscillate over time, and never even reach an equilibrium can be called emergent. This paper will show that thorough analysis, not only through Mathematics but also through Game Theory, can give an explanation for this unexpected behavior of the global predator and prey community.

Lotka-Volterra systems

Volterra [Vol26] first proposed a simple mathematical model for the predation of one species by another to explain the oscillatory levels of certain fish catches in the Adriatic. If $N(t)$ is the prey population and $P(t)$ that of the predator at time t then Volterra's model is

$$\frac{dN}{dt} = N\left(a - bP\right), \frac{dP}{dt} = P\left(cN - d\right)$$

where a, b, c and d are positive constants.

This model is widely known as the Lotka-Volterra model since the equations were also derived by Lotka

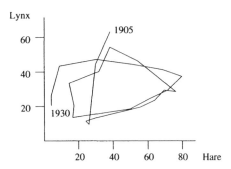

Figure 2: Fluctuations in the number of pelts sold by the Hudson Bay Company. Phase plane plot of the data of the 25 year period starting in 1905. (after Odum [Odu53], Elton and Nicholson [Elt42], Gilpin [Gil73])

[Lot20, Lot25]. [Mur89] is a good source for finding similar mathematical models of biological systems.

The Statistical Approach

There have been many attempts to apply the Lotka-Volterra model to real-world oscillatory phenomena. In view of the system's structural instability, they must essentially all fail to be of quantitative practical use. However, they can be important as vehicles for suggesting relevant questions that should be asked. One particular interesting application is to apply the model to extensive data on the Canadian lynx-snowshoe hare interaction in the fur catch records of the Hudson Bay Company from about 1845 until the 1930's. We assume that the numbers reflect a fixed proportion of the total population of the animals. Although this assumption is of questionable accuracy, the data nevertheless represents one of the very few long-term records available. Fig. 2 reproduces part of this data (pelts gathered from 1905 until 1930). Williamson's book [Wil72] is a good source of population data which exhibit periodic or quasi-periodic behavior.

The availability of pelts shows dramatic, closely linked cycles between the population sizes of the predators and their prey. Unfortunately, as with any field investigation, many variables could influence the relationship between hare and lynx.

The moral of the story is that it is not enough simply to produce a model which exhibits oscillations but rather to provide a proper explanation of the phenomenon which can stand up to ecological and biological scrutiny.

The Artificial Life Approach

While field researchers in Biology don't find uncomplicated examples in nature, Artificial Life researchers try to create these examples through the synthesis of living beings, and their emerging phenomena, in models such as computer simulations. Individuals are programmed according to a couple of rules. A set of these individuals

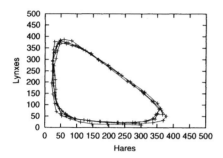

Figure 3: Oscillatory dynamics of the population sizes of grass patches (+), hares (×) and lynxes (⋆) in the simulation as a phase plot. The direction of change is clockwise.

is supplied to a simulation. Unnecessary (and hopefully disturbing) parameters are left out so the population of individuals as a whole - and the hypothesis of predation being at the basis of the oscillatory dynamics in population sizes - can be studied as the biological phenomenon at hand.

For analyzing the predator-prey model we create a set of individuals, each one being either a patch of grass, a hare or a lynx. Repeatedly, a couple of individuals are randomly picked from our artificial life-community [1]. Four rules are applied to this randomly selected couple of individuals: (1) If a grass patch and a hare were picked, the grass patch individual becomes a hare; (2) If a hare and a lynx were taken, the hare individual becomes a lynx; (3) If a lynx and a grass patch were selected, the lynx becomes a grass patch; (4) If two similar individuals are picked from the community, nothing happens. These rules simulate the acquisition of energy through predation, needed for reproduction.

Running the simulation

This system of interaction between individuals, either being grass patches, hares and lynxes, was implemented. The C++ sources of the computer simulations are electronically retrievable from `ftp://ftp.nerdhero.org/pub/alife/`.

An artificial life community consisting of the three different species was initialized and the simulation, based on the implemented rules, was ran. Fig. 3 shows the oscillatory dynamics of the different populations in the simulation.

Can we now call this oscillatory behavior of the system "emergent"? In [Ron99], Ronald, Sipper and Capcarrere propose an "emergence test". The test classifies the phenomenon as emergent if the language of design

---

[1] A life community is a set of populations. A population is a set of similar individuals in the same place at the same time. In this context, the artificial life-community denotes the life community in an artificially created world.

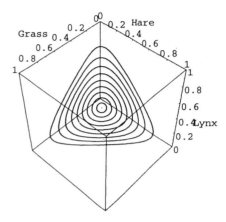

Figure 4: Theoretical phase plane trajectories of grass, hare and lynx population sizes (proportions) in the simulated model. Curves are in the $G + H + L = 1$ plane.

L_1 and the language of observation L_2 are distinct, and the causal link between the elementary interactions programmed in the local rules of L_1 and the behaviors observed in L_2 is *non-obvious* to the observer. In our example, we would expect the population sizes to converge to an equilibrium, but the system keeps oscillating, thereby avoiding the equilibrium.

Although our system seems to have emergent properties, I will show that the surprising oscillatory behavior can be explained by describing L_2 in terms of L_1 mathematically, and that the result of this mathematical analysis has similar properties as the Lotka-Volterra model.

Analyzing the simulation

$G(t)$, $H(t)$ and $L(t)$ are the sizes of the grass patch, hare and lynx populations, at a given time t with $G(t) + H(t) + L(t) = C$, where C is the constant total size of the life community. We can describe the behavior of the population sizes, L_2, mathematically, using our set of rules, L_1. We can write the dynamics of the populations as differential equations:

$$\frac{dG}{d\tau} = G(L - H), \quad \frac{dH}{d\tau} = H(G - L), \quad \frac{dL}{d\tau} = L(H - G)$$

with $\tau \cong t$. By solving this system of differential equations we can find the theoretical trajectories of the population sizes in the $G + H + L = C$ plane:

$$GH(G + H - C) - \left(g_0 h_0^2 + g_0^2 h_0 + C g_0 h_0\right) = 0$$

where $g_0 = G(0)$ and $h_0 = H(0)$. The theoretical trajectories of the population sizes are shown in Fig. 4.

The figure reminds us of the dynamic behavior of the simulated populations in Fig 3 or real life hare-lynx population dynamics as shown in Fig 2. Indeed, if we alter the system of differential equations, and set the size of the grass population, $G(t)$, to a constant g, we get the Lotka-Volterra system for hares (prey) and lynxes (predators), namely

$$\frac{dH}{d\tau} = H(g - L), \; \frac{dL}{d\tau} = L(H - g)$$

The analysis of the set of rules described in L_1 can now be used to study the behavior at the so-called emergent level L_2, without surprise.

The Evolutionary Game Theory Approach

Although the mathematical approach discussed in the previous section explains us in a mathematical way *how* the global behavior in L_2 emerges in function of L_1, there is still no explanation *why* this system of coupled oscillations actually persists, and why the sizes of populations don't converge. Evolutionary Game Theory is the most plausible way to achieve this, complementary to the mathematical approaches discussed in the previous sections.

Standard publications on Evolutionary Game Theory, Evolutionary Stable Strategies (ESS) and Nash equilibria are by Maynard Smith [May74, May82], Maynard Smith and Price [May73] and Nash [Nas50].

The Rock-Scissors-Paper game

In Rock-Scissors-Paper, two players must simultaneously choose either one out of three strategies, Rock (R), Scissors (S) or Paper (P). Rock beats Scissors, Scissors beats Paper, and Paper beats Rock, any other combinations are ties. The payoff matrix for a player is (the player's selected strategy is in the leftmost column, the opponent's choice is in the top row):

	Rock	Scissors	Paper
Rock	0	+1	-1
Scissors	-1	0	+1
Paper	+1	-1	0

None of the pure strategies (consistently playing either Rock, Scissors or Paper) is an Evolutionary Stable Strategy (ESS). If you played Rock-Scissors-Paper as a child, you may remember that you could not win if your opponent knew which strategy you were going to pick. For example, if you pick Rock consistently, all your opponent would need to do is pick Paper and she would win. A child discovers quickly that if it doesn't know what the opponent will pick, then the best strategy is to pick Rock, Paper or Scissors at random, or select Rock, Paper or Scissors with a probability of 1/3, which is formally written as $\{<\frac{1}{3}, R>, <\frac{1}{3}, S>, <\frac{1}{3}, P>\}$. This is the Nash equilibrium mixed strategy of the game.

It should be obvious that if you do know what your opponent is likely to do, then picking a strategy at random with a probability of 1/3 is not the best thing to do (unless this is also your opponent's strategy). Suppose your opponent did not play this Nash equilibrium mixed strategy and preferred S over the other alternative choices. This way he or she adopted the following mixed strategy: $\{<\frac{1}{4}, R>, <\frac{1}{2}, S>, <\frac{1}{4}, P>\}$. We will call this strategy the S-biased strategy. The best reply to the S-biased strategy would be a pure strategy; playing Rock consistently. But while evolving to that strategy, the game comes across the R-biased strategy, namely $\{<\frac{1}{2}, R>, <\frac{1}{4}, S>, <\frac{1}{4}, P>\}$. A P-biased strategy would be the evolved reply to this R-biased strategy, which in turn defeats against the S-biased strategy, causing oscillations in the choice of the three biased strategies, while not converging to the Nash equilibrium. The system is degenerate; it turns out that the steady state Nash equilibrium is surrounded by closed orbits, so that it is a "center," and hence is stable but not asymptotically stable. This was first shown by Zeeman [Zee80]. See Hofbauer and Sigmund [Hof88] and Gauersdorfer and Hofbauer [Gau95] for a discussion of variants of this game. See Boylan [Boy94] for a discussion of the properties and steady states that are robust to perturbations of the dynamics corresponding to deterministic mutations.

The Grass-Hare-Lynx game

Similar to the Rock-Scissor-Paper game we can devise a Grass (G)-Hare (H)-Lynx (L) game with the same rules. Each player repeatedly picks a species, and the player with the species that can 'eat' the other, wins.

Suppose a life community of grass patches, hares and lynxes is playing this game against itself. If the community starts with the H-biased strategy, $\{<\frac{1}{4}, G>, <\frac{1}{2}, H>, <\frac{1}{4}, L>\}$, it can evolve to a L-biased strategy. The G- and H-biased strategies would follow, which makes the cycle complete and would cause the community to have oscillatory population sizes over time, without converging to the Nash equilibrium at $\{<\frac{1}{3}, G>, <\frac{1}{3}, H>, <\frac{1}{3}, L>\}$.

The repeating sequence of strategies is shown schematically in Fig. 5. The game played by the life community shows similar behavior as the Lotka-Volterra or the Artificial Life model. This means that the oscillations are caused through a repetitive choice of a sequence of strategies, due to the life communities' tries to maximize the utility function of the Grass-Hare-Lynx it's playing against itself.

Discussion

Neither of the Mathematical, Artificial Life or Game Theory approach can *exactly* pinpoint how the repetitive oscillations in L_2 emerge from the rules defined in L_1.

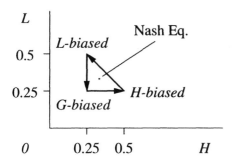

Figure 5: Phase plane trajectory of the cycle of strategies

They are unable to give a general acceptable explanation of the oscillations. A combination of the analytical approaches, however, gives us better insight in the oscillatory dynamics of the system. Also, from this analysis, it is *now* clear that convergence to an equilibrium would be strange and ask for an investigation. Through our analysis, the emergent behavior in the system is now the "obvious" behavior.

For our quest to the origin of oscillatory behavior in predation models only one single hypothesis is formulated and studied rigorously. There might, of course, be other hypotheses that can give a different explanation of the phenomenon, but with similar predictive qualities. As an example, we could introduce a hypothesis based on the flow of energy through the individuals.

Classical science would study the phenomenon in an analytical way, by formulating hypotheses based on the facts surrounding the emergent phenomenon. However, these possibly wrong hypotheses could be able to make perfect predictions about the system. Artificial Life, on the other hand, studies phenomena in a synthetical way. Through derivation of properties of the system (L_1) and synthesizing these properties in a simulation or a model we could again see the emergence of the phenomenon in L_2. But is this way of working satisfactory for other, possibly more complex, phenomena?

The surprise of seeing an illusion, and not understanding how it emerged from a magicians' hands, could sometimes be explained through thorough investigation, analysis and synthesis. But what are the limits of this method? This still remains an open question.

References

[Ark98] Arkin, R. C. 1998. *Behavior-Based Robotics.* Cambridge: MIT Press.

[Boy94] Boylan R. 1994. Evolutionary Equilibria Resistant to Mutation. *Games and Economic Behavior,* 7: 10-34.

[Dac90] Dacosta, N. C. A. and S. French. 1990. The Model-Theoretic Approach in the Philosophy of Science. *Philosophy of Science.* 57: 248-265.

[Gau95] Gaunersdorfer, A. and J. Hofbauer. 1995. Fictituous Play, Shapley Polygons and the Replicator Equation. *Games and Economic Behavior.*

[Gie94] Giere, R. N. 1994. The Cognitive Structure of Scientific Theories. *Philosophy of Science.* 61: 276-296.

[Gun90] Gunsteren, W. F. and H. J. C. Berendsen. 1990. Computational Simulation of Molecular Dynamics: Methodology, Applications and Perspectives in Chemistry. *Angewandte Chemie - International Edition in English.* 29: 992-1023.

[Hof88] Hofbauer J. and K. Sigmund. 1988. *The Theory of Evolution and Dynamical Systems.* Cambridge: Cambridge University Press.

[Lot20] Lotka, A. J. 1920. Undaped oscillations derived from the law of mass action. *Journal of the American Chemistry Society.* 42: 1595-1599.

[Lot25] Lotka, A. J. 1925. *Elements of Physical Biology.* Baltimore: Wilkins and Wilkins.

[May73] Maynard Smith, J. and G.R. Price. 1973. The Logic of Animal Conflict. *Nature.* 246: 15-18.

[May74] Maynard Smith, J. 1974. The Theory of Games and the Evolution of Animal Conflicts. *Journal of Theoretical Biology.* 47: 209-221. Academic Press.

[May82] Maynard Smith, J. 1982. *Evolution and the Theory of Games.* Cambridge: Cambridge University Press.

[Mor99] Morgan, M. and M. Morrison (eds.). Forthcoming. *Models as Mediators.* Cambridge: Cambridge University Press.

[Mur89] Murray, J. D. 1989. Mathematical Biology. Volume 19 from the *Biomathematics Texts* series. Berlin: Springer.

[Nas50] Nash, J. 1950. *Non-Cooperative Games.* PhD thesis, Princeton University.

[Odu53] Odum, E. P. 1953. *Fundamentals of Ecology.* Philadelphia: Saunders.

[Ron99] Ronald, E. M. A., M. Sipper and M. S. Capcarrere. 1999. Testing for Emergence in Artificial Life. In Proc. of *5th European Conference on Artificial Life,* edited by D. Floreano, J.-D. Nicoud and F. Mondado. Berlin: Springer, pp. 13.

[Sch84] Schaffer, W. M. 1984. Stretching and Folding in Lynx Fur Returns: Evidence for a Strange Attractor in Nature? *Amer. Nat.* 124: 798-820.

[Vol26] Volterra, V. 1926. Variations and Fluctuations of a Number of Individuals in Animal Species Living Together. Translation in: R.N. Chapman. 1931. *Animal Ecology.* New York: McGraw Hill pp. 409-448.

[Wil72] Williamson, M. 1972. *The Analysis of Biological Populations.* London: Edward Arnold.

[Zee80] Zeeman, E. 1980. Population Dynamics from Game Theory. *Global Theory of Dynamical Systems,* Lecture Notes in Mathematics. Berlin: Springer. 819: 472-497.

Towards a comprehensive Alife-model of the evolution of the nervous system and adaptive behavior

János Albert

Department of Comparative Physiology, Department of History and Philosophy of Science, Loránd Eötvös University
Pázmány Péter sétány 2. Budapest, H-1117 Hungary
jalbert@mail.matav.hu

Abstract

The potentials and tools that are offered by Alife for biology in modeling the nervous system and animal behavior are mainly unexploited. There is no consistent Alife model of the biological evolution of the nervous system as yet, whereas the modeling tools are at hand and their application for this purpose seems evident. In a biologically grounded model we have to make every possible effort to use principles known from biology, and to minimize the arbitrarily employed organizing rules. The aim of our work is to create a biologically accurate Alife model of the formation and evolution of the nervous system in connection with the adaptive behavior. In this article we concentrate on the structure of the modeled genome, which is the basis of playing a double biological role: to ensure an open-ended evolutionary process, as well as to direct the ontogenesis. The main questions we examined are: what are the basic rules of construction that are sufficient to create a workable nervous system and how can we model them in a biologically realistic way?

Introduction

Adaptive behavior and its neurophysiological background is a complex enough phenomenon to be a serious challenge for the modelers, and the tools of Alife – neural nets, genetic algorithms (GAs), animats and the combinations of them – are excellent for this purpose.

We have to distinguish two ways of modeling adaptive behavior, as they apply these tools in different manners (Wilson 1991; Meyer and Guillot 1994). The aim of models motivated by biology is to study the neural control mechanisms of adaptive behavior of animals (Collett and Land 1975; Beer 1990; Cliff 1992), while models of evolutionary robotics mainly have an "engineering motivation", according to which the most important goal is not biological authenticity but creating an autonomous robot which can perform a special task without human direction (Dorigo and Schnepf 1993; Floreano and Mondada; 1994). The latter ones can apply the biological principles freely and arbitrarily in the interest of success. These models are not those of biological evolution, but they are meant to search for optimal solutions of a special problem with the help of GA. The biological value of these models is doubtful, but it is a problem only if they are supposed to lead to biological conclusions beyond their competence.

The motivations of the two research groups are obvious. On one hand a practical way to understand the organization and development of intelligent systems is to study and to model their existent natural forms (animal behavior and its control mechanisms), because the successful modeling helps us understand the organizing and working principles. On the other hand, if we know the principles of forming intelligence during the biological evolution, we have some chance to create similar systems in an artificial way. The elaboration of successful models has a crucial practical importance, because difficulties of planning grow along with the advance of the complexity of the system. Up to now "real" Artificial Intelligence has successfully resisted the efforts of human designing...

The synthetic approach of Alife has been largely influenced by the results of analytic science, mainly neurobiology and genetics. At the same time we find that the potentials and tools that are offered by Alife for biology in modeling the nervous system are mainly unexploited. Perhaps we can state without exaggeration that neurobiology have not realized the potential that Alife-tools can offer them (Risan 1997). Computational neuroethology is a good example indicating the direction of progress, where there are advanced results by now (Beer 1990; Cliff 1994). But up to the present there is no Alife model of the biological evolution of the nervous system, whereas the modeling tools are at hand and their application for this purpose seems evident. Notwithstanding, there are no signs of applying them to create a consistent model of this evolutionary process. Perhaps this statement is surprising at first hearing, because there really are very interesting and successful models of adaptive behavior and its evolution (Ray 1992, 1994; Harvey, Husband and Cliff 1993; Sims 1994a, 1994b; Nolfi and Parisi 1994a, 1994b, 1995), which illuminate and illustrate important aspects of the biological evolution; and without intending to diminish the importance of the achievement of them we are only saying that they do not lead us any closer to understanding how the nervous system of animals had been created and developed during the evolution – as these models had not been created to study this question. However, we do not have much evidence in connection with the formation and early evolution of the nervous system. In these cases modeling could help us to supply missing links and to test the theories. If we want to create a biologically relevant model of this subject, we have to keep some considerations in view, which normally do not arise in modeling adaptive behavior.

The aim of our work is to create an Alife model of the formation and evolution of the nervous system in connection with adaptive behavior. The following questions stand at the central point of our interest: how came the nervous system and the most primitive forms of intelligence into being? What are the basic rules of construction that are sufficient to create a workable nervous system without specifying the details of the construction? What kind of advantages can be ensured by a primitive protoneural system that is just forming?

The importance of anatomy

An animal – even if we have to simplify it when it is modeled – cannot be viewed as a robot where the body can be handled in separation from the neural network. In the case of an animal body the outputs of the motoneurons cannot be valued by themselves, but only as elements of responses of a more complex system. The role of a certain output unit is not a result of an arbitrary definition, but depends on the anatomy and the position of the muscle that is innervated by that neuron. Moreover, we have to take into account that in a real animal even the simplest movements require the co-operation of many muscle cells, so their innervation has to be carried out in harmony as well. So the behavioral response to an environmental stimulus is influenced not only by the inputs and the outputs of the nervous system, but the anatomical construction of the body as well (Bullock and Horridge 1965; Lawn 1982; Spencer and Arkett, 1984).

If we examine the development of the adaptive behavior, it is not sufficient to model merely the formation and evolution of the neural control mechanisms. We have to consider the whole organism as an evolutionary unit, because keeping contact with the environment and performing the responses are the tasks of the whole living being (Albert 1999a, 1999b). It follows from the foregoing that the bodily construction of a certain animal cannot be a negligible feature in the modeling of adaptive behavior (Mackie 1990). Modeling the evolution of adaptive behavior cannot be reduced to modeling the evolution of the neural network.

A crucial point is to choose the suitable type of animal that can be the basis of the construction of the animat. In our opinion a model of the biological evolution of adaptive behavior and its nervous control cannot be based upon modeling evolved animals with advanced bodily constructions. These animals are too specialized to serve as a starting point for an evolutionary model. Advanced anatomy needs a highly specialized moving control system both in insects and vertebrates. In this case we can only use

GA to test the optimal solutions of the nervous control of this specialized "machinery" (Ijspeert 1999). This task is analogous with evolutionary robotics and not with biological evolution. If we want to model the latter, then we cannot leave out of consideration that the bodily construction, the locomotor organs and their nervous control evolved together, so we cannot model one of them regardless of the others (Sims 1994a, 1994b).

The problem of encoding

In a biologically established model we have to make every effort to avoid or at least to minimize the arbitrarily employed organizing rules. We have to make an attempt to use principles known from biology and not high-handed theories. Therefore in our model the connection of the genotype and phenotype has to differ from that simple relation that is common in models working with neural nets connected to GAs, and we have to bring this relation close to reality. In the case of "traditional" GAs the genome is only a plan containing the solution of a certain problem encoded in it. It is merely the subject of the working of GA, which is able to modify it via mutation and recombination. Then the program reads the data from the genome and creates the phenotype, the solution of a certain problem (Holland 1975). In this case genome is only a passive lump of data that is modifiable and readable, but has no active role creating the phenotype. Consequently GAs use only one aspect of the biological role of the genome, namely the function of a modifiable plan. In this model the flow of information is a one-way process during the life of an individual. Feedback from the phenotype to the genotype is only possible through selection (Balakrishnan and Honavar, 1995) and not during the life of an individual. It is not adequate for biological evolutionary processes, where genome has an active role in forming the phenotype and controlling the development.

Creating the genome, raises the crucial question what properties of the model-organism should be encoded and how to do it, because it determines the working of the model. The higher level properties are encoded in the genes, the easier it is to survey the connection between genotype and phenotype, and the more considerable the effect of the mutations is (Nolfi and Parisi 1995; Albert 1999a, 1999b). At the same time, however, there is no more considerable chance for new properties to arise, only the preprogrammed ones can become better adapted via quantitative changes. If there is no chance for qualitative changes, there will be no chance for new properties to emerge that are not preprogrammed – so the evolutionary process cannot cause a "surprise". This lack of qualitative evolution is a serious problem, because we lose one of the most important aspects of the evolutionary process. We have to make some effort to preserve the possibility of an open-ended evolution, otherwise the model will only mirror the willy-nilly built-in limitations of the programmer. Modelers in evolutionary robotics are also confronted with this problem (Harvey 1992, 1993). The common feature in these attempts is the search for lower level properties encoded in the genome. We have to find the sub-properties or "primitives" that are building elements of higher level properties and take part in building more than one special property (Koza 1992). In real biological systems these encoded units are the amino acids, which build proteins, the real carriers of the properties.

Another problem that can limit the possibilities of a modeled evolution is that the lengths of the genes are predefined in most of the models. It is because of the encoding mechanism, since modeled genes are sequences in which all characters define a certain property of the phenotype. So if somewhere in the genome an additional character gets wedged in or falls out of the genome, all of the following characters shift, and this event can change the inner structures of the genes, causing not only a change in the meaning - as it is in reality - but perhaps making impossible the reading of the genes. In most of the evolutionary models genes are rather rigid structures without any flexibility. A model of an open-ended evolution process would require more flexible, error-tolerant encoding methods that can handle not only the preprogrammed properties, but allow changes of lengths and inner structures of the genes without making impossible the reading of genes. In this case the size of the genome as well as the single genes can vary freely, so this mechanism can serve as a basis for modeling an open-ended evolution.

The structure and the function of the genome

We endeavored to apply the biological principles (Smith-Keary 1991) consistently in our model. The model-genome is built of four types of characters (0, 1, 2, 3), similarly to the four nucleotides of the DNA (adenine, cytosine, guanine and thymine). In most of the Alife models genes consist of a predefined number of characters, and their meaning depends on their position inside the gene. In reality genes are homologous nucleotide-sequences, in which there are no specially positioned characters, so the "meaning" of a gene is not hidden in the single nucleotides, but in the protein, which results from the translation of the whole gene. We applied this principle, therefore the meaning is not assigned to the position of a certain character. In our model the result of the translation of a certain gene is a set of sub-properties. The composition of this set defines the property and its quantitative value encoded in that gene (also, in reality the amino acid sequence defines the property of a certain protein). As the meaning of a gene is not assigned to the positions of the characters inside, the genome can change more flexibly (e.g. the number of the characters of a gene, the length of a single gene and also, the whole genome can vary freely, moreover, the genes can change their position inside the genome) without losing its functionality. This model-genome does not have a strictly organized sequential structure, it can be created as a totally random

sequence, and there is no need to predefine the lengths of the genes and their order in the genome. This structure usually contains less information than it could, but it is also similar to reality, because in the real genomes there are large meaningless sequences among the genes. So this kind of encoding is closer to reality than the common strict way.

The characters of the genome are read in twos, so the codons consist of two characters. Because we have four "nucleotides" (0, 1, 2, 3), the codons are double figures in the base four numerical system, so we have 4 x 4 = 16 different codons. All of them can be connected to a hexadecimal figure (0 – F), which are the results of the translation (the "amino acids"). On the basis of this simple coupling rule we can easily create the "genetic code" of the model (see Fig.1.). In fact real codons consist of three characters, so there are 4 x 4 x 4 = 64 triplets, but because of the degeneracy of the genetic code they encode only 20 amino acids. For this reason mutations in the third letter of the triplets rarely cause real changes. Using two letter codons our encoding system does not have degeneracy, so the ineffective mutations are eliminated. This modification leads to the acceleration of the evolutionary process, but apart from this feature the organizing principles resemble reality, and the number of encoded units is also similar (16 and 20).

	Second character of the codon			
	0	**1**	**2**	**3**
0	STOP	Hex **1**	Hex **2**	Hex **3**
1	Hex **4**	START Hex **5**	Hex **6**	Hex **7**
2	Hex **8**	Hex **9**	Hex **A**	Hex **B**
3	Hex **C**	Hex **D**	Hex **E**	Hex **F**

Fig. 1. The genetic code of the model

The genome does not have special positions in the sequence, all of them are equivalent, regardless of their position. Because of this there must be a "start" and a "stop" codon to sign the forepart and the end of the genes, just like in reality. In our model the "start" codon is "11" (corresponding to AUG codon in reality), while the "end" codon is "00" (corresponding to UAA, UGA and UAG). The "11" in the base four numerical system has the hexadecimal equivalent "5", similarly to the real genetic code where the "start" triplet (AUG) also encodes an amino acid (Methionin). This kind of encoding makes possible the emergence of overlapping sequences of genes, which means that "11" marks the starting point of the genes, but if in a sequence of a gene we find another "11" codon (i.e., before a "00" closes that certain gene), then this new "11" codon has two different meanings. In the original gene that is under translation this codon means a sub-property ("5"), while at the same time this codon serves as a "start" sign for another gene. The sequence of this latter gene is the same as the sequence of the original from this point, and all of them continue until the first "00" codon. So the same part of the genome can encode more than one set of sub-properties (more than one "protein", see Fig. 3). This interesting phenomenon – which increases the efficiency of information storage – is also known in biology, we can find examples of multiple readings for genes in viruses and Prokaryotes.

The process of translation

In our model there are 16 "hexadecimal amino acids" (0 – F) that are the equivalents of the codons. Their determination is inevitably high-handed, since they are not influenced by the laws of physics and chemistry, which determine the properties of real amino acids. The features have to be defined by the modeler, but we have to endeavor not to restrict the potentials of the model.

In reality a certain amino acid does not have a special "meaning", its function depends on the protein which the amino acid molecule is built in. So we cannot order a special function to a single hexadecimal amino acid (and its codon that corresponds to it). The carrier of a function is the whole protein, and a single amino acid can only modify it. Therefore we have to order somehow the functions to the whole proteins (or genes) and not to their separate parts. We connect sub-properties to larger functional units of the proteins (that are called "domains" in real protein molecules). So we have to find these domains in a certain molecule, and they determine its function, but the quantitative features of this function will be modified by the single amino acids of the domains. We defined the repeated hexadecimal amino acid sequences as separators between the domains. If a protein does not contain at least three uniform amino acids successively, then it comprises only one domain. The functions of the domains are determined by the total of the values of its hexadecimals.

Determining the features of the sub-properties we have to keep in view the most general abilities and attributes of a living cell that can be combined with each other and result in different cell-types.

Domains	Type of sub-property	Description of sub-property (The proteins containing these units have the required properties)
0 – 3	**MEMB**	Membrane proteins
4 – 7	**IONCH**	Ion channel proteins (determine the passive and active electric properties of the cell)
8 – B	**CONTR**	Gives the possibility of active contraction in a certain part of the cell
C – F	**SEQ**	Create identifier sequences in a protein molecule

Fig. 2. The sub-properties and their characteristics

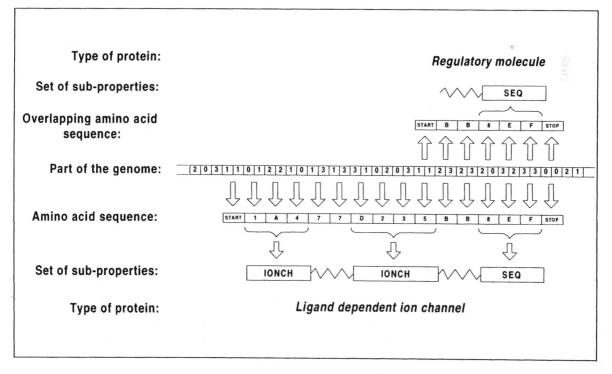

Fig. 3. The process of translation

Regulation of the cells

More or less biologically inspired modelers of neural networks often try to get their models close to reality by planning not an accomplished, well-formed neural net, but an "embryonic-state" one that goes through an ontogenesis and develops by itself (Nolfi and Parisi, 1994a). Some of these models try to simulate ontogenesis without taking into account the active role of the genome in it. In these models turning on and off a certain gene is solved by the mutation of a regulatory gene, so the state of a gene is not influenced directly by the environment, therefore it is not changeable during the ontogenetic process. This solution is not realistic, although it can be suitable for robotic models (Nolfi, Miglino and Parisi 1994; Nolfi and Parisi 1994a, 1994b). Other models try to imitate a more realistic controlling role of regulatory genes inspired by the operon-theory of molecular genetics (Eggenberger 1996, 1997).

Every cell of a living organism contains the same genetic information, but cells are different from each other, because the environmental stimuli activate different genes in them. Most of these stimuli have their origin in the cells themselves. So cells create their own environment inside the body, which influences their further development. In biological systems this regulation is carried out by regulatory genes (Smith-Keary 1991). This needs special molecules called transcription factors, which can be attached to these genes turning them on and off. The number of active genes is changing continually because of the feedback between the environment and the genome, so the phenotype is continuously adapting to this environment. In this model the plasticity has two levels, just like in reality: genotypic and phenotypic plasticity. The former makes possible the evolutionary adaptation of the "species", while the latter makes possible the adaptation during the lifetime of a certain individual.

Chemical regulation has three basic types: intracellular regulation, intercellular regulation directly through the cell membrane and receptor-mediated intercellular regulation. These types of communication are successfully modelled in evolutionary robotics to create and format simple neural nets, and in a model of morphogenetic processes (Eggenberger 1996, 1997). Chemicals have three important features in this context: the type of the molecule (this is important in binding to a receptor), the weight of the molecule (determines the diffusion constant) and the half-life (determines the duration of its effect).

Intracellular regulation (see Fig. 4a.): Some sequences produced by the cell itself can connect to a certain gene and work as a regulatory unit repressing the expression of that gene. In our model, differently from others, there are no predefined regulatory genes. To become a regulatory gene two requirements have to be met. One is a "protein sequence" (product of another gene) that can link to that certain sequence of the genome. But this alone is not enough, because if the gene sequence is a meaningless part of the genome (so it is not between a "start" and a "stop" codon) the linking has no effect. But if this gene-sequence

is a real gene, then the linking prevents it from being expressed.

Intercellular regulation (see Fig. 4b-c): cells can emit chemical signals into the intercellular matrix. Naturally these molecules are also the products of the cells (the features of these chemicals are determined by their sub-properties). If a cell has a proper receptor in its membrane (the identifier sequences of the receptor and the signal correspond to each other), then this cell is sensitive to this signal. The effect of linking is determined by the receptor, and not the signal. There is a different type of intercellular signal that has an effect directly on the genes of another cell without linking to a receptor. (This kind of signal can go through the membrane of the cells and can have an effect on the genes in the same way as the intracellular regulatory molecules.)

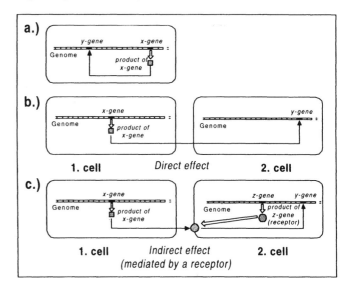

Fig. 4. Levels of the gene regulation

This model of regulation does not define the number of the regulatory genes and the signal molecules in advance; every feature can change by mutation and recombination. So it is not certain whether at the start of a simulation every signal will have a receptor, or if there will be any receptors at all. So in the beginning it is probable that there will be many individuals that are incapable of living, but the fortunate ones have almost unlimited chances to evolve.

Properties of nerve cells

Nerve cells have the most important role in the regulation of the behavior of an animal (or an animat), so in the process of creating a model of adaptive behavior one of the most important steps is to choose a suitable type of model-neuron. There are two restrictions that influence our decision. On the one hand we want to create a biologically relevant model, so the modeled nerve cells cannot be oversimplified. On the other we have to take into consideration that if we want to model an evolutionary process as well, the nerve cells have to have parameters that are well encodable in that kind of genome we outlined above. In consideration of these requirements we created a model neuron which functions based on the Goldman-Hodgkin-Katz equation:

$$E = \frac{RT}{F} \ln \frac{P_{Na}[Na]_{out} + P_K[K]_{out} + P_{Cl}[Cl]_{in}}{P_{Na}[Na]_{in} + P_K[K]_{in} + P_{Cl}[Cl]_{out}}$$

E is the actual value of the membrane potential, while R, T and F have their usual meanings, P is the permeability of the sodium (Na), potassium (K) and the chloride (Cl) ions, in the square brackets there are concentrations of these ions inside (in) and outside (out) the cell. The concentrations have values that are usual in nerve cells, so they are not encoded. The active electric properties of a nerve cell that functions based on this equation depend on the values of three parameters: PNa, PK and PCl. We had to find a solution to encode them in the genome in a way that is biologically reasonable.

The biological subjects of membrane permeability are the ion channel proteins. Their kinetic properties determine the flow of the ions through the membrane, therefore they are the bases of the active electric properties of the nerve cells (Ganong 1987; Prosser 1991). The kinetics of these ion channels can be well described by built-in gating particles. These are hypothetical parts of the ion channel molecules, but in all probability they have equivalents in reality. E.g., in a common voltage-dependent sodium channel there are two types of them, one is an opening (m), while the other is a closing (h) particle. The channel is opened only if they are in a suitable position. As m and h are probability variables, their value is between 0 and 1, and the probability of being open for a sodium channel is m3h, because there are three pieces of m particles connected to a channel.

The values of these probability variables are calculable with the help of some simple equations described by Hodgkin and Huxley (Finkelstein and Mauro 1977; Hille 1984). The equations calculate the values of α (activity constant) and β (inhibitory constant), which are used to calculate τ (time constant), which then is needed to calculate the probability variable. In the general forms of the equations we substituted variables (a, b, c, d) for the numbers connected to a certain ion channel. Knowing their values we can derive P, which determines the kinetics of a modeled ion channel. These parameters seem suitable for encoding the properties of the ion channels, because changing their value by mutation and recombination results in a number of ion channels that have more or less different properties (m is the value of the probability variable after a long term, m(t) is its value at t time).

$$\alpha = a_\alpha \exp\left(\frac{-V}{c_\alpha}\right)$$

$$\beta = a_\beta \frac{b_\beta - V}{\exp\frac{b_\beta - V}{c_\beta} + d_\beta}$$

$$m_\infty = \frac{\alpha}{\alpha + \beta} \qquad \tau_m = \frac{1}{\alpha + \beta}$$

$$m(t) = m_\infty - (m_\infty - m_0)\exp(-t/\tau)$$

This way of encoding is indirect enough to simulate the genetic process, in which a certain biological feature is the result of the functioning of a number of different factors, and not only a single number which directly determines that property. Being the model-genome a randomly generated character-set, the number of domains of a certain protein molecule is not predefined, and so some of these parameters can be missing. These "mutants" can be workable molecules, yet they have different properties. In our model the protein-domain type (or sub-property) that encodes the values of these parameters is called IONCH. So if a protein contains this domain-type, it can become an ion channel, but the resulting function depends on the other domains built in that protein as well.

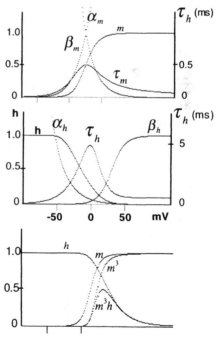

Fig. 5. The values of the parameters in the case of a sodium ion channel.

An ion channel in a real cell can be not only voltage- but ligand dependent as well. In this case the activity depends on the binding of a ligand molecule, and not on a voltage threshold. This type of ion channel needs a receptor sequence that can bind a suitable ligand. If the result of a gene translation is a molecule that has IONCH domain(s) and a SEQ domain (that carries the receptor sequence), then it is defined as a ligand dependent ion channel molecule. Ligands are simpler molecules, which also have a SEQ domain. If the domain of the ion channel corresponds to the sequence of the ligand, they get bound to each other, and it results in the activation of the ion channel molecule

Nerve cells have passive electric properties as well: the time constant (τ) and the space (or length) constant (λ) of the conduction. (This time constant is not the same as the previous one discussed above.) These passive electric properties serve as a basis for the appearance of the active ones. The time constant is the length of time during which the value of membrane potential decreases to its 1/e. The space constant of the membrane is the distance where the membrane potential decreases to its 1/e (Ganong 1987; Prosser 1991). Both of them are derived from the cable-equation, which describes the passive electric properties of biological membranes. These properties play an important role in the formation of behavior during the early evolution of the nervous system (Albert 1999a, 1999b), because in this stage there is no action potential, so passive conduction is the only way to pass electric stimuli. In the model there is a protein-domain type that has the capability of modifying the time- or space constant (they are signed as MEMB in the table, see Fig. 2).

The third important feature of the nerve cells is the ability to move during the early stages of the ontogenesis and to grow processes later. These two activities have similar biological mechanisms. Both of them need contractile molecules built in the cell membrane, and their working is the result of the change of the membrane potential. In our model there is a protein-domain type signed as CONTR (see Fig. 2), which carries this ability of a protein molecule. There is a parameter encoded in this type of domain that determines the voltage threshold of the activity. So the migration or forming processes are not a directly encoded and predefined property of a cell, but the result of its inner state.

The animat

At the recent stage of our model the execution of morphogenesis is not perfect yet (these are only technical problems). So we temporarily applied an animat from our previous model that is well-tried and suitable for testing the behavior resulting from this new control mechanism This animat is a Hydra-like being that is able to perform feeding behavior (Albert 1999a, 1999b; see Fig. 6). The body of the animat consists of three cell-types, epithelial cells, muscle cells and nerve cells. Epithelial cells give the "framework" of the body, defining its shape and boundaries. The layer of these cells provides points for the muscle cells to anchor, and the movements of the animat will result in the change of position of the affected epithelial cells, so these cells serve as a kind of skin. Nerve cells and muscle cells are embedded in the layer of epithelial cells. Muscle cells have fixed positions, while nerve cells can move in the "matrix" of epithelial cells, if they have the capability of movement. Chemical signals emitted by the nerve cells also spread in this matrix by diffusion. The number of nerve cells results from the division of cells, so it can change during the lifetime of an animat. When the "life" of an animat begins, it has only a few "multipotent" cells. If they have the ability of division (it is an encoded property), they start multiplying, but this is not enough to form a nervous system, as it requires the ability to establish connections as well. During the tests the most successful individuals had about 600 – 800 nerve cells, but the fitness also depended on the properties of these cells. In the final version of the recent model the bodily construction will be the result of an ontogenesis, similarly to the nervous system, which directs the behavior.

The environment is modeled as a box. As the animats are sessile, the interaction between the individuals would be minimal, so we perform the tests with only one individual in the box at a time. The stimulus is the appearance of a piece of food, which slowly sinks to the bottom of the box. This is a chemical stimulus for the animat and its strength decreases exponentially with the distance. If the animat is able to catch at least one piece of food, the test is continued. The feeding is successful if the animat not only touches the piece of food, but it also "eats" it, in other words, it gets food particles into its coelenteron through the mouth. The fitness depends on the success of feeding (for details see Albert 1999a, 1999b).

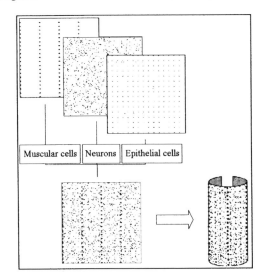

Fig. 6. The bodily construction of the animat

Testing the model

During the tests we examined the success of the feeding behavior of the animat. Then we performed some special tests examining the capabilities of the ontogenetic process (naturally we could only study the ontogenesis of the nervous system and not of the body).

First we created some sets of genomes generating totally random sequences of numbers, only their lengths were defined in advance (There were sets of three different lengths: 100, 200, 300, 400, 500 characters per genome). This was followed by the translation, during which the program read the genes of a certain genome (sequences started with "11" and ended with "00") and determined the encoded sets of sub-properties. At the same time this step was a screening test: we rejected those individuals that were obviously not viable (e.g. they did not have real genes starting with "11"and ending with "00", only meaningless sequences). There were only a few of individuals of this kind. We also rejected those ones that have real genes, but these genes could not produce viable individuals (e.g. they had no receptors or tools to respond to a stimulus). Since genomes were totally random sequences, there were a lot of individuals in this group. The rest of the genotypes were qualified as potentially viable. (They were about 10 % of the total amount of genotypes. The longer the genomes were, the more viable individuals were resulted.) We tested the feeding behavior only of these ones.

Results and discussion

The results of the tests show that only a small part of the potentially viable individuals of the first generation were able to control successfully the feeding behavior of the animat. It is not surprisingly few if we consider the structure of the genome and the way of generating them. Moreover, the successful ones shed light upon the self-organizing ability of the genome structured by this principle, because the genes and their functions arose by themselves from totally random character sets. If there are more or less successful individuals, then the advantages of the flexible genome could manifest themselves during the later phase of the evolutionary process. This is probably because the preprogrammed properties of the "traditional" genomes can evolve easily by using GA, but this evolution has a limit, because when the best genomes are found in a certain environment, there is no other chance for new, better features to appear; the evolution "gets exhausted". The advantage of the flexible genome is that it has the ability to ensure an open-ended evolution including the emergence of new properties. Under these temporary testing circumstances it is the body construction of the animat and the simple environment that limited the possibilities of evolution, and not the genome.

We found that in the beginning several fortunate circumstances are necessary to create a more or less viable individual by generating random genomes. This is the reason of the numerous unviable individuals (~90%) after

generating the starting genomes of the first generation. The members of the second and further generations are the offspring of merely the viable starting individuals (~10%), so the fitness grows rapidly. After about 300 generations the success of feeding behavior stops increasing. We have to stress that the animat we used in this test was originally created for a previous model and it has strong limitations in an evolutionary respect. This is only a stopgap arrangement that is applicable to the first test of this new genetic encoding and the neural control of behavior. During the evolutionary process some characteristic types of nerve cells appeared in the modeled nervous systems (see Fig. 7.)

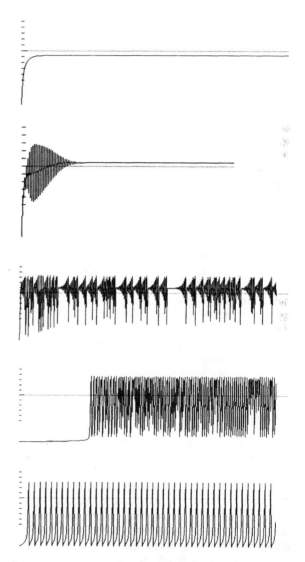

Fig. 7. Some interesting firing patterns of the modeled nerve cells

During the tests we examined the ontogenetic process as well. The model made possible a sophisticated ontogenetic

418

development of the nervous system, but this process was only successful in the "best" individuals. The lives of individuals begin with the ontogenesis and this process lasts throughout the whole of their lifetime, just like in reality. So the ontogenesis is not separated from the "working" of the individuals, which results in the behavior. (Most of the models handle ontogenetic process – if it exists at all – as a separated beginning part of the life, during which the neural network forms, but after this there is no other chance to change the anatomical structure of the network.) The successful animats must have the ability of adaptive behavior from their "embryonic state". In this developmental state there are two successful strategies. One is to develop in a way that continuously ensures the adaptive behavior, the other is to pass rapidly this first period (Nolfi and Parisi 1994a, 1994b).

One of the most interesting qualitative features was that there were individuals which had a "fluid" nervous system, in which the cells were in continuous migration during their lifetime, while their nervous systems preserved the capability of controlling the feeding behavior. The synapses of the cells also changed continuously, but because of the large number of cells they could form new connections instead of the ceased ones. It is very similar to the nervous organization of Hydras, because in these simple animals nerve cells continuously come into being by cell division during the lifetime and they migrate along the body, nevertheless this simple network-like nervous system can preserve its operating ability (Bode 1992). Our previous model (Albert 1999a, 1999b) also showed that in animats at the starting point of nervous evolution with a simple bodily construction (e.g. Hydra and other Cnidarians) and therefore not having special locomotor organs, which would need an advanced nervous control, the large number of nerve cells in a randomly connected network-like nervous system can control the behavior successfully. This result is in harmony with the physiological data (Robson 1975; Spencer 1991).

Another interesting feature that is the result of the lifelong-ontogenesis is the ability of adaptation to the changes of anatomical relations. If we "cut asunder" an animat, in some cases it has the ability to regenerate its neural net, and the resulting "child-animats" can survive this interference and feed more or less successfully. (During the simulation of this process we did not take care of the details of the anatomy, we only pulled apart the tube-like animat at its centre, and the resulting half-animats were considered as two smaller ones.) We only studied the half nerve systems and the behavior produced by them. This is a simple model of the asexual propagation of Hydras, during which an individual produces offspring by gemmation (Bullock and Horridge 1965; Robson 1975).

The "growing together" of two animats had similar results; in some cases this process led to workable animats, which could feed successfully. This surprising regenerating ability is possible because of the simple diffuse structure of the nervous systems in which there are no specialized connections between the cells and there are no special

locomotor organs that would need these connections for their proper working. These simulation results are in agreement with the physiological data, because Hydras and other Cnidarians, which have the most ancestral network-like nervous system, have a very good self-regenerating ability.

Conclusion

All of these early state simulation results demonstrate that our model could be suitable to become a biologically well established model of the evolution of the nervous system. One of the crucial factors of this model is that genome has to exceed its common, passive role and the rigid structure, and has to have the ability to perform the tasks consisting of two parts they have to perform in biological systems. On the one hand it has to make possible developing properties that are not preencoded in the genome, on the other hand it has to ensure the chance to perform the ontogenetic processes.

If we can create a biologically accurate and well-established Alife model of the evolution of the nervous system in relation to the behavior, then it can help Alife to become a useful modeling method for neurobiology. This would help us to make the information flow to a really two-directional process between Alife and biological research. This fruitful relation should accelerate the process that leads to the better understanding of the mysterious phenomenon called intelligence.

References

Albert, J. 1999a. Modeling of an Early Evolutionary Stage of the Cnidarian Nervous System and Behavior. In Proceedings of the 5th European Conference on Artificial Life, ECAL'99, 236-245. Lausanne, Advances in Artificial Life, Springer 1999.

Albert, J. 1999b. Computational modeling of an early evolutionary stage of the nervous system. *BioSystems* 54: 77-90.

Balakrishnan, K.; Honavar, V. 1995. Evolutionary Design of Neural Architectures - A Preliminary Taxonomy and Guide to Literature, Technical Report, CSTR-95-01, Artificial Intelligence Research Group.

Beer, R. 1990. *Intelligence as Adaptive Behavior*. An Experiment in Computational Neuroethology.: Academic Press, Inc.

Bode, H. R. 1992. Continuous conversion of neuron phenotype in hydra. *Trends in Genetics*. 8(8): 279-284.

Bullock, T. H.; Horridge, G. A. 1965. *Structure and Function in the Nervous System of Invertebrates*. 450-506. Freeman, San Francisco, CA.

Cliff, D. 1992. Neural networks for visual tracking in an artificial fly. In Towards a Practice of Autonomous Systems: Proceedings of the First European Conference on Artificial Life, Cambridge MA: MIT Press Bradford Books.

Cliff, D. 1994. Computational Neuroethology. Technical Report, CSRP 338; to appear in: M. A. Arbib (ed.), Handbook of Brain Theory and Neural Networks. MIT Press.

Cliff, D.; Harvey, I.; Husbands, Ph. 1993. Exploration in Evolutionary Robotics. *Adaptive Behavior* 2(1)

Collett, T. S.; Land, M. F. 1975. Visual control of flying behavior in the hoverfly, Syritta pipiens L. *Journal of Comparative Physiology*, 99: 1-66.

Dorigo, M.; Schnepf, U. 1993. Genetics-based Machine Learning and Behaviour Based Robotics: A New Synthesis. In IEEE Transactions on Systems, Man and Cybernetics, 23(1): 141-154.

Eggenberger, P. 1996. Cell Interactions as a Tool of Developmental Processes for Evolutionary Robotics. In 'Animals and Animats' Proceedings of Conference SAB'96.

Eggenberger, P. 1997. Evolving Morphologies of Simulated 3d Organisms Based on Differential Gene Expression. In Proceedings of the Fourth European Conference on Artificial Life. The MIT Press, Bradford Book.

Finkelstein, A.; Mauro, A. 1977. Physical principles and formalisms of electrical excitability. In *Handbook of Physiology*, Section I. The Nervous System vol. 1. Part 1. American Physiological Society, Bethesda.

Floreano, D., Mondada, F. 1994. Automatic creation of an autonomous agent: Genetic Evolution of a neural network driven robot. In From Animals to Animats III: Proceedings of the Third International Conference on Simulation of Adaptive Behavior. MIT Press. Bradford Books. Cambridge, MA.

Guillot, A., Meyer, J. A. 1994. Computer Simulation of Adaptive Behavior in Animats. In Computer Animation '94. IEEE Computer Society Press.

Harvey, I. 1992. Species adaptation genetic algorithms: the basis for a continuing SAGA. In Proceedings of the First European Conference on Artificial Life. The MIT Press, Bradford Books, Cambridge, MA.

Harvey, I. 1993. Evolutionary robotics and SAGA: the case for hill crawling and tournament selection. In *Artificial Life III*, Santa Fe Institute Studies in the Sciences of Complexity, Proc. Vol. XVI. Addison Wesley, 1993.

Harvey, I.; Husbands, P.; Cliff, D. 1993. Issues in Evolutionary Robotics. In Proceedings of the Second International Conference on Simulation of Adaptive Behavior, J. A. Meyer, H. Roitblat and S. Wilson (eds.) The MIT PressBradford Books, Cambridge, MA.

Hille, B. 1984. *Ionic Channels of Excitable Membranes*. Sinauer Associates Inc. Sunderland Mass.

Holland, J. (1975). *Adaptation in Natural and Artificial Systems*. The MIT Press, Bradford Books.

Ijspeert, A. J. 1999. Synthetic Approaches to Neurobiology: Review and Case Study in the Control of Anguiliform Locomotion In Proceedings of the 5th European Conference on Artificial Life, ECAL'99, 236-245. Lausanne, Advances in Artificial Life, Springer 1999.

Koza, J. R. 1992. *Genetic programming: On the programming of computers by means of natural selection.* The MIT Press.

Lawn, I. D. 1982. Porifera, In *Electrical conduction and behaviour in 'simple' invertebrates.* 49-72. G.A.B. Shelton (ed.) Clarendon Press Oxford

Mackie, G. O. 1990. The Elementary Nervous System Revisited. *Amer. Zool.* 30, 907-920.

Nolfi, S.; Miglino, O.; Parisi, D. 1994. Phenotypic Plasticity in Evolving Neural Networks. Technical Report PCIA-94-05. To appear in: Proceeding of the First Conference From Perception to Action, Lausanne

Nolfi, S.; Parisi, D. 1994a. Phylogenetic Recapitulation in the Ontogeny of Artificial Neural Networks. Technical Report 94; Institute of Psychology N.R.C. Rome.

Nolfi, S.; Parisi, D. 1994b. Evolving Artificial Neural Networks that Develop in Time Technical Report 94; Institute of Psychology N.R.C. Rome.

Nolfi, S., Parisi, D. 1995. 'Genotypes' for neural networks. In *Handbook of Brain Theory and Neural Networks*, M. A. Arbib (ed.) MIT Press

Prosser, C. L. 1991. Excitable Membranes; Synaptic Transmission and Modulation 1-21. In *Neural and Integrative Animal Physiology*; Comparative Animal Physiology. C. Ladd Prosser (ed.) Wiley - Liss Publication

Ray, T. S. 1992. Evolution, ecology and optimization of digital organisms. Technical Report Santa Fe Institute 92-08-042

Ray, T. S. 1994. An evolutionary approach to synthetic biology: Zen and the art of creating life. *Artificial Life* 1(1/2): 195-226.

Risan, L. 1997. „Why are there so few biologists here?"-Artificial Life as a theoretical biology of artistry. In Proceedings of the Fourth European Conference on Artificial Life. The MIT Press, Bradford Book

Robson, E. A. 1975. The Nervous System in Coelenterates, 169-209. In *Simple Nervous Systems*, P. N. R. Usherwood, D. R. Newth (eds.) Edward Arnold 1975.

Sims, Karl 1994a. Evolving Virtual Creatures. In Proceedings of the Siggraph '94 Conference. 15-22. New York. ACM Siggraph.

Sims Karl 1994b. Evolving 3D Morphology and Behavior by Competition. In Proceedings of the Artificial Life IV Conference. 28-39. MIT Press.

Smith-Keary, P. 1991. *Molecular Genetics*, Macmillan, London 1991.

Spencer, A. N. 1991. A Protoneuronal System in Sponges; Cnidarian Nervous Systems. 554-567. In *Neural and Integrative Animal Physiology*; Comparative Animal Physiology. C. Ladd Prosser (ed.) Wiley - Liss Publication

Spencer, A. N.; Arkett, S. A. 1984. Radial symmetry and the organization of central neurones in a hydrozoan jellyfish. *J. Exp. Biol* 110. 69-90.

Market-Based Call Routing in Telecommunications Networks using Adaptive Pricing and Real Bidding

M.A.Gibney[1], N.R.Jennings[2], N.J.Vriend[3]

[1]Department of Electronic Engineering, [3] Department of Economics,
Queen Mary and Westfield College, University of London,
London E1 4NS, UK.
[2] Department of Electronics and Computer Science
University of Southampton,
Highfield, Southampton SO17 IBJ, UK.
1 3{M.A.Gibney, N.Vriend}@qmw.ac.uk
2 nrj@ecs.soton.ac.uk

Abstract

We present a market-based approach to call routing in telecommunications networks. A system architecture is described that allows self-interested agents, representing various network resources, potentially owned by different real world enterprises, to co-ordinate their resource allocation decisions without assuming *a priori* co-operation. It is argued that such an architecture has the potential to provide a distributed, robust and efficient means of traffic management. In particular, our architecture uses an adaptive pricing and inventory setting strategy, based on real bidding, to reduce call blocking in a simulated telecommunications network.

1 Introduction

In telecommunications networks, call traffic is typically routed, through the network from source to destination, on the basis of information about the traffic on that path only. Therefore, path routing is carried out without regard to the wider impact of local choices. The main consequence of this myopic behaviour is that under heavy traffic conditions the network is utilised inefficiently: rejecting more traffic than would be necessary if the load were more evenly balanced. One means of performing such load balancing is to centrally compute optimal allocations of traffic across the network's paths using predictions of expected traffic (Bertsekas and Gallager 1987). When such calculations have been completed, the network management function can configure the network's routing plan to make the best use of the available resources given the predicted traffic. However, as networks grow larger and involve more complex elements, the amount of operational data that must be monitored and processed (by the network management function) increases dramatically. Therefore in centralised architectures, management scalability is bounded by the rate at which this data can be processed (Goldszmidt&Yemini1998). In addition, there are a number of well known shortcomings with algorithms to compute optimal network flows; these include progressively poorer performance in heavily loaded networks and unpredictable oscillation between solutions (Kershenbaum 1993). Furthermore, the very centralisation of the network management function provides a single point of failure; thus making the system inherently less robust.

For the above mentioned reasons, a decentralised approach to routing is highly desirable. In such cases, decisions based on more localised information are taken at multiple points in the system. The downside of this, however, is that the local decisions have non-local effects. Thus, decisions at one point in the system affect subsequent decisions elsewhere in the system. Ideally localised control would take place in the presence of complete information about the state of the entire system. Such a state of affairs would enable a localised controller to know the consequences of a choice for the rest of the network. However, there are two main reasons why this cannot be realised in practice. Firstly, the network is dynamic and there is a delay propagating information. This means that a model of the network state held at any one point is prone to error and difficult to keep up to date. Secondly, the scaling issues involved in making flow optimisation computations for the entire network (noted above) would obtain here also. Therefore, a system in which local decision making takes place in the presence of an incomplete view of the wider network is the only feasible solution for providing distributed control.

A promising approach that combines the notion of local decision making with concerns for the wider system context is that of agent technology (Jennings and Wooldridge 1998). Agents address the scaling problem by computing solutions locally, based on limited information about isolated parts of the system, and then using this information in a social way. Such locality enables agents to respond rapidly to changes in the network state; while their sociality can potentially enable the wider impact of their actions to be coordinated to achieve some socially desired effect. Systems designed to exploit the social interactions of groups of agents are called multi-agent systems (MAS). In such systems, each individual agent is able to carry out its tasks through interaction with a small number of acquaintances. Thus, information about the extent of the system is distributed along with whatever functionality the MAS is designed to perform.

One agent-based technique that is becoming increasingly popular as a means of tackling distributed resource allocation tasks is market-based control (Clearwater 1996). In such systems, the producers and consumers of the resources of a distributed system are modelled as the self-interested decision-makers described in standard microeconomic theory (Varian 1992). The individual agents in such an economic model decide upon their demand and supply of resources, and on the basis of this the market is supposed to generate an equilibrium distribution of resources that maximizes social welfare. In market-based control, the metaphor of a market economy as a system computing the behaviour that solves a resource allocation problem is taken literally and distributed computation is implemented as a market price system. That is to say, the agents of the system interact by offering to buy or sell commodities at given prices (Wellman 1996). In our case, such an approach has the advantage that ownership and accountability of resource utilisation are built into the design philosophy. Thus, market-based solutions can be applied to the management of multi-enterprise systems without forcing the sub-system owners to co-operate on matters affecting their own commercial interest.

Within this context, this paper describes a system to balance traffic flow through the paths of a logical network, based on the local action of agent controllers coupled with their social interaction as modelled by a computational market. The approach presented here builds upon the preliminary work reported in (Gibney and Jennings 1998) in that it shares the same architecture, roles and deployment model for the agents. However, to improve upon our earlier results we devised a new approach to the way that agents adapt their pricing and inventory strategies according to the outcome of individual market actions and the profitability of trading in the market. More generally speaking, this paper extends the state of the art in market-based control in the following ways. Firstly, it models a complex two-level economy, in which not only end users but also the internal components of the system compete with one another for resources. The rationale behind using a two-level economic model is to realise call admission control in the same framework as the network management function. This is novel, as market-based control has not previously been used to address two control issues in the same system. Particularly, having two kinds of market within the economy, with agents active in different roles in each of them, provides an elegant way to acquaint agents with one another. This architecture also provides an appropriate way to situate the intelligence of the system in a multi-enterprise network with self-interested enterprises. Secondly, a novel approach is adopted to pricing strategy. Our agents adapt to the outcomes of market interactions which use real bids and offers (i.e. agents state a price in an auction-like market and are then committed to buying or selling at that price in that session). This approach was adopted because it eliminates the lengthy series of interactions between agents that is required to calculate the equilibrium price in the market. Rather, we use real bidding and allow the agents to adapt their bidding behaviour to the outcomes of the auctions over time. Real bidding allows us to use more rapid (one shot) auction protocols as markets.

The remainder of the paper is structured in the following manner. We discuss the background and motivation for this work in Section 2. Section 3 describes the architecture of the system as a whole and the institutional forms of the possible interactions between agents. The design of individual agents is given in Section 4. Section 5 discusses the experiments carried out to evaluate the performance of the system. Finally, Section 6 details our conclusions on the work presented in this paper and discusses the open issues and future work.

2 Background and Motivation

Decentralized approaches to routing, usually in packet switched telecommunications networks, based on the interaction of controllers distributed through the network have existed for some time (Schwartz and Stern 1989). However, agent based approaches extend this idea by modelling the interaction of distributed controllers as a social process. A number of agent based solutions have been proposed to the problem of load balancing in telecommunications networks. (Appleby and Steward 1994) make use of mobile agents roaming the network and updating routing tables to inhibit or activate routing behaviours. (Schoonderwoerd et.al.96) extend and improve this approach by using ant-like mobile agents that deposit "pheremones" on routing tables to promote efficient routes (the majority of ants use the efficient routes and the pheremones re-enforce this behaviour in other ants). (Hayzelden and Bigham 1998) employ a combination of reactive and planning agents in a heterogeneous architecture to reconfigure route topology and capacity assignments in ATM networks. All of these systems exhibit increased robustness and good scaling properties compared to centralised solutions. Indeed, in network environments with symmetric traffic requirements, ant-like agent solutions have even been shown to provide superior load balancing to both statically routed networks and a more conventional mobile agent approach (Schoonderwoerd et.al.96).

However, all the aforementioned approaches model networks as a single resource and therefore act to optimise utilisation of that resource. This makes sense because a poorly managed telecommunications network benefits no one. However, the main disadvantage of such a perspective arises when different telecommunications operators join their networks together (something which is an increasingly common trend). In such cases, if the different sub-network owners agree on a single, unified static network management policy, it is unlikely that this policy will benefit all their interests individually as well as collectively over time. We address this issue by modelling our agents as the resources and groups of resources that enterprises might own or lease in a multi enterprise environment.

Another increasingly common aspect of modern telecommunications deployment is the practice of enterprises in other sectors (banking and other traditional consumers of telecommunications services) leasing bandwidth from telecommunications providers to create virtual private networks in the short, medium and long term (Cisco 1999). Again, this promotes the creation of multi-enterprise networks. In such environments each enterprise clearly has an incentive both to see that overuse does not degrade network performance and to make the greatest possible use of their network ownership. Since these parties cannot agree each traffic policy decision individually, conflicting incentives must be reconciled outside the traffic management domain. Typically this is achieved by allowing sub-network owners to set policy within the remit of their own resources (Stallings 1997). However, the static nature of these policies and the conflict between them at sub-network interfaces often causes institutionalised under-use of the network as a whole.

Both of these trends suggest that telecommunications network management, once a centralised and monolithic undertaking, will increasingly benefit from an open, robust, scalable and inherently multi-enterprise approach. Therefore, one of the aims of this work is to use the multi-agent system paradigm to address the problem of multi-enterprise ownership of the network, while simultaneously addressing the problems of robustness and scalability. Against this background, the resource allocation problem in a network with multiple, non co-operating enterprises can be recast as the problem of reconciling competition between self-interested, information-bound agents. We conjecture that a market economy might be an effective mechanism for achieving this goal. Therefore we decided to implement our telecommunications network management framework using economic concepts and techniques.

3 System Architecture

The overall system architecture consists of three layers The lower layer is the underlying telecommunications infrastructure. The middle layer is the multi-agent system that carries out the network management function. The top layer is the system's interface to the call request software. More details of the rationale to this design are given in (Gibney and Jennings 1998). The remainder of this section concentrates on the agent layer: describing the main components (section 3.1) and how they relate to one another (section 3.2).

3.1 The Agents and their Interactions

The system makes use of three agent types: (1) the *link agents* (section 4.1) that represent the economic interests of the underlying resources of the network, (2) the *path agents* (section 4.2) that represent the economic interests of paths across the network and (3) the *call agents* (section 4.3) that represent the interests of callers using the network to communicate. A *link agent* is used for every link in the network and is deployed at the entry node for that link. A *path agent* is used for each logical path in use across the network and is deployed at the source node for that path. Here we use three path agents for each source destination pair. Three is a reasonable number of alternate paths across which to share a single traffic requirement (alternate static routing systems commonly using three or fewer paths). A *call agent* is used for each source destination pair in the network and is deployed at the source of the traffic requirement that it represents.

The agents communicate by means of a simple set of signals that encapsulate offers, bids, commitments, and payments for resources. We couple the resources and payments with the offers and bids respectively. This reduces the number of steps involved in a transaction (committing agents to their bids and offers ahead of the market outcome), and so increases the speed of the system's decision making (an important consideration in this domain). To enforce these rules the interactions between the different agent types are mediated by means of market institutions (described in section 3.2).

An important notion in agent technology is that agents should be proactive (i.e. be able to anticipate the requirements of the environment and behave accordingly). In our system, we apply this concept to implementing a call routing mechanism that does not need to examine the network state before routing each call. Our path agents proactively determine how many calls they will be able to handle in advance, and seek to obtain the necessary resources to handle them. To be able to offer resources to callers pro-actively, the path agents lease bandwidth from the link agents over a period of time, paying installments on the lease at prices agreed on the link markets.

3.2 Market Institutions

Our system makes use of two types of market institution: At the *link market* (section 3.2.1), slices of bandwidth on single links (the fundamental resources of the system) are sold to path agents. At the *path market* (section 3.2.2), the slices of bandwidth on entire paths across the network are sold to call agents to connect calls.

3.2.1 Link Markets

Link markets are sealed bid double auctions. In this protocol the auctioneer receives two sets of prices in each trading period: bids for resources from buyers and offer prices from sellers, and computes the successful trades according to a set of rules defined for the auction. A sealed bid protocol was chosen because it provides a means to complete institutionally mediated bargaining in one shot. Therefore bargaining that would take an indeterminate time using iterated market institutions such as continuous double auctions can be completed almost instantaneously.

The resources exchanged at the *link markets* are the right to use slices of bandwidth on individual links, which when taken together, provide the necessary bandwidth to connect calls across paths. The link markets use a sealed bid double auction in which buyers and sellers periodically submit bids for individual units of the resource. In our case, buyers and sellers are constrained to their roles in the market by their position in the network. Thus, *path agents* need to buy resources from *link agents* to offer services to callers. The bids and offers are ordered from high to low, and low to high respectively. There will be a range of prices for which the market will clear at which the maximum amount of resources can be traded. Buyers bidding above these prices and sellers offering below it are allowed to trade. The buyers and sellers within this group are matched randomly so that the benefits of trade are assigned between successful agents randomly also. The trading price for each given transaction is determined at random in the range between the buyer's bid and the seller's offer. Notice that this procedure implies that no buyer will pay more than his bid, and no seller will get paid more than his offer. Moreover, the procedure implies that the total surplus realized in the market, a measure of the social welfare, is maximized because the benefits of trade are distributed randomly between all successful agents.

3.2.2 Path Markets

The path market is also a sealed bid auction. This is because it is a critical performance requirement of the system that the allocation of call traffic to paths occur almost instantaneously (so that callers are not kept waiting for calls to be established). This means the auction protocol has to be as short as possible. As before, the most efficient protocols in this respect are the single shot, sealed bid types. Since we have a single caller and multiple path agents offering resources, a single sealed bid auction is appropriate.

A buyer sending a service request message to the market initiates the auction. The auctioneer then broadcasts a request for offers to all agents able to provide the connection. All sellers simultaneously submit offers and the lowest one wins the contract to provide the connection. In this market, we experimented with two protocols: (i) The First Price protocol, in which the price at which the buyer and the seller trade is that of the highest bid submitted. (ii) The Vickrey, or second price auction protocol, in which the price at which the buyer and seller trade is that of the second highest bid submitted. We choose to experiment with two strategies because economic theory predicts that Vickrey auctions provide more competitive market outcomes, doing away with wasteful speculation by encouraging truth telling behaviour on the part of the participants (Varian 1995). However, since we are using simple adaptive agents without speculative bidding strategies, we were unsure as to whether this factor would impact the overall behaviour of the system. To test the impact of this factor on the system as a whole, we implemented the market with both protocols and empirically tested the efficiency of each (section 5).

4 Designing Economically Rational Agents

The range of potential interactions is determined by the market institutions in which the agents participate. In both of our market types, agent communication is restricted to setting a price on a single unit of a known commodity. Therefore, agents set their prices solely on the basis of their implicit perception of supply and demand of that commodity at a given time. When a resource is scarce, buyers have to increase the prices they are willing to bid, just as sellers increase the price at which they are willing to offer the resource (mutatis mutandis when resources are plentiful). Here, agents perceive supply and demand in the market through the success or otherwise of bidding at particular prices.

4.1 Link Agents

A *link agent* has a set of n resources, slices of bandwidth capacity required to connect individual calls, that it can sell on the *link market*. At time t, the price to be asked for each of these units is stored in a vector $p_t = \{p_t^1, ..., p_t^n\}$ with the range of possible prices being zero to infinity, $p_t^i \in [0, \infty>$ for each member of the vector $i = 1,.., n$ and each time period t. At time $t = 0$ the prices for each unit are randomly (uniformly) distributed on $[0, H]$ where H is the initial upper limit on prices asked. When x units have been allocated, the remaining $n - x$ units are offered to the link market for sale simultaneously. Suppose that of the $n - x$ units offered for sale in a given period t, the m units with the lowest prices are successfully sold. The prices in the vector are updated as follows:

$$p_{t+1}^i = p_t^i \, . \, for \, I = 1,..., x \tag{1}$$

$$p_{t+1}^i = p_t^i \times (1 + \varepsilon) \, for \, i = x+1,..., x+m+1 \tag{2}$$

$$p_{t+1}^i = p_t^i \times (1 - \varepsilon) \, , for \, i = x+m+2,...,n \tag{3}$$

$$Where \, \varepsilon = U(0, \sigma) \tag{4}$$

Thus the link agent increases or decreases the price of any unit by a small amount ε after each auction. Here ε is obtained from a uniform random distribution between zero and σ (here 0.1). If previously allocated units are released by the path agent, they join the pool of unallocated units and the price vector is re-ordered to reflect this. This approach was chosen so that the prices of each resource on the link, taken together, should adapt to the demand on the network to carry traffic.

4.2 Path Agents

A *path agent* acts as both a buyer of link resources and a seller of path resources. Their buying behaviour is detailed in section 4.2.1, and their selling behaviour in section 4.2.2. In general, path agents wish to buy resources cheaply from link agents and sell them at a profit to end consumers. To do this, they bid competitively to acquire resources that they then sell on to callers, at a price not less than that paid for them. The path agent tries to maximize its profits by adjusting its inventory and sales behaviour on the basis of the feedback it receives from the market. The mechanism by which the path agent decides what resource level to maintain is described in section 4.2.3.

4.2.1 Buying Behaviour

A *path agent* actively tries to acquire resources (units of link bandwidth needed to connect a call), across the chain of links that it represents. It does this by placing bids at each of the link markets at which the resources it needs are sold. The agent retains a vector of prices that it is willing to pay for resources on each of the links that constitute its path. The agent's strategy is to try to equalise its holding of resources across each of those links; uneven resource holdings have to be paid for but cannot be sold-on or bring in any revenue because they do not constitute complete paths. The *path agent* tries to maintain its resources at a level w that is discovered through hill climbing adaptation (Russell&Norvig 1995) to the behaviour of the market (section 4.2.3). The most profitable value of w is obtained by adjusting it according to changing profit during ongoing buying and selling episodes. When x units have been acquired, the path agent bids for the remaining $w - x$ units on the link market simultaneously. Suppose that, for a given link, of the $w - x$ units that the path agent bids for at time t, the m units with the highest prices are successfully acquired. The prices in the vector are updated as follows (using ε as defined previously in section 4.1).

$$p^i_{t+1} = p^i_t, \text{ for } I = 1, \ldots, x \qquad (5)$$

$$p^i_{t+1} = p^i_t \times (1 - \varepsilon), \text{ for } i = x + 1, \ldots, x + m + 1 \qquad (6)$$

$$p^i_{t+1} = p^i_t \times (1 + \varepsilon), \text{ for } I = x + m + 2, \ldots, w \qquad (7)$$

This price setting mechanism was chosen because it allows the path agents to adaptively determine prices for individual link resources. The price bid for each resource should be as low as possible without failing to win the resource in the auction. Therefore the agent makes a bid for each resource that it needs separately. If a bid fails, the agent increases the price it will bid at the next auction (in order to increase its likelihood of winning the resource). If a bid succeeds, the agent reduces the price it bids for that resource in subsequent auctions (in order to avoid paying over the market price).

4.2.2 Selling Behaviour

A *path agent* will offer to sell a path resource whenever an auction is announced by the path market and it has an appropriate path resource to sell. The price asked is determined by the cost of acquiring the underlying link resources and the outcomes of previous attempts to sell. Let p_t be the price of a path resource at time t (the time of the auction) ranging from the cost to acquire the resource to infinity, $p_t = [cost, \infty >$. The first time the agent offers a resource for sale, it offers it at a price given by $p_t = cost \times (1 + \varepsilon)$ in order to sell at a profit (using ε as defined previously in section 4.1). Subsequently the offer price is given by:

$$p_{t+1} = Max(p_t \times (1 + \varepsilon), cost) \text{ if last offer was successful} \qquad (8)$$

$$p_{t+1} = Max(p_t \times (1 - \varepsilon), cost) \text{ if last offer was not successful} \qquad (9)$$

This price function ensures that the path agent never sells a resource for less than it paid to acquire it in the first place. Given its inventory level (see section 4.2.3), the agent attempts to maximise its income. The price bid for each resource should be as low as possible without failing to win the resource in the auction. Therefore, the agent increases the price it asks whenever it is successful, and decreases it whenever it is unsuccessful. This means the agent adapts its price to the level of competition, as it perceives it from the outcomes of previous auctions.

4.2.3 Inventory Level

The resource levels of the various path agents determine the maximum flows available for traffic on individual paths. In our case, the system design philosophy is to have the individual paths determine their own optimal resource levels. When this is achieved, balancing the load in the network as a whole becomes an emergent property of the social interactions of the agents. Path agents act in response to the economic pressures exerted on them by their consumers, competitors and suppliers. Therefore, we choose to have path agents discover their own optimal flows by adaptation to economic conditions as they perceive them (through interactions in the markets in which they compete). Path agents are both *buyers* and *sellers* that attempt to maximise their profit through trade, where profit is the difference between their revenue (from selling path resources to callers) and their expenditure (cost of acquiring and holding onto resources). In order to maximise profit agents must have an inventory level that is optimal for them, in the competitive environment in which they find themselves. Therefore our path agents adapt their inventory levels to the profits they earn through their interactions with the market. This is implemented by having the inventory level of individual path agents climb the hill of their profits.

In more detail, let R_t be the profit of an agent at time t and I_t be the desired resource inventory of that agent. If profit has increased since the last market interaction ($R_t > R_{t-1}$) and the last change corresponded to an increase in desired inventory level ($\Delta I_t > 0$), the new desired inventory level is increased by one resource unit. If the last change in desired inventory was negative ($\Delta I_t < 0$) then desired inventory is reduced by one unit. However, if profits have fallen, ($R_t < R_{t-1}$) and the last change was positive ($\Delta I_t > 0$) we decrease the desired inventory level; If the last change was negative ($\Delta I_t < 0$) then the desired inventory level will be increased. When decreasing the desired inventory level, the agent chooses to give up the most expensive of its link resources (that are not allocated to a call). This strategy reduces the agents inventory rental cost by the largest amount possible in a single time step.

4.3 Call Agents

Call agents initiate the auctions at which path agents compete by signalling that they wish to buy resources to a given source destination pair. In doing so they set a maximum price that they will not exceed to make a call. This puts downward pressure on the offers made by path agents to provide resources across whole paths, thereby anchoring the system.

5 Experimental Evaluation

Our experiments were designed to test three hypotheses. Firstly, we wanted to know whether market-based systems can compete with static routing algorithms in terms of call routing performance (section 5.1). Secondly, we wanted to know if our system uses the network efficiently (i.e. does it use the best routes) whenever possible, allowing for congestion (section 5.2). Thirdly, we wanted to test if the system discriminates in its choice of routes between paths that would be indistinguishable from one another to a conventional static routing algorithm, without expected traffic predictions (i.e. they differ only in their proneness to congestion) (section 5.3).

5.1 Performance Evaluation

In terms of performance, we sought to address two fundamental questions. Firstly, can our market-based control system perform as well or better than a conventional system? Secondly, what is the effect of using a Vickrey auction protocol rather than a first price auction protocol at the path market?

In a series of experiments we tested the efficiency of our market-based control mechanism (using both first price and Vickrey auctions as path markets) against a static routing mechanism. The static routing mechanism used a number of paths (three) between each source destination pair and routed traffic to these in order of path length, using longer paths when the shorter ones became congested. Here

efficiency was measured as the proportion of calls successfully routed through the network as a percentage of the total number of calls offered. The experiment was configured to simulate a small irregularly meshed network of 8 nodes with link capacities sufficient for 200 channels. Calls arrived on average approximately every 5 seconds, were routed between a randomly chosen source destination pair, and lasted an average of 200 seconds. Call arrival and call duration were determined by a negative exponential time distribution function, with U (0, 1) being a uniform random distribution between 0 and 1. The inter-arrival time between calls and the call duration were calculated using the formula: $f(x) = -\beta \ln U$, where β was the mean inter-arrival time and mean call duration respectively. The simulation was allowed to run for 20,000 seconds in each case. The traffic model was chosen to simulate a realistic call arrival rate and duration. The network dimensions where chosen to reflect a small network under heavy load.

These results show that similar levels of performance are obtained using the market-based control mechanism (95.4% of calls connected) and static routing (94.8% of calls connected) (Figure 1.). It is interesting to note that, contrary to our original hypothesis, using a Vickrey auction for the path market did not improve upon the results obtained using a first-price auction. One possible reason for this is the way in which the path agents in our system adapt their pricing strategy to market outcomes.

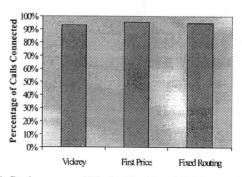

*Fig. 1. **Performance of Market-Based and Static Routing***

Vickrey auctions are designed to make the markets more efficient by making counter speculation between competing agents wasteful. However, with simple adaptive agents being used here such speculation does not occur, so the effect should not be significant.

The ability of our system to perform as well as a static routing mechanism should be taken as a positive result. As well as matching the performance of conventional routing techniques, our market-based approach has a number of distinct advantages for network operators and users. Firstly, it provides an architecture that is open to deployment in multi-enterprise environments without the inefficiency of static internetworking policies at sub-network interfaces. Secondly, our system is scalable in that no agent has to know the address of a significant number of peer agents or possess a map of the entire network. Thirdly, our system allows a much quicker response to call requests because the call routing process does not need to obtain information from the wider network at call set up time. In our case a call request can be processed and the call dispatched to a path (or refused) solely on the basis of information present at its source. This is achieved by having path agents that pro-actively determine their capacity to handle traffic, rather than waiting for call requests before processing.

5.2 Resource Utilisation Efficiency

In addition to the raw connection performance, it is important to know how effectively the network's resources are used. This is important because both the callers and the system benefit from routing calls through the shortest path when one is available. Shorter call routes use fewer system resources than longer ones, and they provide a service with less delay to end users. To assess our system's performance, we analysed the relative percentages of calls that were assigned to first, second and third best paths (by the number of links which make up the path). The data (Figure 2.) shows that the majority of calls were routed through the most efficient routes: 61% using the most efficient route, 25% the second best route and 15% the third best. Thus not only does our system route most of the calls it is presented with, it also makes the most efficient routing choices.

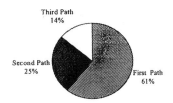

Fig. 2. Utilisation of First, Second and Third Shortest Paths by Market-Based Control

5.3 Congestion Discrimination

One of the claims we made for market-based approaches, is that good system level choices can emerge from local choices, that are influenced by information about the social context (obtained through interaction). To explore this hypothesis, we examined the performance of our system in cases that are indistinguishable from a local perspective. Thus, we focussed on source-destination pairs where all the routes are of equal length. In such cases, an alternate routing mechanism cannot decide between these paths without some notion of congestion through the whole network, which cannot be calculated and propagated in real time (recall the discussion of section 2). Alternate routing mechanisms can assign traffic to paths probabilistically, so that statistically, over time more traffic is routed to less congestion prone paths. However, this method is dependent on the accuracy of past measured traffic as a predictor of future traffic patterns. With our approach, we believe such

discrimination would emerge from the competitive nature of the market place. The reason for this hypothesis is that while path agents for paths of equal route length have to obtain the same number of resource slices, some have to obtain the more congestion prone of those slices. By definition, the more congestion prone resource slices are traded in the more competitive (and hence more highly priced markets). All other things being equal, the profitability of selling these paths will be lower because of the higher costs. With lower profitability comes a lower inventory level and fewer calls being routed via that path.

We wanted to test whether the market-based control mechanism is able to discriminate between paths on the basis of congestion cost in real time. In order to investigate this, we examined the source-destination pairs in our network configuration (nine of them) for which all (three) known paths consisted in an equal number of links. We plotted the percentage of calls routed to each of the three paths in (reverse) order of congestion and took the average of these values across all nine source destination pairs (Figure 3.). Our results clearly show that the market mechanism is able to distinguish between congestion costs entailed in routing across paths of otherwise equal length and assign calls to the least congested path most of the time.

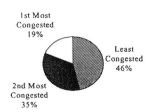

Fig. 3. Utilisation of paths in order of congestion by market-based control

6 Conclusions

We have described the design and implementation of a market-based system for call routing in telecommunications networks. Our system performs comparably with a static routing approach in terms of the percentage of the calls that are connected. However, from an architectural point of view, the market-based approach represents an improvement on static, centralised systems for a number of reasons. Firstly, it provides a platform for implementing network traffic management in a multi-enterprise internetwork. Secondly, it does not rely on a centralised controller to compute network reconfigurations, making the network management function robust to failure. This means that the system described in this work is architecturally more robust than an equivalent network with a centralised control mechanism. It is important to distinguish between this sense of robustness, and robustness as an empirical measure of the performance of the system in the event of the failure of a component of the system (or the agent process managing it). The performance of a decentralised control system degrades with the loss of controlling processes, while a centrally controlled one would be without control if the controlling process were to be lost. The study of the degradation of control efficiency in the system described here has been left to future work. Thirdly, no agent needs to know of the existence of more agents than there are links in the paths of the network (making the agent acquaintance databases compact and the whole system more scalable). Fourthly, there is no requirement to test the network state at call set up time, making the call set up procedure faster and more robust. Fifthly, the cost of each call to the network and the proportion of that revenue owing to each of the enterprises involved in carrying that call can easily be computed from information available to the user terminal equipment when the call is made, thus making call charging more efficient.

The results presented in this paper show that our market-based system performs the call routing and network management tasks adequately. However, the function used to determine the inventory level of path agents is quite simple and responds reactively to burstiness in the call arrival rate; this may be inducing unwanted oscillation in the path inventory parameter which may be adversely affecting performance. We intend to experiment with this function, and the parameters that govern its behaviour, to determine the impact of our choices on the performance of the overall system, and to see if that performance can be improved upon.

Acknowledgement

This work was carried out under EPSRC grant No.GR/L04801.

References

Appleby, S., and Steward, S. 1994. Mobile Software Agents for Control in Telecommunications Networks, *BT Journal of Technology* 12 – 2: 104 - 113

Bertsekas D. & Gallager M. 1987. *Data Networks*, Prentice Hall International , Inc.

Cisco Systems Inc., 1999. A Primer for Implementing a Cisco Virtual Private Network (VPN), Cisco Systems Inc.

Clearwater, S. H. 1996. *Market-Based Control A Paradigm for Distributed Resource Allocation*, World Scientific Press.

Gibney, M..A. & Jennings, N.R, 1998. Dynamic Resource Allocation by Market-Based Routing in Telecommunications Networks, In Proceedings of Intelligent Agents for Telecommunications Applications 1998, Springer Verlag. 102 - 117

Goldszmidt, G. and Yemini, Y., 1998. Delegated Agents for Network Management. *IEEE Communications Magazine* March 1998: 66 -70

Hayzelden, A. and Bigham, J. 1998. A Heterogeneous Multi-Agent Architecture for ATM Virtual Path Network Resource Configuration, In Proceedings of Intelligent Agents for Telecommunications Applications 1998, Springer Verlag 45 - 59

Jennings, N.R. & Wooldridge, M.R. 1998. Agent Technology Foundations, Applications and Markets, Springer Verlag.

Kershenbaum, A. 1993. *Telecommunications Network Design Algorithms,* McGraw-Hill International Editions

Russell, S. J. and Norvig, P. 1995. Artificial Intelligence: A Modern Approach, Prentice Hall, Inc.

Schoonderwoerd, R. Holland, O.E. and Bruten J. 1996. Ant-like agents for load balancing in telecommunications networks, in The First International Conference on Autonomous Agents, ACM Press.

Stallings, W. 1997. *Data and Computer Communications,* Prentice Hall International , Inc.

Schwartz, M. and Stern, T.E. 1989 Routing Protocols, in Computer Network Architectures and Protocols (Second Ed.), ed. Sunshine, C.A, Plenum Press, New York / London

Varian, H.R. 1992. Microeconomic Analysis (Third Ed.), W.W. Norton & Company Inc.

Varian, H.R. 1995. Mechanism Design for Computerised Agents, Proccedings of the 1995 Usenix Workshop on Electronic Commerce.

M. Wellman, M. 1996. Market Oriented Programming: Some Early Lessons, in Market-Based Control a Paradigm for Distributed Resource Allocation, Ed. S. H. Clearwater, World Scientific Press: 74 -95

Evolving Solutions of the Density Classification Task in 1D Cellular Automata, Guided by Parameters that Estimate their Dynamic Behaviour

OLIVEIRA, GINA M. B.

Universidade Presbiteriana Mackenzie
Rua da Consolação, 896, 5° andar, Consolação
01302-907 São Paulo, Brazil
gina@mackenzie.com.br

DE OLIVEIRA, PEDRO P. B.

Instituto Pesquisa e Desenvolvimento, Universidade do Vale do Paraíba
Av. Shishima Hifumi 2911, Urbanova
12244-000 São José dos Campos, Brazil
pedrob@univap.br

OMAR, NIZAM

Divisão de Ciência da Computação, Instituto Tecnológico de Aeronáutica
Praça Marechal Eduardo Gomes 50, Vila das Acácias
12228-901 São José dos Campos, Brazil
omar@comp.ita.cta.br

Abstract

Various studies in the context of one-dimensional cellular automata (CA) have been done on defining parameters directly obtained from their transition rule, which might be able to help forecast their dynamic behaviour. Out of a critique of the most important parameters available for this end, and out of the definition of a set of guidelines that should be followed when defining that kind of parameter, we took two parameters from the literature and defined three new ones, which, jointly provide a good forecasting set. We then used them to define an evolutionary search heuristic to evolve CA that perform a predefined computational task; here the well-known Density Classification Task is used as reference. The results obtained show that the parameters are effective in helping forecast the dynamic behaviour of one-dimensional CA, and can effectively help a genetic algorithm in searching for CA of a predefined kind.

Introduction

Cellular Automata (CA) have the potential of executing complex computations with high degree of efficiency and robustness, as well as modeling the behaviour of complex systems. In spite of their being systems of extremely simple implementation, CA are genuine examples of complex systems with the advantage of being amenable to direct and easy manipulation of their parameters to study their dynamics (Wolfram 1983).

Various pieces of research on the dynamics of CA rely on the study of parameterisations of their rule space, especially static parameters (i.e., those directly obtained from the transition table) designed to help forecast their dynamic behaviour (Li *et al.* 1990; Li and Packard 1990; Li 1991, 1992; Binder 1993).

The present work uses parameters of the latter kind, as heuristics to prune one-dimensional CA rule spaces, thus leading an evolutionary search towards those with a predefined behaviour.

Starting from the analysis of the main published parameters, we conceived guidelines that should be followed in the design of a forecast parameter (Oliveira 1999; Oliveira *et al.* 2000), and selected a set of five parameters, which will be presented below. As an intermediate step, we relied upon a classification scheme of CA dynamic behaviours, which will also be presented as later.

Finally, results are presented in the evolution of CA in the computational task known as the Density Classification Task (Mitchell 1996). In the approach to be described a simple Genetic Algorithm is used, that incorporates a heuristic based on the selected forecast parameters (Oliveira 1999). In the last section, conclusions are drawn and crucial final remarks are made.

Cellular Automata Dynamics

For what follows, the notation adopted is the one used by Mitchell (1996).

Basically, a cellular automaton consists of two parts, the *lattice* where states are defined, and the *transition rule* that establishes how the states will change along time. For one-dimensional CA, the neighbourhood size m is usually written as $m = 2r + 1$, where r is called the *radius*. In the case of binary state (i.e., two-state) CA, the transition rule is given by a Rule Table, which lists each possible neighbourhood with its output bit, that is, the update value of the centre cell of the neighbourhood. The 256 binary, r = 1, one-dimensional CA are called the Elementary Cellular Automata (ECA). Wolfram (1983) proposed an enumeration scheme for ECA, in which the output bits are lexicographically ordered and read from right to left, so as to form a binary number between 0 and 255.

If two CA transition rules are identical, except that one maps a neighbourhood onto a state, while the other maps the same neighbourhood onto a different state, the two rules are said to be neighbors, the distance between them being equal to 1. With this concept of distance, all rules can be thought of as residing in a Rule Space, where each point is a transition rule, and the points arranged in such a way that any two neighboring points are 1 unit of distance apart. The 256 ECA rules form the Elementary Rule Space (Li and Packard 1990).

According to their typical dynamic behaviour (that is, from an arbitrary initial condition), CA can be grouped into classes. A few classification schemes have been used in the literature. Wolfram (1993) proposed a qualitative classification of CA in four dynamic classes. Later, Li and collaborators proposed a series of refinements in the latter classification (Li *et al.* 1990, Li and Packard 1990; Li 1991, 1992). For the present work, we use the classification scheme adopted in (Li 1992):

- Null Rules: the limit configuration is formed only by 0s or only by 1s.

- Fixed Point Rules: the limit configuration is invariable upon reapplication of the cellular automaton´s rule (with a possible spatial shift), the null configurations being excluded.

- Two-Cycle Rules: the limit configuration is invariable upon reapplication of the rule twice (with a possible spatial shift).

- Periodic Rules: the limit configuration is invariable upon reapplication of the cellular automaton´s rule L times, the size L of the cycle being independent or weakly dependent on the system size.

- Complex Rules: although the limit dynamics can be periodic, the transition interval can be extremely long, typically growing more than linearly with the system size.

- Chaotic Rules: they produce non periodic dynamics, which is characterised by an exponential divergence of the length of its cycle with the system size, and by the unstability in respect to perturbations.

Rule Space Parameterisation

The dynamics of a cellular automaton is associated to its transition rule, and it has been proved that the correct forecast of its dynamic behaviour, out of its rule, is an undecidable problem. However, several parameters directly calculated from the rule table have been proposed to help forecast CA dynamic behaviour, notably the λ parameter proposed by Langton (1990). The high cardinality of the CA rule spaces makes their parameterisation a hard task and many studies point to the need of using a group of parameters so as to allow a better characterisation of the dynamics (Li 1991; Binder 1993).

From the study of the parameters published in the literature, and also from others investigated by the authors of this paper, an analysis was made of the main problems found, which led us to propose eight desirable guidelines that should be followed in the design of such a kind of parameter (Oliveira 1999; Oliveira *et al.* 2000). Subsequently, under the light of those guidelines, we revised and analysed four of the main published parameters (Oliveira 1999), namely, λ (Langton 1990); the mean field parameters (Li 1991, 1992); the sensitivity parameter (Binder 1993), and the Z parameter (Wuensche 1999).

One of the main conclusions of that analysis is the fact that λ has a series of conceptual problems that limit its efficacy as a dynamic behaviour forecasting parameter. As a consequence, several researchers suggested that the solution would be to use more parameters together with it, arguing that λ alone would not be sufficient to capture all the singularities of the rule space (Binder 1993; Li 1991; Mitchell 1996). In our critique, we argued that in spite of agreeing with the necessity of using a set of parameters, λ should be discarded completely from that set, and that only parameters more atune with the proposed guidelines should be kept.

Another conclusion was that, just like those guidelines were good to evaluate already proposed parameters, they could also be used to guiding the creation of new parameters, which we then went about. Together with quantitative and qualitative approaches to how to select the parameters, the guidelines provided us with a parameter set which exhibited good performance in helping forecast the dynamic behaviour of CA.

Selected Parameters Set

We selected two parameters among those published: Sensitivity and Z. In addition to them, we conceived and tested several others, following the guidelines, and eventually came up with a set of five parameters. The three new ones in the group were named Absolute Activity, Neighbourhood Dominance and Activity Propagation (Oliveira 1999). All of them have been normalised between 0 and 1.

The set of parameters will not be formally defined here, as this is beyond present purposes; it will suffice to informally describe the three new parameters (Oliveira 1999).

- *Neighbourhood Dominance*: this parameter quantifies how much change is entailed by the CA rule transitions, in the state of the centre cell, in respect to the state that predominates in the neighbourhood as a whole. For example, in the transition $010 \rightarrow 0$, neighbourhood dominance occurs because the state that predominates in the neighbourhood is "0" and the transition maps the centre cell state onto "0" (in this case we say that the transition "follows" the neighbourhood); this is in contrast to transition $100 \rightarrow 1$, where dominance does not occur, as the predominant state in the neighbourhood is "0" but the transition maps the centre cell state onto "1". Accordingly, the parameter value comes from a count (in fact, a weighed sum) of the number of transitions of the CA rule in which neighbourhood dominance occurs, with the additional feature that, the more homogeneous the neighbourhood involved, the higher its weight. For example, in the elementary space, transitions from neighbourhoods 000 and 111 are given weight 3, while the others are given weight 1.
- *Absolute Activity*: the parameter quantifies how much change is entailed by the CA rule transitions, in the state of the centre cell, in relation to two aspects: the state of the centre cell of the neighbourhood, and the states of the pair of cells which are equally apart from the centre cell. Consider, for instance, transition $01011 \rightarrow 1$; since the next state of the centre cell will be 1, the parameter will indicate some activity in relation to the centre cell (which is in state "0"), no activity in relation to the cells which are 1 position away from the centre (those in state "1", underlined), and also some activity in relation to the rightmost and leftmost cells of the neighbourhood (as the latter is in the state "0").
- Activity Propagation: this parameter has a more elaborate definition than the others; roughly, it quantifies, at the same time, both the *neighbourhood dominance* and the *sensitivity* of each CA rule transition.

A formal definition of all the parameters was given in (Oliveira 1999), and will be available soon in a paper currently under preparation.

The main features (Oliveira 1999) that led to the selection of each parameter are summarised next:

Sensitivity and Z help to relatively discriminate null and chaotic behaviours. This characteristic can be observed in the graphics of Figures 1a and 1b. The relative occurrences of all Null and Chaotic rules of the elementary space are plotted there. The Sensitivity parameter is better than Z in performing this task.

Absolute Activity and *Neighbourhood Dominance* help to do the relative discrimination of fixed point and two-cycle behaviours, as can be observed in the charts of Figures 1c

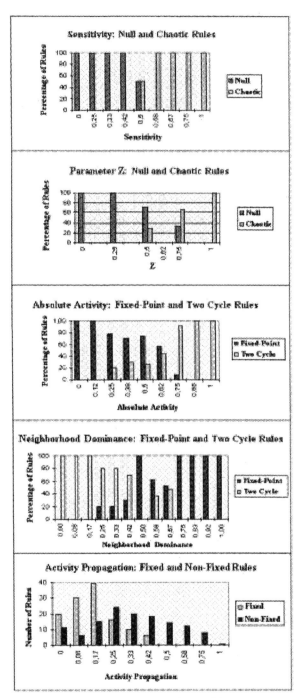

Figure 1. Relative Occurrences of the ECA rules to each value of the parameter: a) Sensitivity b) Z c) Absolute Activity d) Neighbourhood Dominance e) Activity Propagation

and 1d. The relative occurrences of all Fixed Point and Two-Cycle rules of the elementary space are plotted there.

This discrimination is not as perfect as the one obtained for the null and chaotic rules with the Sensitivity and Z parameters. However, it worth remarking that the number of rules involved in the fixed point / two-cycle relative discrimination (176) is much larger than the number of rules involved in the null / chaotic relative discrimination (56). Besides, null and chaotic behaviours are conceptually extremes, while fixed point and two-cycle behaviours are much closer. Therefore, fixed point / two-cycle relative discrimination is undeniably more difficult to achieve.

Activity Propagation helps define the region characterised by Null and Fixed Point rules, as can be observed in the chart of Figure 1e. All Null and Fixed Point rules were grouped as "Fixed" rules, while all the others as "Non-Fixed" rules. It is clear that the region formed by rules with low activity propagation is strongly characterised by the fixed behaviour.

Evidences of the Eficacy of the Parameter Set in High Cardinality Rule Spaces

Once the parameter set had been selected in the elementary space, evidences were searched for that could indicate that it could also be used in rule spaces of one-dimensional binary CA with radius larger than 1. Naturally, due to the high cardinality of these rule spaces, an exhaustive search of the space is just not feasible. However, a few results of classification of dynamic regimes in these spaces do exist, which can then be used as a small testbed, under the light of the selected parameters:

- Elementary Subspaces of CA with radius 2 and 3: In any rule space of one-dimensional binary CA with radius larger than 1, there is a set of 256 rules which are dynamically equivalent to those of the elementary rules, whose classifications are, therefore, known. These rules are the ones whose transitions depend only on the states of the three centremost cells of the neighbourhood, and which can be referred to as the Elementary Subspace with radius r. We calculated the selected parameters for the Elementary Subspaces with both radius 2 and 3.
- Complex Rules with radius 2 and 3: 50 CA rules with radius 2 and 7 CA rules with radius 3 were presented in (Wuensche 1994) and classified, by the author, as rules with Complex behaviour.
- Published rules for the Density Classification Task of CA with radius 3: We extracted from (Mitchell *et al.* 1996), (Andre 1996b), (Juillé and Pollack 1998) and (Cranny 1998), 8 published rules that perform well that computational task. The dynamic behaviour for the Density Classification Task is the Null behaviour.
- Published rules for the Synchronisation Task of CA with radius 2: We extracted from (Mitchell *et al.* 1996) two published rules that perform this computational task. The dynamic behaviour for the Synchronisation Task is the Two-Cycle behaviour.

By calculating the parameter values of the selected set for the 4 groups of known rules mentioned, the values obtained can be shown to fit in the parameter ranges found in the elementary space for each class of behaviour (Oliveira 1999). Therefore, they show there are various evidences that the parameters defined in the elementary space can indeed be used in higher cardinality spaces.

In the following section, we will present an example of the use of the parameter set in the rule space of the binary CA with radius 3.

Incorporating the Parameters in the Evolution of the Density Classification Task

Several researchers have been interested in the relationships between the generic dynamic behaviour of a cellular automaton and its computational abilities as part of the more encompassing theme of the relationships between dynamic systems and computational theories (Wolfram 1984). Various investigations have been carried out on the computational power of CA, with concentrated efforts in the study of one-dimensional CA capable of performing computational tasks (Mitchell 1996, Andre *et al.* 1996a., 1996b; Mitchell *et al.* 1993, 1996; Packard 1988). One of the approaches in this kind of research is the use of Genetic Algorithms (GA) (Goldberg 1989) as a search procedure to find CA with the predefined computational behaviour. Our approach is related to the latter, in that the selected parameter set mentioned earlier was used as an auxiliar metric to guide the processes underlying the GA search.

Calculation of the Parameters for the Density Classification Task

One of the computational tasks we studied was the Density Classification Task (DST) (Mitchell 1996). In this task the objective is to find a binary one-dimensional cellular automaton that can classify the density of 1s in the initial configuration of the 1D lattice, such that: if the initial lattice has more 1s than 0s, the automaton should converge to a null configuration of 1s, after a transient period; otherwise, it should converge to a null configuration of 0s. Various techniques have been described in the literature to find radius 3, two-state, one-dimensional CA with such ability, and the best rules found have been published.

In the current approach, we first calculated the parameters for the best published rules, eight in total, extracted from (Mitchell *et al.* 1996), (Andre 1996b), (Juillé and Pollack 1998) and (Cranny 1998), and summarised in Table 1. In the first column of the table, an acronym is defined for each rule (from their discoverers' initials). In the other columns the following information is provided: the lexicographic specification of the rule in hexadecimal number; the authors of the corresponding piece of work; the year in which the rule was published; the method employed to find the rule (Manual, Genetic Algorithm, Numerical Method, Genetic Programming, or Coevolutionary Genetic Algorithm); and its efficacy for this task. The efficacy of the rule is usually measured in samples of 10^4 initial configurations of different lattices, randomly generated. These cases are the most difficult ones to classify, as they have practically the same density of 0s and 1s.

Table 2 presents the calculated values of the five selected parameters (Sensitivity, Neighbourhood Dominance, Activity Propagation, Absolute Activity and Z) for the eight rules presented from Table 1.

By analyzing Table 2, one can observe that the parameter values for all the rules are distributed in narrow bands (except for the Z parameter). Table 3 presents these bands in its first row.

Rule	Hexadecimal Rule	Authors	Year	Method	Efficacy
GKL	005F005F005F005F005F005FFF5F005FFF5F	Gacs, Kurdyumov, Levin	1978	Manual	81.6%
MHC	050405870500F77037755837BFFB77F	Mitchell, Hraber, Crutchfield	1993	Genetic Algorithm	76.9%
DAV	002F035F001FCF1F002FFC5F001FFF1F	Davis	1995	Manual	81.8%
DAS	070007FF0F000FFF0F0007FF0F310FFF	Das	1995	Manual	82.2%
ABK	050055050500550555FF55FF55FF55FF	Andre, Bennet, Koza	1996	Genetic Programming	82.3%
CRA	00550055005500571F55FF57FF55FF57	Crany	1998	Numerical Methods	82.5%
JP1	011430D7110F395705B4FF17F13DF957	Juillé, Pollack	1998	Coevolutionary G.A.	85.1%
JP2	1451305C0050CE5F1711FF5F0F53CF5F	Juillé, Pollack	1998	Coevolutionary G.A.	86.0%

Table 1 - One-dimensional CA rules with radius 3, published for the DCT.

Rule	Sensitivity	Neighbourhood Dominance	Activity Propagation	Absolute Activity	Z
GKL	0.23	0.91	0.07	0.10	0.25
MHC	0.37	0.91	0.08	0.18	0.54
DAV	0.30	0.88	0.09	0.16	0.24
DAS	0.25	0.87	0.1	0.22	0.38
ABK	0.23	0.88	0.09	0.20	0.50
CRA	0.25	0.88	0.09	0.21	0.47
JP1	0.40	0.85	0.11	0.25	0.36
JP2	0.33	0.84	0.11	0.26	0.43

Table 2 - Parameters of the DCT published rules.

Rule	Sensitivity	Neighbourhood Dominance	Activity Propagation	Absolute Activity	Z
DCT Published Rules (Radius 3)	0.23 to 0.40	0.84 to 0.91	0.07 to 0.11	0.10 to 0.26	0.24 to 0.54
Elementary Null Rules (Radius 1)	0 to 0.50	0.50 to 0.92	0 to 0.25	0.12 to 0.50	0 to 0.75
More than 95% Random Rules (Radius 3)	0.45 to 0.55	0.35 to 0.65	0.15 to 0.35	0.4 to 0.6	0.5 to 0.7

Table 3 - Parameter bands of the DCT published rules, for the elementary null rules and for most of the random rules (more than 95%).

In order for a rule to feature a good performance in the DCT, first of all it has to exhibit a null behaviour. That is why the second row from Table 3 shows the parameter value bands of the null rules of the elementary space, which can then be compared to the parameter values of the published rules. It can be observed that nearly all the latter values do in fact occur in the expected band for the null rules. There is one exception, Absolute Acitivity, but, even for this case, its value range closely matches that of the expected band.

In order to evaluate how characteristic of this task would the parameter value bands be for the published rules, we confronted those values with a sample of 50,000 randomly generated CA rules with radius 3. The third row of Table 3 shows the parameter value ranges of most of the generated samples (at least 95% of the 50,000 rules). It should be observed that the parameter value range of the published DCT rules are not characteristic of the largest portion of the rule space, especially the values of the Sensitivity, Neighbourhood Dominance and Activity Propagation parameters. Furthermore, it is worth noticing that the rules of the random sample have approximately the same density of 1s and of 0s (≈ 0.5) due to the randomly nature of their generation. In spite of this fact, which makes the sample a qualitatively particular one, quantitatively they really represent the largest portion of the rule space.

Therefore, the information from Table 3 leads to the conclusions that, first, the parameter values of published rules for the Density Classification Task are really in the regions where our parameter set forecasts the null dynamic behaviour, and second, that these bands are not trivially found for one-dimensional, two-state CA, with radius 3.

Using a Parameter-Based Heuristic to Guide Evolution of the Density Classification Task

Having obtained the interesting results above, and once information is available on parameter value regions where good rules should be more likely to occur, it is appealing to ask whether would it be possible to use this information in an active way, when searching for CA of a predefined kind.

Once a computational task is defined, it is far from trivial finding CA that perform it. Manual programming is difficult and costly; from another perspective, exhaustive search the rule space becomes impossible, due to its high cardinality. A solution for this problem is the use of search and optimisation methods, particularly evolutionary computation methods.

Packard (1988) was the first to publish results using a Genetic Algorithm as a tool to find CA rules with a specific computational behaviour. He considered one-dimensional CA rules as being individuals in a population and defined their fitness according to their ability to perform the specified task. In this way, the genotype of an automaton is given by its transition rule, and the phenotype by its ability to perform the required task. Crossover among two CA was defined by the generation of two new

transition rules, from segments of two other rules, and mutation by the random flipping of the output bit of one of the transitions of the rule.

We replicated one of the published experiments in which a GA was used to search CA for the DCT (Mitchell *et al.* 1993, 1994). In their experiment, Mitchell and collaborators used radius 3, binary CA, with one-dimensional lattice formed by 149 cells, using a population of 100 individuals, evolving during 100 generations. At each generation, each individual evaluation was obtained out of testing the performance of the automaton in 100 different Initial Configurations (IC). Additionally, elitism (Goldberg 1989) was used at a rate of 20% (that is, the 20 best rules of the population at each generation were always preserved for the next one); the other 80 individuals were obtained through crossover and mutation. Parent selection for the crossover was directly made from the elite without considering each individual fitness. Standard one-point crossover was used at a rate of 80%. Mutation was applied after crossover, in each new generated individual, at a rate of 2% per bit. The results found for our replicated experiment (a series of 100 GA runs) are presented in Table 4.

The efficacy of the GA run was measured by testing the performance of the best rule found, at the end of the run, in the classification of 10^4 different initial configurations. Each row of the Table 4 shows the number of runs in which the efficacy of the best rule found was within the corresponding interval; this result is also present as a chart in Figure 2. In both presentations, the replicated experiment is referred to as "Mitchell". Table 5 presents the efficacy of the two best rules found in the experiment. The results we found are compatible with those reported by Michell *et al.* (1993a, 1993b).

Efficacy Bands (%)	No. of rules found in "Mitchell"	No. of rules found in "Parameters"
≤50	6	0
(50, 55]	4	0
(55, 60]	0	0
(60, 65]	16	14
(65, 70]	69	77
(70, 75]	3	7
>75	2	2
TOTAL	100	100
Average	65%	67.3%

Table 4 – Efficacy achieved for the "Mitchell" and "Parameters" Experiments

Subsequently, experiments were performed where the selected parameter set was used as an auxiliary heuristic in evolutionary searching for CA. The idea was that the introduction of the parameter information would entail improvement in the results obtained, thus providing a clear metric so as to evaluate the ability of the parameters in helping forecasting the dynamic behaviour of CA, and, hence, to gather evidence so as to validate the efficacy of the selected set.

The parameter-based heuristic was coded as a function Fp, which returns a value between 0 and 100 for each transition rule, depending on the values of the rule parameters. For example, in the present experiment, the parameter bands of the published rules were used. Function Fp was defined such that it returned 100 if all the parameters of the cellular automaton rule matched those of the bands of the published rules; otherwise, the value returned by Fp decreased as the parameter values became increasingly away from those bands. All parameters contribute equally in the calculation of function Fp.

The GA was modified so as to incorporate the parameter-based heuristic in two aspects:

- The Fitness Function of a cellular automaton rule was made by the weighted average of the original fitness function (efficacy in 100 different ICs) and the function Fp.

- Biased Reproduction and Mutation: in order to select the crossover point and the rule table bits to be mutated, various attempts were made; among them, only those that generated rules with high Fp value were selected.

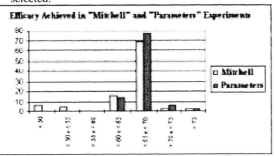

Figure 2 –Efficacy Achieved in "Mitchell" and "Parameters" Experiments

The parameter-based results are presented under the "Parameters" experiment label in Figure 2, and under the corresponding columns in Tables 4 and 5 (Oliveira 1999). Looking at the absolute values found after the insertion of the parameter information, 2 rules were found with higher efficacy than the best ones found by the basic experiment ("Mitchell"). It is also to be remarked that these results are also better than those found by the researchers in the original experiment (76.9% in 300 runs, as reported in (Michell 1993b)). In relative terms, Figure 2 makes it evident that the rules found using the parameter information had a higher efficacy than those found without the information. And finally, the average efficacy of the

"Mitchell" experiment was 65%, while the average of the "Parameters" experiment was 67.3%.

	"Mitchell"	"Parameters"
First Rule	76.1%	80.4%
Second Rule	75.6%	77.2%

Table 5 - Efficacy of the 2 best rules in the "Mitchell" and "Parameters"

Conclusions

Through quantitative and qualitative selection approaches a set of parameters was devised that help characterise the dynamic regions of the elementary space in CA.

A very significant evidence of the efficacy of these parameters is the very good results obtained in the evolution of CA in computational tasks (Oliveira 1999). As shown in Table 1, the insertion of the parameter information managed to improve the performance of the rules found for the Density Classification Task, both in average and in respect to the best rules found.

It should be remarked that the "Parameter" experiment reported here was the best one achieved. However, several others were also performed by varying the function Fp, the relative weight of the function Fp in the total fitness of a rule, and the number of attempts of crossover and mutation. In spite of that, the other results clearly demonstrated an efficacy gain of the rules found when comparing to the corresponding evolution without the parameter information, thus emphasizing the robustness of the parameter-based heuristic.

The issue of replicating the experiments of Mitchell and collaborators was meant to provide the basis upon which the effect of introducing the parameter information could be clearly probed. The implementation of these experiments was well documented and disclosed (Mitchell 1996; Mitchell et al. 1993 1996); consequently, even not having had access to the original code of those experiments, our code seemed to be functionally very close to the original one. This can be checked by the fact that the results reached in the "Mitchell" experiment very closely matched the published ones (Mitchell et al. 1993 and Mitchell 1996). Consequently, the gain obtained in the experiments with the parameter information was definitely, and exclusively due to the insertion of the parameter-based heuristic.

Subsequently to the experiments done by Mitchell et al. (1993), other researchers published significant results on evolving solutions for the Density Classification Task, such as (Andre 1996a, 1996b) and (Juillé and Pollack 1998), and more effective rules to the task were found. However, in these cases the improvement was due to changes in the actual techniques in use (respectively, Genetic Programming and Coevolution) and in the evolution environment itself. As for the latter, it should be remarked that Andre *et al.* (1996a, 1996b) used a population of 51,200 individuals, while Juillé and Pollack (1998) used populations of 400 and 1000 individuals, evolving during more than 1000 generations; undoubtedly, in both cases much more computational resource was allocated to the task than that used in the experiments herein.

The fitness function that guided evolution in all the published experiments was entirely based upon the cellular automaton efficacy in performing the task. The results reported here, as well as many additional aspects we had to omit – see (Oliveira 1999) and (Oliveira *et al.* 2000) – lead us to the clear conclusion that, if the parameter-based heuristic was also incorporated into those more sophisticated, computational intensive experiments, there too a significant improvement in their reported results would be likelier to occur.

Finally, in addition to the Density Classification Task presented in this article, we obtained also excellent results in the Synchronisation Task (Das *et al.* 1995), (Sipper 1998), and in the Grouping Task (Oliveira 1999), a modification of the Ordering Task proposed in (Sipper 1998). These further results are presently available in (Oliveira 1999) but will soon be available in wider published form.

Acknowledgements

P.P.B.O. is grateful to Andy Wuensche and Wentian Li for various conversations, Tim Cranny for helpful information, and CNPq (ProTem-CC: CEVAL project) for funding. G.M.B.O. and N.O. are grateful to ITA for their support, and CAPES for funding.

References

Andre, D.; Bennett III, F. and Koza, J. 1996a. Evolution of Intricate Long-Distance Communication Signals in Cellular Automata Using Programming. In: *Proceedings of Artificial Life V Conference*. Japan, 5.

Andre, D.; Bennett III, F. and Koza, J. 1996b. Discovery by Genetic Programming of a Cellular Automata Rule that is Better than any Known Rule for the Majority Classification Problem. In: *Genetic Programming 96.* Stanford: Stanford University.

Binder, P.M. 1993. A Phase Diagram for Elementary Cellular Automata. *Complex Systems*, 7:.241–247.

Cranny, T. 1998. *Personal Communications.*

Das, R.; Crutchfield, J.; Mitchell, M. and Hanson, J. 1995. Evolving Globally Synchronised Cellular Automata. In: *Proceedings of International Conference on Genetic Algorithms*, San Francisco, 6.

Goldberg, D.E. 1989. *Genetic Algorithms in Search, Optimisation and Machine Learning.* Massachusetts: Addison-Wesley.

Juillé, H. and Pollack, J.B. 1998. Coevolving the "Ideal" Trainer: Application to the Discovery of Cellular Automata Rules. In: *Proceedings of Genetic Programming Conference.* Madison, 3.

Langton, C.G. 1990. Computation at the Edge of Chaos: Phase Transitions and Emergent Computation. *Physica D*, 42:12-37.

Li, W.; Packard, N. and Langton, C.G. 1990. Transition Phenomena in Cellular Automata Rule Space. *Physica D*, 45:77-94.

Li, W. and Packard, N. 1990. The Structure of Elementary Cellular Automata Rule Space. *Complex Systems*, 4:281-297.

Li, W. 1991. Parameterisations of Cellular Automata Rule Space. Santa Fe, NM. Santa Fe Institute. *Preprint.*

Li, W. 1992. Phenomenology of Non-local Cellular Automata. *Journal of Statistical Physics*, 68: 829-882.

Mitchell, M.; Hraber, P. and Crutchfield, J. 1993. Revisiting the Edge of Chaos: Evolving Cellular Automata to Perform Computations. *Complex Systems*, 7: 89-130.

Mitchell, M.; Hraber,P. and Crutchfield J. 1994. Evolving Cellular Automata to Perform Computations: Mechanisms and Impediments. *Physica D*, 75:361-391.

Mitchell, M. 1996. Computation in Cellular Automata: A Selected Review. In: *Nonstandard Computation.* Weinheim: VCH Verlagsgesellschaft.

Mitchell, M.; Crutchfield, J.; and Das, R. 1996. Evolving Cellular Automata with Genetic Algorithms: a Review of Recent Work.. In: *Proceedings of International Conference on Evolutionary Computation and Its Applications.* Moscow, 5.

Oliveira, G.M.B. 1999. *Dinamics and Evolution of One-Dimensional Cellular Automata.* PhD Thesis, Aeronautics Institute of Technology, São José dos Campos, SP, Brazil. (In Portuguese)

Oliveira, G.M.B., de Oliveira, P.P.B., and Nizam, O. 2000. *Guidelines for Parameterizing the Rule Space of One-Dimensional Cellular Automata.* Preprint, under review.

Packard, N. 1988. Adaptation toward the Edge of Chaos. In: *Dynamic Patterns in Complex Systems*. Singapore, 293-301.

Sipper, M. 1998. A simple Cellular Automata that solves the density and ordering problems. *International Journal of Modern Physics*, 9 (7).

Wolfram, S. 1983. Cellular Automata. *Los Alamos Science*, 9: 2-21.

Wolfram, S. 1984. Computation Theory of Cellular Automata – In: *Communication in Mathematical Physics*, 96.

Wolfram, S. 1994. *Cellular Automata and Complexity*. U.S.A.: Addison-Wesley.

Wuensche, A. 1994. "Complexity in One-D Cellular Automata: an Atlas of Basins of Attraction Fields of One-Dimensional Cellular Automata". *Cognitive Science Research Papers: CSRP 321*. Brighton: University of Sussex.

Wuensche, A. 1999. "Classifying Cellular Automata Automatically: Finding gliders, filtering, and relating space-time patterns, attractor basins and the Z parameter". *Complexity*, 4 (3): 47-66.

Cellular Automata Model Of Emergent Collective Bi-Directional Pedestrian Dynamics

V. J. Blue[1]* and J. L. Adler[2]‡

[1] New York State Department of Transportation, Poughkeepsie, NY 12603
[2] Department of Civil Engineering, Rensselaer Polytechnic Institute Troy, NY 12180

ABSTRACT

This paper describes the application of an Artificial Life cellular automata (CA) microsimulation to model the emergent collective behavior of bi-directional pedestrian flows. Since pedestrian flow is inherently complex, even more so than vehicular flow, previous CA models developed for vehicle flow are not directly applicable. It is shown that a relatively small rule set is capable of effectively capturing the collective behaviors of pedestrians who are autonomous at the micro-level. The model provides for simulating three modes of bi-directional pedestrian flow: (a) flows in directionally separated lanes, (b) interspersed flow, and (c) dynamic multi-lane flow. The emergent behavior that arises from the model is consistent with well-established fundamental properties of pedestrian flows.

INTRODUCTION

Cellular Automata (CA) microsimulation is an effective technique for modeling complex emergent collective behavior that is characterized as an Artificial Life approach to simulation modeling (Adami 1998; Levy 1992). CA is named after the principle of *automata* (entities) occupying *cells* according to localized neighborhood rules of occupancy. The CA local rules prescribe the behavior of each automaton creating an approximation of actual individual behavior. Emergent collective behavior is an outgrowth of the interaction of the microsimulation rule set over local neighborhoods.

Traditional simulation models apply equations rather than behavioral rules, but CA behavior-based cellular changes of state determine the emergent results. The self-organization in the collective behavior of Artificial Life modeling stems from decentralized sources of decision making, such as ant colonies, flocks of birds, and vehicles. CA pedestrian simulation, as used here, is a parallel, distributed, bottom-up approach (see Resnick, 1994 for example). By "designing" the CA-based pedestrian from the bottom-up at the interface with one another, higher-level functions, like route selection and trip behavior, can be added later without fundamentally changing the inter-pedestrian dynamics.

CA models are attractive for a number of reasons. The CA interactions of the pedestrians are based on intuitively understandable behavioral rules. They are easily implemented on digital computers, and compared to difference equation-based microsimulation models, run exceedingly fast. CA models function as discrete idealizations of the partial differential equations that describe fluid flows and allow simulation of flows and interactions that are otherwise intractable (Wolfram, 1994). Only the local rules and the sequencing of their use are coded, leaving the many autonomous interactions on the cell matrix to create the emergent macroscopic results. As a result it has been observed in CA simulations that very simple models are capable of capturing essential system features of extraordinary complexity (Bak, 1996).

Over the past several years, researchers have demonstrated the applicability of cellular automata (CA) microsimulation to car-following and vehicular flows. These CA models have included traffic within a single-lane (Nagel and Schreckenberg, 1992), two-lane flow with passing (Rickert, et al., 1995), bi-directional two-lane flow with passing (Simon and Gutowitz, 1998), and network-level vehicle flows in the TRANSIMS model (Nagel, Barrett, and Rickert, 1996). CA traffic models have been shown to provide a good approximation of complex traffic flow patterns over a range of densities (Nagel and Rasmussen, 1994; Paczuski and Nagel, 1995; and Nagel, 1996) including the formation of shock waves in traffic jams.

Though the field of traffic flow modeling is well established, researchers have found the task of modeling pedestrian flows to be somewhat daunting. In several ways, pedestrian movements are more complex than vehicle flows. Pedestrian corridors may have several openings and support movement in several directions. Pedestrian walkways are not regulated as roadways are. Unlike roadways where vehicle flow is separated by direction, bi-directional walkways are the norm rather than the exception. For the most part pedestrian flows are not channeled by direction, leaving pedestrians free to vary speed and occupy any part of a walkway. Pedestrians can form lanes dynamically as the authors have observed at Grand Central Station and on streets in New York City.

* vjblue@gw.dot.state.ny.us
‡ adlerj@rpi.edu

Pedestrians are capable of changing speed more quickly, in one second accelerating to full speed from a standstill or braking to a stop from full speed. Also, since safety and crash avoidance are less of a concern to pedestrians, sidestepping, slight bumping, nudging and exchanging places are often a part of walking through crowded corridors.

Over the past thirty years, researchers have developed several approaches to model pedestrian flows (e.g., Fruin, 1971; AlGhadi and Mahmassani, 1991; Lovas, 1994). Gipps and Marksjo (1985) developed a CA-like model that focused on the use of reverse gravity-based rules to move pedestrians over a grid of hexagonal cells. Helbing and Molnar (1995) advanced a social force model of pedestrian dynamics that captures some properties of bi-directional pedestrian flows, including the formation of dynamic multiple lanes, but is burdened with high computational overhead from floating point calculations. Hoogendoorn and Bovy (2000) have very recently built a gas-kinetic model of pedestrians. Work is underway in the STREETS model (Shelhorn et al., 1999) to create agents with the SWARM simulation system (Langton et al., 2000) that can navigate from behavioral attributes along intended routes and at the same institute a similar path finding approach over spatial systems is underway (Batty, Jiang, and Thurstain-Goodwin, 1998).

The CA model of pedestrian walkways presented here has several advantages, including an intuitively appealing emulation of pedestrian behavior and reliance on integer arithmetic for fast computation. Blue and Adler (1998) demonstrated that unidirectional pedestrian flows emerged from CA simulation experiments that correspond to the fundamental parameters published in a chapter dedicated to pedestrian characteristics in the recent edition of the U.S. Transportation Research Board's Highway Capacity Manual (1994). They further demonstrated the basic framework for the bi-directional pedestrian flow model and examined flows under different exchange probabilities with freer lane changing (Blue and Adler 1999a, 1999b, 1999c).

This bi-directional pedestrian modeling effort is distinctly different from the bi-directional vehicle model developed by Simon and Gutowitz (1998). The bi-directional vehicular flow model focuses on modeling acceleration and passing movements within the framework of two lanes of opposing flow. Due to high speeds of vehicles and the seriousness of collisions, vehicle movements require a more global view of a roadway segment. However, as roadway density increases, the level of vehicular activity (such as lane changing and acceleration) decreases significantly. Pedestrian flow, on the other hand, occurs at lower speeds and collisions are less catastrophic. As a result, pedestrians can adopt a more myopic view of their surroundings. Unlike vehicular flow, as density increases on pedestrian walkways, there can be substantial activity as lane changing, exchanges, and speed variations occur frequently, including the dynamic formation of lanes of various widths.

This paper discusses the application of the bi-directional pedestrian flow model to three modes: (a) separated directional flows (essentially two unidirectional flows), (b) interspersed directional flows, and (c) dynamic multi-lane (DML) flows as important cases that can be treated with the model. DML formation of directional streams is shown in Figure 1 where actual pedestrians can been seen forming six directional lanes, self-organize into this emergent pattern. Also examined is the importance of lane changing (sidestepping) and place exchange in the CA pedestrian model. Simulation results are presented to demonstrate the CA method's ability to capture fundamental properties of pedestrian movements.

Figure 1. Dynamic Multiple Lane formation by pedestrians crossing 35th Street at Fifth Avenue in New York City.

CELLULAR AUTOMATA RULE SET

By incorporating a rule set that eliminates anything but critical behavioral factors, the model facilitates a clear understanding of the underlying fundamental dynamics. There are three fundamental elements of pedestrian movements that a bi-directional microscopic model should account for: side stepping (lane changing), forward movement (braking, acceleration), and conflict mitigation (deadlock avoidance). The basic rule set for the model was developed around these three elements and is designed to work within a framework of parallel updates. As used by Rickert et al. (1995) and Simon and Gutowitz (1998) lane assignment and forward motions change the positions of all pedestrians in two parallel update stages in each time step. Parallel updates avoid succession interdependencies encountered in sequential updates by determining all the new positions before anyone moves. All the entities are then repositioned together. Only the pedestrians in the immediate neighborhood affect the movement of a pedestrian, which, though myopic, is relatively realistic. Each pedestrian is randomly assigned a desired speed of 2, 3, or 4 cells per time step or v_max from a normal distribution of walker speeds (Blue and Adler, 1998). The rule set is presented in Table 1.

Lane change (parallel update 1):

(1) Eliminate conflicts: two walkers that are laterally adjacent may not sidestep into one another

    (a) an empty cell between two walkers is available to one of them with 50/50 random assignment

(2) Identify gaps: same lane or adjacent (left or right) lane is chosen that best advances forward movement up to *v_max* according to the gap computation subsection* that follows the step forward update

    (a) For dynamic multiple lanes (DML):

        (i) step out of lane of a walker from opposite direction by assigning gap = 0 if within 8 cells

        (ii) step behind a same direction walker when avoiding an opposite direction walker by choosing any available lane with gap_same = 1 when gap = 1

    (b) ties of equal maximum gaps ahead are resolved according to:

        (i) 2-way tie between the adjacent lanes: 50/50 random assignment

        (ii) 2-way tie between current lane and single adjacent lane: stay in lane

        (iii) 3-way tie: stay in lane

(3) Move: each pedestrian p_n is moved 0, +1, or –1 lateral sidesteps after (1)-(3) is completed

Step forward (parallel update 2):

(1) Update velocity: Let $v(p_n)$ = gap where gap is from gap computation subsection below*

(2) Exchanges: IF gap = 0 or 1 AND gap = gap_opp (cell occupied by an opposing pedestrian) THEN with probability *p_exchg* $v(p_n)$ = gap + 1 ELSE $v(p_n)$ = 0

(3) Move: each pedestrian p_n is moved $v(p_n)$ cells forward on the lattice.

Subprocedure: Gap Computation

(1) Same direction: Look ahead a max of 8 cells (8 = 2 * largest *v_max*) IF occupied cell found with same direction THEN set gap_same to number of cells between entities ELSE gap_same = 8

(2) Opposite direction: IF occupied cell found with opposite direction THEN set gap_opp to INT (0.5 * number of cells between entities) ELSE gap_opp = 4

(3) Assign gap = MIN (gap_same, gap_opp, *v_max*)

Table 1. Rule Set

In the first parallel update stage, a set of lane changing rules is applied to each pedestrian on a lattice of square cells to determine the next lane of each pedestrian based on current conditions. The lane that best promotes forward movement is chosen from the local decision neighborhood, consisting of the left, same, and right lanes. Once sidesteps are found for everyone, all the pedestrians are moved to the new cells. In the second parallel update, a set of forward movement rules is applied to each pedestrian. The allowable movement (and thus the speed) of each pedestrian is based on the pedestrian's desired speed and the available gap ahead as constrained by the pedestrian in its current position directly ahead. Once speeds are found for everyone, all the pedestrians "hop" forward to new cells.

Pedestrians can change lanes only when an adjacent cell is available. A random number is drawn to designate the lane as free to this pedestrian or to the pedestrian two cells away. If an adjacent lane is free, then lane change is determined by the maximum gap ahead. The base case is interspersed (ISP) bi-directional flow in which pedestrians and exchange places but do not form lanes.

At this point in the lane-change parallel update (Lane Change Rule 2a) dynamic multiple lane (DML) flows will result by assigning a forward gap of 0 if encountering an opposing entity ahead. DML formations are further enhanced when a pedestrian can step behind a same direction walker when avoiding an oncoming pedestrian. This realistic behavioral adjustment helps the pedestrians to move into a same-direction flow lane.

Figure 2 illustrates this emergent DML pattern after 100 seconds of simulation time. In this figure, pedestrians move east-west with the darker cells representing eastbound pedestrians, lighter gray representing westbound pedestrians, and white cells being void. Although the pedestrians were randomly placed at the start of the simulation, after 100 seconds the pedestrians have found their way into same-direction lanes.

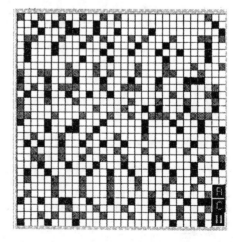

Figure 2. Emerging Dynamic Multiple Lanes

If the maximum gap size is common to two or more lanes, ties are broken to make lane assignments. In this rule set if the current lane is maintained as much as possible, speeds increase over allowing for some lane changes. Evidently, optional lane changes introduce blocks to those who would clearly benefit by changing lanes. Finally, all the pedestrians are moved together into their new lanes.

The second parallel update stage determines the forward movement of the pedestrians. The gap ahead is determined first. The available gap ahead depends on the direction of flow of the next person downstream. From the gap calculation, if the pedestrian ahead is going in the same direction, the new velocity of the follower is the minimum of the desired velocity (v_max) and the available gap ahead. If the pedestrian immediately ahead is going in the opposite direction and within the local neighborhood that both pedestrians could move at maximum speed (i.e., 8 cells is the maximum – 4 in each direction), then the updated velocity is the minimum of v_max and moving halfway forward. Moving halfway forward guards against collisions and hopping over one another.

For opposing pedestrians, place exchange guards against deadlocks by emulating what people actually do. Under constrained conditions opposing pedestrians may slip by one another. People are somewhat elastic and certainly not perfectly square and can thus exchange places. In actuality, temporary standoffs may occur when people guess which way to step past one another. Thus, the simulation contains a probability of a temporary standoff between closely opposing walkers. With probability p_exchg closely opposing pedestrians exchange places in the time step. The opposing entities each move the same number of cells, which is 0, 1, or 2 cells. Finally, the pedestrians go forward based on the gaps.

SIMULATION EXPERIMENTS

Since walking speed varies among pedestrians, a distribution of walking speeds is needed. For this simulation effort, three walking groups were used:

(a) *Fast Walkers* -- maximum speed of 4 cells per time step (about 1.8 m per time step).

(b) *Standard Walkers* – maximum rate of 3 cells per time step (about 1.3 m per time step)

(c) *Slow Walkers* – maximum rate of 2 cells/time step (about 0.85 m per time step)

A 5% fast; 90% standard; 5% slow (5:90:5) distribution of walkers were used to represent the pedestrian population. This distribution had the best realization of the fundamental diagram compared with other distributions in the single-direction case (Blue and Adler, 1998). The distribution also is consistent within ranges of speed and standard deviations used by others (Lovas, 1994).

The pedestrian walkway is modeled as a circular lattice of width W and length G (a rectangular grid that wraps around at the narrow ends). Each cell in the lattice is denoted L(i, j) where $1 \leq i \leq W$ and $1 \leq j \leq G$. Pedestrian densities are predetermined at the start of the simulation and remain constant throughout each run. At the start of each simulation, a density d, where $0.05 \leq d < 1.0$, is generated and N = INT(d*W*G) pedestrians are created and assigned randomly to the lattice. The circular lattice enables the set of pedestrians to interact at constant density and constant space allowance while maintaining strict conservation of flow. Cells in the lattice are considered square at 0.457 meters per side. This cell size, 0.457m on a side, is scaled according to minimal requirements for personal space as described in the Highway Capacity Manual (1994). The scale is also used to generate the speed-flow-density relationships that emerge. A 50x50 lattice of cells is used in the simulation experiments.

One second is the duration of each time step. Each simulation is 1,000 time steps with the first 100 time steps discarded to initiate the simulation and the latter 900 (15 minutes) used to generate performance statistics. The lattice loops back upon itself, allowing the pedestrians to continuously walk on a circular track.

Each set of experiments included runs at 19 densities ranging from 0.05 to 0.95 percent occupancy in increments of 0.05. For statistical accuracy, ten replications at each density level were run and the fundamental parameters were computed as the average over these replications. The emergent fundamental profile of the model is a map of the relationships between speed and flow over the range of densities.

RESULTS

The CA model yields flows that should be expected. As discussed earlier and depicted in Figure 2, dynamic multiple lane formation is generated by the bi-directional model. Figure 3 demonstrates the ability of the model to generate mode locking, a marching effect emerges from unidirectional flow that is very efficient, especially at low density. Mode locking is often seen in complex, self-organizing systems (see Schroeder 1991). This figure depicts the simulation applied to a uni-directional walkway in which all pedestrians are moving eastbound. Several rows in the walkway appear to be almost identically populated with pedestrians; illustrating the mode locking phenomenon.

As further evidence of the model's capabilities, Figures 4-7 show emergent macroscopic measures of performance over a range of densities for balanced (50-50) and unbalanced (90-10) directional splits and place exchange (p_exchg) of 0.5. These figures illustrate the differences between unidirectional, ISP, and DML flows.

Simulations have shown that separated bi-directional flows are essentially equivalent to unidirectional flow (Blue and Adler 1999a, 1999b, 2000). With unidirectional flow and separated flows, very few sidesteps are needed with the restricted lane change rule set used here and generally lane changes are only needed below a density of 0.4 (Figure 6).

Figure 3. Mode Locking

Restricted lane changing (in two- and three-way ties) in unidirectional flow helps aggregate movement rather than inhibits it. In contrast, when tie-breaking rules are applied that allow some choice in lanes rather than staying in lane, people fall out of step and the aggregate speed drops (Blue and Adler 1999b).

While the fastest and most efficient flow is from unidirectional flow with mode locking, those scenarios that come closest to its conditions do next best. Separated flow is an obvious example. DML 50-50 directional splits fare better than 90-10 (Figures 4-5) because the even directional splits allow better lane formation while the minor direction in the 90-10 split can't form lanes. Evidence of this comes from the 90-10 split needs more exchanges than 50-50. Also, DML 50-50 split has fastest drop-off in sidesteps (Figure 6) and the lowest exchange rate above 0.3 density (Figure 7), indicating excellent lane formation. Conversely, the ISP 90-10 split does better than ISP 50-50, because without rules for lane formation the strong major direction defines de facto lanes.

For the ISP scenarios there is a pronounced cusp in the speed-density curve at 0.3 density for the 90-10 split and at 0.35 for the 50-50 split. The transition at these points from low density to high-density performance is evidently a self-organization effect. Below 1/3 density, the grid is relatively sparsely occupied. Above a density of 1/3 at least one in three cells is occupied, a condition where entities will begin to have more adjacent neighbors than not, restricting sidestepping and avoidance of oncoming walkers. The region centered about 1/3 has the highest drop in speed (steepest slope) for all scenarios.

ISP speed and volume improvement from optimal lane changing is effective only up to 0.3 density after which values are essentially the same. However, sidesteps are reduced by 7 per person per minute at the peak. Exchanges are reduced only at low density.

Two to five exchanges per pedestrian per minute do not seem excessive (Figure 7), especially since they are done in pairs. It is unlikely that ISP 50-50 flow could maintain the high levels of position exchange (more than one every 4 time steps) that a p_exchg of 0.5 allows at densities of 0.5 and above (Figure 7). The ISP 50-50 volume curve (Figure 5) levels off at 40 Peds/min/m-of-width that seems unrealistic, but would be an advantage if this type of flow were safe and acceptable. Generally ISP flows are short term, and DML flows or separated flows are prevalent at high density. The model is capable of treating any case.

The Highway Capacity Manual (1994) shows a linear model of speed-density that amounts to a somewhat simplified version of the family of curves shown in Figure 4. Recent work has shown statistical evidence that more closely agree with these CA-based curves (see Blue and Adler, 2000 for a more complete discussion). Though not much empirical data is available on DML and ISP flows, the current edition of the HCM has ascribed to 90-10 flows, where separated lanes do not form, that the peak volumes (see Figure 5) are approximately 85 percent of peak unidirectional or lane separated flow. The bi-directional models show agreement with the HCM in that especially the peak DML volumes (88 percent) and to a lesser extent ISP volumes (77 percent) fall within the HCM's 85 percent range.

DISCUSSION

This bi-directional pedestrian CA model exhibits a range of complex, collective phenomena previously unexplored. This bi-directional model captures formerly intractable flows where discrete automata maneuver at a broad range of densities. The modeled pedestrians appear to exhibit strategic intelligence in choosing the lanes and forward movements with some of the characteristics of actual persons. The self-organization exhibited in the dynamic multiple lane formation (see figure 1) and in the spontaneously arising spatial efficiency of mode locking (see figure 2) reveals that this model exhibits Artificial Life.

While the standard design guide, the Highway Capacity Manual (1994), shows a linear speed-density curve for pedestrians it is generally acknowledged that the unidirectional and separated flow curves would more closely follow the S-curve speed-density relationship realized by the model (see Nagel, 1996 and Blue and Adler, 2000 for discussion). This CA model would further imply that the HCM could indicate that the various types of flow, unidirectional, and bi-directional (a) separated flow, (b) interspersed flow, and (c) dynamic multiple lane flow, have differing speed-flow-density curves.

The model aims at the minimal essential set of rules and parameters to capture bi-directional pedestrian flows. Lane changing and position exchanging are identified as important parameters in modeling ISP and DML flows.

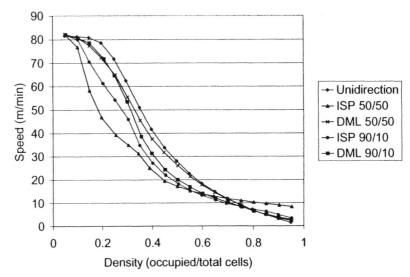

Figure 4. Speed vs density;p_exchg = 0.5

The results of simulations under conditions of (a) unidirectional flow, (b) interspersed bi-directional balanced flows (50-50), (c) dynamic multiple lane bi-directional balanced flows (50-50), (d) interspersed bi-directional unbalanced flows (90-10), (c) dynamic multiple lane bi-directional unbalanced flows (90-10).

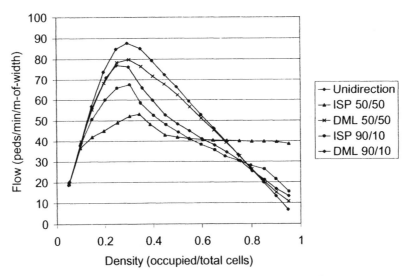

Figure 5. Flow vs density; p_exchg = 0.5

The results of simulations under conditions of (a) unidirectional flow, (b) interspersed bi-directional balanced flows (50-50), (c) dynamic multiple lane bi-directional balanced flows (50-50), (d) interspersed bi-directional unbalanced flows (90-10), (c) dynamic multiple lane bi-directional unbalanced flows (90-10).

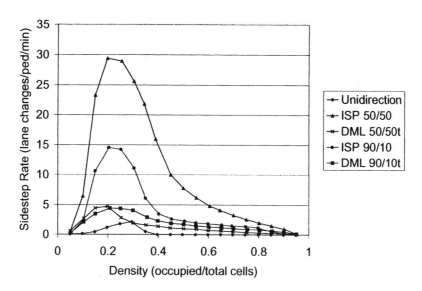

Figure. 6. Sidesteps vs density; p_exchg = 0.5

The results of simulations under conditions of (a) unidirectional flow, (b) interspersed bi-directional balanced flows (50-50), (c) dynamic multiple lane bi-directional balanced flows (50-50), (d) interspersed bi-directional unbalanced flows (90-10), (c) dynamic multiple lane bi-directional unbalanced flows (90-10).

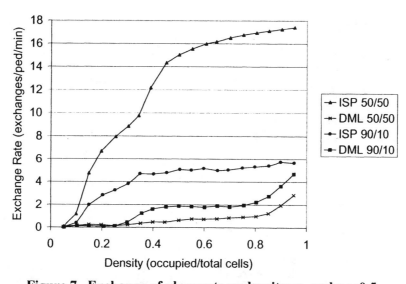

Figure 7. Exchange of place rate vs density; p_exchg = 0.5

The results of simulations under conditions of (a) interspersed bi-directional balanced flows (50-50), (b) dynamic multiple lane bi-directional balanced flows (50-50), (c) interspersed bi-directional unbalanced flows (90-10), (d) dynamic multiple lane bi-directional unbalanced flows (90-10).

Lane change and place exchange variables were set at reasonable values to examine a base case and a mid-range of values. Further calibration from field studies remains to be done.

The CA rules use pedestrian observations of available space and direction of nearby pedestrians. The social forces and gas-kinetic models (see Helbing and Molnar, 1995; Hoogendoorn and Bovy, 2000) imply vectors that give a motivation to act, such as repulsion from others and attractions. In our DML approach persons are repulsed by persons coming from the opposite direction and attracted to following those going in the same direction. However, the CA model is simpler and closer to emulating what pedestrians actually do: evaluate the space nearby, sidestep others blocking the way (change lanes), move forward as much as possible at a desired speed, and exchange places as needed to avert deadlocking.

Due to the vast amount of walking areas for a growing worldwide population, safe design practices would give this model practical purpose. Facility designers may try to avoid high-density situations, but when high-density pedestrian environments are unavoidable, it is especially important to examine the speed-flow consequences. We are next further studying videotapes of pedestrian flows and expanding the model to four-directional flows (Blue and Adler, 2000) and network flows. Simulations can be observed at http://www.ulster.net/~vjblue.

REFERENCES

Adami, C. 1998. *Introduction to Artificial Life*, Springer-Verlag, New York.

AlGadhi, S. A. H. and Mahmassani, H. 1991. Simulation of Crowd Behavior And Movement: Fundamental Relations And Application. *Transportation Research Record* 1320: 260-268.

Bak, P. 1996. How Nature Works: The Science Of Self-Organized Criticality, Springer-Verlag New York, Inc.

Batty, M., Jiang, B., and Thurstain-Goodwin, M. 1998. Local Movement: Agent-Based Models of Pedestrian Flow. Center for Advanced Spatial Analysis Working Paper Series, Paper number 4 (http://www.casa.ucl.ac.uk/working_papers.htm).

Blue, V. J. and Adler, J. L. 1998. Emergent Fundamental Pedestrian Flows From Cellular Automata Microsimulation. *Transportation Research Record* 1644: 29-36.

Blue, V. J. and Adler, J. L. 1999. Bi-Directional Emergent Fundamental Pedestrian Flows From Cellular Automata Microsimulation. in A. Ceder (Ed.), *Transportation and Traffic Theory: Proceedings of the 14th International Symposium on Transportation and Traffic Theory* 235-254, Pergamon, Amsterdam.

Blue, V. J. and Adler, J. L. 1999. Cellular automata microsimulation of bi-directional pedestrian flows. forthcoming in *Transportation Research Record, Journal of the Transportation Research Board).*

Gipps, P. G. and Marksjo, B. 1985. A Micro-Simulation Model For Pedestrian Flows. *Mathematics and Computers in Simulation* 27: 95-105.

Helbing, D. and Molnar, P. 1995. Social Force Model For Pedestrian Dynamics. *Physical Review E* 51: 4282-4286.

Hoogendoorn and Bovy 2000. A Gas-Kinetic Model for Simulating Pedestrian Flows. Proceedings of the 79th Transportation Research Board, Washington D.C.

J. Fruin, 1971. *Pedestrian Planning and Design*. Metropolitan Association of Urban Designers and Environmental Planners, New York, N.Y.

Langton, C., Burkhart, R., Daniels, M., and Lancaster, A. 2000. http://www.santafe.edu/projects/swarm/

Levy, S. 1992. *Artificial Life*. Vintage Books, New York.

Lovas, G. G. 1994. Modeling And Simulation Of Pedestrian Traffic Flow. *Transportation Research* 28B: 429-443.

Nagel, K. 1996. Particle Hopping Models And Traffic Flow Theory. *Physical Review E* 53: 4655-4672.

Nagel, K. and Rasmussen, S. 1994. Traffic At The Edge Of Chaos. *Artificial Life IV: Proceedings of the 4th International Workshop on the Synthesis and Simulation of Living Systems,* 222-225.

Nagel, K. and Schreckenberg, M. 1992. A Cellular Automaton Model For Freeway Traffic. *J. Physique (France)* I 2: 2221-2228.

Nagel, K., Barrett, C. and Rickert, M. 1996. *Parallel Traffic Micro-simulation by Cellular Automata and Application for Large Scale Transportation Modeling.* Los Alamos Unclassified Report 96:0050 Los Alamos National Laboratory, Los Alamos New Mexico.

Paczuski, M. and Nagel, K. 1996. *Self-Organized Criticality and 1/f Noise in Traffic*, in Traffic and Granular Flow, Eds. D. E. Wolf, M. Schreckenberg, and A. Bachem, Singapore: World Scientific.

Resnick, M. 1994. *Turtles, Termites, and Traffic Jams: Explorations in Massively Parallel Microworlds.* MIT Press, Cambridge, Mass.

Rickert, M., Nagel, K., Schreckenberg, M. and Latour, A. 1995. *Two-Lane Traffic Simulations Using Cellular Automata.* Los Alamos Unclassified Report 95:4367 Los Alamos National Laboratory, Los Alamos New Mexico.

Schroeder, M. 1991. *Fractals, Chaos, Power Laws: Minutes from an Infinite Paradise.* W. H. Freeman and Company, New York.

Shelhorn, T., O'Sullivan, D., Haklay, M., and Thurstain-Goodwin, M. 1999. STREETS: An Agent-Based Pedestrian Model. Center for Advanced Spatial Analysis Working Paper Series, Paper number 9 (http://www.casa.ucl.ac.uk/streets.pdf).

Simon, P. M. and Gutowitz, H. A. 1998. Cellular Automaton Model For Bidirectional Traffic. *Physical Review E* 57: 2441-2444.

445

Transportation Research Board 1994. *Special Report 209: Highway Capacity Manual*, National Research Council, Washington, D.C., Chapter 13.

V.J. Blue and Adler, J. L. 2000. Cellular Automata Microsimulation For Modeling Bi-directional Pedestrian Walkways. *Transportation Research B,* forthcoming .

V.J. Blue and Adler, J. L. 2000. Modeling Four-Directional Pedestrian Flows. Proceedings of the 79th Transportation Research Board, Washington D.C.

Wolfram, S. 1994. *Cellular Automata and Complexity*. Addison-Wesley Publishing Company.

Using flocks to drive a Geographical Analysis Engine

James Macgill

Centre for Computational Geography
Department of Geography,University of Leeds, LEEDS LS2 9JT, United Kingdom
Email: j.macgill@geog.leeds.ac.uk
web: www.geog.leeds.ac.uk/pgrads/j.macgill

Abstract

This paper describes a new method for the analysis of spatial data that can be used to solve the NP-hard problem of point pattern analysis in geographic, high-dimensional attribute data. The method builds on an established, highly developed and extensively tested methodology (GAM) and extends and combines it with concepts and methodologies found in the field of Artificial Life (Flocking and Agents). The new methodology is smart in that it is able to adapt to the characteristics of various data sources and makes intelligent use of available computational resources when the problem space becomes so large that brute-force techniques would quickly exhaust all available computer power.

The system is also, by nature, comprised of multiple, discrete computational units that lend themselves to easy parallelization, thus facilitating the use of parallel, or even distributed, architectures. The inspiration for many of the intelligent aspects of the methodology came from existing research into Artificial Intelligence, Artificial Life and Multi-Agent technologies.

Introduction

The well documented 'data explosion' that has resulted since the GIS revolution of the mid 1980s (Openshaw, 1995), has led to the rapid production of large quantities of high quality, spatially-referenced data. This trend shows no sign of slowing. The rate at which new geographic data is being produced is increasing. Indeed, geographic data sets are now being produced many times faster than they can be analyzed (Estivill-Castro and Murray, 1998).

The data that is generated is being linked, to an ever greater degree, to non-spatial attribute data; due partly to the ease with which many new and legacy databases can be linked via their geography, for example through postcodes (Raper *et al*, 1992). This has resulted in a rapid build-up of data that is, in a large number of cases, simply being archived. The under analysis of a number of key spatial data sets has been referred to by Openshaw, 1995 as a crime.

As a result, it became obvious that there was an urgent need for the development of semi or even fully autonomous analysis tools that could start to clear this backlog of under-analyzed data, whilst equipping us with the tools needed to tackle the ever richer sources of data that will arrive in the future.

However, in seeking to meet this challenge it is important not to neglect the unique contribution that intelligent human analysis can make. Additionally, if the aim is the creation of intelligent analysis agents then there are a number of hurdles to be overcome: a demonstration of added value, indications of performance under conditions where manual methods failed to succeed, proof of safety, and preferably more than one good reason for wanting to approach the problem via a route that many would naturally fear and wish to criticize.

One of the most common and important analysis tasks in spatial analysis is having an ability to spot patterns. For many key data sets such as crime and health data, the most important pattern to be identified is a cluster, and it is this capability that has been the main focus for research. Each of the systems that have been developed had to be able to cope with the problem of a varying background population. The problem is that for each location and scale in a study region, there is a different population density. This population-density surface is critical to cluster analysis as a cluster is effectively a region where the incidence of something exceeds the expected rate. It would be unwise to ignore as almost all clustering detected would be as a result of unevenness in the population distribution. Simply put, a large number of cases in a densely populated city may be of less interest than a moderate number of cases in a rural area.

Geographical Analysis Machines

The Geographical Analysis Machine (Openshaw *et al*, 1987) is one of the oldest and most established methods of accurately detecting clusters whilst accounting for the background population. The 'Machine' part of its name gives a clue as to the nature of GAM's approach. GAM uses a brute-force search technique that explores the entire data space at every geographic location and at every scale. In doing so, it is guaranteed not to miss any potential clusters. Unfortunately, as the data sets under investigation increase in both size, and particularly in dimensions, the number of permutations that GAM has to iterate through to complete this exhaustive search explodes.

A single GAM run can generate millions of hypercircles for testing (in multi-attribute data this can escalate to billions). As the size of the data sets increased, it became apparent that alternative approaches were needed that would be able to perform almost as well as GAM but that were smart enough not to require as much computing time. One such approach will be discussed in this paper.

The original concept was to create some form of smart agents that would be able to explore the spatial data in order to find patterns. The first stage of such a system required a mechanism for controlling the agents' movements.

Flocks

One area of research that looked at the movements of multiple interacting entities was that of flocking. The flock algorithm (Reynolds, 1987) was originally devised as a method for mimicking the flocking behavior of birds within a computer both for animation and as a way of studying emergent behavior. The resulting 'artificial birds', or 'boids', exhibited remarkable lifelike behavior as result of a few simple rules such as maintaining separation, alignment and cohesion.

Flocking seemed particularly interesting as a line of attack for two reasons: firstly, because it has a built-in relevance to space: most boid animations take place within a virtual 2d or 3d world, and secondly, because each member of the flock derived its behavior from interactions with all the other members.

The impetus to apply flocking to problem solving came partially from research into the use of ALife as an optimization technology: notably the use of swarms (Eberhart *et al*, 1996) and ant colony optimization (Dorigo and Di Caro, 1999), and partially from an interest in whether the hunter/foraging behaviors modeled in Artificial Life could be used to hunt and forage for patterns and clusters in spatial data.

The Flock-based Geographical Analysis Engine

The flock engine, which builds on the standard flock algorithm, extends the concept of a boid into that of a form of exploratory agent (a geoBoid). As each geoBoid moves around the spatial database, it samples each new location and tests it for signs of interesting pattern.

In a conventional flock, each boid is considered equal, with cohesion, separation and alignment rules applying in the same way to each boid. Within the flock analysis engine, however, each member of the flock is able to evaluate and broadcast its performance based on the data found at its location. Based on this, other flock members can choose to steer towards well-performing geoBoids in interesting areas in order to assist them and steer away from poorly-performing boids in uninteresting regions of the data set. In addition, each geoBoid changes its own behavior based on its performance. For example, a poorly-performing boid will speed up in order to leave an empty or uninteresting part of the geo data space and find an interesting area more quickly. Likewise, a well-performing geoBoid will slow down to investigate an interesting region more carefully by sampling the space at more regular intervals. The following looks at how the basic behavior of a flock was modified in order to develop a behavior targeted at performing an efficient search strategy. The rules were developed a stage at a time as it is difficult, if not impossible, to determine the behavior that will emerge from the rule interactions so after each new rule or modification the behavior of the geoBoids was observed through animation and a decision made as to whether to keep, drop or modify the rule. This process is on-going and new rules, such as ones controlling boid size, still need to be developed further.

Figure 1: The five geoBoid states

The rules used to govern the smart flock used in the majority of analysis to date are as follows:

- Evaluate my current location

  - If there is no population at all then DIE!
  - If there is population but no cases then note poor performance.
  - If there is a non significant number of cases note indifferent performance.
  - If there is a nearly significant number of cases then note good performance.
  - If there is a significant number of cases, note excellent performance and faint (stop).

- Find all detectable (in range) geoBoids then for each one :-

  - If neighbor is too close then feel repelled regardless of its performance.
  - If neighbors performance is indifferent then ignore it.
  - If neighbors performance is good or excellent then move towards it.
  - If neighbor is dead or its performance is poor then avoid it.

- Take weighted average of all target points generated above

- Move towards that point with the following speed rules

  - If I'm performing poorly move faster (this area is dull)
  - If I'm performing well move slower (I don't want to miss anything)
  - If I'm performing excellently, I've fainted so don't move
  - If I'm dead then don't move either.

Figure 1 summarizes the five boid states that arise from these rules.

As can be seen from the above rules, the relative performance of each geoBoid affects both its own and its neighbors' behavior. In the most severe case, where there is no data inside the boid, it dies. The meaning, in this context, is that it stops moving and becomes a warning beacon, discouraging other boids from approaching the area. The net result of this is that, after a short time, a series of 'dead boids' effectively mark off the empty regions of the map, forcing the rest of the flock to concentrate on the areas containing interesting data.

The implementation of the flock that was being used until recently was based on a fairly simplistic implementation of the boids algorithm. Whilst this implementation allowed the first prototypes to be set up quickly, it began to restrict the alterations and improvements that could be made to the flock members' behavior. By following some of the more recent work by Reynolds (Reynolds, 1999), the boids have been re-implemented

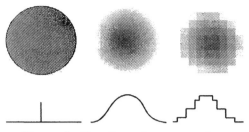

Figure 2: Creation of raster surface

to allow a more complete set of steering behaviors to be used. For example, previous boids would avoid, or 'run away' from, areas of the data set that had been marked as empty or uninteresting. This could mean that a search which was progressing near, but not towards, such an area could be unnecessarily deflected. In the current implementation, a steering behavior capable of 'obstacle avoidance' has been used. This means that only boids in direct danger of entering an empty section of the database will be deflected.

One of the most powerful features of GAM is that it not only explores the entire data set, but also does so at a wide range of scales. To capture this within the flock mechanism, the geoBoids must be able to change size. In the implementation described in this paper, the boids only change size once they have found a particularly interesting result: at which point they attempt to optimize their size for that location. Each boid, however, starts at a randomly determined size, so at any one time each boid is examining the space at a different scale. In addition, the cooperation between geoBoids ensures that interesting areas are studied at a number of scales. It would be beneficial, however, if rules could be imposed, or genetically bred, into the boids to allow them to dynamically adjust their size in response to the densities and distributions of the underlying data.

The rules are only half of the story. Most of the rules require some form of parameterization: blue boids, for example, travel faster, but how much faster should that be? At what point is a result interesting enough to warrant further investigation (red stage) but not interesting enough to mark permanently (yellow stage)?

The second rule states 'Find all detectable (in range) boids...' What is that range? At present, these parameters are being set perhaps somewhat arbitrarily. In the future, however, it seems logical to take the AL-ife paradigm to its logical conclusion and start to breed flock members: perhaps using genetic algorithm techniques, with the most successful pattern hunting boids reproducing to evolve the best parameters.

For the moment, however, there is still scope for humans to set the parameters, watch the differing outcomes and, if necessary, interfere with the search by re-directing attention to a different part of the map.

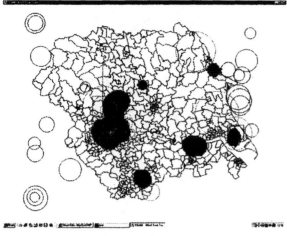

Figure 3: Output from a Flock run

Result Visualization

The aim of GAM, and hence any of its smart decendents, is to distill viable and interesting knowledge from the underlying data. However, given that a GAM run can generate millions of hyper-circles and potentially thousands of interesting results, the end user may find as much difficulty in interpreting the results from a run as they would have in interpreting the underlying data. Early outputs plotted charts of all significant circles, allowing users to see clusters as a result of spotting regions with large numbers of overlapping circles. More recently, GAM and its variants have used a form of result surface built from a raster. In this system, each 'interesting' circle contributes a kernel of values to the result surface in proportion to its score and radius (figure 2).

Despite the radically different driving methodologies between the systematic GAM and the flock-driven approach, the results of both can be plotted in this way, allowing the same visualization tools to be used, as well as making comparisons between the two systems easier. As the size and dimensionality of the data sets being analyzed increases, so too does the complexity of the results to be visualized. As the smart variants of GAM move into data spaces beyond the scalability of GAM, the need arises for visualization systems capable of intuitively conveying the result to end users. To date, efforts have concentrated on the use of animation and interaction to allow users to see numerous result surfaces rather then on trying to devise a visualization which could present all of the result information in a single, potentially overwhelming, display.

Figure 3 shows a flock analysis in mid run. In this diagram, the geoBoids are marked by circles. The dead boids can clearly be seen in regions of the map that lie outside the study region while the more active boids can be seen leaving a trail of significance, as described above, based on their findings at each location.

Method	Speed (per Test)	Thorough- ness	Scalabilty
Random Search	Fast	Low	High
MAPEX (GA)	Average	Average	High
Flock	Average	High	High
GAM	V.Fast	~Total	Low
GAM - T	Slow	Good	V.Low
GEM	Slow	Good	~None

Table 1: Comparision of Multidimensional Geographical Analysis Methods

A multi-engine approach

In addition to the original GAM mentioned in the introduction and the flock method discussed, a number of other approaches have been developed: some based on other AI approaches, such as MAPEX which uses Genetic Algorithms; others designed to explain possible reasons behind the patterns that were found, for example the Geographical Explanation Machine (GEM).

Table 1 summarizes the relative abilities of each of the analysis engines that have been implemented. Given the near total thoroughness of the GAM engine, it is tempting to use it exclusively. Many tests with GAM have shown that it has a high success at locating clusters and, just as importantly, a good tolerance against making false positive detections. (Alexander and Boyle, 1996).

However, GAM owes its efficiency to its highly-optimized, spatial-data retrieval and the fact that it works best on a purely geographical domain. It does not scale to more complex problems involving higher dimensional multi-variate data sets. Although space-attribute and space-time versions have been developed (GEM, GAM-T), they really require HPC levels of performance and, even then, they will be limited to low-dimensional problems because they do not scale well to high dimensional spaces.

The non brute-force systems are attractive for their scalability in terms of dimensions, but using them means sacrificing the level of resolution and completeness that brute-force, GAM-like methods generate in the final results.

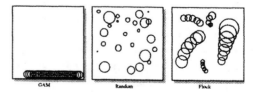

Figure 4: The output from three separate engines

Figure 4 is illustrative of how the search might have progressed for three of the methods shortly after starting. The GAM has been very thorough but has yet to complete one pass of one line at one scale. The random search and flock search have covered more ground and may have found something interesting but it will be some time before either of them has developed a clearer picture of what, if anything, they have found.

For very large, temporal, multi-attribute data sets, the computational time becomes prohibitively large, with CPU time approaching infinity. However, using it to exhaustively search one small part of a spatial data set is less intensive: thus when a less thorough engine finds something of potential interest, it can call on the help of the GAM to exhaustively search that region. In this way, it is possible to narrow down the search by reducing the dimensionality: at which point the GAM style of exploration again becomes technologically viable. Once such an exhaustive search has finished, that section of the data set can be marked as done and avoided by the flock for the rest of the search.

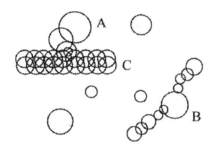

Figure 5: Cooperating Engines

Figure 5 illustrates a potential cooperative scenario in which the random search, flock and GAM are working together. In this scenario, one of the flock methods (A) thinks that it has found something potentially interesting and has called in a miniGAM search (C) to thoroughly examine the area. Meanwhile, one of the random circles (B) has attracted the attention of two other members of the flock for additional study.

Until recently, the database functions, flock search and interactive visualization have been tightly coupled into a single application. Now, however, the process has been successfully split into three tiers to enable such a multi-engine cooperative system to be developed. It is anticipated that the remaining search engines will be similarly implemented over the coming months.

The following section looks at the architecture that is being followed for the implementation of the Multi-Engine Spatial Analysis Tool (MESAT).

The first tier (see figure 6) is some form of spatial-data storage and query facility. In the trials performed here, this is simply a flat file stored on a server. In more complete final applications, it is more likely to be a more advanced online spatial-data warehouse such as that proposed by (Han *et al*, 1989). It should be able to store large amounts of data and respond to spatially-

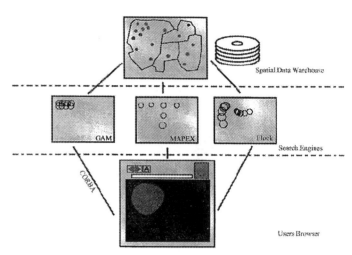

Figure 6: Three tier architecture of the Multi-Engine Spatial Analysis Tool (MESAT)

referenced queries rapidly.

The second tier consists of one or more spatial-exploration engines that will interrogate the spatial data found at the first tier and attempt to find spatial patterns in the data. By developing using the object-oriented paradigm, each engine implements a common set of interfaces that allow them to appear from the outside as identical generic engines. By doing this, additional engines can be implemented and 'plugged-in' to the system, allowing them to instantly communicate with both the first and third tier without having to change either.

The top tier of the system is the visualization engine. This collates the information coming in from the different search tools and displays them in an interactive visualization. This visualization can be built up in real-time, allowing the user of the system to see the progress of the search as it occurs. The major advantage of this is that the user can employ their expertise or intuition to assist and direct the search process.

The visualizer serves a dual role: aside from the obvious displaying of the results, it actually coordinates the activities of the different search engines, starting and stopping them, dividing up and sharing the search space, and setting regions to focus on or ignore. Its controlling role could be entirely human-user driven, autonomous, or a mixture of the two. With human control, the user could drag out boxes to indicate regions that different engines should concentrate on or avoid. In autonomous mode, the application could use the results generated so far to decide on how best to guide the remaining search.

Communication

As stated above, the role of the visualizer is more than that of communicating results to the user: it is also a shared result surface (in AI terms it could be considered a 'blackboard system' (Englemore and Morgan, 1989)). As such, it must also act as the central communication centre for the different agents. Given that the shared result surface is available to all of the analysis engines, what kind of information should be passed back? The first option is that the entire result surface be sent. However, given its size, bandwidth limitations make that impractical. Instead, the shared result surface must provide a useful summary of the search so far. Such information might include:

- The top ten best finds so far (locations and scores)

- Areas of the database that are now known to contain no interesting data.

- Areas of the database that have not yet been studied in depth

- Areas of the database that have been fully explored and need little or no further analysis

- Specific requests for more detailed analysis of a region from another agent or system user.

- Section of database to work in (for division of labour when multiple agents of the same type are running in parallel)

How each engine would take advantage of this information depends very much on the mechanism of the engine itself. For example, a GAM might sit idle on a machine until it receives a specific request to perform a focused search on a particular small region. The flock, on the other hand, employs a number of behavioral patterns to respond to such information.

The generation of these obstacles by the shared result surface quickly marks off uninteresting parts of the database. For example, areas with zero populations such as seas are eliminated early on. With additional information arriving from the shared result surface regarding regions to avoid, the geoBoids would be able to use this behavior to steer around them.

Dealing with duplicate results

Although a single GAM run will generate millions of circles, its systematic nature means that it will never generate two circles of the same size at the same location. As such, every result can be added to the result surface in the knowledge that it is unique.

Members of the flock search, however, have no such restriction and they may well generate results for one section of the map multiple times, each time adding to the surface of evidence. This problem increases as multiple-analysis engines are used as they may, independently, occasionally explore the same areas. As a result, an area of relatively little importance can be added to the result surface enough times to make it seem interesting. This is particularly true for maps with little or no real clustering as the analysis engines will hunt out the closest

Set	Clusters	%Clustered	Radius	Type
1	3	30	5	S
2	3	30	20	S
3	3	60	5	S
4	3	60	20	S
5	3	30	5	S-T
6	3	30	20	S-T
7	3	60	5	S-T
8	3	60	20	S-T
9	3	30	5	S-A
10	3	30	20	S-A
11	3	60	5	S-A
12	3	60	20	S-A
13	3	30	5	S-(T-A)
14	3	30	20	S-(T-A)
15	3	60	5	S-T-A
16	3	60	20	S-T-A

Table 2: Characteristics of the synthetic data used in blind trial S = Space, T = Time, A = Attribute

thing they can find to a real cluster and keep studying it to try and find a scale at which it may become a cluster.

The solution to this was to build two result surfaces. In addition to the normal result surface, the second one records the number of times each part of the map has been examined. This surface is referred to as the search-intensity surface. This result surface can then be divided by the search intensity surface to cancel out the effects of redundant result generation.

As a side effect, this search-intensity surface gives a clear indication of not just where has been searched but, more importantly, where has not yet been searched, providing for the first time a measure of how thorough the flock search is.

Results

In order to compare the capabilites and efectiveness of the flock approach, a trial was set up using both GAM and the geoBoids, as well as a number of other smart techniques (Openshaw, et al, 1999). The data used in the trial was synthetically generated using a system not known to the operators of each technique prior to the trial and included clusters that combined space, time and attribute interactions.

Table 2 summarizes the clusters that were generated for each set, whilst table 3 shows the performance of each system used.

The flock worked well in detecting purely spatial clustering. Its performance was poorest for those data sets where the radius of each cluster was large (20km). Here the boids became susceptible to artifacts from the background population that tended to lead the flock away

Dataset			GAM	GAM/KT		Flock		
	T	A	C	C	T	C	T	A
1			3	1		3		
2			3	1		2		
3			3	2		3		
4			3	2		3		
5	Y		3	3	Y	3	Y	
6	Y		3	3	Y	1	Y	
7	Y		3	3	Y	3	Y	
8	Y		3	3	Y	1		
9		Y	3	3		3		
10		Y	3	3		2		S
11		Y	3	3		3		S
12		Y	3	3		2		S
13	Y	Y	3	3	Y	3	Y	T
14	Y	Y	3	3	Y	3	Y	T
15	Y	Y	3	3	Y	3	Y	ST
16	Y	Y	3	3	Y	3	Y	ST

Table 3: Results of blind trial for GAM and Flock C=Clusters T=Time A=Attribute S=Space

from the real cluster centers.

The system correctly identified all but two of the attribute, space-attribute, time-attribute effects. It should be noted that whilst the flock approach was the only technique able to look for attribute interactions (which, as the table shows, it was successful in achieving) the other systems were still able to detect the clusters in space and time even without taking the attribute information into account.

Conclusions

The flock approach was able to quickly locate indications of clusters within multi-dimensional data. However, the final output describing the located clusters is not as thorough or detailed as the results from a GAM run.

Spatial data sets have special properties not necessarily found in other fields, including high levels of noise, uneven background populations and, most notably, spatial auto-correlation which states that any location in the data set is likely to be similar to any location close to it. By borrowing from behavior observed in natural systems, the geoBoids undertake a search strategy that is particularly well suited to geographic space. The flock-driven approach may well have applications for pattern hunting outside of geography, particularly where data can be represented as a landscape.

Whilst the flock approach is promising in itself, the hybrid approach of combining GAM, flock and other systems offers many potential advantages over using any one of the methods independently. In a worst case scenario, the system should still perform at least as well as the

best contributing component.

This alone may be useful as, given the different capabilities of each engine, it would be difficult to chose which engine to employ.

It is, however, with the introduction of time-attribute clusters that the new system has greatest potential. In this scenario, any individual cluster may be purely spatial, purely temporal, or both, as well as being focused within certain attribute constraints. It is tests such as these which push the computational intensity of the problem upwards and it is here that the hybrid approach is most likely to have greatest effect as the wide exploratory powers of the flock and GA will, it is hoped, find some indication of possible clustering which can then be exhaustively searched by a focused GAM run.

Extensive testing is required, however, to give a full indication of the level of gain offered by combining the different approaches in a cooperative, expandable, distributed architecture. Initial results from such trials, examining each engine separately, have suggested that the combined approach should be capable of producing results of GAM-like quality in considerably less time.

References

Alexander, F. E., Boyle, P. (1996) : Methods for Investigating localised Clustering of Disease ARC Scientific Publications No 135, Lyon, France

Dorigo, M., Di Caro, G. (1999): The Ant Colony Optimization Meta-Heuristic, in : New Ideas in Optimization, eds: Corn, D., Dorigo, M., Glover, F. UK;

Eberhart, R., Simpson, P.K.,Dobbins, R. (1996): Computational Intelligence PC Tools,AP Professional,USA. Computational Intelligence PC Tools,AP Professional,USA

Estivill-Castro, V., Murray, A. (1998): in : Research and development in knowledge discovery and data mining : Second Pacific-Asia Conference, PAKDD-98, eds: Wu, X., Kotagiri, R., Korb, K. Australia;

Englemore, R. and Morgan, T. (1989) eds : Blackboard systems. Addison-Wesley. Reading, MA.

Han,J., Stefanovic, N. and Koperski, K. (1998): Selective Materialization: An Efficient Method for Spatial Data Cube Construction, in: Pacific-Asia Conf. on Knowledge Discovery and Data Mining (PAKDD'98), Australia.

Openshaw, S. (1995): Developing automated and smart spatial pattern exploration tools for geographical information system applications. The Statistician, 44, No 1, pp 3-16.

Openshaw, S., Charlton, M., Wymer, C. and Craft, A.W., 1987, 'A mark I geographical analysis machine for the automated analysis of point datasets', Int. J. GIS, 1, 335-358

Openshaw, S., Perrée, T. (1996): User-centered intelligent spatial analysis of point data. In: Parker, D. (Ed) Innovations in GIS 3, 119-124 ,Taylor and Francis, London

Openshaw, S., Turner, A., Turton, I., Macgill, J. and Brunsdon, C.,(1999), Testing space-time and more complex hyperspace geographical analysis tools paper presented at GISRUK'99

Macgill, J. Openshaw, S. (1998): The use of flocks to drive a Geographic Analysis Machine. In: GeoComputation98, Procedings, Bristol.

Raper, J.F., Rhind, D., Shephers, J.W. (1992): Postcodes: the new geography. Longman, Harlow

Reynolds, C,W. (1987): Flocks, herds, and schools: A distributed behavioral Model. Computer Graphics,21(4):25-34.

Reynolds, C,W (1999): Steering Behaviors for Autonomous Characters, paper presented at the Games Developers Conference http://www.red.com/cwr/steer/

Emergent Design: Artificial Life for Architecture Design

Una-May O'Reilly and **Ian Ross**
Artificial Intelligence Lab, MIT
unamay@ai.mit.edu, ianross@ai.mit.edu

Peter Testa
School of Architecture, MIT
ptesta@mit.edu

Abstract

We report on a software toolbox that is part of an architecture design process we have named "Emergent Design" The toolbox incorporates concepts of artificial life that allow architects to realize conceptual experiments in which the elements of an architectural scenario are endowed with agency and dynamic, spatial interaction. Elements of the scenario combine and interact spatially over time to result in an emergent design. We believe this use of ALife concepts in the design process of architecture to be both novel and powerful.

Introduction

Architecture has predominantly reaped benefits from nascent computational technology in the form of computer-aided design tools. However, the discipline still lacks sufficient tools that actively enrich and extend the design process rather than automate them. The notion that the elements of an architecture scenario (or site) have agency and local interactions determined by both their properties and proximity to other elements is strongly present in both the thought process of a design investigation and the verbal discussions which accompany the investigation. Elements of a design are supposed to emerge from consideration of the non-linear, interdependent factors of a scenario and the non-linear process of design. While this ALife-like perspective is present, to date architects lack adequate tools which would enable them to explore it any further than as a mental exercise. We have coined the term Emergent Design for an architectural design process that emphasizes this perspective. It is characterized in the following ways:

- Given the complexity of a contemporary design scenario, the numerous situational factors of a scenario must be identified and their inter-relationships must be understood as well as possible.

- An effective solution to a complex design scenario is achieved through a non-linear process of bottom-up experimentation involving independent, related or progressive investigations into architectural form and complex organizations. This interactive process increasingly builds up a complex solution that considers the numerous complicated, interdependent relationships of the scenario.

- A complex solution derived in such a bottom-up, investigative style is advantageous because it retains explicability and has the flexibility to be revised in any respect appropriate to a change or new understanding of the design scenario.

- Computer software (perceived as a collection of interacting processes) is an excellent means of performing bottom-up architectural experimentation because, despite a simple specification, a decentralized, emergent software simulation can yield complex behavior, exploit uncomplicated graphics capability to model organization and pattern, can be written flexibly so that alternatives can be quickly examined, and can be controlled and designed by architecture students.

Essential to our notion of Emergent Design is the integration of an effective and powerful software toolbox from the domain of Artificial Life into the process of exploring spatial relationships in an architectural program or among primitive design components. Herein we present our findings that the concepts of Emergent Design and the design of such a software toolbox provide guidance towards informed and innovative designs.

The paper proceeds as follows: First, we describe Emergent Design and discuss the synergy between ALife and architecture. Next, we focus on the Emergent Design software process and software toolbox. The toolbox has been used in a MIT School of Architecture student design studio. To illustrate its capability and range we next describe in detail two investigations that used the toolbox in the studio followed by an appraisal of its performance. Finally we summarize and list future work.

ALife and Emergent Design

Emergent Design is not entirely new. Architects have always sought to identify the elements of a problem scenario and understand them syncretically. Emergent Design is unique in exploiting and emphasizing the role of software designed for self-organizing spatial simulations to do this and to explore solutions. It is a decentralized style of thinking about problems and a way of evolving solutions from the bottom-up. Emergent Design emphasizes appraising and understanding individual behavior (where an architectural component is endowed with agency) as being influenced by other individuals in a system. The system view also incorporates recognition of levels (Resnick 1994; Wilensky and Resnick 1998), and the insight derived from understanding how complex, collective, macroscopic phenomena arises from the simple, local interactions of individuals. Architects continually strive to understand the complexity of a system. The recognition of levels in the system and the phenomena that give rise to the formation of levels provide system level insights that are influential towards arriving at an adaptive design. Examples of such complex adaptive systems can be found in both natural and synthetic architecture environments. These forms may be urban configurations, or spatial and organizational patterns but all are evolved through generative probing and grouping in the space of possibilities.

Emergent Design differs from traditional architecture design approaches which emphasize things in space as fundamental and time as something that happens to them. In Emergent Design things that exist (e.g. rooms, hallways, buildings) are viewed as secondary to the processes through which they evolve and change in time. In this approach the fundamental things in the environment are the processes. The focus of design activity is not simply on events in space time but the processes and Emergent Design seeks to formulate principles of architecture in this space of processes allowing space and time (Architecture) as we know it to emerge only at a secondary level.

There are numerous concepts in the field of ALife (Langton 1989; Langton *et al.* 1992), that are advantageously applicable to an Architecture design process which emphasizes system-level and constituent understanding. In ALife every component of a system, including elements of the environment, is conceptualized as being capable of agency. Thus, a component of the system may or may not act. Acting is not necessarily actually undergoing a change of internal or external state. It also encompasses the way in which a component may exert influence on the state of the environment or other components. Agency between components implies that their interactions create a level of dynamics and organization. Organizations that dynamically define themselves on one level can themselves exhibit agency and, thus, a new level of organization can form as a result of the lower level dynamics. Levels are not necessarily hierarchically organized. They may, in fact, be recognizable by a particular perspective from which the entire system is viewed.

Architectural systems have numerous levels and each level is interdependent with others. For example, an office building can have a level in which components are office furniture, dividers and the floor plan with respect to circulation and windows. Different work groups or work areas emerge as the furniture and dividers are distributed on the floor. Work groups may span different floors. Yet, there is another (totally legitimate and useful, of course) view of the building in terms of floor levels. Office buildings when studied closely turn out to be effectively conceptualized as living systems with a myriad of interdependencies between their components and with numerous levels of interdependent organizations.

As shown by the office building example, in general, architectural systems are very complicated. One type of complication encountered is that small, simple, "local" decisions, when coupled with other similar decisions, have large, complex, global effects on the outcome of a design. For example, a choice about locating a work group solely on one floor has ramifications in the choices of circulation allocation, and work equipment placement. These choices, in turn, influence later choices such as material selection or building infrastructure assignment. ALife studies systems in a manner that highlights and elucidates the consequences of locally defined behavior. ALife models are defined in terms of component agency and decentralized component interactions. The running of an ALife system consists of playing out the local interactions and presenting views of the complex global behavior that emerges. Thus, the Emergent Design version of an ALife system is one in which the environment and components are architectural and spatial in nature. The "running of the system" consists of designating components, defining their agency in local terms and then watching the macroscopic outcome of their interaction.

Both ALife and Architecture find it necessary to consider the influence of non-determinism in the outcome of complicated system behavior. In addition, both ALife and Architecture are also both very aware of how an outcome is sensitive to details of initial conditions. ALife simulations can be defined with parameterized initial conditions and run with different parameter values for the initial conditions. They allow architects to study the impact of initial conditions.

ALife simulations facilitate the investigation of many possible, complicated, spatial outcomes. They model spatial phenomena by using computer visualization that conveys actual movement, changes in a modeled environment or, the influence of spatial proximity. Visualization and spatial studies are also of great interest to architects. For example, an architect may want to investigate the implication of a spatial constraint. How does the area of a plan interplay with the required number of floors, number of window and circulation area? Or, if modular units are chosen for a site, how will these units align with the boundaries of the site?

Emergent Design Steps

Artificial Life software is what truly makes the process of Emergent Design powerful. Without the aid of the software and the speed of simulations that it facilitates, one would not be able to achieve a fraction of the depth, breadth and richness of emergent designs. There are two steps in using the toolbox:

1. Define the goals of an investigation concerning how spatial elements may be configured within a bounded area (i.e. site). Specify initial conditions (e.g. initial quantity of elements, size of size, scale, conditions of site). Identify the elements and the relationships among them. State how elements influence each other and under what conditions they interact. Both the site and its elements can exhibit agency. This identification forms a functional description of the simulation.

2. Using the existing set of Java class methods in the software toolbox, specialize the tool with software that implements the element and site behavior. Run the simulation from initial conditions and investigate the outcome. Usually a simulation has nondeterministic behavior so outcomes from multiple runs are assessed. Based on the outcome, return to either initiative to improve or refine.

The process is practical and not complicated. In the first step, goals and initial conditions are defined. It encourages thinking of elements and site "active" or bestowed with agency. In fact, this style of thinking is very common. The architect says "this is what happens when a comes into contact with b or, when c comes close enough to d or when the corner of any

element of class e touches the boundary, when the bounded area is full", etc. Unfortunately we have noted that our students stop considering agency when they move from thinking to explicitly designing because they can not actualize behavior or manage to "play it out" in complex, non-linear circumstances. We have found that the software toolbox reverses this trend.

The second step involves software programming, i.e. customizing the toolbox for a specific investigation, and running and evaluating the simulations. Multiple runs can be gathered and analyzed for general properties or one particular run may be scrutinized because it shows something unanticipated or interesting. It is possible to find that the initial conditions are too constraining or, conversely, too unconstrained. Often, in view of the results, the architect revises the initial conditions to try out different relationships and actions. If the outcomes are satisfactory, they are now used to take the next step in the larger process of solution definition.

Emergent Design Software

The software toolbox is a collection of Java classes that can be used to generate complex spatial organizations that exhibit emergent behavior. It is an open source, and continually evolving. From the application user's perspective, the application runs in a window in which there are two elements (e.g. Figure 6):

1. The display that graphically represents the current state of the simulation or "run".

2. The Graphical User Interface (GUI) that allows the user to interact with the simulation by altering display options, changing simulation behaviors, and otherwise modifying the simulation.

The toolbox, as software (i.e. from a programming perspective), provides a general framework that can be engaged as is, can be supplemented with new classes, or can act as a group of superclasses that will be specialized. The toolbox software (i.e. classes and methods associated with classes) can be conceptually divided into three parts: **foundation**, **specialization**, and **applet**. The purpose of abstracting foundation, specialization and applet software is to enable the implementation of different applications that, despite their differences, still use a base of common software. The software is conveniently described in terms of the objects and their behavior we conceived as integral to any emergent design application. For clarity, we will italicize Java object classes in the descriptions that follow.

The **foundation** software defines classes that implement the environment of the simulation which is termed a *site*. A *site* represents the two dimensional region of architectural inquiry. Any polygon can define a *site's* extent. A *site* has no inherent scale – the scale is determined by the user. A *site* is defined generally so that it can be conceived in a variety of ways: for example, as a large-scale map of states or counties to investigate transcontinental transportation routes, or, as a dining room to study the layout of furniture.

The *site* can be refined in terms of its sub-parts and in terms of the architectural elements that are imposed on it. We term a sub-part, in general, a *piece* and an architectural element, in general, a *zone*.

A *zone* is an object that is imposed on the site, rather than a component of it. The bulk of an application consists of the applet creating *zones* and enabling their interaction among each other (which results in zone placement, removal or movement) based on different rules, preferences, and relationships. Some *zones* are placed on the site upon initialization. Others are placed during the run, according to the behavior dictated by the *site*, its *pieces* or other *zones*. A *zone* may move during a run or even be removed from the *site*. A *zone* may be represented by a polygon or a line. A *zone* can represent a table, room, building, farm, city, or whatever else the architect wishes to represent place down on the *site*. A zone can be modified: one can change its color, translate, scale, and skew it, or change its type (e.g. "farm"). In terms of its display properties, a *zone* can be transparent (either partially or fully), and can lie on top of or below other *zones*.

The *site* consists of *pieces*. As displayed, the *site* is subdivided into an array of uniformly sized rectangles (much like a checkerboard). The programmer specifies the dimension of these rectangles in terms of pixels (the smallest modifiable units on the computer monitor). Thus, the display may consist of a single large *piece*, or as many *pieces* as there are pixels in that area. By choosing the dimensions of a *piece*, the level of granularity for regarding the simulation is chosen. Resolution has implementation consequences. The finer the resolution, the slower the speed of a run (because of display costs). A *piece* has its private state. That is, it can store information about various local properties of the *site*, as well as information about the uppermost *zone* on it.

Spatial organization occurs on the site via agency of the *site* and *zones*. A *zone* can move itself on the *site* according to criteria based on local conditions such as *zone* s nearby, the properties of a *piece* or conditions of the site. For example, all *zones* may be initially randomly placed on the *site* and then queried in random order to choose to move if they wish. The decision of a *zone* to move will depend on its proximity to other

zones and the properties of the *piece* it sits upon. In a different perspective, but one still generally realizable, the *site* can directs the incremental, successive arrangement or placement of *zones* on its *pieces*. This direction is according to a set of behavioral directives that are likely to take into account emerging, cumulative global conditions (i.e. available free space, current density) as well as local preferences that reflect architectural objectives in terms of relationships.

We have implemented a particular means of property specifications for a *site* which we call a *siteFunction*. A *siteFunction* is a general means of representing various qualities or quantities that change across the Site and which may be dependent on time. It could, for example, be used to model sun cover, wind, pollution, rain, crime, income levels, etc. A *siteFunction* is a function of four variables (x, y, z (3D space which a piece defines), and t (time)). One can create all sorts of interesting relationships between functions and other functions, as well as between zones and sitefunctions or vice versa. By creating feedback loops (where the output of, say, a function is used as input to another function, or to a Zone, whose output could be in turn given as input to the object from which it received input), one can very easily set the stage for emergent behavior between human creations (Zones) and the environment (SiteFunctions).

The foundation software is intentionally very flexible, and can be used to model almost anything of a spatial nature. By defining a group of *zones* and *siteFunctions*, and defining ways for them to interact with each other locally, one can set up the circumstances that lead to extremely complex, unexpected global behavior. The toolbox can thus be employed in investigations that strive for aesthetics, that predict development's effect on the environment (and vice versa), or that have other purposes. The ultimate goal is that the user is only limited by his or her imagination. One could use ideas from ecology, psychology, economics, chemistry, and other disciplines to formulate the ways in which interaction can take place in the simulation.

The **specialized** or **behavioral** software defines aspects that are particular to specific simulations. It facilitates the definition of tools such as models of growth, or different kinds of dynamic relationships between objects in the simulation.

Each software toolbox application is run from its applet. The applet is where the emergent states of the simulation are processed and interpreted. It uses two foundation classes: Display and Site. The Display class implements the graphical display of the site and its updating as the run proceeds and emergent dynamics occur. The perimeter is drawn and then color is used to indicate the properties associated with either pieces or zones of the size.

Case Studies of Emergent Design

We have introduced Emergent Design in a graduate level design studio course. As instructors, we select a thematic project drawn from a real world design scenario that enables students to explore a small number of particular applications of bottom-up, self-organizing or agent-based models that have computational simulation potential. The students themselves extend the initially supplied software toolbox. To assist collaboration, the project software is maintained and shared via the World Wide Web.

A project generally involves architectural problems related to the convergence of several activity programs such as new working environments open to continuous reorganization wherein dynamic systems interact, or, to design problems related to serial systems and pattern structures such as housing and community design. Students model and simulate such factors as patterns of use, development over time, and environmental factors. In each case one objective is to explore how the combinatory dynamics among simple building blocks can lead to emergent complexity.

In a spring 1999 studio course students proposed new models of housing and community for the rapidly expanding and increasingly diverse population of California's Central Valley. Projects investigated emergence as a strategy for generating complex evolutionary and adaptive spatial patterns based on new relationships between dwelling, agriculture, and urbanism.

Student teams developed three sites/morphologies: Field, Enclave, and Rhizome (for details see (Testa *et al.*)). Designs operated at both the community scale (120 dwellings) of pattern formation and detailed design of a prototype dwelling or basic building block. A primary objective was to work toward typological diversity and spatial flexibility planned as combinatorial systems using variable elements.

With respect to the Emergent Design software, each team specified the local spatial relationships that they desired and worked to author a simulation that satisfied local constraints and behaved according to local relationships, and by doing so, exhibited coherent emergent global behavior. Emphasis was placed on procedural knowledge and the dynamics of spatial relationships. The design of these simulations forced the students to thoroughly examine their specifications and the consequences of them. No distinction was made from the "means" and "ends". The nature of the dynamics of the processes were inseparable from the spatial layout of the architectural programs. The students were not trying to create a single static "best" answer to the problems posed to them. Instead, they aimed to explore whole systems of morphology by examining many different simulation runs.

Field: The Field morphology presented the issue of integrating urban housing with farmland. Overall housing density needed to remain fairly low. The existing houses on the site were very widely distributed across it, and the field team wanted to add more residences while preserving this spare agricultural aesthetic. They wanted to minimally disrupt the already established feel of the site while allowing it to support a larger population.

Thus, the group's goal was to determine how such new housing could be aggregated. They wanted to create a distributed housing system, one where there was not a clear dense-empty distinction. Furthermore, they would not be satisfied with simplistic regular patterns of housing. In employing an agent-based, decentralized, emergent strategy, they conceptualized a dynamic simulation in which each newly placed house would be endowed with a set of constraints and preferences and the house would behave by changing its location within the site in order to fulfill its preferences. They wanted to appraise the non-uniform housing aggregations that can arise from easily expressed, naturally suggested interaction constraints. Interaction elements included farms, schools, roads and the size zones.

In the first simulation designed by the field team 150 new houses are placed down at random on the site. The area taken up by these new houses is approximately 1% of the total area of the whole site. Once all of the new houses have been placed, they react with one another according to an attractivity rule that the field team wrote for the houses in order to express their preferences regarding proximity to each other. The rule states that a) a house 3 to 5 units from another moves towards it, b) a house 1-2 units from another moves away from it, and c) a larger collection of houses will move according to the same conditions but as an aggregate.

The simulation was intended to run indefinitely until stopped by the designer. Several patterns of behavior emerged from the application of this attractivity rule:

1. Based on initial conditions, the houses would form small groups, and would travel "in formation" much like a flock of birds. That is, although each house in the group would travel relative to the site, the houses would not move relative to each other.

2. Sometimes, a group of houses would find a steady state, and simply oscillate between two (or more, if

there was more than two houses in the group) orientations.

3. The various groups of houses would sometimes cross paths when travelling across the site. When this happened, the groups would sometimes merge, and sometimes take houses from each other. These collisions would also often alter the trajectories of the groups involved.

4. Groups tended to expand as the simulation went on. Groups sometimes (but not often) broke apart when they encountered the edge of the site, or interacted with another group.

5. Groups sometimes became "anchored" to existing (not newly placed) housing. This was likely because newly placed houses were attracted all types of houses, and since existing houses are stationary, the newly placed houses within the influence of existing houses would not move out of range of the existing houses.

The aim of the attractivity rule was to produce housing patterns that preserved a distributed aesthetic. At the same time the patterns should form mini-communities of small groups of houses. While the houses in a mini-community were expected to be socially grouped, ample space between them was desired. After examining a large number of runs of this simulation, it was apparent that the rule achieved the desired results.

The group's second simulation was a parallel investigation into how new housing could be interestingly distributed in the aforementioned pre-existing farmland site. It explored expressing the proximity preferences with the mechanism of feedback (using a behavior's output as its next input). As a simple experiment, the site was tiled with 144 patches (each of dimension 10 x 10) that were each designated as either a strong attractor, weak attractor, weak repellor, or strong repellor. These designations were assigned randomly with equiprobability. Houses occupying approximately 2% of the site were then randomly placed on top of these patches. At a time step, each house would move according to the strength of the attractor and repellor patches it was close to. Attractor patches made houses go to the center of the patch at varying speeds (depending on the intensity of the patch), while repellor patches made houses go radially outward. In addition, each patch would newly update what type of attractor or repellor it was. This update was the crux of the feedback in the system. The new patch designation was determined by the patch's most recent state and by the ratio of the overall house density across the

site to the density of houses on the specific patch. If a patch had a density that fell some threshold below average, it became less repulsive or more attractive by one designation (that is, strong repellors became weak repellors, weak repellors became weak attractors, etc. Note that nothing would happen to strong attractors in this case). If a patch had a density that was some threshold above average, then the opposite would happen: patches would attract less and repel more.

This case demonstrates negative feedback – this is feedback that damps (or lessens) the behavior of the simulation. Such simulations can usually run forever, with only minimal periodicity (when existent, the periods are long). A relatively uniform density is maintained across all of the patches, due to the aforesaid feedback mechanism. In this case, extremes are usually successfully avoided. Few patches have very small or large densities, and those that do usually take measures to change this. Also, few patches are strong repellors or attractors; they are usually of the weak kind, since their densities usually do not vary by that much.

Negative feedback facilitated the notion of distributed housing, but the condition it created lacked structure; houses did not form coherent groups at all.

Upon seeing multiple instantiations of the "desired result" in the first two investigations, the group decided to refine their goals because the simulations created emergent aggregations that, while non-uniform, were too non-descript. By the aggregations being so amorphous and distributed, they actually disrupted the initial site conditions more than a more structured aggregation. This outcome was very much unanticipated. It was only realized after exploring many runs of the simulations. It was decided that the notion of distributed housing should be combined with some greater notion of form and directionality.

Thus, in a final simulation, (Figures 1...5), the field team chose to restrict the placement of houses to a set of designated farms. After experimenting with a scheme where the designated farms were chosen at random, the team decided the randomness introduced undesirable isolation. That is, they wanted the designated farms to either touch or only be separated by a road. Furthermore, they wanted the non-developable area carved out by the non-designated farms to be somewhat path-like (as opposed to node-like, or central), in order to architecturally recognize a crop rotation scheme. This was accomplished by choosing a farm at random to be non-developable (in contrast to designated as eligible for housing), then having that farm randomly choose one of its neighbors to be non-developable as well. Control would then be passed to the newly chosen neighbor farm which would repeat

460

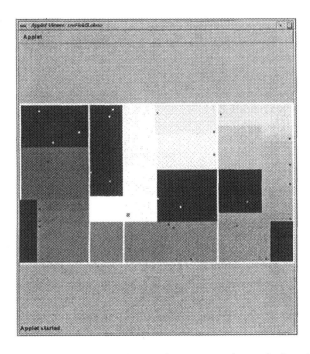

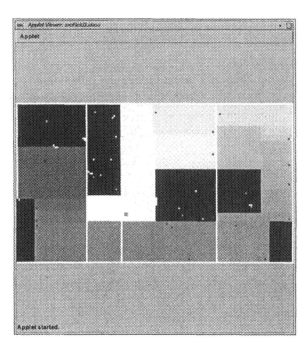

Figure 1: Field 3 simulation depicting 10 housed placed randomly in developable areas. Zones and elements of the site are color-coded.

Figure 2: Field-3 simulation when ten additional houses are placed randomly in the developable area. In a full simulation the process would be iterated 10-20 times.

the process. In the case where a non-developable farm was completely surrounded by other non-developable farms or borders, it passed control back to the farm that chose it. Control was recursively passed backwards until one of these non-developable farms had a neighbor that had not been chosen yet. The first farm would be colored the lightest shade of green, with each successive farm being a darker color of green. The color progression was intended to be suggestive of a possible crop rotation scheme. Once the area threshold has been reached for non-developable area (which is somewhere over 50 % for the field team), this part of the simulation ends.

In the remaining developable area (colored black, for clarity), houses are placed randomly, ten at a time. After they are placed down, an attractivity rule is run, the houses' movement is determined, and they move accordingly. In this particular incarnation of the attractivity rule, houses are attracted to schools (which are stationary) and other houses (both stationary existing ones and movable newly placed ones). When newly placed houses have moved, they become stationary. This rule was enstated so that the simulation did not achieve a predictable result of all newly placed houses being as close as possible to each other and the schools. Instead, the field team wanted attraction to help achieve something more subtle. They wanted houses to tend to gather together in loose communi-

ties, and tend to be located near schools, but not always. Thus, each newly placed house was attracted to other houses within a five piece circular radius (unlike the square radius of their first simulation), and also attracted to the school that was closest to it (other schools would have no influence over it). The attraction is inversely proportional to distance, although any type of attraction could be used.

Attraction independent of distance and attraction proportional to distance were also tried, but attraction inversely proportional to distance yielded the best results. With this sort of attraction, zones form very strong local bonds, and are usually not affected by distant zones. This sort of behavior (strong local interaction, weak global interaction) is often a key component in bottom-up simulations that exhibit emergent behavior.

This simulation yielded the most satisfactory results to date. The housing arrangement was relatively unobtrusive and not formless. Still, the housing collections are not predictable regular patterns, but subtle arrangements, subject to the behavior instilled in each of the houses. It was also recognized that the experiment might be extended by using the negative feedback mechanisms explored in the second simulation. Instead of making houses stationary after they have moved once, one might elicit more complex behavior

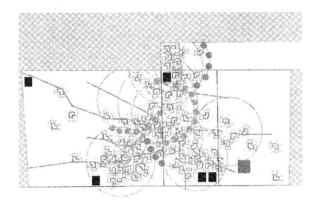

Figure 3: Field project. The local organizations of housing are not predictable regular patterns, but subtle arrangements subject to the behavior instilled in each of the houses.

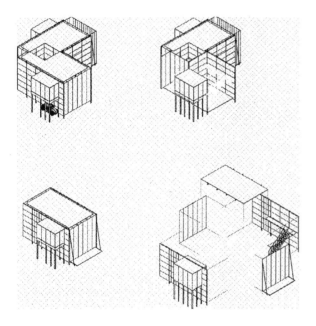

Figure 4: Field project. The individual dwellings were designed as an adaptive assembly of space defining elements.

Figure 5: Field project. Dwellings combine with modeled exterior spaces allowing for expansion over time.

from the simulation by allowing newly placed houses to move indefinitely, with a new rule to handle the inevitable dense clustering of houses around each other and schools. Unfortunately, time did not permit this extension to the investigation. The team felt the result they acquired adequately set them up to proceed with other aspects of the project design.

Enclave: The goal of the enclave team was to generate grids of houses with meaningful reference to the site's borders, each other, and the community space between them. Although the form and arrangement of houses could vary quite dramatically throughout the site, the houses on the site obey several adjacency constraints: to engender the enclave's team idea of community, no house is isolated from all other houses, and a group of houses must number 4 or less. These constraints were satisfied by appropriate placement rules. In the case when satisfaction of one constraint violated another (this was unanticipated, and only noticed after the first simulation was built and ran) a special rule was used. The site was given probabilistic behavior (i.e. a non-deterministic algorithm) to find a satisficing outcome by adding or removing houses from specific locations on the grid. Instead of dictating a specific number of houses to be placed in the site, the team simply chose a density preference. Each particular lot on the site has an approximately 70chance of having a house built on it.

A bifurcated house is used as the atomic housing unit. The prototypical house is composed of a long larger section (the high-density residential part), attached to a square smaller section (the more open recreation/agriculture part). All units are oriented parallel to one another in the site. In the last of an evolving series of simulations (see Figure 6) designed by the enclave team houses are sequentially placed horizontally, from top to bottom along the site's grid. They are placed from right to left (east to west) – this placement order recognizing the existing established town east of the enclave site. All houses on the top row of the site must have their smaller part oriented downward. This establishes another form of directionality for the site, and ensures that these houses' open areas are oriented towards the rest of the community, forming an envelope of sorts along the top border. The remaining placement orientations are determined randomly.

Appraisal

The diversity of the three different investigations (Field, Rhizome, Enclave) showed the flexibility of the toolbox. First, on a simulation level the three teams engaged the tool for investigations of different scale.

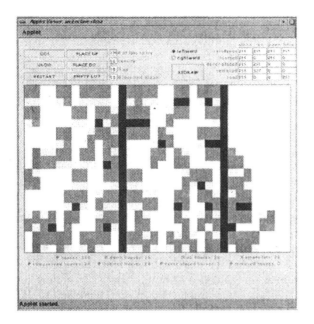

Figure 6: Enclave simulation. Houses are sequentially placed horizontally, from top to bottom along the site's grid. Statistics are noted at the bottom of the page. The simulation accomplished the goal to create medium density housing distribution with non-uniform open space.

The field team used the software to model housing aggregation at the site level. The enclave team studied the possible patterns given an atomic housing unit and the rhizome team investigated the programmatic layout of a single building.

Second, each team found a unique way integrating the tool with the rest of their design process. The field team, which started using the toolbox first, viewed design/conceptual space as an open system that was constantly being modified and added to. They came up with many ideas for different simulations to study different aspects of housing aggregation. Some of the simulations evolved more or less independent of each other, while others were deeper inquiries into subjects that previous simulations had only touched upon. The simulations were modified very little by the field team. Instead of trying to tweak parameters to achieve the desired result, they preferred to try completely new approaches. Many of their simulation ideas were not implemented due to a lack of time. Computation was used in a parallel way, rather than a serial one, with the field team. There was no refinement, so to speak, but rather complete reformulation of the design.

The enclave team chose to use the toolbox for one specific part of their design. Initially, they had a general idea of what behaviors that they wished to model,

but not a full specification. After implementing what they had, it became clear that they had not properly accounted for all of the sorts of interactions that could take place. Little by little, the enclave team fleshed out a full specification of all of the behaviors. Once they finished implementing their specification, they took the results of their simulation (not a particular spatial arrangement of houses, but rather a set of spatial arrangements that satisfied the constraints of their simulation) and incorporated them with the rest of their design. Enclave's use of the tool was far more serial than field's. They had a very narrow domain of investigation, and subjected their tool to many stages of refinement and specialization.

The rhizome team thought extensively about the preferred relationships between the various programmatic elements in their design plan, and encoded them in a "preference table". After collaboration with computer scientists, they arrived at an algorithm which would try to maximize the satisfaction of these preferences. The rhizome team's interaction with the toolbox was the cleanest of the three teams. They had the most concrete idea of what they wanted and how to get it before the implementation started. The complexity and richness of the simulation's results reflected the rhizome team's forethought into the design of the simulation.

Summary

In summary, key concepts of ALife have relevance to Architecture. In particular, because ALife simulations model:

- agency of components and environment

- emergent levels

- local, simple behaviors giving rise to global, complex organization

- sensitivity to initial conditions

- non-determinism affecting outcome

- abstraction using visualization

they are ideal tools for design. The approach to thinking of design elements and their site as dynamic interacting agents is provocative and enlivening to architects. Because global outcomes can not be predicted from the initial conditions, there is an element of suspense and eagerness to observe the outcome of the simulation runs. The hypothetical exploration of this process is engaging and, despite simple questions and initial conditions, it yields complicated, interesting and successful results.

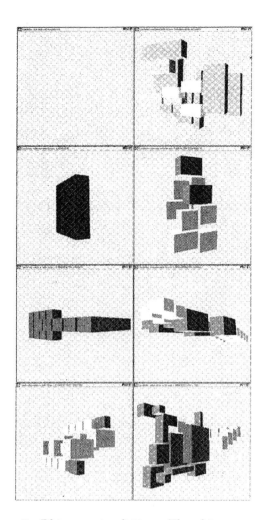

Figure 7: Rhizome simulation. The rhizome genetic algorithm explores different spatial organizations. Its features include portability of the java language, runtime user interaction with the genetic algorithm and a 3d display system.

Future Work

The toolbox is intentionally open-ended for software development. Its object-oriented design was chosen to facilitate the growth of application class libraries. Java was chosen so simulations could run on any platform and from the internet. However, the concepts of Emergent Design are more attractive to us than expending the technical effort required to software engineer the abstracted architectural investigation context. Thus, in the second generation, we will consider using an off-the-shelf agent based software package from which we will re-implement our context. This may be advantageous because the language of such a package may be easier than Java to master and it may be possible to convert the results to Java. We will have to eval-

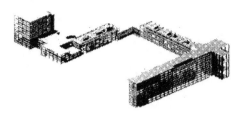

Figure 8: Rhizome project. Axonometric of final proposal combining living and work spaces.

uate whether a package will provide us with sufficient expressibility and flexibility.

In terms of core software extensions we are considering adding an evolutionary algorithm (Holland 1992; Goldberg 1989) component to facilitate a search process during a simulation. We have, in fact, enhanced the rhizome application with a genetic algorithm (Figure 8) since the spring studio course ended. The fitness function of the genetic algorithm measures the outcome's success at fulfilling a preference matrix expressing interaction desires. We would also like to support a project where spatial development and form growth takes place in conjunction with environmental factors.

Acknowledgements We thank Devyn Weiser and the students of 4.144 in Spring 1999.

References

D. E. Goldberg. *Genetic Algorithms in Search, Optimization, and Machine Learning.* Addison-Wesley, Reading, MA, 1989.

J. H. Holland. *Adaptation in Natural and Artificial Systems.* MIT Press, Cambridge, MA, 2nd edition, 1992.

C. G. Langton, C. Taylor, J. D. Farmer, and S. Rasmussen, editors. *Artificial Life II.* Santa Fe Institute Studies in the Sciences of Complexity. Addison-Wesley, Reading, MA, 1992.

C. G. Langton, editor. *Artificial Life.* Santa Fe Institute Studies in the Sciences of Complexity, Proc. Vol. VI. Addison-Wesley, Reading, MA, 1989.

M. Resnick. Learning about life. *Artificial Life Journal,* 1(1,2):229–241, 1994.

P. Testa, U.M. O'Reilly, D. Weiser, and Ian M. Ross. Emergent design: A crosscutting research program and curriculum integrating architecture and artificial intelligence. In submission to Environment and Planning B: Planning and Design.

U. Wilensky and M. Resnick. Thinking in levels: A dynamic systems perspective to making sense of the world. *Journal of Science Education and Technology,* 8(2), 1998.

VII The Broader Context

John von Neumann and the Evolutionary Growth of Complexity: Looking Backwards, Looking Forwards...

Barry McMullin

Artificial Life Laboratory
Research Institute in Networks and Communications Engineering (RINCE)
Dublin City University, Dublin 9, Ireland
Barry.McMullin@dcu.ie

Abstract

In the late 1940's John von Neumann began to work on what he intended as a comprehensive "theory of [complex] automata". He started to develop a book length manuscript on the subject in 1952. However, he put this aside in 1953, apparently due to pressure of other work. Due to his tragically early death in 1957, he was never to return to it. The draft manuscript was eventually edited, and combined for publication with some related lecture transcripts, by Burks (1966). It is clear from the time and effort which von Neumann invested in it that he considered this to be a very significant and substantial piece of work. However: subsequent commentators (beginning even with Burks) have found it surprisingly difficult to articulate this substance. Indeed, it has since been suggested that von Neumann's results in this area are either trivial, or, at the very least, could have been achieved by much simpler means. It is an enigma. In this paper I review the history of this debate (briefly) and then present my own attempt at resolving the issue by focusing on an analysis of von Neumann's *problem situation* (Popper, 1976). I claim that this reveals the true depth of von Neumann's achievement and influence on the subsequent deveopment of this field; and, further, that it generates a whole family of new consequent problems which can still serve to inform—if not actually define—the field of Artificial Life for many years to come.

Burks' Problem: Machine Self-Reproduction

... This result is obviously substantial, but to express its real force we must formulate it in such a way that it cannot be trivialized...

—Burks (1970a, p. 49)

Arthur Burks makes this comment following a 46 page recapitulation of John von Neumann's design for a self-reproducing machine (realised as a configuration embedded in a 29-state, two dimensional, cellular automaton or CA). The comment can be fairly said to have initiated an extended debate about the significance of this work of von Neumann's, which has waxed and waned over time, but still persists to the present day.

Von Neumann's design is large and complex, and relies for its operation on exact and intricate interactions between the many relatively simple parts. In that sense, it is certainly substantial; but Burks is absolutely accurate in pointing out that this intricacy, in itself, is not necessarily interesting or significant. In particular, if the same "results" could be achieved, or the same "problem" solved, with drastically simpler machinery, then the interest of von Neumann's design would be critically undermined.

This is no idle concern on Burks' part. As he himself points out, within the CA framework, one can easily formulate a simple rule whereby a cell in a distinct state (labelled, say, 1) will cause adjacent quiescent cells (labelled, say, 0) to transit to state 1 also. By essentially the same definition or criterion as was applied to von Neumann's system, such a single cell, state 1, configuration, would qualify as a self-reproducing machine—and would seem to render von Neumann's fabulously baroque design completely redundant.

Burks concluded that "... what is needed is a requirement that the self-reproducing automaton have some minimal complexity." And at first sight, this does seem eminently reasonable. Presumably (?) it is relatively easy to construct a "simple" machine, by whatever means; therefore it need not surprise us unduly if we manage to concoct a "simple" machine which can construct other "simple" machines, including ones "like" itself, and which therefore qualifies as self-reproducing. Whereas, it is relatively difficult to construct a "complex" machine, by any means; and therefore it may well be a challenging problem to exhibit a "complex" machine that is capable of self-reproduction. Von Neumann's machine certainly appears "complex", and certainly succeeds in constructing other machines like itself (i.e., in reproducing itself). So if we could just more formally express the precise sense in which von Neumann's machine *is* "complex", then we might indeed be able to clarify the "real force" of his achievement.

However, even at this point, we should be at least somewhat wary—because, while von Neumann himself certainly did introduce and discuss the notion of "complexity" in relation to this work, he did *not* attempt any formalisation of it. Indeed, he described his own concept of complexity as "vague, unscientific and imperfect" (von Neumann, 1949, p. 78). It would therefore seem unlikely that the significance of his eventual results should actually *rely* on such a formalisation.

Nonetheless, Burks went on to propose just such a formal criterion of complexity—namely the ability to carry out universal computation. And by this criterion, von Neumann's design (or, at least, a straightforward derivative of it) would qualify as a "complex" self-reproducing machine, and thus be clearly distinguished from the single cell, "1 state self-reproducer," which would remain merely a "simple" (and thus *trivial*) self-reproducing machine.

This seems a reasonable enough suggestion by Burks; though I would emphasise again that, as far as I have been able to establish, such a thing was never proposed by von Neumann himself—and, indeed, it jars seriously with von Neumann's calculated refusal to formalise complexity.

In any case, it turns out that Burks' suggestion is unsatisfactory and unsustainable. While it is true that von Neumann's machine (suitably formulated) can satisfy Burks' criterion for "complex" self-reproduction, this still represents an interesting result only if this criterion cannot be satisfied by very much simpler means. But in fact—and with hindsight, this now seems unsurprising—Burks' criterion *can* be satisfied by much simpler means than those deployed by von Neumann. This is because universal computation, *per se,* does not actually require particularly complex machinery (see, for example, Minsky, 1967).

This fact was formally demonstrated by Herman (1973), when he essentially showed how the basic single cell, 1 state, self-reproducer described earlier, could be combined with a single cell capable of universal computation. This results in a CA system in which the individual cells are "simpler" than in von Neumann's CA (i.e., have fewer states), and yet there are single cell configurations capable of both self-reproduction and universal computation. Granted, the universal computation ability relies on an adjacent, indefinitely long, "tape" configuration; but that was equally true of the universal computation ability of von Neumann's design, and is not a relevant distinguishing factor.

Herman draws the following conclusion (Herman, 1973, p. 62):

> ...the existence of a self-reproducing universal computer-constructor in itself is not relevant to the problem of biological and machine self-reproduction. Hence, there is a need for new mathematical conditions to insure non-trivial self-reproduction.

So we see that Herman rejects Burks' specific criterion, while still continuing to accept Burks' formulation of the *issue* at stake—namely the identification of a suitable criterion for distinguishing "non-trivial" self-reproduction, albeit in Herman's version this is no longer explicitly tied to the notion of "complexity".

The discussion was taken up again by (Langton, 1984) (though apparently without reference to Herman's work). He presented a rather different analysis of Burks' criterion, but with ultimately complementary results. Langton pointed out that, as a general principle, there is little evidence to suggest that living organisms contain universal computation devices embedded within them. Since the self-reproduction of living organisms is presumably to be regarded as non-trivial (by definition–at least in this context), we should not, therefore, adopt universal computation as a criterion. So in this respect, albeit for different reasons, Langton concurs with Herman.

More importantly, Langton goes on to suggest a specific alternative criterion. He points out that self-reproduction in living organisms relies on a decomposition of the organism into two parts or components playing very distinct roles in the reproductive process:

1. The *genotype*, being an informational pattern stored or recorded in some sort of quasi-static or stable carrier. This information is transcribed or *copied* into a corresponding carrier in the offspring.

2. The *phenotype*, being the visible, active, dynamic, interacting part of the organism. The phenotype of the offspring is created by some sort of *decoding* or *interpretation* of the genotype (rather than by a copying of the parental phenotype).

Langton does not explain just *why* such a decomposition may be important, but rather seems to accept its pervasiveness among biological organisms as reason enough to adopt it as a criterion. And, in some sense it "works", because, indeed, von Neumann's self-reproducing design does have this architecture, whereas (say) Herman's self-declared "trivial" design does not. So it seems like this may be a satisfactory or even illuminating demarcation.

However: Langton did not stop there. With this new criterion in hand, he went on to consider whether it could be satisfied with a design significantly simpler than von Neumann's—and it transpires that it *can*. In fact, Langton was able to present a design for a CA space which itself is rather simpler than von Neumann's (i.e., having fewer states per cell), into which he could embed a self-reproducing automaton which, like von Neumann's, has an explicit decomposition into genotype and phenotype, but which is vastly smaller and simpler, occupying a region of just 150 cells—compared to the several hundred thousand cells required for von Neumann's device!

Langton's automaton is certainly still quite intricate, and its design involved a subtle interplay between designing the CA itself and designing the automaton to be embedded within it. In this sense it is a significant and substantive achievement. But is remains very unsatisfactory from the point of view of evaluating von Neumann's work. If Langton's criterion for non-trivial self-reproduction is accepted as an appropriate measure to judge von Neumann's work by, then we must conclude that the latter's design is vastly more complex than necessary. While this *may* be true, I suggest that we should be reluctant to accept it without some much more substantive rationale for Langton's criterion. Or to put it another way, there may still be more "real force" to von Neumann's achievement than is captured or implied by Langton's criterion.

Von Neumann's Problem: The Evolutionary Growth of Complexity

I propose to resolve this enigma in a rather different way.

Firstly, I fully agree with Burks that to appreciate the full force of von Neumann's work, we must understand what *problem* he was attempting to solve. In particular, if it should turn out that this problem can be solved by trivial (Herman) or at least much simpler (Langton) means then we should have to conclude that it was not such a substantial achievement after all. Where I differ from these, and indeed, most other, commentators, is that I think it is a mistake to view von Neumann's problem as having been wholly, or even largely, concerned with *self-reproduction*!

Of course, this is not to deny that von Neumann did, indeed, present a design for a self-reproducing automaton. I do not dispute that at all. Rather, my claim is that this self-reproducing capability, far from being the object of the design, is actually an incidental—indeed, *trivial*, though highly serendipitous—corollary of von Neumann's having solved at least some aspects of a far deeper problem.

This deeper problem is what I call the *evolutionary growth of complexity*. More specifically, the problem of how, in a general and open-ended way, machines can manage to construct other machines more "complex" that themselves. For if our best theories of biological evolution are correct, and assuming that biological organisms are, in some sense, "machines", then we must hold that such a constructive increase in complexity has happened not just once, but innumerable times in the course of phylogenetic evolution. Note that this claim does not rely on any sophisticated, much less formal, definition of complexity; it requires merely the crudest of qualitative rankings. Nor does it imply any *necessary* or consistent growth in complexity through evolution, but merely an acceptance that complexity has grown dramatically in *some* lineages.

Why is this growth of complexity a problem? Well, put simply, all our pragmatic experience of machines and engineering points in the opposite direction. In general, if we want to construct a machine of any given degree of complexity, we use even more complex machinery in its construction. While this is not definitive, it is surely suggestive of a difficulty.

To make all this a little more concrete, imagine that we could exhibit the following:

- Let there be a large (ideally, infinite) class of machine "types" or "designs"; call this M.

- Now consider those machine types, which we shall call *constructors*, which are capable of constructing *some* other (types of) machine. For any given $m \in M$ let $O(m)$ ("offspring of m") denote that subset of M which m is capable of constructing. m is then a constructor precisely provided $O(m) \neq \phi$; and—incidentally—m is self-reproducing provided $m \in O(m)$. This relation, $O(m)$, induces a directed graph on M; by following arrows on this graph we can potentially see what machine types can, directly or indirectly, be constructed by other machine types.

- Let there be a crude ("vague, unscientific and imperfect") measure of complexity on the elements of M—call it $c(m), m \in M$.

- Let $c(m), m \in M$ span a large (ideally, infinite) range, which is to say that there are relatively simple and relatively complex types of machine, and everything in between, included in M.

Now consider the two (highly schematic) graphs shown in Figure 1. In both cases I have shown the putative set M partitioned more or less coarsely by the complexity measure c. That is, the inner rings or subsets are those of small c (low complexity), while the

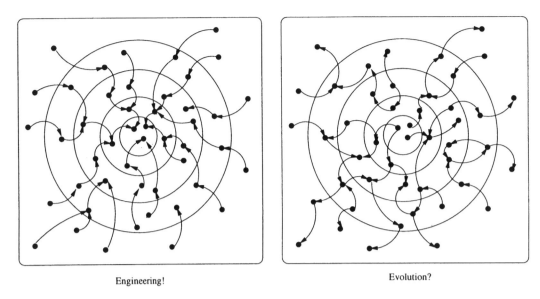

Engineering!

Evolution?

Figure 1: Evolutionary growth of complexity.

outer, more inclusive rings or subsets include machines of progressively greater c (higher complexity). The graph on the left indicates our "typical" engineering experience: all constructional pathways lead inward (from more complex to less complex). As a result, complexity will always, and unconditionally, degenerate in time. Conversely, the graph on the right indicates the abstract situation posited by our best current theories of biological evolution: at least some edges of the graph (constructional pathways) lead from the inner rings to their outer neighbors. Provided there are sufficient such pathways, then, starting from only the very simplest machines (or organisms), there will be potential constructional pathways whereby even the most complex of machines can eventually be constructed (in time). Thus complexity *might* grow in time.[1]

The problem which this presents is to show that the experience and intuition underlying the graph on the left is mistaken; to show, in other words, how the situation on the right might in fact be realised.

What would it take to solve this problem, even in principle? Well, one would need to exhibit a concrete class of machines M, in sufficient detail to satisfy ourselves that they *are* purely mechanistic; one would need to show that they span a significant range of complexity; and finally, one would have to demonstrate that there *are* constructional pathways leading from the simplest to the most complex (there may, or may

not, also be pathways in the other direction, but that is not the point at issue—or at least, not immediately).

I believe that *this* is precisely the problem which von Neumann set out to solve. Furthermore, it seems to me that he did, indeed, solve it; and that it is only by seeing his work in *this* light that its "real force" can be understood.

Von Neumann's Solution

As to the first claim, that this—rather than "self-reproduction" per se—was von Neumann's problem, it is certainly relevant that he introduced his work in essentially these terms in both his recorded public presentations of it—at the Hixon Symposium in September 1948 (von Neumann, 1951, p. 312) and at the University of Illinois in December 1949 (von Neumann, 1949, pp. 78–79). Granted, in both cases he did *also* refer to the issue of self-reproduction. Specifically, he pointed out that self-reproduction is an example of a constructing machine where the output or offspring is just precisely matched in complexity to the parent—complexity neither increasing nor decreasing. But the critical point here—the significance of self-reproduction—is clearly as a special or "watershed" case of the more general situation: it is interesting precisely, and only, because it may mark a transition from strictly-degenerating to potentially-growing complexity. Conversely, if we encounter a form of "self-reproduction" that is *not* associated with such a transition, then it will not be of relevance to von Neumann's problem at all.

Indeed, von Neumann (1949, p. 86) made this point even more explicitly, when he actually noted the trivi-

---

[1]Note that there are still inward, degenerative, pathways shown here; the claim is only that this graph permits the *possibility* of a growth in complexity; whether it actually *will* grow is a different, and an altogether more difficult, question which I do not attempt to address here.

ality of self-reproduction in "growing crystals" (essentially Burks' 1-state reproducer in a CA). But *pace* Burks, *von Neumann's* resolution of this was not at all to impose a criterion of complexity, but rather to stipulate that the reproductive process should be such as to support "inheritable mutations". Taken in context, I think this must be interpreted as supporting at least some "mutations" where the offspring is of *increased* complexity—which again returns us squarely to the problem of the evolutionary growth of complexity, rather than self-reproduction.

My second claim—that von Neumann actually solved this problem—may seem more far fetched; but let us take it in steps.

First note that while von Neumann exhibited the design and operation of only one machine in detail— his self-reproducing machine—he consistently pointed out how his CA space could support an indefinite variety of quite arbitrary machine configurations and behaviours. In this sense, he certainly exhibited not just a single particular machine, but a whole class of machines (which he identified via the infinite set of arbitrary, finite, "initially quiescent" configurations in his CA). This can correspond to the class M in my problem formulation above (though I will refine this somewhat below).

The next question is whether this class of machines spans a significant range of complexity. Given that we are using only the most vague definition of "complexity" here, the answer can be at best informal and qualitative. To my knowledge, von Neumann did not quite comment on this explicitly. He did explain at length that in setting up this sort of "axiomatic" theory of automata, there is necessarily a degree of arbitrariness in selecting the primitive or atomic parts from which the composite automata will be constructed. And indeed, between 1948 and 1953 he considered a number of alternative approaches before settling on a particular CA formulation (the latter concept having been suggested by Ulam—see Burks, 1966, p. 94). But given the highly complicated (co-ordinated, systematic, purposeful) behaviour of the one machine he did design in detail, it certainly seems to me that if we allow, as von Neumann did, for the construction of machines with indefinitely more parts, in arbitrary configurations, then this set surely *might* span a sufficient range of complexity to meet our requirements.

This then leaves what seems by far the most intractable aspect of the problem: to show that there are constructional pathways leading from the simplest to the most complex machines. At first sight, it seems hopeless to demand a *demonstration* of this; firstly because the measure of complexity is so vague; but secondly because it would seem to demand separate analysis or demonstration of the constructional potentialities for most if not all of the elements of the infinite set M. But this, it seems to me, was where von Neumann's crucial insight occurred.

Drawing on the capability for a single universal Turing machine to carry out the computations of any Turing machine at all, given a suitable description of that machine, von Neumann recognised that a single "general constructive automaton" (von Neumann, 1949, p. 85) might be able to construct "any" machine at all (i.e., any element of his specified M), given a "description tape" of that target machine. This immediately suggests the possibility that certain such special, "programmable", constructor (sub-)systems may enable enormous constructive potentialities—and also, incidentally (?) open up a powerful and *generic* route toward self-reproducing capability.

Von Neumann exploited this general idea by making a minor, but subtle, modification of his general constructive automaton (in a manner incidentally having no analog or significance in pure computation theory): as well as requiring it to *decode* a description to produce an offspring, he required that it must also *copy* the description, attaching this as part of the offspring. As Langton surmised, this combination of decoding and copying potentially has quite dramatic consequences—but the challenge is still to make those consequences explicit.

Let me denote a basic general constructive (decoding and copying) machine by u_0. We require that $u_0 \in M$.[2] Let $d(m), m \in M$ denote a "description of m", (relative to u_0). That is, letting $u_0 \oplus d(m)$ denote the composition of u and a tape describing an arbitrary m, this will result in the construction of an instance of m, via *decoding* of the description, itself composed with a *copy* of the original tape, i.e., $m \oplus d(m)$. We write this as:

$$(u_0 \oplus d(m)) \rightsquigarrow (m \oplus d(m))$$

Since this applies for all $m \in M$ it applies to u_0 itself and we have:

$$(u_0 \oplus d(u_0)) \rightsquigarrow (u_0 \oplus d(u_0))$$

This is the single self-reproducing automaton, which has been commonly identified as von Neumann's "result" or "achievement". Certainly at this point we

---

[2]It is neither trivial nor obvious that there exists a general constructive automaton within any given M; this is why the bulk of von Neumann's unfinished manuscript is taken up with the detailed design of a particular example u_0 to establish this result for his particular M.

have a case of construction where complexity has been fully preserved—but the question remains whether this gives us an avenue into the case where complexity can grow.

Now note that u_0 can be combined or augmented with fairly arbitrary ancillary machinery, while still retaining its general constructive ability. That is, we can say that $u_0 \oplus m$ is still a general constructive automaton for "almost all" $m \in M$. The exception would be where m in some sense interferes with or disrupts the normal operation of u_0. Discounting those cases, we can say that the existence of a single general constructive automaton, u_0, actually implies the existence of a whole infinite set of related general constructive automata, $U \subset M$, where $u_m \in U$ has the form $u_m = (u_0 \oplus m)$ (i.e., all the members of this set share the same "general constructive sub-system"). This, in turn, implies the existence of a whole infinite set of self-reproductive automata, S where each $s_m \in S$ has the form $u_m \oplus d(u_m)$. Furthermore, given that U is derived from M just by excluding some exceptional or pathological cases (i.e., where the operation of u_0 would be disrupted), we can say that U, and thus S, will still span essentially the same significant range of complexity as M itself.

Now this is clearly a much stronger result than merely demonstrating the existence of a single self-reproducing machine. In particular, it indicates the possibility of arbitrarily complex machines that are still capable of self-reproduction. This in itself certainly distinguishes von Neumann's result from that of Langton (1984). Although the operation of Langton's machine can be decomposed into copying and decoding activities, it does not incorporate anything like a "general constructive automaton". The decoding capability is extremely impoverished so that, in effect, there is only one description that can be effectively processed—that of the one self-reproducing configuration which Langton exhibited.

But in any case, there is still a further critical step in the analysis of von Neumann's work. The goal is not just to show a class of arbitrarily complex constructors capable of constructing machines of equal complexity (which is what self-reproduction illustrates), but to demonstrate constructional pathways whereby complexity can *grow*. And it turns out that this further result can also be demonstrated by the set S.

To see this, we imagine the possibility of perturbations—"mutations"—to the description tape of a machine $s = (u_m \oplus d(u_m)), \in S$. With reasonable assumptions about the description language, it can be arranged that all, or almost all, tape configurations are "legal" (i.e., can be decoded to *some* target machine),

and that the tape $d(u_m) = d(u_0 \oplus m)$ can be decomposed into a part, $d(u_0)$, coding for u_0, and a part, $d(m)$, coding for the ancillary machinery, m. That being the case, a more or less arbitrary modification of the $d(m)$ portion of the tape will result in a legal description of some other machine, $d(m')$. The reproductive cycle will then result, not in self-reproduction, but in a "mutant" offspring $s' = (u_{m'} \oplus d(u_{m'}))$; but this is still an element of S, and thus self-reproducing in its own right.[3]

Now these mutations essentially open up additional constructional pathways *between* the elements of S. In particular it is now clear that this can allow incremental, bootstrapping, *increases* in complexity. In effect, the density of these mutational pathways reflects the combinatorics of the description code, so that we can be virtually guaranteed, *without any detailed analysis,* that there will be constructional pathways connecting the entire set S. These will include, of course, degenerating pathways, where the offspring are less complex; but will also include vast numbers of pathways of increasing complexity.

In this way, finally, we see how von Neumann's detailed design of a single machine implies at least a schematic solution of the generic problem of the evolutionary growth of machine complexity.

Looking Backwards

While the designation of a distinct field known as *Artificial Life* is comparatively recent (Langton, 1989), I would argue that, looking backwards, it is clear that von Neumann's work properly defines its origin and inspiration. If the formulation above of von Neumann's problem—and his solution to it—is accepted, then a proper assessment of his contribution to the field he effectively founded presents at least three distinct questions:

1. Was von Neumann the *first* one to solve his problem?

2. Has von Neumann presented the *only* known solution to his problem?

3. Is von Neumann's solution the *simplest* known?

My answer to question 1 is clearly an unambiguous "yes"; and it stands as von Neumann's remarkable achievement *both* to have formulated this foundational problem *and* to have actually succeeded in solving it.

Question 2 is slightly more difficult. Given than von Neumann's problem situation has been poorly understood, it follows that subsequent contributors have

---

[3]Indeed, we call such changes "mutations" precisely because they can "breed true" in the offspring.

not generally articulated clearly the relation between their work and this problem.

However, I do not hesitate to say that Thatcher (1970), Codd (1968) and Pesavento (1995) have all offered alternative, or at least significantly refined, solutions. Because these three systems are so very closely related to von Neumann's it follows that they are at least equally satisfactory in solving his problem.

Thatcher provided a somewhat simplified design for the basic general constructive automaton u_0, in the *same* CA space as defined by von Neumann. Codd defined a significantly simpler CA space, which could still accommodate a functionally equivalent general constructive automaton. Much more recently, Pesavento has demonstrated a substantially simplified design for u_0 in a CA space marginally more complicated than von Neumann's; and, further, has actually demonstrated a functioning simulation of this complete u_0.

Conversely, as already indicated, I think we can be clear that Langton's "self-replicating loop" system Langton (1984) does *not* qualify as an alternative solution to von Neumann's. Although it was derived from Codd's earlier system, and although he exhibits a self-reproducing configuration (involving "copying" and "decoding") this does not embody anything like a "general constructive automaton" and therefore has little or no evolutionary potential.

Another possible candidate solution to von Neumann's problem might be that of Berlekamp et al. (1982). There it is claimed that a general constructive automaton, of comparable functionality to von Neumann's u_0, can be realized in Conway's well known "Game of Life" (GOL) CA. This arguably takes Codd's (and thus von Neumann's) result to some kind of limiting case, as the GOL cells have the absolute minimum of only 2 states. On the other hand, in contrast to the fully detailed designs for u_0 presented by Thatcher, Codd and Pesavento, Berlekamp et. al. provide only a very sketchy outline to indicate the *possibility* of an equivalent automaton in GOL. Moreover, because GOL is so radically simplified (as a CA) it would not be quite trivial to establish that it supports embedded automata of comparable range of "complexity" to those of Von Neumann etc.

This last issue affects the comparative evaluation of certain other systems even more seriously. I have in mind particularly `Tierra` (Ray, 1992) and related systems.

`Tierra` is, roughly, a shared memory parallel computer, with fairly simple "von Neumann style" processors.[4] Ray has exhibited processes ("machines" or "au-

tomata" in von Neumann's sense) in this framework which are capable of self-reproduction; and which, moreover, exhibit clear and significant evolutionary change. However, because the `Tierra` system is so very different in style from von Neumann's (or, indeed, *any* CA of 2 or more dimensions), it is very difficult to make even informal and qualitative comparisons of automata "complexity" between these systems.

I also have another quite different, and perhaps more important, misgiving about whether `Tierra` should be regarded as offering an equally satisfactory alternative solution to von Neumann's problem.

I originally expressed this reservation by stating that `Tierra` uses "... a form of self-reproduction based on self-inspection (rather than a properly genetic system in the von Neumann sense)" (McMullin, 1992, Chapter 4). The implication was that the self-reproducing entities in `Tierra` lack the important distinction between "copying" and "decoding"—and that this perhaps affects evolutionary potential in the system. However Taylor (1999) has since persuaded me that this way of presenting the matter is, at best, unclear.

It might be more precise to say that `Tierra` *does* incorporate a distinction between copying and decoding processes: but that the "decoding" is represented by the execution of instructions by the `Tierra` processors. Thus, the decoding is hard-wired, whereas, in von Neumann's system, the decoding is itself a product of the specific design of u_0, and, as such, is at least *potentially* mutable. In this way, von Neumann's system allows for what I have called *Genetic Relativism* (McMullin, 1992, Chapter 4), where the actual "decoding" map can itself evolve. It seems to me that this allows for a more profound form or degree of evolvability in von Neumann's system compared to `Tierra`. Having said that, I should also note that von Neumann himself seems to have discounted this idea. He stated explicitly that mutations affecting that part of a descriptor coding for u_0 would result in the production of "sterile" offspring (von Neumann, 1949, p. 86)—and would thus have no evolutionary potential at all. Clearly, on this specific point, I disagree with von Neumann, and consider that he actually underestimated the force of his own design in this particular respect.

As to my question 3 above—whether von Neumann's solution is the "simplest"—this clearly does not admit of a clearcut answer, given the informality of our notions of simplicity and complexity. Certainly, the alternative solutions of Thatcher, Codd and Pesavento can all be said to be *somewhat* simpler in at

---

[4]Here, of course, I am referring to the so-called "von Neumann computer architecture" (von Neumann, 1945)—

which was not, by any means, solely von Neumann's invention—rather than to his very different work on cellular automata.

least some respects. Similarly, if the outline design of Berlekamp et. al. is accepted then it is *much* simpler in one particular dimension (the number of cell states). Nonetheless, in considering the overall development of the field I would say that none of these "simplifications" either undermine, or significantly extend, von Neumann's seminal results.

Conclusion: Looking Forwards

By re-examining von Neumann's work in the light of his own description of the problem he was working on, I have tried to show that there is much more substance to it than has been generally recognised. However, as Popper (1976) has emphasised, the scientific enterprise is iterative: the solution of any given problem gives rise to a new problem situation, which is to say new problems to be addressed. It seems to me that von Neumann's work is particularly fruitful in this regard, as it poses a number of new and profound questions which need to be addressed by the (still) fledgling field of Artificial Life. Among these are:

1. Can we clarify (or even formalise) the notion of complexity. This of course is not specifically a problem for Artificial Life, but rather underlies the entire discipline of evolutionary biology (Maynard Smith, 1969).

2. What is the precise significance of the self-reproductive capability of von Neumann's machines? Technically, based on the idea of a "general constructive automaton", growth of complexity *per se* could take place in isolation from self-reproduction. One simple scenario here would be to imagine u_0 being made to simply grind through the (countable) infinity of all description tapes, constructing every described automaton in turn. This would require only a trivial extension of the capabilities of u_0. While this would fail in practice due to the essential fragility of von Neumann's automata (discussed further below) it is not clear whether there is any *fundamental* problem with this general idea. Nonetheless, it seems clear that if the growth of complexity is to involve *Darwinian* evolution, then self-reproduction surely is a necessary additional requirement.

3. The problem of identity or individuality. In the alife systems discussion above, the question of what constitutes a distinct individual "machine" or "automaton" is addressed in a completely ad hoc manner. In a real sense, these putative individuals exist as such only in the eye of the human observer. Whereas, the notion of a self-defining identity, which demarcates itself from its ambiance, is arguably essential to the very idea of a biological organism. Some early and provocative work on this question, including simulation models overtly reminiscent of von Neumann's CA, was carried out by Varela et al. (1974). However, there has been relatively little further development of this line since.

4. The evolutionary boot-strapping problem: in the von Neumann framework, at least, u_0 is already a very complicated entity. It certainly *seems* implausible that it could occur spontaneously or by chance. Similarly, in real biology, the modern (self-consistent!) genetic system could not have plausibly arisen by chance (Cairns-Smith, 1982). It seems that we must therefore assume that something like u_0 (or a full blown genetic system) must itself be the product of an extended evolutionary process. Of course, the problem with this—and a major part of von Neumann's own result—is that it seems that something like a genetic system is a *pre-requisite* to any such evolutionary process.

5. Perhaps most importantly of all, what further conditions are required to enable an *actual*, as opposed to merely *potential* growth of complexity? It was well known even to von Neumann himself that his system would not *in practice* exhibit any evolutionary growth of complexity. The proximate reason is that, in his CA framework, all automata of any significant scale are extremely *fragile*: that is, they are very easily disrupted even by minimal perturbation of the external environment. The upshot is after the completion of even a single cycle of self-reproduction the parent and offspring would almost immediately perturb, and thus effectively destroy, each other. This can be avoided by ad hoc mechanisms to prevent all interaction (again, this was suggested by von Neumann, and since shown in practice by Langton and others). However, eliminating interaction eliminates the grist from the darwinian mill, and is thus a non-solution to the substantive problem (growth of complexity). Tierra offers a somewhat different approach, whereby the integrity of the automata (process images) is "protected" by the underlying operating system, while still allowing some forms of significant interaction. This has been very fruitful in allowing the demonstration of *some* significant evolutionary phenomena. However, any serious claim to substantively model real biological organisms will inevitably have to confront their capacity for *self* maintenance and repair in the face of continuous perturbation and material exchange with their environments. This, in turn, is clearly also related to the earlier problem of biological individuality.

In conclusion, it seems to me that von Neumann's work in Artificial Life—properly understood—is as profound and important today as it was half a century ago; and that it should continue to provide structure, insight and inspiration to the field for many years to come.

Acknowledgements

Many of the ideas presented here were first explored in my PhD thesis (McMullin, 1992); I was privileged to carry out that work under the supervision of the late John Kelly. A large number of people have since helped and challenged me in trying to refine the analysis and arguments. I am particularly indebted to Chris Langton, Glen Ropella, Noel Murphy, Tim Taylor, Mark Bedau and Moshe Sipper; the latter also provided critical encouragement to write this particular paper. I am grateful also to the ALife VII reviewers for comprehensive and constructive criticism. Financial support for the work has been provided by the Research Institute in Networks and Communications Engineering (RINCE) at Dublin City University.

References

Berlekamp, E. R., Conway, J. H. and Guy, R. K. (1982), What is Life?, *in* 'Winning Ways for your Mathematical Plays', Vol. 2, Academic Press, London, chapter 25, pp. 817–850.

Burks, A. W. (1970*a*), Von Neumann's Self-Reproducing Automata, in *Essays on Cellular Automata* Burks (1970*b*), pp. 3–64 (Essay One).

Burks, A. W., ed. (1966), *Theory of Self-Reproducing Automata [by] John von Neumann*, University of Illinois Press, Urbana.

Burks, A. W., ed. (1970*b*), *Essays on Cellular Automata*, University of Illinois Press, Urbana.

Cairns-Smith, A. G. (1982), *Genetic Takeover and the Mineral Origins of Life*, Cambridge University Press, Cambridge.

Codd, E. F. (1968), *Cellular Automata*, ACM Monograph Series, Academic Press, Inc., New York.

Herman, G. T. (1973), 'On Universal Computer-Constructors', *Information Processing Letters* **2**, 61–64.

Langton, C. G. (1984), 'Self-Reproduction in Cellular Automata', *Physica* **10D**, 135–144.

Langton, C. G. (1989), Artificial Life, *in* C. G. Langton, ed., 'Artifical Life', Vol. VI of *Series: Sante Fe Institute Studies in the Sciences of Complexity*, Addison-Wesley Publishing Company, Inc., Redwood City, California, pp. 1–47.

Maynard Smith, J. (1969), The Status of Neo-Darwinism, *in* C. H. Waddington, ed., 'Towards a Theoretical Biology, 2: Sketches', Edinburgh University Press, Edinburgh, pp. 82–89. This paper is also accompanied by various addenda and comments (pages 90–105 of the same work).

McMullin, F. B. V. (1992), Artificial Knowledge: An Evolutionary Approach, PhD thesis, Ollscoil na hÉireann, The National University of Ireland, University College Dublin, Department of Computer Science.
http://www.eeng.dcu.ie/~alife/bmcm_phd/

Minsky, M. L. (1967), *Computation: Finite and Infinite Machines*, Prentice-Hall Series in Automatic Computation, Prentice-Hall Inc., Englewood Cliffs, New Jersey.

Pesavento, U. (1995), 'An implementation of von Neumann's self-reproducing machine', *Artificial Life* **2**(4), 337–354.

Popper, K. R. (1976), *Unended Quest*, Fontana/William Collins Sons & Co. Ltd, Glasgow.

Ray, T. S. (1992), An approach to the synthesis of life, *in* C. G. Langton, C. Taylor, J. D. Farmer and S. Rasmussen, eds, 'Artifical Life II', Vol. X of *Series: Sante Fe Institute Studies in the Sciences of Complexity*, Addison-Wesley Publishing Company, Inc., Redwood City, California, pp. 371–408. Proceedings of the workshop on Artificial Life held February, 1990, in Sante Fe, New Mexico.

Taub, A. H., ed. (1961), *John von Neumann: Collected Works. Volume V: Design of Computers, Theory of Automata and Numerical Analysis*, Pergamon Press, Oxford.

Taylor, T. J. (1999), From Artificial Evolution to Artificial Life, PhD thesis, University of Edinburgh.
http://www.dai.ed.ac.uk/daidb/homes/
timt/papers/thesis/html/main.html

Thatcher, J. W. (1970), Universality in the von Neumann Cellular Model, *in* Burks (1970*b*), pp. 132–186 (Essay Five).

Varela, F. J., Maturana, H. R. and Uribe, R. (1974), 'Autopoiesis: The Organization of Living Systems, its Characterization and a Model', *BioSystems* **5**, 187–196.

von Neumann, J. (1945), First Draft of a Report on the EDVAC. The (corrected) version at the URL below has been formally published in the IEEE Annals of the History of Computing, **15**(4), 1993.
ftp://isl.stanford.edu/pub/godfrey/
reports/vonNeumann/vnedvac.pdf

von Neumann, J. (1949), Theory and Organization of Complicated Automata, *in* Burks (1966), pp. 29–87 (Part One). Based on transcripts of lectures delivered at the University of Illinois, in December 1949. Edited for publication by A.W. Burks.

von Neumann, J. (1951), The General and Logical Theory of Automata, *in* Taub (1961), chapter 9, pp. 288–328. Delivered at the Hixon Symposium, September 1948; first published 1951 as *pages 1–41 of:* L. Jeffress, A. (ed), *Cerebral Mechanisms in Behavior*, New York: John Wiley.

What Can We Learn from the First Evolutionary Simulation Model?

Seth Bullock

Informatics Research Institute, School of Computer Studies, University of Leeds
seth@scs.leeds.ac.uk

Abstract

A simple computer program dating from the first half of the nineteenth century is presented as the earliest known example of an *evolutionary simulation model*. The model is described in detail and its status as an evolutionary simulation model is discussed. Three broad issues raised by the model are presented and their significance for modern evolutionary simulation modelling is explored: first, the utility of attending to the character of a system's entire dynamics rather than focusing on the equilibrium states that it admits of; second, the worth of adopting an evolutionary perspective on adaptive systems beyond those addressed by evolutionary biological research; third, the potential for the non-linear character of complex dynamical systems to be explored through an individual-based simulation modelling approach.

With the war-time and post-war development of the first modern computers came a surge of research into computational theory. Seminal work by mathematicians such as McCulloch and Pitts (1943) on the logic of neural circuitry, Turing (1952) on diffusion-reaction models of morphogenesis, Walter (1963) and Ashby (1956) on cybernetics, von Neumann and Burks (1966) on automata theory and self-replication, and later Holland (1975) on the formal properties of adaptation, involved the application of logic, mathematics, robotics, and control theory to essentially biological problems.

The above-cited pieces of research are now recognised as the intellectual precursors to the field that has come to be known as artificial life. Although, more proximally, artificial life can be considered to be the offspring of artificial intelligence (see Brooks, 1991, and Steels, 1994, for accounts of artificial life's relationship to artificial intelligence), it is becoming increasingly apparent that the work published under the artificial life rubric (e.g., models of morphogenesis, cellular automata models, behaviour based robotics, the simulation of adaptive behaviour, etc.) has inherited much of its method, and some would say madness, either directly, or circuitously, from these mid-century pioneers.

However, it will be claimed here that a particular kind of artificial life, the *evolutionary simulation model*, originated far earlier than even the first of these seminal works. Coincidentally, the first evolutionary simulation model holds many lessons that are pertinent today. After introducing the model and discussing its status as an evolutionary simulation model, a series of issues raised by the model will be presented and their implications for modern artificial life explored.

The Ninth Bridgewater Treatise

In 1837, twenty-two years before the publication of Darwin's *On the Origin of Species*, and over a century before the advent of the first modern computer, a piece of speculative work was published as an uninvited *Ninth Bridgewater Treatise*. The previous eight works in the series had been sponsored by the will of Francis Henry Egerton, Earl of Bridgewater, and a member of the English clergy. The will's instructions were to make money available to commission and publish an encyclopedia of natural theology concerning "the Power, Wisdom, and Goodness of God, as manifested in the Creation" (Brock, 1966; Robson, 1990).

The ninth publication in this series is noteworthy in that, unlike typical works of natural theology, it neither sought to draw attention to miraculous states of affairs deemed unlikely to have come about by chance, and thus thought to be the work of a divine hand (e.g., the length of the terrestrial day, which seems miraculously suited to the habits of man and other animals), nor did it seek to reconcile scientific findings with a literal reading of the Old Testament (e.g., disputing evidence that suggested an alarmingly ancient earth, accounting for the existence of dinosaur bones, or promoting evidence for the occurrence of the great flood, etc.). In contrast to these apologetic efforts, the author of the ninth Bridgewater treatise produced what is, to my knowledge, the first instance of an *evolutionary simulation model*.

The author of the *Ninth Bridgewater Treatise* was

Charles Babbage, the designer of the difference engine and analytical engine (the first automatic calculating devices, and thus precursors to the modern computer). Indeed, in 1837 he was one of perhaps a handful of scientists capable of carrying out research involving automated computational modelling. His model stands as a usefully simple case study, and an elegant example of the use to which computers are being put in modern artificial life.

The aim of this paper will not be to draw attention to Babbage's model as an example of a particularly prescient piece of work, anticipating much of modern evolutionary simulation modelling. Rather, this antique evolutionary simulation model will be revived, not only to identify similarities between Victorian science and current artificial life, but in order to highlight important aspects of the contemporary practice of evolutionary simulation modelling. In pursuing this aim, I am well aware of the risks of Whiggism in interpreting historical material (see Hyman, 1990, for discussion of Whiggism in the study of Babbage and his work). Although a modern perspective on the past is unavoidable, the fact that this paper does not primarily seek to re-evaluate Babbage's work but to re-analyse modern evolutionary simulation modelling work in the light of Babbage's model ensures that the risk of mis-treating Babbage's work is slim.

The First Evolutionary Simulation Model

Babbage's (1837) model (see also Babbage, 1864, Chapter XXIX "Miracles" for a rather whimsical account of the model's development) was situated within what was then a controversial debate. It addressed the dispute between *catastrophists* and *uniformitarians*. Prima facie this debate was internal to geology, since it concerned the geological record's potential to show evidence of divine intervention (principally in the form of support for the Old Testament accounts of the Creation and the Deluge). Catastrophists argued for an interventionist interpretation of geological evidence, taking discontinuities in the record to be indicators of the occurrence of miracles (violations of laws of nature). In contrast, uniformitarians insisted that in order to carry out scientific enquiry, the entire geological record must be assumed to be the result of unchanging processes. Allowing a role for divine miracles, the uniformitarians claimed, would render competing explanations equally valid. No theory could be claimed to be more parsimonious or coherent than a competing theory that invoked *necessarily inexplicable* exogenous influences in its account of the phenomena at issue.

Although this dispute had already been dealt some-

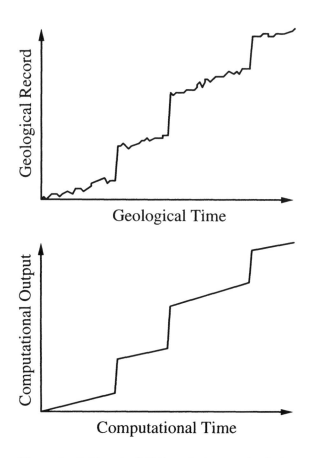

Figure 1: Babbage's (1836) evolutionary simulation model represented the empirically observed history of geological change as evidenced by the geological record (*upper panel*) as the output of a computing machine following a program (*lower panel*). A suitably programmed computing machine could generate sequences of output that exhibited discontinuities without requiring external influence. Hence discontinuities in the actual geological record did not require "catastrophic" divine intervention, but could be the result of "gradualist" processes.

thing of a death blow with Lyell's (1830) publication of his *Principles of Geology*, the publication of the Bridgewater treatises and works like them evidences its slow demise. Only subsequent to the *coup de grâce* provided by Darwin's work on evolution did natural theology texts finally cease to be published (Brock, 1966).

Babbage's response to the catastrophist position was to construct what can now be recognised as a simple evolutionary simulation model (see figure 1). He proposed that a suitably programmed difference engine could be made to output a series of numbers according to some law (e.g., the integers, in order, from 0 onwards), but then at some pre-defined point (e.g, 100,000) begin to output a series of numbers according to some different law (e.g., the integers, in order, from 200,000 onwards). Although the *output* of such a difference engine (an analogue of the geological record) would feature a discontinuity (in our example the jump from 100,000 to 200,000), the *underlying process* responsible for this output would have remained constant (i.e., the general law, or program, that the machine was obeying would not have changed). The discontinuity would have been the result of the naturally unfolding mechanical (and computational) process. No external tinkering analogous to the assumed intervention of a providential deity would have taken place.

Babbage thus tried to show that what might appear to be discontinuities were not necessarily the result of meddling, but could be the natural result of unchanging processes. In doing this he cultivated the image of God as a programmer, engineer, or industrialist, capable of setting a process in motion that would accomplish His intentions without Him intervening repeatedly. In Victorian Britain, the notion of God as draughtsman of an 'automatic' universe, one that would run unassisted, without individual acts of creation, destruction, etc., proved attractive. This conception was subsequently reiterated by several other natural philosophers (e.g., Darwin, Lyell, and Chambers) who argued that it implied "a *grander* view of the Creator — One who operated by general laws" (Young, 1985, p.148).

For the purposes of this paper, what is of interest are not the theological implications of Babbage's work, nor the effect it had on the catastrophist/uniformitarian debate, but the manner in which Babbage mobilised his computational resources to attack a theoretical position. Babbage's computational system was a simple analogue of a natural system (the geology of the planet) implemented mechanically. Babbage did not seek to capture the complexity of real geology in his system. Indeed the analogy between the difference engine's program and geological processes is a crude one. However, the formal resemblance between the computing machine and the geological process is sufficient to enable a point about the latter system's dynamics to be made. Babbage's computing machine is thus clearly being employed as a *model*.

Evolutionary Simulation Models

What grounds do we have for claiming that Babbage's model is an example of an *evolutionary simulation model*? At first glance, the fact that the model involves

no mention of competition, heritable variation, limited resources, etc., would indicate that, even if it could be called a simulation model, Babbage's programmed calculator should not be awarded the status of *evolutionary* simulation model. However, the adjective *evolutionary* is being used here not to invoke the notion of biological evolutionary change, but to draw attention to the fact that Babbage's model was implemented as a dynamic, unfolding, process. For the moment, a few brief observations will serve to give a flavour of what is intended by the phrase[1].

First, the fact that Babbage's model is an unfolding computational process sets it apart from work in which models of dynamic change are constructed as mathematical proofs and are thus not evolutionary simulations. For example, Malthus' (1798) work on population dynamics, in which he demonstrated that population growth would outstrip that of agriculture, was constructed using paper and pencil.

However, it is not merely the computational nature of Babbage's model that ensures its status as an evolutionary simulation model. His reliance on the ongoing dynamic behaviour of his computational model. rather than on any end result it might produce, distinguishes it from modelling work which, although involving computational processes, uses computers as tools for solving what would otherwise prove to be intractable mathematical problems. For example, the use of computers to discover digits of pi, or to iteratively solve the differential equations that might comprise a model of population dynamics, do not count as *evolutionary* simulation modelling since the computational processes involved are merely the means of reaching a particular solution. In contrast, the substantive element of Babbage's model is the evolutionary aspect of the simulation (i.e., the manner in which it changes over time). In the scenario that Babbage considers, his suitably programmed difference engine will, in principle, run forever. Its calculation is not intended to produce some end product, but rather the ongoing calculation is itself the object of interest.

In the following sections particular aspects of Babbage's model will be expanded upon. First, the implications of considering the dynamic behaviour of a model to be central, rather than concentrating on the end-product of some calculation, will be discussed.

Dynamics and Stasis

Artificial life is perhaps exclusively concerned with systems that change over time, and, furthermore, the

---

[1]For a more detailed treatment of the notion of evolutionary simulation modelling see Bullock (1997).

manner in which such systems change over time[2]. Whether the system be a cellular automaton implementing the "game of life", an autonomous robot designed to navigate an extra-terrestrial terrain, or a model of "complexity at the edge of chaos", it is considered as a time-varying system with a certain dynamic character. It is this character that is of interest to the artificial life practitioner. Will the dynamic character of the cellular automaton admit of "universal computation"? Will the dynamic character of the autonomous robot result in robust walking behaviour? Will the dynamic character of the model exhibit "complexity at the edge of chaos"?

We can contrast this interest in dynamic change with the approach taken by game-theoretic models (von Neumann & Morgenstern, 1944). Despite being similarly concerned with systems that change over time (economies, individual economic agents, ecologies, populations of creatures, individual creatures, etc.), game-theoretic accounts of adaptive phenomena typically assume that the systems under consideration are at, or near, equilibria. Once this assumption is in place, the game theorist is faced with the task of specifying a model that admits of stable equilibria with a character that matches that of the observed economic or biological system.

For example, neo-classical economic theory asserts that since the economic agent, *homo economicus*, is an ideal maximiser of expected utility, such agents will clear a market at the equilibrium price. There is thus no sense in asking what behaviour would result from a system comprised of agents who cannot maximise expected utility. Such a system is far from equilibrium, and thus not likely to be found to reflect real economic situations in which markets are either at or near equilibria. Since, from this perspective, economic agents are assumed to be optimal players, one merely needs to identify the equilibrium price analytically in order to describe the behaviour of the market, as this is the price that will be settled upon. Attention to the global dynamics of the model is not necessary since the system will not spend time far from equilibrium.

Whilst these assumptions are, for the most part, eminently reasonable, there are scenarios in which the conditions that real economic agents find themselves in result in their inability to find the equilibrium price, e.g., sellers at an auction who are wary of the existence of cartels amongst their fellow bidders. In order to model these systems, approaches that take into account a richer pallet of dynamic behaviour are nec-

essary (see, e.g., Binmore, 1987, 1988, for critiques of the traditional axiomatic approach to game-theoretic economic modelling).

A similar perspective is evident within game-theoretic accounts of evolutionary systems. Maynard Smith (1982) identifies this problem at the outset of his book, *Evolution and the Theory of Games*,

> "An obvious weakness of the game-theoretic approach to evolution is that it places great emphasis on equilibrium states, whereas evolution is a process of continuous, or at least periodic change. The same criticism can be levelled at the emphasis on equilibria in population genetics. It is of course mathematically easier to analyse equilibria than trajectories of change" (p. 8).

However, unlike economists, evolutionary game theorists have better grounds for pursuing a program of what Frank (1998) terms *comparative statics* than this appeal to the intractability of dynamic models. Identifying equilibria, and exploring their sensitivity to model parameters in order to make predictions about analogous real-world systems is a process with some chance of engaging with empirical biological observations, since, given that the natural systems around us are likely to be at or near equilibrium, we have some chance of collecting appropriate data. The likelihood of obtaining the observations necessary to decide between competing theories invoking trajectories of change is much smaller.

One area in which such data *are* routinely collected is in the construction of phylogenetic histories by systematicians — the bioscience descendents of the geologists that Babbage's model addressed. For our present purposes, these histories are important because differing evolutionary theories often make the same predictions concerning *current* states of affairs. This is because it is present-day phenomena that the theories attempt to account for. However, competing theories may make differing predictions concerning the *prior* states of affairs that have led to the current situation.

For instance, Ryan (1990) attempts to distinguish between theories that compete to account for the character of sensory systems and signalling behaviour extant in the natural world by constructing a phylogenetic tree for several species of frog. From this hypothetical history of speciation events Ryan attempts to discount certain theories whose predictions do not match the historical account he has constructed.

Evolutionary simulation modelling can contribute to this style of hypothesis testing in a way in which modelling methodologies that exclusively attend to equilibria cannot. An evolutionary simulation model provides an account of not only the behaviour of a system

---

[2] Artificial life is clearly not the only field concerned with dynamical systems. Related fields such as cybernetics and control theory, for instance, share similar interests.

at equilibrium, but also the behaviour which that system passed through before it reached this equilibrium. Such accounts of the trajectories followed by evolving populations prior to (potentially) achieving equilibria might be used to distinguish between competing theories.

This is not to say that data from simulations will simply augment data from phylogenetic reconstructions, but that the implications of evolutionary theories for the character of evolutionary trajectories might be clarified through evolutionary simulation modelling. The predictions resulting from such a clarification could then be compared to empirical data in the usual manner.

Although it is perhaps reasonable to expect the natural systems we see around us to be at stable equilibria given the evolutionary timescales involved, as with economic systems, there are situations in which evolving populations may consistently fail to reach equilibria, or in which the equilibria that evolutionary systems do reach are more complicated than point attractors. For example, Maynard Smith follows the passage quoted above with a prediction that *cyclic* attractors will be discovered to characterise much of the behaviour exhibited by players involved in asymmetric games. This prediction has been supported by the discovery of a species of lizard that occurs in three distinct morphs, each of which dominates one other morph, and is dominated by the remaining morph. Such a system is analogous to the parlour game scissors-paper-stone, in which playing one move consistently will never be a lasting strategy since any such strategy can be defeated (Sinervo & Lively, 1996; Maynard Smith, 1996).

In addition, theorists are coming to realise that many interesting games exhibit *multiple equilibria*. Unfortunately, naked game theory is unable to determine, given the existence of more than one equilibrium state, which equilibrium a population will arrive at. Additional criteria for deciding between equilibria (e.g., on grounds of parity, efficiency, etc.) have been offered (Harsanyi & Selten, 1988), but these often appear somewhat arbitrary. The natural solution to this *equilibrium selection problem* is to enquire which equilibria arise from which initial conditions (e.g., Binmore, Gale, & Samuelson, 1995a; Binmore, Samuelson, & Vaughan, 1995b) — a question addressing the dynamic character of the model.

Despite acknowledging that the behaviour of adaptive systems is inherently dynamic, and that attention to these dynamics might enable theorists to distinguish between competing theories, theorists often eschew the study of dynamic change. This accounts for the accent placed on the fixed points of models, whether they be

evolutionary stable strategies in biology, Nash equilibria in economics, or "exit points" in language change (Labov, 1994), rather than the general dynamic behaviour of such models. This is not to say that game theory and other formal approaches cannot tolerate limit sets of a higher order than constant trajectories, or have no consideration of initial conditions or transient behaviours. However, such matters are typically regarded as special cases that require additional analytic techniques if they are to be addressed at all (e.g., Maynard Smith, 1982, devotes an appendix to dealing with cyclic trajectories).

In contrast, evolutionary simulation modelling will principally concern itself with the *character* of a model's evolutionary dynamics rather than some "end product" of these dynamics, whether it be within a population of learning economic agents, a population of evolving creatures, or a population supporting a developing culture or language. From this inherently dynamic perspective, cyclic limit sets and start-up transients, drift and chaos, are on an equal footing with game theory's cardinal limit set, the point attractor.

Subject Matter

Perhaps the aspect of Babbage's model that most clearly distinguishes it from modern evolutionary simulation modelling is its subject matter. Several issues are related to this observation.

First, the overtly theological concerns of the debate Babbage engaged with are for the most part missing from contemporary science. Modern scientists now rarely struggle against matters of faith or religion in print (see Dawkins, 1998, however, for one contemporary example). Despite this, just as certain commentators claim that Babbage's model influenced the conception of God in the 19th century (Young, 1985), some researchers have taken modern artificial life to have implications for the relationship between religion and science (Helmreich, 1997).

Leaving religious matters to one side, it is also the case that whereas Babbage's system models an abstract *geological* process, the majority of current modelling work in this vein addresses *biological* subject matter. Despite the fact that modern evolutionary simulation modelling has tended to address biological questions, commentators have recognised its potential to address problems in other disciplines. However, it is probable that in 1836, Babbage and his contemporaries would not have recognised a difference between, for instance, geology and biology, since these fields and many others under the umbrella of natural philosophy had yet to part company and begin to specialise. Now that we understand that the mechanism by which organismic evolution proceeds (the differential transmission of genetic material) is not present in geological, economic, linguistic or psychological systems, should we be wary of the recent trend within artificial life of applying simulation models of adaptation to an ever broader class of topics and problems?[3]

Some of the non-biological disciplines that are beginning to adopt this approach have been exploring questions of change for some considerable time. For instance, both anthropology and linguistics involve a historical, diachronic or developmental element. The study of language change, for instance, predates Darwin's theory of evolution; indeed Darwin made use of research into the history and relatedness of languages in formulating his theory of natural selection.

In contrast, other disciplines have taken up the challenge of understanding the dynamics of their phenomena more recently. For example, economics (despite the prompting of Veblen, 1898, at the turn of the century) has only recently begun to consider the processes that might underly the evolution of economic systems (c.f., the inception of the Journal of Evolutionary Economics in 1991, that builds on the pioneering work of Joseph Schumpeter, e.g., 1934). Previously such matters were the preserve of historians of economics.

In the most extreme cases, novel scientific programs must be developed in order to pursue the implications of an adaptive systems perspective. Memetics, the study of the evolution of ideas, is one example of such a neonatal paradigm.

What links this rather disparate group of disciplines is their concern with the *dynamics of adaptation*. These dynamics are often studied by modellers using techniques developed specifically to deal with their indiginous problems, with no reference to evolution, or even with explicit rejection of evolutionary thinking. However, increasingly theorists are coming to see parallels between the dynamics underlying many different adaptive systems (c.f., the proliferation of journals, meetings and book titles involving *Evolutionary* as a leading adjective). Although, the systems that they study do not involve adaptation in the form of orthodox organismic evolution, nevertheless, the necessary ingredients for adaptation can be identified: competition for limited resources and heritable variation. For economic systems, the limiting resource is utility, the

variation exists at the level of economic strategy and inheritance occurs socially through some kind of learning mechanism. For linguistics, the limiting resource is language users, the variation exists at the level of language structure and inheritance again occurs socially through language transmission.

Evolutionary biologists enjoy an advantage over evolutionary simulation modellers dealing with non-biological systems in that they possess a detailed understanding of the mechanisms underlying biological evolution. As yet, comparable understanding of economic, linguistic, cultural, or psychological adaptation is relatively lacking. Non-biological adaptationists have made progress by appropriating mechanisms from biological evolution. However, the extent to which these adaptive mechanisms are merely being used metaphorically is often unclear. For example, the epidemiological notions of contagion used by Cavalli-Sforza and Feldman (1981) to model the spread of innovations may be considered to apply literally or metaphorically, depending, perhaps, on one's perspective. In contrast, it has been argued (Crick, 1989) that the concept of competitive exclusion that underlies "Neural Darwinism" (Edelman, 1987) does not constitute a literal example of natural selection in the brain due to crucial disimilarities between Darwinism and "Edelmanism". Fundamental research into the unique character of these non-biological adaptive systems must eventually reify or replace these placeholders (e.g. Gatherer, 1998).

The prospect of multiple levels of adaptation interacting with one another further complicates the picture. The learning mechanisms invoked by economists, linguists and cultural anthropologists are not fixed entities, but are themselves the results of adaptive evolutionary processes. Some effort has been made to model the interaction between learning and evolution (e.g., Hinton & Nowlan, 1987) and to apply the insights thus gained to non-biological adaptationist research programs (e.g., Kirby & Hurford, 1997). But until theoretical approaches to parallel, interacting adaptive systems (e.g., Laland, Odling-Smee, & Feldman, in press) can be shown to be sound, these modelling enterprises will be less secure than their biological forebears.

In summary, any research paradigm that studies the behaviour of systems of entities which interact with each other and their environment over time such that they change in an adaptive fashion is amenable to the evolutionary simulation modelling approach. While Babbage's application of a simulation model to what is now considered to be a non-biological topic presaged modern simulations of non-biological adaptive systems, his model is lacking in a sophisticated notion of geo-

---

[3]For example, Miller (1997) has recently appealed to the notion of self-organization in an attempt to resolve the apparent paradox presented by, on the one hand, the discovery of Hitler's indolence, and, on the other, the intensely structured order of the Third Reich. Miller quotes evidence from artificial life simulations that suggest that complex order may arise from simple local interactions, rather than requiring global co-ordination from some central executive.

logical adaptation. Indeed it is unclear whether geological systems can be classed as adaptive in the sense described above. It is apparent that modern descendents of Babbage's model are more concerned with the details of *how* adaptation occurs in natural systems, rather than merely demonstrating that some phenomena can be replicated on a machine. This notion will be pursued further in the next section.

Emergence and Individuals

The phenomenon at the heart of Babbage's model — discontinuity — is emblematic of the concerns of present-day researchers employing evolutionary simulation models. However, it is in its approach to non-linearity that Babbage's model departs most significantly from modern models.

There is a superficial resemblance between the catastrophist debate of the 19th century and the more recent dispute over the theory of punctuated equilibria introduced by Eldredge and Gould (1973). Both arguments revolved around the significance of what appear to be abrupt changes at geological time scales. However, while Babbage's dispute centered on whether change could be explained by one continuously operating process or must involve two different mechanisms (the first being geological processes, the second Divine intervention), Gould and Eldridge take pains to point out that their theory does not supercede phylogenetic gradualism, but augments it. They wish to explain the two apparent modes of action evidenced by the fossil record (long periods of stasis, short bursts of change), not by invoking two processes, but by explaining the unevenness of evolutionary change. In this respect, the theory that Eldridge and Gould supply attempts to meet a modern challenge: that of explaining non-linearity, rather than merely accommodating it.

Whereas Babbage's aim was merely to demonstrate that a certain kind of non-linearity was logically possible in the absence of exogenous interference, modern researchers are probing questions of how and why non-linearities arise from the homogeneous action of low-level entities, and what implications these non-linearities may have for the systems under examination. In order to achieve this, modern simulation models have had to move beyond the elegant but simplistic form of Babbage's demonstration.

This increased sophistication stems, in part, from the use of an explanatory strategy that invokes a "constructive" relationship between at least two relevant levels of description: a level of explicitly modelled individual atomic entities and a higher level of aggregate phenomena. The notion that the possibly complex and non-linear behaviour of higher-level phenomena emerges from the action of lower-level entities allows that an understanding of the constructive relationship that links them may be all that is required to account for the behaviour at the aggregate level of description. For example, a simulation model of traffic dynamics may involve routines that deal with the individual vehicles comprising the traffic without at any point explicitly invoking the higher-level phenomenon of "jams". Despite this, the simulation may be a useful way of modelling how jams behave if, through analysis of the simulation, an understanding of how the "emergent" phenomena derive from the atomic entities can be achieved.

It is important to explicate the differences between this kind of explanatory project and the task met by Babbage's model. Babbage did not need to model a system of entities at some atomic geological level of abstraction and then simulate the emergence of discontinuities in the geological record since his project was merely to demonstrate that a class of phenomena could exist in the absence of an element that had previously been considered necessary — external intervention. As such nothing hinged on the manner in which the natural phenomena actually did arise. This kind of explanation is a proof of concept of the type: "it is commonly thought that M is needed to generate P, but here is a model in which M is missing, but something that looks like P is exhibited". One of the challenges for modern evolutionary simulation models is to move beyond this kind of explanation, and reveal the constructive relationships that hold between atomic and emergent levels of description (see, e.g., Di Paolo, Noble & Bullock, this volume, for further discussion).

Babbage himself was not satisfied with merely demonstrating through simulation modelling that apparent discontinuities could be the result of unchanging mechanical processes. He also spent some time developing theories with which he sought to explain how specific examples of geological discontinuity could have arisen as the result of physical geological processes. One example of apparently rapid geological change that had figured prominently in geological debate since being depicted on the frontispiece of Lyell's *Principles of Geology* was the appearence of the Temple of Seraphis on the edge of the Bay of Baiae in Pozzuoli, Italy. The surface of the 42-foot pillars of the temple are characterised by three regimes. The lower portions of the pillars are smooth, their central portions have been attacked by marine creatures, while above this region the pillars are weathered but otherwise undamaged. These abrupt changes in the character of the surface of the pillars were taken by geologists to be evidence that the temple had been partially submerged for a considerable period of time.

For Lyell (1830), an explanation could be found in the considerable seismic activity which, historically, had characterised the area. It was well known that eruptions could cover land in considerable amounts of volcanic material and that earthquakes could suddenly raise or lower tracts of land. Lyell reasoned, that a volcanic eruption could have buried the lower portion of the pillars before an earthquake lowered the land upon which the temple stood into the sea. Thus the lower portion would have been preserved from erosion, while a middle portion would have been subjected to marine perforations, and an upper section to the weathering associated with wind and rain.

Recent work by Dolan (1998) has uncovered the impact that Babbage's own thoughts on the puzzle of the pillars had on this debate. Babbage, while visiting the temple, noted an aspect of the pillars which had hitherto gone undetected: a patch of calciated stone located between the central, perforated, section, and the lower, smooth, portion. Babbage inferred that this calciation had been caused, over considerable time, by calcium bearing spring waters which had gradually flooded the temple, as the land upon which it stood sank lower and lower. Eventually this subsidence caused the temple pillars to sink below sea-level and resulted in the marine erosion evident on the middle portion of the columns.

Thus Babbage's explanation invoked gradual processes of cumulative change, rather than abrupt episodes of discontinuous change, despite the fact that the evidence presented by the pillars is that of sharply seperated regimes. Babbage's account of this gradual change relied on the notion that a central, variable source of heat, below the earth's crust, caused expansion and contraction of the land masses above it. This expansion or contraction would lead to subsidence or elevation of the land masses involved. Babbage exploited the power of his new calculating machine in attempting to prove his theory, but not in the form of a simulation model. Instead, he used the engine to calculate tables of values that represented the expansion of granite under various temperature regimes. With these tables, Babbage could estimate the temperature changes that would have been necessary to cause the effects manifested by the Temple of Seraphis.

Here, Babbage is using a computer, and is moving beyond a gradualist account that merely tolerates discontinuities (i.e., his Bridgewater Treatise) to one that attemptes top explain them. However, his engine is not being employed as an evolutionary simulation model, but as a prosthetic calculating device. The complex, repetitive, computations involved in producing and compiling these tables of figures would normally have been carried out by "computers", people employed to make calculations manually. In replacing this error-prone, slow and costly manual calculation with his mechanical reckoning device, Babbage demonstrates the application of computing power to solving problems that are otherwise intractable. This use of computers has become widespread in modern science. Numerical and iterative techniques for calculating (or at least approximating) the results of what would be extremely taxing or tedious problems has become a mainstay of much academic practice.

In contrast, evolutionary simulation models of the kind discussed in this paper offer a new role for powerful computers. Where Babbage employed his machine to either (i) demonstrate that some natural phenomenon could be simulated in the absence of an element that had previously been deemed a necessary pre-requisite, or (ii) perform otherwise intractable calculations in order to support theory building/testing, modern simulation modellers attempt to move beyond these uses in pursuing an explanatory strategy in which simulations are used to directly explore explanatory theories proposed as ways of understanding how complex, aggregate behaviour might arise from the homogeneous action of lower-level entities.

A successful example of completing this explanatory task can be found in Di Paolo's (2000) model of the evolution of co-ordinated communication. Using an individual-based evolutionary simulation model featuring a spatially distributed population of agents and playing an action-response game, Di Paolo shows that even in situations for which game-theoretic considerations predict that co-ordination will be unstable, co-ordination may arise and persist. At this point the model has fulfilled the same explanatory role as that of Babbage's Bridgewater Treatise: an aggregate phenomena (co-ordinated communication) has been demonstrated in the absence of an element previously deemed necessary (i.e., an appropriate equilibrium in the underlying game). Di Paolo accounts for the presence and character of this co-ordinated communication by first drawing attention to the manner in which the spatial structure of the medium gives rise to clusters of individuals. Subsequent exploration reveals that individuals in the center of such clusters face a scenario that differs significantly from those at the periphery. Analysis demonstrates that the strategic asymmetry induced by this spatial organisation is sufficient to enable co-ordinated communication to persist as a stable strategy.

Notice that merely appealing to the existence of spatial clustering, and invoking the idea that co-ordination "emerges" from the interactions of game-players that exist in a spatially-structured medium would be to fall short of such a successful explanation. Such an appeal would in fact be closer to the apologeticist use of "miracles" as explanations for phenomena that would otherwise be inexplicable. Indeed construing emergent phenomena to be those aggregate phenomena for which, as yet, we have no reductionist explanation (Ronald, Sipper, & Capcarrère, 1999) would seem to invite this comparison.

In summary Babbage's model usefully demonstrates a simple explanatory strategy which modern evolutionary simulation modelling hopes to move beyond. In order to do so, evolutionary simulation models must do more than invoke the notion of emergence. The relationship between atomic and aggregate phenomena must be explicated successfully.

Conclusions

In many respects Babbage's model second-guesses aspects of modern evolutionary simulation modelling. His use of an ongoing computational program to model the dynamics of a natural system, his attention to a debate that would now be regarded as external to evolutionary biology, and his concentration on high-level non-linear phenomena and the ability of low-level processes to give rise to them, are all prominent features of modern individual-based simulations of adaptive systems.

However, Babbage's model does not address the manner in which exploring the dynamics of a simulation can explicate the constructive relationships that account for high-level aggregate phenomena in terms of the behaviour of systems of low-level atomic entities. The model does provide an example of a more simple explanatory strategy, that of demonstrating that some high-level phenomenon survives the removal of an element previously deemed to be a necessary prerequisite for it. Moving beyond this class of explanation has been identified here as an important challenge for modern evolutionary simulation modelling.

Babbage's model may resemble current artificial life in one final respect that is perhaps worth noting. Upon publication of the *Ninth Bridgewater Treatise*, the work was treated with some disdain. His demonstration of the power of automatic computing was generally regarded as impressive. But although his machine was a remarkable feat of engineering it was perceived to be a tool ill-suited to the job of natural philosophy and his model gained little credibility as a result. It was generally agreed to have overstepped some boundary. In contrast, Dolan's (1998) recent work has shown

that Babbage's empirically-driven theories on the geological processes responsible for the appearence of the Temple of Seraphis were readily taken on board by the eminent uniformitartian geologists of the time. Like Babbage's Bridgewater treatise, contemporary evolutionary simulation models may also be "uninvited". Perhaps the lessons we can learn from Babbage's work will ensure that current modelling efforts have more chance of gaining a better reception.

Acknowledgments

Thanks to Ezequiel Di Paolo, Henrietta Wilson and two anonymous referees for helpful comments.

References

Ashby, W. R. (1956). *An Introduction to Cybernetics.* Chapman & Hall, London.

Babbage, C. (1837). *Ninth Bridgewater Treatise: A Fragment* (2nd edition). John Murray.

Babbage, C. (1864). *Passages from the Life of a Philosopher.* Longman, London.

Binmore, K. (1987). Modelling rational players, part I. *Economics and Philosophy, 3*, 179–214.

Binmore, K. (1988). Modelling rational players, part II. *Economics and Philosophy, 4*, 9–55.

Binmore, K., Gale, J., & Samuelson, L. (1995a). Learning to be imperfect: The ultimatum game. *Games and Economic Behavior, 8*, 56–90.

Binmore, K., Samuelson, L., & Vaughan, R. (1995b). Musical chairs: Modelling noisy evolution. *Games and Economic Behavior, 11*, 1–35.

Brock, W. H. (1966). The selection of the authors of the Bridgewater Treatises. *Notes and Records of the Royal Society of London, 21*, 162–179.

Brooks, R. A. (1991). Intelligence without representation. *Artificial Intelligence, 47*, 139–159.

Bullock, S. (1997). *Evolutionary Simulation Models: On their Character, and Application to Problems Concerning the Evolution of Natural Signalling Systems.* Ph.D. thesis, School of Cognitive and Computing Sciences, University of Sussex, Brighton, UK.

Cavalli-Sforza, L., & Feldman, M. (1981). *Cultural Transmission and Evolution: A Quantitative Approach.* Princeton University Press, Princeton, NJ.

Crick, F. (1989). Neural edelmanism. *Trends in Neurosciences, 12*(7), 240–248.

Darwin, C. (1859). *The Origin of Species by Means of Natural Selection.* John Murray, London.

Dawkins, R. (1998). *Unweaving the Rainbow: Science, Delusion and the Appetite for Wonder.* Allen Lane, London.

Di Paolo, E. A. (2000). Ecological symmetry breaking can favour the evolution of altruism in an action-response game. *Journal of Theoretical Biology, 203*, 135–152.

Di Paolo, E. A., Noble, J., & Bullock, S. (2000). Simulation models as opaque thought experiments. This volume.

Dolan, B. P. (1998). Representing novelty: Charles Babbage, Charles Lyell, and experiments in early Victorian geology. *History of Science, 36*, 299–327.

Edelman, G. M. (Ed.). (1987). *Neural Darwinism: The Theory of Neuronal Group Selection*. Basic Books, New York.

Eldredge, N., & Gould, S. J. (1973). Punctuated equilibria: An alternative to phyletic gradualism. In Schopf, T. J. M. (Ed.), *Models in Paleobiology*, pp. 82–115. Freeman, Cooper and Co, San Francisco.

Frank, S. A. (1998). *Foundations of Social Evolution*. Princeton University Press, Princeton, NJ.

Gatherer, D. (1998). Why the thought contagion metaphor is retarding the progress of memetics. *Journal of Memetics — Evolutionary Models of Information Transmission, 2*. http://www.cpm.mmu.ac.uk/jom-emit/1998/vol2/. Last accessed 24.1.2000.

Harsanyi, J., & Selten, R. (1988). *A General Theory of Equilibrium Selection in Games*. MIT Press, Cambridge, MA.

Helmreich, S. (1997). The spiritual in artificial life: Recombining science and religion in a computational culture medium. *Science as Culture, 6*(3), 363–395.

Hinton, G. E., & Nowlan, S. J. (1987). How learning can guide evolution. *Complex Systems, 1*, 495–502.

Holland, J. H. (1975). *Adaptation in Natural and Artificial Systems*. University of Michigan Press, Ann Arbour. Reprinted by MIT Press, 1992.

Hyman, R. A. (1990). Whiggism in the history of science and the study of the life and work of Charles Babbage. *Annals of the History of Computing, 12*(1), 62–67.

Kirby, S., & Hurford, J. (1997). Learning, culture and evolution in the origin of linguistic constraints. In Husbands, P., & Harvey, I. (Eds.), *Proceedings of the Fourth European Conference on Artificial Life (ECAL'97)*, pp. 493–502. MIT Press / Bradford Books, Cambridge, MA.

Labov, W. (1994). *Principles of Linguistic Change*, Vol. Volume 1: Internal Factors. Blackwell, Oxford.

Laland, K. N., Odling-Smee, F. J., & Feldman, M. W. (in press). Niche construction, biological evolution and cultural change. To appear in Behavioral and Brain Sciences.

Lyell, C. (1830/1970). *Principles of Geology*. John Murray, London.

Malthus, T. (1798). *An Essay on the Principle of Population*. Patricia James (Ed.), 1989, Cambridge University Press, Cambridge.

Maynard Smith, J. (1982). *Evolution and the Theory of Games*. Cambridge University Press, Cambridge.

Maynard Smith, J. (1996). The games lizards play. *Nature, 380*, 198–199.

McCulloch, W. S., & Pitts, W. (1943). A logical calculus of the ideas immanent in nervous activity. *Bulletin of Mathematical Biophysics, 5*, 115–133.

Miller, J. (1997). Start The Week. BBC Radio 4. Discussion of the television documentary 'Nazis: A Warning from History".

Robson, J. M. (1990). The fiat and the finger of God: The Bridgewater Treatises. In Helmstadter, R. J., & Lightman, B. (Eds.), *Victorian Crisis in Faith: Essays on Continuity and Change in 19th Century Religious Belief*. Macmillan, Basingstoke.

Ronald, E. M. A., Sipper, M., & Capcarrère, M. S. (1999). Testing for emergence in artificial life. In Floreano, D., Nicoud, J.-D., & Mondada, F. (Eds.), *Advances in Artificial Life: Fifth European Conference on Artificial Life (ECAL'99)*, pp. 13–20. Springer, Berlin.

Ryan, M. J. (1990). Sexual selection, sensory systems, and sensory exploitation. *Oxford Survey of Evolutionary Biology, 7*, 157–195.

Schumpeter, J. A. (1934). *The Theory of Economic Development*. Harvard University Press, Cambridge, MA.

Sinervo, B., & Lively, C. M. (1996). The rock-paper-scissors game and the evolution of alternative male strategies. *Nature, 380*, 240.

Steels, L. (1994). The artificial life roots of artificial intelligence. *Artificial Life, 1*(1/2), 75–110.

Turing, A. (1952). The chemical basis of morphogenesis. *Philosophical Transactions of the Royal Society of London, Series B, 237*, 37–72.

Veblen, T. (1898). Why is economics not an evolutionary science? In Lerner, M. (Ed.), *The Portable Veblen*, pp. 215–240. Viking Press, New York. Collection published 1948.

von Neumann, J., & Burks, A. W. (1966). *Theory of Self-Reproducing Automata*. University of Illinois Press, Urbana, IL.

von Neumann, J., & Morgenstern, O. (1944). *Theory of Games and Economic Behavior* (Princeton, NJ edition). Princeton University Press.

Walter, W. G. (1963). *The Living Brain*. W. W. Norton, New York.

Young, R. M. (1985). *Darwin's Metaphor: Nature's Place in Victorian Culture*. Cambridge University Press, Cambridge.

Real artificial life: Where we may be

David H. Ackley

Department of Computer Science
University of New Mexico

Abstract

Artificial life research typically employs digital computers to *implement models* of living systems. However, there is now a growing if pre-theoretical feeling that computers, or perhaps the software running on them, *are themselves* some kind of living systems. Such a possibility can impact artificial life research in at least two ways: By highlighting that computers and communications networks can be *subjects*, as well as tools, for artificial life modelling, and by highlighting that insights, tools, and models from the life sciences can have explanatory, predictive, and design consequences for the construction of future computation and communications systems. This paper seeks perspective on such 'real artificial life', looking backwards and forwards at the rise of living systems in manufactured computer and communications systems.

Real artificial life

Strong notions of artificial life, as discussed in (Sober, 1992) for example, claim that computers and computer programs may not merely *simulate*, but actually *instantiate* living systems. When considering the small, closed-world models typical of most of our work in artificial life, the strong claim is often dismissed with variants of the 'confusion-of-levels' objection, pointing out that 'a model of a hurricane won't get you wet' and so forth. Supporters of the strong claim might argue that life and hurricanes are qualitatively different, or that a hurricane model *will* drench a *model of you*, or that some simulations might be different from but every bit as useful as "the real thing" (Dennett, 1978), and the discussions continue.

Here I want only to recognize that such unsettled issues exist, then set them aside and veer erratically towards the real and the concrete, accepting uncertain philosophical footings below. I suspect that if you asked the millions of computer-using people today to give an example of 'artificial life', a common answer would be 'computer viruses'. Whatever the ontological status of a simulation of a forest fire, a computer virus is as real as the computer programs and data files it infects. The Melissa virus invaded over 80,000 reported computers in under two weeks in 1999 (Vatis, 1999). Even with

no 'viral payload' whatsoever, Melissa clogged networks and incapacitated servers around the world; it had direct effects on the real people in the real world, it is not 'merely' a model. Much as we may want to distance ourselves from the ethical and moral questions, do we really want to argue that the Melissa virus is *not* artificial life?

We might dismiss Melissa on the grounds that viruses shouldn't be considered alive anyway. The natural world supports some reasonably clear distinctions between hardy 'living' systems that flourish in wide-ranging environments, versus mere 'parasitic replicators' that require a 'virus-friendly' environment, one willing not only to copy nearly any information at hand but then also to interpret the copied program regardless of what it does (Dawkins, 1991). But in that sense there is a strong kinship between living cells and manufactured computers in that both cell and computer interiors present tremendously virus-friendly environments. Moreover, the actual physicality of a computer itself may support richer notions of life, compared to the apparent insubstantiality of a computer program and the resulting sometimes anemic quality of purely software candidates for artificial life.

Though we may prefer to work entirely with small, manageable programs—executable models that are 'close to theory' in some sense—do we really want to argue that the rapidly expanding world of internetworked computer systems is *not* artificial life? Here, my aim is to run with the naive view to see what insights it may afford, beginning with the position that the connections between computation and life are genuine and the ramifications of that linkage are manifest not merely in small-scale closed-world computer models of natural world phenomena, but in the past, present, and likely future of manufactured computing devices on Earth.

There have been approaches to 'real world' computing issues from explicitly alife perspectives—the enterprise I am here calling 'real artificial life'—for example (Cohen, 1987; Spafford, 1992; Kephart, 1994; Ackley, 1996). Either from the outset or over the course of their development, all of those efforts had a pronounced emphasis on computer security. Indeed, one of the largest areas of impact of the artificial life mindset upon current and emerging computing practices is in the area of 'computer immune systems' for improving security and robustness, e.g. (Forrest et al., 1996; Kephart et al., 1995; Forrest et al., 1998). Later we suggest why this has been so.

In this exploration we seek understanding of what is happening around us as the internet grows. We seek leverage from the idea that the stunning explosion of computing power on earth over the last few decades is not unprecedented, but has antecedents in the long development of life of earth. We seek better approaches to the design of manufactured computing, possibilities of a different relationship between us and our computers, and between computers and each other.

This endeavor turns the original charter of artificial life almost precisely on its head. Rather than seeking to understand natural life-as-it-is through the computational lens of artificial life-as-it-could-be (Langton, 1989), we seek to understand artificial computation-as-it-could-be through the living lens of natural computation-as-it-is. The endeavor can fail; there is no *a priori* assurance the connections between life and computation are bidirectional, or that any identified points of contact will be substantial and specific enough to be usefully predictive.

Still, given the tremendous current and future impacts of the computer and communications hardware and software that we choose to design, and the current paucity of a systematic basis underlying computational design for robust security and privacy, and the relentlessly myopic market-driven approach that most often dominates deployment decisions, it seems worth some struggle to uncover new perspectives.

Outline

In keeping with the 'Looking backwards, looking forwards' theme of this Artificial Life conference, this paper contains takes on the past, present, and future of life in manufactured computing. The next section sets the stage with a familiar tale carried into these new circumstances, bringing us from the more or less the beginning up to more or less the present. Following that, we draw out of the present state of affairs some basic 'living systems' principles and guidelines as they increasingly seem to apply to networked computer systems, and present a few instantiations of such principles in research software development. We characterize computer source code as a principal genotypic basis for living computation, and consider methods of applying tools and techniques from biology to the understanding of computer systems. Finally we speculate briefly on possible futures for living computation and consider some possible implications of artificial life 'in the real'.

Living Computation: The Past

Depending on how inclusive one chooses to be, the history of manufactured computing can stretch back hundreds and even thousands of years. To keep things manageable, we open in the recent past, with machines that are in some sense directly traceable ancestors of the machines surrounding us in the world today.

An origins story

In five decades, manufactured computer technology has come from nowhere to account for 10% of the world's industrial economies (Dertouzos, 1997), a spectacular growth process that, along the way, has been playing out a tale as old as storytelling:

1940's-1950's: The age of innocence Though there are many plausible candidates for the title of 'first manufactured computer', it's fair enough to call the Colossus machine built at Bletchley Park an 'early computer'. In January of 1944, the Colossus Mk 1 began breaking messages encoded with the Nazi's Enigma cipher within hours of their interception. By the end of the war, 63 million characters of German messages had been recovered by ten Colossus machines (Sale, 1998).

This war-torn period is the age of innocence in this story because the computer itself knew nothing of the conflict; it was unaware of allies or axis, of friends, enemies, or spies. It was completely open and at the mercy of whoever could reach the main plug panel next to the paper tape reader. Of course, buried deep in F Block on the ground of the top-secret Bletchley Park operation, such naive trust was both reasonable and efficient.

1950's-1970's: The knowledge of good and evil By the debut of the Digital Equipment Corporation PDP-10 in 1967, computers had found many viable niches and were spreading, but they were still major capital investments. The rise of time-sharing operating systems such as TOPS-10—with the ability to provide useful services to many users at once—sent communications links snaking across campuses and through office buildings. Mere physical control over the central hardware was no longer sufficient for system security.

User accounts, passwords, the distinction between ordinary 'user mode' and privileged 'supervisor mode' operation—all sorts of mechanisms for creating fences, enforcing separations, and permitting limited sharing—date from this period. Now for the first time, the internal design and operation of the computer begins to reflect the divisions and separations of the world outside it. Trust is no longer implicit and automatic; now it is explicit and conditional. Now for the first time, both in hardware and sofware the computer itself manifests a distinction between *self* and *other*, and the system administrator appears explicitly in the design, playing a third role, that of the 'trusted other'.

1970's-1980's: You can't go home again In August of 1981, the IBM 'personal computer' was launched. At an astoundingly low price, compared to mainframe and minicomputers, the PC offered computing power, inexpensive and convenient floppy disks for bulk storage, a monitor and keyboard for interactive use, printers and peripherals. What it didn't offer was any of the complex and resource-consuming trust management mechanisms of the time-sharing systems. Gone were the user ids, gone the passwords, gone the protected memory, gone the distinction between user and supervisor modes of operation. "This is not a time-sharing machine!" we can imagine somebody arguing, "This is a computer for *one* person! Who's to protect *from?*"

And then of course, the PC had scarcely hit the market when the first PC computer viruses appeared and began to spread via those same convenient floppy disks, without the least bit of immunological defense by the PC.

1980's-2000's: The big big world Belatedly comes the realization that 'personal computing' does not imply 'isolated computing'—if anything just the reverse, compared to the dedicated-function mainframes of old—but the genie is out of the bottle. Now we find ourselves in the odd situation of having myriads of these so-called personal computers connecting to the internet with essentially no systematic defenses, no immune system, and precious little sense of self. The scale of it is quite staggering, with the internet growing from thousands to tens of millions of hosts just in the last decade. Epidemic waves of infections flash through the networked population faster and farther than any natural pathogen—constrained to transmission vectors involving slow-poke matter—ever did, even as the exploding size and thus value of the global network makes remaining unconnected increasingly untenable for many purposes.

It is no wonder that the computer security industry is booming. Widely underappreciated is the fact that current commercial security systems, constrained to deal with the deployed base of hardware and software, are for the most part exactly as ad hoc, awkward, and unreliable as a plastic bubble is as substitute for an immune system—it works better than nothing, but surely you wouldn't trust your life to it if there was a more integrated and robust alternative.

The missing element

The story so far is one of machines designed in the image of the *conscious minds* of their creators—a single strictly serial process proceeding step by deliberate step, with nothing changing except as directed by the processor, with no need for coordination or communication, no peripheral or preconscious awareness, and so forth. Boundary conditions could be applied only before the computation began; when started it simply ran until finished. The machine had no stimulus-response ability, no

interrupts, no hardware monitoring; as far as its functional repertoire was concerned, the machine not only could not control its 'body', it wasn't even aware it had one.

Living computation: The present

Thus has the evolution of manufactured computing systems to the present been backward to the evolution of natural computing systems such as the brain. The brain appeared only recently in the scope of the history of life on earth—and is no great shakes as a general-purpose algorithmic computing device—but from the beginning it has been richly interconnected to a body possessing sensory-motor apparatus that beggars anything we are currently able to manufacture at any price. The body has an extensive array of active and passive defense mechanisms, and the brain has extensive hardware support for threat assessment, triage, extrapolation, and rapid response.

The living computation perspective predicts that we are nearing childhood's end for computers, that current designs—still steeped in their innocent, safe, and externally-protected origins—will give way to designs possessing significant kinesthetic senses, defense and security at many levels, and with rich and persistently paranoid internal models representing the body and the stance of the body within the larger computational and physical environment.[1]

Manufactured computers began as 'pure mind designs' but were not thereby excused from the demands of existing in the physical world. With the rise of networked computing, and the rise of computers intended to survive in consumer environments and without the benefit of an expert human system administrator, the '*IOU: A body*' notes issued fifty years ago and more are now rapidly coming due.

Life principles for computation design

Many aspects, problems, and developments in current and emerging computing can be understood in this context. Here, we highlight several computational aspects of living systems that stand in contrast to traditional approaches to computing, then offer a few examples from current work illustrating some ways of that such strategies can apply in current and near-future systems.

Termination considered harmful In the fundamentals of computer science, an *algorithm* is typically defined as a *finite effective procedure* (Horowitz et al., 1997). A 'procedure'—description or plan of action to accomplish something—that is 'effective'—so it can actually be performed, step-by-step—and is 'finite'—so it will definitely

---

[1] I would have said this prediction was too obvious to be worth making, except that so much computer research, development, and deployment continues to ignore even the rudiments of robust system design.

stop eventually. What are we to make of, say, an operating system, whose number one mission in life is to never ever stop?

(Horowitz et al., 1997) acknowledge the importance of such non-algorithmic 'computational procedures', but only in the process of excluding them from further consideration. Here, in contrast, they are our central focus—so much so that calling a system 'living' may in some sense *mean* the system is running an *infinite effective procedure*. Such an approach, though certainly unconventional, is compatible with the 'Computation as interaction' approach to redefining introductory Computer Science (Stein, 1998), and more generally with the ongoing shift from algorithmic and procedural computation to object-based open-ended computation. Living systems potentially offer an epistemological framework surrounding and motivating these newer characterizations of computation.

Programs are physical Given the mathematical and algorithmic emphasis underlying computers and computation, it is unsurprising that most computer scientists and many other computing professionals tend to think of a computer program fundamentally in terms of the algorithm it implements, and to carry that sense of abstraction over to the computer program itself. While that is often a helpful, or at least harmless, way of thinking, it is at root a categorical error.

An actual, functioning computer program is literally a physical entity. It occupies actual physical space, in RAM, disk, or other media; while one computer program occupies some particular space, nothing else can be there. A functioning computer program consumes actual energy as it executes, producing waste heat that must be dissipated by a cooling system. It doesn't matter if the same amount of waste heat is produced by an operating CPU regardless of what program happens to be running, what is essential is that when some particular program is running on some particular CPU, the energy that is consumed is consumed at the behest of that program.

What distinguishes digital software from most other organizations of matter is that in a computer it can be copied so quickly and easily at high fidelity; DNA molecules in a cell of course have the same property. The flip side in both cases is that either can also be easily erased. Computers provide an additional feature for software that cells at least in principle could provide for DNA but to my knowledge do not: The ability to transduce losslessly between the relatively stable matter-based representation and ephemeral, fast-moving wave forms.

For internetworked computers only the physical boundaries between separately owned and administered systems are truly fundamental. Despite the much-touted *non*-spatiality of 'cyberspace', driven by precisely that lossless transduction, in fact each piece of hardware—each computer and disk, each wireline and switch—is localized in space, and each piece of hardware has an owner. For small personal computers no less than huge corporate computing facilities, physical access and legal ownership are the two key elements defining a player in the game.

The evolutionary mess Establishing and maintaining those boundaries, today, is a disaster. Modern computer security amounts to a porous hodge-podge of corporate firewalls, third-party virus scanners, hastily written and sporadically applied operating system patches, and a myriad of woefully inadequate password authentication schemes. Dozens or hundreds of break-ins occur daily, computer virus infections are everyday life for millions of computer users, and new virus detections are booming, even as more and more businesses and business transactions move to the internet, and as software companies race with each other to deliver new ways of embedding code inside data.

Though this situation is far from what one would expect to find in a thoughtfully engineered, deployed, and maintained system-of-systems, it is precisely what one expects to find in systems produced by blindly reactive evolutionary processes. For example, it is too easy simply to lay blame for computer viruses on the early mass-market computer designers, even considering the body of knowledge available from the era of time-sharing. On the contrary, the tale of the PC and the virus is one of evolution in action: When the machine was designed, there were essentially no viruses in the wild—there was no wild to speak of—and code exchanges were either in large system-administrator-managed mainframe environments or in the tiny computer hobbyist community. Why would anybody waste design and manufacturing resources, increase costs greatly, and sacrifice time-to-market, just to defend against a non-existent problem?

Having humans in the loop, with all our marvelous cognitive and predictive abilities, with all our philosophical ability to frame intentions, does not necessarily change the qualitative nature of the evolutionary process in the least. Market forces are in effect regulated evolutionary forces; in any sufficiently large and distributed system, nobody is in charge, and evolutionary forces are constantly at work.

Living computation: Examples

This section provides a few examples drawn from the author's work, illustrating ways that living computation principles can be applied in the implementation of near-term software systems, and how some of the analytical tools of the life sciences can find applications in understanding the expanding software systems around us.

For	Primary purpose
End users	Peer-to-peer security-aware chat system / graphical MUD
'Agents' and artificial life	Scalable distributed environment with human interactions
Real artificial life	Investigate life-like computation and communications strategies

Select version dates		v0.4+ algorithms	
v0.4.0=	3/30/00	Authentication:	El Gamal
v0.2.6=	8/ 4/98	Encryption:	Twofish
v0.1.91=	9/15/97	PKI:	included
v0.1.80=	4/11/95	Transport:	TCP/UDP
v0.1.24=	10/26/92	Addressing:	IPv4+PKI

	File counts and code size (by wc)			
Content	Files	KLines	%tot	%new
C code	721	187.3	49.1	57.0
history	19	50.4	13.2	92.9
documentation	80	34.4	9.0	60.7
Perl code	73	33.0	8.6	100
development	153	24.0	6.3	41.5
Tcl/Tk code	37	18.1	4.7	100
assembly code	145	15.2	4.0	0.0
C++ code	43	10.4	2.7	93.4
uncategorized	24	6.7	1.7	1.2
ccrl code	19	2.3	0.6	100
empty	1	–	0.0	–
Total	1315	381.7	100.0	64.9

Table 1: Views of the **ccr** research prototype: Purpose, history, technology, and 'genetic' content. '%new' refers to code developed within the project rather than acquired from the environment.

Living computation by design

For several years, we have been building a series of research prototypes to explore 'life-like' design strategies for networked computations. This **ccr** project overall has been introduced previously (Ackley, 1996); here we provide only the briefest overview, then draw a few examples from the current system design, and then use the software itself as a sample object of study.

At its core **ccr** is a code library for peer-to-peer networking with research emphases on security, robust operation, object persistence, and run-time extensibility. Built upon the core libraries are the graphical user interface world **ccrTk**, and the text-only world **ccrt**. Table 1 provides views of the system along several dimensions: Its major functions, history, and technology, along with a breakdown of the current contents of the system software. Figure 1 visualizes the system in the linear 'tar file' format in which it normally reproduces. The 14Mbytes in the 'genome' are depicted in terms of 'coding' regions that are the actual source code; 'promoter' regions contain metadata guiding the migration of code segments during early development; 'garbage' regions are simply wasted by the tar file format. Successive 'zooms'

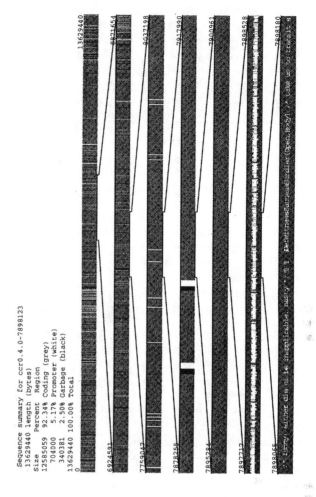

Figure 1: Views of the ccr v0.4.0= genome

in Figure 1 across some five orders of magnitude ultimately resolve individual bytes, illustrating the scale of a modest-sized system like **ccr**.

A fundamental design element of **ccr**, arguably a *sine qua non* for an independent living system, is its peer-to-peer communications architecture: A **ccr** 'world' *can* communicate with other **ccr** worlds, but it doesn't *require* any other worlds to function; each **ccr** world is both 'client' and 'server' in interactions with other worlds.

Self-reliance Taking seriously the independent living system approach implies that we must make as few assumptions about the 'outside world' as we reasonably can, and be as self-reliant as we can. This caution extends through all levels of the system and along multiple dimensions. One approach would simply avoid all external influences, but when it comes to communications and communications risk, fundamentally we are damned if we do and damned if we don't: We might have had warning of a threat in time to avoid it, had we been allowing external factors to affect system behavior.

Most work in computer-based communication has focused on efficiency and power rather than safety. High-speed communications protocols, for example, can now supply data as fast or faster than most processes are ready to digest it, and even the ubiquitous TCP/IP internet protocol requires reacting to a connection attempt even just to ignore it, leading in part to the 'denial of service' attacks currently occurring on the internet. Similarly, nearly all software for personal computers focuses heavily on adding ever more programmed abilities and 'features,' even as automated network access is woven more deeply into the system rather than being more isolated, with the predictable results that your own computer's processing and data can be stolen out from underneath you by anyone simply by 'speaking' to your machine in particular ways.

Natural living systems have a large variety of mechanisms for evaluating interactions, assigning degrees of trust, and allowing only limited influences in proportion to estimated risk. Complex chemical signals and hard-to-duplicate bird songs, for example, increase confidence that messages are genuine; between **ccr** worlds cryptography serves that purpose, among others, helping on the one hand to establish identity and increasing (in particular) the sender's cost to generate a valid message on the other hand.

Ritualized interactions such as mating behaviors allow gradual and mutual stepping-up of trust and acceptable risk as confidence in identities and intentions grows. A **ccr** world wishing to communicate, likewise, engages in protocols designed to capture as much of the value of communication as possible while exposing the world to as little risk as possible. Here, we describe a few of those protocols and mechanisms, to provide concrete examples of living computation design strategies in artificial systems.

A cautious "Hello World" The protocol by which **ccr** worlds establish a communications link with each other moves through several stages, with a gradually increasing 'message size limit' allocated to the connection as the stages are successfully negotiated. Note that all of the strategies discussed here are in addition to the mechanisms provided by the TCP and IP version 4 transport mechanisms. Initially a **ccr** world will read only small messages, of no more than 128 bytes, from an incoming connection. Such messages are sufficient to exchange version information and establish a cryptographic 'session key' for the connection, which both establishes identities and insulates the channel from eavesdroppers and intruders. Any attempt to send a larger message causes the connection to be cut at the receiving end.

If this initial stage succeeds, more trust is warranted, and the incoming message size limit is raised to 1K-byte, which is enough to complete the connection establishment protocol. At the successful conclusion of the 'greeting ritual', the message size limit is raised to 100K-bytes. Note that while that number is high enough for most typical channel uses, it is much less than it could be. Higher limits, if desired, can be set by deliberate act of the **ccr** world owner. This strategy is typical of **ccr**'s self-protection mechanisms. Even once a remote **ccr** world is identified and the communications channel secured, still only limited trust is granted to the channel because inconvenient or dangerous things still may happen, either due to a user's mistake, or to malicious intent, or to bugs in the code.

Friction as friend To help guard against such possibilities, in addition to the message size limit, a **ccr** world also places rate limiters on every established network connection, to reduce the risk of accepting and acting on network communications, whatever they may be. The *inbound bandwidth limiter* specifies the maximum average rate in bytes/second that one world is willing to read data from another world, with an effective default of 2Kb/s, which is enough to allow most normal channel usage to flow unimpeded. If more than 2Kb/s is supplied, the receiving world simply *doesn't read it* until enough time has passed that the overall bandwidth doesn't exceed the set bandwidth limit.

In addition to the communication bandwidth control, a **ccr** world also maintains a *processing bandwidth limiter* for each world that is in contact. For two worlds to communicate meaningfully, it is necessary that data sent from one world *somehow* affect the processing that occurs on the other world; therein also lies the risk. Although **ccr** uses several mechanisms to control what operations remote worlds can perform locally—for example, by controlling the language used to express the messages the receiving world will choose to read—here we focus only the processing time control. Each operation a **ccr** world can perform has an associated cost in terms of "work units". Each computation request received from another world is tagged locally with the identity of the requesting world, and as processing proceeds, work units are logged against the remote world. As with the bandwidth limiter, if processing on behalf of a remote world exceeds a specified rate, then the local world delays accepting further input from that world until the overall processing rate drops within established limits.

With a default of 20 work units/world/second, once again most normal inter-**ccr** operations are at most minimally impacted by the limiter. In pathological situations, however, the protection they afford can be significant. Figure 2 illustrates their effect via two simulated denial-of-service attacks launched by World 'B' against World 'A'. World 'A's data appears in the upper graph

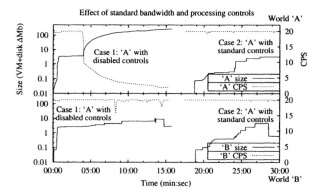

Figure 2: Using rate-limiters on bandwidth and processing time to mitigate denial-of-service events. *Case 1:* World 'B' floods world 'A'—with defenses disabled—starting at 00:00. *Case 2:* World 'B' floods world 'A'—with defenses in place—starting at 18:00. See text for details.

and World 'B's in the lower graph; in both cases the solid lines represent memory usage and are measured in megabytes of growth (the left side y-axis labels), and the dotted lines represent the currently main loop processing rate in cycles/second (the right side y-axis labels). A rate of 20CPS is the target 'heartbeat' rate of a **ccr** world; it will 'sleep' if it doesn't require the full 50ms to complete all scheduled tasks.

Before the first attack, which starts at zero minutes into the displayed Case 1 data, World 'A's limiters were effectively disabled by setting the acceptable rates to extremely large values. 'A' grows by several megabytes quickly and then stabilizes for several minutes, and then shortly before the 5:00 mark it begins growing exponentially and its cycle rate crashes. The simulated attack was stopped shortly after 15:00 minutes, at which point 'A's size had exceeded 200Mb and it had nearly exhausted the swap space on its machine. Both worlds were then restarted and the attack was repeated, this time with the normal values for the limiters. Now in Case 2 'A's growth rate is slower, and remains under 10Mb, and it turns a steady 20CPS throughout the event: The controls are performing effectively.

The growth behavior displayed by World 'B' through the two events was somewhat unexpected: It was much *less* different between the two cases than anticipated. The 'attack' was performed by instructing a character in world 'B' to 'speak' 10Kb strings of random numbers approximately 50 times per second. In both cases 'B's size gradually grows. In Case 2 'B' grows because 'A' is deliberately delaying reading from 'B', to protect itself, while 'B' continues to speak, causing 'B's 'pending

output' buffers to expand. In Case 1 the same effect occurs, but there it is because 'A' is in severe distress from memory thrashing and processing overload, and its cycle rate has crashed, so there also it is not reading from 'B' as frequently. In both cases 'B's size eventually drops, when the communication channel was closed by yet another watchdog within the system. That mechanism injects 'Are you alive?' messages into communications streams at random intervals and times how long the response takes; if no response is received after several minutes the connection is killed.

Software genetics

The amount of code in the world is exploding, as is the amount of code in any given program. Today, essentially all application programs take advantage of prewritten libraries of code—at the very least the runtime library of the chosen programming language(s), and usually many other existing components as well, for graphical interfaces, database access, parsing data formats, and so forth. The analogy to natural genetic recombination is quite strong: Computer source code as genome; the software build process as embryological development; the resulting executable binary as phenotype. The unit of selection is generally at the phenotypic level, or sometimes at the level an entire operating system/applications environment.

A main place where the analogy breaks down is that in manufactured computers, but not in the natural world, there are two distinct routes to producing a phenotype. The extreme 'copy anything' ability of digital computers means that source code is not required for to produce a duplicate of a phenotype. Source code *is* a requirement, in practical terms, for significant evolution via mutation and recombination.

Commercial software is traditionally distributed by direct copying of precompiled binary programs while guarding access to the 'germ line' source code, largely to ensure that nobody else has the ability to evolve the line. In that context, the rapidly-growing corpus of 'open source' software is of particular interest. With source code always available and reusable by virtue of the free software licensing terms, an environment supporting much more rapid evolution is created. The traditional closed-source 'protect the germ line at all cost' model is reminiscent of, say, mammalian evolution; by contrast the free software movement is more like anything-goes bacterial evolution, with the possibility of acquiring code from the surrounding environment and in any event displaying a surprising range of 'gene mobility', as when genes for antibiotic drug resistance jump between species. There is therefore reason to expect open source code, on average, to evolve at a faster rate than closed source, at least up to some level of complexity depending on design where the chances of new code being useful rather than disruptive become negligible.

As software systems grow, and software components swallow each other and are in turn swallowed, and older 'legacy systems' are wrapped with new interface layers and kept in place, we are arriving at the situation where actually *reading* fragments of source code tells us less and less about how—if at all—that code ever affects the aggregate system behavior. As this trend accelerates, tools and techniques from biological analysis are likely to be increasingly useful.

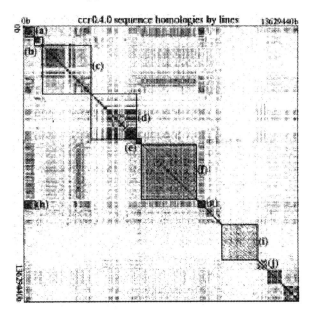

Figure 3: ccr genome vs itself. Darker represents greater homologies. See text for discussion.

Figure 3 presents another view of the **ccr** genome, using the 'dotplot' program (Helfman, 1996) for visualizing large data sets. This view shows the **ccr** genome plotted against itself, using lines of code as the fundamental u-nit of similarity; there is a black line representing perfect overlap down the main diagonal. Dark areas significant-ly off the main diagonal represent similarities between widely-separated code regions; squares on the diagonal represent 'cohesive' regions with more similarity within than without. 'Looking under the hood', we find that often such regions either are or are components of larger functional units—'genes'—within the genome. Several such genes have been highlighted with black outlines: Region *(a)* codes for **ccr**'s web server/client program; *(b)* is the configuration system that guides the overall system ontogeny; regions *(c)–(e)* are separately-evolved code segments (for JPEG images, long integer manipu-lation, and regular expressions, respectively) that have become incorporated, essentially unchanged, into **ccr**. The large region near the middle of the genome *(f)* con-tains the **ccr** core components themselves. Region *(g)* deals with processing animated images; interestingly, it

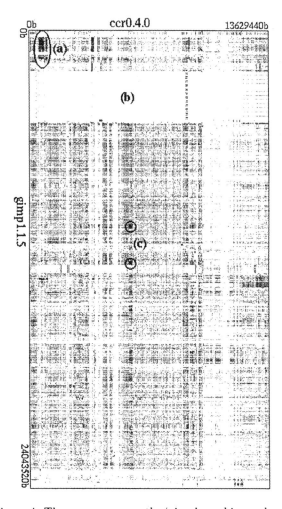

Figure 4: The **ccr** genome vs the 'gimp' graphics package genome. See text for discussion.

shows a perceptible overlap *(h)* with the web code *(a)* even though their functions are very different and in fac-t they are expressed in different languages (C++ *(a)* vs C *(g)*)—but, it turns out, both were created by the same programmer. Semi-automated project historical notes *(i)* display a distinctive pattern, as does program-generated Postscript documentation *(j)*.

Figure 4 compares **ccr** to a different 'species'—the 'GNU Image Manipulation Program' (GIMP) (Mattis and Kimball, 2000). There are fewer dark regions, re-flecting a generally lower degree of similarity between the code sequences. The dark region *(a)* is aligned with Figure 3*(b)* and reveals that both systems use evolved variants of the same 'autoconf' developmental control system—though the region is rectangular indicating that GIMP's instantiation of the code is bigger than **ccr**'s. A large 'internationalization' segment *(b)*—allowing GIMP to operate in some eleven natural languages—is strikingly different than almost everything in **ccr**.

A surprising element in this comparison are two short lines *(c)*—black diagonals indicating identical sequences. The relevant sequence is the GNU regular expression package, which is used in both **ccr** (Figure 3*(e)*) and the GIMP, and is a good example of a highly useful gene incorporated into multiple different applications out of the free software environment. *Two* short black lines, aligned vertically, show that the GIMP contains two identical copies of the GNU regex gene. Rather than being wasteful—as traditional software practices might have construed it—such gene duplication reduces epistasis and increases evolvability; one copy is deep in the application core and the other in a relatively peripheral 'plug-in scripting' segment. Furthermore, from one point of view it's odd that this gene appears at all, because regular expression support is actually a required part of any POSIX-conforming operating system. Yet, like complex living systems, both **ccr** and the GIMP acquired captive regex 'organelles' instead, reducing their vulnerability to environmental variations and increasing their ecological range.

Real artificial life: The Future

I have argued that the similarities between living systems and actual computational systems are too overwhelming to dismiss. I've suggested that many of the differences between manufactured computation and natural living systems, both superficial and substantive, have arisen from the complementary circumstances surrounding the origins of the two technologies, but that both approaches must address the same imperatives and are therefore on converging evolutionary paths.

If these arguments even mostly hold up, then we can predict major changes in future architectures of manufactured computation. A recurring theme here has been that many of the defining claims for digital computation and communication—ranging from 'instant communication' to 'frictionless commerce' to 'location transparency', and possibly even the notion of 'general-purpose computing'—simply are too good to be true, having been purchased at the expense of utterly ignoring the basic tenets of self-versus-other and local self-reliance. It will not continue this way. Even though many parties would like to have control over individual hardware systems—ranging from software and hardware manufacturers to internet and application service providers to governments and regulatory agencies—in the end the geometry of physical space will assert itself over 'cyberspace' as computing systems become aware of themselves and their universe.

What we can do The hypothesis is that mass-market computer communications systems, at least, will become more and more like natural living systems. The scope and nature of that evolution is far from clear, and ar-

tificial life research and researchers can contribute significantly to the process, bringing fresh technical, biological and philosophical perspectives to the growth and development of the network, a process which only with great naiveté can be regarded simply as engineering. Alife models developed for natural systems can be and will increasingly need to be applied to the hardware, software and data of the internet. Biological principles, hypotheses, and scaling laws may find analogs in the growing computational ecosystem. There is much to be done.

Where we may be Over sixty years of development, computer programs have grown from a few bytes to hundreds of megabytes, from a few lines of assembler source code to tens of millions of lines of complex programming language code. We have been living with the 'software crisis'—which usually means rapidly increasing software development and maintenance time and cost, often with decreasing reliability—now for several decades, and a number of proposed solutions have come and gone.

Especially over the last fifteen years, 'object-oriented programming' (Meyer, 1988; Booch, 1994, and many others) has emerged in various forms as a durable programming methodology. There are debates over technical details, and factionalism surrounding specific object-oriented programming languages, but the overall approach continues to gain design wins for more and larger projects when significant new code is needed.

From the living computation perspective, one interpretation of that history is difficult to resist. In coarsest outline the arc of software development paralleling the evolution of living architectures: From early proteins and autocatalytic sets amounting to direct coding on bare hardware; to the emergence of higher level programming languages such as RNA and DNA, and associated interpreters; to single-celled organisms as complex applications running monolithic codes; to simple, largely undifferentiated multicellular creatures like SIMD parallel computers. Then, apparently, progress seems to stall for a billion years give or take—the software crisis.

Some half a billion years ago all that changed, with the 'Cambrian explosion' of differentiated multicellular organisms, giving rise to all the major groups of modern animals (Gould, 1989, for example). Living computation hypothesizes that it was primarily a programming breakthrough—combining what we might today view as object-oriented programming with plentiful MIMD parallel hardware—that enabled that epochal change.

Where we may be is in the leading edge of the Cambrian explosion for real artificial life. If so, there is of course no certainty, from our vantage point today, how or how quickly the process will play out. On the other hand, in this interpretation we are aligning perhaps three billion years of natural evolution with perhaps a century of artificial evolution.

We are living in interesting times.

Acknowledgments

Many of the ideas in this paper have been developed and elaborated in collaboration with Stephanie Forrest; but for exigencies of time and deadlines she would have been a co-author. The **ccr** development group at UNM has included Adam Messinger, Brian Clark, Neal Fachan, Nathan Wallwork, Suresh Madhu, Peter Neuss, Steve Coltrin, and Jeff Moser. This work was supported in part by an NSF research infrastructure grant, award number CDA-95-03064, and in part by the DARPA Intelligent Collaboration and Visualization program, contract number N66001-96-C-8509.

References

Ackley, D. H. (1996). ccr: A network of worlds for research. In Langton, C. and Shimohara, K., editors, *Artificial Life V. (Proceedings of the Fifth International Workshop on the Synthesis and Simulation of Living Systems)*, pages 116–123, Cambridge, MA. The MIT Press.

Booch, G. (1994). *Object-Oriented Analysis and Design WIth Applications*. Addison-Wesley Object Technology Series. Addison-Wesley, 2nd edition.

Cohen, F. (1987). Computer viruses. *Computers & Security*, 6:22–35.

Dawkins, R. (1991). Viruses of the mind. On the web 1/10/2000 at http://www.santafe.edu/ shalizi-/Dawkins/viruses-of-the-mind.html.

Dennett, D. C. (1978). Why you can't make a computer that feels pain. *Synthese*, 38(3):415–456. Also in *Brainstorms: Philosophical Essays on Mind an Psychology*, Bradford Books.

Dertouzos, M. L. (1997). *What Will Be: How the New World of Information Will Change Our Lives*. Harper Edge, San Francisco, USA.

Forrest, S., Hofmeyr, S., Somayaji, A., and Longstaff, T. (1996). A sense of self for unix processes. In *Proceedings of the 1996 IEEE Symposium on Computer Security and Privacy*. IEEE Press.

Forrest, S., Somayaji, A., and Ackley, D. H. (1998). Building diverse computer systems. In *Sixth Workshop on Hot Topics in Operating Systems*.

Gould, S. (1989). *Wonderful life: The Burgess Shale and the nature of history*. W.W. Norton.

Helfman, J. (1996). Dotplot patterns: A literal look at pattern languages. *Theory and Practice of Object Systems*, 2(1):31–41.

Horowitz, E., Sahni, S., and Rajasekaran, S. (1997). *Computer algorithms/C++*. W. H. Freeman Press.

Kephart, J. O. (1994). A biologically inspired immune system for computers. In Brooks, R. and Maes, P., editors, *Proceedings of Artificial Life IV*, Cambridge, MA. MIT Press.

Kephart, J. O., Sorkin, G. B., Arnold, W. C., Chess, D. M., Tesauro, G. J., and White, S. R. (1995). Biologically inspired defenses against computer viruses. In *IJCAI '95. International Joint Conference on Artificial Intelligence*.

Langton, C. G. (1989). Artificial life. In Langton, C. G., editor, *Artificial Life (Santa Fe Institute Studies in the Sciences of Complexity, Vol VI)*, pages 1–47, Reading, MA. Addison-Wesley.

Mattis, P. and Kimball, S. (2000). The GNU Image Manipulation Program home page. At http://www.gimp.org/.

Meyer, B. (1988). *Object-oriented Software Construction*. Prentice Hall.

Sale, A. E. (1998). The colossus of Bletchley Park: The German cipher system. On the web 10/31/1999 at http://www.inf.fu-berlin.de-/~widiger/ICHC/papers/Sale.html. Presented at the International Conference on the History of Computing, Paderborn, Germany.

Sober, E. (1992). Learning from functionalism: Prospects for strong artificial life. In Langton, C. G., Taylor, C., Farmer, J. D., and Rasmussen, S., editors, *Artificial Life II*, pages 749–765, Reading, MA. Addison-Wesley.

Spafford, E. H. (1992). Computer viruses—a form of artificial life? In Langton, C. G., Taylor, C., Farmer, J. D., and Rasmussen, S., editors, *Artificial Life II*, pages 727–745. Addison-Wesley, Redwood City, CA.

Stein, L. A. (1998). What we've swept under the rug: Radically rethinking CS1. *Computer Science Education*, 8(2):118–129.

Vatis, M. A. (1999). Statement for the record before the House Science Committee, Subcommitte on Technology. Comments by the Director of the National Infrastructure Protection Center. On the web at http://www.house.gov/science/vatis_041599.htm.

Simulation Models as Opaque Thought Experiments

Ezequiel A. Di Paolo[1], Jason Noble[2] and Seth Bullock[3]

[1]GMD—German National Research Center for Information Technology (AiS)
[2]Center for Adaptive Behavior and Cognition, MPI für Bildungsforschung, Berlin
[3]Informatics Research Institute, School of Computer Studies, University of Leeds
Ezequiel.Di-Paolo@gmd.de, noble@mpib-berlin.mpg.de, seth@scs.leeds.ac.uk

Abstract

We review and critique a range of perspectives on the scientific role of individual-based evolutionary simulation models as they are used within artificial life. We find that such models have the potential to enrich existing modelling enterprises through their strength in modelling systems of interacting entities. Furthermore, simulation techniques promise to provide theoreticians in various fields with entirely new conceptual, as well as methodological, approaches. However, the precise manner in which simulations can be used as models is not clear. We present two apparently opposed perspectives on this issue: simulation models as "emergent computational thought experiments" and simulation models as realistic simulacra. Through analysing the role that armchair thought experiments play in science, we develop a role for simulation models as *opaque thought experiments*, that is, thought experiments in which the consequences follow from the premises, but in a non-obvious manner which must be revealed through systematic enquiry. Like their better-known transparent cousins, opaque thought experiments, when understood, result in new insights and conceptual reorganisations. These may stress the current theoretical position of the thought experimenter and engender empirical predictions which must be tested in reality. As such, simulation models, like all thought experiments, are tools with which to explore the consequences of a theoretical position.

Introduction

Imagine that you have constructed an artificial life system in which the interactions of many simple agents give rise to complex patterns at the global level. Suppose that these complex patterns remind you of some real-world phenomenon, such as termite nest construction or the behaviour of human investors on the stock market. How do you go about demonstrating the scientific value of your work? There exists a range of answers to this question. At one extreme is the "strong artificial life" position, which suggests that your work is not a *model* of nest construction or investment behaviour, but an *instantiation* of the phenomenon (and

hence more than just a simulation). Accepting this view means seeing your piece of work as a new data point to be added to those found in the natural world; scientific investigation proceeds as a search for common features across natural and artificial versions of the phenomenon. An opposing position states that what you have done can have no scientific value, as it is ultimately just a computer program that rearranges symbols in a logical fashion, and as such cannot arrive at new knowledge. This is the idea that if the premises or input are already known, then the conclusions or output cannot constitute an empirical discovery.

We are unhappy with both of these extremes. However, as will be argued below, we are also not content with some of the intermediate positions that have been advanced by artificial life researchers over the past decade. Our goal in this paper is to clearly spell out one way in which the type of systems characteristic of artificial life can make a contribution to science as simulation models — note that we are concerned only with artificial life research conducted in a scientific mode, and will have nothing to say about work directed towards other goals, such as engineering or education.

After reviewing previous attempts to describe the scientific role of artificial life simulations, we will develop our own position through an extended comparison with thought experiments. Our view, in brief, is that although simulations can never substitute for empirical data collection, they are valuable tools for re-organising and probing the internal consistency of a theoretical position. Because simulations are complex, their internal workings are opaque: it is not immediately obvious what is going on or why. This opacity means that researchers must spend time developing and testing a theory of the simulation's operation, before relating this internal theory back to theories about the world, and, ultimately, to the world itself through empirical investigation. Links in this chain are often missing in current artificial life research.

Previous Suggestions:
Babies and Bathwater

Several authors have attempted to carve out a niche for the style of simulation modelling pioneered within artificial life (Bonabeau & Theraulaz, 1994; Fontana, Wagner, & Buss, 1994; Ray, 1994; Taylor & Jefferson, 1994; Miller, 1995; Sober, 1996; Bedau, 1999; Maley, 1999). Two basic approaches can be identified. First, the worth of artificial life models is sometimes located in their unique ability to explore some important class of subject matter. Second, artificial life modelling is sometimes claimed to offer new and perhaps superior techniques with which to attack problems that would traditionally be dealt with using existing formal logico-mathematical approaches. These two lines of argument (which are put forward by strongly overlapping groups of authors) will be reviewed in caricature below. Subsequently, two perspectives on the role of evolutionary simulation models in scientific enquiry will be presented. The first takes such models to be "emergent thought experiments" (Bedau, 1999), whilst the second considers their use to form part of a conventional cycle of hypothesis generation and testing dubbed the "physics model" (Kitano, Hamahashi, Kitazawa, Takao, & Imai, 1997).

Unique Object of Enquiry

Artificial life researchers have sometimes claimed that the simulation models which they construct enable them to explore phenomena which lie beyond the ambit of more traditional modelling techniques. The argument runs that since simulations are built from low-level mechanisms (e.g., simulated organisms) which instantiate low-level behaviours (e.g., locomotion), they have the potential to explore the nature of phenomena which although not straightforwardly instantiated by these low-level mechanisms, are nonetheless robust aspects of their high-level, aggregate behaviour (e.g., flocking). This intuition underlies the assertion by Miller (1995), Bonabeau and Theraulaz (1994) and Taylor and Jefferson (1994) that the strength of artificial life simulation models lies in their ability to model natural phenomena which are complex, emergent, and/or self-organising, and that it is through such modelling that artificial life simulations will prove to be most useful since these phenomena are hard to model using previously existing techniques. Indeed some claim that "analytic approaches are certainly doomed" (Bonabeau & Theraulaz, 1994, p. 315).

However, there are potential dangers involved in closely associating the utility of a modelling technique with its application to a specific set of new concepts (e.g., the role of self-organisation in evolution). Such an association may lead to the growing conviction that an important class of phenomena can only be modelled using one particular approach (other approaches being "doomed"), and a reduced tendency to engage with alternative modelling enterprises will be the result. This could seriously impede our ability to construct unifying explanations of the phenomena, since doing so requires that we reconcile the conflicting suggestions of alternative modelling approaches and not merely dismiss them.

Furthermore, the validity of a modelling practice should not stand or fall on the worth of a particular theoretical idea to which it is applied. Recently this issue has been the topic of debate in ecology: to what degree is the methodology of individual-based modelling wedded to a philosophical position regarding the nature of the systems being modelled? Whereas Judson (1994) asserts that a growing awareness of ecosystems as chaotic systems has led directly to the adoption of individual-based simulation modelling practices (which are more able to capture the complexities of such systems), James Bullock (1994) has countered that accepting the utility of these modelling techniques is in no way dependent on accepting this particular perspective on ecosystems. He offers an alternative benefit to this kind of model by suggesting that individual-based modelling can augment more traditional modelling techniques within orthodox theoretical frameworks (see below).

In summary, by claiming that evolutionary simulation modelling is a new technique which should properly be applied exclusively to a new class of problems, modellers incur a risk of scientific isolation and an attendant lack of rigour.

Unique Method of Enquiry

A parallel route has been to claim not that evolutionary simulation models are associated with particular novel phenomena, but that they have properties which make them better than or at least different from existing modelling techniques (e.g., Taylor & Jefferson, 1994; Miller, 1995) in their application to existing phenomena of interest.

We agree with Miller (1995) and Taylor and Jefferson (1994) when they note that mathematical assumptions which are made in the construction of tractable equational models may be relaxed under a simulation-based regime. The infinite, random-mating, unstructured populations often assumed within evolutionary models based upon differential equations may be replaced with finite, structured populations in order to highlight effects of genetic drift, frequency dependent selection, extinction, and other evolutionary phenomena.

In addition, the difficulties faced by equational models in capturing non-linearities, or increasingly complex inter-dependencies between the actions of agents, are largely absent from simulation-based models. Further characteristics of natural phenomena which prove difficult to incorporate within equational models include the representation of spatially distributed phenotypes, and repeated interactions between individuals (Nowak & May, 1992; Lindgren & Nordahl, 1994). The difficulty in constructing equational models of many verbal arguments is highlighted by Miller (1995) and Di Paolo (1996). Both suggest that simulation models of such arguments might prove easier to construct.

However, some authors go further, claiming that equational models and individual-based simulation models can be distinguished on the basis of general considerations such as clarity, explicitness, and inter-subjectivity. For example, Miller (1995) claims that simulations are more explicit models than those built from differential equations, since, during simulation design, the processes which govern the evolution of the system must be rendered as particular pieces of computer code. Miller also claims that simulations may be passed easily between researchers allowing more effective peer-validation of simulations than of equational models. Similar claims are made by Taylor and Jefferson (1994) who state that the "explicit" representation of behaviour within a simulation compares favourably with the "implicit" representation of an organism's behaviour within equational models. The authors also maintain that simulations are a more direct "encoding" of behaviour, and that this facilitates their design, use, and modification, to such an extent that these processes are necessarily easier to carry out for simulations than equational models.

These unqualified claims are misleading. For example, it is equally admissible to claim of equational models that they capture theoretical assumptions *more* explicitly than simulations since they do not involve extraneous processes which are necessary in order to implement the model as an unfolding, automated process, but which are not spoken to by the theory being tested, and are thus the source of potential artefacts. Similarly, the claim that simulations can be exchanged by modellers in order for their validity to be checked, can be made more forcefully for equational models, which can be presented in their entirety within an academic paper, rather than requiring an additional exchange of computer code. In general, unqualified claims of the superiority of one style of modelling over another are not compelling. Clarity, ease of design, ease of presentation, etc., will vary from model to model to a greater extent than they vary from modelling style to

modelling style.

Whilst we agree that simulation models sometimes offer advantages over equational models, the view of individual-based evolutionary simulation models as merely augmenting existing modelling efforts is also overly limiting. It fails to acknowledge that a new tool does not merely increase the number of ways to attack old problems, but also changes the nature of these existing problems, and, in an extreme case, may reveal whole new classes of problem to systematic enquiry. Evolutionary simulation models are not merely a trivial addition to the arsenal of modelling techniques at the disposal of, for instance, the theoretical biologist, but offer a chance to reconsider and explore the theoretical commitments made within existing modelling paradigms and compare them to those made within the new modelling paradigm (Di Paolo, 1996).

In summary, evolutionary simulation models may sometimes offer advantages over traditional methods. However, this is not true as a general rule, and must be assessed on a case-by-case basis. Furthermore, in claiming that, when they are advantageous, evolutionary simulation models are merely a new tool for an old job, there is a risk of undue conservatism and a failure to fully exploit the potential of a novel modelling paradigm.

Emergent Thought Experiments vs. The Physics Model

The positions outlined in the previous two sections address the issue of what challenge artificial life models are best suited to meet — extending old models or modelling new phenomena. We argue that artificial life simulation models have the potential to meet both of these challenges. However, there is a more fundamental question — how can simulations be models at all? In this section we consider two perspectives on *how* simulation models of the kind developed within artificial life can be made to subserve either of the scientific projects identified in the last two sections.

Maynard Smith (1974) maintains that, in the context of ecology, the difference between models and simulations is that whereas models strive for a minimum of detail, simulations strive for a maximum — models are general whereas simulations are specific, gaining validity and scientific worth to the extent that they accurately capture as much about a particular real system as possible. Does the notion of a simulation model imply a departure from this understanding of the role of a simulation? If so does this departure mean that the standards by which we judge the worth of a simulation must also change? Should a simulation model be judged on the same considerations as a model (i.e., generality, parsimony, coherence, etc.) or a simulation (i.e., fidelity, realism, resolution, etc.)?

Two extreme positions on this issue are apparent in the artificial life literature. One position takes the role of simulation models to be essentially in line with that proposed by Maynard Smith for simulations — they are maximally faithful replicas — whereas the other understands simulation models to be more like thought experiments: unrealistic fantasies which nevertheless shed light on our theories of reality.

Kitano et al. (1997) propose that detailed simulations of particular biological systems can serve as the source of novel hypotheses. As such the use of simulation models fits into what they term the "physics model" of scientific enquiry (see figure 1c), in which theories give rise to predictions, which are cast as hypotheses and are tested through experimentation, the results of which have implications for the generation of new theories, giving rise to new predictions, and so on. Under this reading, the attraction of simulation models lies in the claim that within disciplines studying systems comprised of many interacting components, such as those concerning biologists, particular kinds of hypotheses may be unattainable through conventional mathematical analysis.

Proponents of this position claim that in order for a simulation to be a useful source of hypotheses it must be "valid", that is, the behaviour of the simulation must square with data available from real experiments. Only once a simulation has been validated in this manner can one have any interest in any novel insights that might be suggested by it. Kitano et al. liken the process of obtaining these insights to "virtual experiments" designed to provide "decisive evidence" on specific biological questions. Once these virtual experiments have been carried out, the resulting hypothetical answers to these questions can be corroborated, or challenged, by experimentation in the real world. The results of these real experiments will add to biological knowledge and may require accommodating changes in the design of the simulation, new "virtual experiments", and so on.

In contrast, Bedau (1998, 1999) has presented a radically different perspective on artificial life simulation models, claiming that they are best understood as "emergent, computational thought experiments". Far from being authentic reproductions of existing natural systems, simulation models may provide explanations possessing both "simplicity and universality" if their designers "*abstract away* from the micro-details in the real system" in order to construct "models which are as *unrealistic* as possible", but suffice, nevertheless, to provide instances of the phenomena of interest (Bedau,

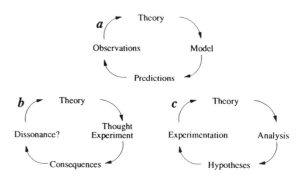

Figure 1: The thought experiment (*b*) and the physics model (*c*) can be understood as examples of a more general cycle of scientific enquiry (*a*).

1999, p. 20).

With the advent of such simulation models, some claim that philosophers have gained a valuable tool which can be brought to bear on contentious thought experiments (e.g., Dennett, 1994; Bedau, 1998, 1999). For instance, those thought experiments involving complex biological systems are often hard to apprehend unaided. For example, if we rewound the geological clock and started evolution from scratch again, would the same history unfold (Gould, 1989)? Would any differences caused by random events be minor ones, or would radically different states of affairs come about as this imaginary terrestrial history diverged from our own? Bedau claims that until a computer version of this thought experiment is constructed and run "all guesses about its outcome — including Gould's [(1989) guess that the evolution of some intelligent lifeform like ourselves would be very unlikely] — will remain inconclusive" (Bedau, 1998, p. 145). What is required in order to settle this matter, Bedau claims, is a powerful computer simulation to aid us in discovering the implications of the premises postulated in the original thought experiment. The prefix *emergent* is appended to the label *thought experiment* in such a situation presumably because it is precisely the emergent properties of the complex systems implicated in some thought experiments which are hard to intuit about. For these situations, in which our natural reasoning apparatus stumbles, Bedau proposes the computer simulation as a philosophical crutch.

How much credence should we give to each of these two visions of individual-based artificial life simulation models, and how might their differences perhaps be reconciled? In the next section, we consider these two proposals and how they contrast with our own.

Simulations as Opaque Thought Experiments

One of the key methodological questions about the scientific use of simulations — what knowledge can be gained from one if all that is 'fed into it' is already known — mirrors exactly the question of what use a thought experiment can be if, unlike a real experiment, it cannot bring to light any new information about a natural phenomenon. It may be of some use to examine briefly how this question has been answered for thought experiments and then see if the same answer works for abstract simulation models as well. In doing so we will uncover one possible use of simulation models and reveal the presence of internal tensions within the roles that Bedau and Kitano et al. attribute to them. However, it should be borne in mind that the analogy between thought experiments as practiced in the armchair and simulation modelling as practiced in the computer laboratory is not complete. There are important differences between the two which demand discussion.

Kuhn (1977, p. 241) poses the following questions about thought experiments. First, to what conditions of verisimilitude must a thought experiment be subject? Second, given that a successful thought experiment involves the use of prior information which is not itself being questioned, how "can a thought experiment lead to new knowledge"? Finally, "what sort of knowledge can be so produced"? Kuhn says that these questions have a set of rather straightforward answers which are important but "not quite right". These answers suggest that all the understanding that can be gained from thought experiments will be understanding about the researcher's *conceptual apparatus*; eliminating previous confusions or revealing inconsistencies within a theory, for instance. If one employs thought experiments to this end, the only requirement of verisimilitude that must be fulfilled is that "the imagined situation must be one to which the scientist can apply his concepts in the way he has normally employed them before" (ibid., p.242)[1].

Kuhn describes a well known thought experiment in which Galileo demonstrates the Aristotelian concept of speed (something similar to the present day idea of average speed) to be paradoxical. An immediate interpretation of this thought experiment can be given along the lines mentioned above. However, by carefully analysing how the Aristotelian concept of speed was used by Aristotle and his followers, Kuhn finds that it "displayed no *intrinsic* confusion" (ibid., p. 261). Contradictions arise when the scientist tries to apply this concept to previously unassimilated experience, as occurred when the 'corrected' Galilean concept of speed was itself confronted with situations where its application proved inappropriate (such as the additivity of the velocities of electromagnetic waves). In this way, Kuhn argues, the thought experiment is indirectly saying something about nature and has a historical role similar to empirical observation. But how is this possible when it was assumed that no new empirical information was 'fed' into the thought experiment? Kuhn answers: "If the two can have such similar roles, that must be because, on occasions, thought experiments give the scientist access to information which is simultaneously at hand and yet somehow inaccessible to him" (ibid., p. 261).

Scientists decide to pay attention to "problems defined by the conceptual and instrumental techniques already at hand" (ibid., p. 262). Therefore, some facts, although known, are pushed to the periphery of scientific investigation, either because they are thought not to be relevant, or because their study would demand unavailable techniques. A thought experiment will, on occasions, bring the relevance of these facts into focus, and therefore catalyse a re-conceptualization which may involve anything from an undramatic re-organisation of relationships between existing concepts to a scientific revolution.

This understanding of thought experiments suggests that they question a theoretical framework in the way depicted in figure 1b. It is important to contrast this view with a commonly held attitude towards artificial life simulation models as synthetic sources of empirical data. As we saw in the previous section, Bedau regards "emergent thought experiments" in a way that seems to imply the latter attitude. According to him, "[it] is worth emphasizing that a model can explain how some phenomenon occurs only if it produces *actual examples of the phenomena in question*, it is not sufficient to produce something that represents the phenomenon but lacks its essential properties", (Bedau, 1999, p. 21, our emphasis). Simulation models that, like mathematical or pen-and-paper variants, merely represent the phenomena of interest cannot serve Bedau's purposes. This is a courageously different understanding of thought experiments, and models in general, and the burden of proof falls on its proponents who must show how a simulation, which always starts from a previously agreed upon theoretical stance, could ever work like a source of new empirical data about natural phenomena.

---

[1] To this one could add Popper's further requirement that, in the case of an argumentative thought experiment, idealizations should always work in favour of the position that the experimenter is trying to debunk, (Popper, 1959, p. 444).

Bedau is right when he notes that the conclusions of armchair thought experiments may not be justified in the case of complex systems of many interacting elements. But, by suggesting that "emergent thought experiments" can provide the evidence that will settle such matters, he is making a category error which implicitly raises emergent computational thought experiments to the status of empirical, rather than conceptual, enquiries. Gould's argument concerning "replaying evolution's tape" would perhaps be better construed as a speculation about an empirical (albeit hypothetical) state of affairs rather than a thought experiment[2]. Even if scores of "emergent thought experiments" supported this speculation, Gould might still be wrong since there may always exist undiscovered phenomena that would invalidate his reasoning. But, such phenomena can never be *discovered* through building simulations because these are always based on existing theoretical knowledge and, as we have seen from Kuhn's arguments, can only reshuffle existing theoretical ideas, not deliver new facts. Thus Bedau's "emergent computational thought experiments" fall between two stools. If they are sources of empirical data then they are not thought experiments (only real empirical experiments are sources of empirical data). On the other hand, if they are not sources of empirical data then they cannot do the job he requires of them, that is, to provide decisive evidence one way or the other about the validity of our intuitions concerning an armchair thought experiment.

To see how a simulation model *could* function as a thought experiment, consider, as an example, a famous paper by Hinton and Nowlan (1987) in which a clear demonstration of the Baldwin effect is provided using an elegant evolutionary simulation scenario. The Baldwin effect, which stipulates that phenotypic plasticity can speed up an evolutionary process, had existed as a theoretical idea for nine decades at the time the paper appeared. According to Maynard Smith, the Baldwin Effect "has not always been well received by biologists, partly because they have suspected it of being lamarckist [...], and partly because it was not obvious that it would work. What Hinton and Nowlan have

done is to answer these objections", (Maynard Smith, 1987, pp. 761–762). In other words, they have discovered nothing new, but have helped in changing an attitude towards an already known piece of information. This change is evidenced by the amount of literature related to the Baldwin Effect that followed, both in theoretical biology and evolutionary computing. Hinton and Nowlan's simulation model thus plays the role of a successful thought experiment, demanding a reorganisation of an existing theoretical framework.

The use of thought experiments sketched above parallels that of abstract models in general. There is, however, an important difference between thought experiments and abstract *simulation* models. A thought experiment has a conclusion that follows logically and clearly, so that the experiment constitutes in itself an *explanation* of its own conclusion and its implications. If this is not the case, then it is a fruitless thought experiment. In contrast, a simulation can be much more powerful and versatile, but at a price. This price is one of *explanatory opacity*: the behaviour of a simulation is not understandable by simple inspection; on the contrary, effort towards explaining the results of a simulation must be expended, since there is no guarantee that what goes on in it is going to be obvious.[3]

This difficulty in achieving an adequate *understanding* of a simulation model threatens to nullify the advantage that simulation models enjoy in terms of the ease with which they may be designed. Whereas many authors have claimed that in important situations the construction of a simulation model tends to be far less of a chore than devising an equivalent formal mathematical model, few have reported the flipside of this advantage — that effort must be made to reconstruct the *relationships between classes* (which mathematical treatments get for free and utilise in explaining the behaviour of analytically derived models) from the *instances* which the simulation model generates. Thus, although, under certain conditions, the construction of simulation models might prove more tractable than the construction of analogous equational models, the analysis of such simulation models often requires an additional effort which threatens to more than compensate for any increased ease of design. This situation has been dubbed the "law of uphill analysis and downhill invention" (Braitenberg, 1984), and has been given a more thorough exposition by Clark (1990), in terms of a general distinction between automatic and manual models.

---

[2] If the conclusions prompted by an armchair thought experiment are unclear or tendentious — if different thought experimenters reach different conclusions from the same premises — then it is just not a very good thought experiment. Thought experiments are successful to the extent that the conclusions follow trivially from the premises. This unanalysed straightforwardness may of course disguise general reasoning biases which, although shared by all thought experimenters, are in fact fallacious (for example holding that humans are superior to other animals, or that nothing can be smaller than an atom, and so on). These biases drive troublesome *intuition pumps*.

---

[3] However, although opaque, simulations are explicit and manipulable, and hence may be cognitively penetrable to a greater extent than thought experiments carried out 'in the head' where hidden assumptions may be harder to uncover.

To return to Hinton and Nowlan's simulation model, it is possible to say that, despite its clarity as a simulation, the model did not achieve the logical 'closure' of a good thought experiment. As a matter of fact, the model posed open questions that other researchers have investigated in subsequent work, and which have also turned out to be of theoretical importance (e.g., Harvey, 1993; Mayley, 1996). For instance, Harvey (1993) has investigated a subsidiary aspect of the model: the number of genotypic loci that fixate as non-plastic. The number of non-plastic alleles increases very rapidly during the first stages of the simulation, but then tends to stabilise at some high, but sub-optimal value. Harvey shows that genetic drift (random fluctuations in finite populations) is the cause of this phenomenon, thus discovering further richness in the original model, which did not consider this factor, and at the same time demonstrating that unlike clear thought experiments, even simple and elegant simulation models may have a hidden explanatory structure.

Intuitively, we can expect this difference between simulations and thought experiments to become more accentuated as the complexity of the phenomena of interest increases. It is in precisely such cases, according to Kitano et al. (1997), that simulations are deemed most useful. However, lack of transparency in a simulation signifies an important problem for the strategy of validation against empirical data that Kitano et al. propose (see also Maley, 1998). Their "physics model" relies heavily on simulations being implementations of current theory, otherwise the truth or falsity of predictions generated by the model will have no implications for the theory being modelled, but only for the validity of the simulation being built. Specifically, in such a situation, a prediction made by the simulation model which fails to be borne out by experimentation in reality, may be attributed to the simulation's failure to adequately implement the theory, rather than a failure of the theory to adequately account for the relevant natural phenomena. But, despite one's best efforts, empirical validation cannot guarantee that the simulation will implement the theory one intends it to, since different theories could be neutral with respect to the data that one is validating it against. (Particularly if the validation involves some sort of parametric adjustment.) The explanatory opacity of simulations is a disarming problem for their proposed strategy. The lack of a priori certainty about what happens in a simulation may be something that we will have to learn to live with, if they are to be applied to the understanding of complex systems involving many interacting parts[4].

---

[4]Notice that this criticism does not invalidate the

A Workable Methodology

We now turn to the question of how this difference between simulations and thought experiments can be reconciled within a workable methodology if we want them to perform a similar scientific role. There is no general answer to this question. The following paragraphs describe a possible way of using simulations as scientific tools in which the difference between them and thought experiments is made methodologically evident. However, this description is no prescription. In particular, not much will be said about how a simulation should be built, or when it is adequate for a particular job. Rather, some landmarks in its use will be pointed to which ultimately will help in ensuring that the simulation plays a scientific role without undermining its potential[5].

The first preconception that must be challenged is the idea that all that is required from a simulation model is the choice of a plausible mechanism and the replication of a certain pattern in order to claim that an explanation of a similar natural pattern has been achieved. This idea is based on the premise that successful replication implies understanding of how the pattern arises in the simulation (figure 2). It does not always work like that; in fact, it rarely does. Only in a limited number of cases will the researcher be concerned with just the basic mechanisms of the model — typically when she wants to present a proof of concept of the type: "it is commonly thought that M is needed to generate P, but here is a model in which M', which is simpler (more plausible, nicer, etc.) than M, reproduces something that looks like P".

Whether explanations are couched in terms of the atomic properties of the simulation or involve higher-level entities as well, it need not be directly obvious how the patterns of interest arise or which aspects of the model are involved and which are inconsequential. Simulations are opaque and must be explored. Relevant observables must be chosen and may lead to the discovery of non-obvious patterns, some of which may not have been suspected initially. That this may happen is a further consequence of the difference between thought experiments and simulations. In the former all relevant entities are already defined whereas in a

---

"physics model" when other formal approaches are used instead of simulations. In general, in a good mathematical model, everything is (or should be) spelled out, so that if it is not a good implementation of a theory this should eventually become apparent, if not to the researcher, then to critics of the model.

[5]There is nothing particularly new about this description. This methodology has been applied successfully in many instances (Boerlijst & Hogeweg, 1991; Fontana et al., 1994, and others).

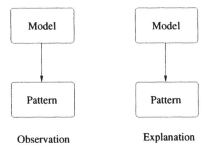

Observation   Explanation

Figure 2: Direct explanation. Observed patterns are explained exclusively in terms of modelled entities and processes. This is to be contrasted with the indirect explanation shown in Figure 3.

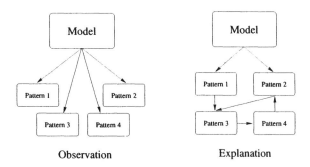

Observation   Explanation

Figure 3: Indirect explanation. Some observed patterns are explained in terms of modelled entities and processes *and* by other observed patterns.

simulation some of the entities that are interesting, at least from a descriptive (but also possibly explanatory) perspective, are not modelled explicitly and are discovered only after the simulation has been observed. An explanation of these entities is often germane to the task that the simulation sets out to achieve.

Notice that simply treating these non-obvious patterns or entities as 'emergent' is not an explanation at all, but rather the statement of a problem. For a simulation model to be of any use, both obvious and non-obvious patterns must be explained and not brushed under the carpet of emergence as this amounts to an admission of failure[6]. This is not an advocation of reductionism, since we are not implying that the explanation must proceed only from the micro- to the macro-structure of observed entities. Some observations can be explained in terms of the basic constituents of the simulation model but others may have to be explained in terms of higher order structures and patterns. Consequently, different observations have to be related through an *explanatory organisation* which, in general, can be more complex than that shown in figure 2 and more like the one depicted in figure 3 where the observed patterns play different explanatory roles,

and only some patterns are explained exclusively at the micro-level. Ultimately, this explanatory organisation (which explains what happens *within* the simulation) must then be related to the corresponding theoretical terms which describe analogous phenomena in the natural world.

Consider as an example the model of spatially distributed catalytic reactions presented by Boerlijst and Hogeweg (1991). In this simulation model different chemical species in a two-dimensional lattice spontaneously form dynamic macroscopic patterns in the form of rotating spiral waves. This is one observation. It is also observed that hypercycles (closed loops of catalysing relations between species and reactions) are resistant to the invasion of parasites (chemical species that take advantage of catalysis without themselves catalysing any other reaction in the loop). This is a second observation. This latter observation is important because such resistance is not observed in mathematical models of catalytic reactions taking place in a mixed medium (i.e., those which do not consider the possible effects of spatial structure). Based on this evidence and crucial experiments to test relevant hypotheses, the authors conclude that the first pattern provides an explanation for the second one. They show that the rotational dynamics of spiral waves do not allow parasites to invade. In this simulation model, then, one high level pattern explains another. Now, the authors must relate this explanatory structure to the real case. Since spiral waves have also been observed in the analogous natural systems, it is possible for Boerlijst and Hogeweg to recast the explanatory relevance of this known natural phenomenon using the explanatory organisation developed to account for the behaviour of their simulation model. Whether the resulting explanation holds water will be discovered through actual

---

[6] Marr (1977) makes a similar point using his distinction between theories of Type-1 (essentially explanations) and Type-2 (essentially descriptions) when he argues that "failure to find [a Type-1 theory] does not mean that it does not exist" (p. 135). Acceptance of a Type-2 solution to a problem in the belief that a more principled understanding of the nature of the problem is made impossible by its emergent, chaotic, non-linear, or context-sensitive nature is at best premature. Although some problems no doubt are of this type (Marr offers protein folding as one possibility), merely assuming that this is the nature of the beast is an unproductive strategy. "Such pieces of research have to be judged very harshly, because their lasting contribution is negligible." (ibid.).

empirical experimentation on the natural systems concerned.

In general, we can distinguish three different phases for *using* simulation models in the way we propose:

1. *Exploratory phase:* After the initial simulation model is built, explore different cases of interest, define relevant observables, record patterns, re-define observables or alter model if necessary.

2. *Experimental phase:* Formulate hypotheses that organise observations, undertake crucial "experiments" to test these hypotheses, explain what goes on in the simulation in these terms.

3. *Explanatory phase:* Relate the organisation of observations to the *theories* about natural phenomena and the hypotheses that motivated the construction of the model in the first place, make explicit the theoretical consequences.

The first two phases concern the simulation itself. Here the practitioner is dealing with her own created system. The organisational hypotheses formulated during the second phase prevent random fact-gathering and provide a theoretical perspective proper to the simulation, laying the groundwork for the third phase in which those hypotheses that were developed and supported in the experimental phase can be meaningfully compared with existing theories or hypotheses about natural phenomena.

This final comparison involves a 'backward metaphorical step'. The first, forward use of metaphor occurs when the model is built. Entities in the model represent theoretical entities metaphorically or analogically. However, nothing guarantees that this same set of metaphors will be sufficient when one wants to project the observations made after running the simulation back onto existing theoretical entities relating to the natural world. This may be a trivial step if the observed patterns, or relationships between patterns, have corresponding counterparts in the existing theory. But it is possible to discover relationships between observations that are not easily accommodated by an *existing* theoretical framework. This is a tricky but interesting situation, because this tension may mean that the simulation model is flawed — that it is not modelling what it is supposed to. Alternatively, it may be the current theories that are at fault — the model may be pointing to genuinely new theoretical constructs, which perhaps deserve new names.

The organising theory achieved by the modeller in order to understand what is observed in the simulation may provide a new perspective with which to understand the analogous, existing theory of the natural

phenomena being modelled. Conversely, this existing theory may prompt a re-consideration of the organising theory developed during the three phases of simulation modelling, prompting the modeller to explore where the crucial differences lie and make a possibly useful conclusion about them[7].

These possible outcomes of simulation modelling are directly analogous to the results of armchair thought experiments. When a thought experiment generates dissonance (i.e., the consequences of the thought experiment are not easily accommodated by our current understanding of the phenomena involved) we must question both the integrity of our current theories, and the validity of the intuitions which guided our thoughts during the thought experiment. A dialectic process must be invoked in order to reconcile this kind of theoretical impasse.

Conclusion

Thus it is reasonable to understand the use of computer simulations as a kind of thought experimentation: by using the relationships between patterns in the simulation to explore the relationships between theoretical terms corresponding to analogous natural patterns. Through this practice, theoretical terms may be shown to stand in different relationships than previously thought. However, this is an unusual kind of thought experiment. Due to their explanatory opacity, computer simulations must be observed and systematically explored before they are understood, and this understanding can be fed back into existing theoretical frameworks. The necessity of this systematic enquiry into the workings of computer simulations is not part of armchair thought experimentation. The irony here is that, although we advocate an understanding of simulations as tools of *theoretical* enquiry, working with simulations in the way proposed above does have an 'empirical' flavour precisely because complex simulations are not obvious; hence the aptness of the phrase 'computer experiment'.

An additional difference lies in the fact that it may indeed be possible to make a stronger case with simulations than with a 'naked' thought experiment since a simulation can also provide insights that could not be arrived at by thinking alone. As with traditional thought experiments, the information 'fed' into the computer model may not be controversial but, in the end, the researcher may be forced to focus on facts or processes that were at the periphery of her conceptual

---

[7]See Hesse (1980) for further discussion on the use of metaphors in science and particularly on the two-way conceptual dynamics which is generated when two domains are related metaphorically.

structure and place them in novel relationships with other theoretical terms.

Acknowledgments

Thanks to Walter Fontana, Mike Wheeler and three anonymous reviewers for their helpful comments.

References

Bedau, M. A. (1998). Philosophical content and method of artificial life. In Bynum, T. W., & Moor, J. H. (Eds.), *The Digital Phoenix: How Computers are Changing Philosophy*, pp. 135–152. Basil Blackwell, Oxford.

Bedau, M. A. (1999). Can unrealistic computer models illuminate theoretical biology? In Wu, A. S. (Ed.), *Proceedings of the 1999 Genetic and Evolutionary Computation Conference Workshop Program*, pp. 20–23. Morgan Kaufmann, San Francisco.

Boerlijst, M. C., & Hogeweg, P. (1991). Spiral wave structure in pre-biotic evolution: Hypercycles stable against parasites. *Physica D, 48*, 17–28.

Bonabeau, E. W., & Theraulaz, G. (1994). Why do we need artificial life? *Artificial Life, 1*(3), 303–325.

Braitenberg, V. (1984). *Vehicles: Experiments in Synthetic Psychology*. MIT Press, Cambridge, MA.

Bullock, J. (1994). Letter. *Trends in Ecology and Evolution, 9*(8), 299.

Clark, A. (1990). Connectionism, competence and explanation. In Boden, M. A. (Ed.), *The Philosophy of Artificial Intelligence*, pp. 281–308. Oxford University Press.

Dennett, D. (1994). Artificial life as philosophy. *Artificial Life, 1*(3), 291–292.

Di Paolo, E. A. (1996). Some false starts in the construction of a research methodology for artificial life. In Noble, J., & Parsowith, S. R. (Eds.), *The Ninth White House Papers*. Cognitive Science Research Paper 440, School of Cognitive and Computing Sciences, University of Sussex.

Fontana, W., Wagner, G., & Buss, L. W. (1994). Beyond digital naturalism. *Artificial Life, 1*(1/2), 211–227.

Gould, S. J. (1989). *Wonderful Life: The Burgess Shale and the Nature of History*. W. W. Norton, New York.

Harvey, I. (1993). The puzzle of the persistent question marks: A case study in genetic drift. In Forrest, S. (Ed.), *Genetic Algorithms: Proceedings of the Fifth International Conference on Genetic Algorithms*, pp. 15–22. Morgan Kaufman, San Mateo, CA.

Hesse, M. B. (1980). The explanatory function of metaphor. In *Revolutions and Reconstructions in the Philosophy of Science*. Harvester Press, Brighton, UK.

Hinton, G. E., & Nowlan, S. J. (1987). How learning can guide evolution. *Complex Systems, 1*, 495–502.

Judson, O. P. (1994). The rise of the individual-based model in ecology. *Trends in Ecology and Evolution, 9*(1), 9–14.

Kitano, H., Hamahashi, S., Kitazawa, J., Takao, K., & Imai, S.-i. (1997). The virtual biology laboratories: A new approach of computational biology. In Husbands, P., & Harvey, I. (Eds.), *Proceedings of the Fourth European Conference on Artificial Life*, pp. 274–283. MIT Press, Cambridge, MA.

Kuhn, T. (1977). A function for thought experiments. In *The Essential Tension: Selected Studies in Scientific Tradition and Change*. Chicago University Press.

Lindgren, K., & Nordahl, M. G. (1994). Evolutionary dynamics of spatial games. *Physica D, 75*, 292–309.

Maley, C. C. (1998). Models of evolutionary ecology and the validation problem. In Adami, C., Belew, R. K., Kitano, H., & Taylor, C. E. (Eds.), *Artificial Life VI*, pp. 423–427. MIT Press, Cambridge, MA.

Maley, C. C. (1999). Methodologies in the use of computational models for theoretical biology. In Wu, A. S. (Ed.), *Proceedings of the 1999 Genetic and Evolutionary Computation Conference Workshop Program*, pp. 16–19. Morgan Kaufmann, San Francisco.

Marr, D. (1977). Artificial intelligence — A personal view. In Boden, M. A. (Ed.), *The Philosophy of Artificial Intelligence*, pp. 133–147. Oxford University Press. Collection published in 1990.

Mayley, G. (1996). Landscapes, learning costs and genetic assimilation. *Evolutionary Computation, 4*(3), 213–234.

Maynard Smith, J. (1974). *Models in Ecology*. Cambridge University Press.

Maynard Smith, J. (1987). When learning guides evolution. *Nature, 329*, 761–762.

Miller, G. F. (1995). Artificial life as theoretical biology: How to do real science with computer simulation. Cognitive Science Research Paper 378, School of Cognitive and Computing Sciences, University of Sussex.

Nowak, M. A., & May, R. M. (1992). Evolutionary games and spatial chaos. *Nature, 359*, 826–829.

Popper, K. R. (1959). *The Logic of Scientific Discovery*. Hutchinson, London.

Ray, T. S. (1994). An evolutionary approach to synthetic biology: Zen and the art of creating life. *Artificial Life, 1*(1/2), 179–209.

Sober, E. (1996). Learning from functionalism — Prospects for strong artificial life. In Boden, M. A. (Ed.), *The Philosophy of Artificial Life*, pp. 361–378. Oxford University Press.

Taylor, C., & Jefferson, D. (1994). Artificial life as a tool for biological inquiry. *Artificial Life, 1*(1/2), 1–13.

Artificial Life as a bridge between Science and Philosophy

Alvaro Moreno

Department of Logic and Ph. of Science
University of the Basque Country
Post Box 1249
20080 San Sebastian -- Donostia
ylpmobea@sf.ehu.es

Abstract

Artificial Life is developing into a peculiar type of discipline, claiming the principle of computational construction as the main avenue to explore and produce a new science of life "as it could be". In this research program, the generation of complex virtual systems may become the actual object of the theories, somehow substituting the usual empirical domain. This brings along not only a deep change in the traditional relationship between the ontological, epistemological and methodological levels, but also the appearance of a new relationship between the theoretical and technical realms, in what constitutes a relevant epistemic problem. Such a state of affairs forces us to reconsider the solid differences apparently established between science and philosophy. Even if the frontiers between these two kinds of knowledge do not completely disappear, new, very dynamic and complex, technologically mediated ways of interaction are being developed between them.

Introduction

Artificial Life (AL) was originally conceived as a generalization of biology, since it was meant to become the study of "all possible life" (Langton, 89). However, like in the case of Columbus, who discovered the American continent when trying to reach India, in the course of its brief history as an independent discipline, AL has developed into something rather different from that foundational claim. This is partly because it is at stake whether it may actually reach that goal, and partly because it has generated a new set of problems and a new way of approaching the living phenomenon.

AL can be defined as the study of "lifelike" systems created by human action. But there are several elements in this definition that need to be clarified and stated more precisely. First, the very sense of the term "study", for AL is not only an epistemic process but also a technical activity; in AL the objects of study are literally created through technological action. Second, the actual meaning of "lifelike systems" (or systems that show "lifelike" behavior) becomes much complex (and controversial) than in traditional biology, since this concept has been understood in a much wider sense than that of empirically real biological systems. Thus, as C. Emmeche (94, p. 161)

has pointed out, AL is founded as a *modal* discipline, establishing as its own objective the study of life "as it could be" and not simply "as we know it" (even if we include here extraterrestrial forms of life that might be discovered in the future). Last but not least, the idea of "artificiality" should be subject to examination. In addition to its generic sense of human construction, the term *artificial* has a double-sided peculiarity in the context of AL. On the one hand, it has a paradoxical meaning that comes from the idea that such humanly constructed systems be capable -like the actual natural living beings- of exhibiting creativity. The so-called "emergence" (in the behavior, capabilities, morphology, etc.) must be understood precisely as a form of indirect human creation (Boden, 96; Risan, 97) what in other words Langton describes as "getting the humans out of the loop", and thus endowing the actual machine with creativity. That is to say, one of the essential features of AL is that the artificially created system displays some type of agency, which allows us to speak (without falling into contradiction, although somewhat paradoxically) about autonomy in those cases. And on the other hand, *artificial* has the peculiar meaning (not exclusive but prevailing), like in AI, of virtual "construction" as opposed to physical realization.

The complex combination of all these elements has provided AL with a proper identity as a discipline, distinct not only from traditional biology, but also from the whole set of traditional empirical sciences. The idea that the living entity subject to study will be generated by a human agent -and, on top of that, in a computational universe- is truly suggestive; but, nevertheless, it also brings about novel issues and challenges in many respects.

AL as a computational research project

As it was mentioned above, the most usual meaning of the term *artificial* in AL research does not denote that the systems under study be the result of human action generically (i.e., as material realizations), but specifically refers to the generation of virtual systems in the computational universe. It is in fact necessary to take into consideration the strength that, since half a century ago, the tendency to separate organizational and informational aspects from material and energetic ones has acquired in the study of systems, if we are to understand even the development (and specificity) of a discipline like AL at the end of this 20th century. Both AI and AL (following in this regard the precedent set up by cybernetics) share a common approach based on the idea that all of the material and energetic aspects of an organization do not affect its logical essence. Therefore, although a wide range of research enterprises (like robotics or some new branches of bioengineering) may be included within the field of AL, the main interest and area of activity is focused on the study of virtual systems generated in a computational environment.

Philosophically speaking, the computational versions of both IA and AL have been associated to functionalism. This position assumes and defends that the specific materiality that sustains a certain capability (mental, biological or of some other nature) is not relevant (Block, 80; Sober, 92). In the specific case of AL, it is claimed that biological phenomenology is exclusively the result of some organizational arrangement, rather than of a particular material implementation of it (Langton, 89). In fact, the question of whether those organizational arrangements are sustained by carbon or silicon molecules or by patterns of electrons in a computer is considered as completely irrelevant. Accordingly, and given the enormous possibilities to create and manipulate formal processes in the computational realm, it is obvious that the major line of research in AL be the purely computational one[1].

This huge potential for exploration of virtual organizations becomes in fact a way of "experimenting" with formal entities in formal environments, which should be therefore empirically interpreted. Hence, it is only at the end of the whole process that the analysis of the result can proceed; a result that will probably be difficult to compare or extrapolate to the real domain, since it is the consequence of pure abstractions of processes taking place in empirical environments. Nevertheless, even if this new strategy is highly promising with regard to the problem of the universalization of biology, it poses also new intriguing questions.

In particular, we want to refer here to three issues that we consider most interesting. The first one, which is basically epistemological (although it has ontological implications, as well), refers to the confusion between object and model; the second one, of methodological nature, is the problem of evaluating hypotheses (and, thereafter, research programs, too); and the third one concerns the interrelation between science, philosophy and technology.

---

[1] This is the reason why Langton has defended the idea of universalizing classical biology through abstracting the materiality of biological phenomena and studying such phenomena assuming it may take place in a purely formal organizational domain.

The confusion between "object" and "model"

In biology (like in all the other traditional empirical sciences) conceptual models are elaborated to represent operatively a certain type of empirical systems, natural living systems, in this case. Such empirical systems thus constitute the objects of reference of the models. Regardless of whether these models bring about computational simulations or not, the latter involve the pre-existence of a reference system, whose behavior is trying to be, total or partially, reproduced. However, computational "models" of AL are elaborated without direct and precise reference to the empirical biological reality[2]. C. Emmeche (op. cit., p. 163) regards them as "second-order *simulacra*, that is, copies of the copies themselves", generated not as abstractions on empirical biological systems (like in the case of concepts and theories of biology, which would be "first-order simulacra") but on the theories themselves. Their main goal is to allow a new way of "computational experimentation" that enables us to "discover" the universal principles of living systems.

Sometimes, even the "model" or "simulacrum" is literally considered as a realization, that is to say, as an object whose phenomenology would make it equivalent to any other natural system of the corresponding empirical domain. This occurs when, due to the functionalist theses predominant in AL, the actual systems created in the computer are conceived not as representations of natural biological systems, but literally as artificial creations of the same type of systems[3]. Nevertheless, among those authors that regard these systems as true realizations, the use of the term "model" is quite widespread (even being aware of the differences with the classical meaning), probably because in the course of research and experimentation subsequent virtual tentative structures are generated (that is, for methodological reasons).

As a matter of fact, then, depending on circumstances, sometimes a computational system is interpreted as an

---

[2] Indeed, there are also some intermediate cases of computational models based on standard AL methodologies. However, the design of such models aims only at the study of certain phenomenologies of living beings. In my view, the epistemological status of these models, rather than constituting a special differentiated case, is somewhere between models in traditional biological sciences and in AL Usually these models are closely related to certain empirical biological systems, and typically elaborated and developed according to the synthetic and 'bottom-up' criteria of AL. This is why they potentially aim to find principles that not only govern the structure of those known systems taken as a reference, but of all possible realizations of the same type.

[3] For instance, in the so-called "strong AL" it is claimed that computational simulations of living systems may really come to be living systems. Whereas "weak" AL considers that models represent certain aspects of living phenomena, strong AL would be ready to defend that the phenomenology that takes place in the actual computational environment is life in a proper sense.

object (ultimately, what is interpreted as an object is the model associated to the physical structure of the machine that sustains the execution of such a model[4]), some other times as a proper model, and some other even as a tool or methodological technique mediating between the "true object" (natural biological systems) and the "true theory" (theories of empirical life sciences)[5].

There is another source of confusion, which must be located in the tendency to regard the computational system as a representation of a particular type of empirical biological system, i.e., as the realization of a universal that would include all possible systems of its own kind; or in other words, as they could be and not only as we happen to know them. In fact, there is a broad range of systems where the known phenomenology only constitutes a subset of all possible empirical systems of that type.

In short, the solution to these dilemmas involves determining which may be, ultimately, the ontological status of such artificial systems. Are they real or virtual? Which is the empirical reference of a computational "simulacrum"?[6] Whichever answer may be given to these questions, it will certainly have methodological implications, since all those problems are tightly related to that of establishing the criteria for evaluation of the hypotheses that the very design of such systems attempts to test.

The problem of "empirical" evaluation

The fundamental problem in the methodology of AL research programs is that the ways of evaluating models are not empirically conclusive, for, by definition, the hypothetical empirical references of these models belong to a domain broader than the already known and even than the effectively existent. Despite the fact that "virtual experimentation" provides some formal rigor to the methodology, there is always a difficult problem of global empirical interpretation (Casti, 97; 99). This is one of the main differences with the classical scientific methodology, where experiments and measurements, no matter how idealized, are always performed or stated within the general framework of models that can be given a rigorous empirical interpretation.

Typically, an AL model is designed taking as a starting point some basic principles inspired in the general theories of a certain area of empirical biology. Then, these principles are introduced in the design of a computer program in which only low level instructions (local rules) are made explicit, so that new patterns (which play the role of new rules at higher levels) may appear. The characteristics of these patterns, however, are not previously known by the designer In this process, the emergence of new structures in the computer (clearly distinguishable in spatial and/or temporal terms) is fundamental. Some of these new patterns get organized hierarchical and functionally so that, eventually, they become subject to empirical interpretation. As a result, the methodology consists in a continuous revision of the values of the parameters (and even, of some of the actual principles according to which the original design of the model was made), depending on the degree of coherence that the results being produced keep with the phenomenology of the empirical domain under study.

Therefore, the problem is manifest firstly because in AL computational systems evaluation is not discernible from the difficulties of interpreting "emergent" patterns or processes generated in such systems, and neither from the establishment of the epistemological criteria to select the primitives that define the models. So we find the initial problem of how to set up the main features of the model. Here, apart from the basic criterion for simplicity, the usual way to proceed is to search for (sub)models that, somehow, implicitly ensure that the assumed abstraction be "valid"; i.e., that the new provided information be coherent with what happens in well-known biological systems. The limits on simplification are set so as to avoid that the results produced appear trivial in the selected phenomenological domain[7]. The aim is, thus, double-sided: simplicity is pursued, but provided that, at the same time, the model is able to generate enough complexity in the course of the simulation.

Thus, the design of models ought to fulfil two basic requirements. On the one hand, it is important that emergent processes or structures, which may be interpreted as new pieces of knowledge in the empirical domain under investigation, are obtained. But on the other hand (and in order to avoid the problem of how to determine when the empirical interpretation of these new structures and/or

---

[4] E.T. Olson (97), for instance, has argued against the objections of the "immateriality" of those "virtual organizations", defending that one could actually take them as real physical systems if, instead of looking at the computer program in abstract terms, one considers the patterns of electrons which materially support the execution of that program. Yet, this argument misses the point, because the material structures which support the *operational level of computer simulations* are entirely passive, since their intrinsic dynamics is completely constrained (Moreno & Ruiz-Mirazo, 99).

[5] H. Pattee (89) has criticized this confusion between computational simulations and realizations of material biological systems, arguing that the former are mere symbolic systems, whereas in the latter their capacities derive precisely from the symbolic constraint (autonomously generated) in material systems.

[6] B. Smith (96), for instance, claims that the computational world, in fact, does not constitute a proper subject matter (of study); rather, if should be considered as a complex social practice that involves the design, construction, maintenance and use of intentional artifacts (pp. 75 and 359).

[7] In fact, one of the main objectives of a model in AL is to show something new not only in that phenomenological domain but also in the domain of the internal correlations of the systems under study.

processes is objectively justified), it is necessary to design the computational experiment so that at least the type of expected results can be forecasted, and these are specific enough within each particular domain. If only the first condition is met, the problem then becomes how to justify its interpretation; and if it is only the second one, the risk is that the result becomes trivial. Thus, depending on the results given by running the program it will be assessed whether the model is valid or not, although there are some other elements that take part in the evaluation, like in traditional sciences (for instance: auxiliary theories, the adequacy of the computational tools employed, or of the set-up values for the initial parameters, etc.)

Nevertheless, the problem of the evaluation of models goes further beyond this, since, depending on the epistemological status given to computational systems, the actual research program may vary quite radically. Should we try to design "plausible" models (understanding plausible in the sense that computational models of AL ought to "resemble" the behavior of the corresponding natural systems) or, on the contrary, the main goal should be oriented in some other direction, like in the search for models that fulfil primarily certain formal criteria (for example, the capacity to generate computational complexity, or similar ones).

The problem with a purely formalist conception of AL (i.e., that the task of AL models is mainly to "unfold" and make explicit the logic consequences of the starting premises) is that AL moves into the risk of becoming a discipline closer to mathematics than to any empirical science. Several authors have formulated serious doubts and wariness about this issue. Maynard Smith, for instance, has objected that AL is a "science without facts", referring to the problem of how to assess a set of computational models whose (potential) empirical references are imprecise and generic (quoted in Horgan, 95). However, this criticism is too strong, for most Alifers, both in the design of the models and in the evaluation of their results, usually take quite seriously into account the theories as well as the behaviors present in the respective empirical domain. At the same time, the worries to escape mathematization might well put at stake the actual autonomy of AL as a discipline[8].

The attempt to answer these questions takes us well beyond the strictly methodological realm, as it requires determining the actual nature of that which the model discerns (sometimes, rather than a particular phenomenon in a particular empirical domain, what the model addresses and hopefully elucidates are certain computational issues...). Nevertheless, this is related at the same time to the role that meta-theoretical conceptual elements play in the design of models, as well as to questions that have to do with the nature of the technical objects employed.

---

[8] Miller (95), for example, has claimed that AL can only become a fully scientific discipline through its integration in the field of theoretical biology.

The problem of the interrelation between science, philosophy and technology

AL constitutes a discipline that, due to the nature of the problems it addresses, has progressively moved into an intermediate area between philosophy and science. AL allows to face, analyze and sometimes solve (or at least re-formulate) in an entirely new way several important philosophical issues in biology, such as the debates between reductionism and vitalism, the problem of emergence, or the matter-form relationship. This is so as a result of the deep entanglement between the AL methodology and the meta-theoretical problems of biology. Regarding this question, J. Casti (97, pp. 187-8) and D. Lane (95) hold that the construction of "models" in AL (and other computational sciences) requires theories about particular empirical domains; namely, about the kind of objects that are found in each of them, and about how these objects relate to each other, or the type of processes that change their nature (including creation and destruction processes). The relevance of computational models with regard to their corresponding empirical domains would be mediated by these theories, which constrain the interpretation of the former (i.e., the models). In turn, such models have a significant advantage, for they allow the analysis of these theories with the aid of more powerful tools than ordinary language.

Yet, this idea poses some problems. Firstly, as it has been broadly argued by the post-positivist philosophy of science, the concepts on which models in traditional empirical sciences are constructed, are also mediated by theories, in a similar sense to that pointed by Lane (95). This is why the differences between one and the other type of models are not stated in such terms, although it is quite true that the use of this kind of virtual models does indeed involve, and rather directly, the aforementioned meta-theoretical problems. The question to address is why it does so.

Secondly, these models are, in principle, purely formal, even though the formulation of its basic principles is inspired in a given biological domain. As we mentioned above, the difficulties of interpreting and testing AL models in terms of the current biological knowledge lie on the meta-theoretical implications that are found implicit in the epistemological status of the entities and relations that constitute such models. Nevertheless, the main criterion that we will use to evaluate the usefulness of these models is their capacity to improve our understanding of the theories about the empirical biological world. Apparently, this is a vicious circle.

However, according to Lane (95), these problems could be progressively solved. On the one hand, the models of AL and other computational sciences can improve our theories about the corresponding empirical domains; and on the other hand, these theories become essential for testing the relevance of virtual models. In this way, a kind of "hermeneutic" circularity is established between models and theories (Kleindorfer et al, 98), since the models are used for the elucidation of the theories (about the empirical world) and these for the interpretation of the models. A highly significant feature of this idea is that the interaction is not established merely between a philosophical or meta-scientific level and a proper scientific one, but a new technological level (constituted by the computational tools, with their double-sided nature, as software and hardware) plays also a crucial role in this hermeneutic process. This is because computers[9] make possible the exploration and visualization of a given set of work premises through the creation of virtual worlds of indefinite complexity.

D. Dennett (94) holds a similar position when he points out that AL can be conceived as a special sort of philosophy[10] which allows the creation and testing of complex thought experiments, "kept honest by requirements that could never be imposed on the naked mind of a human thinker alone". That is why this author considers that the research program in AL consists basically in the creation of prosthetically controlled thought experiments of indefinite complexity. By increasing the capacity and precision of the human mind through these prostheses that are the computers, we would be ready to increase indefinitely the complexity of such experiments, as well.

Anyway, it is not just a question of the computer acting as an extension of our conceptual world, as Dennett asserts; the novelty lies on the fact that the blending between concepts and computational technology creates a new domain with its own ontological claims. Somehow, in AL a set of systems generated as a result of putting together human concepts and artifacts, gets constituted as a set of objects. In other words, an ontology is founded from an epistemological standpoint. Yet, this happens in a way that such concepts, once embodied in the machine, are transformed (since they have acquired some ontological dimension) and establish, in turn, a new epistemological relationship with their creators and users.

This is why AL, on the one hand, constitutes an engineering activity whose starting point is located at a set of basic biological intuitions about the empirical biological domain; it attempts to build systems capable of producing by themselves emergent and functional processes. In order to do so, it proceeds by integrating creatively different technological resources, together with theoretical developments in computer science and physics. And on the other hand, it involves a process of interpretation of the behavior of such systems with the aim of broadening the phenomenology of the empirical biological domain and of developing new concepts coherent with the body of knowledge of traditional biology, which properly studies that domain. Both processes -construction and interpretation- are deeply entangled, since the AL research

program is a continuous concatenation of constructions, interpretations, modifications, new interpretations and re-constructions.

Conclusion

If things are displayed and understood in this way, AL would come to be a bridge between empirical science and philosophy of science, as each of them has been traditionally conceived. The methodological realm appears deeply intertwined with the properly theoretical one, so as to make it very difficult to discriminate among the possible contributions of a certain model and state neatly which are the heuristic, the epistemological, the theoretical and even the "empirical" aspects involved. Whereas in traditional philosophy of science the interaction under analysis is that between the empirical domain, the empirical theories and the meta-theories (a relationship in which technology only plays a significant role in the interactions between the first two), in AL a new way of interrelating empirical theories and meta-theories arises, a way that involves the mediation of technological devices. This new connection is, on the one hand, more limited than the established in the traditional philosophy of biology, since AL must be restricted to deal with conceptual issues that come out in the context of implemented (computational) systems; but on the other hand, AL is more rigorous and powerful, because it uses quasi-experimental methods of validation (computational experimentation) and also because it can, in principle, make the traditional range of problems of biological meta-theories even wider. Hence, research in AL is opening up radically new perspectives, not only in the sense of bringing about profound changes in the meaning of concepts like experimentation, model or evaluation (of theories); but also because the status of the philosophy of biology is clearly modified and deeply interwoven with technology. All this puts at stake the classical differences made between science and philosophy and demands a more elaborate framework to give proper accounts of the more and more complex and dialectical bonds established between them.

Acknowledgments.

The author acknowledges funding from the Research Project Number PB95-0502 from the DGICYT-MEC, the EX-1998-146 and HU-1998-142 from the Basque Governement, and the UPV 003.230-HA079/99 from the University of Basque Country. K. Ruiz-Mirazo read a first draft of this paper and contributed to make complete its final version.

References

Block; N. 1980. What is functionalism? In N. Bolck, (ed) *Readings in the Philosophy of Psychology*. Vol 1. 171-184 Cambridge, Mass.. Harvard University Press

---

[9] Although there are, of course, various kinds of technological tools within AL (like robots or other devices), the most important and usual way to develop and test models in this research field is through computers. Therefore, in this paper, the discussion will be focused on them.

[10] Some other authors showing enthusiasm for the future of AL as a philosophy are Bonabeau & Theraulaz (94)

Boden, M 1996. Autonoimy and Artificiality in Boden, M. (ed) *The Philosophy of Artificial Life*. 95-108 Oxford University Press

Bonabeau, E.W, & Theraulaz, G. 1994. Why do we need Artificial Life. *Artificial Life* 1(3): 303-325.

Casti, J. 1997. *Would-be Worlds. How Simulation is Changing the Frontiers of Science*.New York.: John Wiley.

Casti, J. 1999. The Computer as a Laboratory. *Complexity* 4(5):12-14.

Dennett, D. 1994. Artificial Life as Philosophy. *Artificial Life* 1(3): 291-292.

Emmeche, C. 1994. *The Garden in the Machine* Princeton University Press.

Horgan, J. 1995. From Complexity to Perplexity. *Scientific American*, June :104-.109.

Kleindorfer,, G. B, O´Neil L & Ganeshan, R. 1998. Validation in Simulation: Various Positions in the Philosophy of Science. *Management Science*, 44 (8):1087-1099.

Lane, D. 1995. Models and Aphorisms. *Complexity* 1(2): 9-13.

Langton, C. 1989. Artificial Life. In Langton, C. (Ed.) *Artificial Life*. 1-47 Redwood City CA.: Addison-Wesley.

Miller, G. 1995. Artificial Life as Theoretical Biology. How to do real science with Computer Simulation. Technical Report,CSRP 378, COGS. University of Sussex.

Moreno, A & Ruiz-Mirazo, K. 1999. Metabolism and the problem of its universalizati on. *BioSystems* 49 (1):45-61.

Olson, E.T., 1997. The ontological basis of strong artificial life. *Artificial Life* 3(1): 29-39.

Pattee, H. H. 1989. Simulations, Realizations and Theories of Life. In Langton, C. (Ed.) *Artificial Life*. 63-77.Redwood City CA. Addison-Wesley.

Risan, L. 1987. Why are there so few biologists here? In Husbands, P & Harvey, I.(eds) Proceedings of the Fourth European Conference on Artificial Life. 28-35. MIT Press.

Smith, B.C. 1996. *On the Origin of Objects*. MIT Press.

Sober, E. 1992. Learning from Functionalism-Prospects for Strong Artificial Life. In C. Langton, C. Taylor, D. Farmer & S. Rassmusen (Eds.) *Artificial Life II*. 749-765. Redwood City CA. Addison-Wesley.

Artificial Justice

J. McKenzie Alexander
Logic & Philosophy of Science
University of California, Irvine
Irvine, CA 96297

Abstract

Recently, there has been an attempt among some philosophers to provide cultural evolutionary grounds for certain norms of distributive justice. The most noteworthy attempt (Skyrms, 1996) uses a simple evolutionary model based upon the replicator dynamics. I argue that the replicator dynamics is not the most appropriate model of interactive human behavior in societies, and present an alternative agent-based model. I demonstrate how the conditions under which norms of distributive justice arise depend on how we construe the underlying evolutionary dynamics.

0. Introduction

The presence of a concept of justice is a fundamental property of human societies. Intuitions over what exactly this concept includes vary widely, but, generally speaking, a concept of justice can be thought of as a set of principles identifying the appropriate relationship between individuals and the society to which they belong. Two topics of natural interest to philosophers concern *descriptive* and *normative* questions of the concept of justice: What concept of justice do people actually have? Why should people hold one concept of justice instead of another?

Constructing a theory of justice which is both descriptively and normatively adequate is a daunting task, given that it must accurately describe the concept of justice individuals have, it explains why individuals should have *that* particular concept of justice instead of any other, and it explains what we mean when we say we *ought* to follow certain principles of justice. In this paper I shall concentrate on a very minor question about distributive justice which properly belongs to a general theory of justice. This concentration will not prove problematic for my purposes, though, as presently I wish to argue that philosophers interested in the sorts of questions mentioned above can benefit by approaching these questions employing techniques used within the artificial life community. To argue this, it suffices to show how certain fundamental questions of distributive justice can be profitably studied within the artificial life framework. It must be remembered that the models discussed here address only a very small area of the full concept of justice most people possess; a complete account of our concept of justice would need to expand upon the treatment given here to arrive at a more empirically adequate account.

1. The distribution problem

In its most general form, the distribution problem consists of a set of goods G to be distributed among the members of a population P, with situational considerations S, subject to two constraints:

1. No good is assigned to two members of the population.[1]

2. Each good is assigned to some member of the population.[2]

---

[1] It may seem that this restriction limits the generality of distribution problems considered, but it really does not. To handle problems involving public goods which may be shared among people, such as highways, parks, and baseball fields, we simply need to adjust our conception of the good to be distributed. Instead of conceiving of the good (say, the highway) as a single item which must be assigned to one and only one agent, we do not assign the highway itself but rather *time-shares* in sections of the highway, one time-share to each person who desires a part of it. After all, although the highway, as a whole, may be used by more than one person at the same time, no single part of the highway (one hopes) will be used by more than one person at the same time. This approach allows us to make the simplifying assumption that no good can be assigned to two members of the population without a loss of generality.

[2] This restriction simply requires any solution to be Pareto-optimal. This captures the commonly held belief that, if the position of any person may be improved without negatively affecting the position of anyone else, then that person's position *should* be so improved. Of course, if the set G we are distributing over the population contains undesirable items (say, diseases, debts, or unpleasant duties), it might be appropriate to drop this requirement.

The situational considerations S allow us to incorporate various relevant facts concerning needs, rights, prior claims to some of the goods in G, and so on, into the distribution problem.[3]

A *solution* to the distribution problem is an assignment of sets of goods to each member of the population subject to the above constraints. (By a *good* I mean any object of value which is capable of being assigned to any arbitrary member of the population. The point of this restriction is to eliminate from consideration those objects judged to be of value which cannot be, for whatever reason, objects of exchange.) Clearly the "problem" here is not with finding a solution, for many solutions exist: any function $f_S : G \to P$ gives a solution satisfying our two constraints. The problem concerns *selecting* a particular solution (or solutions) out of the many possible ones available to receive the label "fair" or "just."

Two particular distribution problems are of particular philosophical interest because their simple structures seem to correspond to primitive principles, or norms, of distributive justice. These distribution problems have the additional virtue of being extensively studied by economists, resulting in a large body of theoretical and empirical results which may be brought to bear on the problem. In the rest of this section, I shall describe the two problems which shall receive our attention for the rest of this paper, with brief summaries of some of the relevant experimental work on the subject. Throughout I endeavor to explain why these particular problems deserve our attention, their simple forms notwithstanding.

The Nash bargaining game

The simplest distribution problem of interest involves allocating a divisible good G, which I shall refer to as "cake," between two people.[4] For simplicity, we assume the measure of the amount of cake is such that the total amount of cake available for distribution is 10, where the units of measurement correspond to a natural quantity, e.g., slices of cake. Furthermore, let us

assume that the two individuals are perfectly symmetric in all relevant respects. This assumption insures that the good is equally useful to each person, each person has the same need, and that no person has a prior claim on the good that would trump the other person's claim, among other things.

In this symmetric case, we have strong intuitions urging that the "just" or "fair" distribution allocates exactly half of the cake to each person. Our common intuitions suggest that the relevant principle of distributive justice almost everyone holds is the one singling out the equal split as the (uniquely) just solution. Although few would object to saying that the equal split is the correct principle of distributive justice to hold, at least for the perfectly symmetric case of divide-the-cake, studies indicate that this intuition is, in fact, widely shared.

In 1974, Nydegger and Owen conducted an experimental test of people's behavior for the game of divide-the-cake (although they had subjects divide a dollar instead of a cake). Not surprisingly, they found that *all* pairs of subjects agreed on the 50–50 split. While some doubt over the generality of the results may be warranted given the small, biased sample size, their claim seems correct that, "The outcome of this study is quite impressive if for no other reason than the consistency of its results." (Nydegger and Owen, 1974, page 244) seems correct.

That people *do* ask for half of the cake in a perfectly symmetric situation is indisputable. Explaining *why* people always ask for half of the cake is more difficult. The equal split in the game of divide-the-cake is an equilibrium in informed, rational self-interest (also known as a Nash equilibrium) in that each player's request is optimal given the other player's request.[5] However, given the particular structure of divide-the-cake, *every* solution which does not give all of the cake to one player is a Nash equilibrium. Yet the common game-theoretic solution concept of a Nash equilibrium does not help us identify *why* the equal-split ought to be favored over any other alternative.

It should be noted that Nash (1950) presents an argument which singles out, in certain cases, the equal-split outcome from the many other Nash equilibria possible. Chronicling the objections made against, and virtues of, Nash's approach would take me too far afield; however, the interested reader may consult Luce and Raiffa (1957, pp. 128–134) for an excellent criti-

---

[3]For a discussion of the effects situational considerations have on the favored solution to particular distribution problems, see Yaari and Bar-Hillel (1984).

[4]The one-person distribution problem has an obvious solution—the person receives everything—that I take to be unproblematic. Conceivably, though, one could argue that if the person were incapable of using all of the good, giving her more than she could use would be unjust since it eliminates the possibility of another individual using the remainder of the good. In all of the cases of the distribution problem I consider, these sorts of complicating contextual factors are assumed to be impossible, simply because the additional complexity introduced obscures the basic problem.

[5]If both A and B ask for half of the cake, neither player can improve her situation by changing her request, provided the other player's request remains fixed. For example, if A decides to ask for 60% of the cake when B still requests 50%, the total amount they ask for overshoots the amount of cake available, and each player receives nothing.

cism of Nash's approach. Skyrms (1996) contains a good discussion on why explaining the equal-split outcome of divide-the-cake should be so difficult.

One might suspect that the difficulty in explaining the equal-split outcome is primarily due to certain artificialities in our framing of the problem: since agents only play the game once, there is no opportunity for them to communicate their preferences to the other player. In an *iterated* distribution problem, the iteration creates a kind of communication between the agents, allowing one agent to, in effect, tell the other that she will not accept an offer below a certain threshold, possibly allowing the other to coordinate on the "fair" solution. Consider the following variation of the Nash game: as before, the two players must agree on how to divide a cake but, unlike before, we do not require them to arrive at an agreement by the end of the first round of play. At any particular time it is one player's turn to suggest a possible division of the cake. If the other player accepts the proposed division, the game ends; if the other player does not accept the proposal, then the two swap roles and the other player may suggest a division. This alternation continues until a decision is reached. The catch is that the total amount of cake available for dividing decreases as time passes—after t minutes, only $100\delta^t\%$ remains, for some $\delta \in [0, 1]$.

In this new setting, there are still an infinite number of Nash equilibria, but the majority of them fail to be subgame perfect. Rubenstein (1982) showed that if we require the agents to follow subgame perfect strategies, there is a single equilibrium in which the first player proposes slightly more than half of the cake and the second player agrees immediately to this distribution. Initially, this result may seem to settle the question as to why people tend to ask for half in divide-the-cake type games, but closer examination reveals that the matter is not so simple. Kreps (1990) notes that if the cake is not infinitely divisible, the problem of multiple equilibria reappears. Moreover, if the two players differ in their response time, the subgame-perfect equilibrium selected distributes the good proportionally to the response rate of individual agents; and if each agent incurs a cost when making a proposal, then the agent incurring the smallest cost receives the majority of the cake. Since all three phenomena (discrete granularity of the good, variable response time, and variable cost of interaction) are present in real bargaining situations, we see that considering the iterated game as a way to settle the question of why people ask for half of the cake does not suffice.

The ultimatum game

A slightly more complicated distribution problem arises when we lift the requirement (present in divide-the-cake) that the two agents are perfectly symmetric. The *ultimatum* game provides a useful example of such a game. Here we again have two agents A and B who need to determine how best to share some good (say a cake) between them. In the ultimatum game we assume that one party, say A, has initial possession of the cake, and presents B with an offer of how much of the cake A is willing to give B (this is the ultimatum). B may either accept or reject the offer. If B rejects the offer, A and B each receive nothing (if you want a reason for this, suppose they begin arguing and the cake spoils). If B accepts the offer, each person receives the appropriate amount of cake. The fact that each player in this game has distinct (and different) roles makes the extended-form representation shown in figure 1 the most natural one.

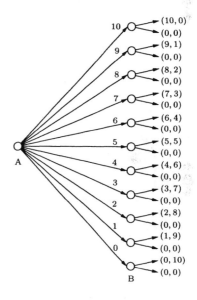

Figure 1: Extended form representation of the ultimatum game

Intuitions as to the "just" resolution to this particular distribution problem seem to vary more than for the Nash bargaining game. A fairly strong intuition exists which says that, since neither player has any special need or entitlement to the cake (the fact that A has possession of the cake is seen as a historical accident), the equal split, or at least something relatively close to the equal split, is again perceived as the only just outcome.

Between 1982 and 1990, many experiments were conducted to determine people's actual behavior in the ultimatum game, as the original results reported by Güth (1982) were quite surprising, reporting that people's actual behavior did not conform to the game-theoretic predictions. (Traditional game theory predicts player B will accept any nonzero offer; after all, some of the cake is better than none of the cake, and so as long as player A leaves player B with some cake, B should take it.) Although some people behaved in accordance with the "unjust" solution predicted by traditional game theory, the modal offer was the equal-split. Numerous follow-up experiments attempted to isolate the factors eliciting this response.[6] Ultimately, though, the experimental evidence suggests that there exists a principle of distributive justice, guiding people's behavior in the ultimatum game, which favors the equal split (or something close to it).

3. Evolutionary Explanations of the Equal Split

The Nash bargaining game

Skyrms (1996) offers an evolutionary explanation of why the equal split may be so widely followed, using the replicator dynamics of Taylor and Jonker (1978). As detailed discussions of his model may be found elsewhere (Skyrms, 1996), I shall only sketch the details here. For simplicity, assume that the cake is sliced into 10 pieces, and that individual requests are restricted to an integer number of slices. We represent the total state of the population by a vector $\vec{s} = (s_0, s_1, \ldots, s_{10})$ where s_i denotes the proportion of the population asking for i slices. Assume that the current "growth" rate of the types of individuals asking for i slices of the cake is approximately equal to the expected fitness of the demand i request in the population $\vec{s}$ [denote this by $F(i|\vec{s})$]. According to the replicator dynamics, the rate of change of the proportion of individuals requesting i slices of the cake is given by $\frac{ds_i}{dt} = s_i\big(F(i|\vec{s}) - F(\vec{s}|\vec{s})\big)$, where $F(\vec{s}|\vec{s})$ denotes the average fitness of the population. One should note that we are describing a cultural evolutionary process here, where the evolutionary dynamics describe changes in *beliefs*, rather than an evolutionary process operating on the biological level.

In a series of 100,000 trials, each trial beginning from a randomly selected point in the state space of the population, the final (converged) state of the population is distributed as follows:

Polymorphism	Count
Fair division	60822
4–6	28131
3–7	9324
2–8	1663
1–9	60
0–10	0

Notice that the population evolves to a state where fair division is the predominant norm approximately 60% of the time. The other rows in the table correspond to polymorphic states of the population where some fraction of the population requests i slices of the cake and the remainder requests $10 - i$ slices.

This provides the start of an evolutionary explanation of the norm of fair division in the Nash bargaining game, but it has the unfortunate consequence that it depends, rather heavily, on the initial conditions of the population. In terms of an explanation of why we think the equal split is *just*, this account appears a bit wanting. Although it offers some explanation of why the norm of fair division proves so widespread (namely, the initial conditions of our population lay within the basin of attraction for fair division), it does not explain why we think we *ought* to ask for half of the cake: had the initial conditions been otherwise, the evolutionary dynamics would have carried the population to (say) the 4–6 polymorphism, and people's beliefs as to what the appropriate sort of division was would be very different.

Skyrms does show that if interaction in the model is correlated (that is, people are more likely to interact with people following the same strategy than another), then once the degree of correlation exceeds a certain value, the basin of attraction for fair division expands to the interior of the state space. This seems to provide a plausible beginning to an evolutionary explanation of (certain) norms of distributive justice.

The ultimatum game

Skyrms also constructs a replicator dynamic model of the ultimatum game in order to see whether a similar evolutionary explanation of the norm of fair division existing in that game can be provided as well. In order to make such a model tractable, one needs to restrict the set of possible strategies, for even if we assume that the cake divides into only ten slices, there are $11 \cdot 2^{11}$ possible strategies an individual may follow.[7] One particularily interesting group of strategies to study is as follows:

---

[6]See, for example, Binmore *et al.* (1985), Güth and Tietz (1985), Neelin *et al.* (1988), Ochs and Roth (1989), Roth *et al.* (1991). For comprehensive surveys of the relevant experimental results, see Thaler (1988), Roth (1995), and Güth and Tietz (1990).

[7]Each strategy consists of two parts: the amount of cake one offers when one has possession of the cake, and the offers one is willing to accept. There are 11 possible offers one may make (offer 0 slices,..., offer 10 slices) and 2^{11} possible acceptance strategies.

Strategy	Offer	Accept
Gamesman	1	anything
S2	1	nothing
S3	1	accept 5, reject 1
Mad Dog	1	accept 1, reject 5
Easy Rider	5	anything
S6	5	nothing
Fairman	5	accept 5, reject 1
S8	5	reject 5, accept 1

Not every initial state of the population converges to a state where a "fair" or "just" strategy (Fairman or Easy Rider) dominates. A population in which all strategies appear equally likely converges to a state containing (roughly) 87% Gamesman and 13% Mad Dogs (Skyrms, 1996, pg. 31). However, certain initial population proportions *do* lead to states where only the strategies of Fairman and Easy Rider are present. This account still falls prey to the previous criticism of unacceptable dependence on the initial conditions.

Criticisms of Skyrms's model

Several criticisms of Skyrms's project of providing an evolutionary account of distributive justice have appeared in the philosophical literature. Since covering these criticisms in detail would take me too far from my present purpose, the interested reader may wish to consult Barrett (1999), Kitcher (1999), and D'Arms *et al.* (1998) for further discussion. I shall concentrate here on two criticisms:

1. The appropriateness of using the replicator dynamics to model human populations.

2. Concerns regarding Skyrms's introduction of correlation in his model of the Nash bargaining game.

First, using the replicator dynamics to model human populations requires that one make two assumptions which, taken together, seem highly implausible. When deriving the replicator dynamics, one needs to assume that the size of the population is sufficiently large to warrant identifying individual fitness with expected fitness. (This allows one to keep track of the evolution of the proportions of each type of strategy in the population.) Unfortunately, one also needs to assume that any two members of the population are equally likely to interact. While it may be true for sufficiently small populations of humans that any two interactions are equally likely, the plausibility of this decreases as the population size increases. By the time one reaches a population of the size of, say, New York, it certainly is no longer true that any two members are equally likely to interact.

Second, in his model of the Nash bargaining game, Skyrms only considers the effect of positive correlation

among strategies. D'arms *et al.* point out that, while this makes sense for strategies which request at most half of the cake, this does not make sense for strategies asking for *more* than half of the cake. If one is going to introduce correlation into the model, one needs to allow for both positive and negative correlation. Negative correlation, in the Nash bargaining game, would correspond to some sort of avoidance behavior, where, say, individuals asking for six slices of the cake would try to steer clear of individuals asking for the same amount. D'Arms *et al.* construct a model, similar to Skyrms's, finding that when one allows for negative as well as positive correlation, the unfair polymorphisms reappear.

4. An agent-based, social network model

Description

In this section, I describe an agent-based social network model which improves upon Skyrms's replicator dynamic model, in the sense that it gives more robust results concerning the emergence of fair division for the Nash bargaining game. As before, for sake of simplicity, we assume that the cake is sliced into 10 pieces, and that individual requests are restricted to an integer number of slices. We replace the replicator dynamic assumption that we have an essentially infinite population by the assumption that the population P under consideration has only finitely many finitely many agents. Each agent in the population has a particular belief (or strategy) determining her behavior in the game of interest. Furthermore, we assume that an individual interacts only with those people who stand in some appropriate social relation to her. In general, these relations could be given by any connected graph whose nodes are the individual agents in the population. In this paper, I assume that the underlying social network has the form of a square lattice, where each agent is connected to some subset of the Moore 24 neighborhood. These relations are considered fixed since we assume individual beliefs change much faster than an agent's social relations. Finally, we assume that individuals follow some sort of imitative rule which determines how they change their beliefs over time.

The evolutionary dynamics used in this model are relatively common: at the start of each generation, each player receives a score equal to the number of slices of cake she receives when playing the appropriate game (either the Nash bargaining game or the ultimatum game) with her neighbors. At the end of each generation, an agent will change her strategy if some other agent in her neighborhood earned a higher score.

We allow agents to use one of four different update rules, each rule having a certain degree of plausibility.

The first update rule considered is "imitate the best neighbor." A very common update rule [see Nowak and May (1992; 1993), Lindgren and Nordahl (1994), Huberman and Glance (1993), and Epstein (1998)]. Each agent looks at her neighbors and mimics the strategy of the neighbor who did the best, where "best" means "earned the highest score." If ties occur, agents choose a random strategy by essentially flipping a coin.

The second update rule is "imitate with probability proportional to success." As before, each agent compares her score with those of her neighbors, modifying her strategy only if at least one neighbor did strictly better. However, instead of ignoring those neighbors who did better but not well enough to include their strategy in the set of highest-scoring ones, this update rule assigns to every neighbor who did better than an agent a nonzero probability that the agent will adopt her strategy. [For a formal description of this update rule, and the others, see Alexander (1999).]

The last two update rules are "imitate the strategy with the best expected payoff" and "adopt the best response strategy." Under the former rule, agents calculate the expected payoff of each strategy in their neighborhood, selecting the one with the highest value. With the latter rule, agents compute the best response strategy assuming that, in the next generation, none of their neighbors will change their strategies.

Obviously all of these update rules provide only rough approximations of the sort of rules real human agents would use. However, they are reasonable approximations in that they assume individual agents will use some rough-and-ready heuristic when determining what strategy to use in the next generation, instead of a computationally intensive optimization algorithm which only the most cognitively sophisticated could employ. Most of the time we find that the *simpler* update rules are the ones conducive to the evolution of norms. Perhaps our norms of fair division arose because they are the sorts of behaviors most beneficial to boundedly rational agents whose interactions are constrained by social networks.

The Nash bargaining game

Figure 2 illustrates the evolution of a 200×200 world in which all strategies are equally likely. Each block represents a single agent, and different shades of gray indicate how many slices of the cake each agent requests. Each slide portrays the state of the model after one generation, so the entire evolutionary process shown in figure 2 represents only nine generations. The dark gray color dominating the last few images corresponds

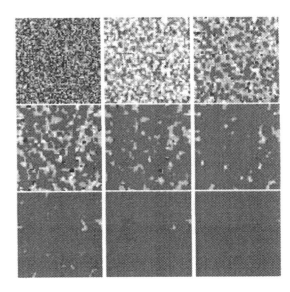

Figure 2: Fair division emerging from uniform random conditions

to the strategy of fair division.

Figure 3 shows the ability of fair division to invade a world starting at one of the unfair Nash equilibria. A single agent adopting the strategy of fair division is sufficiently successful to initiate the spread of fair division throughout the population. Figure 4 illustrates the robustness of fair division in the presence of "mutation." In a world randomly initialized (with all strategies equally likely) with a mutation rate of 5%, fair division still becomes the dominant strategy with only the expected amount of mutational noise occurring in the background.

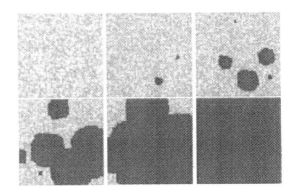

Figure 3: Fair division emerging from 4–6 polymorphism

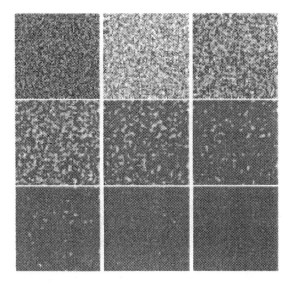

Figure 4: Fair division persisting in the face of mutations ($\mu = 0.05$)

Table 1 lists convergence results based on neighborhood and dynamic. In the column labeled "Dyn," "1" indicates the "imitate with probability proportional to success" update rule, "2" the "imitate best neighbor" update rule, and "3" the "imitate the strategy with best expected payoff." One can see that virtually no dependence on the neighborhood size or update rule exists, provided that we only consider agents using one of these three update rules. Figure 5 suggests, though, that using the "best response" update rule can give quite different results.

Nbhd	Dyn	Polymorphism						
		0–10	1–9	2–8	3–7	4–6	5	Other
VN	1	0	0	0	0	29	9970	1
	2	0	0	0	0	26	9966	8
	3	0	0	0	0	13	9984	3
M(8)	1	0	0	0	0	26	9973	1
	2	0	0	0	0	26	9908	66
	3	0	0	0	0	24	9970	6
M(24)	1	0	0	0	8	110	9879	3
	2	0	0	0	21	220	9721	38
	3	0	0	0	0	62	9934	4

Table 1: Convergence results based on neighborhood and dynamic

The reason for this odd behavior under the best response rule can be easily appreciated. Agents surrounded by a majority of neighbors who demand four slices of the cake will compute that the best-response strategy in the next generation is to request six slices of the cake. In a region consisting primarily of agents who request four slices, in the next generation all of those

agents will request six slices. Of course, the best response strategy will then be to request four slices of the cake, and so on. Given the specification of the Nash bargaining game, this oscillatory behavior leads to a horribly *suboptimal* result for all of the agents, with a long-run average payoff of only two slices of cake; this is considerably less than what they would receive if they used a less sophisticated update rule (such as "imitate the best neighbor") and converged to a state where everyone asked for half of the cake.

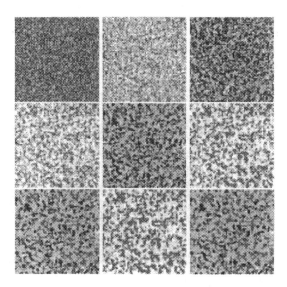

Figure 5: The disadvantageous best response update rule

The ultimatum game

The agent-based, social network model of the ultimatum game considered in this section is virtually the same as the model of the Nash bargaining game discussed in the previous section. The only relevant difference is that, on account of the slightly more complicated nature of the ultimatum game, we need to introduce an additional calculation in each generation. Before calculating each agent's score, for each interaction between two agents we randomly assign one agent the role of ultimatum giver and the other agent the role of ultimatum receiver. An agent's score equals the sum of the individual payoffs earned when that agent plays the ultimatum game with each of her neighbors, where each pairwise ultimatum game uses the assignment of roles made at the start of the generation.

Given the extent to which the agent-based, social network model increased the probability of fair division emerging for the Nash bargaining game, one might expect a similar effect to occur for the ultimatum game. (In the ultimatum game, the analogous effect would be to have the evolutionary dynamics cause the Fairman and/or Easy Rider strategy to acheive predominance.) However, it turns out that in the agent-based model described above the Fairman strategy *fails* to dominate in the vast majority of cases.[8]

Figure 6 illustrates how, under the "imitate best neighbor" update rule, the Gamesman and Mad Dog strategies dominate. Since we are now considering the underlying game to be the ultimatum game, rather than the Nash bargaining game, it should be noted that the color scheme used in figures 6, 7, and 8 has a different meaning than before. In these three figures, black represents the Gamesman strategy, and the darker gray color, Mad Dog. In the final image of figure 6, the only surviving strategies are Gamesman and Mad Dog.

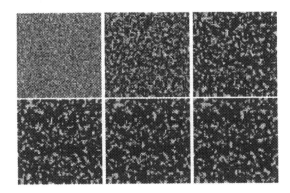

Figure 6: The emergence of gamesman and mad-dogs

Under the update rule of "imitate the best neighbor," the Fairman strategy lacks the robustness properties possessed by the demand-half strategy in the Nash bargaining game. For example, figure 7 shows how a pure population of Fairman (indicated by light gray) can eventually be invaded and overwhelmed by Gamesmen if we allow a very small amount of mutation. Since the sequence of figure 7 was not sampled at constant time intervals (unlike the rest of the figures in this paper), I indicate the exact generation of each image explicitly.

[8]In the context of the ultimatum game, by a "fair-playing" strategy I mean either Fairman or Easy Rider, since these are the only two self-consistent strategies which offer half of the cake. I take the other two strategies which offer half of the cake to be ones which we would not expect rational agents to adopt, since they are not self-consistent—they refuse the very offers they make! (See the table in section .)

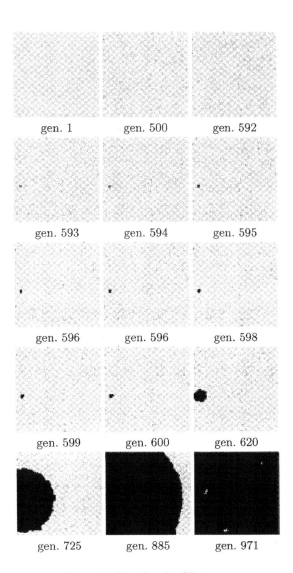

gen. 1   gen. 500   gen. 592

gen. 593   gen. 594   gen. 595

gen. 596   gen. 596   gen. 598

gen. 599   gen. 600   gen. 620

gen. 725   gen. 885   gen. 971

Figure 7: The death of Fairman

During the first 500 generations, the only mutant strategies which survive in a pure Fairman population are Easy Riders. By the 500th generation, though, enough Easy Riders mutants have appeared to allow a Gamesman mutant to flourish. Once a reasonably sized Gamesman cluster has appeared (which occurs in the sequence of figure 7 by generation 593), Gamesmen may spread into regions occupied by Fairman without needing to piggyback on the presence of Easy Riders.[9] Over time, the population will eventually arrive at a state consisting primarily of Gamesmen, with a few Mad Dogs.[10]

However, if we modify the game slightly, allowing Fairmen to "punish" agents who do not make fair offers, we find that Fairmen, who previously became extinct in a few generations, may persist in the hostile environment created by the presence of Gamesmen. Figure 8 illustrates the evolution of a population in which the Fairman strategy punishes greedy neighbors (which, in this context, means merely that when a "greedy" strategy interacts with the Fairman strategy, they receive a negative payoff). The presence of the Fairman strategy, represented by light gray, shows how the strategy corresponding to our norm of fairness persists when the parameter controlling the severity of the punishment exceeds a certain value.

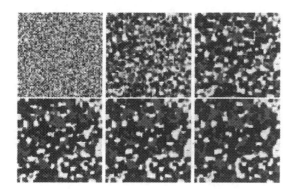

Figure 8: The emergence of fair play in the ultimatum game

---

[9]Since Easy Riders accept any offer, a Gamesman mutant appearing in a region with a high concentration of Easy Riders may exploit their presence to obtain a critical foothold in the world. The random assignment of roles may fall in favor of the Gamesman mutant, meaning that he will make offers to Easy Riders and receive offers from Fairman. When this happens, the Gamesman mutant will typically earn a score high enough to persist into the next generation, spreading his strategy to neighboring agents.

[10]Mad Dogs appear by mutation and may survive in a pure Gamesman population since, when no other strategies are present, they are indistinguishable from Gamesmen.

6. Conclusion

There exists a rich philosophical tradition which seeks to ground norms of moral and political obligation in the self-interested actions of individual agents. The models of this paper demonstrate that the emergence of norms can depend on dynamical considerations which are not immediately apparent. In particular, these models suggest that our norm of fair division in games having the structure of the Nash bargaining game depends on the constraint of some underlying social network on our interactions. In addition to this constraint, in the ultimatum game the models suggest that acquiring the ability to punish "deviant" strategies plays an essential role in the development of our concept of justice. The primary point is that evolutionary accounts of norms which neglect to take seriously the underlying dynamics, and the structure of the underlying game, do so upon risk of descriptive inaccuracy and hence do not provide a plausible account of normativity.

References

Jason M. Alexander. The (spatial) evolution of the equal split. Technical report, Institute for Mathematical Behavioral Sciences, U.C. Irvine, 1999.

Martin Barrett, Ellery Eells, Branden Fitelson, and Elliott Sober. Models and reality—a review of Brian Skyrms' *Evolution of the Social Contract*. *Philosophy and Phenomenological Research*, 59(1):237–241, March 1999.

K. Binmore, A. Shaked, and J. Sutton. Testing noncooperative bargaining theory: A preliminary study. *The American Economic Review*, 75(5):1178–1180, December 1985.

Justin D'Arms, Robert Batterman, and Krzyzstof Górny. Game theoretic explanations and the evolution of justice. *Philosophy of Science*, 65:76–102, March 1998.

Joshua A. Epstein. Zones of cooperation in demographic prisoner's dilemma. *Complexity*, 4(2):36–48, 1998.

W. Güth and R. Tietz. Strategic power versus distributive justice. An experimental analysis of ultimatum bargaining. In H. Brandstätter and E. Kirchler, editors, *Economic Psychology. Proceedings of the 10th IAREP Annual Colloquium*, pages 129–137. Rudolf Trauner Verlag, Linz, 1985.

Werner Güth and Reinhard Tietz. Ultimatum bargaining behavior: A survey and comparision of experimental results. *Journal of Economic Psychology*, 11:417–449, 1990.

Werner Güth, Rolf Schmittberger, and Bernd Schwarze. An experimental analysis of ultimatum bargaining. *Journal of Economic Behavior and Organization*, 3:367–388, 1982.

Bernardo A. Huberman and Natalie S. Glance. Evolutionary games and computer simulations. *Proc. Natl. Acad. Sci.*, 90:7716–7718, August 1993.

Philip Kitcher. Games social animals play: Commentary on Brian Skyrms' *Evolution of the Social Contract*. *Philosophy and Phenomenological Research*, 59(1):221–228, March 1999.

David M. Kreps. *Game Theory and Economic Modelling*. Oxford University Press, 1990.

Kristian Lindgren and Mats G. Nordahl. Evolutionary dynamics of spatial games. *Physica D*, 75:292–309, 1994.

R. Duncan Luce and Howard Raiffa. *Games and Decisions: Introduction and Critical Survey*. John Wiley and Sons, Inc., 1957.

John F. Nash. The bargaining problem. *Econometrica*, 18:155–162, 1950.

Janet Neelin, Jugo Sonnenschien, and Matthew Spiegel. A further test of noncooperative bargaining theory: Comment. *The American Economic Review*, 78(4):824–836, September 1988.

Martin A. Nowak and Robert M. May. Evolutionary games and spatial chaos. *Nature*, 359:826–829, October 1992.

Martin A. Nowak and Robert M. May. The spatial dilemmas of evolution. *International Journal of Bifurcation and Chaos*, 3(1):35–78, 1993.

R. V. Nydegger and G. Owen. Two-person bargaining: An experimental test of the Nash axioms. *International Journal of Game Theory*, 3(4):239–249, 1974.

Jack Ochs and Alvin E. Roth. An experimental study of sequential bargaining. *The American Economic Review*, 79(3):355–384, June 1989.

Alvin E. Roth, Vesna Prasnikar, Masahiro Okuno-Fujiwara, and Shmuel Zamir. Bargaining and market behavior in Jerusalem, Ljubljana, Pittsburgh, and Toyko: An experimental study. *The American Economic Review*, 81(5):1068–1095, December 1991.

Alvin E. Roth. Bargaining experiments. In J. Kagel and A. Roth, editors, *The Handbook of Experimental Economics*, chapter 4, pages 253–348. Princeton University Press, 1995.

A. Rubinstein. Perfect equilibrium in a bargaining model. *Econometrica*, 50:1123–42, September 1982.

Brian Skyrms. *Evolution of the Social Contract*. Cambridge University Press, 1996.

Peter D. Taylor and Leo B. Jonker. Evolutionary stable strategies and game dynamics. *Mathematical Biosciences*, 40:145–156, 1978.

Richard H. Thaler. Anomalies: The ultimatum game. *Journal of Economic Perspectives*, 2(4):195–206, 1988.

Menachem E. Yaari and Maya Bar-Hillel. On dividing justly. *Social Choice and Welfare*, 1:1–24, 1984.

Engineering, Emergent Engineering, and Artificial Life: Unsurprise, Unsurprising Surprise, and Surprising Surprise

Edmund M. A. Ronald[1], and **Moshe Sipper[2]**

[1]Centre de Mathématiques Appliquées, Ecole Polytechnique, 91128 Palaiseau, France. eronald@cmapx.polytechnique.fr.
[2]Logic Systems Laboratory, Swiss Federal Institute of Technology, 1015 Lausanne, Switzerland. Moshe.Sipper@epfl.ch.

Abstract

We examine the eventual role of surprise in three domains of human endeavor: classical engineering, what we call "emergent engineering," and the general unrestricted field of artificial life. Our study takes place within the formal framework of the recently proposed "emergence test." We argue that the element of surprise, central in the test, serves to illuminate the fundamental differences between these three fields. This we achieve by distinguishing between three different forms of surprise: unsurprise, unsurprising surprise, and surprising surprise.

Technofright

Given the choice, would you care to entrust your one and only living body, even fleetingly, to the structurally stable bridge depicted below, which was designed by evolution [4, 5]?

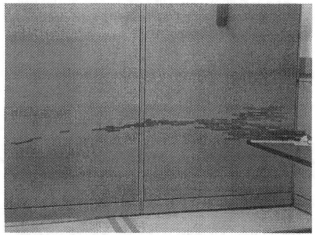

Copyright (c) 1999 Pablo Funes & Jordan Pollack. Used by permission.

We think not.

Introduction

The above bridge design appears amusing when presented as a toy, yet irrationally frightening when envisaged as a life-or-death experience. On reflection, "irrationally frightening" boils down to "what is this weird new thing?": evolutionary computation has generated an inchoate structure, totally lacking recognizable patterns such as pillars, arches, frames, stays, or cables. This bizarre creation fails altogether the test of trust-by-familiarity, a test each of us applies routinely, e.g., to food (is it healthy?), or to airplanes (are two engines really enough?). Like a new airplane design, the evolved structure evokes in us a sense of amazement, or surprise, but a rather uneasy one at that, when we think of entrusting our lives to it.

Thus, the acceptance or rejection of an engineering application is conditioned by factors which can be qualified as emotional, but rest on some quite real criteria. In particular, novelty and risk mix badly in the public eye, while widespread understanding of a method's theoretical underpinning will confirm its soundness.

Engineers commonly employ nowadays techniques classified as "emergent," and this tendency will most likely increase as artificial life (Alife) moves from the laboratory to the field. At least two bio-inspired emergent methodologies are already in widespread engineering use: evolutionary algorithms [3] and artificial neural networks [8]. Newer emergent methods will also mature from the science to the engineering stage, e.g., cellular computing [12] and ant algorithms [2].

When engineers use technologies which exhibit (putative) emergent properties, rather than classical design techniques, they are faced with a problem: how can you use an emergent technique and still guarantee the customer that the supplied product is foolproof? Cars come with guarantees; artificial neural networks should too, yet usually do not: who would sign a guarantee certificate stating that the neural network-based handwriting recognizer will always work perfectly with *your* handwriting, or stating that the voice-activated computer will reliably follow *your* commands?

A widespread sense of uneasiness undoubtedly accompanies the introduction of emergence in engineering. Although not as extreme as the jitters induced by sight of the experimental bridge reproduced above, doubts are commonly expressed about electronic devices being employed to control heavy machinery (cars, airplanes)— and which develop a will of their own. In our exploration of the roots of such misgivings we have been led to reflect on the role of surprise in engineering.

In this paper we examine three domains of human endeavor: classical engineering, what we call "emergent engineering," and artificial life. Our study takes place within the framework of the recently proposed "emergence test" [9, 10]. We argue that the element of surprise, inherent in the test, serves to illuminate the fundamental difference in the confidence we accord to products of the aforementioned three fields. We will reach a distinction between three different degrees of surprise implied in the design process: unsurprise, unsurprising surprise, and surprising surprise.

The plan of the paper is as follows: in the next section we briefly summarize the emergence test, and use it to justify conferring the emergence label on neural-network technology. Next, we present the three forms of surprise. In the subsequent three sections we discuss, respectively, classical engineering, emergent engineering, and the differences between the two. We then focus on the enterprise of artificial life and the various forms of surprise it involves, ending with some concluding remarks.

The Emergence Test

In this section we shall first recapitulate the setup associated with our emergence test, and then use the test to confer the emergence label on neural-network classifiers.

The test is an operant definition in the spirit of Turing's intelligence test, with the aim of tagging a given construction as emergent. Originally (along with our colleague Mathieu Capcarrère), we presented the *emergence test* as a sort of emergence certification mark which would garner approval from the Alife community [9, 10]. Herein, we employ the test in the manner of a definition, namely, as a tool for reasoning about the properties of emergent and non-emergent phenomena.

As noted in our previous paper [9], the test is aimed at what Herbert Simon called the "sciences of the artificial" [11], of which artificial life is a quintessential example. The test consists of three criteria—design, observation, and surprise—for conferring the emergence label.

Assume that the scientists attendant upon an Alife experiment are just two: a system designer and a system observer (both of whom can in fact be one and the same), and that the following three conditions hold:

(i) **Design.** The system has been constructed by the designer, by describing *local* elementary interactions between components (e.g, artificial creatures and elements of the environment) in a language $\mathcal{L}_1$.

(ii) **Observation.** The observer is *fully aware* of the design, but describes *global* behaviors and properties of the running system, over a period of time, using a language $\mathcal{L}_2$.

(iii) **Surprise.** The language of design $\mathcal{L}_1$ and the language of observation $\mathcal{L}_2$ are distinct, and the causal link between the elementary interactions programmed in $\mathcal{L}_1$ and the behaviors observed in $\mathcal{L}_2$ is *non-obvious* to the observer—who therefore experiences surprise. In other words, there is a cognitive dissonance between the observer's mental image of the system's design stated in $\mathcal{L}_1$ and his contemporaneous observation of the system's behavior stated in $\mathcal{L}_2$.

When assessing this clause of our test one should bear in mind that as human beings we are quite easily surprised (as any novice magician will attest). The question reposes rather on how *evanescent* the surprise effect is, i.e., how easy (or strenuous) it is for the observer to bridge the $\mathcal{L}_1$–$\mathcal{L}_2$ gap, thus reconciling his global view of the system with his awareness of the underlying elementary interactions.

The above three clauses, relating design, observation, and surprise, describe our conditions for diagnosing emergence, i.e., for accepting that a system is displaying emergent behavior [9, 10].

We will now use the emergence test to justify our conferring the emergence label on artificial neural-network (ANN) classifiers. These are artificial neural networks which take the description of a pattern as input, and assign this input pattern to one of a number of predetermined classes. Handwritten character recognizers fall into this category, outputting a character value for each input gesture.

- **Design.** The design language $\mathcal{L}_1$ is that of artificial neuron transfer-function definitions, network topologies, and synaptic weights.

- **Observation.** The observation language $\mathcal{L}_2$ is that of input-output behavior, i.e., input patterns and class-membership assignments.

- **Surprise.** While fully aware of the underlying neuronal definitions, of the topological connections, and of the synaptic weights, the observer nonetheless marvels at the performance of the network, in particular its ability to generalize and classify novel inputs, previously unseen patterns—a behavior which he *cannot* fully explain.

? **Diagnosis:** emergent behavior is displayed by ANN classifiers.

Thus, by conferring the *emergence label* on the classifier networks, we formally acknowledge that a significant degree of surprise accompanies any thoughtful consideration of their behavior.

Surprise

In his book *Scientific Literacy and the Myth of the Scientific Method*, Henry H. Bauer wrote [1]: *"To make sense of the tension between innovation and conservatism in science, more helpful than the banal distinction between what is known and what is not known is the discrimination of three categories: the known, the known unknown, and the unknown unknown."* In the same vein, we hold that there are three categories of surprise: (1) unsurprise (i.e., no surprise); (2) unsurprising surprising, where our surprise is confined within well-defined bounds; and (3) surprising surprise, where we are totally and utterly taken aback. We shall show below that classical engineering, emergent engineering, and artificial life, are fundamentally different, in large part owing to the different categories of surprise involved.

Classical Engineering: Unsurprise

Engineering is *"the application of science and mathematics by which the properties of matter and the sources of energy in nature are made useful to people"* (Merriam-Webster online dictionary at www.m-w.com). Viewed through emergence-test spectacles, the engineer's *modus operandi* can be seen as a continual shuffling between $\mathcal{L}_1$ and $\mathcal{L}_2$. Let us demonstrate this via the following scenario:

The motel scenario. Your company has won the tender to build a building in Smallville, for a well-known national chain of motels. As a civil engineer, your boss has assigned you the task of drawing up the plans. The specifications you are given consist essentially of the number of rooms of the building, the standardized motel units that must be used, and the land plot's shape (the plot is supposed large).

The motel company has spoken in the observation language $\mathcal{L}_2$, describing observable (functional and behavioral) properties: the hotel's location, its intended size, and the global expense. Now you, the engineer, will select a layout, e.g., a number of wings and the number of rooms to each wing, and draw up plans.

From $\mathcal{L}_2$ to $\mathcal{L}_1$ you need to translate the *requirement specifications* couched in $\mathcal{L}_2$ terminology into *design specifications* couched in $\mathcal{L}_1$, bricks-and-mortar terminology, i.e., actual plans for the builder. The components you will specify are prefabricated room units which are mostly trucked in, dropped into place and then hooked up. You also need to place corridors and ducting.

Your work in going from $\mathcal{L}_2$ to $\mathcal{L}_1$ is not a single drive down a one-way street: once you have an $\mathcal{L}_1$ design in hand—a layout—you will need to traverse in the opposite conceptual direction, from $\mathcal{L}_1$ to $\mathcal{L}_2$, to check that the cost requirements are met. Of course, you can see immediately whether the complete construction fits in the alloted space; in this respect the design and specification languages $\mathcal{L}_1$ and $\mathcal{L}_2$ overlap.

The $\mathcal{L}_2$ specifications of the client may be altered while the engineer's work is in progress, for instance the hotel chain may add a few rooms. Such a change might be accommodated by merely lengthening a wing. Or cost projections might cause the client to amend his initial specs. The shuffling back and forth between $\mathcal{L}_1$ and $\mathcal{L}_2$, amending things on one side and checking the effects on the other side, is at the heart of the engineering enterprise.

There are no surprises, there should be no surprises in the scenario described above: in classical engineering we always seek *unsurprise* (no surprise). The engineer in these non-emergent classical domains has at his disposal a set of scientific theories, which supply him with a satisfactory model of how his $\mathcal{L}_1$ constructions will behave when observed in $\mathcal{L}_2$ by the "client."

Regrettably though, the real-world will *always* provide some variation to the models which scientists elaborate. Concerning this, Albert Einstein said: *"As far as the laws of mathematics refer to reality, they are not certain, and as far as they are certain, they do not refer to reality."* But apart from the unavoidable gaps which will inevitably develop between the predictions of science-based models and reality, the classical engineer expects no surprises. He will expect his constructions to be used *only* within conceptual areas where the models apply. If surprises arise he will not ignore them, but rather attempt to stamp them out. Classical engineering is a conservative discipline.

Emergent Engineering: Unsurprising Surprise

Whereas in the preceding section we focused on engineering with the help of a theory, which would be labeled as non-emergent, we now wish to move our attention to the case where *emergence* strides onto the scene.

The scanner scenario. The mayor of your town appreciates your engineering skills, and calls you to his office one bright morning to discuss a project. There have recently been several infiltrations to city hall, the mayor explains, and as a result the council has decided to install an automatic scanner at the entrance to the building. The scanner will scan the incoming crowd, and will identify potential troublemakers, whose faces are stored in a database.

To solve the assigned face-recognition task you decide to use an emergent technology: artificial neural networks, whose emergent behavior we discussed earlier. Despite their emergent nature, artificial neural networks are used routinely nowadays by engineers. In particular, they represent an *a priori* promising choice for your scanner problem.

The scanner scenario (cont'd). Having implemented the neural network-based scanner, you invite the mayor and his council to your lab and demonstrate proudly the operation of the device. You report to them that tests have shown the network to recognize faces with a success rate of 98.5%. The mayor is very happy with this figure and reaches immediately for the official city checkbook.

On its first day of effective operation, at the entrance to city hall, the scanner attains a recognition rate of 2.3%.

You are surprised, of course, by this abysmally low performance rate and eventually figure out the $\mathcal{L}_2$ cause: a change in the illumination at city hall is to blame. But you *cannot* bridge the $\mathcal{L}_1$-$\mathcal{L}_2$ gap! You have no idea whatsoever how to explain the failure in $\mathcal{L}_1$ terms: neurons, synapses, synaptic weights. All you can do is retrain the network on location; and training is a process whose goals and procedures are fully described in the $\mathcal{L}_2$ language. But, in all honesty, the emergent nature of neural networks means that you—the designer—are surprised every time you think about them hard, *even when they are working as designed!*

The more you think of "engineering with emergence," or *emergent engineering* as we call it, the more it comes to resound oxymoronically. Emergent engineering, while inherently containing a non-evanescent element of surprise, seeks to restrict itself to what we call *unsurprising surprise*: though there is a persistent $\mathcal{L}_1$-$\mathcal{L}_2$ understanding gap, and thus the element of surprise does not fade into oblivion, we wish, as it were, to take in this surprise in our stride. Yes, the neural network works (surprise), but it is in some oxymoronic sense *expected* surprise: as though you were planning your own surprise birthday party.

Emergent vs. Non-emergent Engineering

Before moving on to examine the wider field of artificial life and its immanent forms of surprise we ask the following: how can the engineer, in commercial practice, decide to deploy emergent approaches in engineering applications if the (unsurprising) surprise effect is omnipresent in the method employed?

During the design phase of a device, which embodies an aspect of emergence, two distinct modalities of unease can arise, and these we wish to distinguish rather than conflate:

1. The engineer's task of *creating a design* may be rendered difficult by the emergent aspects. For example, neural-network training is still a somewhat black art rather than a perfectly predictable process.

2. The behavior of the deployed application may manifest surprising aspects. Thus handwriting recognizers have been known to develop surprising allergies after hitting the market.

Despite these difficulties, emergent engineering is not impossible, as evidenced by multitudinous examples of real-world applications using neural-network techniques, and the gradual adoption of evolutionary algorithms in industry. However, we would like to consider more in detail the practical implications of these caveats which are due to emergence.

- Point (1) above, as impinging mainly on the engineer, requires essentially mental adjustment to a less systematic engineering process. For instance, when employing neural networks or evolutionary optimization, the designer will have to adapt to performing multiple runs of the corresponding stochastic algorithms. These runs may or may not converge to yield the desired quality of solution. Manual tuning of some parameters may prove necessary, and supervising the process can be both expensive in computing resources and psychologically frustrating. With time (maybe years) a process becomes more well understood, emergence evanesces, and the engineer's task becomes a more predictable routine.

- Point (2) above impacts the company selling the product as a whole, as surprises in the real world may generate costly product liability suits. ABS car brakes, for example, are typical of an application which would benefit from the most advanced control algorithms but where the associated legal risks are high (indeed ABS brakes have been known to fail dangerously after hitting the market).

To combat the unease described in Point (2), we would argue that the only acceptable way to gain confidence in such a product is to specify an extremely rigorous testing regime: classical, non-emergent methods induce trust because they rest on well-understood theoretical models, and the envelopes of confidence of those models are known. Emergent methods are always pushing the envelope—yet there is none! Hence, a methodology is needed whereby the engineer can confidently represent to management (and ultimately to clients) that a design has been adequately tested.

Indeed, a practical application of the emergence test might be in the review of risks to which companies may subject their products before launch: when a product conforms to classical engineering practice it will require a minimal amount of due-diligence testing. However, the presence of any technology diagnosed as emergent should trigger a stringent review of the testing process before a design is signed off for production. At the very least such a procedure might prevent embarrassment (as was the case of the Apple Newton PDA whose handwriting recognizer was lampooned nationwide in the Doonesbury comic strip) and in some applications the review of testing may save lives.

Because they are distinct, the two mentioned difficulties induced by emergence surprise can surface independently. An example of a situation where the engineer faces (1) and not (2) is in, say, a product palette packing problem, where each geometric solution to the packing is perfectly understandable and usable, even though a stochastic evolutionary algorithm needs to be invoked to find it. Conversely, an adaptive load-balancing algorithm on a cluster of web servers may be perfectly implemented and effective, but may sometimes suffer such brittle degradation of performance under saturation load that its adoption would pose a risk to the sites that run it.

Artificial Life: Unsurprising and Surprising Surprise

Alife is a constructive endeavor: some researchers aim at evolving patterns in a computer, some seek to elicit social behaviors in real-world robots, others wish to study life-related phenomena in a more controllable setting, while still others are interested in the synthesis of novel lifelike systems in chemical, electronic, mechanical, and other artificial media. Alife is an experimental discipline, fundamentally consisting of the observation of run-time behaviors, those complex interactions generated when populations of man-made, artificial creatures are immersed in real or simulated environments. Published work in the field usually relates the conception of a model, its instantiation into real-world or simulated objects, and the observed behavior of these objects in a collection of experiments [9].

Perusing the Alife literature one can discern three underlying motivations driving practitioners in the field. We argue that each motivational category goes hand in hand with a different category of surprise.

1. Alife as a tool for investigating natural phenomena of interest (what Langton called *life-as-we-know-it* [6]). This includes, for example, studying the relations between learning and evolution, or investigating insect behavior using robots. This view of Alife often goes by the name of "weak artificial life."

   Within this category of Alife practice we might put forward a theory and try to confirm it using Alife methods; in this case we would hope for complete and utter unsurprise: confirmation of our theory. Or, at most, we would accept facing an unsurprising surprise: some new explanation of the phenomenon in question, to emerge out of our Alife experiment, but which—though surprising—is not "too" surprising.

2. Alife as a method for tackling vaguely defined problems. When we use an evolutionary algorithm to evolve a bridge or an artificial neural network that recognizes faces we have a very precise problem definition—we are dealing with *engineering*. Using an evolutionary setup to study flocking behavior, however, involves the study of a vaguely defined problem: we are not necessarily interested in the natural phenomenon of flocking (as in item 1), nor do we have a precise definition (e.g., as embodied by a fitness function) of the exact behavior we expect.

   In this category we are using Alife as a metaphorical lunar rover to explore new, unfamiliar terrain. Though we aim to find new data or novel phenomena, we wish to stay firmly seated in our comfortable rover, that is, we are seeking unsurprising surprise. If our flocking experiment suddenly produces totally bewildering behavior (perhaps not even flocking) then we shall tumble off our virtual rover: we have just experienced surprising surprise.

3. This last category of Alife practice is usually called "strong artificial life": seeking to bring about new forms of life, or *life-as-it-could-be* as dubbed by Langton [6]. This form of Alife (as yet unattained) involves *surprising surprise*: the new form of life is, *ipso facto*, entirely novel, hence producing a strong, lingering sense of amazement.

Concluding Remarks

Surprising surprise in engineering is almost invariably a nasty surprise, from a neural network's low performance to a bridge's collapse. This kind of surprise is what we feel when we view the evolved bridge shown at the beginning of this article—hence our reluctance to traverse it. Engineering and surprising surprise do not go hand in glove. So why, when we wish to abolish surprise, would we let emergence enter engineering at all?

An insight to this paradox may be given by a trivially extreme case of the emergence test: in [9] we applied the test to Minsky's putative Society of Mind according to which mind emerges from a society of myriad, mindless components [7]. We concluded that:

> Mind is an emergent phenomenon, *par excellence*, since the observer always marvels at its appearance. [9]

Now, we are led to reflect that given a complex design task which can only be vaguely specified, *any solution to that task will be surprising* and will thus pass the emergence test!

Hence the presence of emergence in engineering may be a natural consequence of the modern trend which is leading engineering into areas where we expect machines to do things which we cannot really specify, but, like intelligence and life, can only say "I will know it when I see it!"

MISSION CONTROL: If the computer should turn out to be wrong, the situation is still not alarming. The type of obsessional error he may be guilty of is not unknown among the latest generation of HAL 9000 computers...

No one is certain of the cause of this kind of malfunctioning. It may be over-programming, but it could also be any number of reasons.

2001: A SPACE ODYSSEY

References

[1] H. H. Bauer. *Scientific Literacy and the Myth of the Scientific Method.* University of Illinois Press, Urbana, 1994.

[2] M. Dorigo, G. Di Caro, and L. M. Gambardella. Ant algorithms for discrete optimization. *Artificial Life,* 5(2):137–172, Spring 1999.

[3] D. B. Fogel. *Evolutionary Computation: Toward a New Philosophy of Machine Intelligence.* IEEE Press, Piscataway, NJ, second edition, 1999.

[4] P. Funes and J. Pollack. Computer evolution of buildable objects. In P. Husbands and I. Harvey, editors, *Proceedings of Fourth European Conference on Artificial Life,* pages 358–367. The MIT Press, Cambridge, MA, 1997.

[5] P. Funes and J. Pollack. Evolutionary body building: Adaptive physical designs for robots. *Artificial Life,* 4(4):337–357, Fall 1998.

[6] C. G. Langton. Preface. In C. G. Langton, C. Taylor, J. D. Farmer, and S. Rasmussen, editors, *Artificial Life II,* volume X of *SFI Studies in the Sciences of Complexity,* pages xiii–xviii, Redwood City, CA, 1992. Addison-Wesley.

[7] M. Minsky. *The Society of Mind.* Simon and Schuster, New York, 1986.

[8] R. Rojas. *Neural Networks: A Systematic Introduction.* Springer-Verlag, Berlin, 1996.

[9] E. M. A. Ronald, M. Sipper, and M. S. Capcarrère. Design, observation, surprise! A test of emergence. *Artificial Life,* 5(3):225–239, Summer 1999.

[10] E. M. A. Ronald, M. Sipper, and M. S. Capcarrère. Testing for emergence in artificial life. In D. Floreano, J.-D. Nicoud, and F. Mondada, editors, *Advances in Artificial Life: Proceedings of the 5th European Conference on Artificial Life (ECAL'99),* volume 1674 of *Lecture Notes in Artificial Intelligence,* pages 13–20. Springer-Verlag, Heidelberg, 1999.

[11] H. A. Simon. *The Sciences of the Artificial.* The MIT Press, Cambridge, Massachusetts, second edition, 1981.

[12] M. Sipper. The emergence of cellular computing. *IEEE Computer,* 32(7):18–26, July 1999.

Art and Artificial Life — A Coevolutionary Approach

Gary R. Greenfield
Department of Mathematics and Computer Science
The University of Richmond
Richmond, VA 23173
E-mail: ggreenfi@richmond.edu

Abstract

Looking backwards, we recall art of Sommerer and Mignonneau, Sims, and Latham that was inspired by artificial life principles. Assessing current artificial life inspired art, we examine the methods of fitness by aesthetics and user-guided evolution of evolving expressions as practiced by Rooke, Ibrahim, Musgrave, Unemi and the author. Looking forwards, we consider autonomously evolved artistic works using algorithmic aesthetics. We survey what little is known about this topic and proceed to describe our new coevolutionary approach based on hosts and parasites.

Introduction

It would be prohibitive to attempt a comprehensive survey of all the artistic endeavors that have been influenced or inspired by artificial life principles, both for reasons of space and because of the difficulty of referencing many of the works that have been exhibited. But in looking backwards we are struck by two themes: the incorporation of *emergent behaviors* into artistic works, and the exploration of *simulated evolution* for artistic purposes. Emergent behavior forms the cornerstone for many interactive works including, for example, the installations of Sommerer and Mignonneau(39; 40) and Allen(1). Possibly such works trace their origins to the MIT Media Lab *ALIVE* project(24). Spurred on by autonomous robotics, the gaming industry, and the "Furbies" craze, behavior engines and emergent behaviors continue to make their presence felt in the world of fine art. On the other hand, the use of simulated evolution for artistic purposes is perhaps less widely recognized or accepted. We will begin our discussion of this topic by investigating more carefully the origins of simulated evolution in the fine arts.

Michael Tolson, co-founder of the digital effects company *Xaos Tools*, won the prestigious 1993 *Prix Ars Electronica* award for his series of still images titled "Founder's Series." The series was generated with the aid of evolved neural nets. Since Tolson's software was proprietary, details of precisely how this was done are fragmentary. In print, Tolson described his method as applying the genetic algorithm to populations of neural nets to breed intelligent brushes(31). His neural nets were then released onto a specially prepared image where they could sense cues introduced by the artist. By responding to the cues, the image was modified according to the brush *procedures* the neural nets had been bred to implement. As a Siggraph panelist, Tolson showed videotape of the breeding stages of a population of neural nets that were trained to be photosensitive. When patches of pure white were added to an image as cues, the photosensitive neural nets, which were moving at random on the image surface, would streak toward these patches *dragging along the underlying image colors*. Tolson's efforts seem not to have been duplicated[1]. A second example involving neural nets is Lund's *Artificial Painter*(22; 28). It has a much different flavor: Each neural net is coded as a bit string so that the genetic algorithm can be applied, but the computer generated image is obtained by mapping the net's output response at each cell (i.e. pixel) of an environment to a color. Fitness and evolution of the neural nets are accomplished by using the method of Dawkins which we describe next.

In a seminal Artificial Life proceedings paper written by Dawkins(9), the fundamental concept of *user-guided* evolution was introduced. Dawkins implemented, and later marketed commercially, his *Biomorphs* program which allowed users to guide the evolution of a population of two dimensional forms by interactively assigning their fitness values. The forms were recursive Pascal drawing routines whose drawing parameters constituted the form's genotype. Karl Sims combined Dawkin's user-guided evolution technique with an imaging method called evolving expressions for the purpose of providing an artificial life inspired art *medium* in his

---

[1]Research on intelligent brushes is not easy to characterize. A popular and widely used "intelligent brush" technique, first introduced by Haeberli(17), is based on a drawing program's interactive response to user input through a mouse. Another algorithmically based approach to intelligent brushes is the "computational" brushes recently described by Berzowska(5). Her brushes are Java applets. One intriguing aspect of her work is that her brushes are dynamic and can, for example, erase themselves over time.

Siggraph paper(34), "Artificial evolution for computer graphics." Subsequently, Sims published several variations on this theme(36; 37). For Sims, an image — the *phenotype* — was generated from a LISP expression — the *genotype*. The user, viewing the population of phenotypes, interactively assigned an aesthetic fitness value to the phenotypes so that mating and mutation of the genotypes of the most fit individuals could take place in accordance with the rules of the underlying artificial genetics. During this period, William Latham, with technical assistance from programmer Steven Todd, began exhibiting computer generated synthetic three dimensional organic sculptures also created using the Dawkins' evolutionary paradigm which requires the user to assign fitness based on aesthetics(41). The difference between the two approaches is that Sim's genotypes were expressions implemented as trees, while Latham's were gene sequences implemented as bit strings. The use of evolving expressions is currently widely recognized because of the fundamental role it plays in Koza's optimization method known as Genetic Programming(21).

This takes us to the present. While Sims and Latham have gone on to form companies for developing and promoting their endeavors (*Genetic Arts Inc.* and *Computer Artworks Ltd.* respectively) others have followed in their footsteps. Since Sims' original design was too computationally demanding to be of general use, Sims' successors have proceeded to refine his methods for designing and implementing image generating systems based on evolving expressions and fitness by aesthetics. Most well known are the works of Rooke(18; 45). However the author was an early convert(12; 13; 14; 15) and Unemi(42) made available a very restrictive X windows version of a Sims' style system. Papka et al built a Sims' style system to evolve three dimensional polygonal isosurfaces as an application to help test the immersive CAVE environment(8). McGuire described a Sims' style three dimensional polygonal modeling system(25). Ibrahim invoked Sims' technique to evolve Renderman shaders(19). Musgrave developed a Sims' style prototype for MetaCreations(26). Bedwell and Ebert investigated the possibility of using Sims' method to evolve implicit surfaces(3). Mount has a web page (http://www.cs.cmu.edu/ jmount/g3.html) which offers a Sims' style system based on quaternion maps. It is a successor to earlier systems designed for the web by Witbrock and Neil-Reilly(47). Additional examples are described by Rowbottom(33). Undoubtedly there are many more examples of which we are not aware.

Looking forwards, what does the future hold? One vision for the future is that virtual reality will be populated by artists (populations of image producing agents) and art critics (populations of image consuming agents) who will decide what is to be viewed by the humans visiting the virtual environment. For this to occur, however,

some means must be found to make simulated evolution automatic so that the user need n completely guide the evolution. There are three initial forays that have been made into this area. They deserve our careful attention and will serve as our introduction to simulated aesthetics.

Simulated Aesthetics

In 1994, three CMU graduate students — Baluja, Pommerleau, and Jochem — published the results of their efforts to *fully* automate the Sims' process(2). They first designed and implemented a bare-bones Sims' style image generation system and then logged images from users' sessions in order to create a database of images. The images in the database were then numerically rated for their aesthetic value. Images from the database were resolved to 48×48 pixels, and then training and testing sets of images were drawn with equal representation from low, medium, and high ranked images. A neural net was trained and tested, and then asked to guide the interactive evolution without human assistance. The effort expended, together with the number of different experiments the authors performed, is impressive. According to the authors, the results were "somewhat disappointing" and "mixed and very difficult to quantify." It was concluded that it was difficult for the neural nets to learn or discern any aesthetic principles. The authors also noted that the neural net's exploration of image space was "very limited and largely uninteresting." They pointed out that the greatest potential for using such automated approaches may be to prune away uninteresting images and direct the human (assistant) to more promising ones.

A short time later Rooke undertook efforts of a quite different nature to evolve his own art critics which he called "art commentators" to perform aesthetic evaluations of the images his system generated, and which he could then use to guide the image evolution process(32). Rooke's goal was to present his critics with seeded, as opposed to random, starting populations of genotypes. Thus unlike Baluja et al his critics were not forced to start from scratch. Each critic — itself an expression, but one with the capability of examining selected portions of the image phenotype — assigned an aesthetic fitness value to each image in the population. The training set Rooke used consisted of one hundred evolved images together with Rooke's own aesthetic rankings. As populations of expressions, albeit expressions including image processing functions, "tesselation" functions, and statistical measurements, Rooke's critics could also be evolved. Rooke evolved his critics until they could duplicate his fitness rankings to within an acceptable tolerance. To put his critics to work, Rooke gave them his top ranked images from twenty successive generations of an evolutionary run. After each subsequent generation, the oldest of these images would be removed and the image from the current population with the best aesthetic fitness as judged by the critics would be kept. Thus, after twenty generations the critics were in complete control. Rooke let the critics guide the evolution for three hundred generations. Rooke judged his art critics to have been capable of learning *his* aesthetics, but once again, they seemed incapable of using it to explore new areas of image space. One plausible explanation is that Rooke's critics were being caught in eddies of image space. Rooke suggested that it might be necessary to work side by side with his critics and intervene every so often to put them back on track by reassigning aesthetic fitness rankings to the current image set, and then re-evolving the critics — a human assisted coevolution scenario. Following the 1997 Digital Burgess Conference, Rooke and Steve Grand initiated an on-line discussion[2] about the viability of using artificial life coevolution techniques for aesthetics. One idea that emerged was that one needed a "physics" for aesthetics, by which they meant a theoretical framework from which aesthetic principles could be derived and/or tested.

A tangential development which we find significant for understanding the aesthetics of visual images is the recent effort of Belpaeme(4) to evolve expressions whose internal nodes consist solely of image processing primitives in an attempt to discover new and useful digital filters. The aesthetic or fitness value used for the experiment was how successful the filter was at *distinguishing* between the various images of a test set. One intriguing outcome from Belpaeme's experiments was how small the evolved filters turned out to be. One might think that there was a hidden bias towards computational efficiency incorporated into the fitness metric. The explanation Belpaeme offered was that the chaining of image processing functions caused a significant loss of image information content.

Such prior work helps motivate why we think the automation problem for evolving images is a difficult one, and why we see coevolution based on the Sims' method as a significant challenge for the future. Before taking up this challenge, we must review some of the developments of artificial life coevolution research. The first artificial life coevolution simulation was published by Hillis(16), who applied coevolutionary techniques to an optimization problem involving sorting. In the visual arena, Sims gave a stunning example of artificial *learning* based on coevolution(35; 38). Using directed graphs for genotypes, Sims constructed virtual "creatures" to compete in virtual contests of "capture the flag." Sims made mesmerizing videos of the evolved behavior of his creatures. Since then, by linking creature evolution with environments supporting artificial physics, other impressive behaviors

have been evolved, including artificial walking and swimming(43).

There are several obstacles to overcome in trying to adapt coevolutionary research to evolving images based on fitness by aesthetics. First, Hillis' sorters are easy to assign a fitness too. Clear optima are recognizable. The sorters either sort or they don't. Similarly his coevolving population of parasites are either successful at invading the sorters by finding examples of difficult lists to sort or they are not. Second, the coevolutionary behavior generators of Sims and his successors seem to depend on competition between individuals within a single population for a resource, or success at completing a task. Further complicating matters is the fact that recently the fundamental nature of coevolution and its underlying principles have begun to be reconsidered. Cliff and Miller point to the difficulty of recognizing and measuring the so-called Red Queen effect that results as two coevolvers change each other's fitness landscapes(7), while Ficici and Pollack question the sustaining power of the coevolutionary arms race by analyzing mediocre stable states and the prevalence of cycling(10).

Now we are able to frame the fundamental problem we are trying to solve: Given a Hillis coevolutionary framework, where now one species consists of host images and another species consists of image parasites, in order for the parasites to prey upon the host images based on aesthetics, how will the parasites be judged? In other words, what will the *algorithmic* assessment of aesthetic fitness be? Nake in commenting on early attempts by Max Bense and Abraham Moles to use the Shannon concept of information as the guiding principle for an analysis of the aesthetic processes concludes:

> Although some exciting insight into the nature of aesthetic processes was gained this way, the attempt failed miserably. Nothing really remains today of their theory that would arouse any interest for other than historical reasons(27).

In a recent issue of the Journal of Consciousness Studies, Ramachandran and Hirstein sparked considerable debate by offering a series of computational aesthetic principles, the primary one being *exaggeration*(29). The framework for exaggeration that they propose is strikingly similar to artificial life sexual selection experiments of Werner and Todd(46). In considering the challenge of evolving populations of images and populations of aesthetic observers, one is heartened by the words of one of the visionaries of artificial life, Thomas S. Ray, who writes,

> We do not know yet, if we can ever expect evolution in the digital medium to express a level of creativity comparable to what we have seen in the organic medium. However, it is likely that evolution can only reach its full creative potential, in any medium,

[2]http://www.biota.org/conf97/reviews.html.

Figure 1: The phenotype from the genotype of a binary basis function (after Maeda) defined on the unit square. Its postfix expression is V0 V1 B19.

when it is free to operate entirely by natural selection, in the context of an ecological community of co-evolving replicators(30).

Inspiring words. In the sequel, we shall take up the co-evolutionary challenge based on aesthetic fitness.

Images from Expressions

In this section we will describe how we generate phenotypes (images) from genotypes (expresssions). A genotype is an expression tree E written in postfix form. The leaves of the tree are chosen from a set consisting of constants with values ranging from 0.000 to 0.999 in increments of 0.001 together with the variables $V0$ and $V1$. The internal nodes are chosen from sets of unary and binary primitives or *basis* functions. A unary primitive is a function from the unit interval to itself, and a binary primitive is a function from the unit square to the unit interval. A left to right stack evaluation procedure assigns to each point $(V0, V1)$ of the unit square a value $E(V0, V1)$ in the unit interval which is then mapped to a color. For convenience, we shall resolve phenotypes at a resolution of 100×100. The nodes of the genotype, sometimes referred to as alleles or nucleotides, possess *arity* — technically, the number of arguments each basis function requires — zero for terminals, one for unary basis funcions, two for binary basis functions. A sample binary basis function, which we adapted from Maeda(23), is shown in Fig. 1.

Parasites for Images

Our motivation is as follows: A visually "interesting" image is one that causes our filtering apparatus — our eyes — to generate anomalies for our brain to process. Recent findings of Vogt (44) offer evidence in favor of this hypothesis. We want our images to evolve in such a way that our filtering apparatus will be affected by them. Thus we attach simple digital filters to fixed locations on the image, convolve local portions of the image with the filters, and then compare the convolved image with

the original image. We are seeking images for which the convolved image is significantly different from the original. The filter is parasitic upon the image, attempting to blend with the image, while the host image attempts to repel the parasite by making it visible as a blemish. We provide the necessary details.

Given a 100×100 host image with values $h_{i,j}$ in the interval $[0, 1]$, at location L we extract a 10×10 patch $p_{i,j}$ with $1 \leq i, j \leq 10$. A parasite is represented as a 3×3 matrix of integers $(f_{i,j})$ with $0 \leq i, j \leq 2$ whose values are restricted to lie in the interval $[-P_{\max}, P_{\max}]$. The neighborhood of the patch is the 12×12 region of the image consisting of the original patch surrounded by a one pixel wide border. When we pass the filter over the neighborhood we obtain a convolved patch $v_{i,j}$ defined as

$$v_{i,j} = \frac{\sum_{r=0}^{2} \sum_{c=0}^{2} p_{i+(r-1),j+(c-1)} \tilde{f}_{i+(r-1),j+(c-1)}}{S} , \quad (1)$$

where

$$S = 1 + |\sum_{i,j} f_{i,j}|. \quad (2)$$

To make precise the comparison between the original patch and the evolved patch we assign a fitness to the host image via

$$h_{\text{fitness}} = \sum_{i,j} \delta_{i,j} , \quad (3)$$

where

$$\delta_{i,j} = \begin{cases} 1 & \text{if } |v_{i,j} - p_{i,j}| > \varepsilon \\ 0 & \text{otherwise} , \end{cases} \quad (4)$$

and ε is the host's *exposure* threshold. Since the patch is 10×10 we define the fitness for the parasite to be

$$p_{\text{fitness}} = 100 - h_{\text{fitness}}. \quad (5)$$

When *multiple* parasites are attached to a host, the host's fitness is the average fitness taken over all parasites.

Artificial Genetics for Hosts and Parasites

We use the standard genetic operators for the host expression genotypes: The host crossover operator exchanges subtrees between two host genotypes, and the host point mutation operator causes every node of the host genotype to have a small probability of being substituted for using a different basis function selected from among the set of basis functions of the same arity. We are not aware of any artificial genetics having been previously implemented for (3×3) image filters. Because we felt that filters/parasites should be viewed as exceedingly primitive organisms we did not use any mating operators. Instead reproduction was accomplished by cloning the parasite and subjecting the clone, with small probability, to a small number of transcription operators (e.g., exchange of two rows or columns, shifts of a row or column, or exchange of two entries) before passing it to the parasite point mutation operator which, with small probability, allows each entry of the array to be perturbed.

Coevolution of Evolved Images

Our coevolutionary scenario is now straight forward to describe. During initialization we fix a number of locations for parasites to attach to. We generate a random host population and attach a randomly generated parasite to each of the fixed locations on each host. The parasite populations are managed according to the location they are specific to, analogous to the way a species of fish might have wholly different parasites for specific internal organs. At each time step, fitness updates are calculated and the least fit hosts are removed from the population. Random matings between the survivors are used for replacements. Similarly, for each *location*, the least fit parasites are removed and their replacements are determined by cloning and mutating the most fit survivors from that location's population. A new host inherits the parasites that were attached to the host it is replacing[3]. Since a host's parasites only act by filtering a small patch on the host, and since the phenotype does *not* have to be generated in visual form, the coevolution implementation is fast. Of course to monitor the coevolution we must *cull* the host population and examine the phenotypes. Typically we cull one or two hosts with the highest fitness every two hundred time steps.

Some Results

Since our goal is to obtain visually interesting images, we found it necessary to impose one additional constraint on the genetics of our system. Before describing it, we remark that by having individual parasite populations irritate the hosts locally while the hosts can only react globally *viz., the basis functions used as nodes in the genotype are globally defined,* a tension is set up between these local irritations and the global response. Unfortunately, there are two obvious and uninteresting ways for hosts to fight the local invasion. The first is to evolve thin vertical bands in the phenotype that have many contrasting values so that a parasite can not adjust to

---

[3] Here our biological motivation is weak. While it is easy to argue that clones from the most fit parasites would be the ones most successful at attaching themselves to a location where a host has just successfully repelled a parasite, it would take a stranger scenario to justify re-attaching existing parasites to newly bred hosts. Presumably space limitations would need to dictate that a newly bred host must be "deposited" at a spot being vacated by a host that is to be removed.

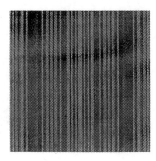

Figure 2: Degenerate images, presumably local optima, that were coevolved by exploiting a flaw in the aesthetic measure of fitness. These cloud or striped hosts present a rugged fitness landscape to the parasites because of the many local discontinuities.

the resulting global discontinuities, and the second is to create tiny islands of such discontinuities resembling a cloud of droplets. Examples of such degenerate images are given in Fig. 2. As is so often the case when using the genetic algorithm, our hosts quickly found this flaw in our physics. We countered by restricting the percentage of *unary* primitives that could appear in a genotype. The purpose of using such primitves is to allow visual "smoothing." Since our binary primitives offer much more better visual contrast, forcing our hosts to incorporate sufficiently many of them into their genotypes helped sustain the diversity we needed in the population's gene pool.

We have tested our coevolutionary simulation using up to thirty hosts and by assigning up to five locations per host giving rise to populations of up to one hundred and fifty parasites. During one representative simulation run using these parameters and lasting 1000 time steps, 10,030 hosts and 79,505 parasites were considered though only ten were culled. No human could examine this many host images. Given our upper bound on the size of the host genotype (up to 100 nodes) and our restriction on the number of unary primitives allowed in the host genotype, it is remarkable that only 24 times during this run was a mating attempt unsuccessful in the ten tries alloted for achieving a valid host crossover.

Does cycling ever occur during our coevolution? Yes, at least to some extent. Hybrids and variants of the degenerate images like the kind shown in Fig 2 do appear and re-appear during the course of coevolution. However, we have visual evidence that, even though we start with random populations, subsequent evolutionary trajectories escape from these uninteresting degeneracies. Our explanation is that since parasites in some sense "chase" their hosts over the fitness landscape, or to put it another way, hosts ward off parasites by fleeing from them, under evolutionary pressure hosts flee parasites following *different* trajectories in image space, hence

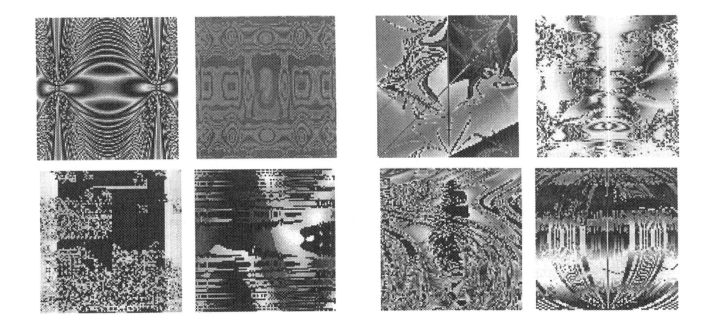

Figure 3: Coevolved images from four different runs. All were evolved starting with small random populations. Noteworthy is the ability of the system to produce diverse imagery by exploring different evolutionary trajectories in image space.

Figure 4: Coevolved images from four runs using larger genome sizes but with the simulation run for only 1500 generations. Host population size is thirty with three parasites per host. The most fit images are culled for inspection every 200 generations.

newly emerging fitter hosts will tend to be quite different than fitter hosts from earlier epochs. (See Fig 3.)

Future Work

Additional benchmarking of the coevolutionary simulation's capabilities is needed. To make reasonable comparisons with images evolved by humans and images evolved from other systems of the type we have described will require considerable additional effort, which is further complicated by the practice of seeding runs from archived gene banks and by the wide disparity of basis function sets that artists and researchers use. Because we did not use seeding, images such as those shown in Fig. 3 are probably best thought of as organisms from the "primordial ooze." More indicative of the image *complexity* that can arise using coevolution is revealed by the representative images shown in Fig. fig4 obtained from runs which permitted hosts to use larger genomes, but restricted the number of generations the simulation was allowed to run for. Work is currently in progress on extending our model so that it is coevolutionary in the sense of Hillis, incorporating a network topology of demes which at intervals share hosts and parasites with neighboring demes(6).

References

Allen, R., Emergence, *Siggraph Conference Abstracts and Applications, Computer Graphics Annual Confer-ence Series*, 1999, ACM Publications, New York, NY, 182.

Baluja, S., D. Pomerleau, and T. Jochem, Towards automated artificial evolution for computer-generated images, *Connection Science*, **6** (1994), 325–354.

Bedwell, E., and D. Ebert, Artificial evolution for implicit surfaces, *Conference Abstracts and Applications, Computer Graphics Annual Conference Series, 1998*, Association for Computing Machinery, New York, NY, 1998, 261.

Belpaeme, T., Evolving visual feature detectors, *Advances in Artificial Life, Proceedings of the Fifth European Conference*, 1999, D. Floreano et al (eds.), Lecture Notes in Artificial Intelligence, Vol. 1674, Springer Verlag, 266–270.

Berzowska, J., Computational expressionism: a model for drawing with computation, *Siggraph Conference Abstracts and Applications, Computer Graphics Annual Conference Series*, 1999, ACM Publications, New York, NY, 210.

Clement, D., Coevolution of evolved expressions, *Honor's Thesis*, University of Richmond, Richmond, VA, 2000.

Cliff, D., and G. Miller, Tracking the Red Queen: measurements of adaptive progress in co-evolutionary simulations, *Advances in Artificial Life: Third European Conference on Artificial Life, Granada Spain, June 1995, Proceedings*, Lecture Notes in Artificial Intelligence 929, Springer-Verlag, Berlin, 1995, 200–218.

Das, S., T. Franguiadakis, M. Papka, T. DeFanti, and D. Sandin, A genetic programming application in virtual reality, *IEEE World Congress on Computational Intelligence, Proceedings of the First IEEE Conference on Evolutionary Computation June 1994*, IEEE Press, 1994, 480–484.

Dawkins, R., The evolution of evolvability, *Artificial Life*, C. Langton (ed.), Addison-Wesley, Reading, MA, 1989, 201–220.

Ficici, S., and J. Pollack, Challenges in co-evolutionary learning; arms-race dynamics, open-endedness, and mediocre stable states, *Artificial Life VI*, C. Adami et al (eds.), MIT Press, Cambridge, MA, 1998, 238–247.

Graf, J., and W. Banzhaf, Interactive evolution of images, *Genetic Programming IV: Proceedings of the Fourth Annual Conference on Evolutionary Programming*, J. McDonnell et al (eds.), MIT Press, 1995, 53–65.

Greenfield, G., An algorithmic palette tool, University of Richmond Technical Report TR-94-02, 1994.

Greenfield, G., Evolving expressions and art by choice, *Leoonardo*, Vol 33. No. 2, 2000, to appear.

Greenfield, G., New directions for evolving expressions, *1998 Bridges Conference Proceedings*, R. Sarhangi (ed.), Gilliland Publishing, Arkansas City, KS, 1998, 29–36.

Greenfield, G., On understanding the search problem for image spaces, *1999 Bridges Conference Proceedings*, R. Sarhangi (ed.), Gilliland Publishing, White Plains, MD, 1999, 41–54.

Hillis, D., Co-evolving parasites improves simulated evolution as an optimization procedure, *Artificial Life II*, C. Langton et al (eds.), Addison-Wesley, Reading, MA, 1991, 313–324.

Haeberli, P., Paint by numbers : abstract image representations, *Computer Graphics (Proceedings SIGGRAPH '90)*, 1990, 207–214.

Hitchcock, N., Painting pictures through time, *Computer Artist*, December/January 1996, 9–9.

Ibrahim, A., GenShade, *Ph.D. Dissertation*, Texas A&M University, 1998.

Keith, M., and M. Martin, Genetic programming in C++: implementation issues, *Advances in Genetic Programming*, K. Kinnear, Jr. (ed.), MIT Press, Cambridge, MA, 1994, 285–310.

Koza, J., *Genetic Programming III : Darwinian Invention and Problem Solving*, Morgan Kaufmann, San Francisco, CA, 1999.

Lund, H., L. Pagliarini, O. Miglino, Artificial painter, *Abstract Book of the Third European Conference on Artificial Life (Ecal '95)*, 1995.

Maeda, J., *Design by Numbers*, MIT Press, Cambridge, MA, 1999.

Maes, P., ALIVE : Artificial Life Interactive Video Environment, in *Computer Graphics Visual Proceedings, Annual Conference Series, ACM Siggraph 1993*, ACM, New York, NY, 1993, 189–190.

McGuire, F., The origins of sculpture: evolutionary 3D design, *IEEE Computer Graphics & Applications*, January, 1993, 9–12.

Musgrave, K., http://www.wizardnet.com/musgrave/mutatis.html.

Nake, F., Art in the time of the artificial, *Leonardo*, Vol. 31, No. 3, 1998, 163–164.

Pagliarini, L., H. Lund, O. Miglino, and D. Parisi, Artificial life: a new way to build educational and therapeutic games, *Artificial Life V*, C. Langton and K. Shimohara (eds.), MIT Press, Cambridge, MA, 1997, 152–156.

Ramachandran, V., and W. Hirstein, The science of art: a neurological theory of aesthetic experience, *Journal of Consciousness Studies*, Vol. 6, No. 6-7, June/July 1999, 15–52.

Ray, T., Evolution as artist, in *Art@Science*, C. Sommerer & L. Mignonneau (eds.), Springer-Verlag/Wien, 1998, 81–91.

Robertson, B., Computer artist Michael Tolson, *Computer Artist*, August/September 1993, 20–23.

Rooke, S., personal communication.

Rowbottom, A., Evolutionary art and form, in *Evolutionary Design by Computers* (ed. P. Bentley), Morgan Kaufmann Publishers, San Francisco, CA, 1999, 261–277.

Sims, K., Artificial evolution for computer graphics, *Computer Graphics*, **25** (1991), 319–328.

Sims, K., Evolving 3D morphology and behavior by competition, *Artificial Life IV*, R. Books and P. Maes (ed.), MIT Press, Cambridge, MA, 1994, 40–48,

Sims, K., Interactive evolution of dynamical systems, *Toward a Practice of Autonomous Systems: Proceedings of the First European Conference on Artificial Life*, MIT Press, 1991, 171–178.

Sims, K., Interactive evolution of equations for procedural models, *The Visual Computer*, **9** (1993), 466–476.

Sims, K., Virtual creatures, *Computer Graphics, Annual Conference Proceedings, 1994*, ACM SIGGRAPH, 1994, 15–23.

Sommerer, C., and A. Mignonneau, Art as a living system, in *Art @ Science*, C. Sommerer & A. Mignonneau (eds), Springer-Verlag/Wien, 1998, 148–161.

Sommerer, C., and A. Mignonneau, Art as a living system : interactive computer art works, *Leonardo*, **32** No. 3 (1999), 165–173.

Todd, S., and W. Latham, *Evolutionary Art and Computers*, Academic Press, San Diego, CA 1992.

Unemi, T., *sbart* - Interactive Art Using Artificial Evolution, User's Manual, 1994.

Ventrella, J., Attractiveness vs. efficiency - how mate

preference affects locomotion in the evolution of artificial swimming organisms, *Artificial Life VI*, C. Adami et al (eds.), MIT Press, Cambridge, MA, 1998, 178–186.

Vogt, S., Looking at paintings: patterns of eye movement in artistically naive and sophisticated subjects, Extended Abstract, *Leonardo*, Vol. 32, No. 4, 1999, 325.

Voss, D., Sex is best, *WIRED*, December, 1995, 156–157.

Werner, G., and P. Todd, Too many love songs: sexual selection and the evolution of communication, in *Fourth European Conference on Artificial Life*, P. Husbands & I. Harvey (eds.), The MIT Press, Cambridge, MA, 1997, 434–443.

Witbrock, M., and S. Neil-Reilly, Evolving genetic art, in *Evolutionary Design by Computers*, P. Bentley (ed.), Morgan Kaufmann Publishers, San Francisco, CA, 1999, 251–259.

Synthetic Harmonies: an approach to musical semiosis by means of cellular automata

Eleonora Bilotta, Pietro Pantano and **Valerio Talarico**

Centro Interdipartimentale della Comunicazione,
Università degli Studi della Calabria,
Arcavacata di Rende, Cosenza, Italy
{eleb|piepa|tvalerio}@abramo.it

Abstract

This paper deals with a software environment based on cellular automata devoted to musical experimentation, realised through a methodology by which, mathematical structures, produced by AL models, the general theory of signs, as proposed by Charles Peirce and music, which consists of acoustic and perceptual relationships are connected. The main features of this environment are the following:

1. semiotics and musical language as tools for reading and interpreting mathematical configurations generated by cellular automata and other AL models;

2. musical expression as creative artefacts;

3. artificial universes as contexts in which to detect perceptual patterns and the correlated emotions music produces;

4. experimentations in aural perception in humans as a method for evolving musical artefacts.

We can know the real world which is near us and the artificial world which is in the computer only by means of thought. Artificial worlds can be equivalent to the phenomenological world and both could be manipulated and organised by thought. We have to apply to artificial worlds the same method humans have developed in organising and giving meanings to the physical world. It'll be necessary to detect patterns generated by Artificial Life machines and to give them meanings. We are trying to apply this methodology through mathematics and music, combining them in a semiotic approach.

Introduction

Cellular automata and AL models have extraordinary capacity for representing some of the biological characteristic of life [24], utilising mathematical structures. This potential (which has not yet been fully explored) seems to have some basic peculiarities in common with language and natural languages, and with semiotics on the one hand. On the other "music is a sign in itself, and the various ways of organising musical material can be viewed as forms of semiosis" [19]. This relationship realises a *triangle of signification* where, at one vertex

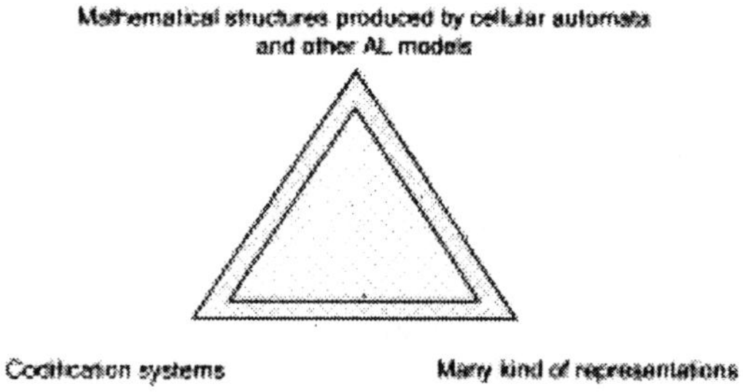

Figure 1: A triangle of signification.

we can find mathematical structures produced by cellular automata systems. At the second vertex, there is the codification system we can use. At the third vertex, there are the various kind of representations (visual, aural, but also space-temporal, multidimensional and multimodal) we can obtain, according to the kind of codification systems we have chosen. One interpreter reads these mathematical structures, chooses an appropriate codification and translates them into other languages, producing *artificial artefacts* of a different kind, whose artistic potential has to be further exploited. We have chosen music as one of the codification systems to be used in the conceptual framework we propose and have translated the mathematical structures produced by cellular automata in musical compositions. Human subjects have listened to these compositions, interpreting them.

The computer is the context in which these semiotic structures live, develop and are represented. As Wolfram [25] points out the computer is a virtual environment in which experimentation and computation allow us to re-create natural phenomena.

The computer is the place where Artificial Life is growing up, or, as Langton [16] says: "a field of study devoted to understanding life by attempting to abstract the fundamental dynamical principles underlying biological phenomena, and recreating these dynamics in other physical media - such as computers - making them accessible to new kinds of experimental manipulation and testing".

It is possible to create *artificial universes* we can understand through naturalistic and artistic mimesis, giving life to new forms of reality, which have the following characteristics:

1. productivity and many production rules;

2. an infinite number of productions;

3. general and abstract character of productions;

4. arbitrariness of codification and representation;

5. many kind of semantics related to representations;

6. many kind of readings it is possible to do in the mathematical configuration spaces;

7. many patterns, both local and global, which have relationships with other patterns;

8. many behaviours of evolution.

There are many semantic tools we can utilise in order to know and to describe the worlds generated by artificial systems. In this paper we'll utilise mathematics and music, through a double process of codification: first by translating cellular automata space-time dynamical patterns into musical variables and then reading them by using the tonal scale of musical notation. The computer allows us to produce musical phenomena through a correlation between a numerical set associated with cellular automata and some physical variables. A computer programme can simulate music, like some other natural event. Computer programmes developed in the musical sector have algorithms and procedures, which reproduce some types of musical composition. Many researchers are studying *fugues, canons* and everything which is related to mathematical formalisation of musical composition. The computer has played an important role in developing digital music and computer music [18], [17].

The need for improving musical creativity has pushed some researchers to use new concepts that scientific theories have developed. Evolving music is based on cellular automata, genetic algorithms, L-systems and many other tools, which allow the artists to generate music in a non-traditional manner [10].

Music and Mathematics are expressions of human creative thought. They can be studied from the point of view of their relationships, as many philosophers, both ancient and modern, have pointed out. They can be analysed by means of the perceptual patterns the human mind utilises in hearing behaviour. They can produce emotions, creating what Leibniz calls the *arithmetic of soul*. They can be analysed by means of the mental patterns artists utilise in their compositions, a topic which is interconnected with some scientific models such as chaos, complexity, recursion and so on [22]. The close connection between Music and Mathematics had already been recognised by Pythagoras, who found that particular relationships allowed sounds to fit well together, realising beautiful melodies for the human ear. According to Pythagoras, universe is made up of these numerical properties, and man has only to read them. The synthetic worlds elaborated by Artificial Life machines can be interpreted as harmonies. In fact, a key concept in Pythagorean philosophy was harmony. Harmony is the principal quality of numbers by means of which opposites (like form and substance) are combined. Philolao (fragment D44B6), one of Pythagoras's disciples, points out that it is necessary to have godlike abilities to understand the substance of the universe, which is made of different things. For this reason an ordered process, which is harmony, has combined different elements of the universe. If the elements of the universe were the same, harmony wouldn't be necessary. Only harmony is able to unify the universe.

Nothing would be comprehensible, neither things nor their relationships, if there weren't number and its substance. But this, harmonising in the soul everything through perception, made things and their relations knowable (Philolao, fragment D44B4).

In this fragment Philolao stresses the connection between number and the sensory world, which has been apprehended in the Pythagorean philosophy and will be later used by modern science to give a mathematical explanation of physical world and of some of the sensory and perceptual processes. This interpretation of the world made by Philolao is a characteristic of gnosiological thought and could be a modality to use in experimenting in artificial worlds. Is it possible to provide a semantic for artificial worlds? The answer could be yes, since we give meanings to every thing. We know real world which is outer and the artificial world which is in the computer only by means of thought. Artificial worlds can be equivalent to the phenomenological world and both could be manipulated and organised by thought. We have to apply to artificial worlds the same methods humans have developed in organising and giving meanings to the physical world. It'll be necessary to detect patterns and give them meanings. We are trying to apply this methodology through mathematics and music, combining them in a semiotic approach.

Does a semiotics of Artificial Life exist?

Semantics is the science of meaning. The scientific study of meaning is one of the more complex things in the Human Sciences. Psychology, Linguistics, Anthropology, Arts, Literature, Information Theory and so on deal with meaning. According to Linguistics, Semantics is also the study of the signification process; of how people give meaning to things and what relationhip meaning has with linguistic signs. People often superimpose Semantics upon Semiotics, since these topics are closely interconnected: both deal with the process of giving meanings to things. For Semiotics, semantics is one of its topics, since Semiotics is the science of signs [21], or the theory of signs. It involves the study not only of what we refer to as signs in everyday speech, but also to anything which stands for something else. In the semiotic approach signs include words, images, sounds, gestures and objects. Such signs are studied not in isolation but as part of a semiotic sign system. Morris [20] divides the subject into three branches:

1. semantics, the science of meaning (the relationship of signs to what they stand for);

2. syntactics (or syntax), the relationship between signs;

3. pragmatics, the way in which signs are used and interpreted.

Could Semiotics be applied to the analysis of cellular automata signs? Which methodology might be necessary? In the following, we'll try to explain, in a comparative manner, some of the methodological assumptions Semiotics uses and how it is possible to translate them into the context of Artificial Life worlds. Many researchers study meaning in relationship to the contexts in which it occurs. The difference between common thought and mathematical or logical thought resides in the fact that context is very relevant to the first way of thinking whereas context is not so influential in the second. An artistic form is acceptable or unpleasant according to the different relationship it has with all the other forms in a context. A mathematical expression isn't influenced by a context. It is not possible to deal with meaning by extrapolating it from the context in which it occurs. Similarly, the cellular automaton plays its role in the relationship it activates within its neighbourhood to develop its configurations. Different neighbourhoods lead to different configurations and so to different kinds of signs and meanings. Wuensche [27] writes: "Processes consisting of concurrent networks of interacting elements which affect each other's state over time occur in a wide variety of natural systems, the dynamics depending both on the pattern of connections (wiring) and on the update rules for each element."

In fact, cellular automata, like signs, are organised within a *syntagmatic structure* (the horizontal dimension of a net), which involves studying their structure and the relationships between their parts. Signs and their correlate meanings have a pragmatic aspect, that is, the way in which signs are used and interpreted. This characteristic translates meanings into social behaviour. According to the concept of *semantic potential*, every time a speaker has to behave linguistically, he/she makes a choice from a wide array of items. This array of choices represents the paradigmatic dimension of a language. Cellular automata have also a *paradigmatic structure*, which means studying the patterns of evolution of a net. In particular, it is possible to identify the signification of the evolution of a pattern and that particular occurrence rather than another pattern. It is also possible to detect the underlying meaning of that pattern, such as one pattern with certain characteristics as opposed to another one with different characteristics. We could use Wolfram's classification of the cellular automata dynamics to operate these kinds of discriminations/oppositions. In this way, we'll arrive at individuating of the invariants of artificial worlds. Could an interpreter, through the medium of music notice this invariant? And if these models of intensity, frequency and separation are pointed out by an observer, what could be the learning process or the notational system the observer uses to communicate to other people how he/she has worked on the artificial world? This idea leads us to reflect on how the recognition process of musical models occurs. We can use the same methods Psychology uses in discrimination tasks with humans [5].). To notice a pattern we can use templates, or global representations of a pattern; features description, or the conditions or characteristics useful in detecting a pattern and structural description of a pattern [4]. It is also possible to describe a musical pattern produced by artificial worlds, utilising the visual medium, by linking computer graphics with evolving music. Pattern recognition based on template matching has some problems: a note occurring on the first beat of a measure will sound different and carry a different musical meaning from the same note occurring on the second beat, even if both are played in exactly the same way. And if we want that musical composition to function as a sign, or to be recognised as a meaningful musical object, it must be apprehended as a *gestalt*, i.e. a form or a structure taken as a whole, and not as the sum of its components, a form that can be perceived against a ground [7]. One problem is that the template has to have the same position, orientation and size as the pattern to be recognised. Furthermore, any object of perception can signify (take on meaning) only in relation to the space within which certain types of activity have the potential to take place.

A second problem is the extreme variability of patterns (a slight different pattern could be produced by the same rule of evolution). A third problem is the grade of difference amongst patterns. A matching between two patterns could indicate that they are similar because one is superimposed upon the other. But this matching says nothing about how they are different, because to have a measure of the difference it is necessary to take note of the specific properties of each pattern. Moreover, each pattern could be described by means of a minute analysis. By codifying the artificial world into music, it is possible to study the perceptual organisation of hearing phenomena, moving from simple to complex patterns, from local to global dynamics. There is another important characteristic of meaning, which distinguishes between logic and emotional meaning. This dichotomy is based on the difference between the communicative function of language, intended as a means of transmission of thought, and the expressive function of language, as an expression of emotions. Music generates emotions [9]. If cellular automata could be translated into music and music, like voice [23], generates emotions, then cellular automata can produce emotions. Music is, for the human brain, a dynamical organisation of time. It seems that the quality we perceive immediately in a musical composition is the quantity of rhythm (quantity of notes in the time unit). The perceptual aural models we perceive are sequences of notes (or sounds), with some expressive qualities directly linked to the emotions. There are as many musical perceptual models as there are principal emotions (or their variants). This is certainly the most complex and broad field of musical/perceptual semiotic inquiry. In fact, if we think of every emotion we can experience and their potential combinations and if we make an analogy with the musical models which can produce emotions, we gain an idea of the great creativity of the musical medium. In our conceptual framework, we'll examine the relationship between mathematical structures, music, and emotions.

Related Works

The idea that a sequence of numbers could generate music has been used in the literature since modern technologies, Artificial Intelligence and psychological research on cognitive abilities have shown that a musicians compositional selections imply mathematical parallels. Combinatorics, fundamental constants, functions, prime numbers and so on are some of the

kind of mathematics practised. Recursion, iteration and complex mathematics can be seen as an extension of traditional music compositional practice. Algorithms generate sequences of numbers which, appropriately coded and transformed into musical parameters, such as frequency and duration, can produce pleasant harmonies for the humans. The results depend on both algorithms and musical rendering techniques.

Using recursive functions that work on a feedback process produces generative music in analogy with the ways musicians unconsciously scaled, shifted, flipped, expanded and compressed melodies in their mental musical spaces. The resulting algorithms are organised on the mathematical models science has produced, such as chaos [6], fractals [1], the theory of complexity [26] L-systems [8] and so on.

Many musical systems have been realised using techniques related to the mathematical models of Artificial Life. By analysing the literature present on the Internet, it is possible to organise the following musical systems' taxonomy:

- Fractal music;
- Music produced by chaotic structures;
  - Chua's circuit;
- L-systems;
  - Used to generate Midi files;
- Generative music based on grammars;
- Evolutionary music based on genetic algorithms;
  - Lee Spector has used genetic algorithms to produce interactive jazz music;
  - GenJam is a GAs based program, which learns to play jazz songs;
- Genetic music;
  - DNA sequences are used to generate MIDI music;
- Music generated by Cellular Automata.

For example, Chaosynth [11] is a system that uses the ideas of granular synthesis to generate music and is based on cellular automata. Each granular sound produced by Chaosynth is made by many components, using three parameters: frequency, amplitude, and duration. Chaosynth checks frequency and duration values, while amplitude values are configured by hand, before starting the process of music generation.

Camus and Camus 3D [12] [13] [14] use *Life* and *Demon Cyclic Space* cellular automata to generate musical compositions. The first space is necessary to determine a triad that will be played at a certain time in the composition, while the *Demon Cyclic Space*'s state will be used to fix the orchestration of the song. In this system, Cellular Automata drive the composition processes, utilising well-known forms of pattern propagation, while stochastic selection routines are used as they constitute a method of specifying long term structure to be used in compositions.

The *Isle Ex system* also uses cellular automata to generate sequences or numbers that later will be rendered by musification maps. These maps are rather complex and operate both on cycles and transient structures. Some related internet address are given in Appendix A.

The system which has been realised

To exemplify the conceptual framework we have designed, we developed *Musical Dreams*. It is a software environment based on cellular automata, devoted to musical experimentation. The main features of this environment are the following:

1. semiotics and musical language as tools for reading and interpreting mathematical configurations generated by cellular automata;

2. musical expressions as creative artefacts;

3. artificial universes as contexts in which to detect perceptual patterns and the correlated emotions music produces;

4. experiments in aural perception in humans as a method for evolving musical artefacts.

Our idea is to let the system become an environment in which to investigate the different features related to AL models and some different kind of codification systems in order to get different production rule, to be utilised in musical compositions.

From the technical point of view the system is made up of three main workspaces (Figure 2). Ideally, these environments correspond to the *signification triangle*. There is a *pattern generator*, to produce cellular automata structures or other kind of AL tools. A *codification system*, which contains a directory of codes, the user can choose, like different types of grammars, to vary the kind of music he/she can obtain. A *musical meanings system* (or representational workspace), with a rendering engine, which transforms the codification system utilised in a MIDI file, giving to the output the particular features users have chosen.

In the patterns generation workspace, a given input is inserted into the cellular automata space and the resulting numbers, once an appropriate codification system has been chosen, are rendered to produce an output, which represents the musical composition.

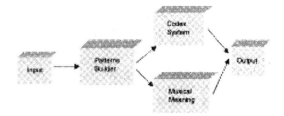

Figure 2: Musification process obtaining different types of musical meanings.

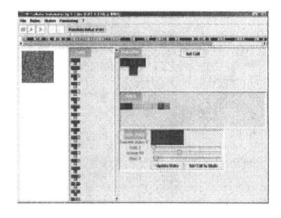

Figure 3: The aspect of the simulation environment and the system's input window.

We preferred to distinguish clearly between the pattern generation task and music production (also at the interface level of the tasks' organisation) because both of them use arbitrary mechanisms that can give different results, depending on the choices made by the users.

The first element of the system is a classical multiple states cellular automata space. In fact, the first version of the system is actually limited to one-dimensional networks. At the interface level, the window shows three main frames (Figure 3). The left-hand highlights the network evolution in a graphical form, the central one shows the transition rules and in the right-hand, users can work with various tools, in order to vary the graphical aspect, modify the transition rules and generate the initial state condition.

The network has a dimension of 100; the CA has 8 states and both the initial state and the transition rules have been generated randomly.

Simply interacting with the system by means of the mouse, users can specify particular sequences of the initial data. Moreover, rules can be inserted one by one. This is not advantageous for automata with many states, since it is impossible to obtain significant patterns, if the user doesn't define (by hand) a high number of rules (for example the entire rule set for a four states automaton is made up of many rules). Another possibility is to generate the entire sequence of transition rules on which to operate randomly, or by hand, directly on a specific rule. The resulting pattern can be zoomed in and out by the magnifying lens tool.

In the *codification system* workspace there is a directory of the codification systems it is possible to use.

1. *Simple musification.* Let the user control only the frequency parameter. In a network, a column corresponds to a note. If the state for that position is equal to 1 or greater, the note is played. If the state is zero, no note is played. The user can choose which octaves to play on a given sequence. No control of the tempo is given. This kind of codification produces very simple sounds, which are not perceived as melodies by human subjects.

2. *Random musification.* This is analogous to the former, but random choices are made between the columns and notes, from row to row. This process changes from one row in the pattern to another. The user can choose which parts of the pattern to render. This kind of codification gives different kinds of sounds which have the characteristics of not being well fitted to human subjects since they cannot identify a melody. Random choices seem to make worse the musical production we can obtain.

3. *Musification with evolutionary functions.* It is possible to think that every note undergoes an evolution process, produced by a certain generative function. This function can be of various types: predefined by the user or externally selected, like a logistic map where the user can choose the value of some parameters. In this case, the results become extremely complex, allowing every note to evolve in different ways. The presence of cycles, both in the cellular automata space and in such functions, leads to interesting results from the musical point of view.

4. *Complex musification.* As in Camus, we can think of using other cellular automata spaces to render the patterns. We can also operate as in Isle Ex to start the musification processes, based on triads. The possibilities offered by this codification system are multiple and produce extremely diversified results.

At the moment, the codification process we have chosen has the following conventions. Each column of the cellular automata pattern corresponds to a specific note, the octaves increasing from left to right. The

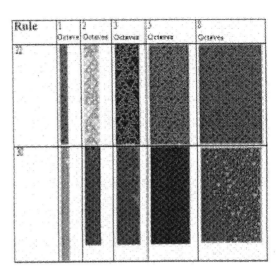

Figure 4: Resulting patterns.

Cell State represents the starting offset for that note with respect to the beginning of the bar fraction that the current row represents. The notes belonging to a row are all played in the time fraction assigned to the row itself.

The other parameters, such as tempo, instrument assignments and other variables are chosen in the MIDI rendering window.

The parameters the user can change are:

- the number of cellular automata state transitions per musical bar;

- the musical tempo model (i.e. 4/4, 3/4 etc.);

- the number of beats per minute;

- the instruments split definition and assignment.

In the *Representational workspace*, once the user has provided the initial data and generated the pattern using the selected rules, the rendering process can be started. The data generated by the cellular automata are transferred to the rendering engine for the MIDIfication process. The rendering process is also extremely arbitrary and the results obtained depend strongly on the choices made for the various parameters. The rendering process output is a MIDI file.

The idea behind these arbitrary choices is to have a system which is flexible enough to allow us to investigate both the perceptual features related to the different process of musification, and the analysis of results based on the patterns typically produced by cellular automata. The system has been completely developed

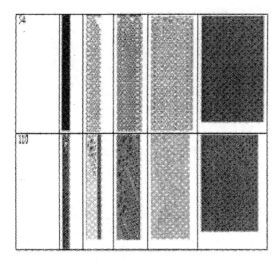

Figure 5: Resulting patterns.

in Java, in order to be portable on many platforms and to build a set of classes that will be easily used in applet, available on the web.

In particular, we used the Java Development Kit 1.2.2 with the integration of the Java Media Framework 2.0 beta, in order to provide the Java platform with MIDI functionality.

The need for flexibility, completeness, ease of use and effective visual representation, has guided system development.

The case of Cellular Automata with Boolean states

We have used the system to generate MIDI files using two-state cellular automata networks and we have collected all data. They have been generated for various octaves. Figures 4 and 5 represent the patterns which result for four rules and various octaves. At a first glance, the data present some interesting results.

If we analyse the networks' behaviour in general, using for example DDLab, we can see different situations. In the case of the first octave, the network dimension is 12, and for rule 22 we will have 5 typologies of basins of attraction and 12 basins in total. For rule 54, we will have 8 typologies of basins of attraction and 33 basins in total. For rule 110, we will have 5 typologies of basins of attraction and 11 basins in total. In practice, as the dimension of the network increases, it is impossible to exhaustively analyse every possibility. From the musical point of view, or, more precisely, from the perceptual side, important results of this analysis are the following. The difference between the various melodies,

contained in the MIDI files collected, does not change in a relevant manner as the network dimension changes (the most significant result is the growth of cacophony which makes difficult discrimination between the various emergent structures), while it is rather relevant when different rules are used.

Multi state system and genetic algorithms

In the previous section we hypothesised that cellular automaton could assume just two values (0 and 1) for the state. If we allow it to have different values (ranging from 0 to 3, 0 to 7, 0 to 15 and so on) the results become immediately more interesting. Non-Boolean cellular automata are scarcely studied and there is a lack of classifying rules analogous to those of Boolean networks.

We have used our system to generate music with a four state cellular automaton and a collection of examples, obtained by generating randomly both the initial state and the transition rules, has been collected and stored. The melodies obtained with this method don't seem to be really significant and a listening group has expressed conflicting opinions in reply to their pleasantness, interest, and acceptability. The subjects opinions about musical compositions are however discordant. Some were inclined to interpret the melodies as horror movie soundtracks, considering the musical compositions to be neither agreeable nor acceptable. Many reported that there were no melodies and that that kind of music was very different from traditional music. It is interesting to point out that music generated by other complex systems, such as Chua's circuit, was used in movies with special effects and "alien" soundtracks.

The number of possible transition rules for a cellular automaton with four states for every cell and a neighbourhood of three cells is very high (4^{4^3}) and thus it is not possible to examine them all. Furthermore, the results obtained are strongly dependent on the initial state, while the dimension of the network seems to be scarcely relevant.

Thus, for simplicity, we have considered a network of dimension 12-24-36.

We have produced a set of 100 MIDI files using 100 different rules and we have selected the rules that seem to produce the most acceptable files. As said before, initial data significantly influence the results, while the presence of cycles allows identification of structures inside the melodies, which make these melodies more pleasant.

Few values, different from zero, allow us to better detect these rhythms and self-organising structures.

In Figure 6, the results obtained for some values of the

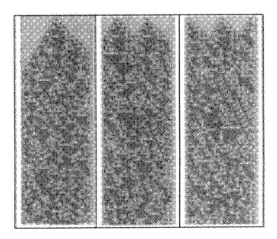

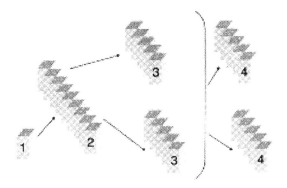

Figure 7: Evolution after various generations.

Figure 6: Resulting patterns with different initial states.

initial state are given. In the next step, we started an evolutionary process in the following way. We wrote the above mentioned rules as a sequence of 64 characters (each character could assume one value amongst four), which we will call from now on the genotype.

Later, we generated ten different genotypes, which were variations of the first and differs from the first only by one character.

We generated five MIDI files (phenotypes) from each genotype, with the same initial states shown in the figures, using two random initial states. We thus obtained ten families of five MIDI files each.

We proposed the files to a listening group made up of five elements, asking them to assign a score ranging from 1 to 10 to each song. By summing the scores obtained by each file in a family, we obtained the family's fitness. The families that obtained the greatest fitness were allowed to evolve, in a fashion analogous to that formerly exposed. We produced five new sequences for every family allowed to evolve, thus keeping the number of new phenotype families to ten (Figure 7).

This process was repeated for ten generations. The results obtained can be found on the authors' reference site (http://uni.abramo.it/esg). We verified that the music varies but not in a significant way, even if the average fitness shows a growth trend in the direction of complexity.

We think that the process needs to be repeated for a very high number of generations, in order that natural selection be effective and in order to surpass the effects related to the initial state and the rendering process. We started with other processes of musification. In fact, there are two important directions to-

wards which we are moving in exploring musical context. The first is a scientific experimentation on the kind of artefacts we can obtain to detect a musical theory that can be utilised to write new kinds of composition in connection with the experimentation on the underlying musical cognition [3]. The second concern musical fitness, which in our hypothesis, can provide us with some global characteristics of a melody and can help us in reconstructing the evolution of Western musical notation. While in evolutionary music, musical fitness is generally operated intuitively by the listeners, we have elaborated a fitness function which allows selection amongst the produced musical pieces. Such a fitness function is based on the theory of Pythagorean consonance [2], since it has been demonstrated that there is a biological basis for consonance [28] [29].

Conclusions

New keys of reading into the global dynamics of Artificial Life machines could be generated utilising the semiotic approach this paper presents. By means of a codification system, represented by musical language, it is possible to give meaning to many characteristics of the patterns global dynamics produce and to use the results in the Computer Arts dominion. At the moment of writing this paper, only a part of the proposed methodology has been developed in the software Musical Dreams, but the authors idea is that the system will become an environment for analysing music generated by cellular automata and other AL models. The resulting structures are complex enough to effect self-organisation and evolutive processes, while the patterns generated give us new insights for improving the semiotic approach. The system is already complex and even insignificant variations on one element or a different kind of codification produce important variations in the music we can obtain. The music it is possible to produce depends on the kind of analysis we have organised. It'll be possible to analyse how different kinds of cellular automata contexts will produce different kind of music. It'll be possible to make an analysis of cellular automata patterns by means of *syntagmatic* and *paradigmatic* analysis. It'll be necessary to carry out exhaustive experiments with human subjects in order to individuate perceptual organisation of dynamical patterns on the one hand and the continuum of emotions cellular automata music can produce from the other. Furthermore, many studies of musical tone systems established in different historical and ethnic cultures suggest that their evolution is mainly dependent on perception of the audible frequency range, and harmonic consonance. It'll be possible to search for fitness functions in music through genetic algorithms to ascertain if the strong relationship between the historical evolution of tone systems and their cultural refinement by emphasis on consonant aspect of music is true or not.

Appendix

Fractal music:

- http://hometown.aol.com/strohbeen/fml.html
- http://www-ks.rus.uni-stuttgart.de/people/schulz/fmusic/

Music produced by chaotic structures

- http://www.ccsr.uiuc.edu/People/gmk/Projects/ChuaSoundMusic/ChuaSoundMusic.html
- http://www.organised-chaos.com/java/newindex.html

L-systems:

- http://www.geocities.com/Athens/Academy/8764/lmuse/lmuse.html

Generative music based on grammars:

- http://cs.eou.edu/ jefu/Grammidity.html

Evolutionary music based on genetic algorithms:

- http://hampshire.edu/ lasCCS/genbebop.html
- http://www.it.rit.edu/ jab/genjam.html

Genetic music:

- http://algoart.com/

Use of neural networks:

- http://www.btinternet.com/ scroberts/thesis/
- http://www.nici.kun.nl/mmm/

References

[1] Barnsley M. 1988. *Fractals Everywhere*, New York, Academic Press.

[2] Bilotta E., P. Pantano and V. Talarico. *Evolutionary Music Based on Cellular Automata*, in press.

[3] Bilotta E. and P. Pantano. *In Search for Musical Fitness on Consonance*, in press.

[4] Bregman A. S. and A.I. Rudnicky. 1975. Auditory segregation: Stream or streams? *Journal of Experimental Psychology:Human Perception and Performance*, 1, 263- 267.

[5] Bregman A. S. 1990. *Auditory Scene Analysis: The Perceptual Organisation of Sound*, Cambridge, Massachusetts, The MIT Press.

[6] Field, M. and M. Golubitsky. 1992. *Symmetry in Chaos*. Oxford, Oxford University Press.

[7] Koehler W. 1947. *Gestalt Psychology*, New York, Liveright Publishing Corporation.

[8] Lindenmeyer A. and P. Prusinkiewicz. 1990. *The Algorithmic Beauty of Plants*. New York, Springer-Verlag.

[9] Meyer L. 1956. *Emotion and Meaning in Music*, Chicago, University of Chicago Press.

[10] Miranda E. R. 1999. Modelling the dynamic evolution of synthesised sounds and the propagation of musical forms using cellular automata. *Musikometrika*, 9, Bochum (Germany), Universitatsverlag Norbert Brockmeyer.

[11] Miranda E. R. 1999. Modelling the Evolution of Complex Sounds: A Cellular Automata approach. In *Proceedings of the AISB'99 Symposium on Creative Evolutionary Systems*, Edinburgh College of Art and Division of Informatics, University of Edinburgh.

[12] McAlpine K. B., E. R. Miranda and S. G. Hoggart. 1997. A Cellular Automata Based Music Algorithm: A Research Report. In *Proceedings of IV Brazilian Symposium on Computer Music*, 7-17 .

[13] McAlpine K. B., E. R. Miranda and S. G. Hoggart. 1997. Dynamical Systems and Applications to Music Composition: A Research Report. In *Proceedings of Journees d'Informatique Musicale*, 106-113.

[14] McAlpine K. B., E. R. Miranda and S. G. Hoggart. 1998. Music Composition by Means of Pattern Propagation. In *Preprint Series: paper 98/25*, Dept. Of Mathematics, University of Glasgow.

[15] McAlpine K. B., E. R., Miranda and S. G. Hoggart. 1999. Making Music with Algorithms: A Case-Study System, *Computer Music Journal*, 23 (2), 19-30.

[16] Langton C. G. 1992. Preface. In C. G. Langton, C. Taylor, J. D. Farmer, and S. Rasmussen (Eds.), *Artificial Life II*, Volume X of *SFI Studies in the Sciences of Complexity*, xiii-xviii, Addison-Wesley, Redwood City, CA.

[17] Loy G. 1989. Composing with computers. A survey of some compositional formalism and Music Programming Languages. In M. Mathews and J. R. Pierce (Eds.) *Current Directions in Computer Music Research*. Cambridge, MIT Press, 292-396.

[18] Manning P. 1993. *Electronic and Computer Music*, Oxford, Claredon Press.

[19] Martinez J. L. 2000. Semiotics and the Art Music of India, *Music Theory Online, The Online Journal of the Society for Music Theory*, vol. 6, 1.

[20] Morris C. 1971. Writings on the General Theory of signs, The Hague, Mouton.

[21] Peirce, Charles S. 1938-1956. *The Collected Papers* 8 vols. Eds. Charles Hartshorne, Paul Weiss, and Arthur W. Burks. Cambridge, Harvard University Press.

[22] Pickover, C. A. 1990. *Computers, Pattern, Chaos, and Beauty*. New York, St Martins Press.

[23] Scherer K. R. 1995. Expression of emotion in voice and music. *Journal of Voice*, 9(3), 235-48.

[24] Toffoli T. and N. Margolus. 1987. *Cellular Automata Machines: A New Environment for Modeling*, Cambridge, MIT Press.

[25] Wolfram S. 1984. Computer Software in Science and Mathematics. In *Scientific American*, 251,188-203.

[26] Wolfram S. 1994. *Cellular Automata and Complexity*, Reading, Addison Wesley.

[27] Wuensche A. 1998. Genomic regulation modelled as a network with basins of attraction. In *Proceedings of the 1998 Pacific Symposium on Biocomputing*, 89-102.

[28] Zentner M. R. and J. Kagan. 1996. Perception of music by infants, *Nature*, 383, p.29.

[29] Zentner M. R. and J. Kagan. 1998. Infants, perception of consonance and dissonance in music, *Infant Behavior and Development*, 21, pp. 483-492.

Modeling Emergence of Complexity: the Application of Complex System and Origin of Life Theory to Interactive Art on the Internet

Christa SOMMERER & Laurent MIGNONNEAU

ATR Media Integration and Communications Research Lab
2-2 Hikaridai, Seika-cho, Soraku-gun
61902 Kyoto, Japan
christa@mic.atr.co.jp, laurent@mic.atr.co.jp

Abstract

The origin of this paper lies in the fundamental question of how complexity arose in the development of life and how one could construct an artistic interactive system that can model and simulate this emergence of complexity. Based on the idea that interaction and communication between entities of a system are the driving forces for the emergence of higher and more complex structures than its mere parts, we propose to apply principles of Complex System Theory to the creation of an interactive, computer generated and audience participatory artwork on the Internet and to test whether complexity within the system can emerge.

1. Introduction

The Internet seems to be especially capable of dealing with interactions and transformations of data and information. Users on the Internet can be considered entities or particles who transport information, such as written texts or images. As these data of information or entities are carried from location to location they could, in principle, change their status and value. We could imagine a system that can increase its internal complexity as more and more users interact with its information. Just as a genetic string or "meme" (Blackmoore & Dawkins, 1999), these strings of information would change and mutate as they are transmitted by the users; they eventually could create an interconnected system that features, similar to the models presented by Stewart Kauffman (1995), a phase transition toward more complex structures. Based on these considerations, we propose a first prototype system for modeling a complex system for the Internet, introduce its construction principles and translation mechanisms, and analyze how the data of information have changed over time:

2. Conceptual Objective

The aim of this research is to construct an Internet based interactive artwork that applies and tests principles of Complex System Theory and Origin of Life theories to the creation of a computer generated and audience participatory networked system on the Internet. Complex Systems Theory is a field of research that allows simpler subsystems to increase in complexity by using phase transitions. These phase transitions take place when a network of particles is given and these particles can switch one another on or off to catalyze or inhibit their production. The proposal of this paper is to test the principle of phase transition for an interconnected web of people who can transmit visual and written information over the Internet. As the information is transported from location to location it will be transformed, creating an interconnected open-ended system that features phase transitions toward more complex structures. Before investigating how to actually build the system, a short summary is given of the theories that ground this research proposal.

3. Origin of Life Theories

The search for "laws of form" to explain the patterns of order and complexity seen in nature has intrigued researchers and philosophers since the Age of Enlightenment. These searchers have included famous scholars such as William Bateson (1894), Richard Owen (1861), Hans Driesch (1914), D'Arcy Wentworth Thompson (1942), and Conrad Waddington (1966). Their quest could generally be subsumed under the term Rational Morphology, a counterpart to the functionalistic approach of the Natural Theology promoted by Charles Darwin (1859, 1959) and Neo-Darwinist Richard Dawkins (1986). Whereas Natural Theology considers form mainly a function of natural selection and adaptation, Rational Morphologists emphasize the creative principle of emergence that accounts for the order of structures found in nature. The quest for the "laws of form" is closely linked to the question of the Emergence of Life. The discussion on how life emerged has a long tradition and basically involves two opposing views: the Aristotelian and the Platonic. These two views of the natural world have dominated science over the past two millennia (Lewin, 1993). Baltscheffsky (1997) notes that "Fundamental to a deeper understanding of complex biological functions are ideas about how life originated and evolved. They include questions about how the first compounds, essential to life, appeared on Earth; how the first replicating molecules came into being; how RNA and DNA were formed; how prokaryotes and the earliest eukaryotes emerged; how different species, with traits like susceptibility, sentience, perception, cognition, and self-consciousness, and with various patterns of behaviors, evolved; and how with these developments, the environment and the ecological systems changed. "

Speculations of how life on earth might have originated have a long history, perhaps as long as the history of humanity. The widely accepted hypothesis that life originated from chemical processes largely derives from the work of Russian biochemist Alexander I. Oparin (1924, translated to English, 1938). In the 1930s, Alexander I. Oparin and J.B.S. Haldane (1932) suggested that life on earth could have emerged by natural means in an early atmosphere filled with different gases such as methane, ammonia, hydrogen and water vapor. Oparin and Haldane called this early atmosphere the Primordial Soup. In their Primordial Soup Theory, life would have originated in the sea as a reaction of these chemical gases triggered by the energy of lightning, ultraviolet radiation, volcanic heat and natural radioactivity.

In the early 1950s, Stanley Miller (1953) of the University of Chicago's Chemistry Department simulated such a primordial atmosphere and was able to synthesize significant amounts of amino acids, main components of all life forms, from methane, ammonia, water vapor and hydrogen. This experiment gave credence to the belief that the chemical building blocks of life could be created by natural physical processes in the primordial environment. Modern proponents of the Primordial Soup Theory now think that the first living things were random replicators that assembled themselves from components floating around in the primordial soup (Miller, 1953). Based on experiments by Sol Spiegelman (1967), who was able to create self-replicating RNA strings in an environment filled with a primitive "seed" virus and a constant supply of replicase enzymes, Manfred Eigen (1992) went a step further by omitting the initial "seed" virus. Eigen succeeded in showing that self-replicating RNA strands can assemble themselves from only replicase enzymes. In Eigen's theory of the origin of life, RNA molecules can evolve self-replicating patterns and finally develop a primitive genetic code. As the molecules specify and take on different functions, complex and cooperative interactions take place: Eigen calls these the "hypercycles" (Eigen, 1992). Mutation and competition among these hypercycles finally create prototypes of modern cells and the earlier chemical evolution is finally replaced by biological evolution. A similar theory on the origin of life was also presented by Walter Gilbert (1986).

Even though the "RNA world" model seems very convincing, the question of where RNA came from in the first place remains open. L. Orgel (1987), C. Böhler (1995), and P. Nielsen (1991) found that a peptide nucleic acid, called PNA, could be a pre-form of RNA because it can act to transcribe its detailed genetic information directly to RNA; consequently, PNA could have initiated the RNA world. Another scientist, Hendrik Tiedemann suggests that the nucleotide bases and sugars needed in RNA could have been built from hydrogen cyanide and formaldehyde, both available in the early atmosphere of the Earth.

Completely opposite to the "RNA world" theories on the origin of life is the Dual-Origin Theory of A.G. Cairns-Smith (1982). According to Cairns-Smith, the starting point in early crystallization of life was not "high-tech" carbon but "low-tech" silicon, a component of clay. In his theory clay has the capacity to grow and re-assemble itself by exchanging its ion components through mutation and mechanical imperfections. More recent proponents of the mineral and early molecular based theories on the molecular evolution of metabolism subscribe to the "iron-sulphur world" theory of Wächtershäuser (1997), the "thioester world" theory of deDuve (1991), and the "inorganic pyrophosphate world" or "PPi world' theory of Baltscheffsky (1991). Wächtershäuser (1994) proposes a model where early evolution of life as a process begins with chemical necessity and winds up in genetic exchange.

Somewhat related to the question of how life occurred in the first place, whether the first stages of life were metabolic or genetic, is the question of how to draw the line between life and non-life. While generally it is agreed that the RNA world (Gilbert, Eigen, Böhler, Nielsen, Orgel) is a first stage of life, Wächtershäuser (1997) and others believe that rather primitive entities on mineral surfaces can also be called alive; however he calls them "two-dimensional life. " On the other hand, Maynard Smith and Szathmáry (1995) stress that a living organism needs to possess at minimum a reproduction mechanism, and Gánti (1979) proposes that a minimum requirement for a living organism is that it possesses three essential subsystems: a genetic system, a functioning unit synthesizing the components, and a membrane part.

Another big question in understanding life's origin is to determine the origin of the translation apparatus and the genetic code (Crick, 1968, Crick et al. 1976, Woese, 1967). Clas Blomberg (1994) claims that the only way to get a stable translation mechanism is a feedback between the code and the proteins that were synthesized by the mechanisms they controlled. Furthermore, Maynard Smith and Szathmáry (1995) suggest that the relations between amino acids and nucleic acid sequences were established before the translation apparatus, serving as an improved catalyst in the RNA world.

It would exceed the scope of this report to describe all the other theories on the origin of life in detail; however some of them should be mentioned here briefly: the "Membrane First" theory of Harold Morowitz (1992) and the "Self-replicating protein" theory of Ghadiri et al. (1996). Theories that life was first introduced by meteorites that came from other planets or stars include the "Radiopanspermia" theory of Hoyle and Wickramasinghe (1979) and the "handedness of the solar system" theory and its influence on the origin of life of Carl Chyba (1997) as well as the "Chirality" theories of Yoshihisa Inoue (1992).

John Casti notes in the manuscript of his forthcoming book "Paradigms Regained" (Casti, 2000) (from which much of the above information is taken), that "when it comes to defining what it means to be alive, there are as many answers as there are biologists." While the numerous theories about the origin of life suggest that scientists today are still in the dark about the details of life's beginnings and have not been able to create it from scratch, Richard Dawkins (1986) argues that this is rather to be expected. "If the spontaneous origin of life turned out to be a probable enough event to have occurred during a few man-decades in which chemists have done their experiments, then life should have arisen many times on Earth and many times on planets within the radio range of Earth. "

4. Complex System Theory

Closely related to the question of how life on earth originated is the question of how complexity arises. Complex System Theory, as a field of research, has emerged in the past decade. It approaches the question of how life on earth could have appeared by searching for inherent structures in living systems and trying to define common patterns within these structures. Among others, researchers at the Santa Fe Institute in New Mexico, USA have been looking at emergent structures in nature and have called this approach the new science of Complex System Theory. Stuart Kauffman is one of the most prominent proponents of this new theory. According to Kauffman (1995), the pure evolutionary view of nature in the Darwinian sense fails to explain the vast structures of order found in nature. By stressing only natural selection, patterns of spontaneous order cannot be sufficiently described or predicted. In Kauffman's view, this order arises naturally as an "order for free." As a consequence, life is an expected phenomenon deeply rooted in the possibilities of the structures themselves. Kauffman argues that, considering how unlikely it is for life to have occurred by chance, there must be a simpler and more probable underlying principle. He hypothesizes that life actually is a natural property of complex chemical systems and that if the number of different kinds of molecules in a chemical soup passes a certain threshold, a self-sustaining network of reactions - an autocatalytic metabolism - will suddenly appear. It is thus the interaction between these molecules that enables the system to become more complex than its mere components taken by themselves.

4.1. Complexity through Phase Transition

Kauffman and other researchers at the Santa Fe Institute for Complex Systems Research call the transition between the areas of simple activity patterns and complex activity patterns a phase transition. Kauffman (1995) has modeled a hypothetical circuitry of molecules that can switch each other on or off to catalyze or inhibit one of their production. As a consequence of this collective and interconnected catalysis or closure, more complex molecules are catalyzed, which again function as catalyzers for even more complex molecules. Kaufmann argues that, given that a critical molecular diversity of molecules has appeared, life can occur as catalytic closure itself crystallizes. A model built by Kauffman is the Boolean network model, which basically describes the connections and relations between three elements (Kauffman, 1995). The networks described by Kauffman in the Boolean network model show stablility, homeostatis, and the ability to cope with minor modifications when mutated; they are stable as well as flexible. The poised state between stability and flexibility is commonly referred to as the "edge of chaos."

Other researchers have also analyzed this phase transition between order and chaos. Brian Goodwin (1994) describes this transition phase as a kind of biological attractor: "For complex non-linear dynamic systems with rich networks of interacting elements, there is an attractor that lies between a region of chaotic behaviour and one that is 'frozen' in the ordered regime, with little spontaneous activity. Then any such system, be it a developing organism, a brain, an insect colony, or an ecosystem will tend to settle dynamically at the edge of chaos. If it moves into the chaotic regime it will come out again of its own accord; and if it strays too far into the ordered regime it will tend to "melt" back into dynamic fluidity where there is a rich but labile order, one that is inherently unstable and open to change."

4.2. Life at the Edge of Chaos

Two of the first scientists to describe the idea of complex patterns and the ones who defined the term "life at the edge of chaos" were Christopher Langton (1992) and Norman Packard. They discovered that in a simulation of cellular automata there exists a transition region that separates the domains of chaos and order. Cellular automata were invented in the 1950s by John Von Neumann (1966). They form a complex dynamical system of squares or cells that can change their inner states from black to white according to the general rules of the system and the states of the neighboring cells. When Langton and Packard observed the behaviour of cellular automata, they found that although the cellular automata obey simple rules of interaction of the type described by Stephen Wolfram (1986), they can develop complex patterns of activity. As these complex dynamic patterns develop and roam across the entire system, global structures emerge from local activity rules, which is a typical feature of complex systems. Langton and

Packard's automata indeed show some kind of phase transition between three states. Langton and Packard hypothesize that the third stage of high communication is also the best place for adaptation and change and in fact would be the best place to provide maximum opportunities for the system to evolve dynamic strategies of survival. They furthermore suggest that this stage is an attractor for evolving systems. Subsequently, they called the transition phase of this third stage "life at the edge of chaos" (Langton, 1992).

Other researchers at the Santa Fe Institute have extended this idea of life found in this transition phase and applied it to chemistry. In 1992, Walter Fontana developed a logical calculus that can explore the emergence of catalytic closure in networks of polymers (Fontana, 1992). A related approach is seen in the models of physicist Per Bak (1991), who sees a connection between the idea of phase transition, or "life at the edge of chaos," and the physical world, in this case a sand pile onto which sand is added at a constant rate (Bak, 1991).

To summarize, we can see that the various observations and models of Kauffman, Langton, Packard, Fontana and Bak describe complex adaptive systems, systems at the "edge of chaos" where internal changes can be described by a power law distribution. These systems are at the point of maximum computational ability, maximum fitness and maximum evolvability. It is hypothesized that these models could indeed function to explain the emergence of life and complexity in nature. While Kauffman's concept of phase transition is not the only model for creating complexity (many more approaches are currently being discussed on-line, see: www.comdig.org, http://necsi.org/ or published in recent conference proceedings, see: Bar-Yam, 2000), it does however provide an advantageous starting point for creating an artistic system that tries to incorporate some of the features of complex adaptive systems.

5. VERBARIUM - Modeling Emergence of Complexity for Interactive Art on the Internet

Based on the above objective and the literature search in Origin of Life and Complex System Theories, with special focus on the concept of phase transitions, we have developed a first prototype to model a complex system for the Internet (Sommerer and Mignonneau, 1999).

Artists have been working with the potential of user interaction on the Internet over the past several years, and some of the pioneering artworks include works by Anzai (1994), Fujihata (1996), Amerika (1997), and Goldberg (1998). A good overview of this work is also provided by the on-line exhibition "Net-Condition" at the ZKM Center in Karlsruhe, Germany (ZKM, 1999). While many of the above works feature a significant amount of user interaction, their main interest does not seem to be based

on the objective of modeling complexity as described in Chapter 2 of this paper.

Our system, called VERBARIUM, is an interactive web site where users can choose to write email messages that are immediately translated into visual 3-D shapes. As the on-line users write various messages to the VERBARIUM's web site, these messages are translated by our in-house Text-to-Form editor into various 3D shapes. By accumulation, these collective shapes can create more complex image structures than the initial input elements. It is anticipated that through the users increased interaction with the system increasingly complex image structures will emerge over time.

5.1. VERBARIUM System Over View

VERBARIUM is available on-line at the following web page: http://www.fondation.cartier.fr/verbarium.html.
The on-line user of VERBARIUM can create 3-D shapes in real-time by writing a text message within the interactive text input editor in the lower-left window of the web site. Within seconds the server receives this message and translates it into a 3-D shape that appears on the upper-left window of the web site. Additionally, this shape is integrated into the upper-right window of the site, where all messages transformed into shapes are stored in a collective image. An example screenshot of the VERBARIUM web site is shown in Fig. 1.

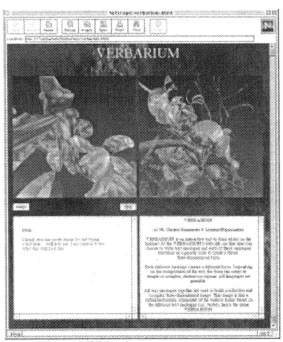

Fig. 1 VERBARIUM web page

VERBARIUM consists of the following elements:

1. a JAVA based web site (Fig.1)
2. an interactive text input editor
   (lower-left window in Fig.1)
3. a graphical display window to display the 3-D forms
   (upper-left window in Fig.1)
4. a collective display window to display the collective 3-D
   forms (upper-right window in Fig.1))
5. a genetic Text-to-Form editor to translate text characters
   into design functions

5.2. VERBARIUM's Text-to-Form Editor

We have set up a system that uses the simplest possible component for a 3-D form that can subsequently model and assemble more complex structures. The simplest possible form we constructed is a ring composed of 8 vertices. This ring can be extruded in x, y and z axes, and during the extrusion process the rings' vertices can be modified in x, y and z axes as well. Through addition and constant modification of the ring parameters, the entire structure can grow, branch and develop. Different possible manipulations, such as scaling, translating, stretching, rotating and branching of the ring and segment parameters, creates diverse and constantly growing structures, such as those shown in Fig. 2.

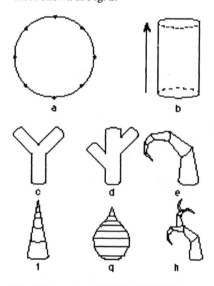

Fig. 2 Example of VERBARIUMS's growing structures

Figure 2a shows the basic ring with 8 vertices, and Fig. 2b shows the extruded ring that forms a segment. Figures 2c and 2d show branching possibilities, with branching taking place on the same place (=internodium) (2c) or on different internodiums (2d). There can be several branches attached to one internodium. Figure 2e shows an example of segment rotation, and Fig. 2h shows the combination of rotation and branching. Figures 2f and 2g are different examples of scaling. In total, there are about 50 different

design functions, which are organized into the design function look-up table (Fig. 3). These functions are responsible for "sculpting" the default ring through modifications of its vertex parameters.

function1 translate ring for certain amount (a) in x
function2 translate ring for certain amount (a) in y
function3 translate ring for certain amount (a) in z
function4 rotate ring for certain amount (b) in x
function5 rotate ring for certain amount (b) in y
function6 rotate ring for certain amount (b) in z
function7 scale ring for certain amount (c) in x
function8 scale ring for certain amount (c) in y
function9 scale ring for certain amount (c) in z
function10 copy whole segment(s)
function11 compose a new texture for segment(s)
function12 copy texture of segment(s)
function13 change parameters of RED in segment(s)texture
function14 change parameters of GREEN insegment(s)texture
function15 change parameters of BLUE in segment(s)texture
function16 change patterns of segment(s)texture
function17 exchange positions of segments
function18 add segment vertices
function19 divide segment in x to create branch
function20 divide segment in y to create branch
function21 divide segment in z to create branch
function22 create new internodium(s) for branch(es)
function23 add or replace some of the above functions
function24 randomize the next parameters
function25 copy parts of the previous operation
function26 add the new parameter to previous parameter
function27 ignore the current parameter
function28 ignore the next parameter
function29 replace the previous parameter by new parameter
....
function50

Fig. 3 VERBARIUM's design function table

The translation of the actual text characters of the user's email message into design function values is done by assigning ASCII values to each text character according to the standard ASCII table shown in Fig. 4.

33 !	34 "	35 #	36 $	37 %	38 &	39 '
40 (| 41)	42 *	43 +	44 ,	45 -	46 .	47 /
48 0	49 1	50 2	51 3	52 4	53 5	54 6	55 7
56 8	57 9	58 :	59 ;	60 <	61 =	62 >	63 ?64
@ 65 A	66 B	67 C	68 D	69 E	70 F	71 G
72 H	73 I	74 J	75 K	76 L	77 M	78 N	79 O
80 P	81 Q	82 R	83 S	84 T	85 U	86 V	87 W
88 X	89 Y	90 Z	91 [	92 \	93]	94 ^	95 _
96 `	97 a	98 b	99 c	100 d	101 e	102 f	103 g
104 h	105 i	106 j	107 k	108 l	109 m	110 n	111 o
112 p	113 q	114 r	115 s	116 t	117 u	118 v	119 w

| 120 x | 121 y | 122 z | 123 { | 124 | | 125 } | 126 ~ |

Fig. 4 ASCII table

Each text character refers to an integer. We can now proceed by assigning this value to a random seed function *rseed*. In our text example from Fig. 5, *T* of *This* has the ASCII value 84, hence the assigned random seed function for *T* becomes *rseed(84)*. This random seed function now defines an infinite sequence of linearly distributed random numbers with a floating point precision of 4 bytes (float values are between 0.0 and 1.0). These random numbers for the first character of the word *This* will become the actual values for the modification parameters in the design function table. Note that the random number we use is a so-called "pseudo random," generated by an algorithm with 48-bit precision, meaning that if the same *rseed* is called once more, the same sequence of linearly distributed random numbers will be called. Which of the design functions in the design function table are actually updated is determined by the following characters of the text, i.e., *his*; we then assign their ASCII values (104 for *h*, 105 for *i*, 115 for *s* ...), which again provide us with random seed functions *rseed(104)*, *rseed(105)*, *rseed(115)*. These random seed functions are then used to update and modify the corresponding design functions in the design function look-up table, between design function1 and function50. For example, by multiplying the first random number of *rseed(104)* by 10, we get the integer that assigns the amount of functions that will be updated. Which of the 50 functions are precisely updated is decided by the following random numbers of *rseed(104)* (as there are 50 different functions available, the following floats are multiplied by 50 to create integers). Figure 5 shows in detail how the entire assignment of random numbers to design functions operates. As mentioned above, the actual float values for the update parameters come from the random seed function of the first character of the word, *rseed(84)*.

Example word: *This*

T => *rseed(84)* => {0.36784, 0.553688, 0.100701,...}
  (actual values for the update parameters)

h => *rseed(104)* => {0.52244, 0.67612, 0.90101,...}
  # 0.52244 * 10 => get integer 5 => 5 different
functions are called within design function table

  # 0.67612 * 50 => get integer 33 => function 33
within design function table will be updated by   value
0.36784 from 1. value of *rseed(84)*

  # 0.90101 * 50 => get integer 45 => function 45
within design function table will be updated by   value
0.553688 from 2. value *rseed(84)*
  .... until 5. value

Fig. 5 Example of assignment between random functions and design functions

As explained earlier, the basic "module" is a ring that can grow and assemble into segments that can then grow and branch to create more complex structures as the incoming text messages modify and "sculpt" the basic module by the design functions available in the design function table in Fig. 3.

5.3. VERBARIUM´s Complexity Potential

Depending on the complexity of the incoming text messages, the 3-D forms become increasingly shaped, modulated and varied. As there is usually great variation among the texts, the forms themselves also vary greatly in appearance. As a result, each individual text message creates a very specific three-dimensional structure that can at times look like an organic tree or at other times look more like an abstract form. All forms together build a collective image displayed in the upper-right window of the web site: it is proposed that a complex image structure could emerge that represents a new type of structure that is not solely an accumulation of its parts but instead represents the amount and type of interactions of the users with the system. Another example of forms created by a different text message is shown in Fig. 6, this time the text was written in French.

Fig. 6 VERBARIUM web page - example

6. Conclusions and Outlook

We have introduced an interactive system for the Internet that enables on-line users to create 3-D shapes by sending text messages to the VERBARIUM web site. Using our text-to-form editor, this system translates the text parameters into design parameters for the creation and modulation of 3-D shapes. These shapes can become increasingly complex as the users interact with the system. A collective image hosts and integrates all of the incoming messages that have been transformed into 3-D images, and as users increasingly interact with the system it is anticipated that an increasingly complex structure will emerge. As it will no longer be possible to deconstruct the collective image into its initial parts, some of the features of complex systems are thought to have emerged. However, it remains to be tested whether one can call this system a truly complex and emerging system. Future versions of the system should address the current shortcomings such as the limited amount of design functions as well as the somewhat nontransparent translation process. Furthermore, we plan to expand the capacity of the system to simultaneously display all messages in the browser's window; this should make it possible for users to retrieve all messages ever sent. Finally, another crucial aspect will be the ability of the forms to start to interact with each other more actively; this could be done by using the genetic exchange of information (text characters) between forms, creating off-spring forms through standard genetic cross-over, and mutation operations as we have used them in the past (Sommerer and Mignonneau, 1997).

References

Amerika, M. 1997. *Gammatron*, http://www.grammatron.com/

Anzai, T. and Nakamura, R. 1994. *RENGA*, http://www.renga.com/

Baas, N., Olesen, M. W. and Rasmussen, S. 1999. Generation of Higher Order Emergent Structures. *Physica D*. North Holland Publishing Company.

Bak, P. and Chen, K. 1991. Self-Organized Criticality. *Scientific American*. January 1991, pp. 46-54.

Baltschefsky, H. 1997. Major "Anastrophes" in the Origin and Early Evolution of Biological Energy Conversion. *J. Theoretical Biology*, 187, pp.495-501.

Bar-Yam, Y (ed.), 2000. Unifying Themes in Complex Systems: Proceedings of the First Necsi International Conference on Complex Systems, New York: Perseus Books.

Bateson, W. 1894. *Materials for the Study of Variation*. Cambridge: University Press.

Blackmoore, S. and Dawkins, R. 1999. *The Meme Machine*. New York: Oxford University Press

Blomberg, C., Liljenström, H. Lindahl, B.I.B. and Århem, P. eds. 1994. Mind and matter: essays from biology, physics and philosophy. *J. Theoretical Biology*, 171, pp.1-122.

Böhler, C., Nielsen, P.E. and Orgel, L.E. 1995. *Nature* Vol.376. pp. 578-581.

Cairns-Smith, A.G. 1982. *Genetic Takeover*. Cambridge: Cambridge University Press.

Cairns-Smith, A.G. 1986. *Clay Minerals and the Origin of Life*. Cambridge: Cambridge University Press

Casti J. 2000. *Paradigms Regained*. New York: William Morrow & Co. forthcoming.

Chyba, C. 1997. A Left-Handed Solar System? *Nature*, Vol.389, pp. 234-235.

Clancey, W. J. 1979b. Transfer of Rule-Based Expertise through a Tutorial Dialogue. Ph.D. diss., Dept. of Computer Science, Stanford Univ.

Crick, F.H.C. 1968. The origin of the genetic code. *J. Molecular Biology*, Vol. 38, pp. 367-379.

Crick, F.H.C., Brenner, S., Klug, A. and Pieczenik, G. 1976. A speculation of the origin of protein synthesis. *Origins Life*, Vol. 7, pp.389-397.

Darwin, C. 1959 [1859]. *The Origin of Species*. Reprinted. London: Penguin.

Dawkins, R. 1986. *The Blind Watchmaker: why the Evidence of Evolution Reveals a Universe without Design*. London: Longman.

De Duve, C. 1991. The Beginnings of Life on Earth. *American Scientist*, Vol. 83, pp. 428-437.

Driesch H. 1914. *The History and Theory of Vitalism*. London: Macmillan.

Eigen, M. and Winler-Oswatitch, R. 1992. *Steps Towards Life: A Perspective on Evolution*. New York: Oxford University Press.

Fontana, W. 1992. Algorithmic Chemistry. *In:* C.G. Langton, J.D. Farmer, S. Rasmussen and C. Taylor (eds.) *Artificial Life II: Santa Fe Institute Studies in the Sciences of Complexity*. Vol. 10. Reading, MA: Addison-Wesely, pp. 159-209.

Fujihata M. 1996. *Light on the Net Project*, http://www.flab.mag.keio.ac.jp/light/

Gánti, T. 1979. *A Theory of Biochemical Supersystems and its Application to Problems of Natural and Artificial Biogenesis.* Baltimore, MD: University Park Press.

Gilbert, W. 1986. The RNA World. *Nature,* Vol. 319, p. 618.

Goldberg, K. 1998. *Memento Mori: an Interface to the Earth,* http://memento.ieor.berkeley.edu/

Goodwin, B. 1994. *How the Leopard Changed its Spots.* London: Orion Books (Phoenix), p. 169.

Haldane, J.B.S. 1932. *The Forces of Evolution.* London: Longmans Green.

Hoyle, F. and Wickramasinghe, N.C. 1979. *Diseases from Space.* New York: Harper and Row.

Inoue, J. 1992. Asymmetric Photochemical Reactions in Solution. *Chemical Reviews,* Vol.92, pp.741-770.

Kauffman, S. 1995. *At Home in the Universe. The Search for Laws of Complexity.* New York: Oxford University Press, pp. 23 - 112.

Langton, C. 1992. Life at the edge of chaos. *In :* C.G. Langton, J.D. Farmer, S. Rasmussen and C. Taylor (eds.) *Artificial Life II: Santa Fe Institute Studies in the Sciences of Complexity .* Vol. 10. Reading, MA: Addison-Wesely, pp. 41-91.

Lee, D.H., Granja, J.R.; Martinez, J.A., Severin, K., Ghadiri, M. R. A Self-Replicating Peptide. 1996. *Nature ,* Vol. 382, pp. 525-528.

Lewin, R. 1993. *Complexity: Life at the Edge of Chaos.* London: Macmillan Publishing, pp. 178.

Maynard Smith, J. and Szathmár, E. 1995. *The Major Transitions in Evolution.* Oxford: W.E. Freeman.

Miller, S.L. 1953. Production of amino acids under possible primitive earth conditions. *Science.* 117: 528.

Morowitz, H. 1992. *Beginnings of Cellular Life.* New Haven, CT: Yale University Press

Nielsen P. E., Egholm, M., Berg, R.H. and Buchardt, O. 1991. *Science.* Vol. 254, pp. 1497-1500.

Orgel, L.E. 1987. The Origin of Self-Replicating Molecules. In*: Self-Organizing Stysems: The Emergence of Order.* (F.E. Yates, ed.), New York: Plenum Press, pp. 65-74.

Oparin, A. I. 1924 [translated to English 1938, new edition 1957]. *The Origin of Life on Earth.* New York: Academic Press.

Owen, R. 1980 [1861]. *Paleontology or a Systematic Summary of Extinct Animals and Their Geological Relation.* London: Ayer Co. Publisher

Ruse, M. 1997. The Origin of Life: Philosophical Perspectives. In: *J. Theoretical Biology,* 187, pp.473-482.

Sommerer, C. and Mignonneau, L. 1997. Interacting with Artificial Life: A-Volve, In *Complexity Journal.* New York: Wiley, Vol. 2, No. 6, pp. 13-21.

Sommerer, C. and Mignonneau, L. 1999. VERBARIUM and LIFE SPACIES: Creating a Visual Language by Transcoding Text into Form on the Internet, In: *1999 IEEE Symposium on Visual Languages,* Tokyo, Japan, pp. 90-95.

Spiegelman, S. 1967. An in Vitro Analysis of a Replicating Molecule. *American Scientist,* Vol. 55, pp: 3-68.

Thompson, D'A. W. 1942. *On Growth and Form.* Cambridge: University Press.

Von Neumann, J. 1966. *Theory of Self-Reproducing Automata.:*Completed and edited by A.W. Burks, Champaign-Urbana: University of Illinois Press.

Waddington, C. H. 1966. *Principles of Development and Differentiation.* New York: Macmillan.

Wächterhäuser, G. 1997. The Origin of Life and its Methodological Challenge. In: *J. Theoretical Biology,* 187, pp.483-494.

Whitehead, A. N. 1971 [1920]. *The Concept of Nature.* Reprinted. Cambridge: Cambridge University Press.

Woese, C.R. 1967. *The Genetic Code - The Molecular Basis for Genetic Expression.* New York: Harper and Row.

Wolfram, S. 1986. *Theory and Applications of Cellular Automata.* Singapore: World Scientific Press

ZKM Center Karlsruhe, Germany, 2000. *Net-Condition,* http://on1.zkm.de/netCondition.root/netcondition

Artificial Evolution: creativity and the possible

Arantza Etxeberria

Logika eta Zientziaren Filosofia Saila (Dept. of Logic and Philosophy of Science)

Euskal Herriko Unibertsitatea (University of the Basque Country)

1249 Posta Kutxa, 20.080 Donostia-San Sebastian, Spain

ylpetaga@sf.ehu.es

Abstract

In this paper the aims and goals of artificial evolution are discussed in relation to two of the founding features of Alife: how to characterize the domain of the possible and the criterion of lifelikeness. It is argued that artificial evolution should aim to understand the evolution of organizations and that this will bring about a better understanding of possible evolutions.

Introduction

Artificial Life (Alife) emerged as a research programme to study living phenomena in artificial media so as to exempt the notion of life from its dependence on a single example. Three main features characterize the goals of this science as it was conceived by Langton (1989). First, its object is to explore the domain between the existent life-as-we-know-it and the possible life-as-it could-be. In relation to biology, which studies life on Earth, Alife is a science of universal living phenomena. Second, this universality will be achieved by studying artificially generated invariants of form or organization, and not of matter. The systems under study are synthetized, instead of (or, ideally, before) analysing them, and it is assumed that the relevant phenomena have more to do with the organization of the parts than with the nature of the components. Third, the extension of the systems under inquiry is open, but characterized by the notion of lifelikeness; thus, the question of defining life is left aside and substituted by a criterion based on experience and intuition. Research may emulate different aspects: complex organization, stability in a given medium, capacity to absorb, transform and use energy, self-reproduction, evolution, development and growth, adaptation, etc. and those aspects will be interesting according to how lifelike they are. All the three –the domain of the possible, universal organizing principles independent of matter, and lifelikeness– are interesting start points, but deserve careful discussion.

In previous work several authors have pointed out the problems of analyzing universality by means of organizations that do not emerge from a realistic material dynamics (Pattee 1989, 1995; Cariani 1992, Moreno et al 1994). In this paper I intend to discuss artificial evolution in relation to the other two: how to characterize the domain of the possible and the criterion of lifelikeness.

The distance between existent life and possible life is manifest in that different possible evolutions could have taken place on Earth starting from the same or similar initial conditions, because the history of evolution contains both contingent, or fortuitous, events and necessary ones, determined by the properties of evolving matter. An interesting line of research for a discipline interested on the universal aspects of possible life is then to tell the necessary and the contingent of the history of evolution. In this sense, and presuming that laws are universal, the domain of the possible has, at least, two orders of magnitude: starting with the same initial conditions would give us the scope of non-realized possibilities of life on Earth, whereas different initial conditions would generally describe the scope of other possible lives. This would be the counterfactual approach to evolution.

However, another way of considering the domains of possibility that artificial evolution may unfold has to do with evolutionary theory itself and the way it conceives creativity or production of novelty. The theory of evolution is epistemologically challenging because it introduces creativity in the realm of science. Unlike other fields, like physics, that try to discover the laws of nature underlying the behavior of all systems, evolutionary theory may be read as saying that almost anything is possible, because it describes a procedure by which novelty appears and develops in nature.

It is in this sense that it is important to find a criterion of lifelikeness as a foundation for research in artificial evolution. The creativity and open-ended nature of life has been characterized as "supple adaptation", as a hypothesis that life can be defined in terms of a system that exhibits "lifelike" evolution (Bedau 1998). This is an interesting idea, but somehow developed at the cost of renouncing to consider lifelikeness at the level of individual living beings and preferring, instead, to focus on the whole of life as a process. In this paper my interest is directed to the problem of how to conceive the evolution of embodied agents, a perspective that obliges to take into account the nature of living organization and its relation to evolution. An adequate understanding of the evolution of organizations is still lacking: evolution and organization are difficult terms or perspectives to bring together. This is probably the reason why researchers who have worked on the problem of biological organization, like Varela and Rosen, have somehow left the problem of evolution aside as secondary. Rosen, for example says: "We cannot answer the question (...) "Why is a machine alive? with the answer "Because its ancestors were alive". Pedigrees, lineages, genealogies and the like, are quite irrelevant to the basic question. Ever more insistently over the past century, and never more so than today, we hear the argument that biology *is* evolution; that living systems instantiate evolutionary processes rather than life; and ironically, that these processes are devoid of entailment, immune to natural law, and hence outside of science completely. To me it is easy to conceive of life, and hence biology, without evolution." (Rosen, 1991, pp. 254-55). A similar feeling may be found in Varela (1979): "I maintain that evolutionary thought, through its emphasis on diversity, reproduction, and the species in order to explain the dynamics of change, has obscured the necessity of looking at the autonomous nature of living units for the understanding of biological phenomenology. Also I think that the maintenance of identity and the invariance of defining relations in the living unities are at the base of all possible ontogenic and evolutionary transformation in biological systems" (p. 5). Although these positions might sound extreme, they reveal that there is a real problem for artificial evolution to be a process of evolving organizations, given the use of evolutionary theory done by many researchers. Thus, a perspective like the one I want to suggest requires to work on what I call diachronic embodiment. This would constitute an original contribution of Alife beyond actual biology.

Three perspectives of artificial evolution

Artificial Evolution has been explored according to different interests by the participants in this interdisciplinary research area. Very roughly we could distinguish three main tendencies of research:

- In a first one evolutionary theory is used as a problem solving strategy, convenient for certain purposes, but which does not require any biological plausibility as it is not being applied to generate life or lifelike phenomena. As the main goal of this case is the efficiency of the method used to achieve certain practical goals, both theoretical and empirical knowledge can be violated (lamarckian evolution or very high mutation rates) in order to achieve certain goals such as optimization.

- A second group uses artificial models to study biological phenomena for which no adequate model or theory has been developed so far. The aim in this case is to develop modelling strategies to enhance our capabilities of explaining biological phenomena, by developing theory or analytical tools based on artificial systems.

- A third division is constituted by those artificial systems in which the frontiers or boundaries between the artificial and the biological are diffuse, and instances of alternative universes are constructed that, although do not correspond literally with the facts of real life, provide an intuitive grasp of lifelikeness.

These three categories correspond roughly to the production of three different *things*: tools, models and instantiations (Etxeberria 1995). *Tools* are clearly methodological innovations that may or may not apply to life, although inspired by it; whereas models and instantiations are distinguished by their respective different relation to a target real phenomenon to be

explained or grasped. The term *model* is reserved for the case usually developed by the natural sciences: as a way of representing certain relevant aspects of an empirical domain by means of data collected through measurements. In this case the natural systems are, so to say, the referents of the artificial models built to explain or emulate them. This reference relation is looser in *instantiations*, more related to the way models are conceived in formal sciences: as an instance or a system built after a set of axioms. As they do not have a concrete empirical referent, their value resides in their being new creations of ontology that, although inspired by natural phenomena, provide as an end result something different from life on Earth, but with a strong lifelikeness.

The three kinds of artificial systems have different goals. Tools enhance our epistemological capacities in the sense that they provide information processing ways that, although inspired by the biological theory, do not operate in a transparent way for the human mind. They are directed to fulfil human or technological goals. Models may allow us to study the distance between realized and non-realized life, the conditions under which other possibilities of our life could have happened and counterfactuals based on different parameters. They may provide theoretical contributions if the data were set to run the appropriate experiments. Instantiations may be much more abstract and their only connection with reality depends on the way we apply the theory we have. Although it may be the case that instantiations will provide the most radical ways of exploring life-as-it-could-be, present day work is strongly limited by its dependence on difficult to overcome theoretical assumptions.

Intuitive notions of evolution and evolutionary theory

Models and instantiations are, in principle, more likely to produce a notion of lifelike artificial evolution than tools. At the moment most of the research in artificial evolution is based either on an intuition or on a theory, or on both. The intuition is that evolution, whatever it is, must reveal itself as the growth of something: complexity, adaptation, optimization, or some other global property. Even if this view of evolution as growth tends to be rejected by the latest evolutionary theory, where the usual perspective is neutral with respect to this, it remains an important guiding notion for a science that has to provide artificial examples of something so poorly understood as evolution. Evolutionary Biology considers that evolution takes place when there is a change in the gene frequencies of a population (the Hardy-Weinberg Law), change driven by different "evolutionary forces": natural selection, sexual selection, genetic drift, etc. This notion of evolution assumes a fairly neutral situation with respect to what is expected from evolution, whether there is or not an increase in complexity in evolution, or whether evolution has a direction. Most artificial evolution models are not that neutral.

The motivation of researchers trying to study or to

implement artificial evolution might in fact preclude that neutrality. For instance, when von Neumann (1966) studied the problem of artificial self-reproduction, he took into account the possibility of an increase of complexity of the reproduced system. In his view, the capacity living systems have to evolve derives from the possibility that they, unlike machines, have to produce other systems which are of equal or superior complexity; and many of the constraints he imposed to the self-reproducing logic, like the necessity of a universal constructor, where implied by this conception of open-ended evolution. Similarly, evolutionary computation, as a general problem solving tool, assumes that this methodology will bring along some sort of "improvement" or optimisation of the proposed solutions.

Thus, Alife experimenters trying to develop artificial evolution face the disjunctive option of either providing an external goal to their systems, or to have no natural "goal" at all. Many artificial systems are indeed supplied with an external goal: this is the case of most tools, and also of experiments of artificial evolution that explore a kind of creativity which is in fact evaluated by humans (an example could be art created by this procedure). However the case of an artificial evolution system that wants to be true to nature or lifelikeness is more tricky: on the one hand, the evolving system is not supposed to have any intrinsic direction and, on the other, it is assumed that under the appropriate evolutionary dynamics (and the appropriate encoding of the problem) anything is possible. The reason for considering that anything is possible is that the kind of abstraction of natural evolution used by artificial evolution must be able to derive any interesting system or organization as its evolution under the appropriate variety and the selection pressure. This is why in general a procedure of artificial evolution, such as the genetic algorithm, gives the impression that anything is possible. Yet, a situation in which anything is possible is equivalent to one in which nothing interesting happens. The problem is that this is not the case of natural evolution and the reason for it is that there are intrinsic trends that have not been sufficiently investigated by artificial systems.

This general idea can be expressed in another way by saying that the evolution should be based on synchronically complex or organized systems. The study of complex systems presents a dichotomy of methodologies: the approach taken is either synchronic or diachronic. The synchronic one (sometimes also called *vertical or emergent*) studies the relation of components with the aggregates or totalities they form, the target is complexity or interesting behaviours at a global level, starting from simple components or interactions at a lower level. It is a bottom-up perspective, largely inspired by Thermodynamics. Examples of this approach are auto-catalytic sets, swarm organisations, neural nets and embodied robotics. The diachronic one (also *horizontal or transformational*) studies how complex systems arise and develop in time, as a substitution of subsequent generations. The system is considered in terms of the appearance of novelty through time, and not in relation to levels, selection being the main mechanism or force acting in the system.

Work in artificial evolution, like the genetic algorithm, is an example of this.

Hence, in many cases artificial evolution seems to rely in an intuitive idea of evolution, instead of trying to develop a new, productive notion to explore what evolutionary change would be like for artificial systems. This kind of new notion will only be developed by an appropriate notion of the organizations –as opposed to bit strings– able to evolve.

Organization, functionality, and design

The lack of a well developed notion of diachronic embodiment is evident in the treatment of functionality. The functional analysis of complex systems is a hard issue for biology and Alife and this can be explained, even if in a sketchy way, by referring to the difference between two traditions of thought, the Kantian and the Darwinian, through the way each of them compares the living organism with a watch.

For the Darwinian tradition this comparison poses the problem of the "argument from design" (developed, among others, by Aquinas, Hume and Paley). Paley said that if in crossing a heath I found a watch I will not think that (like a stone) it had lain there for ever, but the inference will be that the watch must have had a maker: "Arrangement, disposition of parts, subserviency of means to an end, relation of instruments to a use, imply the presence of intelligence and mind" (Paley, in Ruse 1998, p. 38). In the same way, the design of organisms takes us to necessarily accept the existence of a creator: "every manifestation of design which existed in the watch, exists in the works of nature, with the difference on the side of nature of being greater and more, and that in a degree that exceeds computation" (idem, p. 39). When the Darwinian tradition responds to this argument, a natural explanation of design is produced that requires no divine intervention –the principle of natural selection–, but it lets a likening of the watch and the organism in, maybe accepting it as a good one.

Kant had already used the same comparison in the *Critique of judgement*, but in a rather different way. He notices a fundamental difference between the two of them, whereas the watch is formed by fixed components, fabricated before hand and later ensembled, in the organism the parts are formed for the others, some parts produce the others. Kant accepts an internal teleology in the living system.

Many authors (for example, Mayr) have said that the later development of the theory of evolution corrects the Kantian summon to teleology and makes it possible to explain this in another way. Maybe this is true, but Kant points to a problem that the Darwinian tradition has not collected: the relation among the parts to form an organization. Actually, like in the case of the watch, for the Darwinian tradition the assimilation of watch and organism is not problematic, whereas the Kantian tradition feels that distinction must be set and explained.

This distinction implies that there is a difference between explaining the function of components in relation to organization or to adaptation, either in the historical or a-historical way, like it is usual in the evolutionary biology. However, both traditions coincide in considering that the problem of organization must be approached in an analytical way that derives the whole from the properties –decomposable or not– of the parts. The question is whether this should continue being this way, whether it would not be much better to produce a complementary "synthetic" way of looking at this problem.

This is the reason why embodiment –both in a synchronic and a diachronic perspective– is such an important issue for Alife. Now the field of "embodied cognition" seeks to understand how physical properties contribute to the processing of information needed for the interactions involved in the autonomous behaviour of robots or artificial agents in general (Mataric 1997). The hint is that when the physical properties of bodies (shape, orientation, degrees of freedom, etc.) are taken into account, a great deal of the explicit information processing becomes superfluous. Probably taking into account more aspects, like metabolism, would induce more radical changes (Moreno & Ruiz-Mirazo 1999). Traditionally this phenomenon could be conceived as an imperfection, as some sort of handicap or constraint that limits information processing capabilities, but embodiment precludes the search for "clean" functions and this has serious consequences for the task of designing embodied agents. In other words, if evolution were a process of creation of functionalities (that is to say, a process in which matter is shaped to produce –and reproduce– certain functions), in most cases these would be non-detachable properties of the physical substratum.

The problem of functions brings us directly to the one of design. Every theory of evolution attempts to provide natural mechanisms for the origins and diversification of living forms, so that no designing agent has to be proposed. Evolution is intuitively different from design in its non-intervention, evolutionary explanations are naturalistic because form or function emerge from matter and interactions. Naturalistic explanations of design are, nevertheless, difficult; even if it is agreed that the process of evolution is not designed and, moreover, that it does not itself design. For machines, the case is easier, but the aim of Alife is to produce lifelikeness. Let's turn for a moment to Polanyi´s ideas about machines. He described them as working under two distinct principles: a higher one of design, and a lower one harnessed by the former, which consists in the physical-chemical processes on which the machine relies (Polanyi 1968). This analysis may hold for living beings as well, but, for them, the higher level of design (at any scalar level) is not easy to characterise. Then the attempt to substitute design for artificial evolution in the construction of complex machines, such as adaptive robots, should imply to start with no functions, to let functions emerge.

Artificial evolution and natural selection

In Alife artificial evolution has been often used to build artificial creatures with features adapted to their artificial environments by a procedure similar to evolution by natural selection (for example, the genetic algorithm).

The methodologies developed with this inspiration generate (at random or hand crafted) a population of bit strings which are interpreted as possible solutions for the problem; the "fitness" of each of them is scored using an evaluation function; there is a process of selection of individuals according to these scores (usually the best are taken to be the "parents" of the subsequent generation, but, sometimes, samples of medium or even bad individuals are also preserved); and individuals are modified using the genetic operators (mutation and/or recombination) to produce a new generation. This is basically the genetic algorithm; by letting the system "evolve" in this way, the population usually converges into a situation of mutual environment/creatures adaptation. In "artificial worlds" this kind of artificial evolution governs the change of certain features of simulated agents and environment, it introduces a higher scale temporal change than the one of processes taking place at the scale of the life time of individuals, for example, learning. In the most interesting cases, the interaction between both scales produces interesting phenomena. In "evolutionary robotics" this method has been conceived either as an alternative to the explicit design of the morphology and/or the cognitive systems of artificial creatures, or as a design methodology (there is some ambiguity about which is the case).

In the first kind of work, the artificial world is usually a two-dimensional array of cells (though it can be n-dimensional, and, sometimes, continuous) where a population of agents coexists with similar ones and other living (predators, parasites, etc.) or/and non-living systems (food, shelter, geographical barriers, etc.). Agents are typically represented by a control system for behaviour specification (such as a neural network, a finite state machine, or an abstract grammar), and, very seldom, with a simulated physical body. They are endowed with a sensorimotor apparatus to perceive the relevant features of the world and with a set of behaviours to act in it. The sensorimotor system is, in general, as complex as the world: sometimes both are relatively realistic, while, in others, perception consists merely in a detection of the state of the nearby cells. The metabolism of agents is usually represented as an energy storage, whose level determines whether they are in good shape to perform several actions (mate, escape, etc.) or close to death. The generational replacement will result in a modification of the control system (and/or the body) of the agents: like in evolution by natural selection, agents will be able to survive and reproduce according to their abilities in the world they inhabit.

Most of these models try to simulate the relevant features considered by evolutionary models (morphological, functional or behavioural traits that influence the fitness of individuals, size of populations, selection pressures, etc.) to be able to set up experiments. These are rather close to the phenomena studied by population genetics. The problem is that the change (that is to say, the evolution) likely to arise in simulations is, in general, already known from the theoretical results and the creativity or exploration of potential novelty expected from an evolutionary process is generally lacking.

These works bring about little new about interesting

ways in which evolution can influence complex (vertical) organizations. Within the field of Artificial Life the main criticisms they have received have to do with their being simulations. As such, agents cannot present the physical organization that makes it possible to study embodied cognition in the simulated domain, therefore research should be directed towards physical realizations. I think this criticism is correct, but insufficient. The problem of the kind of artificial evolution has been overlooked, and it is as important as realism at the time of implementing the dynamics of cognitive interactions.

In the work on robot evolution, the evaluation function depends on how well the robot behaves in the real world, instead of in a simulation (Cliff, Harvey, & Husbands 1993). Because this methodology requires a combination of simulations (encoding of the evolving structure, genetic operators applied to selected individuals) and physical building (performance, test and scores), different technical problems arise in transferring the evolved individuals to the robot for evaluation. The rationale for adopting the methodology of artificial evolution is related to the problems encountered with an external functional decomposition and the complexity and big number of interactions among sub-systems, which sometimes are even mediated through the environment. Thus, agents whose interaction with the environment is very sophisticated need also a very complex sensorimotor system that is difficult for hand design. Artificial evolution will be used as the search method and also the heuristic to handle the intractable complexity that up to certain levels arises in other hand design incremental procedures (Brooks 1986, Harvey 1996). There is an in principle interest in avoiding functional decomposition and to evolve whole architectures that can adopt non decomposable functions within the system. This is, also, linked with the ideal goal of obtaining via evolution an incremental complexification of the behavioural capacities of the robots in a non-foreseen way. Nevertheless, the separation of the diachronic and synchronic aspects of the process (performed respectively in simulation and realisation), and the correspondent diachronic disembodiment, demand further investigation of artificial evolution.

If the artificial evolution procedure applied in these systems is analysed, then it is clear that getting rid of the task of designing is not that easy after all, many problems arise as a difficulty not to intervene in the process from the outside. Questions like how the genotypes of the population are encoded (and decoded), the relation between evolving populations and their environments, and how the artificial models exhibit "natural" selection make apparent that there is a fundamental lack of understanding of how a spontaneous, not externally directed evolutionary process takes place.

The first problem, how to encode the evolving traits of agents in the genotype (only the encoded ones are affected by the GA) is still poorly understood in most work. The genotype-phenotype relation tends to be very simple (in most cases, a direct mapping) and, normally, only a few genes are allowed to evolve and in a fixed-length genotype. Besides, genes evolve individually, there is no epistatic interaction between them, and this reveals what we may call a "diachronic disembodiment" of the evolved structures. There is a growing awareness that the process of development has consequences for the way genotypes evolve, and there have been interesting attempts to overcome the difficulties (for example, using variable length genotypes and, more important, developing more complex ways to represent the genotype-phenotype mapping, so that some aspects of development are taken into account). Nevertheless important background questions are still underdeveloped. Von Neumann, in his work on self-reproduction, already said that only systems endowed with a self-description were capable of open-ended evolution. It was important to establish this threshold, but it still requires deeper understanding in the light of new research. In fact. the way von Neumann conceived the relation between the description and the constructed system is not satisfactory (Etxeberria & Ibañez 1999) and should be a matter of concern and study for artificial evolution.

The second problem, the relation between the agents and the environment as a common history, is also difficult to implement, in such a way that conditions appear in which agents and environment inform one another and intervene in the change of the other (Lewontin 1983). Simulated environments are too simple, fixed and external, and this makes it difficult to observe the action of agents to produce their own environments. Yet, even the real environments in which robots operate are usually very controlled and not at all constructed by the creatures themselves.

Finally, it is very difficult to model the conditions in which selection can arise from the characteristics of the whole system, that is to say, to define the possible adaptive landscapes and intrinsic evaluation or fitness functions according to them. In fact, some models use straightforward (human) artificial selection, using theoretical or aesthetic criteria for it and others define conditions that will be automatically evaluated, but with an absolute criterion, that is to say, they pre-define a "perfection" to be achieved. Natural selection in artificial evolution is external and designed, when a simple evaluation function scores individuals of the population. But it is not a lot more "natural" when creatures live, reproduce or die in this world according to their abilities, because the poor genotype-phenotype encoding and the simple agent-environment relation makes it rather obvious which of the parameters will be that define the best "fitness" in that world. An interesting solution for this has been sought in co-evolution, so trying that the fitness of a population becomes evaluated in relation to changing conditions (Hillis 1992, Sims 1994).

For evolutionary biology, artificial selection is the process of change induced in a population by a human agent selecting something for some purpose. Natural selection, in turn, is blind, and even if some can consider it a process that produces adaptations, or even further, design, these are not foreseen in advance: they depend on many changing conditions that shape the interaction between the evolving entity and the environment. For artificial evolution the situation is confusing because the concept of natural selection is, in some ways, epistemologically very close to that of artificial selection. In fact, the metaphor of artificial selection in breeding was used by Darwin himself to propose natural selection as a explanatory mechanism for evolution. Perhaps the word "selection" cannot but suggest a selective agent, or it might be that natural selection needs further examples (as suggested, for example, by Depew & Weber 1995).

Artificial Evolution's contribution to science

Some issues presented in the last section are a warning about the achievements of artificial methodologies whose main criterion is biological plausibility reduced to a standard textbook theory. This is an important problem for Artificial Life because, since organisms are a clear example of complex systems and the phenomena under study are complex, the obvious way to tackle it appears to appeal for biological plausibility. However, an excessive tribute to the biological might be a handicap, because to obtain insight of some phenomena being studied it is important to acknowledge the characteristics of the medium in which the model is produced. For example, embodiment, emergence, or complexity itself, are difficult notions to understand and copying biology is not necessarily the best way. In fact, the strategies of biology and cognitive science are often reductionist and if research in the artificial domain is expected to overcome some of the problems encountered by these approaches, then it should look for other ways.

In biology there are theoretical problems to integrate two apparently opposed perspectives on organisms, the main being the functional and the structural. The notion of self-organisation is problematic for the standard theory of evolution which has problems to elaborate systemic points of view. Self-organisation, unlike the notion of natural selection, can be characterised as a *systemic* property, in the sense that the entity that organises itself is composed of parts whose configuration and interaction determine the whole they form (which cannot be reduced to those parts). It is also a property that has to do with the production of *spontaneous order* in the system, which is a result of the dynamics of interaction among the components of the system and has nothing to do with an external organising agent or with external design. Finally, it is a capacity that expands the explanatory domain of classical Physics and Chemistry; hence, even if it is not a uniquely biological (it also appears in inanimate systems), self-organization provides a view of nature that suggests a *continuity between the inanimate and the animate*. These characteristics contrast with the atomistic, externalist and non-physicalist perspective of evolution by natural selection.

Then, one possibility for artificial evolution is to explore the integration of both perspectives. This way, it would do more than provide computational versions of the analytical models already developed, it would contribute to the unsolved problems of the field. In fact, artificial models of biological systems have already contributed before; for example auto-catalytic sets (Farmer et al. 1986, Kauffman 1986), Turing's reaction-diffusion model of morphogenesis, or models of Random Boolean Automata have been relevant to reconsider the

biological theory or at least to open interesting debates within it. An application of the new ideas proposed in evolutionary biology to the domain of the artificial is important in order to expand the often too narrow application of evolutionary principles found in some artificial models of evolution, like for example, in genetic algorithms.

An idea that is particularly relevant is the notion of *developmental constraint* as a "bias on the production of variant phenotypes or a limitation on phenotypic variability caused by the structure, characteristics, composition, or dynamics of the developmental system" (Maynard Smith et al 1985, p. 266). The possible sources of those constraints are varied, but they all introduce intrinsic limits to the action of natural selection, both universal or local (only in certain taxa). If developmental constraints can be characterised, then they contrast with ideals of perfection: natural selection is not a mechanism of unlimited optimisation, evolution does not produce "perfect design". Thus, developmental constraints are material limits to perfection.

Among the universal developmental constraints, those studied by Kauffman (1993) are conceived as generic properties of organised matter which do not depend upon natural selection, even if they could influence the conditions in which it takes place. These generic properties are based on the capacity for spontaneous order upon which natural selection acts. Self-organisation is previous to the constitution of the system itself, it prepares the conditions in which natural selection can take place. Kauffman has constructed mathematical models of genetic regulatory systems, as logical networks of connections. By changing the connectivity parameters, it is possible to study the conditions in which attractors and limit cycles appear. Intermediate systems, those found between order and chaos, have the most propitious landscape to evolve. This set of conditions define a new null hypothesis to determine whether there is evolution, it may act as a substitute for the Hardy-Weinberg equation. (Burian & Richardson 1996)

Yet, evolution can also be understood in terms of stages, like the model of *generative entrenchment* suggests (Shank & Wimsatt 1986). The starting standpoint is systemic: nature is divided into systems which are only partially decomposable and form several levels and unities. The idea is that natural systems are locked in stable ontogenetic paths which, once formed, can not be reshaped again, except when there is a general reorganisation of the great phylogenetic taxa. Development, then, makes evolution a quasi irreversible process in which each stage strongly determines subsequent evolution.

Also, from a thermodynamic perspective, Brooks & Wiley (1986) present what they call a Unified Theory of Evolution. A consideration of this may be useful to compare whether the generic properties are best expressed in terms of dynamics or of thermodynamics. Dynamic descriptions are deterministic, reversible and require a detailed knowledge of the initial conditions, while thermodynamic descriptions are stochastic, irreversible and require a selective description of initial data. A thermodynamic approach of evolution can bring together the diachronic and the synchronic perspectives.

If artificial evolution is elaborated in this way (in which if the evolving system is conceived as a complex material entity) both the change of embodied systems and the study of the generic properties of evolutionary change may be reunited. Many of the works where Artificial Evolution is used to design robots or artificial autonomous agents recognise and try to advance in the study of synchronic embodiment, but diachronic embodiment is not questioned. They combine a self-organising perspective of the structure and behaviour of the agent, with an limited selective perspective. Hence, elements which are disembodied in the diachronic sense are automatically introduced (adaptationism, optimality models and evolutionary stable strategies, externalism, poor genotype-phenotype mappings). Because of these reasons, a natural expansion of the notion of embodiment is to develop new forms of artificial evolution that can find a unified perspective.

The unified perspective contributes to embodiment because it can develop a different way of understanding functions as emergent and embodied constraints. Moreover, the philosophical discussion of whether they have ontological reality may be converted into more concrete ones. For some, functional or symbolic explanations only consider secondary properties, largely related to human observation and categorization, whereas others think that some natural systems (for example, the cell) produce their own functional or symbolic structures, that "stand for" longer reactions (or information processing) and can be transmitted as such in evolution (Pattee 1977). The search of generic properties is not an alternative between these positions, what is actually contended is that, underlying the property, there will be a certain specific relation between levels and it might be a non-detachable unit of evolution for the system. Thus, the investigation of functions as generic properties can integrate the diachronic and synchronic perspectives of embodiment, because it involves a further freeing from design (or a naturalistic understanding of it).

Counterfactual evolutions become also interesting. Instead of focusing on higher order functional properties as such, these are attached to the evolution of matter in the universe. The kind of properties and questions proposed for research in this sense have been of the type of why there are only two sexes in most taxa. Probably it is insufficient to investigate this by looking at different selection pressures that could have produced a variety of sexes. However, there might be insights to find if the question is posed in terms of self-organising properties and cohesion.

Conclusion: evolution, creativity and the possible

Artificial Evolution and research in autonomous systems inherit ideas from sources of a very different epistemological style. The first has mainly followed standard evolutionary biology, while the second is based on self-organisation. An effort to expand the theoretical basis of artificial evolution should be made so as to achieve a better understanding of the evolution of organizations. This would have interesting consequences for the way we understand evolutionary creativity and the domain of the possible opened by it. When creativity is conceived in a combinatorial way, almost anything is possible, but no interesting phenomena occur. The reason for this is that organizations are based on emergent properties that confer some kind of physical cohesion to the system that makes it something different from a conjunction of parts or elements (Collier & Muller 1998).

The different purposes driving artificial evolution seek creativity in different ways but all of them –tools, models and instantiations– would benefit from a notion of constrained creativity that the evolution of organizations suggests. Non-intervening dynamical models are important to advance towards a notion of natural selection that is indeed natural, especially in the case of models of evolutionary phenomena. It is not so important to preclude conscious intervention in art, where perhaps the emphasis is placed more in using the evolutionary inspiration as a way to explore new forms of creativity, rather than in making the process biological. Yet, probably in both cases, biology and art, there is a similarity in the effort to understand the sources of creativity, which, also in the two of them, are not completely free or combinatorial, but constrained in ways we would like to understand better. A constrained creativity not only limits variety, it also enables novelty to appear in the system.

Acknowledgement

I thank two Alife7 referees for comments. Funding for this work was provided by the DGICYT (MEC, Madrid) Grant PB95-0502, Basque Government Grants HU-1998-142 and EX-1998-146, and University of the Basque Country (UPV) Grant 003.230-HA079/99.

References

Bedau, M. A. 1998. Four Puzzles about Life. *Artificial Life* 4 (2): 125-140

Brooks, R. A. 1986. A Robust layered control system for a mobile robot, *IEEE J. Rob. Autom.* 2: 14-23.

Brooks, D. & Wiley, E.O. 1986 *Evolution as entropy: Toward a unified Theory of Biology*. Chicago: Chicago University Press, 2nd edition.

Burian, R. & Richardson, R. 1996. Form and order in Evolutionary Biology. In M. Boden Ed. *The Philosophy of Artificial Life*. Oxford University Press.

Cariani, P. 1992. Emergence and Artificial Life. In C.G. Langton, C. Taylor, J.D. Farmer & S. Rasmussen Eds. *Artificial Life II*, Redwood City CA: Addison Wesley, 775-797.

Cliff, D., Harvey, I. & Husbands, P. 1993. Explorations in Evolutionary Robotics. *Adaptive Behavior* 2 (1): 73-110.

Collier, J. & Muller, S. 1998. The Dynamical Basis of Emergence in Natural Hierarchies. In George Farre and Tarko Oksala (eds) *Emergence, Complexity,*

Hierarchy and Organization, Selected and Edited Papers from the ECHO III Conference, Acta Polytechnica Scandinavica, MA91 (Finish Academy of Technology, Espoo, 1998).

Depew, D. & Weber, B. (1995) *Darwinism Evolving. Systems Dynamics and the Genealogy of Natural Selection.* Cambridge, MA & London: MIT Press.

Etxeberria, A. 1995. Vida Artificial. Bases teóricas y metodología. In I. Ramos, M. A. Fernandez, P. Gonzalez, G. Moreno & T. Rojo (eds) *Vida Artificial,* Albacete: Ed. Universidad de Albacete, Colección Estudios, pp. 15-35.

Etxeberria, A. 1998. Embodiment of Natural and artificial Agents. In G. van der Vijver, S. Salthe & M. Delpos (eds.) *Evolutionary Systems,* Kluwer, pp. 397-412.

Etxeberria, A. & Ibañez, J. 1999. Semiotics of the Artificial: The self of self-reproducing systems in cellular automata. *Semiotica* 127 (1/4): 295-320.

Farmer, J.D., Kauffman, S.A. & Packard, N.H. 1986. "Autocatalytic Replication of Polymers", *Physica D,* **22**: 50-67.

Gould, S.J. & Lewontin, R. 1979. The spandrels of San Marco and the Panglossian Paradigm: A critique of the Adaptationist Program, *Proceedings of the Royal Society of London,* 205: 581-98.

Harvey, I. 1996. Artificial Evolution and Real Robots. In Proceedings of International Symposium on Artificial Life and Robotics (AROB) Beppu, Japan Feb 18-20 1996. Masanori Sugisaka (ed.). ISBN4-9900462-6-9 (1996), pp. 138--141.

Hillis, W. D. 1992. Co-evolving parasites improve simulated evolution as an optimization procedure, En Langton, C.G., Taylor, C., Farmer, D. & Rassmusen, S. Eds. *Artificial Life II,* Addison-Wesley, Redwood City CA, 313-324.

Husbands, P., Harvey, I., Cliff, D. & Miller, G. 1997. Artificial Evolution: A new Path for Artificial Intelligence? *Brain and Cognition,* 34 (1): 130-159.

Kauffman, S.A. 1993. *The Origins of Order: Self-organization and Selection in Evolution.,* Oxford: Oxford University Press.

Langton, C.G. 1989. Artificial Life. In Langton, C.G. Ed. *Artificial Life,* Addison-Wesley, Redwood City CA, pp. 1-47.

Lewontin, R. 1983. The organism as the subject and object of evolution, *Scientia,* 118, pp 65-82.

Mataric, M. 1997. Studying the Role of Embodiment in Cognition, *Cybernetics & Systems,* 28 (6). Special issue on Epistemological Aspects of Embodied AI

Maynard Smith, J., Burian, R., Kauffman, S. Alberch, P., Campbell, J, Goodwin, B. Laude, R., Raup, D. & Wolpert, L. 1985. Developmental constraints and evolution: A perspective from the Mountain Lake Conference on development and evolution, *Quaterly Reviews of Biology,* 60: 265-287.

Moreno, A., Etxeberria, A. & Umerez, J. 1994. Universality without matter? In R. Brooks, R.& P. Maes Eds. *Artificial Life IV,* Cambridge (MA): MIT Press, 406-410.

Moreno, A. & Ruiz-Mirazo, K. 1999. Metabolism and the Problem of its Universalization, *Biosystems.* 49 (1): 45-61

Pattee, H.H. 1977. Dynamic and Linguistic Modes of Complex Systems, *Int. J. General Systems,* **3**: 259-266.

Pattee, H.H. 1989. Simulations, Realizations and Theories of Life, In C. Langton Ed. *Artificial Life,* Redwood City CA: Addison Wesley, 63-77.

Pattee, H. H. 1995. Artificial life needs a real epistemology. In F. Moran, et al Eds. *Advances in Artificial Life,* Springer-Verlag. pp 23-38.

Polanyi, M. 1968. Life's Irreducible Structure, *Science,* **160**: 1308-1312.

Ray, T. 1992. An approach to the synthesis of life, En Langton, C.G., Taylor, C., Farmer, D. & Rassmusen, S. Eds. *Artificial Life II,* Addison-Wesley, Redwood City CA, 211-254.

Rosen, R. (1991) *Life itself,* New York: Columbia University Press.

Ruse, M. 1998. *Philosophy of Biology,* Prometheus Books.

Shank, J.C. & Wimsatt, W. C. 1986. Generative Entrenchement and Evolution, PSA 1986, Vol. 2, pags. 33-60.

Sims, K. 1994. Evolving 3D morphology and Behavior by Competition, En Brooks, R. & Maes, P. Eds. *Artificial Life IV,* Cambridge, MA: MIT Press, 28-39.

Varela, F. (1979) *Principles of Biological Autonomy,* New York: Elsevier North Holland.

von Neumann, J. 1966. *Theory of Self-reproducing Automata,* (A. Burks Ed.) Urbana: University of Illinois Press.

Umerez, J. 1998. The Evolution of the Symbolic Domain in Living Systems and Artificial Life. In G. Van der Vijver, S. Salthe & M. Delpos Eds. *Evolutionary Systems,* Dordrecht: Kluwer, pp. 377-396.

Author Index

Printed in the United States
by Baker & Taylor Publisher Services